Santa Cruz Summer Workshops in Astronomy and Astrophysics

S.E. Woosley
Editor

Supernovae

The Tenth Santa Cruz Workshop
in Astronomy and Astrophysics
July 9 to 21, 1989, Lick Observatory

With 311 Illustrations

Springer-Verlag
New York Berlin Heidelberg London
Paris Tokyo Hong Kong Barcelona

S.E. Woosley
Lick Observatory
University of California
Santa Cruz, CA 95064
USA

Printed on acid-free paper.

Softcover reprint of the hardcover 1st edition 1991

Camera-ready copy prepared by the authors.

9 8 7 6 5 4 3 2 1

ISBN-13: 978-1-4612-7748-4 e-ISBN-13: 978-1-4612-2988-9
DOI: 10.1007/978-1-4612-2988-9

Preface

Twelve years ago, in July 1977, the Santa Cruz Summer Workshop series began with a three week meeting on "Supernovae." Over the years these workshops have treated a variety of subjects ranging from the Big Bang to star formation to astronomical instrumentation, but with this one, the tenth, we returned to the original theme. It is interesting to contrast this year's meeting with the one twelve years ago. The original workshop lasted three weeks with fifty participants and had no written proceedings. This one lasted two weeks, had 177 participants from 17 nations, and resulted in the proceedings which you now see of over 110 contributions. Certainly progress is not measured in numbers alone, but from the talks summarized here I hope that the reader will be able to sense the excitement that presently surrounds the study of supernovae, excitement generated not only by the recent occurrence of the brightest event since before the invention of the telescope, but also by progress on many fronts on a problem that arguably involves more physics than any other phenomenon in astronomy. This progress reflects the efforts of many people, a substantial fraction whom were not working on supernovae ten years ago; the rapid development of computer technology in the last decade; and access to instrumentation that was previously unavailable, especially space and balloon borne detectors of UV, x-ray, and γ–radiation, sensitive infrared detectors on the Earth and on the Kuiper airplane, and, of course, neutrino telescopes.

True progress is measured not only by those questions which are answered but by the deeper insight provided, insight that invariably leads to further questions. Thus we now know with certainty that massive stars explode and make Type II supernovae, but were surprised by the light curve of SN 1987A and its occurrence in a blue supergiant. Two years later everyone understands that the two are related, but still not all agree why the star Sk 202 —69 was blue (mixing? Metallicity? rotation?) and until we do it is difficult to say how common such events as SN 1987A are.

Similarly, we now know that Type II supernovae are accompanied by bursts of neutrino radiation and that a neutron star forms inside the supernovae, but still unresolved, or at least controversial, are the exact mechanism for the explosion (though substantial progress was made on this at the workshop, see §7), and the properties of the neutron star left behind — its mass, rotation rate, and magnetic field strength.

Whatever their energy source, we are certain now that supernova explosions create new elements, some of them in the form of radioactive progenitors that decay and make γ–radiation. It was surprising though to find that elements made at the bottom of the supernova had somehow made their way out into material that had once formed the hydrogen envelope. Now we know that *mixing* occurs in the explosion and many new multidimensional calculations are being done (see §4), but we remain a bit fuzzy about just what instabilities and initial seed perturbations led to the mixing and how common mixing is in supernovae of other types. We are also just beginning to explore the role of rotation in these events and have a long way to go in addressing observational evidence for deformation or, worse still, jets in the explosion.

Observations too seem always to point out the need for more observations. Theorists and observers alike have now learned the value of panchromatic studies in keeping track of the supernova energy budget and for indicating the formation of dust, but, so far, we only have such observations for one of the 600 and some odd supernovae that have occurred. Similarly, though we recognize the importance of the radius of the presupernova star and its recent mass loss history in shaping the light curve and spectrum of the supernova, we know the identification of just one presupernova star.

With larger amounts of data of increasing precision we have categorized supernovae into several groups, not just I and II, but Ia, Ib, IIp, II-L, 87A, with probably more to come. But still, in all cases except IIp and 87A, we are unclear as to the evolution of the progenitor star (or stars). What makes a Type Ia supernova and what would a progenitor system look like? Are the progenitors of Type Ib all Wolf-Rayet stars? Also of continued great interest is the interaction of supernovae with their surroundings, their role in stirring and heating the interstellar medium and, perhaps, in triggering the formation of new stars. Do they really do that? We would also like to use supernovae, certainly among the best and brightest of standard candles, to determine Hubble's constant.

Finally one area where enormous progress has clearly been made in the last twelve years is in the theoretical modeling of supernova spectra (§8). It is by way of the spectrum (and the less informative light curve) that the models make most direct contact with the observations. The detail and physical realism of the model spectra, now routinely calculated from the infrared through γ-rays, motivate the theorists to provide a commensurate level of detail in the models themselves and will surely guide us to a more complete understanding of the supernova phenomenon.

The Santa Cruz Workshop discussed and these proceedings present our progress in addressing all of the above issues and much, much more. The dates of the meeting were July 9–12, 1989 and the Scientific Organizing Committee included, besides myself, John Danziger (ESO); Mike Dopita (MSSSO); Dave Helfand (Columbia); Mark Phillips (CTIO); Tom Prince (CIT); Bob Kirshner (Harvard); Ken Nomoto (Univ. Tokyo); Al Mann (Univ. Pennsylvania); and Craig Wheeler (Univ. Texas). The meeting consisted of two weeks of invited morning talks followed by from one to three technical workshops in the afternoon. The first half of the meeting centered on observations and models of Supernova 1987A. During the second week other supernovae and supernova remnants were discussed. The Table of Contents of this book follows roughly the time order in which the sessions occurred.

The grand success of the meeting was chiefly a reflection of the enthusiasm (and endurance!) of the participants and the fun we all have working on supernovae, but the Organizing Committee laid out a coherent and challenging schedule and we are grateful for their help. Furthermore the Workshop would not have been possible and certainly not as much fun without the personal effort and attention of Sue Robinson, our Workshop Coordinator. She was assisted by Pat Shand who is additionally responsible for assembling the book you now hold. In as much as it is my prerogative as editor, I dedicate this book to those two ladies. We also gratefully acknowledge the support of the National Science Foundation, the National Aeronautics and Space Administration, Cal Space, and the Institute for Geophysics and Planetary Physics at Lawrence Livermore National Laboratory. Without their support there also would have been no Workshop.

Finally, I must mention that not all of nature's disasters are as fun as supernovae. In the late afternoon of October 17, 1989 an earthquake of magnitude 7.1 occurred in Santa Cruz which indirectly delayed the publication of this book by about two months. I apologize for the delay. Though the ground shook ferociously, so far as I am aware there was no accompanying neutrino burst.

Stan Woosley
February 1990

Contents

VI. THE NEUTRON STAR IN SN 1987A AND OTHER SUPERNOVAE

VII. EXPLOSION MECHANISMS AND NEUTRINO BURSTS

XIII. SUPERNOVA RATES, SEARCHES, AND USE AS STANDARD CANDLES

Identification Chart

1. Noam Sack
2. Greg Shields
3. Friedrich-K. Thielemann
4. Bruno Leibundgut
5. Jay Boisseau
6. Peter Sutherland
7. Russell Cannon
8. Nicholas Suntzeff
9. Jeruska Suntzeff
10. Mark Phillips
11. Mario Hamuy
12. Douglas Swartz
13. Jerry Cooperstein
14. Alak Ray
15. Ralph Mönchmeyer
16. Unidentified
17. Hans-Thomas Janka
18. Jordi Isern
19. Robert Mochkovitch
20. Maurice Aufderheide
21. Enrico Cappellaro
22. Robert Harkness
23. Craig Wheeler
24. Itamar Lichtenstadt
25. Sid Bludman
26. Ken Van Riper
27. Dave Arnett
28. Norbert Langer
29. Zalman Barkat
30. Robert Kirshner
31. Stirling Colgate
32. George Sonneborn
33. Nikos Prantzos
34. Tom Prince
35. Fred Witteborn
36. Dick McCray
37. Charles Petit
38. Nino Panagia
39. Francesca Matteucci
40. Robert Gilmozzi
41. Peter Meikle
42. Unidentified
43. Albert Petschek
44. M. Kirshner
45. Bonnard Teegarden
46. Mounib El Eid
47. Reinhard Hanuschik
48. Juliana Hsu
49. D.K. Nadezhin
50. Angela Bragaglia
51. Anurag Shankar
52. Marcos Montes
53. Ron Eastman
54. Unidentified
55. Roger Chevalier
56. Nikolai Chugai
57. Kurt Weiler
58. Unidentified
59. Ken'ichi Nomoto
60. David Branch
61. John Danziger
62. Lisa Ensman
63. Leon Lucy
64. Mark Leising
65. Hank Donnelly
66. Robert Wagoner
67. John Cowan
68. Ed Baron
69. Alfred Mann
70. Massimo Turatto
71. Tom Weaver
72. Almas Chalabaev
73. James Wilson
74. David Jeffery
75. Timothy Young
76. Claes Fransson
77. Jack Hughes
78. Dieter Hartmann
79. Yasuo Tanaka
80. Rino Bandiera
81. Valery Chechetkin
82. Mike Howard
83. Brad Meyer
84. Christian Pollas
85. Jason Spyromilio
86. Ramon Canal
87. Neil Gehrels
88. Jack Tueller
89. Hong-wei Li
90. L.S. Lubimkov
91. Peter Höflich
92. Alexei Filyukov
93. Yueming Xu
94. Robin Catchpole
95. Alain Porter
96. Cecilia Kozma
97. Peter Lundqvist
98. Mikio Nagasawa
99. Toshikazu Shigeyama
100. Mitsue Den
101. Rosario López
102. Tatsuo Yoshida
103. Patrice Bouchet
104. Christian Gouiffes
105. Mike Dopita
106. Donald Osterbrock
107. Adam Burrows
108. Ikko Fushiki
109. Toshikazu Ebisuzaki
110. Hideyuki Suzuki
111. Chris Smith
112. Alan Uomoto
113. Brad Peterson
114. Philip Pinto
115. Diane Wooden
116. Patricia Whitelock
117. Joseph Wampler
118. Jim Peters
119. Zhen-Ru Wang
120. Stan Woosley
121. Eric Schlegel
122. Joe Shields
123. Kai-wing Chan

List of Participants

ARNETT, David	University of Arizona
AUFDERHEIDE, Maurice	SUNY at Stony Brook
BANDIERA, Rino	Osservatorio Astrofisico di Arcetri, Italy
BARKAT, Zalman	Hebrew University of Jerusalem, Israel
BARON, Edward	SUNY at Stony Brook
BARTEL, Norbert	Harvard-Smithsonian Center for Astrophysics
BLUDMAN, Sidney	University of Pennsylvania
BLUMENTHAL, George	University of California, Santa Cruz
BODENHEIMER, Peter	University of California, Santa Cruz
BOISSEAU, Jay	University of Texas at Austin
BOUCHET, Patrice	European Southern Observatory, Chile
BRAGAGLIA, Angela	Universita di Bologna, Italy
BRANCH, David	University of Oklahoma
BREGMAN, Jesse	NASA-Ames Research Center
BURROWS, Adam	University of Arizona
CANAL, Ramon	Universidad de Barcelona, Spain
CANNON, Russell	Anglo-Australian Observatory, Australia
CAPPELLARO, Enrico	Osservatorio Astronomico di Padova, Italy
CATCHPOLE, Robin	SA Astronomical Observatory, South Africa
CHALABAEV, Almas	Observatoire de Haute Provence, France
CHAN, Kai-Wing	University of California, San Diego
CHECHETKIN, Valery	Keldysh Inst. of Applied Math., USSR
CHEVALIER, Roger	University of Virginia
CHUGAI, Nikolai	Astronomical Council, Ac.Sci., USSR
CLINE, David	University of California, Los Angeles
COLGATE, Stirling	Los Alamos National Laboratory
COOPERSTEIN, Jerry	Brookhaven National Laboratory
COWAN, John	University of Oklahoma
DANZIGER, John	European Southern Observatory, FRG
DELLA VALLE, Massimo	University of Padova, Italy
DEN, Mitsue	Kyoto University, Japan
DICKEL, John	University of Illinois
DOANE, Jay	University of California, Santa Cruz
DONNELLY, Hank	University of California, Santa Cruz
DOPITA, Michael	Mt. Stromlo & Siding Spring Obs., Australia
DRAKE, Frank	University of California Observatories/Lick Observatory
DUROUCHOUX, Philippe	CEN-Saclay, France
DWEK, Eli	NASA-Goddard Space Flight Center
EASTMAN, Ronald	Harvard-Smithsonian Center for Astrophysics
EBISUZAKI, Toshikazu	University of Tokyo, Japan
EL EID, Mounib	Universitäts-Sternwarte Göttingen, FRG
ELLMAN, Nancy	University of California, Santa Cruz
ENSMAN, Lisa	University of California, Santa Cruz

FESEN, Robert	University of Colorado
FILIPPENKO, Alexei	University of California, Berkeley
FILYUKOV, Alexander	Keldysh Institute, Academy of Sciences, USSR
FRANSSON, Claes	Stockholm Observatory, Sweden
FRYXELL, Bruce	University of Arizona
FULLER, George	University of California, San Diego
FUSHIKI, Ikko	University of California, Santa Cruz
GEHRELS, Neil	NASA-Goddard Space Flight Center
GILMOZZI, Roberto	Space Telescope Science Institute
GOUIFFES, Christian	European Southern Observatory, Chile
HAMILTON, Andrew	JILA, University of Colorado
HAMUY, Mario	Cerro-Tololo InterAmerican Obs., Chile
HANUSCHIK, Reinhard	Ruhr-Universität, FRG
HARKNESS, Robert	University of Texas at Austin
HARTMANN, Dieter	University of California, Santa Cruz
HELFAND, David	Columbia University
HOFFMAN, Robert	University of California, Santa Cruz
HÖFLICH, Peter	Max-Planck-Institut für Astrophysik, FRG
HOWARD, Mike	Lawrence Livermore National Laboratory
HSU, Juliana	Massachusetts Institute of Technology
HUGHES, John	Harvard-Smithsonian Center for Astrophysics
ILLINGWORTH, Garth	University of California Observatories/Lick Observatory
ISERN, Jordi	Centre d'Estudis Avangats Blanes, Spain
JANKA, Hans-Thomas	Max-Planck-Institut für Astrophysik, FRG
JEFFERY, David	University of Oklahoma
KAY, Laura	University of California, Santa Cruz
KELLER, Jeffrey	University of California, Santa Cruz
KIM, Soo Bong	University of Pennsylvania
KIRSHNER, Robert	Harvard-Smithsonian Center for Astrophysics
KLEIN, Richard	Lawrence Livermore National Laboratory
KOO, David	University of California Observatories/Lick Observatory
KOZMA, Cecilia	Stockholms Observatorium, Sweden
LANGER, Norbert	University of California, Santa Cruz
LATTIMER, James	SUNY at Stony Brook
LEIBUNDGUT, Bruno	Astronomical Institute Basel, Switzerland
LEISING, Mark	Naval Research Laboratory
LI, Hong-Wei	JILA, University of Colorado
LI, Zongwei	Beijing Normal University, PR China
LIANG, Edison	Lawrence Livermore National Laboratory
LICHTENSTADT, Itamar	Hebrew University of Jerusalem, Israel
LINGENFELTER, Richard	University of California, San Diego
LÓPEZ, Rosario	Universidad de Barcelona, Spain
LOREDO, Tom	University of Chicago
LOSECCO, John	University of Notre Dame
LUBIMKOV, L.S.	Crimean Astrophysical Observatory, USSR

LUCY, Leon	European Southern Observatory, FRG
LUNDQVIST, Peter	Lund Observatory, Sweden
MANN, Alfred	University of Pennsylvania
MARVIN, Heidi	University of California, Berkeley
MATTEUCCI, Francesca	Max Planck Institut für Astrophysik, FRG
MAYLE, Ronald	Lawrence Livermore National Laboratory
MEIKLE, Peter	Imperial College, England
MEYER, Brad	Lawrence Livermore National Laboratory
MOCHKOVITCH, Robert	Institut d'Astrophysique de Paris, France
MÖNCHMEYER, Ralph	Max-Planck-Institut für Astrophysik, FRG
MONTES, Marcos	Stanford University
MURRAY, Stephen	University of California, Santa Cruz
McCRAY, Richard	JILA, University of Colorado
McKEE, Chris	University of California, Berkeley
NADYOZHIN, D. K.	Institute of Theoretical & Experimental Physics, USSR
NAGASAWA, Mikio	National Lab for High Energy Physics, Japan
NAKANO, George	Lockheed Palo Alto Research Lab
NOMOTO, Ken'ichi	University of Tokyo, Japan
O'CONNELL, Bob	University of California, Santa Cruz
OLIVA, Ernesto	Osservatorio Astrofisico di Arcetri, Italy
OSTERBROCK, Donald	University of California Observatories/Lick Observatory
PACINI, Franco	Osservatorio Astrofisico di Arcetri, Italy
PANAGIA, Nino	Space Telescope Science Institute
PERLMUTTER, Saul	University of California, Berkeley
PETERS, James	San Francisco State University
PETSCHEK, Albert	New Mexico Institute of Mining & Technology
PHILLIPS, Mark	Cerro-Tololo InterAmerican Observatory, Chile
PINTO, Philip	Harvard-Smithsonian Center for Astrophysics
POLLAS, Christian	Observatoire de la Côte d'Azur, France
PORTER, Alain	National Optical Astronomy Observatories
PRANTZOS, Nikos	Institut d'Astrophysique de Paris, France
PRINCE, Thomas	California Institute of Technology
PRIMACK, Joel	University of California, Santa Cruz
RANK, David	University of California Observatories/Lick Observatory
RAWLINGS, Steven	Mt Stromlo & Siding Spring Obs., Australia
RAY, Alak	Tata Institute of Fundamental Research, India
REED, Jeri	Institute of Astronomy, England
RICHMOND, Michael	University of California, Berkeley
ROBINSON, Lloyd	University of California Observatories/Lick Observatory
SACK, Noam	Hebrew University of Jerusalem, Israel
SCHLEGEL, Eric	Harvard-Smithsonian Center for Astrophysics
SHIELDS, Joseph	University of California, Berkeley
SHIGEYAMA, Toshikazu	University of Tokyo, Japan
SIEVERLING, Eric	University of California, Santa Cruz
SMITH, Chris	Harvard-Smithsonian Center for Astrophysics

SOKER, Noam	University of Virginia
SONNEBORN, George	NASA-Goddard Space Flight Center
SPYROMILIO, Jason	Imperial College, England
STARRFIELD, Sumner	Arizona State University/LANL
STATHAKIS, Raylee	Anglo-Australian Obs./Univ. Sydney, Australia
STRINGFELLOW, Guy	University of California, Santa Cruz
SUNTZEFF, Nicholas	Cerro-Tololo InterAmerican Observatory, Chile
SUTHERLAND, Peter	McMaster University, Canada
SUZUKI, Hideyuki	University of Tokyo, Japan
SWARTZ, Douglas	University of Texas at Austin
TANAKA, Yasuo	Inst. Space & Astronautical Science, Japan
TEEGARDEN, Bonnard	NASA-Goddard Space Flight Center
THIELEMANN, Friedrich-K.	Harvard-Smithsonian Center for Astrophysics
TIMMES, Frank	University of California, Santa Cruz
TRAN, Hien	University of California, Santa Cruz
TUELLER, Jack	NASA/Goddard Space Flight Center
TURATTO, Massimo	Osservatorio Astronomico di Padova, Italy
UOMOTO, Alan	Johns Hopkins University
van den BERGH, Sidney	Dominion Astrophysical Observatory, Canada
Van RIPER, Ken	Los Alamos National Laboratory
VEILLEUX, Sylvain	University of California, Santa Cruz
WAGONER, Robert	Stanford University
WALKER, Merle	University of California, Santa Cruz
WAMPLER, Joseph	European Southern Observatory, FRG
WANG, Zhen-Ru	JILA/Nanjing University, PR China
WEAVER, Tom	Lawrence Livermore National Laboratory
WEHINGER, Peter	Arizona State University
WEILER, Kurt	Naval Research Laboratory
WHEELER, Craig	University of Texas at Austin
WHITELOCK, Patricia	SA Astronomical Observatory, South Africa
WHITFORD, Albert	University of California Observatories/Lick Observatory
WILSON, James	Lawrence Livermore National Laboratory
WINKLER, Frank	Middlebury College
WITTEBORN, Fred	NASA-Ames Research Center
WOODEN, Diane	NASA-Ames Research Center
WOOSLEY, Stan	University of California, Santa Cruz
XU, Yueming	JILA, University of Colorado
YAMADA, Yoshiyuki	Kyoto University, Japan
YOSHIDA, Tatsuo	Hokkaido University, Japan
YOUNG, Timothy	University of Oklahoma

SECTION I

SN 1987A — THE LIGHT CURVE AND SPECTROPHOTOMETRY

The Bolometric Light Curve of SN 1987A

Nicholas B. Suntzeff & Patrice Bouchet

1. Introduction

Perhaps the most fundamental quantity needed for understanding the physics of SN 1987A is the temporal evolution of the bolometric luminosity. The calculation of the bolometric luminosity must sum together the all the flux, from the gamma rays to the microwave that is associated with the energy generated within the supernova nebula yet exclude the flux which is not part of the prompt energy release from these sources, such as the flux from an echo or stars which lie along the line-of-sight. The bolometric luminosity can be divided into two disjoint ranges, the ultraviolet-optical-infrared luminosity (the "uvoir" luminosity or L_{uvoir}), and the "high-energy" luminosity of gamma and X-rays observed from space. The latter represents the energy as it is produced from the radioactive decay of the nuclides produced in the supernova explosion or the Compton scatterings associated with the high-energy radiation. L_{uvoir} is produced by that part of the energy from the shock that goes into radiation (as opposed to mechanical energy) for the first few weeks after outburst, and the thermalized photons from the radioactive decays for the late-time evolution. Other energy sources, such as a buried pulsar may contribute to either the high-energy or uvoir flux. The comparison of the observed L_{uvoir} and the predictions of the models can provide measures for the type and amount of radioactive nuclides synthesized in the explosion, and can constrain the existence of other sources of energy.

The observed L_{uvoir} can be calculated by integrating the flux distribution of SN 1987A. We have two estimates for the flux distribution – the broad-band photometry and the spectrophotometry. The spectrophotometry is a more accurate means of integrating the flux but is observationally more difficult to obtain. The broad-band photometry, converted to equivalent monochromatic fluxes at the effective wavelengths of the filter response functions, provide a much more complete picture of the time evolution of the bolometric luminosity, but at the expense of accuracy. The following discussion is a summary of the work presented by SUNTZEFF and BOUCHET [1] and BOUCHET, *et al.*[2].

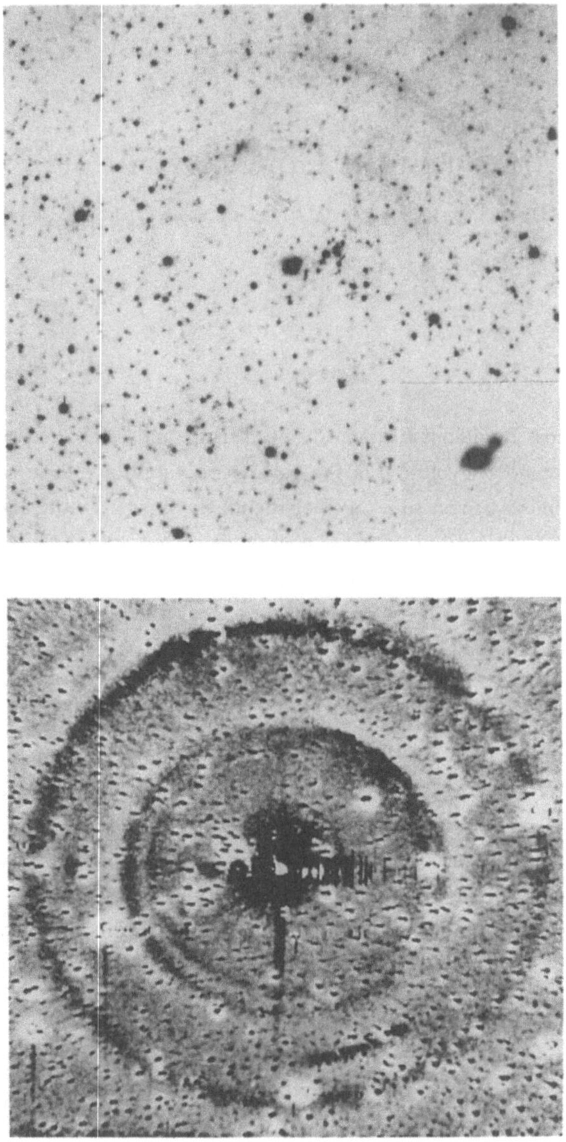

Figure 1. CCD images of SN 1987A. Fig. 1a (top) is a V image taken on 14 March 1989 in 0.75[″] seeing with the CTIO 4m telescope. North is up and east to the left. The field is 3.2[′] across. The inset is an expanded image in Strömgren u, where stars 2 (p.a.=110[°]) and 3 (p.a.=310[°]) are clearly visible. Fig. 1b (bottom) is an image of the echo from SN 1987A in R light, taken on the 0.9m telescope on 16 Dec 1988. North is up and east is to the *right.* The structure within 20[″] of SN 1987A and the spikes are effects from the overexposure of SN 1987A. The stars have been removed to show the echo. Four rings or partial rings are visible.

2. Observations

In Fig. 1, I present CCD frames from CTIO of SN 1987A taken at the beginning of 1989 to give you a feel of what we are trying to observe. Figure 1a presents U and V exposures when SN 1987A was at V=12. The field is extremely crowded with blue main sequence stars which are members of the LMC field population (there are 38 stars brighter than V=20 within 30["] of SN 1987A!). Stars 2 and 3, in the numeration of WEST, *et al.* [3] are clearly visible in the inset to Fig. 1a. Figure 1b shows the optical echo complex (in R_{KC} light) with the stars subtracted, where the two main dust sheets at 115 and 305[pc] in front of SN 1987A (SUNTZEFF, *et al.* [4]) are seen to break up into multiple components, which are expanding at ~20c.

The photometric data set used in my review is optical photometry from CTIO (UBVRI photoelectric and CCD photometry; HAMUY, *et al.* [5], SUNTZEFF, *et al.* [6], HAMUY and SUNTZEFF [7]) and infrared aperture photometry from ESO (JHKLM, and N1, N2, N3, Q0; BOUCHET, *et al.* [8], BOUCHET and MONETI [9]). The ESO/CTIO data set has roughly 200 nights of U to M (0.32[μm] to 5[μm]) photometry, and 65 nights of N1 to Q0 (8[μm] to 20[μm]) photometry, from day 1.1 to 903. Another large data set is from SAAO and is reviewed in this volume by WHITELOCK, *et al.* [10]. The SAAO data set contains roughly 200 nights of U to M photometry through day 800, but does not contain the critical far-infrared flux points (longward of 5[μm]) that dominate the uvoir spectral energy distribution starting around day 500. After that time, any discussion of L_{uvoir} based only on U to M interpolations is meaningless, unless one assumes that essentially the *entire* far-infrared flux is not associated with the prompt energy release from SN 1987A. The SAAO group [10,31] have attempted to add on the thermal component to the U to M data by adding black-body *model* fits to the thermal component. Since the thermal component dominates the U to M flux after day 600, the L_{uvoir}based on the SAAO data is esentially just the *model* fits, and is not really a direct observational derivation of L_{uvoir}. In particular, the black-body models do not include any of the flux in emission lines since they were based on continuum points, and the total flux predicted by the model fits is a sensitive function of the T_{eff} chosen.

The spectrophotometry is a combination of CTIO optical CCD spectra [PHILLIPS, *et al.* [11]), CTIO Infrared Spectrometer (IRS) spectra, ESO CVF infrared spectra [2,8], and KAO infrared spectra (WOODEN [12]). The spectral range is 0.31[μm] to 13.5[μm] with a few small gaps at 1.4 and 1.9[μm]. We have added the Q0 flux point to complete the spectral coverage out to 19[μm]. The CTIO IRS spectra, which were not always taken under photometric conditions, were scaled by a nightly constant to match the ESO CVF data. Since the various pieces of spectra were not always taken on the same night, the data were logarithmically interpolated to the same date. The spectrophotometry covers days 134 to 725.

3. Measurement of the UVOIR Bolometric Luminosity

The broad-band magnitudes were converted to equivalent monochromatic fluxes, dereddened by E(B-V)=0.15 assuming a Galactic reddening of 0.04 and 30 Doradus reddening of 0.11 (see [5]) using a logarithmic, linear, or spline fit, corrected for the missed flux shortward of 3200[Å] or longward of 19[μm], integrated, and scaled for a true distance modulus of 18.5. The extrapolation from 19[μm] to zero frequency was done assuming a Rayleigh-Jeans law, which after day 700, accounted for more than 10% of L_{uvoir}. The extrapolation shortward of 3200[Å] was made by forcing the flux to go to zero at 3000[Å]. Except for the first 10 days, this correction was always less than 3% of the total flux. For these first days, we added in the IUE flux (KIRSHNER, *et al.* [13]) and interpolated from the longest wavelength of the IUE flux to 3600[Å]. We performed two integrations, one from U to Q0, and the other from U to M to compare to the SAAO data. The spectrophotometry was dereddened, extrapolated, and scaled to the distance of the LMC in the same fashion. In Fig 2. I plot the spectrophotometry and broad-band photometry converted to monochromatic fluxes for 28 Oct 1988 (day 615) where it can be seen that the broad-band magnitudes agree with the spectrophotometry, and that a very strong thermal component was present that can be fit by a scaled black body.

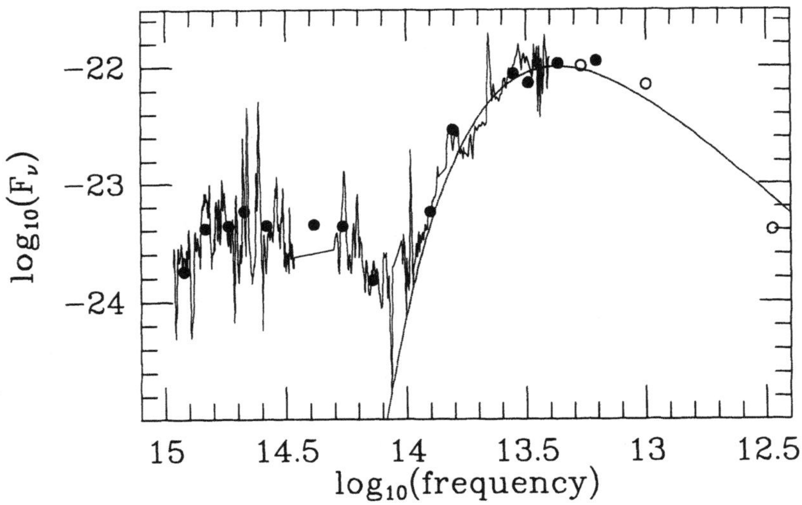

Figure 2. A comparison of the ESO/CTIO spectrophotometry (solid curve) with the broad-band photometry (filled circles) for day 615. The open circles are from MOSLEY, *et al.*[14] and LESTER [15]. The dotted line is the best-fit black body (400[°K]) to the L through Q0 photometry.

3.1 Uncertainties in the Bolometric Luminosity

Uncertainties to L_{uvoir} are dominated by the errors associated with the distance modulus to the LMC and the reddening to SN 1987A. The true distance modulus to the LMC lies in the range of 18.3 to 18.6 (FEAST and WALKER [16]) where the lower distance corresponds to the fits to the color-magnitude diagrams and the longer distance to the Cepheid and RR Lyrae distance scale. Recent work (VANDENBERG and POLL [17]) indicated that the new theoretical isochrones may favor the longer distance. The reddening to SN 1987A lies between 0.15 and 0.25 and more likely between 0.15 and 0.20 based on the Balmer decrement nearby the supernova (WAMPLER, *et al.* [18], DANZIGER, *et al.* [19]), the colors of the Sk -69 202 coupled with a spectral type of B3 ([3], WALBORN, *et al.* [20], PANAGIA, *et al.* [21], FITZPATRICK [22]), or the color evolution over the first few days [21]. WALKER and SUNTZEFF [23] have measured E(B-V)=0.17±0.02 based on the colors of 23 main sequence stars within 1.5['] of SN 1987A.

Another large source of error in L_{uvoir} is due to the broad-band photometry. The broad-band photometry is essentially spectrophotometry done at a spectral resolution of ~5, with large non-rectangular (in wavelength) pixels, which can have significant overlap or be quite non-contiguous. The conversion of the broad-band magnitudes to monochromatic magnitudes uses a calibration based on A stars, which is clearly not a very good assumption. We can explore the uncertainties in the photometric measurement of L_{uvoir} by trying different interpolation schemes, by using the two different data sets, and by adding artificial noise to the data at the level of the photometric uncertainties. I summarize the estimated uncertainties in Table 1. The typical uncertainty is 0.02-0.05[dex] for the various intrinsic uncertainties. Note that the change in assumed reddening does not produce a simple shift because the bulk of the flux changes from ~1[μm] to longward of 5[μm] by day 600. Table 1 shows that the estimate of L_{uvoir} based on the broad-band photometry is in reasonable agreement with the spectrophotometry, with perhaps a small systematic difference between the two estimates.

Table 1. Uncertainties in $Log_{10}(L_{uvoir})$

days since outburst	E(B-V) 0.15 to 0.20	ESO/CTIO - SAAO U to Q0	logarithm - linear interpolation	U to Q0 - spectrophot.	photometric errors
134	-0.033	-0.016	-0.010	0.00	0.002
262	-0.037	-0.039	-0.013	-0.03	0.002
455	-0.040	-0.030	-0.015	-0.05	0.003
526	-0.039	-0.029	-0.016		0.005
635	-0.022	-0.016	-0.018	-0.01	0.017
729	-0.016	-0.008	-0.021	-0.07	0.028
813	-0.024		-0.011		0.022

On day 126, when SN 1987A just entered the exponential decline phase, the L_{uvoir} was 41.514[$\log_{10}$ erg s^{-1}]. This corresponds to 0.071[$M_{\odot}$] of ^{56}Co (and therefore ^{56}Ni) produced in the outburst (WOOSLEY [24]). Given the range in the distance modulus and reddening, and a "typical" 0.03[dex] error from the other sources, I find that , *the actual allowed range in the amount of nickel synthesized in the explosion is 0.054[$M_\odot$] to 0.090[$M_\odot$].* For those who think that this is an overly pessimistic calculation, I would only note that if you multiply these numbers by 1000, you have roughly the range of values for H_0, whose main uncertainty is due also to the distances to the local calibrators, of which the LMC is perhaps the most important.

One final correction needs to be applied to the data; the correction for the field star contamination. From CCD photometry, we [23] have measured the optical magnitudes for the stars in the vicinity of SN 1987A. Stars 2 and 3 as numbered by [3], which are 1.6["] and 2.9["] from SN 1987A, have (V,B-V,U-B) of (14.20,0.06,-0.74) and (15.60,0.46,-0.89). The aperture photometry will include these and other stars. For the CTIO photometry, our *preliminary* value for the UBVRI colors of all the flux inside a 19["] aperture excluding SN 1987A is (13.71,14.45,14.23,14.14,13.99). This corresponds roughly to a main sequence star of spectral type A2. We have added on the infrared colors of an A2 star to estimate that the contamination adds 37.362[$\log_{10}$ erg s^{-1}] to L_{uvoir}. We have subtracted this contamination from the photometry and the spectrophotometry, since the latter was also calibrated through a 20["] slit.

3.2 The Temporal Evolution of the UVOIR Bolometric Luminosity

In Fig. 3, I show the L_{uvoir} evolution of SN 1987A based on the U to Q0 integration, the U to M intergration, and the spectrophotometry. I have plotted the U to M integration as a line since the points are so dense. The U to M estimate was terminated at day 500 because the interpolation did not include most of the flux, which by then shifted to the far infrared. The light curve can be divided into three parts. From day 1 to day 10, the luminosity and temperature dropped as the nebula expanded after the shock passed through. From day 20 to 125 the hydrogen recombination front propagated inward (in mass) into the nebula which had been heated from the energy released from the radioactive decay of ^{56}Ni and ^{56}Co . From day 126 to 903, L_{uvoir} of SN 1987A has fallen exponentially with a remarkably constant e-folding time of 96.3 days.

Some details of the light curve need to be noted. Around day 25, satellite emission lines appeared in the spectrum of SN 1987A (PHILLIPS and HEATHCOTE [25], HANUSCHIK and DACHS [26]). PHILLIPS [27] noted that the optical color evolution also slowed its rapid reddening at this time and interpreted the "bluing" to the appearance of radiation from the diffusion wave. In Fig. 3, it can be seen that L_{uvoir} went through an inflection point around day 30, which may be a manifestation of this event. The transition from maximum light to the "radioactive tail" was extremely fast. From day 105 to 120, the luminosity declined at a rate of 0.016[dex day^{-1}]. By day 128, the rate had changed to 0.004[dex day^{-1}]. This changeover happened in less than a week, implying a remarkable coherence to the process which has not been theoretically explained.

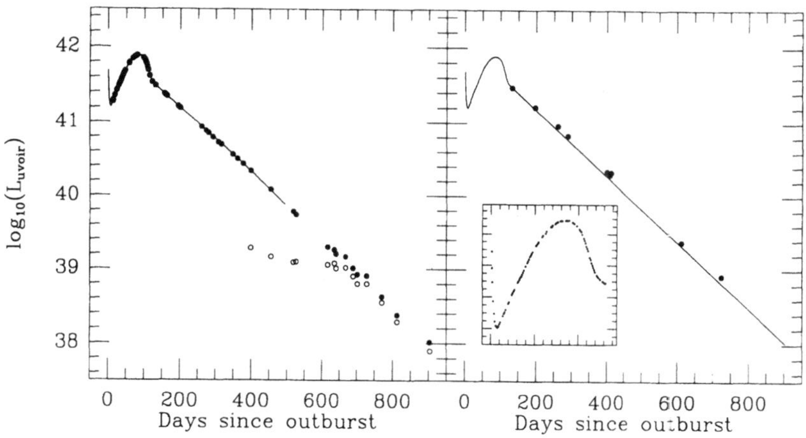

Figure 3. UVOIR bolometric luminosity evolution of SN 1987A. The left panel shows L_{uvoir} based on the U to M interpolation (solid line), U to Q0 interpolation (filled circles), and the luminosity of the thermal component (open circles). The right panel shows L_{uvoir} based on the spectrophotometry compared to a smoothed version of the U to Q0 interpolation. The inset shows the detail of the first 140 days. The units of L_{uvoir} are erg s^{-1} .

Finally, the ESO/CTIO data show that during the exponential decline from day 126 to 300, the observed e-folding time of 101 days was *faster* than the e-folding time of 111.3 days predicted from the decay of ^{56}Co . This contradicts the conclusion of the SAAO group (WHITELOCK, *et al.* [28]) where they found that L_{uvoir} fell at *exactly* the decay rate of ^{56}Co , and only after day 270, did the decline steepen. However, a reanalysis of the SAAO data [6] found that the L_{uvoir} based on SAAO data indeed declined more rapidly (105 days) than the ^{56}Co decay rate over this time period. We interpreted this result to mean that the diffusion wave did not pass out discontinuously, but left a tail that died out in time, steepening the exponential decline. The uncertainty of the e-folding time of 0.6 days quoted by the SAAO group [28] for the exponential decline ignores the uncertainties listed in Table 1, and the true error is an *order of magnitude greater* than they claimed.

Figure 4 compares the observed L_{uvoir} to the theoretical models of the luminosity evolution of SN 1987A by PINTO, *et al.* [29] and KUGAMAI, *et al.* [30]. These models fit the observed gamma-ray line fluxes and X-ray continua to predict the total energy escaping as high-energy photons, and when subtracted from the total energy liberated by 0.075$[M_\odot]$ of ^{56}Co , the resulting curve is the predicted L_{uvoir}. These models are reasonably well constrained through the latest gamma ray observations on day 631 (see WHITELOCK, *et al.* [31]). The agreement of theory and observation as seen in Fig. 4 though day 800 is remarkably good.

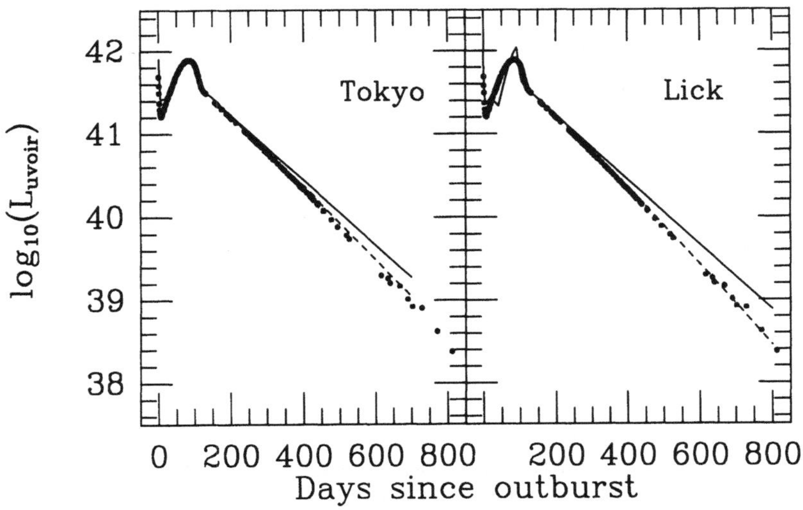

Figure 4. A comparison of the observed and theoretical L_{uvoir}. The filled circles are the U to Q0 interpolation. The solid line represents the total energy produced in SN 1987A (mainly from ^{56}Co). The dotted line is the predicted L_{uvoir} based on the gamma and X-ray fluxes.

4. Dust or Infrared Echo?

The preceding discussion has assumed that the spectral energy distribution from 0.30[μm] to 20[μm], with appropriate extrapolations to zero frequency should be integrated to give the bolometric luminosity of SN 1987A. Indeed, the remarkable constancy of exponential decline from day 126 to 903 argues strongly for a simple physical explanation, which has been provided by the Lick and Tokyo groups. They explain that the observed L_{uvoir} is exactly the difference between the energy produced by the decay of ~0.07[$M_\odot$] of ^{56}Co and the fraction of that energy which is not thermalized and leaks out as high-energy photons. This simple mechanism is contradicted by the observations reported by ROCHE, *et al.* [32], who found that on day 580, the 10[μm] image of SN 1987A was resolved at 1.6[″]. This is much larger than the expected size of the supernova nebula, which should be only tens of milli-arcseconds. They conclude that the thermal component must be due to a dust cloud 0.2[pc] *behind* the supernova along the line of sight, which is reradiating the energy absorbed when the SN 1987A was at maximum light. If the thermal component is due to an echo, it must be subtracted from L_{uvoir}.

There can be no doubt that the spectral energy distribution shifted from the optical and near-infrared to longward of 5[μm] starting about day 500. To study the nature of the thermal component, I have fit scaled black bodies to the far-infrared broad-band flux points starting around day 400. There was clearly an infrared excess before this date, but it did not add more than 5% to the total flux. A typical fit is shown in Fig.

2, where it can be seen that a simple scaled black body fits the thermal component extremely well. The total luminosity in the thermal component is plotted in Fig. 3, and the T_{eff} Fig. 5. The black-body evolved from 600°[K] on day 400 to 225°[K] by day 900. The ratio of the flux in the thermal component to the uvoir flux is also shown in Fig. 5 where it can be seen that by day 600, more than half of L_{uvoir} is due to the thermal source and by day 680, 80% is being emitted in the far infrared.

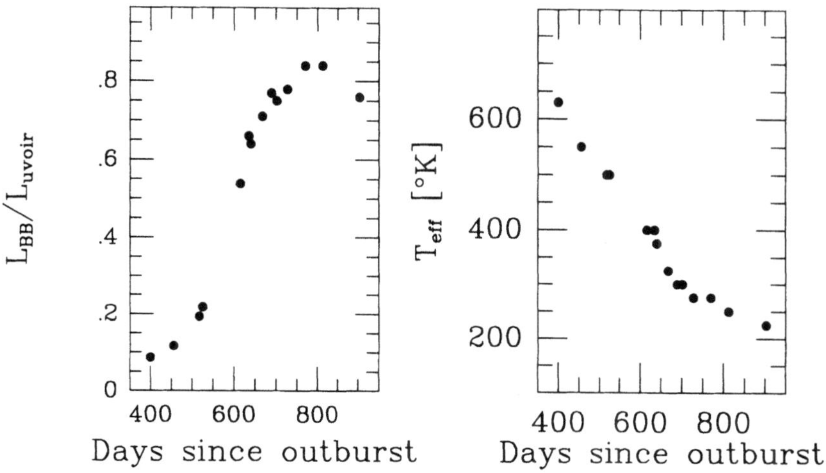

Figure 5. The ratio of the luminosity in the thermal component to L_{uvoir}(5a; left panel), and the temperature evolution of the thermal component (5b; right panel). By day 600, more than half of L_{uvoir} was due to the thermal component.

The thermal component is obviously dust; but has it formed in the SN 1987A nebula or is it an "infrared echo" as envisioned by DWEK [33]? I feel there are a number of compelling reasons to support the idea that dust has formed in the nebula. Perhaps the most compelling reason is that noted above; *the observed L_{uvoir} followed a simple exponential law from day 126 to day 903, even though the spectral energy distribution shifted dramatically to the far infrared on day 600. At the same time, the sum of observered L_{uvoir} and the high-energy luminosity predicted by the the models fitted the energy released from ~0.07[$M_\odot$] of ^{56}Co* . Another way of stating this argument is that the thermal component, as plotted in Fig. 3, after day 600, declined at the *same rate* as did L_{uvoir} before the formation of the thermal component, which in both cases, was close to the decay rate of ^{56}Co . Since the echo is due to the re-radiation of light from the time of maximum of SN 1987A, the decline rates of the echo and the radiative tail should have no connection. Also, if this thermal component were an echo, we would have to subtract off this flux from our estimate of L_{uvoir}, in which case, starting around day 500, the sum of L_{uvoir} and the high-energy flux would fall well short of the energy released by 0.07[$M_\odot$] of ^{56}Co .

Another indication that dust formed is that the optical colors began to decline more rapidly starting on day 500, at the same time the thermal component strengthened. In Fig. 6 I plot the temporal evolution of U-V (the U-B or B-V show the same trends) and the V-K colors, and the individual UBVK magnitudes. Compared to the evolution from day 125 to day 400, the UBV magnitudes began to decline more rapidly starting around day 500. The U-V color evolved monotonically from day 150 to day 800, but the V-K color reddened suddenly around day 500. This points to the formation of large (0.1[μm]) grains, since the reddening is not strongly peaked towards the ultraviolet. The optical reddening concurrent with the formation of a thermal component is seen in many novae, and has also been interpreted as dust formation (GERTZ [34]). The slight wiggle at day 550 in the U-V curve has been interpreted as a "blueing" event also associated with the dust formation (LUCY [35]). Such small wiggles could easily be due to changes in the emission-line spectrum as [FeII] began to dominate UBV. Until we understand in detail the reason why, for example, U-V fell by *4 magnitudes in 500 days*, we should be cautious about interpreting second order effects in the color curves.

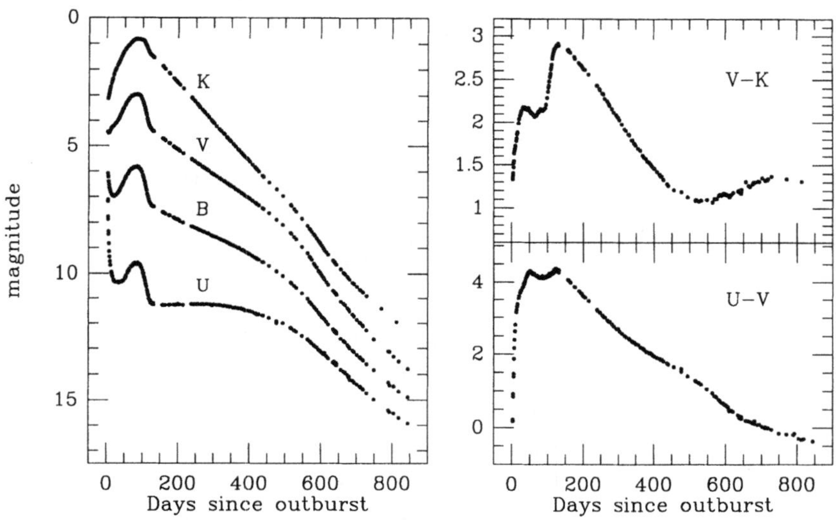

Figure 6. The color evolution of SN 1987A in UBVK (left panel; 6a) and U-V, V-K colors (right panel; 6b). Note the sudden dimming in UBV and reddening in V-K around day 600, at the time when L_{uvoir} shifted to the thermal component.

Other evidence for the formation of dust local to SN 1987A comes from spectral data where it has been shown that the emission lines suddenly blue-shifted at the time the thermal component appeared (DANZIGER, *et al.* [36]). LUCY, *et al.* [37] have interpreted this blue-shift as the formation of dust in the line-forming regions. Finally, the effective size of the thermal component is consistent with the kinematic size of the supernova nebula. Figure 7 shows the evolution of the angular radius of the black-body emission, which by day 650, is ~0.015["]. The terminal velocity of the absorption lines is about -2250[km s^{-1}] [11] which, if integrated over 700 days gives a kinematic size of the nebula of ~0.018[″] in close agreement with the size of the thermal component.

Besides the observation of a resolved source on day 580, there are some aspects of the echo model which fit the observed data well. The temperature of the thermal component is quite cool (400°[K]) compared to most regions of dust formation, such as in novae. The cool temperature is naturally explained by the simple dilution of light out to 0.2[pc]. Also, the size of the resolved source (0.36[pc]) is consistent with the minimum distance ($c\Delta t/2 = 0.2$[pc] where Δt is the difference in time between maximum luminosity and the observation) from the supernova to the ellipsoid of reflection.

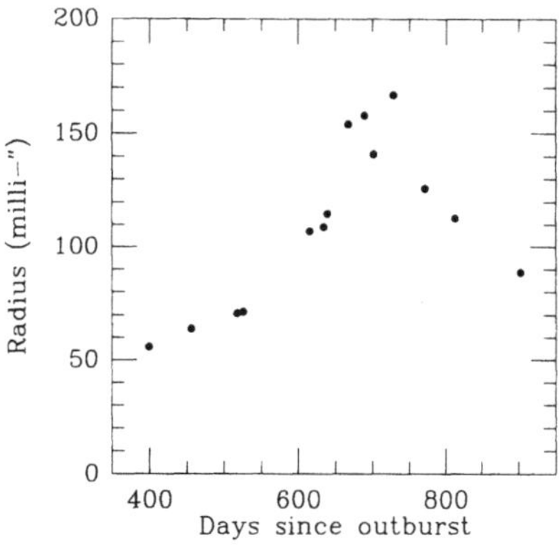

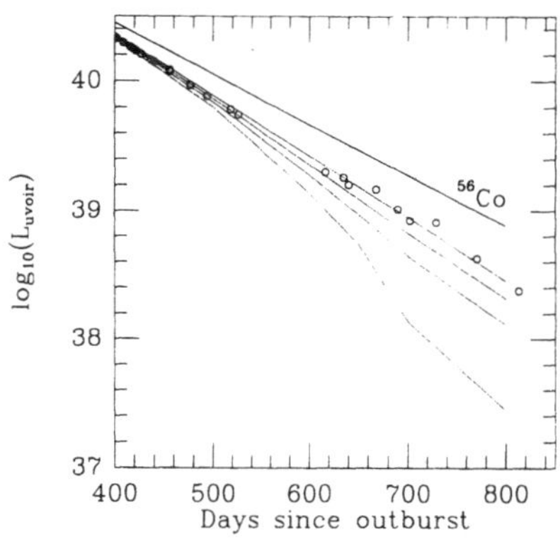

Figure 7. (left panel). The temporal evolution of the angular size of the thermal component.
Figure 8. (right panel). The predicted L_{uvoir} for different contributions to the thermal component from the infrared echo. The open circles are the observed L_{uvoir}. The solid line (at the top) is the energy released by 0.07[$M_\odot$] of ^{56}Co. The four lower lines were calculated under the assumption that 100% (bottom line), 60%, 30%, or 0% of the black-body luminosity was due to an echo.

In Fig. 8 I present a comparison of the L_{uvoir} data and the Lick models, where I have subtracted 0%, 30%, 60%, and 100% of the thermal component from the prediction of L_{uvoir} by the Lick group. No more than 30% of the thermal component can be due to an echo (or other sources) before the energy budget is not met. With the same sort of calculation (not shown), I find that the upper limit to the amount of energy I can add to L_{uvoir} before it begins to level out and thus deviate from the model predictions is $\sim 2x10^{38}$ [erg s^{-1}]. This is the upper limit to the amount of energy input from other sources such as a buried pulsar or other radioactive nuclides, provided that this energy is thermalized.

This work has been done in collaboration with Mario Hamuy, Mark Phillips, and Alistair Walker. I would like to thank Diane Wooden and Dan Lester for communicating their data prior to publication.

References

1. N. B. Suntzeff, and P. Bouchet: A. J. submitted (1990).
2. P. Bouchet, C. Gouiffes, M. M. Phillips, and N. B. Suntzeff: Astr. Ap. in preparation (1990).
3. R. M. West, *et al.*: Astr. Ap. 177, L1 (1987).
4. N. B. Suntzeff, *et al.*: Nature 334, 135 (1988).
5. M. Hamuy, *et al.*: A. J. 95, 63 (1988).
6. N. B. Suntzeff, *et al.*: A. J. 96, 1864 (1988).
7. M. Hamuy, and N.B. Suntzeff: A. J. submitted, (1990).
8. P. Bouchet, *et al.*: Astr. Ap. accepted (1989).
9. P. Bouchet, and A. Moneti: Astr. Ap. submitted (1990).
10. P. Whitelock, R. Catchpole, and M. Feast: In Supernovae, ed. S. E. Woosley (Springer-Verlag, New York, Berlin) (1990).
11. M. M. Phillips, *et al.*: A. J. 95, 1087 (1988).
12. D. Wooden, *et al.*: In Supernovae, ed. S. E. Woosley (Springer-Verlag, New York, Berlin) (1990).
13. R. P. Kirshner, *et al.*: Ap. J. 320, 602 (1987).
14. H. Mosley, *et al.*: IAU Circ. No. 4689 (1988).
15. D. Lester: private communication.
16. M. W. Feast, and A. R. Walker: Ann. Rev. Astr. Ap. 25, 345 (1987).
17. D. A. VandenBerg, and H. E. Poll: preprint (1989).
18. E. J. Wampler, *et al.*: preprint (1989).
19. I. J. Danziger, *et al.*: Astr. Ap. 177, L13 (1987).
20. N. R. Walborn, *et al.*: Astr. Ap. 219, 229 (1989).
21. N. Panagia, *et al.*: Astr. Ap. 177, L25 (1987).
22. E. L. Fitzpatrick: private communication (1989).
23. A. R. Walker, and N. B. Suntzeff: P. A. S. P. accepted (1990).
24. S. E. Woosley: Ap. J. 330, 218 (1988).
25. M. M. Phillips, and S. R. Heathcote: P. A. S. P. 101, 137 (1989).
26. R. W. Hanuschik, and J. Dachs: Astr. Ap. 182, L29 (1987).
27. M. M. Phillips: In Supernova 1987A in the Large Magellanic Cloud , ed. by M. Kafatos, and A. G. Michalitsianos, (Springer-Verlag, New York, Berlin 1988) (p. 248).
28. P. A. Whitelock, *et al.*: M. N. R. A. S. 234, 5P (1988).
29. P. A. Pinto, *et al.*: Ap. J. Lett. 331, L101 (1988).
30. S. Kugamai, *et al.*: Ap. J. in press (1989).
31. P. A. Whitelock, *et al.*: M. N. R. A. S. submitted (1989).
32. P. F. Roche, *et al.*: Nature 331, 533.
33. E. Dwek: Ap. J. 274, 175 (1988).
34. R. D. Gertz: Ann. Rev. Astr. Ap. 26, 377 (1988).
35. L. B. Lucy:In Supernovae, ed. S. E. Woosley (Springer-Verlag, New York, Berlin) (1990).
36. I. J. Danziger, *et al.*: IAU Circ. No.4746 (1989).
37. L. B. Lucy, *et al.*: In Supernova 1987A in the Large Magellanic Cloud , ed. by M. Kafatos, and A. G. Michalitsianos, (Springer-Verlag, New York, Berlin 1988) (p. 323).

SN 1987A Light Curves

Patricia Whitelock, Robin Catchpole, &
Michael Feast

1. Introduction

The observation of the light curves of SN1987A over a wide range of wavelengths is perhaps the most straightforward way of studying the energy generation and dissipation within the supernova. The measurements discussed here were made largely at the SAAO (MENZIES *et al.* [1], CATCHPOLE *et al.* [2, 3, 4], WHITELOCK *et al.* [5, 6] and the AAO (AITKEN and ROCHE *private communication*). These observations provide a coherent body of data covering the period from day 1.5 to 880, and comprise broad–band optical and infrared photometry. To the best of our knowledge photometry obtained in Chile and Australia provides very similar results if account is taken of the different filter transmissions and procedures for determining bolometric magnitudes in use at the various observatories.

While there is some interest in examining the light curves as measured through individual filters it is the bolometric light curve which is of particular importance in understanding the underlying physical processes and for comparison with the models. Although strictly speaking the term "bolometric luminosity" should include all contributions from γ–rays through radio waves, we will follow the common practice of referring to the ultraviolet, optical and infrared flux (0.35 to 5 μm)as bolometric and discuss separately contributions from other spectral regions when they are important to the total energy balance.

2. Bolometric Light Curve

The calculation of the bolometric luminosity from broad–band photometry has been discussed in detail elsewhere [1–6] and the resulting light curve is shown in Fig 1. The curve can be seen as going through a number of distinct phases which are discussed in turn below.

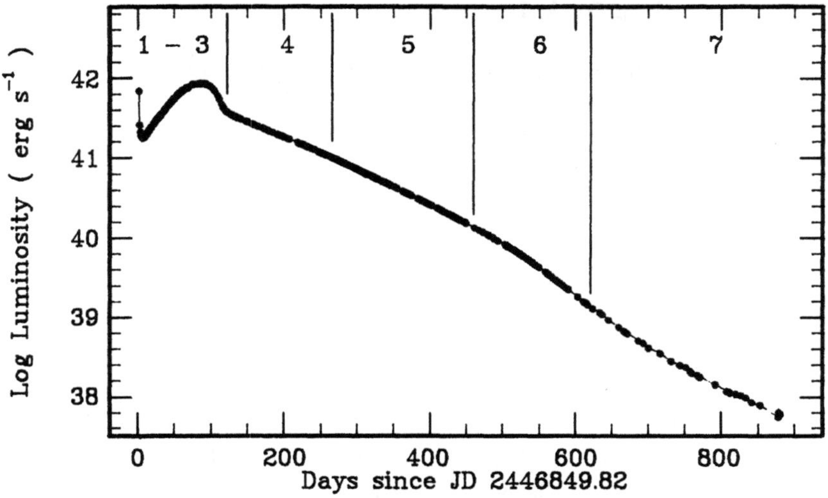

Figure 1. Bolometric light curve from U to M photometry [6].

2.1 Early Stages (phases 1 to 3)

The first 200 days of the curve are illustrated in more detail in Fig 2 (see also CATCHPOLE [7]). Phase 1 is the initial decline in brightness following the ultraviolet flash and lasting until day 7. This behaviour is the result of a very rapidly declining photospheric temperature. As photospheric expansion compensates for the temperature drop the luminosity rises and phase 2 is entered. According to the theory (e.g. ARNETT *et al.* [8]) the first, approximately four weeks of the light curve are powered by energy deposited in the envelope by the initial shock wave. Subsequently the release of trapped energy from the radioactive decay of ^{56}Co to ^{56}Fe dominates, and the luminosity reaches a maximum around day 88. Phase 3 is the steep drop preceding the linear decline phase. The overall form of the phase 2 and 3 maximum (plateau) is strongly influenced by hydrogen recombination as trapped ^{56}Co decay energy diffuses out. Phase 4 is entered when the energy generated by ^{56}Co decay is balanced by that radiated and an exponential decline in luminosity results.

The rest of the light curve has a number of interesting features, which are discussed below, but it is worth noting at this stage that the radioactive decay of cobalt remains the sole significant source of energy up to day 850 and possibly beyond.

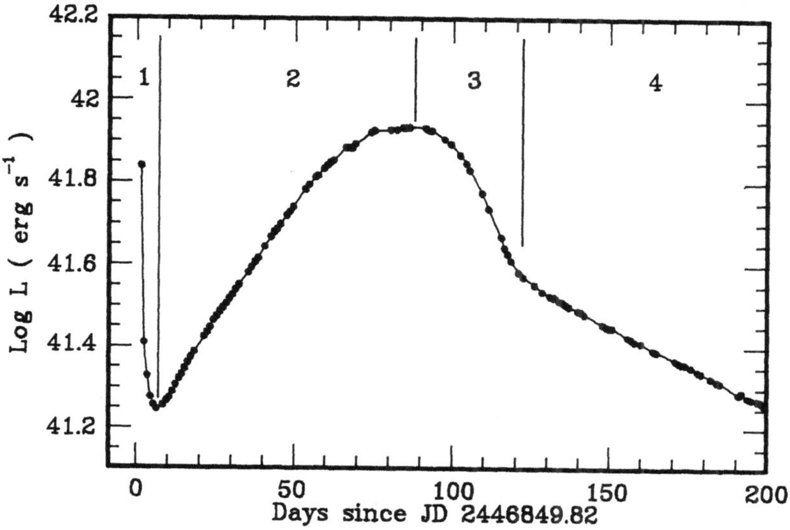

Figure 2. The first 200 days of Fig 1.

2.2 Radioactive Decline (phases 4 and 5)

The SAAO data indicate that between day 147 and 260 (phase 4) the decay rate was closely linear (FEAST [9], [3]). If the luminosity is calculated by spline fitting U to M photometry after correcting for an interstellar extinction of A_V = 0.6 mag then an e–folding time of 110.4 ± 0.6 day is found for phase 4. This is very close to the 111.26 day e–folding time for the radioactive decay of ^{56}Co, and provided direct evidence that type II supernovae are powered by this type of radioactive decay. It was during phase 4 that X– and γ–rays were first detected from SN1987A, indicating that some degree of mixing of the ejecta must have occurred (SUNYAEV *et al.* [10], DOTANI *et al.* [11], MATZ *et al.* [12]). The measured high energy fluxes were very weak at this stage and there is considerable uncertainty associated with the flux values. Nevertheless if we use the published models (e.g. KUMAGAI *et al.* [13]) to estimate the total high energy contribution and re–examine the e–folding times then values between 112 and 113 day are found. Thus the good agreement with the expectations from the decay of ^{56}Co is retained.

From the linear decline of the logarithmic luminosity it was possible to establish that 0.08 $M_{\odot}$ of ^{56}Ni was produced during the supernova explosion. This figure is important for a variety of reasons, but in the context of this discussion its significance is that it establishes a lower limit to the total luminosity at later times. This minimum luminosity is provided by the decay of ^{56}Co. Excesses above this limit can come from many sources, e.g. the radioactive decay of other isotopes, light echoes, direct input from a neutron star etc.; but a deficit from the ^{56}Co limit can only arise if the SN emission is non–isotopic so that the radiation impinging on the Earth is less than that emitted in some other direction.

After day 460 the decline rate of the bolometric luminosity increased (phase 5) and the energy emitted from U to M fell below that predicted for ^{56}Co decay. Various supernovae theories predicted that as the ejecta expanded, and the optical depth to γ–rays decreased, an increasing fraction of the radioactive decay energy would be emitted as γ–rays and an ever smaller fraction would be downgraded into what we are calling bolometric luminosity. That γ–ray emission was detected considerably before these theories predicted is generally interpreted as the effects of mixing and clumping of the ejecta. It has been shown that the difference between the ^{56}Co decline rate and the measured bolometric one was compensated for by an increasing flux of high energy photons [5].

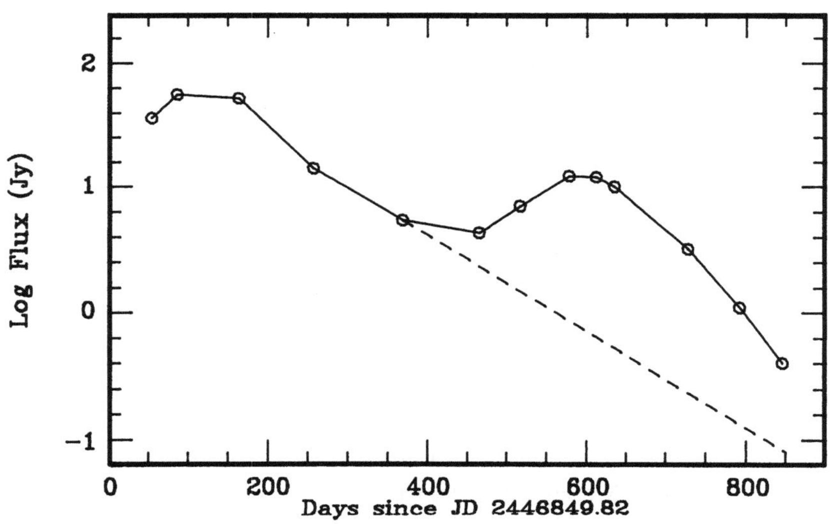

Figure 3. 10 μm light curve from AAO and KAO data, references in [6].

2.3 Dust Formation (phases 6 and 7)

From approximately day 500 to 600 (phase 6) the bolometric light curve went through a period of much steeper decline than during phase 5. At first it seemed likely that this was a continuation of the earlier trend and represented an increase in the release of high energy radiation. However, adding the X- and γ-rays to the bolometric luminosity calculated from U to M photometry left a clear deficit from the predictions for radioactive decay of ^{56}Co [6, Fig 6]. It was therefore necessary to look for this "lost energy".

We were aware of a significant flux at wavelengths beyond those dealt with in the SAAO monitoring program, i.e. in the mid- and far-infrared. Figure 3 shows the light curve for 10 μm based on AAO and KAO observations, and it is clear that an excess developed round about day 500. A detailed discussion of this excess is given by others (e.g. LUCY [14], DWEK [15]) and it suffices to say here that it is clearly produced by dust emission. The critical question for the energy balance concerns whether that emission originated from material condensing within the ejecta or was a light echo from pre-existing grains. If the dust lies within the ejecta then its energy must be added to the U to M flux to derive the true bolometric luminosity. If the infrared flux is an echo then the dust is reprocessing flux emitted from the SN at early times and can be neglected from considerations of the current luminosity. Measurements by ROCHE *et al.* [16] seemed to provide a clear answer: the emission was extended with an FWHM of 1.5 $\pm$ 0.2 arcsec, and therefore must be a light echo. For this reason it was not initially included in our calculations of the total luminosity. However, having failed to find sufficient flux from other spectral regions and following the suggestion by DANZIGER *et al.* [17] that line profile changes in the optical spectra could be understood if dust had formed within the ejecta it seemed worth examining the dust contribution in more detail.

The total luminosity from the mid- and far-infrared can be estimated with the simplifying assumption that the dust particles emit as blackbodies (see also WOODEN [18]) at the colour temperatures derived from the infrared spectra. Figure 4 shows the calculated dust luminosity as a function of time. Adding this luminosity to the previously calculated U to M luminosity (Fig 1) gives the dotted line in Fig 5. The steep decline disappears and the light curve

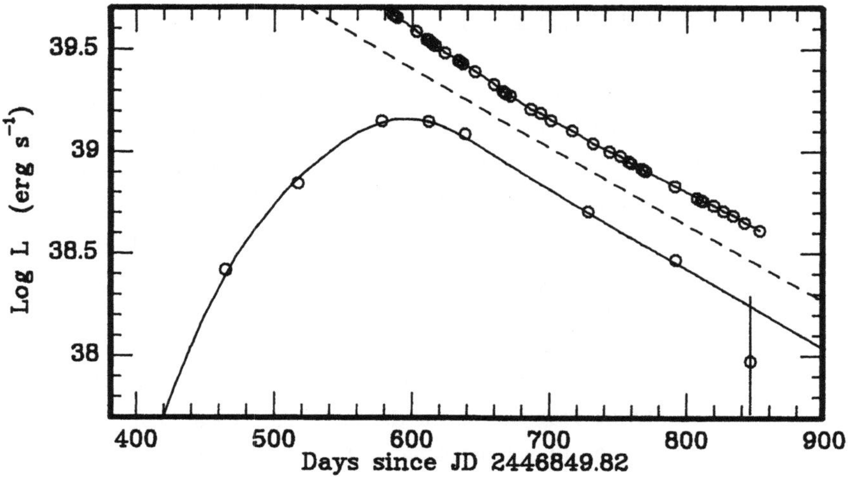

Figure 4. Total dust luminosity (lower curve) and U to M plus X– and γ–ray luminosity (upper curve). The dotted line has the slope for decay of ^{56}Co and ^{57}Co [6].

continues the trend of phase 5. Adding the modelled high energy flux to this curve results in remarkably good agreement with the cobalt radioactive decay energy (Fig 5). The contributions from both ^{56}Co and ^{57}Co are included in this illustration as the contribution from ^{57}Co is significant towards the end of the period under consideration (see [6] for more detail).

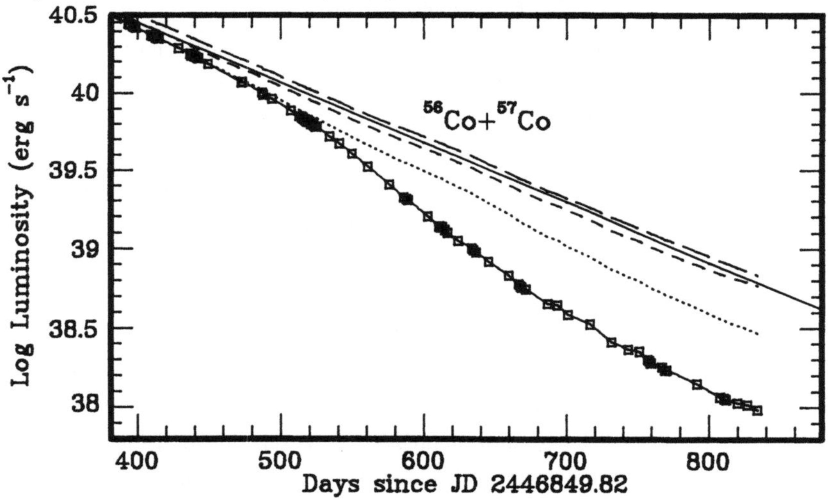

Figure 5. U to M luminosity (squares), plus dust (dotted line), U to M plus dust plus X– and γ–rays from 14E1 (short dashes) and 10HMM (long dashes) [6].

Returning to the light curve (Fig 1) it is clear that following the steep drop (phase 6) which indicated dust formation there was a phase of less rapid decline, labelled as phase 7. It was this slowing down of the decline rate from about day 620 which was responsible for the spate of suggestions that the long awaited effects of the neutron star might be manifesting themselves in the light curve. What was not understood at the time was that the previous steep drop was due to dust extinction and that the slackening off represented a return to what might be regarded as the more normal decay rate.

It is now clear from the bolometric light curve that after about 4 weeks the supernova was powered by the radioactive decay of cobalt (^{56}Co + ^{57}Co), with no significant input from any other source, up to at least day 850. In coming to this conclusion it is necessary to include the dust luminosity within the energy budget. It would therefore seem that the dust must have condensed within the ejecta. This picture does not explain the extended nature of the 10 μm source [16]. It is however, possible that more precise modelling of the dust properties will show that some fraction of the infrared flux originated from an extended light echo.

3. The Dust

The final datum on Fig 4, that for day 847, was obtained by AITKEN *et al.* (*private communication*) under poor conditions and the error bars represent their estimates of the uncertainty. A remarkable characteristic of the dust emission is that between day 600 and at least 800 (depending on what we make of this last point) it decays at the same rate as radioactive cobalt. This can be seen in Fig 4 by comparison with the dashed line which has the slope expected for cobalt decay. At the same time the rest of the SN emission, i.e. the near–IR through γ–rays, also parallels the cobalt decay. We know from the IR spectra that the colour temperature of the dust decreased from about 400 K on day 600 to around 200 K on day 800. It is therefore simple to show that in order to reproduce the dust light curve in Fig 4 the emitting volume must have increased. In fact the radius of the assumed blackbody must have increased linearly with time, as would be expected from uniform expansion. It thus appears that after day 600 the dust maintains a constant fraction of the total emitting volume. An alternative illustration of this is shown in Fig 6 which compares the percentage of the total luminosity emitted by the dust and by everything else. It can be seen that the dust maintains an approximately constant 30% of the total emission after day 600.

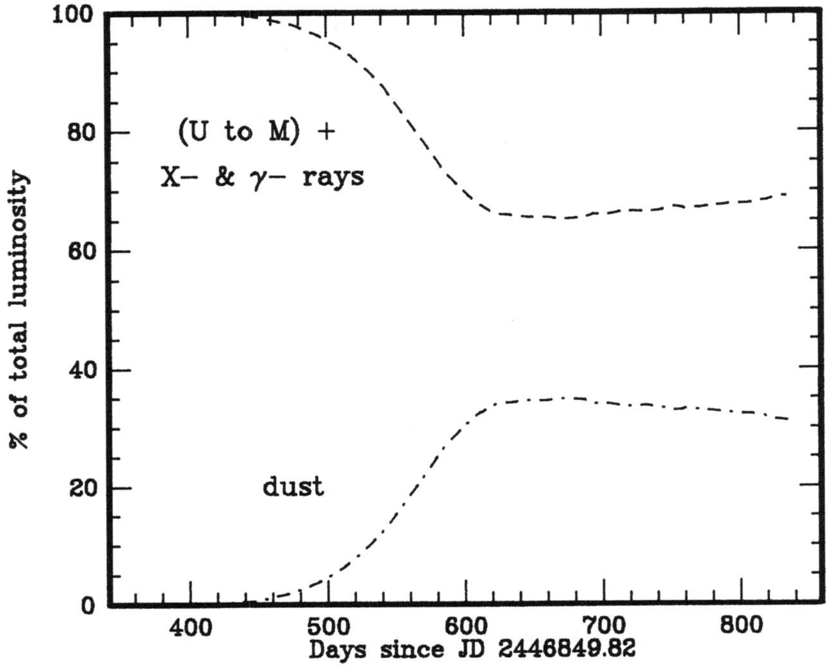

Figure 6. The apparent decoupling of dust emission from the shorter wavelength luminosity after day 600.

It is not easy to explain the above result if the dust is optically thin after day 600. However a simple interpretation is possible if optically thick dust is formed in clumped ejecta. Let us suppose that the dust forms in or around the high density clumps which were still optically thick to X– and γ–rays. As the dust thickens it becomes optically thick to visual, ultraviolet and near–IR radiation. Thus within the clumps all of the radioactive decay energy is reprocessed into thermal dust emission. This would result in the decoupling of the dust emission evident in Fig 4 and 6. It might also go some way towards explaining the colour changes during phase 6 which are discussed below.

4. Colours

Any detailed discussion of the colour changes seen in SN1987A should obviously be made with reference to the spectrum. However colour curves, such as those illustrated in Fig 7 and 8 do show some interesting characteristics. First, certain colours are completely dominated for long periods of time by individual spectral features. The most obvious example of such a feature is the CO fundamental vibration–rotation band which falls in the M (4.8 μm) window. Thus K–M monitors the changing strength of CO emission except for a short period between days 500 and 600 when dust also made a significant contribution to the M flux. The rise in K–L between day 400 and 600 was due to the growth of the [NiI] line in the L band at 3.12 μm. Similarly V–I_c and V–R_c are strongly influenced by the CaII and Hα lines respectively.

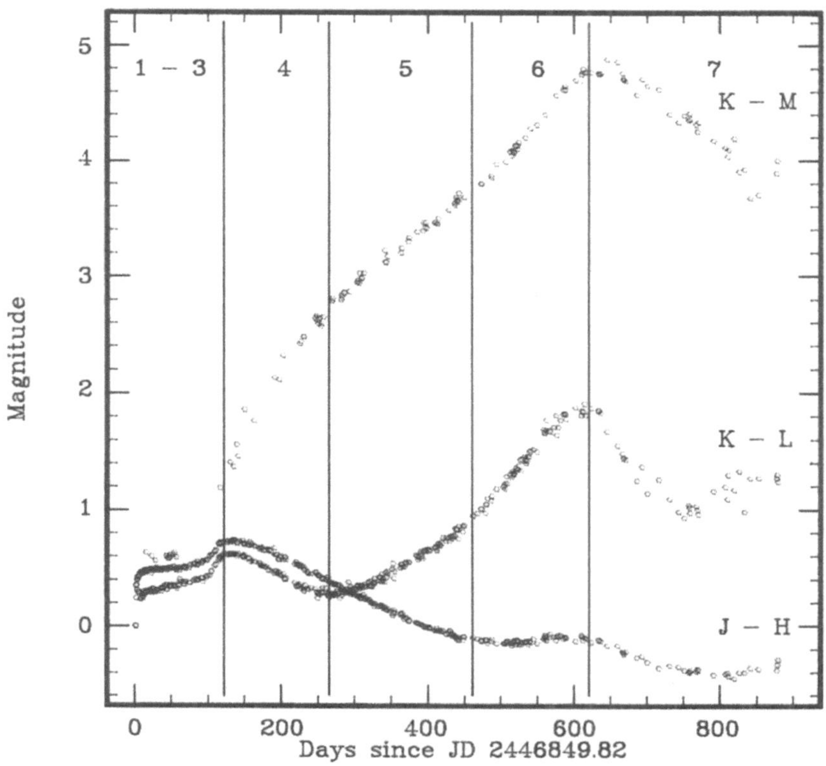

Figure 7. The infrared colours.

Another point of potential interest is that some kind of colour change has been associated with the onset of each phase of the bolometric light curve, which can be seen by comparing Fig 7 and 8 with Fig 1. While we do not yet know how typical this is of the general development of type II supernovae, it seems possible that when only limited wavelength coverage is available for a particular supernova, a close examination of the colours may provide some clue as to the behaviour of the bolometric light curve.

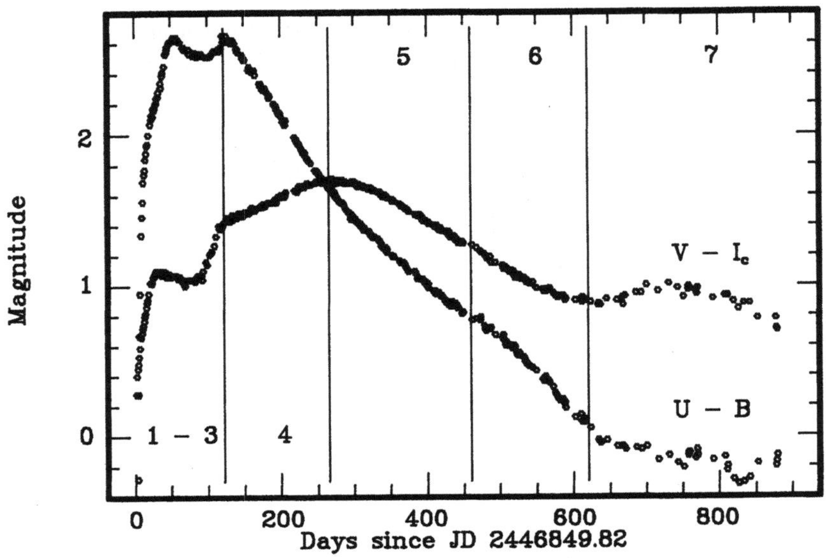

Figure 8. The optical colours. Note in particular the increase in slope of (U–B) around day 500, coinciding with the dust formation phase.

Note that the only change in colour trend accompanying the dust formation of phase 6 was an increase in the rate of blueing of U–B and B–V. The absence of any reddening at this stage is possibly why the dust formation went unnoted for some time. It might be possible to understand the blueing with the optically thick clump model outlined above. Prior to phase 6 the observed light must have originated from regions of both high and low density. When the dust formed the high density regions were preferentially obscured. At this stage the optical light was originating from, on average, deeper in the ejecta. Thus there was a sudden acceleration of the blueing which had been occurring gradually for some time. There is a very obvious change in all colours around day 600 as phase 7 is entered.

Acknowledgements

We are extremely grateful to Dave Aitken and Pat Roche for communicating their observations in advance of publication. This paper reviews observations by a large number of people [1–6], and we particularly thank our colleagues at SAAO for their enthusiasm.

References

1. J.W. Menzies *et al.*: MNRAS, 227, 39P (1987).
2. R.M. Catchpole *et al.*: MNRAS, 229, 15P (1987).
3. R.M. Catchpole *et al.*: MNRAS, 231, 75P (1988).
4. R.M. Catchpole *et al.*: MNRAS, 237, 55P (1989)
5. P.A. Whitelock *et al.*: MNRAS, 234, 5P (1988).
6. P.A. Whitelock *et al.*: MNRAS, in press (1989).
7. R.M. Catchpole: In Highlights of Astronomy vol. 8, p 185, ed. by D. McNally (IAU 1989).
8. W.D. Arnett, J.D. Bahcall, R.P. Kirshner & S.E. Woosley: Ann. Rev. A.A., 27, 629 (1989).
9. M.W. Feast: In SN1987A in the LMC p 51, ed. by M. Kafatos & A.G. Michalitsianos (Cambridge Univ. press 1988).
10. R. Sunyaev *et al.*: Nature, 330, 227 (1987).
11. T. Dotani *et al.*: Nature, 330, 230 (1987).
12. S.M. Matz *et al.*: Nature, 331, 416 (1988).
13. S Kumagai *et al.*: Ap.J. 345, in press (1989).
14. L.B. Lucy: In this volume (1989).
15. E. Dwek: In this volume (1989).
16. P.F. Roche, D.K. Aitken, C.H. Smith & S.D. James: Nature, 337, 533 (1989)
17. I.J. Danziger, C. Gouiffes, P. Bouchet & L.B. Lucy: IAU Circ. no 4746 (1989).
18. D. Wooden: In this volume (1989).

Optical Spectrophotometry of Supernova 1987A: The First 133 Days

Reinhard W. Hanuschik

1. Introduction

Optical spectrophotometry constitutes the most important kind of data from supernovae. At effective temperatures in the range of 5 - 10000 Kelvin, the expanding envelope emits the bulk of electromagnetic energy in the optical and near-infrared range. A record of this energy therefore provides key information about the physical conditions at the photosphere (temperature, radius, and density), as well as in the layers above it (geometry, composition, and ionization state from the line spectrum and from detailed line profiles). Thus, optical spectrophotometry provides an essential key for the understanding of the physical processes in the debris from the fatal explosion.

However, achieving spectrophotometric data, i.e. flux-calibrated spectroscopic information at photometric accuracy, is not an easy task, especially when a detailed record is desired. Therefore, only very few supernovae could be studied spectrophotometrically in some detail before 1987. A nearby supernova like SN 1987A offers the unique opportunity to gain such data at unprecedented detail and time coverage, provided the following is available: (1) the optimum telescope and detector, (2) a well-calibrated system of standard stars, and (3) last but not least: lots of observing time.

The brightest extragalactic supernova in historical times, Supernova 1987A in the LMC occurred just in time to face an armada of well-equipped telescopes and eager observers. Spectrophotometric work was undertaken at virtually any larger southern observatory (for a list see [1]). Five of these campaigns were at photoelectric accuracy, namely the campaigns at AAO, at the Bochum telescope, at CTIO, at ESO, and at SAAO. By now, three of these observatories have published atlases of their data: the Bochum group [2, 3], CTIO [4], and SAAO [5 and references therein]. Only the first two groups have performed truly spectrophotometric work, i.e. direct calibration by spectrophotometric standards instead by broad-band photometry.

2. The Bochum Telescope Campaign on SN 1987A

The 0.61m telescope of the Ruhr University of Bochum is located at the ESO site at La Silla. It is equipped with a rapid spectrum scanner (for a description see [6]) and a single-channel photomultiplier, thus combining low efficiency (rather essential for a naked-eye object like SN 1987A) with high linearity. The entrance slit can be opened as wide as 20 - 30", essential for an accurate flux record. A system of bright southern standard stars, connected to α Lyrae and the equatorial standard stars of HAYES [7], has been constructed by TÜG [8,9]. For SN 1987A, the Bochum telescope thus was the right telescope at the right time.

The scanner spectrograph was used until day 315 (1988 Jan. 5); we further continued observations with a different detector (OMA, a multi-channel detector), until on day 450 the signal from the supernova finally faded below reasonable level at our telescope. At present, data for the first 315 days are fully reduced. They cover the wavelength range 3200 – 8700 Å (at some dates the range starts at 3900 Å) at 10 Å, after day 200 at 15 Å resolution. A number of wavelength bands enclosing interesting line features were measured at higher (3 – 5 Å) resolution and flux-calibrated. Absolute calibration was performed by comparison to several flux standards from the TÜG [8,9] list. Accuracy of the flux calibration is $\approx$3 %, and slightly worse in the UV and near-IR range. The average signal-to-noise ratio is 100 – 200. For days 2 - 20, we had to employ Walraven and Cousins photometry to achieve absolute flux calibration as the entrance slit at that epoch was chosen too narrow.

3. Evolution of the Continuum Radiation

A multiple scan plot of all data until day 133 is shown in Fig. 1. Only every second day is shown, slightly shifted by a constant amount. A more detailed overview is depicted in Fig. 2.

The evolution of the continuum and the line spectrum of SN 1987A occurred initially very fast, subsequently much slower. The UV flux (at 3500 – 4000 Å) decreased by almost a factor of 100 within the first 10 days, while the flux longwards of 5000 Å continuously increased towards the rather flat brightness maximum which occurred on day 82 ± 2 (day 87 at shorter wavelengths).

Integrating the optical fluxes yields the *optical* bolometric lightcurve, L_{opt} (Fig. 3a), which differs from the total bolometric lightcurve mainly by neglecting the IR and UV flux. Effective temperatures have been determined from our data by black-body fitting for the first 5 days and by determination of the approximate continuum maximum thereafter (Fig. 3c). Photospheric radii, r_{ph}, have been calculated from these temperatures and the SAAO [13] bolometric lightcurve (Fig. 3b).

The optical lightcurve demonstrates that the flux evolution during the first 133 days can quite naturally be divided into four phases:

(1) The phase of *rapid cooling* (lasting from day 0 to day 5) is characterized by the adiabatic cooling of the envelope. The UV radiation, as a result, falls off rapidly. The spectrum shows hydrogen and helium lines, Ca II λ8600 starts to develop. (2) Phase II, the *brightening phase*, is defined for days 5 – 82 and exhibits a slow and steady brightness increase until maximum. The temperature stays at an almost constant value, T $\approx$ 5100 K. A rich absorption and P Cygni-type spectrum

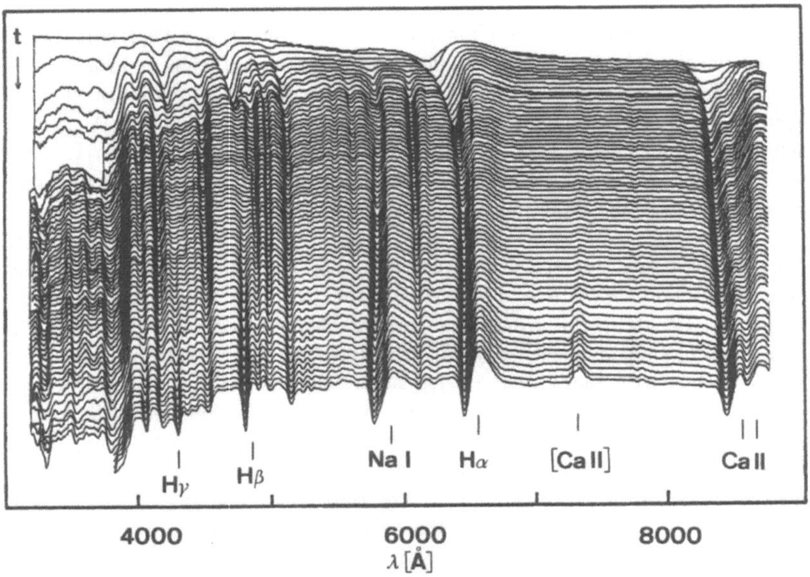

Figure 1: Evolution of the optical flux in SN 1987A. Flux is shown on a logarithmic scale for every second day, starting with day 2 (= 1987 Feb. 25) on top. Spectra for days without measurements have been interpolated. This plot is best suited for following the *evolution of spectral lines*. The *lightcurve* can be better traced in intensity-coded plots as in [2,3,4,31]

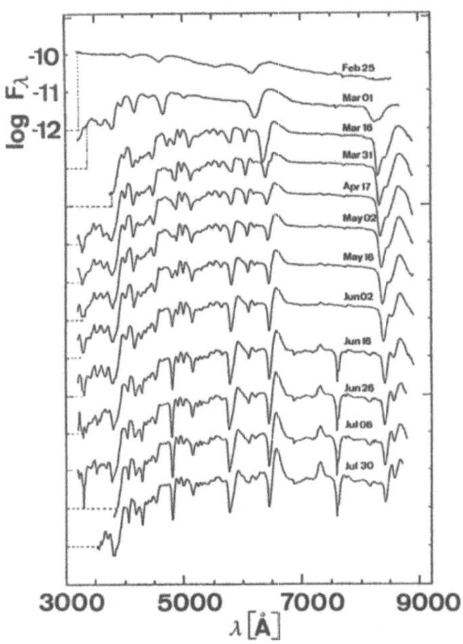

Figure 2: Selected spectra for SN 1987A. The logarithmic flux scale in erg s^{-1} cm^{-2} $Å^{-1}$ is for the first Bochum spectrum (Feb. 25.0, 1987). All other spectra are shifted by $\Delta \log F_\lambda$ = 1.0 each; the broken lines indicate log F_λ = -12

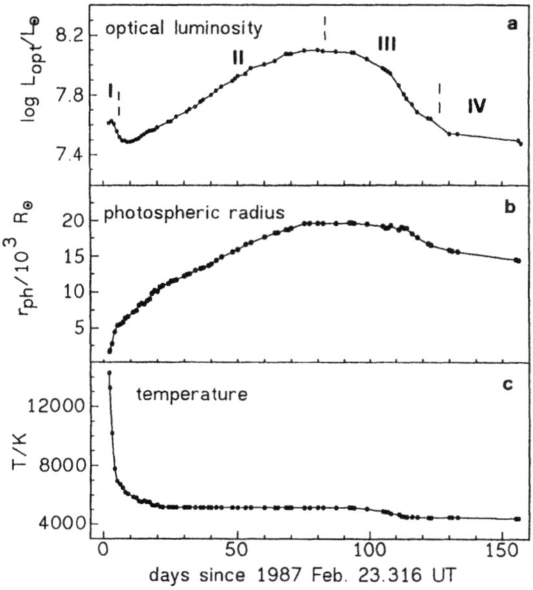

Figure 3a: The optical bolometric lightcurve for SN 1987A derived from the Bochum fluxes. Integration of fluxes has been carried out between 3200 Å and 8700 Å. A reddening of 0.16 mag and a distance of 50 kpc have been assumed.

Figure 3b: Photospheric radius in units of 10^3 $R_\odot$ as calulated from the temperatures in Figure 3c and the SAAO bolometric lightcurve [13].

Figure 3c: Effective temperatures for SN 1987A, derived from the Bochum flux record by black-body fitting (days 2 - 5) and by estimation of the Wien flux maximum thereafter

develops. Forbidden emission of [Ca II] starts around day 36 [ref. 3]. (3) The *maximum phase*, phase III (days 82 – 126), exhibits a steep continuum decline, a slightly decreasing temperature and the transition into a nebular-type spectrum. (4) Finally, phase IV as the *exponential decay phase* shows a less steep, exponential flux decline and a well-developed nebular spectrum with strong allowed and forbidden emission lines, among them [O I], and fading absorption troughs.

4. Comparison of Spectrophotometry

As optical spectrophotometry provides key data for any envelope modelling in supernovae, it is worthwile to compare the published fluxes for SN 1987A. As is visible in Fig. 4a, the *integral fluxes* $\int F_\lambda d\lambda$ of the Bochum and the CTIO record show excellent agreement. Some systematic differences, however, show up in the red wavelength range (Fig. 4b) around and after the maximum (CTIO excess by 5 – 10 %), and in the blue after about day 20 (CTIO deficit by ≥10%). A revision of the calibration of δ Doradus, the primary comparison star for the CTIO fluxes, is presently undertaken at CTIO [10]. The revised flux now shows good agreement with the Bochum flux for this star, so that there is confidence that most of the observed blue discrepancy will disappear when a careful recalibration of the CTIO fluxes is performed. The red discrepanies are probably due to the fact that here the CTIO response curves are near to their red end and therefore less reliable than the Bochum calibration.

The SAAO fluxes seem to deviate much more severely from both flux sets. A comparison of the (arbitrarily chosen) Apr. 4 = day 40 spectrum of SN 1987A from all three observatories shows that the SAAO flux deviates by as much as 30 % at certain wavelengths from both Bochum and CTIO calibrations. The cause for this is clearly the B and V filter photometric calibration of these fluxes with rather uncertain transformation coefficients.

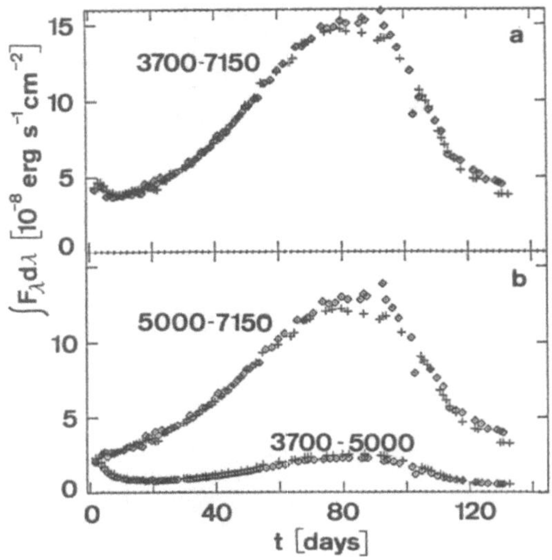

Figure 4: Comparison of Bochum and CTIO fluxes.

Figure 4a: Integral optical fluxes (λλ3700 – 7150) from the Bochum (plusses) and the CTIO (diamonds) atlases.

Figure 4b: Same as Figure 4a, for the red and blue wavelength ranges separately

5. The Bolometric Lightcurve

All presently existing bolometric lightcurves for SN 1987A [11 – 13] are derived from fits to optical and infrared photometry, thus mainly suffering from the notorious uncertainties in the transformation of broad-band into monochromatic fluxes. The optimal solution to this problem would be the integration of optical and infrared (and, in the first few days, UV) spectrophotometry. As a first step towards this final goal, the Bochum fluxes were combined with ESO infrared photometry [14] for a number of selected days, thus covering the wavelength range 0.32 μ – 20 μ.

Figure 5 shows our results calculated with two values of E_{B-V}: with 0.20, the value chosen by the SAAO observers, and with 0.15, corresponding to the CTIO choice (a true distance modulus of 18.5, or D = 50 kpc, is adopted in either case). The agreement of the SAAO and the Bochum lightcurve is excellent until maximum light. After day 100 when the relative strength of emission lines increases dramatically, photometric lightcurves become more clearly inadequate. The CTIO lightcurve seems to underestimate the total flux from SN 1987A: only half of the downshift towards the SAAO and the Bochum lightcurves is due to the lower E_{B-V}. A more thorough interpretation of Fig. 5 will be undertaken when a final Bochum lightcurve has been finished.

6. The Line Spectrum

A rich line spectrum developed very quickly, one of the many surprising facts in SN 1987A. Line identifications can be found, e.g., in [15] and [2]. A very helpful tool for following and identifying individual features is the calculation of ΔF/F, essentially equivalent to the (normalized) first spectral derivative of the fluxes

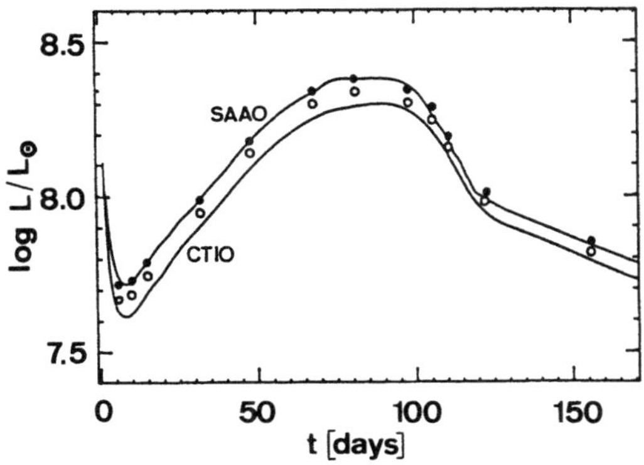

Figure 5: The Bochum bolometric lightcurve, obtained by integrating the Bochum fluxes and IR photometry from ESO [14]. A distance of 50 kpc is assumed; calculations were carried out for E_{B-V} = 0.20 (filled circles) and 0.15 (open circles). The SAAO and CTIO lightcurves are shown as solid lines

$F(\lambda,t)$, a representation which is depicted in Fig. 6 of HANUSCHIK et al. [3].

Absorption trough velocities, as documented by HANUSCHIK and DACHS [16], PHILLIPS et al. [4], and HANUSCHIK & SCHMIDT-KALER [17], show a rather complex evolution, characterized by: (1) a steep decrease within the first ≈20 days and a rather flat fall-off thereafter, and (2) a stratification indicated by the fact that Balmer line, as well as Ca II λ8600, velocities are larger by a factor of two or more than those of "weaker" lines, e.g. of Ba II and Fe II. The Na I-D line complicates this picture by the fact that its trough velocity *increases* after about day 30 (as does Hα after day ≈85). Finally, fine-structure in some of the lines (see next chapter) makes the definition of the trough minimum rather doubtful between days ≈20 and 70.

The picture of line opacity decreasing by geometrical dilution and thus revealing *peu à peu* layers of smaller radial velocity is obviously a good first-order approximation. However, detailed spectral line information as available for the first time from SN 1987A, reveals that this simple picture needs further refinement e.g. by NLTE calculations of the ionization equilibrium.

Both line and photospheric velocities show distinct phases of evolution. Fig. 6 demonstrates that the velocity v_{ph} of the matter at the photosphere (calculated from r_{ph} in Fig. 3b assuming homologous expansion and t_o = Feb. 23.316 UT, 1987) shows three distinct phases within the first 111 days. Assuming a power law, $v_{ph} \sim t^n$, the power law exponents *n* are +0.16, -0.49 ± 0.02 and -0.93 ± 0.07 [17]. In electron-scattering dominated photospheres, this power law is related to the electron density law at the photosphere,

$$v_{ph} \sim t^{2/(1-\alpha)} \quad , \tag{1}$$

(derived from BRANCH [18]), with the power law exponent α for the "initial" density distribution, when homologous expansion started. Consequently, $\alpha = 5.1\pm0.2$ before day 73 and 3.2 ± 0.2 thereafter. Before day 5, the density law must have been very steep [18,19].

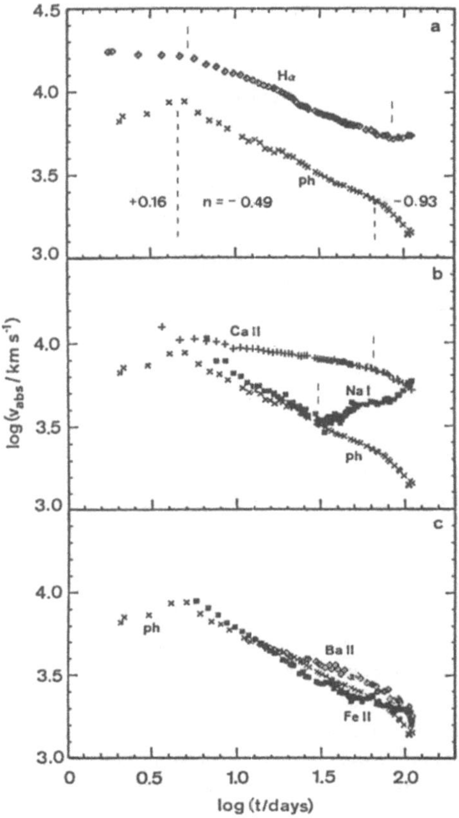

Figure 6: Logarithms of spectroscopic and photospheric velocities vs. logarithm of time.

Figure 6a: Hα and photospheric (ph) velocities. Phase transition in the evolution are marked by broken lines

Figure 6b: Ca II λ8542 and Na I-D lines.

Figure 6c: Ba II λ6141 and Fe II λ5018

Interestingly enough, some of the absorption lines also show distinct velocity phases, i.e. straight lines in the log v — log t plot of Fig. 6. This is especially remarkable because relating trough velocities to geometric radii and physical conditions there (namely density and ionization state) is to some degree model-dependent. These conditions are obviously constant over longer periods and then change fairly quickly (cf. Hα in Fig. 6a, Ca II and Na I-D in Fig. 6b). This supports the suspicion that the trough velocity evolution is mainly governed by the local density law.

A comparison of effective radii of maximum absorption, $r_{abs} = v_{abs}\,(t-t_0)$, for 9 optical lines, selected for their weakness and for the high certainty of their identification, and of the photospheric radius yields a Baade-Wesselink distance of D_{BW} = 50.0 kpc (distance modulus 18.5) for SN 1987A, if spherical symmetry, black-body radiation field and a reddening of 0.16 mag are assumed [20]. Correcting for ellipticity with a true axis ratio of $\mu = 0.7$ on day 20 which is implied by combining speckle and polarimetric data [21, 22], yields

$$D_{ell} = 46.9\ \varepsilon^{1/2} \pm 1.3\ \text{kpc},\ (m-M)_0 = 18.36 + 2.5\log\varepsilon. \qquad (2)$$

Here is ε the model-dependent ratio of the true emergent flux, πF, and the black-body flux at temperature T, σT^4. For a discussion of ε see, e.g., BRANCH [32].

7. Fine-Structure in Spectral Lines: the Bochum Event

One of the most interesting spectroscopic surprises during the first 130 days of SN 1987A was the detection of spectroscopic fine-structure in Hα and other spectral lines, the so-called Bochum event [16, 23 - 26]. It was followed up by the Bochum observers in virtually each night at full resolution (3 Å) and high signal-to-noise ratio (> 100). Fig. 7 shows detailed plots of the Hα, Hβ and Na I-D lines all showing different aspects of this event (discussed in detail by PHILLIPS & HEATHCOTE [25] and HANUSCHIK & THIMM [26]).

The fine-structure shows up in the absorption trough, *as well as in the emission hump* when the latter is visible (in Hα and Na I-D). Its typical radial velocity is declining from ±5000 to less than ±3000 km s^{-1}, close to the photospheric value. Another characteristic is its sudden appearance as a peak-like trough structure (on day 19 in Hα and on day 17 in Hβ [26]) and its gradual fading until it disappears by days 70 — 100. The phenomenon is obviously not restricted to the optical [26] and IR [25] hydrogen lines, but is also visible in Na I-D and, very faintly, in Ca II λ8600. Thus no doubt is left that we are dealing with an *intrinsic line phenomenon*, rather than with a transient line blend.

A number of possible origins for the Bochum event has been discussed extensively [see 24 - 26]. The width of the peak-like trough fine-structure, as well as its radial velocity, time evolution and symmetry, rule out the suggestion that these satellite features arise in the circumstellar environment of SN 1987A. An origin within the expanding envelope, namely closely above the photosphere, seems much more likely: the layers causing the fine-structure then would be hidden inside the photosphere before day 17 and suddenly become visible around that date, thus providing an attractive explanation for the observed temporal behaviour [30].

There is observational evidence [25, 27] that the net (= emitted minus absorbed) flux in Hα vanishes on day 20, but is larger than zero before and afterwards. We thus observe the onset of a new Hα heating source around day 20, after the shock-deposited energy is released by hydrogen recombination. This energy source presumably is heating by clumps or fingers of synthesized, mixed-up and now decaying ^{56}Co. The coincidence of the Bochum event and the sudden change in the Hα energy budget then would imply that this event is the earliest evidence for ^{56}Co mixing in SN 1987A.

Type-II SNe 1985*l* [28] and 1985*p* [29] show some fine-structure in their Hα profiles which resembles the Bochum event in SN 1987A. This provides some evidence that mixing is not unique to SN 1987A, but instead might be of general relevance in type-II supernovae.

Availability. The already published Bochum spectrophotometric atlas of SN 1987A [2, 3; days 158 - 315 are in preparation] can be obtained as a FITS formatted magnetic tape copy from the author.

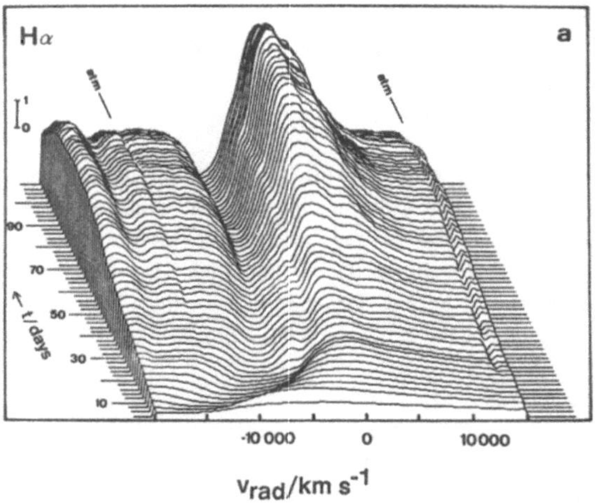

Figure 7: Detailed flux records at 3 Å resolution. Scans were averaged for every two days; the few observational gaps were filled by interpolation. The first scan is from day 2. Elapsed time in days is indicated. The radial velocity scale has been corrected for the LMC system velocity of +280 km s^{-1}. The vertical bar indicates 10^{-10} erg s^{-1} cm^{-2} $Å^{-1}$.

Figure 7a: The Hα line. Fine-structure is visible as blueshifted, peak-like flux excess in the trough and as redshifted flux deficit in the emission hump.

Figure 7b: The Hβ line. Only the trough is visible, redshifted Hβ photons are suppressed by the Fe II lines at λλ4924/5018. Fine-structure shows up as blueshifted flux excess.

Figure 7c: Na I-D λλ5890/5896. The depression at v ≈ 0 km s^{-1} is caused by the interstellar galactic and LMC components. Fine-structure appears as a depression in the red side and as a weak flux excess in the trough

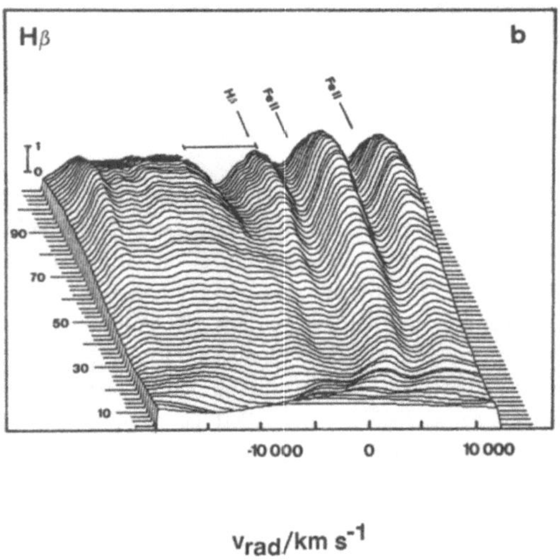

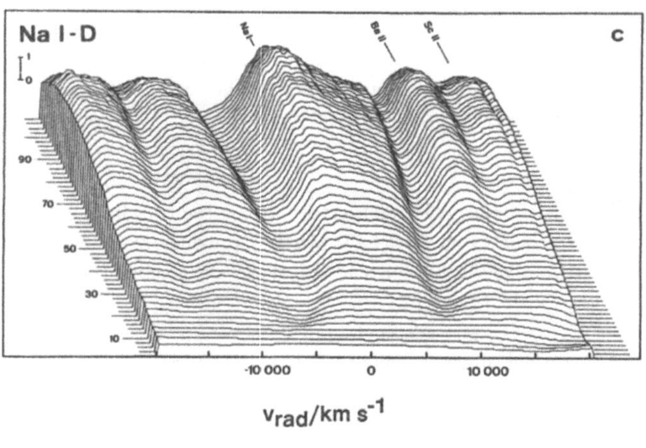

Acknowledgements. This review is based on the work of many co-investigators at the Bochum Institute. Especially worthwile to mention are the dedicated and hard efforts of more than half a dozen observers at the Bochum telescope, and the personal engagement of our Director, Prof. Th. Schmidt-Kaler, who provided ideal and financial support for this project.

References

1. J.B. Hearnshaw & V.J. McIntyre: Proc. Astr. Soc. Australia 7, 424 (1988)
2. R.W. Hanuschik & J. Dachs: Astr. Astrophys. **205**, 135 (1988)
3. R.W. Hanuschik, G. Thimm & K.J. Seidensticker: Astr. Astrophys. **220**, 153 (1989)
4. M.M. Phillips, S.R. Heathcote, M. Hamuy & M. Navarette: Astr. J. **95**, 1087 (1988)
5. R.M. Catchpole et al.: MNRAS **237**, 55P (1989)
6. W. Haupt et al.: Astr. Astrophys. **50**, 85 (1976)
7. D.S. Hayes: Ap. J. **197**, 593 (1970)
8. H. Tüg: Astr. Ap. Suppl. **39**, 67 (1980)
9. H. Tüg: Astr. Ap. **82**, 195 (1980)
10. M.M. Phillips: private communication (1989)
11. N.B. Suntzeff, M. Hamuy, G. Martin, A. Gomez, & R. Gonzalez: Astr. J. **96**, 1864 (1988)
12. M.A. Dopita et al.: Astr. J. **95**, 1717 (1988)
13. R.M. Catchpole et al.: MNRAS **229**, 15P (1987)
14. P. Bouchet, A. Moneti, E. Slezak, T. Le Bertre, & J. Manfroid: Astr. Astrophys., *in press* (1989)
15. R.E. Williams: Ap. J. **320**, L117 (1987)
16. R.W. Hanuschik & J. Dachs: in *ESO Workshop on the SN 1987A*, ed. by I.J. Danziger (ESO, Garching: 1987), p. 153
17. R.W. Hanuschik & Th. Schmidt-Kaler: MNRAS *in press* (1989)
18. D. Branch: Ap. J. **320**, L23 (1987)
19. M.A. Dopita: Space Sci. Rev. **46**, 225 (1988)
20. R.W. Hanuschik & Th. Schmidt-Kaler: *in preparation* (1989)
21. M. Karovska, L. Koechlin, P. Nisenson, C. Papaliolios, & C. Standley: in Highlights in Astronomy **8**, ed. by W.C. Liller (Reidel, Dordrecht: 1989)
22. H.E. Schwarz & R. Mundt: Astr. Astrophys. **177**, L4 (1987)
23. V.M. Blanco et al.: Ap.J. **320**, 589 (1987)
24. R.W. Hanuschik, G.J. Thimm & J. Dachs: MNRAS **234**, 41P (1988)
25. M.M. Phillips & S.R. Heathcote: PASP **101**, 137 (1989)
26. R.W. Hanuschik & G.J. Thimm: Astr. Astrophys. *in press* (1989)
27. G.J. Thimm, R.W. Hanuschik & Th. Schmidt-Kaler: MNRAS **238**, 15P (1989)
28. A. Filippenko & W.L.W. Sargent: Astr. J. **91**, 691 (1986)
29. A.A. Chalabaev & S. Cristiani: in *ESO Workshop on the SN 1987A*, ed. by I.J. Danziger (ESO Garching: 1987), p. 655
30. L. Lucy: in 4th *George Mason University Workshop in Astrophysics "Supernova 1987A in the LMC"*, ed. by M. Kafatos & A. Michalitsianos (Cambridge University Press: 1988), p. 323
31. R.W. Hanuschik, G.J. Thimm & J. Dachs: ESO Messenger No. 51, p.7 (1988)
32. D. Branch: in *Proc. ASP* 100th *Ann. Symp. "The Extragalactic Distance Scale"* (ASP Conf. Ser., San Francisco: 1988), p. 146

SN 1987A: Optical Spectrophotometry 130–900 Days After Core Collapse

M. M. Phillips & R. E. Williams

1. Introduction

SN 1987A in the Large Magellanic Cloud has provided an unprecedented opportunity to study the spectral characteristics and evolution of a Type II supernova at epochs >6 months after core collapse. The initial brilliance of type II supernovae at optical wavelengths is dictated in great measure by the compactness of the hydrogen envelope of the progenitor. Thus, the unusually low luminosity of SN 1987A during the first month or so after outburst was due primarily to the fact that its progenitor, Sk -69°202, was a blue rather than a red supergiant. For the same reason, the spectral evolution of SN 1987A at these early epochs was much faster than normal. However, within a few weeks of outburst, energy released from the radioactive decay of ^{56}Co began to power a long steady rise in the bolometric light curve, culminating in a broad maximum which was reached around day 90. By the beginning of the exponential decline phase in the bolometric light curve that followed, SN 1987A was no longer subluminous, indicating that the amount of mass 56 material synthesized in the explosion of Sk -69°202 must have been comparable to that produced in "normal" type II supernovae. Thus, in spite of the peculiarities early on, it is reasonable to expect that the late-time spectral evolution of SN 1987A should be representative of type II supernovae in general.

This paper presents a preliminary discussion of optical spectrophotometry of SN1987A obtained 130-900 days after outburst during the exponential decline phase of the bolometric light curve. We shall concentrate on several aspects of the emission-line spectrum, including line identifications, flux measurements, and profile comparisons, from which we hope to derive basic information on the physical conditions in the expanding ejecta.

2. Observations

Except where otherwise stated, the observations reported in this paper were obtained at the Cerro Tololo Inter-American Observatory (CTIO). The "2D-Frutti", a two-dimensional photon-counting detector, was employed on the Cassegrain spectrograph of the 1.0 m telescope at approximately biweekly intervals to record spectra covering the wavelength range 3700-7100 Å at a full-width at half-maximum (FWHM) resolution of 5 Å. A GEC CCD was used with the Cassegrain spectrographs of the 1.5 m and 4.0 m telescopes to obtain lower-dispersion (11-16 Å FWHM) spectrophotometry over the wavelength range 3200-11000 Å on a less-frequent (1-3 month) basis. Details concerning the acquisition and calibration of these data sets are given by PHILLIPS *et al.* [1, 2].

3. Line Identifications

The development of the optical spectrum of SN 1987A from day 198 to 907 is illustrated in Fig. 1. In general, these data closely resemble the few late-time spectra that are available for other type II supernovae. In particular, we call attention to the striking similarity of the top spectrum in Fig. 1 to

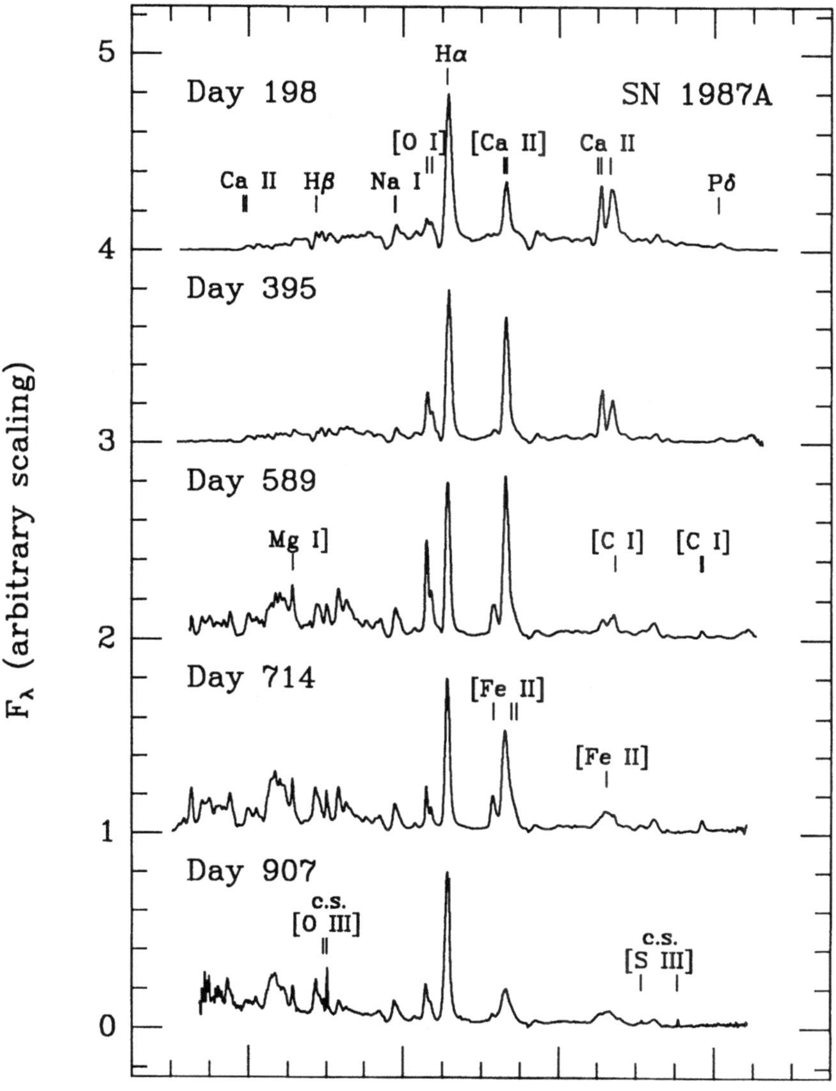

Figure 1. Optical spectral evolution of SN 1987A from days 198-907. Identifications are given for selected emission lines. The narrow [O III] λλ4959,5007 and [S III] λλ9069,9532 lines visible in the spectrum for day 907 are due to the circumstellar material that surrounded the progenitor Sk -69°202.

spectra of the two *plateau* type II supernovae 1985P and 1986I obtained approximately six and nine months, respectively, after outburst [3, 4]. Observations of the *linear* type II supernova 1980K obtained between 240-670 days past maximum are also reasonably similar, although this supernova never showed the strong P Cygni absorption in Hα that was visible in SN 1987A through day 600 [5].

A preliminary discussion of emission-line identifications in the optical spectrum of SN 1987A during these late epochs has been given by TERNDRUP *et al.* [6]. An updated summary is given in the remainder of this section. As the supernova spectrum slowly faded, narrow emission lines due to circumstellar material associated with the progenitor, Sk -69°202, became increasingly visible. This accounts for the narrow [O III] λλ4959,5007 and [S III] λλ9069,9532 emission present in the spectrum for day 907 shown in Fig. 1. Similar emission due to Hα and [N II] λλ6548,6583 is also faintly visible in this spectrum atop the broad Hα emission of the supernova.

3.1 H I

During virtually the entire period covered by this paper, Hα was the strongest emission line in the optical spectrum. Until day 600 or so, weak P Cygni absorption was also visible. As may be seen in Fig. 1, other Balmer lines such as Hβ and Hγ showed similar absorption through days 500-600, but without any significant emission components. However, in the latest spectra, an emission feature which we feel is most likely identified with Hβ has been increasing slowly in intensity with respect to Hα.

3.2 Ca II, [Ca II]

The Ca II H and K lines were observed strongly in absorption within a few days of outburst (BLANCO *et al.* [7]) and have remained visible in the spectrum since this time. The Ca II λλ8498,8542,8662 "triplet" appeared shortly after outburst as well, but with a complicated P Cygni emission-line profile [8]. Over the period covered by this paper, the strength of the triplet emission slowly decreased relative to Hα. Forbidden-line emission due to [Ca II] λλ7291,7323 was first observed around day 35 (HANUSCHIK *et al.* [9]) and grew progressively stronger with respect to Hα until around day 475, at which point these lines were the strongest feature in the optical spectrum. After day 475, however, the [Ca II] emission began to weaken in strength with respect to Hα, with this trend having continued to the present (see Fig. 1).

3.3 [O I]

The [O I] λλ6300,6364 lines first became clearly visible in the spectrum of SN 1987A around day 150, although close inspection of the high signal-to-noise spectra obtained by the Bochum group (HANUSCHIK *et al.* [9]) shows that such emission was actually present as early as day 115 (see Fig. 2). As in the case of the [Ca II] λλ7291,7323 lines, the intensity of the [O I] emission gradually increased with respect to Hα until around day 475, at which point this trend was reversed. Interestingly, however, the spectrum obtained on day 907 shows that the [O I] lines have once again begun to increase in strength with respect to Hα, whereas the [Ca II] lines continue to decline in intensity (see Fig. 1).

3.4 Mg I]

Mg I] λ4571 emission first became detectable around day 275 and steadily increased in strength with respect to Hα through day 575. As Fig. 1 shows, this behavior is somewhat different from that of the

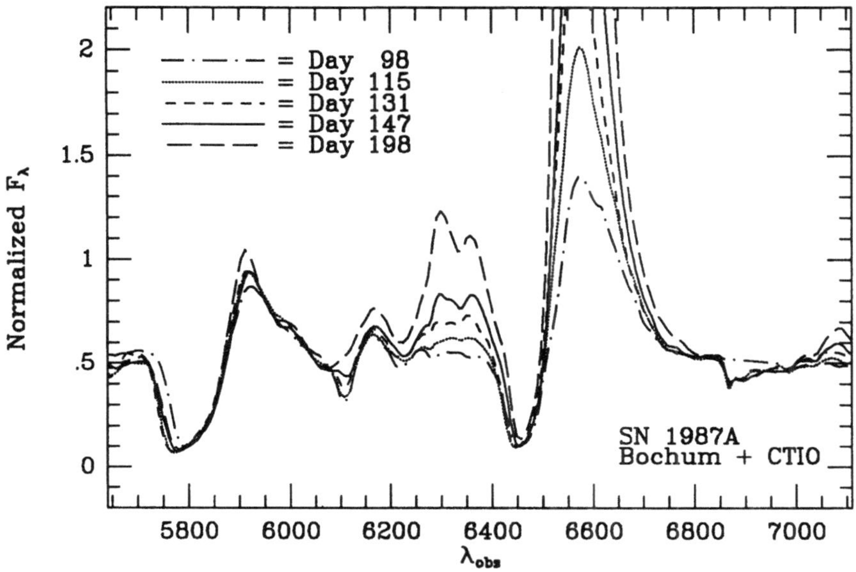

Figure 2. Spectra obtained by the Bochum group and at CTIO are plotted to show the rise of [O I] λλ6300,6364 emission in SN 1987A between days 98-198. The spectra have been normalized to the same continuum intensity at a wavelength of 6825 Å.

[Ca II] λλ7291,7323 and [O I] λλ6300,6364 lines which, as mentioned above, experienced a fairly sudden decrease in strength starting around day 475.

3.5 [C I]

An emission line which TERNDRUP *et al.* [6] tentatively identified with the [C I] λλ9824,9850 blend appeared in the spectra around day 350 and slowly increased in strength with respect to Hα. Beginning on day 589, a small "bump" appeared in the red wing of the Ca II λ8662 line which appears most likely to have been due to [C I] λ8730. The λλ9824,9850 emission peaked in strength with respect to Hα around day 700 and then declined sharply in intensity, falling below the limit of detectability in the spectrum obtained on day 907.

3.6 [Fe II]

TERNDRUP *et al.* [6] identified several weak lines at wavelengths between 7000-9500 Å with multiplets 1F, 13F, and 14F of [Fe II]. CTIO spectra show that one of the strongest of these [Fe II] lines, λ7155 of multiplet 14F became clearly visible by day 198 and steadily increased in strength relative to Hα through day 650 or so, at which point it began to decline in intensity. Beginning around day 500, a multitude of blended lines became discernable in the wavelength range 3200-5500 Å, growing steadily brighter until by day 805 they were stronger than all over emission lines in the spectrum except for Hα and the [Ca II] λλ7291,7323 lines. Comparison with the spectrum of the emission-line star η Carina shows that these lines are also mostly due to [Fe II] (see Fig. 3). These blue [Fe II] multiplets seem to have reached a peak in strength with respect to Hα between days 800-900, and may now be declining.

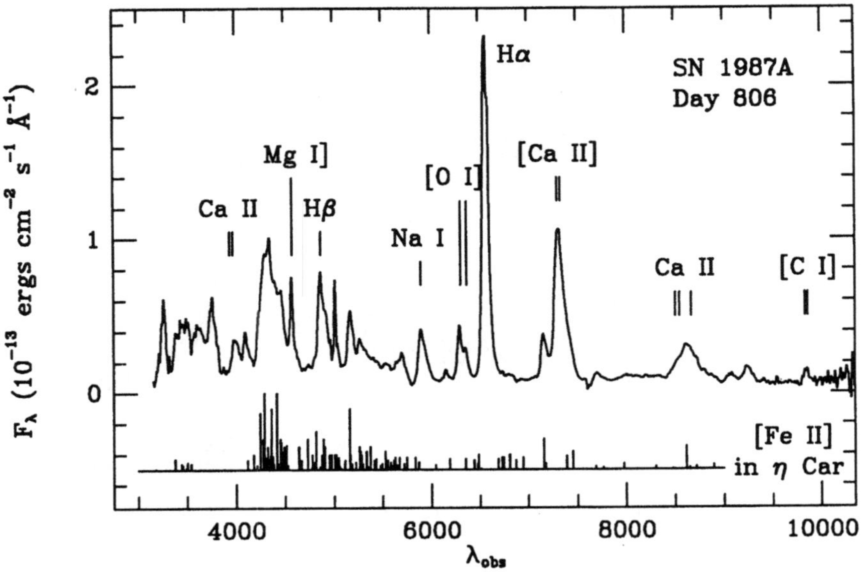

Figure 3. Comparison of an optical spectrum of SN 1987A obtained on day 806 with the positions and relative intensities of the [Fe II] emission lines observed in the spectrum of the peculiar star η Carina (THACKERAY [10]).

4. Emission-Line Luminosities

Fig. 4 shows a comparison of the Hα and [O I] $\lambda\lambda$6300,6364 emission-line luminosities and the ultraviolet-optical-infrared (UVOIR) bolometric light curve from days 200-900. For reference, the luminosity generated by 0.071 $M_\odot$ of ^{56}Co is also plotted. On average, the luminosities of these and the other strong emission lines approximately followed the UVOIR bolometric light curve during this period, providing direct evidence that the dominant source of ionization and heating was the γ-rays produced by the radioactive decay of ^{56}Co. UOMOTO and KIRSHNER [5] concluded the same for the linear type II SN 1980K, whose Hα emission-line flux declined with a mean lifetime of 108 days between days 240-670.

In Fig. 5a, we have replotted the Hα luminosities as a fraction of the UVOIR bolometric light curve so as to factor out the time dependence of the energy deposition of the ^{56}Co γ-rays. We see that the fractional luminosity of Hα reached a broad peak around day 300, slowly declined until day 700, and then began to climb again. Measurements for Paβ, Pfγ, and H7-6 obtained from infrared observations carried out at CTIO, the European Southern Observatory, and the Anglo-Australian Observatory (AAO) are plotted in the same figure for comparison. Both Paβ and Pfγ appear to have peaked somewhat earlier than Hα, whereas H7-6 reached a maximum slightly later.

Fractional emission-line luminosities of the [Ca II] $\lambda\lambda$7291,7323, [O I] $\lambda\lambda$6300,6364, [Fe II] λ7155, Mg I] λ4571, and [C I] $\lambda\lambda$9824,9850 emission lines between days 130-900 are plotted in Fig. 5b. Included in this diagram are measurements of the [Co II] 10.52 μm and [Ne II] 12.81 μm lines obtained by AITKEN [13] and AITKEN *et al.* [14] at the AAO. The behavior of the [Ca II] $\lambda\lambda$7291,7323 and

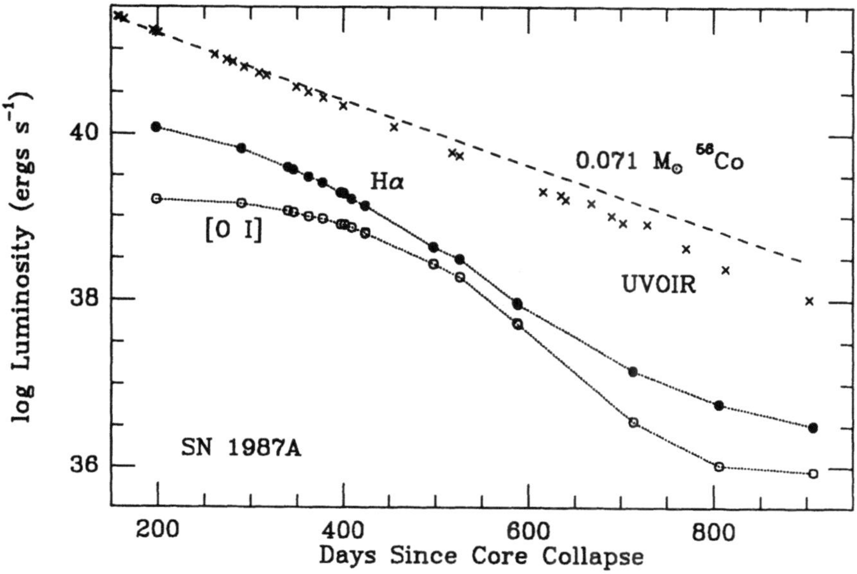

Figure 4. Comparison of the emission-line luminosities of Hα and [O I] λλ6300,6364 with the UVOIR bolometric light curve (SUNTZEFF and BOUCHET [11]) between days 200-900. The dashed line shows the radioactive decay energy corresponding to 0.071 $M_\odot$ of ^{56}Co.

[O I] λλ6300,6364 lines was very similar over most of this period, with the fractional luminosities of both lines having reached a maximum around day 400. The Mg I] λ4571 and [Fe II] λ7155 lines peaked approximately 100 days later, whereas the fractional luminosity of the [C I] λλ9824,9850 blend did not start to turn down until day 700 or so. The behavior of the [Ne II] 12.81 μm and [Co II] 10.52 μm lines was very similar, with both apparently having peaked between days 400-500.

5. Emission-Line Profiles

The evolution of the Hα, Paβ, and Brγ emission-line profiles from days 198-431 is shown in Fig. 6. Although Paβ and Brγ were somewhat narrower than Hα, the overall shapes of all three of these hydrogen lines were clearly very similar. The profiles were distinguished by redshifted peaks and asymmetric red wings similar to those reported by WITTEBORN *et al.* [15] for the [Ar II] λ6.983 μm and [Ni II] 6.634 μm lines from spectra obtained on day 415 with the Kuiper Airborne Observatory. Indeed, close examination shows that the profiles of the [Ar II] and [Ne II] lines agree very closely with those of the hydrogen lines shown in Fig. 6. Other non-hydrogen lines showing redshifted peaks during this same time period were Na I D, Ca II λ8662, He I 1.083 μm, O I 1.129 μm, [Fe II] 1.257 μm, [Fe II] 17.93 μm, [Ni I] 3.119 μm, and [Co II] 10.52 μm (TERNDRUP *et al.* [6]; DANZIGER *et al.* [16]; SPYROMILIO *et al.* [17]; JENNINGS *et al.* [18]). The peak of Mg I] λ4571 also appears to have been redshifted, although the weakness of this line and the complicated structure of the underlying spectrum of the supernova at this wavelength make this conclusion somewhat tentative (see PHILLIPS *et al.* [2]).

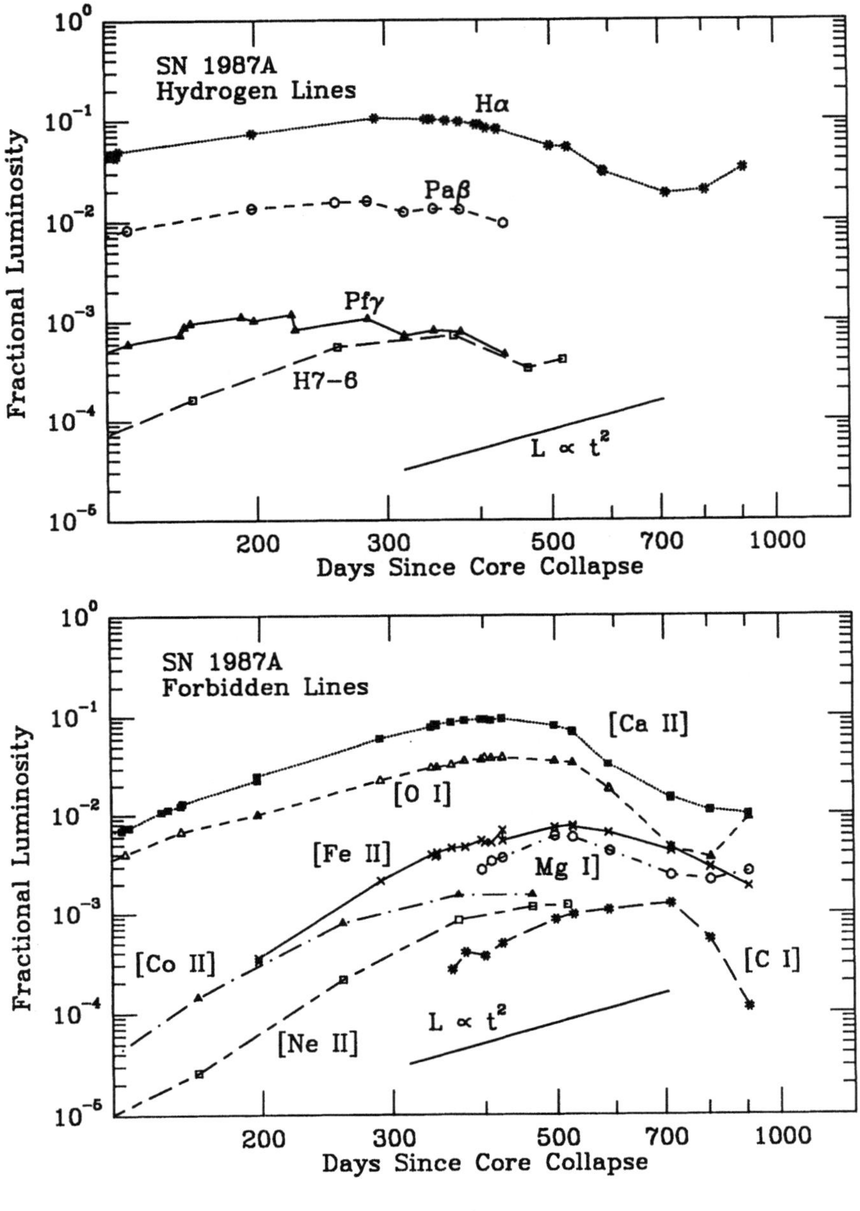

Figure 5. Time evolution of the luminosities of selected hydrogen (5a; top panel) and optical/infrared forbidden (5b; bottom panel) emission lines plotted as a fraction of the UVOIR bolometric light curve to remove the time dependence of the energy deposition of the ^{56}Co γ-rays. The forbidden line identifications are [Ca II] $\lambda\lambda$7291,7323, [O I] $\lambda\lambda$6300,6364, [Fe II] λ7155, Mg I] λ4571, [Co II] 10.52 μm, [Ne II] 12.81 μm, and [C I] $\lambda\lambda$9824,9850. The H7-6, [Co II] and [Ne II] luminosities were calculated from flux measurements made by AITKEN [13] and AITKEN *et al.* [14], while some of the Pfγ observations are due to BOUCHET *et al.* [12]. All other measurements were made from optical and infrared spectra obtained at CTIO.

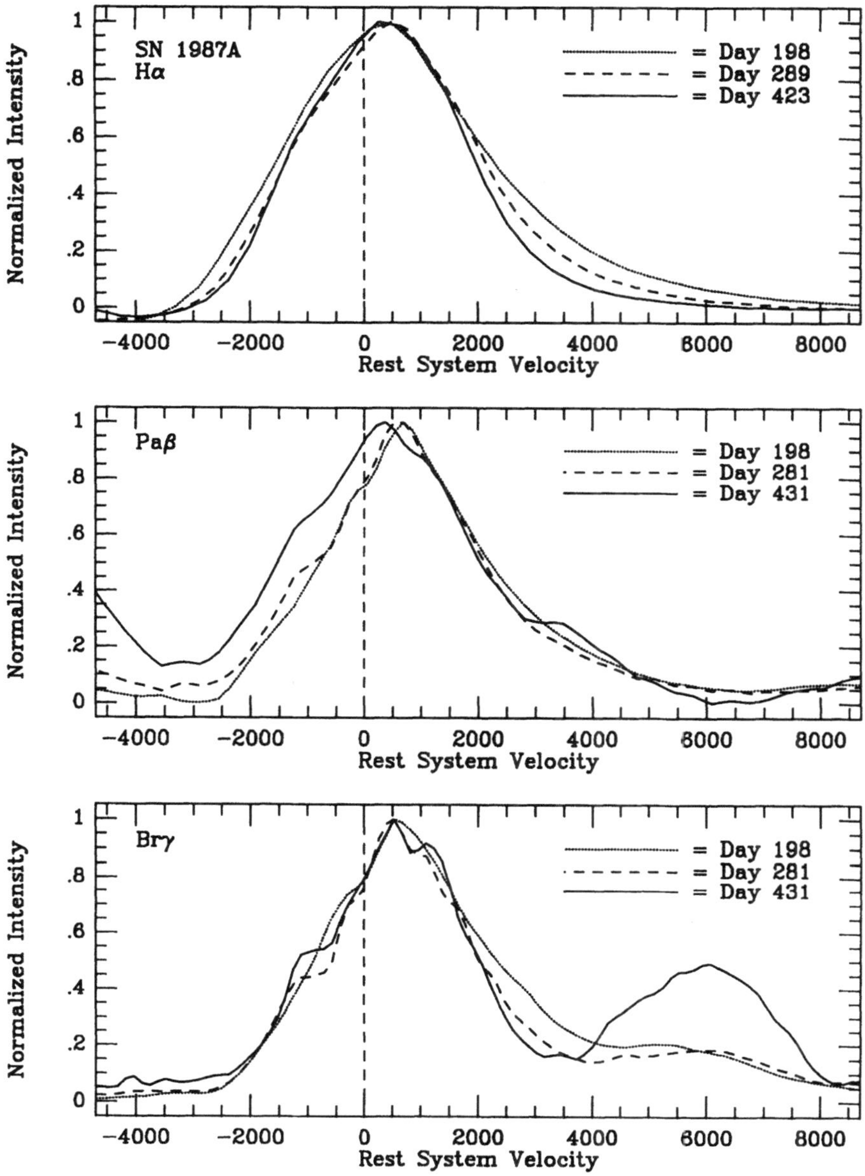

Figure 6. Evolution of the Hα (6a; top panel), Paβ (6b; middle panel), and Brγ (6c; bottom panel) emission-line profiles between days 198-431.

WITTEBORN *et al.* [15] interpreted the redshifted peak and asymmetric red wing of [Ar II] λ6.983 μm as resulting from scattering by electrons in the hydrogen envelope. However, SPYROMILIO *et al.* [17] have argued against this explanation in the case of the [Fe II] 1.256 μm and [Ni I] 3.119 μm lines, suggesting instead that bulk asymmetries in the density or temperature distribution of the ejecta were responsible.

Between days 200-500, the [O I] $\lambda\lambda$6300,6364 emission-line profiles differed strikingly from the above-mentioned lines, being characterized by a narrower width and with the peaks never having developed a significant redshift. This latter difference was first pointed out by DANZIGER *et al.* [16] who suggested that the peaks of the [Ca II] $\lambda\lambda$7291,7323 lines also lacked a redshift. We have examined the [Ca II] blend in a spectrum obtained on day 290 and conclude that the profiles of the two components were, indeed, very similar to that of the [O I] λ6300 profile from the same epoch. The only other emission lines which clearly lacked a significant redshift during this period were the [C I] $\lambda\lambda$9824,9850 blend (PHILLIPS *et al.* [2]) and Mg I 1.503 μm (SPYROMILIO *et al.* [17]).

The evolution of the Hα and [O I] $\lambda\lambda$6300 line profiles from days 198-713 is illustrated in Figs. 7a and b. The above-mentioned profile differences are clearly evident in this diagram. Even more striking, however, is the sudden blueshifts of the peaks of both lines which occurred between days 526-589. This corresponds almost precisely in time with a pronounced steepening that took place in the V light curve which was mimicked at other wavelengths in the optical and near-infrared. LUCY *et al.* [19] have argued convincingly that these changes were produced by absorption from dust grains forming in the ejecta.

6. Physical Conditions

FRANSSON [20] and CHUGAI [21] have pointed out that, even for forbidden lines, optical depth effects can be important in the late-time spectra of supernovae, and the optical spectra of SN 1987A certainly confirm this. Perhaps the most obvious example is the [O I] $\lambda\lambda$6300,6364 lines. Since both of these transitions arise from the same upper level, their relative intensities under optically-thin conditions are set by the ratio of their transition probabilities, which happens to be 3:1. However, as Fig. 2 clearly shows, the observed I(λ6300)/I(λ6364) ratio was initially much closer to 1:1 (see also CHUGAI [21]). Between days 198 and 588, we estimate that the I(λ6300)/I(λ6364) ratio changed smoothly from a value of 1.25 to 2.57. If we assume that the escape probability in each line is given by $\varepsilon = (1-e^{-\tau})/\tau$ and use the Sobolev approximation for the optical depth, τ, we calculate that the optical depth in λ6300 decreased over the same period from $\tau_{\lambda 6300}$ = 5.0-0.5.

Optical depth effects can probably also account for some of the behavior of the emission-line luminosities plotted in Figs. 5a and b. Under conditions of constant electron temperature, the luminosity of an optically-thick emission line should increase in proportion to t^2 due to simple expansion. As Fig. 5b shows, the fractional luminosities of the [Ca II] $\lambda\lambda$7291,7323 and [O I] $\lambda\lambda$6300,6364 lines rose at nearly exactly this rate until approximately day 250. It is tempting to interpret the turnover in luminosity which followed as the transition from optically-thick to optically-thin emission, particularly in view of the direct evidence provided by the changing [O I] I(λ6300)/I(λ6364) ratio. However, it is important to note that both the electron density and temperature were likely to be decreasing during this period, which could also effect the observed emission-line intensities. FRANSSON and CHEVALIER [22] have predicted that a fairly sudden transition from a spectrum dominated by optical emission lines to one dominated by infrared and far-infrared lines should take place when the electron temperature falls to 2000-3000 K. For models of both a 15 $M_{\odot}$ and 25 $M_{\odot}$ star, this so-called "IR-catastrophe" was found to occur around day 700 which, as Fig. 5b shows, matches reasonably well the timing of the fall-offs in luminosity suffered by the major optical lines. Dust

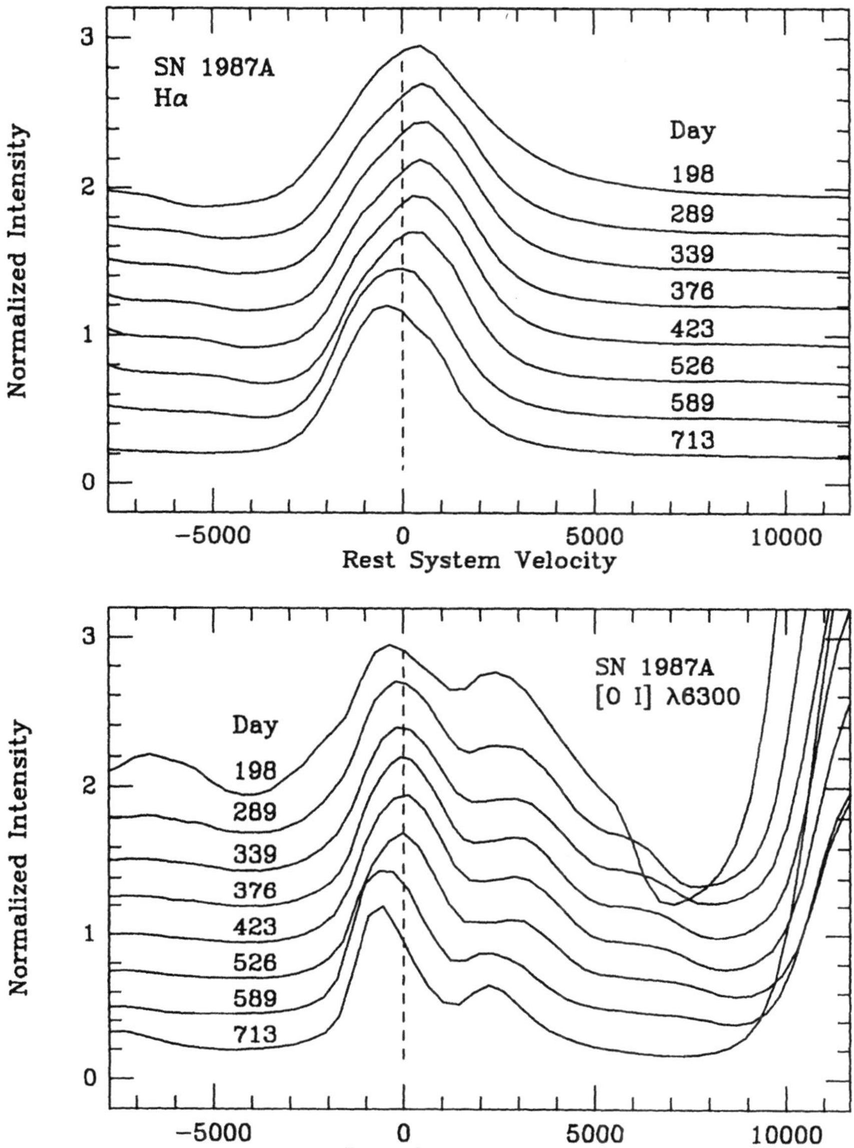

Figure 7. Evolution of the Hα (7a; top panel) and [O I] λλ6300 (7b; bottom panel) line profiles between days 198-713. The bump on the red shoulder of λλ6300 is due to [O I] λ6364.

absorption may have also had a strong effect on the observed line intensities. We note in Fig. 5b that decreases in the fractional luminosities of the [Ca II] λλ7291,7323, [O I] λλ6300,6364, Mg I] λ4571, and [Fe II] λ7155 lines occurred at the same time (between days 526-589) that the peaks of these lines underwent sudden blueshifts.

MEIKLE *et al.* [23] have pointed out that the relative strengths of the infrared hydrogen emission lines in SN 1987A during the first year after outburst were inconsistent with case B recombination. An even more glaring discrepancy observed in the optical spectra was the huge Hα/Hβ intensity ratio that existed during much of the period covered by this paper. Xu (this meeting) has suggested that line trapping may be the explanation for the latter, with Hβ photons being scattered and converted to Hα and Pβ photons. Between days 20-90, the Hα luminosity increased nearly exactly as t^2, suggesting that this line was initially optically-thick as well. The peaks in the fractional luminosities of Hα, Paβ, Pfγ, and H7-6 which occurred between days 200-350 (see Fig. 5a) may thus correspond to the transition from optically-thick to optically-thin conditions. Optical depth effects may also explain the dramatic change that took place in the I(Pfγ)/I(H7-6) ratio, which remained constant at ~4.5 from day 50-100, but which had decreased by day 350 to the expected Case B value of ~1. Further evidence of the gradual thinning of the hydrogen envelope is found in the apparent increase of Hβ emission (and corresponding decrease in the Hα/Hβ ratio) observed in the latest spectra (see Fig. 1).

The profile differences summarized in §5 provide direct evidence for the existence of different emitting zones within the ejecta of SN 1987A. The case of the Ca^+ lines is particularly interesting. Several years ago, KIRSHNER and KWAN [24] suggested that the strong Ca II "triplet" $\lambda\lambda$8498,8542,8662 and [Ca II] $\lambda\lambda$7291,7323 emission lines observed in the late-time spectra of type II supernovae were collisionally-excited. However, the fact that the triplet line profiles in SN 1987A displayed a significant redshift between days 200-500 whereas the profiles of the forbidden lines during the same period lacked such a redshift tells us that these lines could not have arisen from the same emitting region. The absence of Ca II H and K emission in the supernova spectra is also noteworthy. Indeed, these lines have appeared as *absorption* features during nearly the entire evolution of the supernova. Such a large intensity ratio of triplet to H and K emission can only be explained if the optical depths in H and K were very large (*e.g.*, see FERLAND and PERSSON [25]). This observation along with the different profiles of the triplet and forbidden lines leads us to wonder if a significant fraction of the observed triplet emission might be produced by resonance fluorescence rather than collisional excitation.

One of the major goals of studying the emission-line spectrum of SN 1987A is to determine the abundances of the heavy elements produced through explosive nucleosynthesis and compare these to theoretical predictions. We have attempted to use the observed [O I] λ6300 luminosity to estimate the oxygen mass in SN 1987A since this element is particularly sensitive to model assumptions (see WOOSLEY *et al.* [26]). To estimate the electron temperature, T_e, we tried measuring the [O I] $\lambda5577/(\lambda6300+\lambda6364)$ intensity ratio. However, the upper limit of 0.02 which we obtained from a spectrum taken on day 198 is relatively uninteresting, corresponding to T_e < 8000 K for an electron density $N_e > 10^6$ cm^{-3}. A potentially more useful line ratio is [C I] I(λ8730)/I($\lambda9824+\lambda9850$). Unfortunately, the blending of λ8730 with the Ca II triplet lines (see Fig. 1) makes a precise measurement of this line ratio very difficult. Our best value is for day 589, where we estimate an upper limit of 0.6 which implies T_e < 4000 K for $N_e > 10^6$ cm^{-3}. A third method for estimating T_e is to assume that the peak intensity of [O I] λ6300, which we have shown to have been optically-thick during the first year after outburst, must have been less than or equal to the Planck function. For day 198, this implies a lower limit of $T_e \geq 3000$ K assuming that the characteristic maximum velocity of the emitting region corresponds to the half-width at half-intensity of 1400 km s^{-1} measured for this date.

If, instead, we take the half-width at zero-intensity as the characteristic velocity, a lower limit of $T_e \geq$ 2250 K is deduced.

To derive an abundance, we must also assume a value for the electron density. As discussed by SCHLEGEL and KIRSHNER [27], a lower limit of $N_e \geq 10^6$ cm^{-3} is quite plausible within the first year after core collapse. Since this value is greater than the critical density of [O I] λ6300, it reasonable to further assume that the population of the upper level of the λ6300 transition is closely approximated by the Boltzmann equation. The mass of neutral oxygen is then given by

$$M_{O^\circ} = (6.69 \times 10^{-43} L_{\lambda 6300}) \bullet \exp(22816/T_e) \; M_\odot. \tag{1}$$

For day 198, we derive $M_{O^\circ} = 1.4\ M_\odot$ for $T_e = 3000$ K, which must be taken as a lower limit since λ6300 was optically-thick on this date. This value agrees quite well with the WOOSLEY *et al.* [26] nucleosynthesis calculations for a 20 $M_\odot$ star. However, it must be emphasized that the derived oxygen mass is *highly* dependent on the assumed value of the electron temperature. For $T_e = 4000$ K, the lower limit to the neutral oxygen mass drops to $M_{O^\circ} \geq 0.2\ M_\odot$, which is no longer very interesting. Clearly, a better way to estimate masses would be to use fine structure lines in the far-infrared such as [O I] 63.15 μm which are much less sensitive to temperature. Barring such observations, it will probably be necessary to develop detailed radiative transfer models to provide further constraints on the abundances of elements in the ejecta.

This work has been carried out with the help of Jay Elias, Brooke Gregory, Mario Hamuy, Steve Heathcote, Mauricio Navarrete, and Lisa Wells.

References

1. M. M. Phillips, S. R. Heathcote, M. Hamuy, and M. Navarrete: A. J. 95, 1087 (1988).
2. M. M. Phillips, M. Hamuy, S. R. Heathcote, N. B. Suntzeff, and S. Kirhakos: A. J. submitted (1989).
3. A. A. Chalabaev, and S. Cristiani: in ESO Workshop on the SN 1987A, ed. by I. J. Danziger, (ESO, Garching bei München 1987) p. 655.
4. A. V. Filippenko: in Supernova 1987A in the Large Magellanic Cloud, ed. by M. Kafatos, and A. G. Michalitsianos, (Cambridge University Press, Cambridge 1988) p. 106.
5. A. Uomoto, and R. P. Kirshner: Ap. J. 308, 685 (1986).
6. D. M. Terndrup, *et al.*: Proc. Astron. Soc. Australia 7, 412 (1988).
7. V. M. Blanco, *et al.*: Ap. J. 320, 589 (1987).
8. I. J. Danziger, *et al.*: Astr. Ap. 177, L13 (1987).
9. R. W. Hanuschik, G. Thimm, and K. J. Seidensticker: Astr. Ap. in press (1989).
10. A. D. Thackeray: M. N. R. A. S. 135, 51 (1967).
11. N. B. Suntzeff, and P. Bouchet: A. J. submitted (1989).
12. P. Bouchet, *et al.*: Astr. Ap. in press (1989).
13. D. K. Aitken: Proc. Astron. Soc. Australia 7, 462 (1988).
14. D. K. Aitken, *et al.*: M. N. R. A. S. in press (1989).
15. F. C. Witteborn, *et al.*: Ap. J. 338, L9 (1989).
16. I. J. Danziger, *et al.*: in Supernova 1987A in the Large Magellanic Cloud, ed. by M. Kafatos, and

A. G. Michalitsianos, (Cambridge University Press, Cambridge 1988) p. 37.
17. J. Spyromilio, W. P. S. Meikle, and D. A. Allen: M. N. R. A. S. in press (1989).
18. D. E. Jennings, R. J. Boyle, G. R. Wiedemann, and S. H. Moseley: IAU Circ. No. 4671 (1988).
19. L. B. Lucy, I. J. Danziger, C Gouiffes, and P. Bouchet: in IAU Colloquium No. 120, Structure and Dynamics of the Interstellar Medium, ed. by G. Tenorio-Tagle, M. Moles, and J. Melnick, (Springer-Verlag, Berlin 1989) in press.
20. C. Fransson: in ESO Workshop on the SN 1987A, ed. by I. J. Danziger, (ESO, Garching bei München 1987) p. 467.
21. N. N. Chugai: Astro. Zircular No. 1525 (1988).
22. C. Fransson, and R. A. Chevalier: Ap. J. 322, L15 (1987).
23. W. P. S. Meikle, D. A. Allen, J. Spyromilio, and G. -F. Varani: M. N. R. A. S. 238, 193 (1988).
24. R. P. Kirshner, and J. Kwan: Ap. J. 197, 415 (1975).
25. G. J. Ferland, and S. E. Persson: preprint (1989).
26. S. E. Woosley, P. A. Pinto, and T. A. Weaver: Proc. Astron. Soc. Australia 7, (1988).
27. E. M. Schlegel, and R. P. Kirshner: A. J. 98, 577 (1989).

The ESO Infrared Data Set

P. Bouchet, I. J. Danziger, & L. B. Lucy

Several talks during this conference are based partly on infrared data obtained at ESO (SUNTZEFF *et al.*, DANZIGER *et al.*, LUCY *et al.*, these proceedings). These data have been published by BOUCHET *et al.* [1] and BOUCHET *et al.* [2], and discussed, for instance, in DANZIGER *et al.* [3] and LUCY *et al.* [4]. Broad-band photometry from $1.2\mu m$ to $20\mu m$, as well as narrow band photometry with a resolution of $\lambda/\Delta\lambda \sim 80$ (CVF) have been obtained at ESO, La Silla, with the 3.60m, 2.2m and 1m telescopes, continuously since day 4 after outburst. Some higher resolution IR spectroscopy ($\lambda/\Delta\lambda \sim 1500$) has also been obtained with the 3.6m telescope.

An illustration of our data is given in Fig. 1, which displays the evolution of the spectral energy distribution of SN 1987A from November 1987 until April 1989. In this figure, our CVF data are superimposed on the broad-band photometry. The identifications of the main features present in the 1–$20\mu m$ range are given in Fig. 2. Note the strength of the fundamental vibration band of CO at $4.6\mu m$, which contributes significant flux to the M band, and of the first overtone band at $2.3\mu m$. Figure 2 also shows an excess near $8\mu m$ due to the long wavelength part of the fundamental band of SiO, and the development of the [Ni I] emission line at $3.12\mu m$, which became stronger than Brα by July 1988. However, the most striking point in Fig. 1 is the evolution of the flux distribution longward of $5\mu m$. The significance of this as an indicator of dust in the ejecta has been pointed out by DANZIGER *et al.* [5] and elaborated by LUCY *et al.* [4] and SUNTZEFF *et al.* [6].

To elaborate on earlier presentations at this conference, we would like to emphasize the following two points:

1) LUCY *et al.* [4] (and these proceedings) have shown that since day ~ 530 after outburst spectroscopic indicators suggested dust condensation in the ejecta of SN 1987A. This model has been supported by quantitative modelling of optical and infrared photometry. We present here in more detail the light-curves of the 8 to $20\mu m$ photometry shown in Fig. 3. At the onset of dust formation (points at day 520), the N2 light-curve ($\lambda_{eff} = 9.69\mu m$) displays an

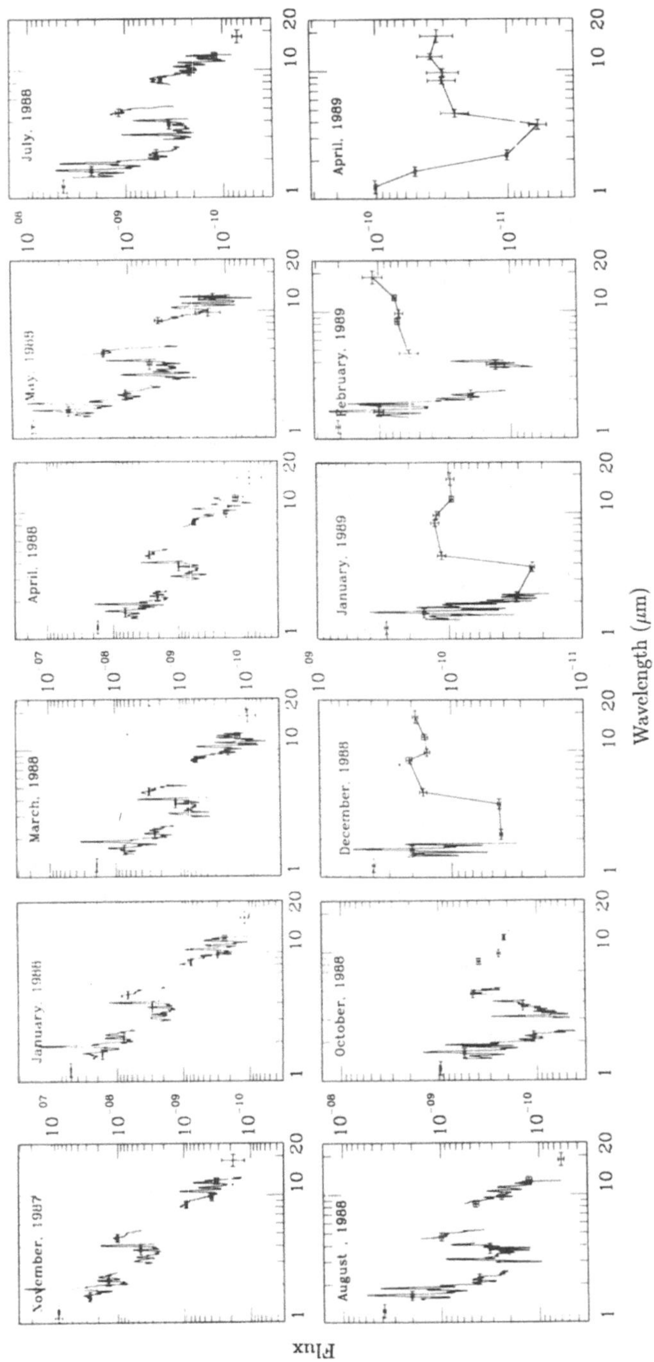

Fig. 1. Spectral energy distribution of SN 1987A between 1 and 20μm at the epochs indicated. When available, the CVF photometry has been superimposed on the broad-band photometry.

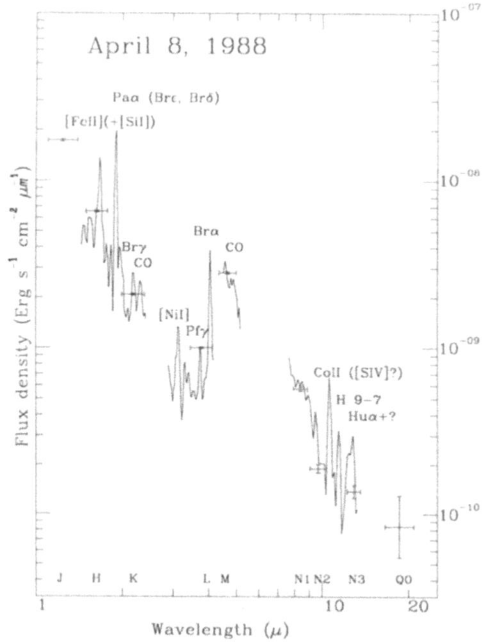

Fig. 2. Broad-band photometry and CVF spectra obtained on April 8, 1988.

increase more pronounced than in the other light-curves. As this effect is less obvious in the N1 filter ($\lambda = 8.38\mu$m), it cannot be attributed to a continuous reddening of the supernova. Since the N2 filter ($\Delta\lambda = 1.65\mu$m) is centered by design on the peak of the silicate emission band, the behaviour of the light-curve in this filter at this early stage may be best interpreted as enhanced emission from diffuse silicate dust. This effect relative to that of the neighbouring photometric bands would be reduced later when black body radiation from the clumped dust became dominant. This process of clumping may have proceeded rapidly. Although Fig. 4, showing the evolution of the 8–13μm spectrum, does not reveal any peaked emission at this wavelength on the spectra obtained on May 31, 1988 ($t = 463d$) and August 3, 1988 ($t = 526d$) (dates when the broad-band photometric data have been obtained), it should be noted that this region of the spectrum is quite noisy, due to a poor atmospheric transmission and a weak signal from the supernova. The continuum (within the noise) is remarkably constant at 9.69μm while it is not in the other regions of the spectrum. Furthermore, Fig. 4 shows that the relative energy distribution in the N1, N2 and N3 ($\lambda_{eff} = 12.89$) filters experienced a sudden change between August 1988 and October 1988.

2) Of special interest also is the behaviour of the line at 10.52μm, which results from a transition between fine structure levels in the ground state (a^3F) of Co II. This line is the

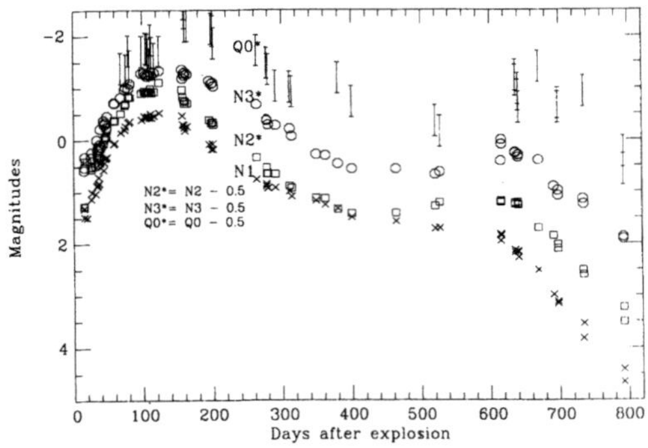

Fig. 3. The light-curves in the N1 ($\lambda_{eff} = 8.38\mu$m), N2 ($\lambda_{eff} = 9.69\mu$m), N3 ($\lambda_{eff} = 12.89\mu$m) and Q_0 ($\lambda_{eff} = 18.56\mu$m) filters.

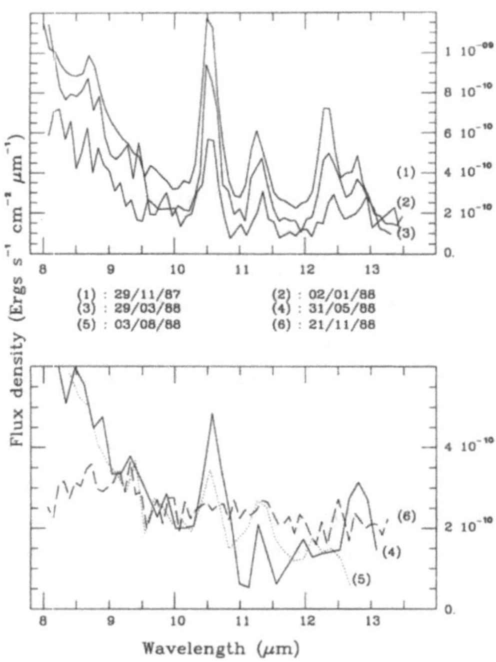

Fig. 4. Evolution of the 8–13μm spectrum. Note how the 10.52μ emission line has vanished on the spectrum taken on November 21, 1988.

most convincing evidence for the production of cobalt and its role in powering the light-curve. The quantitative temporal behaviour of this line has been followed and the results are presented in [2], by DANZIGER and BOUCHET [7] and by Danziger *et al.* (this conference). The line passed through maximum luminosity at day 280 ± 20 days, and then its strength decreased at an exponential rate as a result of the radioactive decay and, within the observational uncertainties, in good accord with the bolometric light curve and with the models requiring an initial production of 0.075 $M_\odot$ of ^{56}Co. Figure 4 shows qualitatively how the line vanished between August 3, 1988 ($t = 526d$) and November 21, 1988 ($t = 637d$). This effect is significantly larger than would be expected from radioactive decay alone and has been attributed to extinction by silicate dust in the ejecta [4].

The completeness of the ESO IR data set should help in sorting out many difficulties which have arisen during this conference.

References

1. P. Bouchet, A. Moneti, E. Slezak, T. le Bertre, J. Manfroid: Astr. Ap. Supp. Series, in press (1989) (ESO Scientific Prep. No. 592).
2. P. Bouchet, I.J. Danziger, A. Moneti, T. le Bertre: in preparation (1989).
3. I.J. Danziger, P. Bouchet, R.A.E. Fosbury, C. Gouiffes, L.B. Lucy, A.F.M. Moorwood, E. Oliva, F. Rufener: In Supernova 1987A in the LMC, ed. by M. Kafatos and A.G. Michalitsianos, (Cambridge University Press 1988).
4. L.B. Lucy, I.J. Danziger, C. Gouiffes, P. Bouchet: In Structure and Dynamics of the Interstellar Medium, ed. by G. Tenorio-Tagle, M. Moles and J. Melnick, IAU Colloquium No. 120 (Springer-Verlag 1989).
5. I.J. Danziger, P. Bouchet, C. Gouiffes, L. Lucy: IAU Circ. 4746 (1989).
6. N.B. Suntzeff, P. Bouchet: Astronom. J., submitted (1989).
7. I.J. Danziger, P. Bouchet: In *Evolutionary Phenomena in Galaxies*, ed. by J.E. Beckman and B.E.J. Pagel (Cambridge University Press 1989).

Infrared Emission from SN 1987A: Light Echoes or Dust Formation?

Eli Dwek

1. Introduction

The presence of an infrared (IR) continuum in the spectrum of SN 1987A was discovered in the early stages of its evolution by various photometric and spectral observations in the 5 – 20 μm wavelength regime [1, 2]. This IR continuum component represented an IR excess over that expected from a Planck spectrum fitted to the SN emission at optical wavelengths. Subsequent observations with the Kuiper Airborne Observatory (KAO) showed this continuum to be quite flat out to 100 μm with a value of ≈ 5 –10 Jy around day 265 after core collapse [3, 4]. AITKEN *et al.* [5], were first to interpret this IR excess as free–free emission from the expanding ionized gas. As expected from an expanding cooling gas, the emission from this component has been continuously declining since its first discovery, its 10 μm emissivity reaching a minimum of about 1 Jy around day 400 [6]. Subsequent observations by ROCHE *et al.* [6] revealed an upturn in the 10 μm continuum light curve of the SN, indicating either the delayed arrival of thermal emission from pre–existing circumstellar/interstellar dust heated by the UV–visual output of the SN (i.e. an IR echo), or the onset of dust formation in the SN ejecta.

2. The Case for an Infrared Echo

ROCHE *et al.* [6] advanced several argument in favor of the echo interpretation for the evolution of the IR light curve. First, the observed profile appeared to be broader than that of a point source, with a FWHM of ≈ 2.2 arcsec centered around the SN (a point source has a FWHM of ≈ 1.6 arcsec). Second, if attributed to emission from a dust shell travelling at velocities typical of the SN ejecta (i.e. < 10,000 km s^{-1}), a shell optically thick at 10 μm should completely block out the visible output from the SN, contrary to observations. Finally, east–west scans across the SN taken on day 750 [7], showed that the centroid of the 10 μm emission shifted from its previous position in the westward direction. Such movement is expected from an echo which follows the distribution of matter around the SN, but not from a dusty ejecta.

3. The Case for Dust Formation

This seemingly straightforward interpretation of the nature of the IR emission was, however, complicated by DANZIGER *et al.* [8], who presented optical evidence for the presence of dust in the SN ejecta. The observations (presented in more detail by DANZIGER [9]) show that during the period of Aug. – Oct. 1988, the [OI] 6300 Å, [CI] 9844 Å, and MgI] 4571 Å lines became asymmetric, with their peak emission blueshifted by 500–600 km s^{-1}. This development in the spectral line shape can most readily be understood if dust formed in the SN, preferentially obscuring lines emitted from the receding portions of the ejecta. The lack of total obscuration of the SN can be attributed to the fact that the dust resides in clumps in the ejecta. LUCY [10, 11], presented detailed models in which the dust giving rise to the shift in the line profiles resides in ejecta with velocities less than about 2,000 km s^{-1}. In these models, the dust is radiatively heated by the optical output of the SN, and can account for all the observed IR emission longwards of about 8 μm. The onset of the dust formation epoch (around day 450) agrees well with that predicted by DWEK [12] for dust with a condensation temperature of ≈ 1000 K.

4. Optical Echoes From Within 2 Arcsec Around the SN

To discriminate between the two interpretations for the origin of the IR emission FELTEN and DWEK [13] considered the implication of the presence of an echoing dust cloud on the optical light curve, and on the existence of diffuse optical emission around the SN. They concluded that the expected optical echo from the cloud should be resolvable, and could be very bright with an integrated visual brightness of ≈ 10.3 mag around day 650. However, further optical observations, as expressed in SN light curve, showed no inflection in the light curve at the predicted level.

Nevertheless, CROTTS, KUNKEL and McCARTHY [14] searched and found a complex structure of diffuse optical emission within 2 arcsec of the SN, albeit at a level significantly lower than predicted. These observations, taken on day 750, provided the first definite proof of the existence of a dust cloud at a distance of about 1 lyr behind the SN. Figure 1 is a schematic presentation of the morphology of the visual arcs around SN 1987A, presented in more detail in [14]. The cross indicates the approximate location of the SN, and the two circles represent the brightness distribution from stars 2 and 3 in the field. The solid line represents the location of the peak visual intensity of the diffuse emission from the arcs. It is clear that the dust cloud responsible for the diffuse visual emission should also give rise to an IR echo.

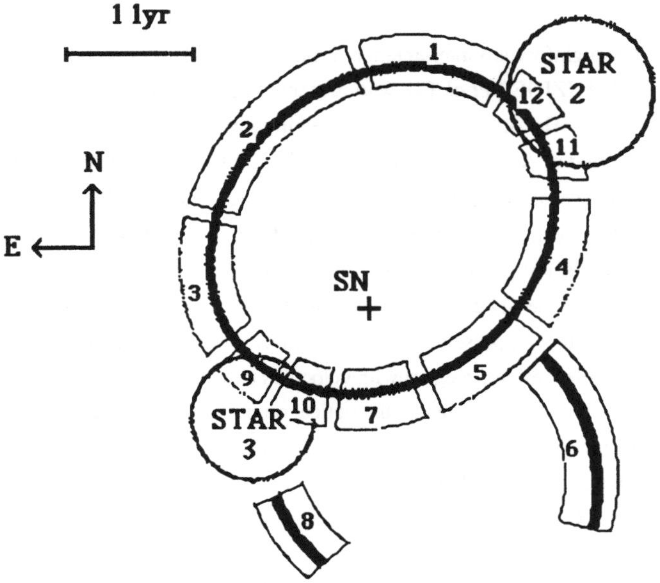

Figure 1. A schematic drawing of the position of the visual arcs, stars 2 and 3, and the SN. The regions used for modeling the IR echo are also shown in the figure.

5. The Infrared Light Curve of SN 1987A: Echo or Dust Formation?

The ambiguity in the interpretation of the origin of the IR emission resembles that surrounding the IR emission from SN 1980k (DWEK *et al.* [15], DWEK [16]). In contrast to that event, where both explanations are still equally valid, the nature of the IR emission from SN 1987A can be resolved by examining the energy output of the SN over all wavelengths, and by studying the temporal and spatial evolution of the IR emission.

5.1 The Energy Distribution

One clue to the nature of the IR excess is provided by the energy output of the SN [17]. The total bolometric output of the SN is determined by the available energy from the Co decay. Most of this energy is converted to optical light, with a fraction escaping in the form of hard X–rays and gamma radiation. In the echo model, the IR emission is reprocessed visual light from a previous epoch, and should not be added to the current energy budget of the SN. In the dust formation model, the IR emission is reprocessed visual light from the current epoch, and is part of the energy budget of the SN.

The detailed energy budgets of the SN for days 616 and 768 after core collapse are presented in Table 1. The table clearly shows that the optical and γ–ray emission alone cannot account for the available Co–decay energy, and the IR must be part of the energy budget of the SN. From the energetics of the SN we therefore conclude that the IR emission must be from newly–formed dust in the SN ejecta. The same conclusion has been also been reached by WHITELOCK *et al.* [18], WHITELOCK [19], and WOODEN [20].

The above conclusions conflict with the reported spatial resolution and movement of the centroid of the IR emission from the SN. There are difficulties in attaining the spatial distribution of the emission from the SN, since the SN and the reference 10 μm point source may have been observed through a different airmass. Furthermore, seeing conditions may vary between the observations of the SN and the reference source. Independent confirmation of the spatial extent of the IR emission from the SN would have been useful in resolving this conflict.

Table 1. The Energy Budget of SN 1987A (1)

	Day 613(2)	Day 768
Co–decay (3)	4.9	1.3
UV–visual ([18])	1.4	0.2
X– and γ–rays ([21,22])	2.0	0.7
Infrared ([17, 23])	1.7	0.3
Total	5.1	1.2

(1) Energies calculated for an LMC distance of 50 kpc, and presented in 10^{39} [erg s^{-1}]. Numbers have been rounded off to nearest significant digit.
(2) The energy budget for day 613 was presented by MOSELEY *et al.* [17]
(3) Assuming 0.75 $M_{\odot}$ of Cobalt

5.2 The Temporal Evolution of the IR Spectrum

The infrared spectrum of an optically thin dust cloud is solely determined by the dust temperature distribution in the emitting region. In the echo model, the dust is heated by the SN output at peak visual light (about day 85 after core collapse). Changes in the dust temperature and IR flux are therefore determined by the spatial distribution of the emitting dust clouds around the SN.

Figure 2 shows the IR spectrum of the SN taken with the KAO on days 638 [17] and 768 [23] after core collapse. At both epochs the IR spectrum was quite flat, at a level of ≈ 10 Jy and ≈ 4 Jy on day 638 and 768, respectively. The solid lines in the figure represent the spectrum of 0.1 μm graphite particles with a temperature chosen to give the best visual fit to the continuum emission component in the spectrum. The dust temperature was found to decrease from about 400 K on day 638 to about 250 K on day 768.

In the echo model, the emitting dust cloud must be confined to a region defined by scattering angles $\theta > 112^o$ (FELTEN & DWEK [13]), where θ is the angle subtended between the radius vector from the SN to the dust cloud and the line–of–sight to the observer. At time t, the paraboloid of maximum visible light could have therefore travelled a distance of 0.5ct to ≈ 0.73ct from the SN. The maximum increase in the distance of the emitting region from the SN between days 638 and 768 is therefore by a factor of ≈ 1.46. For dust with a λ^{-1} emissivity law, the dust temperature dependence on radial distance is $r^{-2/5}$. The dust temperature is therefore expected to change at most by a factor of $\approx (1.46)^{-0.4} = 0.86$ (i.e. from 400 to 340 K) between days 638 and 768. This change is clearly smaller than the observed one.

In the dust formation model, changes in the dust temperature are not solely determined by geometrical considerations. Additional factors, such as the decay in the optical light, the increased transparency of the ejecta, and the growth of the dust particles, play an important role in determining the spectrum of the IR emission. All these factors may have changed considerably between days 638 and 768, so that the temporal behavior of the IR spectrum can be more readily explained if it is due to dust formation in the SN ejecta.

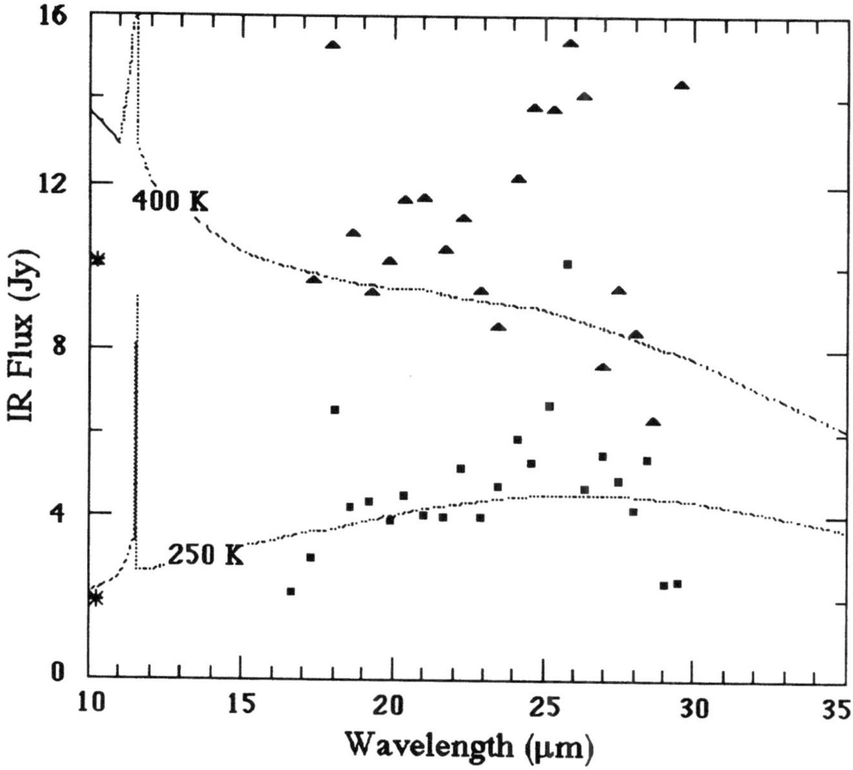

Figure 2. The IR spectra of SN1987A on days 638 (triangles) and 768 (squares) [17, 23]. Data from day 768 are preliminary. The 10 [μm] continuum data from [6, 7] are given by stars. The solid lines are visual fits of 0.1 [μm] graphite particles 400 and 250 [K] to the continuum of the spectra.

6. Infrared Echoes From Within 2 Arcsec Around the SN

6.1 General Consideration

Since energetics arguments have shown that a significant fraction (if not all) of the IR emission must be due to dust formation in the SN ejecta, it is interesting to determine what fraction (if any) can be attributed to an IR echo. It may be possible, that the IR spectrum contains two distinct components: a short–wavelength component centered around 10 μm from hot ejecta dust, and a second component from cooler circumstellar dust emitting at wavelength longer than $\lambda \approx 15$ μm. For example, the first contribution may be due to Al_2O_3

dust which has a strong and narrow emission peak at ≈ 10 μm, and is among the first minerals expected to form in the ejecta (KOZASA, HASEGAWA, and NOMOTO [24]).

First, it is interesting to examine the various factors that determine the relative strength of the reflected visual and thermal IR emission from an echoing dust cloud. The intensity, F_{sca}, of the reflected light is proportional to $P(a,\lambda,\theta)Q_{sca}(a,\lambda)$, where Q_{sca} is scattering efficiency of a dust particle of radius a at wavelength λ, and P is its phase function at scattering angle θ. The intensity of the thermal IR emission, F_{IR}, is proportional to $Q_{abs}(a,\lambda{\approx}V)$, i.e. the dust absorption efficiency at visual (V) wavelengths. For particles with sizes smaller than the incident wavelength, $Q_{sca}(a,\lambda) \sim (a/\lambda)^4$, and $Q_{abs}(a,\lambda) \sim (a/\lambda)$. Furthermore, grains with sizes below ≈ 0.2 μm are isotropic scatterers, so that $P \approx 1$ (e.g. [25]). The ratio of the reflected visual–to–thermal IR is therefore strongly dependent on grain size, and for sizes less than ~ 0.2 μm is given by:

$$\frac{F_{sca}}{F_{IR}} \sim \frac{P(a,\lambda{=}V,\theta)Q_{sca}(a,\lambda{=}V)}{Q_{abs}(a,\lambda{\approx}V)} \sim a^3. \qquad (1)$$

Given an IR flux, the intensity of the scattered light is proportional to a^3. If all the IR emission is attributed to an echo, then the low brightness of the visual arcs (compared to the predictions of [13]) can be understood if the radius of the dust particles is much smaller than the value of 0.1 μm assumed by [13]. Conversely, for a given visual echo, the intensity of the associated thermal IR echo is proportional to a^{-3}. Therefore, to calculate the IR echo from the reflecting dust clouds behind the SN we need to know the detailed size distribution as well as the spatial distribution and composition of the dust around the SN.

6.2 Detailed Models of the IR Echo from SN 1987A

To model the IR echo from the reflecting dust cloud, the visual arcs around the SN were divided into 12 segments, each representing the intensity of scattered light from a distinct cloud around the SN projected on the plane of the sky (see Fig. 1). The clouds were assumed to be optically thin in the visual, and to have identical dust properties. The only distinguishing factor between the clouds is therefore their optical depth, distance to the SN, and the intensity of their reflected visual light. Adopting a coordinate system in which the SN is at the origin, the radial distance, R, of a cloud from the SN is given by

$$R = \{x^2 + y^2 + z^2\}^{1/2}, \qquad (2)$$

where (x, y) are the coordinates of the cloud on the plane of the sky at the time of observations t, and z is line–of–sight distance of the dust cloud from the SN. The distance z is given by:

$$z = \frac{x^2+y^2}{2c\Delta t} - \frac{c\Delta t}{2} \qquad (3)$$

where Δt is the delay in the arrival time of the reflected (or reemitted) light, and is given by $\Delta t = t - t_{SN}$, where $t_{SN} \approx 85$ days is the epoch of maximum visual light.

The time behavior of the SN light curve was approximated by a flat top–hat function with a luminosity of 8.5×10^{41} erg s^{-1} between days 80 and 90, and zero at all other times. The dust was assumed to consist of pure silicate grains with an $a^{-3.5}$ size distribution between a radius of 10 Å and an upper cutoff radius, a_{max}, which was taken to be a free parameter of the model. The choice of silicates was motivated by the fact that graphite particles placed at the same distance as silicates would be hotter, and when normalized to fit the $\approx$ 20 μm emission would account for most of the observed 10 μm flux as well, leaving very little residual emission to originate from the SN ejecta. The intensity of the reflected light was calculated for a Henyey–Greenstein phase function (e.g. MARTIN [25]), and dust optical properties were taken from DRAINE and LEE [26]. Equilibrium dust temperatures were determined by equating the dust heating rate (the SN spectrum at maximum light was assumed to be that of a 6000 K blackbody) to its cooling rate by IR emission (e.g. DWEK [27]). Then, for a given visual intensity and dust composition, the associated IR echo is only a function of the grain size distribution, or a_{max}. The resulting IR spectrum which gives a best visual fit to the observed IR emission above ~ 15 μm is shown in Fig. 3. To attain this spectral fit, a_{max} had to be finely tuned to a value of 0.047 μm. This value is significantly smaller than that encountered in the general interstellar medium of our galaxy, where $a_{max} \approx 0.25$ μm (e.g. [26]). If the upper limit of the grain size distribution in the clouds around SN 1987A is increased to that value, the resulting 20 μm echo flux would be reduced by a factor of 3.7×10^{-3}, which is of the order of $(0.047/0.25)^3$.

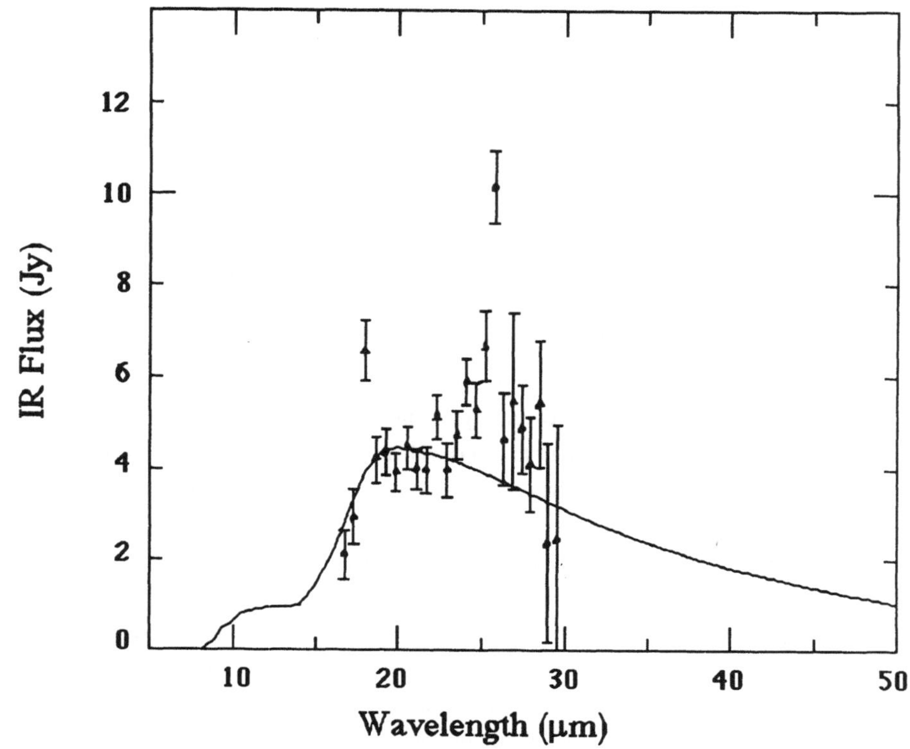

Figure 3. The spectrum of the IR echo. The grain size distribution was chosen to match the observations at wavelengths > 15 [μm].

6.3 The Spatial Evolution of the IR Echo

We have shown that it is possible to construct, by a contrived choice of dust parameters, an echo model for the long wavelength (> 15 μm) IR emission from the SN. In this model, two–thirds of the 8 – 15 μm emission is radiated by newly–condensed dust, and approximately accounts for the "missing" energy in the total energy budget of the SN on day 768. The remaining one–third of the observed 10 μm emission can be attributed to an IR echo. By construction, the spatial distribution of this 10 μm emission follows that of the visual arcs. To compare this distribution with the scans of ROCHE [7], the 10 μm emission was projected onto the x–axis running in the east–west direction through the SN. The projected emission was then convolved with a gaussian beam with a FWHM of 1.6 arcsec, which represents the spatial response of their instrument to a point source. The resulting one–

dimensional spatial distribution of the 10 μm emission is depicted in Fig. 4. The bold solid line presents the spatial distribution of a point source, centered on the SN. The dotted line represents the distribution of the emission from the echo. The figure clearly shows that the emission from the echo is broader than that of a point source, and that its centroid is displaced to the west. However, the offset from the origin is considerably smaller than the observed ≈ 0.6 arcsec [7]. Furthermore, the offset is smaller than the error in the calculations produced by uncertainties in the brightness and location of the arcs in the vicinity of stars 2 and 3. The actual distribution of the 10 μm emission must, however, be further modified to take into account the 10 μm emission emanating from the SN ejecta. From Fig. 3 we see that only one–third of the observed IR flux is attributed to an echo. The thin solid line in the figure represents the distribution of the 10 μm emission from the ejecta and the echo, in which two–thirds of the observed flux is attributed to a point source centered on the SN. The resulting distribution of the IR emission shows hardly any spatial extension compared to a point source.

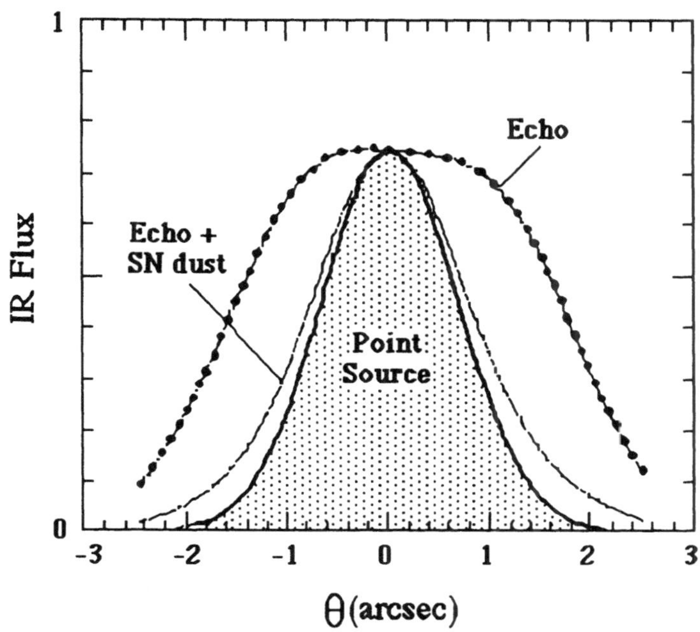

Figure 4. Theoretically calculated one–dimensional spatial distribution of the 10 [μm] emission in the east–west direction with the SN at the origin. Details are given in the text.

7. Conclusions

The rise in the IR flux from SN1987A, which occurred about 400 days after core collapse, can most readily be interpreted as infrared emission from dust that formed in the cooling ejecta of the SN, instead of an echo from a dust cloud in the vicinity of the SN. The dust formation model accounts for the distribution of the Co decay energy at the visual, X– and gamma–ray, and IR wavelengths. It also provides a better explanation for the temporal evolution of the IR spectrum.

Optical images of the SN revealed a complex structure of arcs within 2 arcsec from the SN. The diffuse emission from these arcs originates from a dust cloud located about a light year behind the SN. This dust cloud should also give rise to an IR echo. However, model calculations show that for any reasonable grain size distribution (such as that encountered in the interstellar medium of our galaxy), the IR flux from this cloud is negligibly small compared with the observations.

Acknowledgements

Arlin Crotts provided the visual continuum fluxes from the various segments of the arcs used in modeling the IR echo, and Jim Felten made useful comments on an earlier version of the manuscript. This work was supported in part by NASA Astrophysics Division RTOP No. 188–44–23–55.

References

1. I. J. Danziger *et al.*: In Supernova 1987A in the Large Magellanic Cloud, ed. by M. Kafatos and A. Michalitsianos (Cambridge University Press, Cambridge 1988), page 37.
2. W. J. Couch: In Supernova 1987A in the Large Magellanic Cloud, ed. by M. Kafatos and A. Michalitsianos (Cambridge University Press, Cambridge 1988), page 60.
3. S. H. Moseley, E. Dwek, W. Glaccum, J. R. Graham, R. F. Loewenstein, R. F. Silverberg: Ap. J. in press.
4. P. M. Harvey, D. Lester, M. Joy: IAU Circular No. 4518 (1987).
5. D. K. Aitken, C. H. *et al*: M. N. R. A. S. 231, 7p (1988).
6. P. F. Roche, D. K. Aitken, C. H. Smith, S. D. James: Nature 337, 533 (1989).
7. P. F. Roche: private communication (1989).
8. I. J. Danziger, C. Gouiffes, P. Bouchet, L. B. Lucy: IAU Circular No. 4746 (1989)

9. I. J. Danziger: In this volume.
10. L. B. Lucy, I. J. Danziger, C. Gouiffes, P. Bouchet: In Structure and Dynamics of Interstellar Medium IAU Colloquium No. 120, ed. by G. Tenorio–Tagle, M. Moles, and J. Melnick (Springer Verlag;) in press.
11. L. B. Lucy: In this volume.
12. E. Dwek: Ap. J. 329, 814 (1988).
13. J. E. Felten, E. Dwek: Nature 339, 123 (1989).
14. A. P. S. Crotts, W. E. Kunkel, P. J. McCarthy: Ap. J. (Letters), in press.
15. E. Dwek *et al.*: Ap. J. 274, 168 (1983).
16. E. Dwek: Ap. J. 274, 175 (1983).
17. S. H. Moseley, E. Dwek, W. Glaccum, J. R. Graham, R. F. Loewenstein, R. F. Silverberg: Nature 340, 697 (1989).
18. P. A. Whitelock *et al.*: M. N. R. A. S., in press.
19. P. A. Whitelock: In this volume.
20. D. Wooden: In this volume.
21. S. Kumagai, T. Shigeyama, K. Nomoto, M. Itoh, J. Nishimura, S. Tsuruta: Ap. J. 345, 000 (1989).
22. S. E. Woosley, P. A. Pinto, D. Hartmann: Ap. J., submitted.
23. S. H. Moseley, E. Dwek, W. Glaccum, J. R. Graham, R. F. Loewenstein, R. F. Silverberg: in preparation.
24. T. Kozasa, H. Hasegawa, K. Nomoto: Ap. J. (Letters), in press.
25. P. G. Martin: In Cosmic Dust (Clarendon, Oxford 1978), chapter 4.
26. B. T. Draine, H. M. Lee: Ap. J. 285, 89 (1984).
27. E. Dwek: Ap. J. 297, 719 (1985).

SECTION II

SN 1987A — SPECTROSCOPY

Molecules, Dust and Ionic Abundances in SN 1987A

I. J. Danziger, L. B. Lucy, P. Bouchet, & C. Gouiffes

I. Introduction

The continuing evolution of SN 1987A has brought new developments even if some of them had been predicted at various epochs in the near or distant past. Formation of molecules, CO and SiO, in the expanding envelope at a very early stage focussed some attention on the possibility of subsequent formation of dust. From observations made at ESO, La Silla, irrefutable evidence for the formation of dust near day 530 will be presented together with other spectroscopic and photometric data that seem to be associated with this event. This qualitative interpretation will be elaborated more fully in a quantitative treatment by LUCY *et al.* [1]. Some observational data concerning the strength and temporal behaviour of the vibration-rotation bands of CO is discussed because theoretical attempts to define the molecular chemistry are now appearing.

Although there are numerous caveats to be added to any determination of abundances of ionic species in the ejecta of SN 1987A, we have attempted to provide a first order attempt at abundance determinations. These are compared with recent nucleosynthetic models to determine whether any major discrepancies exist. The observational data for this are now available in some detail since the optimum epoch falls between the time when the envelope became optically thin ($\sim$ day 300 but depending on the lines in question), and the time when dust formation occurred (day 530).

II. Line Shifts

In the first part of this presentation we show some of the observational data that pointed to the formation of dust in the envelope of SN 1987A near day 530 (DANZIGER *et al.* [2]; LUCY *et al.* [3]). Figure 1 shows a section of low resolution spectra centered on the [O I] 6300,63 lines. One immediately recognizes that the line peaks have shifted bluewards by several hundred kilometres/sec in this seven-month period. In fact a detailed study shows that the

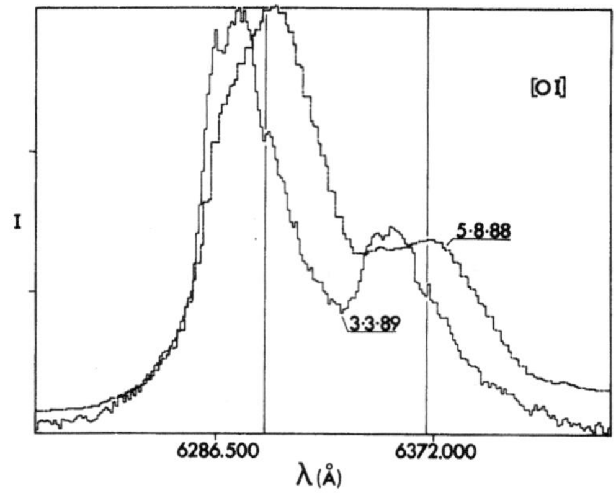

Fig. 1. Spectra of the region of [O I] 6300,64 showing the emission line profiles prior to and subsequent to the formation of dust.

shift of the maximum was monotonically increasing with time up to day 730, with a tendency to slow down or cease entirely at this later time. Note that the short wavelength side of the line profile has not shifted significantly.

This is the spectroscopic signature of dust distributed over a finite region in the envelope where the far (receding) side would be more obscured by the greater path length through dust than the near (approaching) side. The effect in varying degrees is apparent in all emission lines in the wavelength interval 3700–10000 Å. The following table shows the range of velocities recorded for a selection of lines.

Table 1: Velocity of Emission Line Peaks

	Velocity (km/s)	Day	Velocity	Day
Mg I] 4571	+100	400	−750	830
[O I] 6300	0	400	−650	830
Hα 6562	+421	375	−216	634
[Fe II] 7155,72	+700	400	−750	830
[C I] 9823,49	0	400	−500	830

With the exception of the [Fe II] 7155,72 lines the amplitude of the shift decreases with increasing wavelength, a detail of some importance in determining the nature of the dust grains [1,3].

The behaviour of the [Fe II] 7155,72 feature differs from that of the other lines in having a significant redshift before the onset of dust formation. There are uncertainties involved with this feature due to blending with strong nearby features and possibly due to unknown contributors to the feature. Its behaviour however is somewhat akin to that reported for forbidden lines in the IR spectrum, where the redshift for the emission lines prior to day 530 reported by DANZIGER *et al.* [4] is now accepted to be real.

These differences before the onset of dust are providing information about the structure of the envelope, but the decoding of these into temperature and abundance dependences on radius remains for the future. At present the total amplitude of the shift of the [Fe II] 7155,72 line, 1450 km/sec, is also inconsistent with that for the other lines.

III. Line Strengths

(a) Atomic Lines

In the following series of plots are shown the temporal behaviour of the observed flux of emission lines from different atomic species. The behaviour of the Hα emission line flux is shown in Fig. 2. It reached maximum strength between days 80 and 100 and has decreased monotonically since that time. During no reasonable time interval after maximum has the decay followed an e-folding time scale of 111.3 days (that for radioactive decay of ^{56}Co) as

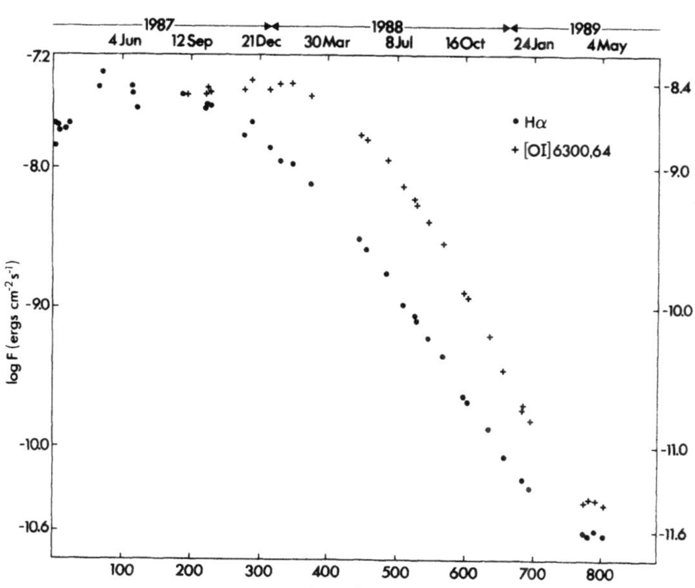

Fig. 2. The temporal behaviour of the flux of Hα and [O I] 6300,64. The scale on the left refers to Hα and that on the right to [O I] 6300,64.

had been reported for the Type II SN 1980k by UOMOTO and KIRSHNER [5] from observations extending over 650 days. At day 530 there is not convincing evidence for an increased rate of decrease of the flux which might be expected if the hydrogen emitting region were mixed with the dust. Since we might expect most of the hydrogen to be in an outer extended region of the envelope relatively free of the dust our observed result is hardly surprising.

Although it is not very obvious in Fig. 2 as plotted, there is an increased rate of decrease of the [O I] 6300,64 flux at day 530 which would be expected if dust were mixed with the oxygen emitting region. It is not so obvious because it is already imposed on a steep decline. This is intuitively in accord with the idea of dust condensations in the deeper metal-rich parts of the ejecta absorbing and scattering photons traversing the same region.

In Figs. 3, 4 and 5 we show analogous plots for Mg I] 4571, Si I (+ [Fe II]) 1.644 μ and [Ni I] 3.12 μ. The quantitative analysis of some of this data will be given by LUCY *et al.* [1]. Here we draw attention to the following possible interpretations. The onset at day 530 of the increased decay rates of these line strengths suggests that they may be intimately related to dust formation at the same epoch. To first order we might expect oxygen and magnesium to be mixed together in the expanding envelope, and therefore the line emission to be affected in the same way by dust. Due allowance must be made for the wavelength dependence of the absorption and scattering process of the dust particles.

The amplitude of the increased decay beginning at day 530 is comparable for the Mg I] 4571 and Si I 1.644 μ lines. Thus any wavelength dependence seems almost insignificant when compared to the measurable differences in velocities referred to above. Note also that the continuum (at 1.66 μ and 2.1 μ) decay rate does not increase nearly as much at day 530. One possibility to explain the difference between line and continuum behaviour is that we are seeing the effect of depletion of silicon into grains (of silicate dust). Another additional possibility is that the continuum is formed in regions less affected by dust. Since silicon is expected to be a minor component of the nuclear composition but a major component of the dust if silicates are formed, efficient condensation into dust might result in this effect.

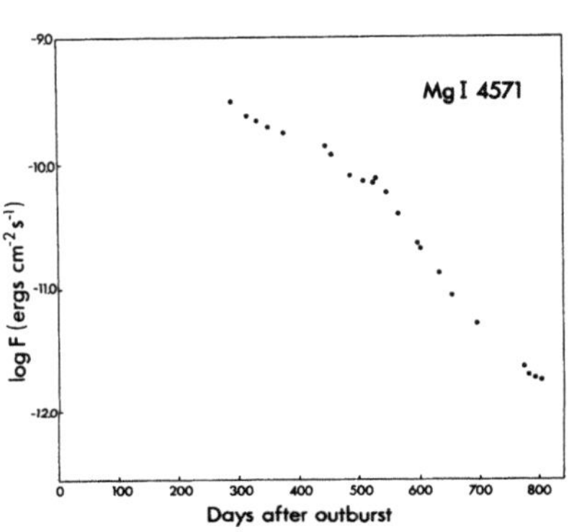

Fig. 3. The temporal behaviour of the flux of Mg I] 4571. Note the sudden increase in the rate of decline at day 530.

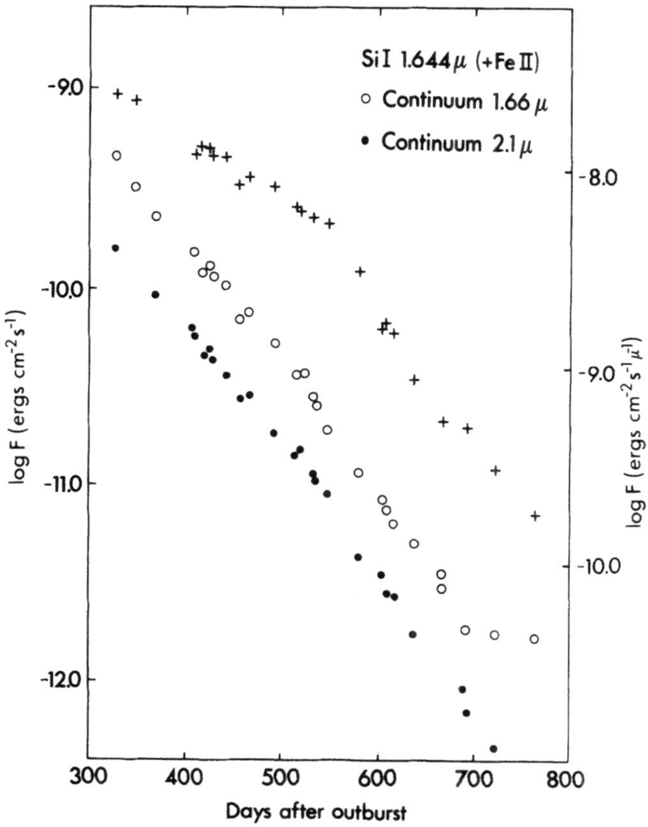

Fig. 4. The temporal behaviour of the flux of Si I (+ [Fe II] 1.644 μ) (left hand scale) and nearby continuum points (right hand scale). The increased rate of decline of the Si I 1.644μ line at day 530 is not so pronounced for the continuum points.

The fact that the [Ni I] 3.12 μ line also behaves in such a way suggests that other processes may be also operating. If we ignore the possibility of nickel participating in significant grain formation (we know little about nickel in this respect), then another explanation is that a dramatic extra cooling effect began near day 530, and this may be a prelude to the infra-red catastrophe.

In summary the following effects need to be quantified to explain all of the observations: type and quantity of dust, distribution of dust, distribution of relevant nuclear species and line and continuum forming regions, depletion of nuclear species, cooling as a function of time.

(b) Molecular Bands

The only authentically identified molecular species in the envelope of SN 1987A are carbon

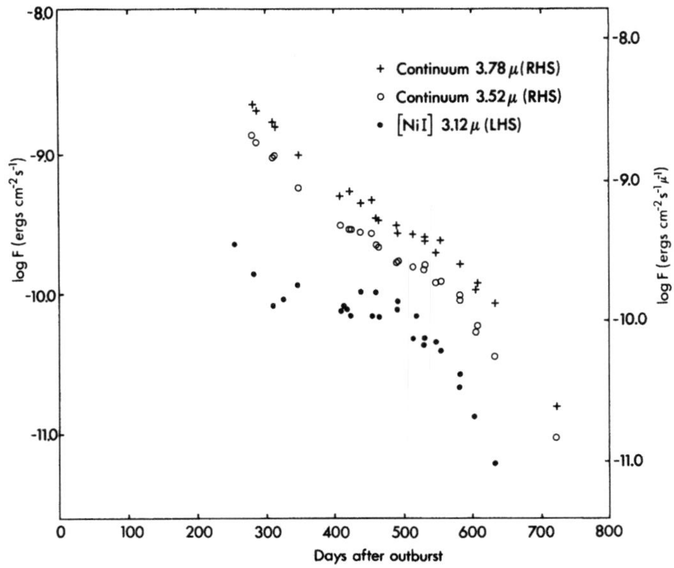

Fig. 5. The temporal behaviour of the flux of the [Ni I] 3.12 μ line and neighbouring continuum points. The decrease in the [Ni I] 3.12 μ line seems stronger after day 530 than that for the continuum points.

monoxide (CO) and silicon monoxide (SiO). Claims for the presence of molecular bands of other species (CO^+, CS) rest upon inappropriate wavelength coincidences. Here we discuss only CO because the study of SiO can be adequately achieved only with KAO data.

In Fig. 6 we show the temporal behaviour of the strength of the first overtone band of CO at 2.3 μ which has been measured on CVF spectra at ESO, La Silla. This behaviour must be eventually interpreted through modelling because of the temporal variation of temperature, density, optical depth, and CO formation. Nevertheless it is interesting to note that the peak emission occurs some 150–200 days from outburst in good agreement with predictions of Latter and Black [6] for the formation of CO in such an expanding envelope. The subsequent decrease in strength may not be reflecting a change in the mass of CO present, but more a decrease in temperature. In the discovery paper [4] it was pointed out that the fundamental band at 4.6 μ was almost certainly optically thick, and that therefore the two bands would change in strength at different rates as the temperature dropped. That this has indeed occurred is exemplified in Fig. 7, where we show spectra taken at days 287 and 492. The ratio of the strengths of fundamental to first overtone bands has increased dramatically meaning that the decrease in absolute strength of the fundamental band has been much slower than that of the first overtone band shown in the figure. Whether there has been a significant change in the actual mass of CO remains to be modelled. Work is in progress on this question.

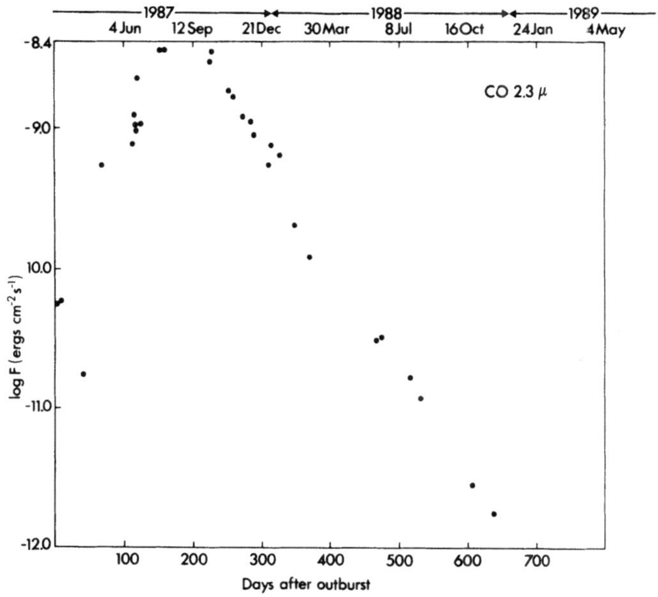

Fig. 6. The temporal behaviour of the flux in the CO 2.3 μ first overtone band of carbon monoxide.

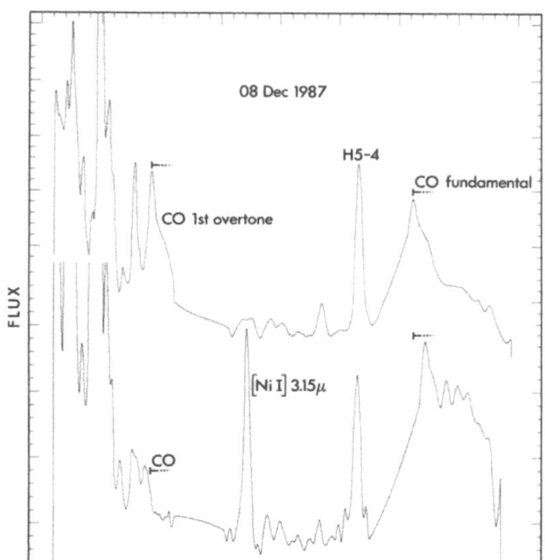

Fig. 7. CVF spectra from 1.4–5.4 μ of SN 1987A at 2 epochs. Note the temporal evolution of the strength of the fundamental and first overtone bands of CO, and the way in which the strength of the [Ni I] 3.12 μ has changed.

IV. Abundances

(a) Stable Species

In order to obtain a first order estimate of abundances in the envelope, a code has been developed that synthesizes the emission-line spectrum of a statistically uniform, expanding spherical mass. Levels are assumed to be populated according to the Boltzmann formula; ionization is not treated but must be specified; self absorption is treated in the Sobolev approximation; and clumpiness is incorporated via a filling factor. Currently this code includes $\simeq 1000$ forbidden lines from $\simeq 20$ ions. This code reproduces rather well the late time emission features of Type 1 SNe shown by KIRSHNER and OKE [7] and MEYEROTT [8] to be due to overlapping forbidden lines of [Fe II] and [Fe III].

Using this code to determine abundances we present results for SN 1987A at day 410. We rationalize the choice of this day as follows: there are reasons for supposing that the envelope was optically thin at both optical and infrared wavelengths; there was complete wavelength coverage at La Silla on this date as well as some extra data from the Kuiper Airborne Observatory; this date preceded the onset of dust formation at day 530 so corrections for absorption are not necessary. See Fig. 8 for the temporal development of the optical spectrum and Fig. 9 for the IR spectrum. The temperature indicators that we have available particularly for iron suggest that, if we are forced to choose an optimum single temperature, 2700 °K is the best choice. A single temperature for all of the ions must be a gross approximation to reality. Indeed since clumping and stratification must occur a single temperature for even one ion must be a simplification. Naturally, imprecise definition of the appropriate temperature adversely affects abundances determined from lines resulting from levels of high excitation. Abundances involving fine structure transitions will be much less severely affected. Our oxygen results fall in the first category and results for cobalt in the second. An earlier discussion of ionization conditions and abundances is given by OLIVA, MOORWOOD and DANZIGER [9].

The results are summarized in Table 2 where we give abundances in solar masses of a particular ion (using the emission lines shown) with an assumed distance of 50 kpc. We also give in brackets a factor by which the result should be multiplied in order to obtain a mass of the element by allowing for other ions. These factors are probably reasonably secure for Fe, Co, and Ni, but much less certain for the others. The result for oxygen is uncertain as can be seen by the result, 0.2 $M_\odot$ in brackets, obtained at an earlier epoch when it seemed possible that a temperature for O I had been obtained through observing the O I 5577 auroral transition. Thus there continues to be a large uncertainty in the measured mass of oxygen. It is particularly encouraging to note that estimates of the mass of iron have increased with time in accord with the predictions of radioactive decay of ^{56}Co. The silicon result was obtained in the following manner. The 1.644 μ emission line is a close blend of [Fe II] and Si I lines. We assumed that 0.075 $M_\odot$ of iron was present in the ionic proportions given in our table. This

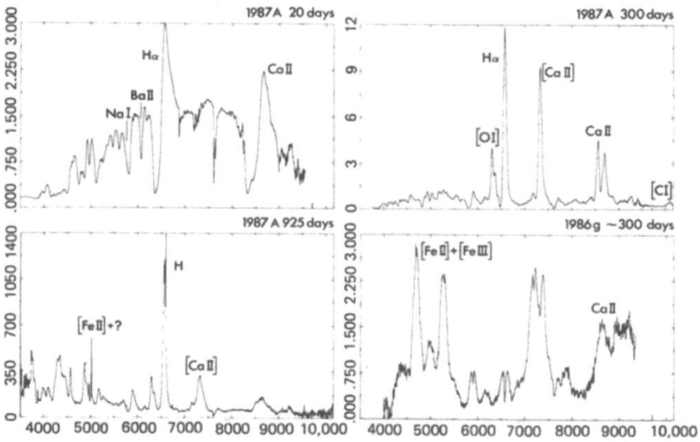

Fig. 8. Optical spectra of SN 1987A demonstrating its development from an optically thick extended envelope to an optically thin one. The spectrum of the Type Ia SN 1986g at 300 days is completely different from that of 1987A.

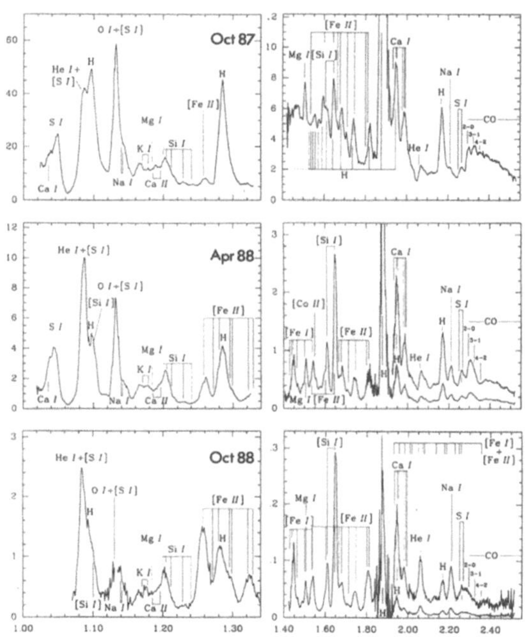

Fig. 9. IR spectrum of SN 1987A showing the temporal development.

Table 2: SN 1987A

	ABUNDANCES DAY 410 APRIL 1988			
	OBSERVED $M/M_\odot$		RECENT MODEL* $M/M_\odot$	
C I (9823,49)	0.036	(×2)	0.114	
O I (6300,64)	3.0 (0.2)	(× ?)	1.48	
Si I (1.644 μ)	0.034	(×3)	0.1	
Ar II (6.98 μ)	0.0005	(× ≥ 1.5)	0.004	(3)
Ca II (7291,7324)	< 0.007	(×1.5)	0.0033	
Fe I (1.423–1.462 μ)	0.043	(×3)	0.075	
Fe II (1.248–1.27 μ)	0.098	(×1.5)		(1)
(17.93 μ)	0.055			(4)
Co II (10.52 μ)	0.00193	(×1.5)	0.00188	^{56}Co only
Ni I (3.12 μ)	0.00074	(×3)	0.0177	(2)
Ni II (6.63 μ)	0.0014	(×1.5)		(2,3)

Isothermal LTE = 2700°K
Expansion velocity = 2000 km/s
* Thielemann, Hashimoto, Nomoto [10]

(1) Consistent also with [Fe II] 7155,72.
(2) Consistent also with [Ni I] 8203, [Ni II] 7380,7410,8300.
(3) Line strengths from Witteborn *et al.*, IAU Circular # 4592.
(4) Line strengths from Moseley *et al.*, IAU Circular # 4576.

provided a predicted line strength at 1.644 μ which was subtracted from the observed one. The remainder was ascribed to Si I 1.644 μ. An upper limit is implied for Ca II because the upper metastable level of the transition for the [Ca II] 7291,7324 lines (3^2D) is not populated in LTE if scattering of Ca II H and K photons succeeds in populating the upper level for the Ca II IR triplet emission lines (4^2P^0) with subsequent decay to the metastable level.

In this table we also present for comparison the results of nucleosynthesis in the model of Thielemann, Hashimoto and Nomoto [10]. Within the uncertainties, which can be considerable in some cases, there appears to be reasonable agreement. At this point the only believable difference is that for nickel where our abundances are lower than that of the model. All of the nickel lines available to us suggest this same discrepant result.

(b) Radioactive Species

The case of cobalt requires special discussion because of its temporal variation due to radioactive decay and because we obtained observations of the [Co II] 10.52 μ line over a considerable period.

Flux calibrated CVF spectra of SN 1987A in the 8–13 μ region have been obtained at regular intervals at ESO, La Silla. An example of such a spectrum taken on 2 Jan 1988 is shown in Fig. 10. In Fig. 11 we show the temporal behaviour of the intensity of the [Co II] 10.52 μ line. Points plotted near day 100 and earlier should be treated with caution because the intensity measurement of an emission feature near the expected wavelength was extremely difficult owing to its very small equivalent width. The array of spectra presented by AITKEN [11] demonstrate this effect.

The line strength has reached a maximum near day 250 and has decreased since that time until it became undetectable at day 625. We have used the program described above to calculate the expected line strength assuming that (a) 0.075 $M_\odot$ of ^{56}Co was originally produced and (b) 0.075 $M_\odot$ of ^{56}Co plus 0.0017 $M_\odot$ of ^{57}Co were produced. This latter amount is taken from the suggestion of WOOSLEY and PINTO [12] that the ratio of ^{57}Co/^{56}Co might be the same as the ratio of ^{57}Fe/^{56}Fe observed in the solar system. The dashed points connected by the vertical bars represent these theoretical points assuming only ^{56}Co production. The upper of the 2 bars in all cases results from assuming all cobalt is in the form of Co II; the lower bar allows for the possibility of some significant amount of Co I. We have already indicated that at day 410 there is good reason to believe from observations of iron and nickel in 2 stages of ionization that Co I / Co II = 1/2. The plus signs for ^{56}Co + ^{57}Co do not include a correction for the existence of Co I, and therefore a more realistic calculation would bring those points down by the same amount indicated by the vertical lines.

It is worth noting that at day 410 the Sobolev optical depth in the [Co II] 10.52 μ is approximately 0.1, whereas at day 225 it is 1.4. At about this time there is some indication of a real discrepancy between the computed strength and the observed strength. While some of this discrepancy may be due to the inadequacy of the treatment of the line transfer at large optical depths, it is also likely that at these earlier epochs not all of the cobalt has appeared above the region where the photosphere is formed. Obviously an understanding of this point is important in constructing models which include the mixing of radioactive material into the outer regions of the envelope.

The sudden decrease in strength of the 10.52 μ line after day 529 has been ascribed to absorption by dust by LUCY *et al.* [3]. Combined with extinction at shorter wavelengths, it pointed to silicates being the major component of the dust. Unfortunately however this effect has interfered with our expectation that after day $\sim$ 600 we could anticipate that ^{57}Co would become a major contributor to the line strength, because its decay time was 3.5 times longer than that for ^{56}Co. Nevertheless, if we concentrate on observed points near days 410, 476 and 520–529 we see that the best fit to the data requires a contribution of ^{57}Co approximately 1.5 times the amount suggested by WOOSLEY and PINTO [12]. However Fig. 9 contains no error bars on the points, observed or theoretical, and they may not be insignificant. Possible contamination of the observed line by other unidentified lines would decrease the implied

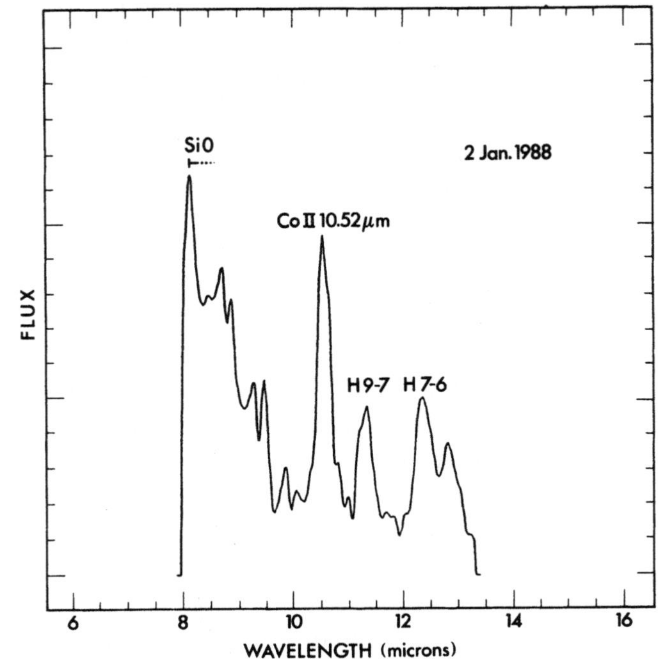

Fig. 10. A CVF spectrum of SN 1987A in the 8–13 μ region highlighting at this epoch the strength of the [Co II] 10.52 μ line, and the fundamental band of SiO.

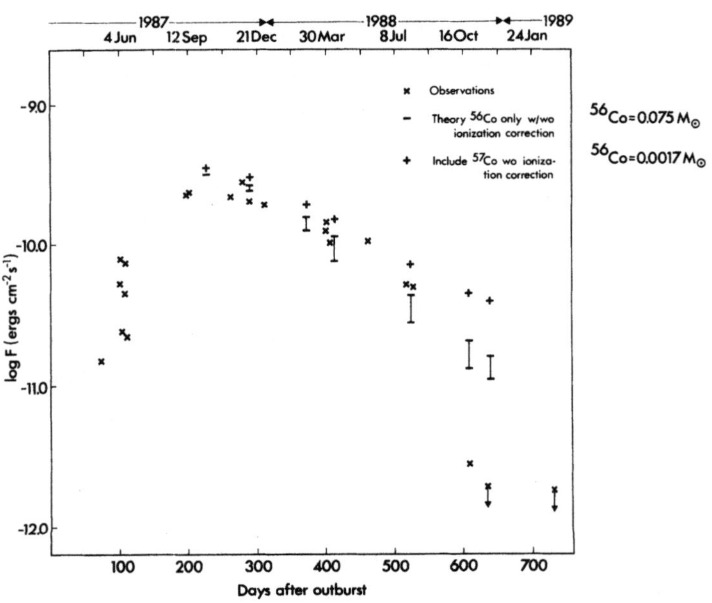

Fig. 11. The temporal behaviour of the strength of the [Co II] 10.52 μ line together with the predicted strengths under assumptions discussed in the text.

amount of ^{57}Co. For the purposes of discrimination at these epochs there is still a significant uncertainty in the Co I / Co II ratio, and in the oscillator strength for the transition producing the 10.52 μ line.

Finally we note that at later epochs there has been an increasing flux in the blue region ascribed to the increasing predominance of [Fe II] lines (Phillips, this conference). See Fig. 8. We have used the above program to calculate the [Fe II] spectrum over the complete wavelength range 0.35–20 μ. We have included all the known forbidden lines with their published oscillator strengths. In the wavelength region 3500–5500 Å we have compared our observed spectra with the computed one. While it is true that there are a sufficient number of coincidences of line peaks to create the impression that [Fe II] emission is a significant contributor to the observed features, there are also many major features in the observed spectrum that have no counterpart in the computed spectrum. There is also a mismatch between the computed flux level of the spectrum in this wavelength region and the observed one which is much higher. This reinforces the conclusion that there is significant flux in the observed spectrum not accounted for by the [Fe II] lines. It remains to be demonstrated whether much of this remaining emission originates from allowed transitions of Fe II or from other ionic species. From the point of view of nucleosynthesis alone it is important to solve this problem.

References

1. L.B. Lucy *et al.*: this conference.
2. I.J. Danziger, P. Bouchet, C. Gouiffes, L.B. Lucy: IAU Circ. # 4746 (1989).
3. L.B. Lucy, I.J. Danziger, P. Bouchet, C. Gouiffes: In *Structure and Dynamics of the ISM*, ed. by G. Tenorio-Tagle, M. Moles and J. Melnick (Lecture Notes in Physics, Springer-Verlag 1989).
4. I.J. Danziger, P. Bouchet, R.A.E. Fosbury, C. Gouiffes, L.B. Lucy, A.F.M. Moorwood, E. Oliva, F. Rufener: In Supernova 1987A in the LMC, ed. by M. Kafatos and A. Michalitsianos, (Cambridge University Press 1988).
5. A. Uomoto, R.P. Kirshner: Astrophys. J. **308**, 685 (1986).
6. W.B. Latter, J.H. Black: Astrophys. J. submitted (1989).
7. R.P. Kirshner, J.B. Oke: Astrophys. J. **200**, 574 (1975).
8. R.E. Meyerott: Astrophys. J. **239**, 257.
9. E. Oliva, A.F.M. Moorwood, I.J. Danziger: In *Proc. 22nd ESLAB Symposium on Infrared Spectroscopy in Astronomy*, Salamanca, Spain, 7–9 December 1988, ESA SP-290, p. 375 (1989).
10. F.-K. Thielemann, M. Hashimoto, K. Nomoto: to appear in Astrophys. J. (1989).
11. D.K. Aitken: Proc. A.S.A. **7**, 462 (1988).
12. S.E. Woosley, S. Pinto: In *Workshop on γ-ray Spectroscopy*, ed. by N. Gehrels and G. Share (AIP 1988).

Dust Condensation in the Ejecta of SN 1987A, II

L. B. Lucy, I. J. Danziger, C. Gouiffes, & P. Bouchet

1. Introduction

Work at ESO on the formation of dust from newly synthesized elements in the ejecta of SN 1987A began in late January when blue shifts of $\simeq$ 600 km s^{-1} were found for various optical emission lines in spectra taken after September 1988. It was quickly realized that these shifts could be due to the attenuation of emission from the far, and therefore receding, side of the ejecta caused by dust within the ejecta. However, dust formation — if it occurred — was expected to be discovered photometrically through a precipitous fading in the UV and optical (DWEK [1], KOSAZA *et al.* [2]) with concomitant brightening in the IR. But no such fading had occurred and the brightening in the IR after day $\simeq$ 450 (SMITH *et al.* [3]) had been attributed to a thermal echo by pre-existing circumstellar dust, an interpretation apparently supported decisively by the measurement of a small, but finite angular size of $\simeq$ 1.5 arcsec at 8-10 μ (ROCHE *et al.* [4]).

Nevertheless, on inspection, the photometric record did reveal support for our spectroscopic diagnostic of dust condensation. With respect to optical fading, the wonderfully precise and densely sampled V light curve of the Geneva observers (BURKI *et al.* [5]) showed that an increase in the rate of fading occurred on $\simeq$ day 530 (= 6 Aug. 1988), which we therefore identified (DANZIGER *et al.* [6]) as the onset of the dust condensation episode that later gave rise to the blue shifts. The dust's relatively subtle effect on the optical light curves was attributed to low condensation efficiency or, more probably, to its clumpiness [6]; but contributing also is the extended duration ($\simeq$ 150 days) of the condensation episode.

With respect to IR brightening, we pointed out in Paper I (LUCY *et al.* [7]) that the absence in the ROCHE *et al.* [4] data of an offset between the SN and the photocentre of the resolved 8-10 μ "echo" allowed a substantial fraction of this radiation to be attributed to dust in the ejecta. Moreover, the absence in optical light curves of the scattering analogue of this putative IR echo supported this reattribution [6] — see also FELTEN and DWEK [8].

Note also that, since the echo interpretation is inconsistent [4] with a spherically symmetric distribution of circumstellar dust, the absence of an offset then implies a special configuration of Earth-SN-dust. Finally, and perhaps most tellingly, the comparison in Paper I (§8) of the predicted bolometric light curve (PINTO *et al.* [9]) with two extreme empirical estimates derived by assuming all or none of the far-IR flux to be promptly thermalized input from the radioactive isotopes decisively favoured the former assumption. Taken together, these arguments constitute a strong case that the bulk of the far-IR flux originates within the ejecta, favouring therefore its interpretation as thermal emission by newly formed dust.

In this paper, we briefly review and comment on our subsequent quantitative analysis [7] of these spectroscopic and photometric consequences of dust condensation as well as reporting our recent attempt to construct and interpret an extinction curve for supernova dust. In addition, since it complicates the measurement of dust-induced blue shifts, we comment on the probably unrelated phenomenon of time-dependent red shifts for selected lines.

2. A Simple Model

In Paper I, a better fit to observed line profiles was achieved with the assumption of uniformly distributed dust than with dust confined to a thin shell. In addition, a spectroscopic diagnostic for the dust's albedo was identified theoretically but not confirmed observationally. Accordingly, the quantitative analysis in Paper I was largely based on the simplest possible model for the dusty ejecta at elapsed time t: a spherically symmetric distribution of matter expanding with velocity law $v = r/t$ and within which the volume emissivity for line photons is independent of r as also is the volume absorption coefficient $k_\lambda \rho$ of zero-albedo dust.

With these assumptions, line profiles can be derived analytically and have a scale-free shape depending only on the optical depth

$$\tau_\lambda = k_\lambda \rho R_E \ , \qquad (1)$$

where $R_E = V_E t$ is the radius of the ejecta at time t. These analytic profiles show the expected shift to the blue as τ_λ increases — see Fig. 4 in Paper I — and yield the useful formula

$$v_\lambda = -1 + \tau_\lambda^{-1} ln(1 + \tau_\lambda) \qquad (2)$$

for the velocity shift of the peak of a line at wavelength λ in units of V_E, the expansion velocity at the surface of the ejecta.

Equation (2) allows $\tau_\lambda(t)$ to be determined from the time variation of the blue shift of a particular line. In Fig. 7 of Paper I, τ_λ's thus derived are plotted for the lines Mg I] λ4571 Å, [O I] λ6300 Å, and [C I] λ9844 Å. The [O I] data were the most accurate and showed that the optical depth of the dust at $\lambda 6300$ Å had been a smooth, monotonically increasing function of t, being negligible prior to $t \simeq 530d$, increasing rapidly after $t \simeq 580d$, and then only slowly after $t \simeq 670d$. Thus, to the accuracy permitted by the spectroscopic sampling, the onset of dust condensation determined spectroscopically agreed with that determined photometrically ($t \simeq 530d$ [6]).

Under the assumptions of the adopted model, the mean escape probability $p(\tau_\lambda)$ for photons emitted within the ejecta is given by OSTERBROCK'S [10] formula. This then allows the dust-free magnitude $m_\lambda^\circ = m_\lambda - A_\lambda$ to be calculated since the mean extinction is simply

$$A_\lambda = -2.5 \log p(\tau_\lambda) \ . \tag{3}$$

In Paper I, Table 1 these spectroscopically-derived extinction corrections were tabulated for the V-band and the resulting corrected V light curve plotted in Fig. 8 of that paper. In the corrected light curve, the accelerated decline after day 530 is largely removed, but an increasing departure from the earlier (130–400d) exponential decay remains and is of course attributable to the increasing transparency to X- and γ-rays. This exercise indicates that $\tau_V(t)$ determined from the blue shifts is quantitatively consistent with the amount of extinction A_V suggested by the morphology of the light curve.

This same simple model can be used to compute the equilibrium temperatures of dust in the ejecta's ambient radiation field as well as the resulting IR spectrum. Such calculations were reported in Paper I (§7) for various grain species with, in each case, the maximum of amount of dust permitted by the photometry being determined as a function of t. When the grains were small particles of "astronomical" silicate (DRAINE and LEE [11]) or grey absorbers, the resulting maximum $\tau_V(t)$ curves agreed rather well with $\tau_V(t)$ determined from the blue shifts. But poor agreement was found with small particles of graphite, amorphous carbon or iron. Accordingly, small grains of these substances can only be minor constituents of supernova dust — unless in opaque clouds (see below).

On the basis of these dust-emission calculations, a crude attempt was made to construct a composite grain model for supernova dust by imposing both spectroscopic and photometric constraints. Specifically, the spectroscopically-derived τ_V had to be satisfied while still allowing a modest fraction ($\sim 20\%$) of the 8–10$\,\mu$ flux to be assigned to the thermal echo of Roche *et al.* [4]. This latter constraint limits the amount of silicate dust, which emits efficiently at 10$\,\mu$, but the resulting shortfall in τ_V can be made up with small graphite or amorphous carbon grains since these have higher equilibrium temperatures and radiate at $\sim 4\,\mu$. Although the resulting mixture is not implausible [1,2], this argument for a contribution from carbon grains is predicated on the correctness of ROCHE *et al.*'s [4] claim to have resolved the 8–10$\,\mu$ emission. But some doubt on this point follows from informal remarks at the Workshop by A. Chalabaev, who stressed the difficulty that the time-dependence of seeing (SARAZIN [12]) poses for such differential-scanning experiments. Accordingly, the presence of carbon grains should be regarded as not proven until supported by modelling of continuum fluxes in the L and M bands.

A further calculation readily made with this simple model is the condensation efficiency — i.e., the dust mass derived from τ_V expressed as a fraction of the maximum value permitted by the element abundances in the ejecta. For the abundances predicted by HASHIMOTO *et al.* [13], the condensation efficiency up to day 775 — by which time grain formation had greatly slowed, if not stopped — was $\lesssim 10^{-3}$, a surprisingly low value. However, an efficiency of ~ 1 is permitted if the dust is clumped — i.e., if one of the model's basic assumptions is abandoned.

3. Extinction Curve

The interstellar extinction curve is unquestionably a basic datum in studies of the nature of local Galactic dust. The corresponding curve for supernova dust is therefore much to be desired.

In principle, the theory of dust-induced blue shifts allows the extinction curve of supernova dust to be constructed: measurements of these shifts for numerous lines covering a wide wavelength range would give, via (2) and (3), the dependence of A_λ on λ. Moreover, this procedure would give the extinction curve as a function of time, offering thereby the prospect of determining changes in composition and grain size as the dust condensation episode developed. Unfortunately, the number of unblended lines with good S/N data are few. Moreover, the derived A_λ's are model dependent, and systematic errors due to a model's failures are likely to depend on the emitting element and its ionization stage. A further problem for some lines is the uncertainty as to how much of the blue shift is due to dust and how much to the weakening of an independent effect that caused redshifts at earlier epochs.

Because of these difficulties, at present we use this spectroscopic technique to derive extinctions only at the wavelengths of the [O I] and Mg I] lines and at the representative epoch $t = 625d$. These A_λ values, derived from the τ_λ's in Fig. 7 of Paper I, are plotted as asterisks in Fig. 1. Despite the difficulties, further spectroscopic points on this extinction diagram would be useful, especially in the IR.

In view of the presently meagre spectroscopic data, an attempt has been made to determine A_λ's photometrically. This possibility follows from the successful identification of the condensation episode by inspection of the morphology of the V light curve (see above). This success suggests that useful estimates of $A_\lambda(t)$ are given by $m_\lambda(t) - \tilde{m}^\circ_\lambda(t)$, where m_λ is the observed magnitude and $\tilde{m}^\circ_\lambda$ is an estimate of the dust-free light curve obtained by extrapolating the trend shown by $m_\lambda(t)$ prior to dust condensation. Specifically, $\tilde{m}^\circ_\lambda(t)$ is given by the least squares straight line fitted to $m_\lambda(t)$ in the interval 450–525d. The A_λ's at $t = 625d$ thus derived from the broad-band magnitudes UBV... M are plotted as filled circles in Fig. 1. Note that this procedure gives $A_V = 0.65$ at $t = 625d$, in reasonable agreement

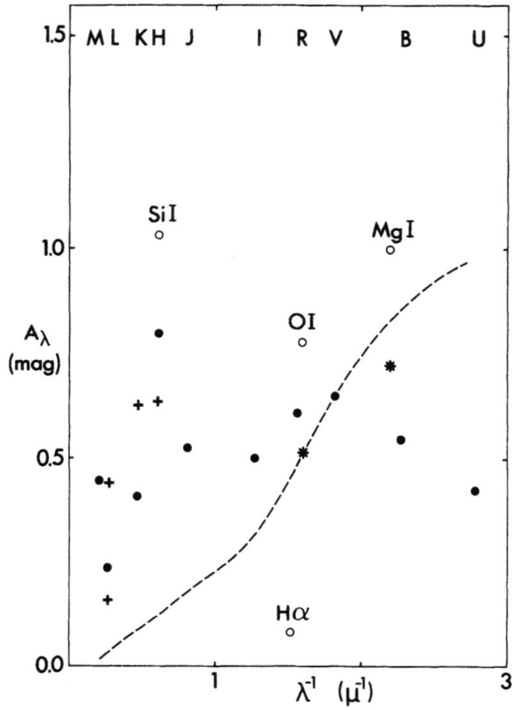

Fig. 1. Estimates of A_λ, the mean extinction within the ejecta at wavelength λ. Data are from blue shifts (*asterisks*), broad-band photometry (*filled circles*), line fluxes (*open circles*), and continuum fluxes (*crosses*). The interstellar extinction curve scaled to $A_V = 0.65$ is also shown.

with the value 0.58 found from the blue shifts (Table 1, Paper I).

Further points on the extinction diagram can be obtained by applying the above technique to line- and continuum fluxes measured from calibrated spectra. Examples of such data and the linear fits are shown in Figs. 2–4 for the lines Mg I] λ4571 Å, [O I] $\lambda\lambda$6300,6364 Å and [Si I] $\lambda 1.65\,\mu$, respectively. Fig. 4 also includes (with arbitrary vertical displacement) the measurements and fits for the neighbouring continuum. As with the photometry, differences in the fits are evaluated at $t = 625d$ to yield the A_λ's plotted as open circles (line fluxes) and crosses (continuum fluxes) in Fig. 1. Note that this diagram includes points additional to those derived from Figs. 2–4.

In addition to illustrating the derivation of A_λ's, Figs. 2–4 — especially the Mg I] and [Si I] data — provide striking support for $t \simeq 530d$ as the onset of a major dust condensation episode. The Mg I] data also hint at an earlier, minor episode at $t \simeq 450d$ that would coincide with the initial IR brightening [3] and with the onset of fluctuations in U and B light curves [5,7].

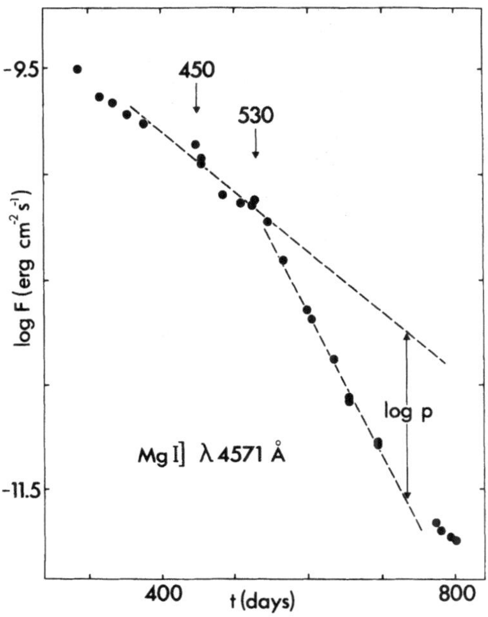

Fig. 2. Measured fluxes for the Mg I] $\lambda 4571$ Å emission line. Least-squares linear fits are shown and their use to estimate the escape probability and hence A_λ illustrated. The straight line segments intersect at $t = 530d$.

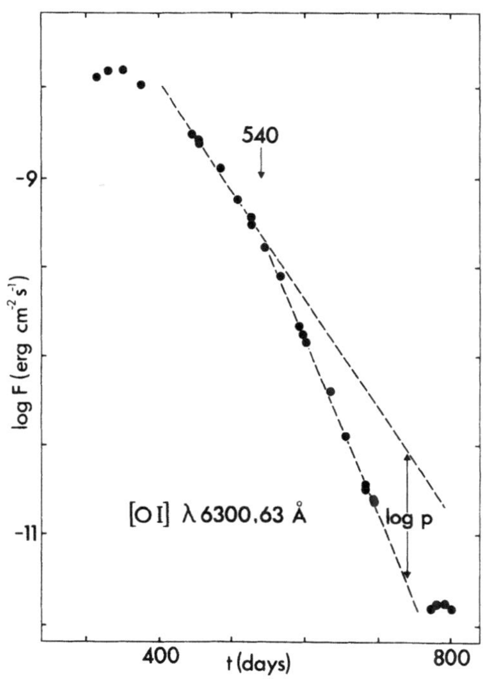

Fig. 3. As Fig. 2 but for the [O I] doublet.

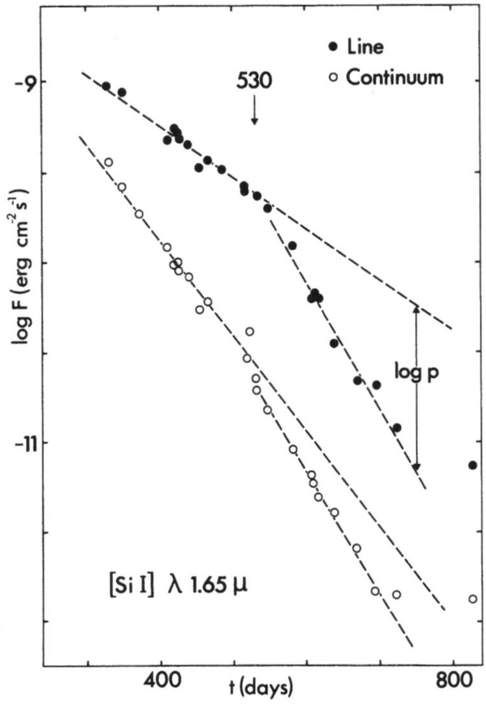

Fig. 4. As Fig. 2 but for the [Si I] $\lambda 1.65\,\mu$ line. The continuum fluxes (open circles) are plotted with arbitrary vertical displacement.

A further point of interest in Fig. 4 is the markedly stronger fading of the [Si I] $\lambda 1.65\,\mu$ line flux relative to the neighbouring continuum. This is strongly suggestive of depletion of Si from the gas phase in consequence of the formation of silicate grains, an interpretation that requires a condensation efficiency greatly exceeding the estimate of $\lesssim 10^{-3}$ derived with the simple model (§2). Of course, this data could also be interpreted as a temperature effect, but this would imply a coincidence between the onset of dust condensation and an increase in the rate of cooling in the Si zone. Physical modelling of this zone should settle this question. If the stronger line fading is in fact due solely to depletion, then $\simeq 27\%$ of silicon had condensed by day 625 and this rises to $\simeq 56\%$ by day 700.

4. Interpretation

To guide the interpretation of the extinction data collected in Fig. 1, the interstellar extinction curve (Savage and Mathis [14]) scaled to $A_v = 0.65$ is also plotted.

Blueward of V, interstellar extinction increases and a similar rise is evident for supernova dust when the two Mg I] points (from blue shifts and line fluxes) are compared with the corresponding [O I] points. However, this trend is not confirmed by the A_λ's derived from broad band photometry; indeed, surprisingly, we find $A_V > A_B > A_U$, so that the SN in fact got bluer both in B–V and U–B as dust condensation proceeded, a result requiring explanation (WHITELOCK *et al.* [15]). Interestingly, this counter-trend defined by the broad-band magnitudes is confirmed by IUE data. From plots published by KIRSHNER and GILMOZZI [16], we estimate that, on day 625, $A_\lambda = 0.35$ at 3100 Å, 0.23 at 2750 Å, and is < 0.1 at shorter wavelengths.

One possibility for explaining this blueing despite increasing V-band extinction is that, notwithstanding the grains' increasing extinction efficiency as λ decreases, the expected reddening is offset by the increasing albedo — in contrast with extinction of starlight, a scattering in the SN's ejecta does not preclude a photon from eventually contributing to the measured magnitude. This effect has been investigated by computing the mean escape probability $p(\tau_\lambda; \omega_\lambda)$ from a uniform sphere when the absorbing matter has albedo ω_λ. (In fact, these calculations use the excellent analytical approximation

$$p(\tau;\omega) = p(\tau;0)[1-\omega+\omega p(\tau;0)]^{-1} , \tag{4}$$

derived on the assumption that scattered photons are uniformly distributed in the sphere. Here $p(\tau;0)$ denotes OSTERBROCK'S [10] formula.) For a given grain model and with τ_V chosen to give $A_V = 0.65$, such calculations yield a theoretical extinction curve for supernova dust at $t = 625d$ with scattering accurately taken into account. Fig. 5 shows the curves for astronomical silicate [11] with various grain sizes. These calculations show that increasing the grain size to $1\,\mu$ does indeed eliminate the reddening we associate with extinction. Nevertheless, the effect falls far short of accounting for the observed blueing, and the same holds for graphite grains.

A second possibility for explaining the blueing is that, with increasing transparency to γ-rays, those that are thermalized suffer this fate at increasing mass coordinate. Accordingly, an increasing fraction of the (declining) UV and optical radiation is emitted by matter exterior to the ejecta and much of this radiation escapes without traversing the dust in the ejecta. An example of this effect is the low A_λ derived from H_α line fluxes — see Fig. 1.

This second suggestion was criticized at the Workshop by M. Phillips on the grounds that much of the radiation in the U and B bands is from [Fe II] and so certainly originates in the ejecta. This is a valid point and prompts the further suggestion that an anti-correlation between dust clumps and Co/Fe clumps has allowed the increasing relative importance of Fe II emission to weight the spectrum towards the blue despite increasing V-band extinction. Further analyses of the kind exhibited in Figs. 2–4 but now choosing points that include and exclude [Fe II] features could perhaps resolve this issue.

Redward of V, interstellar extinction falls because most interstellar grains are then small

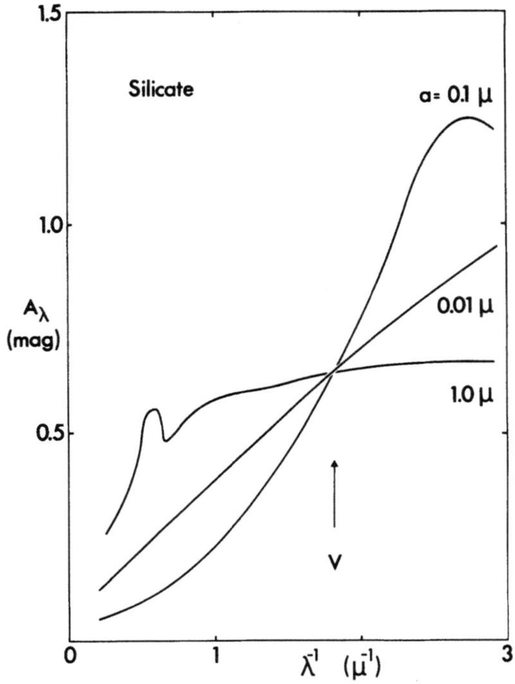

Fig. 5. Theoretical mean extinction curves. Silicate particles are uniformly distributed in the ejecta and have the indicated radii. Curves have $A_V = 0.65$, corresponding to day 625.

(i.e., $2\pi a \ll \lambda$) and therefore have extinction efficiencies that (apart from resonances) decrease with increasing λ. Surprisingly, no corresponding fall is evident in the extinction data (Fig. 1) for supernova dust; indeed, to the accuracy of the data, A_λ remains constant at $\simeq$0.4–0.6 mag. out to $\lambda \simeq 4\,\mu$. (Note that the technique used to derive the observed A_λ's allows negative values, so that the data could have scattered about an extinction curve that $\rightarrow 0$ as $\lambda \rightarrow \infty$.)

An obvious suggestion to explain this continuing extinction into the mid-IR is that supernova dust is dominated by large particles — i.e., with a $\gtrsim 1\,\mu$ — see Fig. 5. But this is inconsistent with the evidence of selective extinction at short wavelengths provided by the [O I] – Mg I] comparison. In addition, for silicates, large grains are excluded by our non-detection [7] of the spectroscopic diagnostic of dust albedo.

A second and preferred suggestion is that neutral (i.e., grey) extinction is found in the IR because the bulk of the dust is in opaque clouds which therefore geometrically occult background emission. This can be made consistent with selective extinction at short wavelengths if there is a diffuse distribution of small dust particles between the opaque clouds.

In addition to explaining the extinction data, this opaque cloud model for the dusty ejecta is consistent with our expectation that the condensation efficiency should be high (§2) as well as with the indications of silicon depletion (§3). It is also consistent with the absence of dust emission features in IR spectra — e.g., the silicate peak at $\simeq 9.7\,\mu$. Strong support also comes from the finding that the spectroscopic diagnostic of dust — i.e., the blue shift — is found also for IR lines — e.g., [Fe II] $\lambda 1.26\,\mu$ (SPYROMILIO *et al.* [17]) and [Ni II] $\lambda 6.65\,\mu$ (RANK [18]).

5. Infrared Spectrum

For a given grain model and with the dust content (i.e., τ_V) maximized, the infrared spectrum predicted by the simple model of §2 is a function of t only with no free parameters. But now a parameter quantifying the population of opaque clouds is necessary. This can be taken to be

$$\tau_* = \tfrac{3}{4} n r_b^2 \,, \tag{5}$$

where n is the number of identical opaque spherical clouds in the ejecta and $r_b = R_b/R_E$ is their fractional radius. This occultation optical depth is such that $e^{-\tau_*}$ is the probability that a line segment of length R_E does not intersect an opaque cloud.

With τ_* specified, the amount of diffusely distributed dust may be maximized exactly as for the simple model. In such calculations, the generally unequal temperatures of the diffuse dust and of the black-body emitting opaque clouds are determined, in a procedure analogous to that described in Paper I (§7), from their separate conditions of global thermal equilibrium in the ambient radiation field.

An IR spectrum predicted by this model for day 625 is plotted in Fig. 6 (curve c). In this case, the diffuse component comprises silicate grains with $a = 0.01\,\mu$ and the occultation optical depth $\tau_* = 0.4$, corresponding to minimum extinction $A_* = 0.31$ — i.e., that due solely to occultation by opaque clouds. In addition, IR spectra obtained with the simple model (§2) — i.e., without opaque clouds — are also plotted for different grain species — silicate with $a = 0.01\,\mu$ (spectrum b) and graphite with $a = 0.1\,\mu$ (spectrum a). Comparison of spectra b and c reveals how effectively the opaque clouds model with realistic τ_* reduces the prominence of the silicate peak at $9.7\,\mu$. Spectrum a, on the other hand, illustrates again (see Paper I) that, if the dust is not clumped, grain species that radiate inefficiently in the IR cannot be dominant constituents of supernova dust.

Also plotted in Fig. 6 are the fluxes derived from ESO broad band photometry (K–Q_0) as well as the nearly contemporaneous continuum flux of $\simeq 10$ Jy at 20–30 μ obtained from the Kuiper Airborne Observatory (MOSELEY *et al.* [19]). Strong excesses relative to the cloud model are seen in the M and N_1 fluxes, but these are due to CO and SiO emission, respectively, and so are not failures of the model. Of most significance, given the

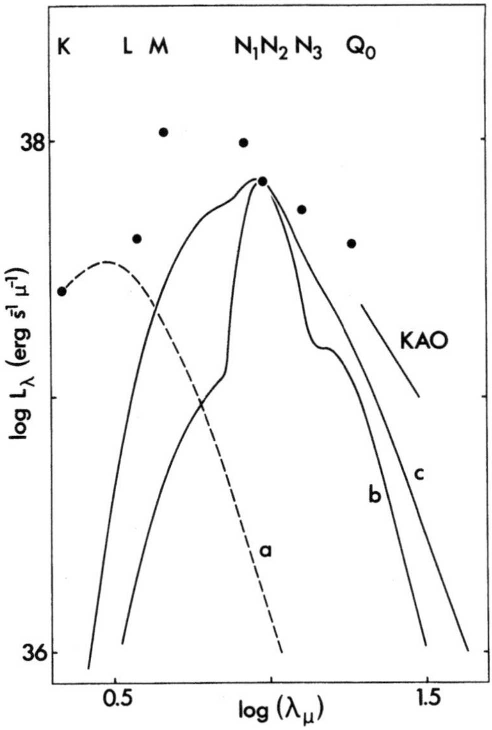

Fig. 6. Theoretical IR spectra for day 625. Spectra *a* and *b* are for uniformly distributed graphite ($a = 0.1\mu$) and silicate ($a = 0.01\mu$) grains, respectively. Spectrum *c* is for the opaque clouds – diffuse dust model, with $\tau_* = 0.4$ and silicate grains ($a = 0.01\mu$) as the diffuse component. Observed fluxes are from ESO (K–Q_0) and KAO (20–30 μ).

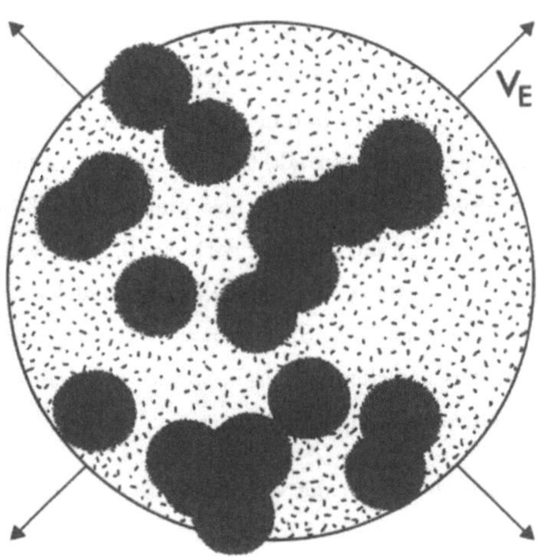

Fig. 7. Opaque clouds – diffuse dust model. Here the number of clouds $n = 20$ and have radii such that $\tau_* = 0.4$.

declining emission efficiencies of single grains as $\lambda \to \infty$, is the quality of the fit at the longest wavelengths — i.e., to the KAO flux, which is matched to within a factor two. This degree of success with such a simple modelling exercise together with the arguments of (§4) constitute a strong case for the existence of opaque dust clouds in the ejecta of SN 1987A.

Fig. 7 illustrates the opaque clouds — diffuse dust model used to compute the IR spectrum just discussed. In this diagram, the number of clouds $n = 20$ and their radii $r_b = 0.163$ are such that $\tau_* = 0.4$. The corresponding area covering factor of the clouds is $C_* = 0.402$, which is derived from the formula

$$C(\tau) = \tfrac{4}{3}\tau p(\tau) , \qquad (6)$$

where once again OSTERBROCK'S [10] formula appears — here denoted by $p(\tau)$. The corresponding volume filling factor is $\alpha_* = 0.084$ and is derived from the formula

$$\alpha(r,n) = 1 - (1 - r^3)^n . \qquad (7)$$

6. Redshifts

As noted earlier (§1), certain lines that developed blue shifts after day 530 had earlier exhibited redshifts, and so there is some uncertainty as to what fraction of the net shift to the blue should be attributed to dust. The favoured explanation is that these redshifts are due to electron scatterings in the expanding hydrogen envelope (WITTEBORN *et al.* [20]). But this cannot be entirely correct in view of the absence noted by DANZIGER *et al.* [21] of a redshift for the [O I] $\lambda\lambda$6300,6364 Å doublet. A revision of this model suggests itself when one notes that the lines that did show a redshift — see PHILLIPS and WILLIAMS [22] for a review of the data — are by and large from ions (Fe II, Ni II, Co II) that inevitably have close proximity to the ^{56}Co power source or are lines (H, He I, Ar II) that require close proximity because of high ionization or excitation potentials. Conversely, lines that did not show redshifts (O I, Mg I, C I, Ca II) involve low ionization and excitation potentials. This division could be explained if the ^{56}Co is in discrete clumps, each of which maintains around it a zone of enhanced ionization, excitation and electron density. Line photons emitted from within these zones would then have a higher probability of undergoing electron scattering than line photons emitted elsewhere.

References

1. E. Dwek: Ap. J. **329**, 814 (1988).
2. T. Kosaza, H. Hasegawa, K. Nomoto: Ap. J. **344**, 325 (1989).
3. C. Smith, S. James, D. Aitken, G. Orton, P. Roche: IAU Circ. No. **4645** (1988).
4. P.F. Roche, D.K. Aitken, C.H. Smith, S.D. James: Nature **337**, 533 (1989).
5. G. Burki, N. Cramer, M. Burnet, F. Rufener, B. Pernier, C. Richard: Astr. Ap. **213**, L26 (1989).
6. I.J. Danziger, C. Gouiffes, P. Bouchet, L.B. Lucy: IAU Circ. No. **4746** (1989).

7. L.B. Lucy, I.J. Danziger, C. Gouiffes, P. Bouchet: in *IAU Colloquium No. 120, Structure and Dynamics of the Interstellar Medium*, ed. by G. Tenorio-Tagle, M. Moles, and J. Melnick, in press (1989) (Paper I).
8. J.E. Felten, E. Dwek: Nature **339**, 123 (1989).
9. P.A. Pinto, S.E. Woosley, L.M. Ensam: Ap. J. **331**, L101 (1988).
10. D.E. Osterbrock: in *Astrophysics of Gaseous Nebulae* (W.H. Freeman; San Francisco 1974).
11. B.T. Draine, H.M. Lee: Ap. J. **285**, 89 (1984).
12. M. Sarazin: ESO Messenger **49**, 37 (1987).
13. M. Hashimoto, K. Nomoto, T. Shigeyama: Astr. Ap. **210**, L5 (1989).
14. B.D. Savage, J.S. Mathis: Ann. Rev. Astr. Ap. **17**, 73 (1979).
15. P.A. Whitelock, R.M. Catchpole, J.W. Menzies, M.W. Feast, S.E. Woosley, D.A. Allen, F. van Wyk, F. Marang, C.D. Laney, H. Winkler, K. Sekiguchi, L.A. Balona, B.S. Carter, J.H. Spencer Jones, J.D. Laing, T. Lloyd Evans, A.P. Fairall, D.A.H. Buckley, I.S. Glass, M.V. Penston, L.N. da Costa, S.A. Bell, C. Hellier, M. Shara, A.F.J. Moffat: M.N.R.A.S., in press (1989).
16. R.P. Kirshner, R. Gilmozzi: in *Exploring the Universe with the IUE Satellite* (2nd ed.), ed. by Y. Kondo (Kluwer, Dordrecht), in press (1989).
17. S.H. Moseley, E. Dwek, W. Glaccum, J.R. Graham, R.F. Lowenstein, R.F. Silverberg: Nature **340**, 697 (1989).
18. J. Spyromilio, W.P.S. Meikle: M.N.R.A.S., in press (1989).
19. D.M. Rank: This volume (1989).
20. F.C. Witteborn, J.D. Bregman, D.H. Wooden, P.A. Pinto, D.M. Rank, S.E. Woosley, M. Cohen: Ap. J. **338**, L9 (1989).
21. I.J. Danziger, P. Bouchet, R.A.E. Fosbury, C. Gouiffes, L.B. Lucy, A.F.M. Moorwood, E. Oliva, F. Rufener: in *Supernova 1987A in the Large Magellanic Cloud*, ed. by M. Kafatos and A.G. Michalitsianos; p. 37 (Cambridge U. Press 1988).
22. M.M. Phillips, R.E. Williams: This volume (1989).

AAT Observations of SN 1987A & Other Supernovae

Raylee A. Stathakis, Michael A. Dopita, Russell D. Cannon, & Elaine M. Sadler

1. Introduction

Extensive observations of nearby supernova SN1987A have been made at the 3.9m Anglo–Australian Telescope (AAT). Most of these are spectra, with a wide range of wavelengths and resolutions, which are being processed and collated into an open–access archive. Here we present two subsets of this archive: a set of low–dispersion optical spectra taken from 10 to 755 days after core collapse, and a study of emission line profiles made at very high dispersion and signal–to–noise ratio. SN1987A has also sparked interest in other supernovae and we report on the progress of a program to follow the spectral evolution of SNe at late times.

2. SN1987A at Low Dispersion

SN1987A was observed by RAS, RDC and MAD on 49 occasions between March 1987 and March 1989 with the Faint Object Red Spectrograph (FORS) on the AAT. FORS is a fixed–format low dispersion spectrograph covering the wavelength range 5500 – 11000Å at a resolution of 20Å. It is an ideal device for forming a uniform data set, and can be used in parallel with most other AAO instruments.

Since SN1987A was so bright in the first few months, various devices were used to attenuate the light. Most frequently we used a Fibre Image Slicer (FIS) which feeds the slit with a 2mm diameter bundle of 100μ optical fibres. The image was defocused across the bundle, which thus spread the light over a large percentage of the GEC CCD detector. Unfortunately FIS was found to distort the instrument response, so the FORS data were corrected to match the flux distribution in the Bochum data set ([1]) up to day 100. Fluxes redward of 8500Å were calibrated using a low–order extrapolation since our data extended further into the red than the Bochum observations. At later times the object had faded considerably and we were able to observe it directly and to take wide–slit exposures. Consequently, the continuum shape of these data did not require correction, but since the weather was often non–photometric we used the SAAO I–band photometry ([2,3,4,5,6]) to scale the spectra to absolute spectrophotometry. The FORS data set is now available on request.

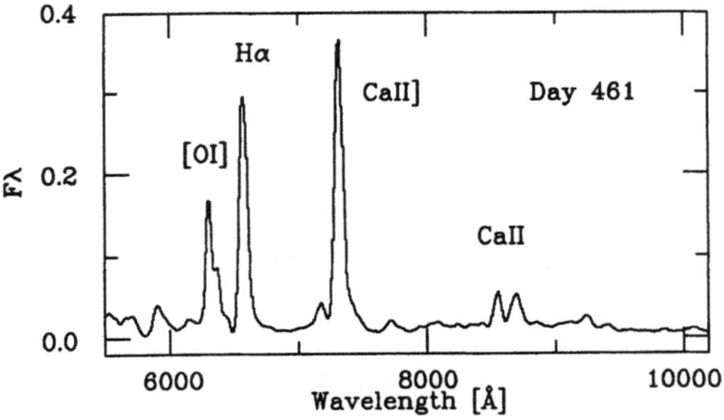

Figure 1: FORS spectrum of SN1987A on 29 May 1988. Flux is in 10^{-10} ergs/cm^2/s.

Between days 200 and 600 the spectrum of SN1987A was characterised by strong nebular emission lines. An example is shown in Fig.1, where the stronger lines are identified. The fluxes of these lines were measured from the FORS spectra, and the evolution was compared to predictions made by FRANSSON and CHEVALIER in mid-1987, [7,8]. In Fig.2 the measured flux evolution of CaII] 7291–7324Å and [OI] 6300,6363Å is plotted together with the predicted values from the $15M_\odot$ and $20M_\odot$ models from [7] and [8] respectively. We find that the shape of the CaII] flux curve is well reproduced in both models, with a slightly better fit from the $20M_\odot$ model. The shape of the [OI] flux curve agrees with the $20M_\odot$ model, with some discrepancy before day 300. However, the $15M_\odot$ model results in a far better match to the CaII]/[OI] ratio, especially from day 400. These results may indicate that the mass of the progenitor of SN1987A was closer to $15M_\odot$, but that the overestimate of [OI] flux line at early times is due to the simplicity of the model. A more sophisticated treatment is required incorporating the effect of mixing and considering the whole of the supernova envelope in order to understand any discrepancies.

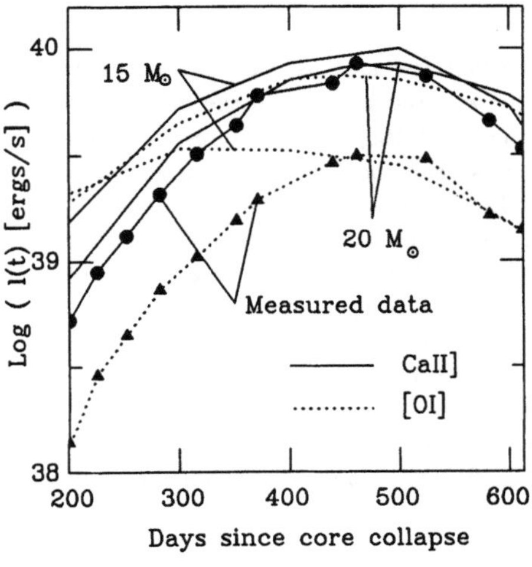

Figure 2: Measured fluxes of CaII] and [OI], between days 200 and 612, compared to predicted fluxes from [7,8]. l(t) is defined in [7,8] as luminosity corrected for Co^{56} decay and scaled to day 300.

3. Line Profiles of SN1987A

Spectra of the strongest optical emission lines were observed on several occasions during the supernebular period between days 200 and 600. These spectra, taken on the RGO Spectrograph with a GEC CCD detector, have resolutions of 9 – 60 km/sec, and S/N of 100 – 1300. All of the lines studied displayed evolution in the overall shape of their profiles and in structure with a characteristic scale of a few hundred km/sec. In addition, the [OI] 6300,6363Å doublet and the [CI] 8727Å line contained structure with a scale length of ~ 80km/sec and amplitudes of 3 – 10 percent of the average line profiles. Figure 3 shows the 6300 and 6363Å lines overlaid on a velocity scale relative to the velocity of the circumstellar lines. The correspondence between the features in the two lines is undeniable. [CI] was more difficult to observe as it was severely blended with CaII 8662Å at all times. Nevertheless many of the features also match with [OI] in velocity space. These features remained stationary in velocity throughout the period covered and changes in amplitude, if any, were small.

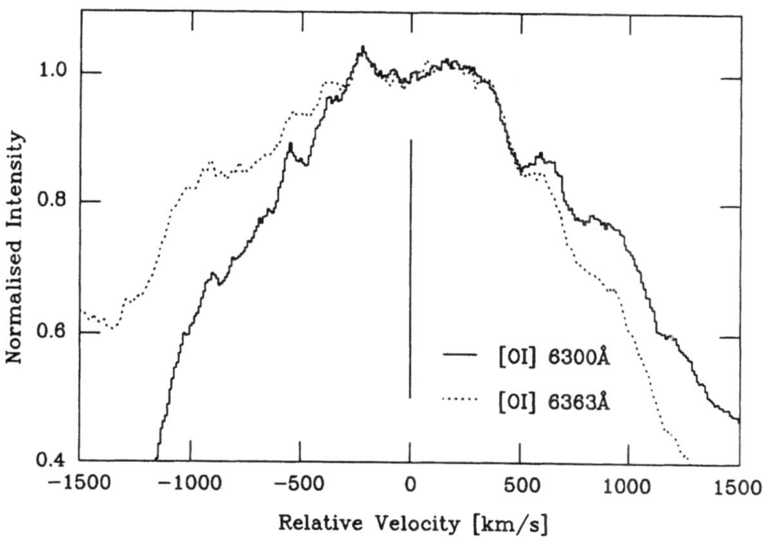

Figure 3: The line profiles of [OI] 6300,6363Å in SN1987A on 10 December 1987, on a velocity scale relative to +286 km/s (the velocity of the CSM lines). Both profiles were normalised to one at zero velocity. The small features throughout the lines match up very well, confirming their physical nature.

These features can be attributed to relatively dense clouds within the emitting region. About 50 such clumps are observed, which implies that ~ 5% of the emitting volume is occupied by clump material. These features are well resolved in our data, so we are confident that 60 km/sec is a physical lower limit to the size of these clouds. From Fourier transform analysis we find that the population of features does not increase towards smaller scale lengths. This evidence argues for clumps formed by Rayleigh–Taylor instabilities rather than thermal instabilities since the latter process

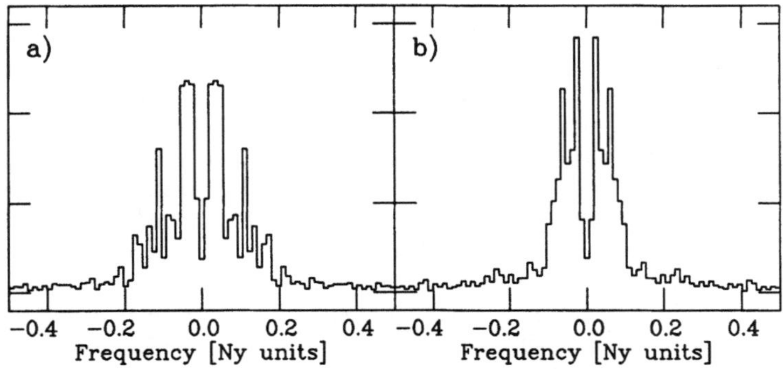

Figure 4: The power spectra of a) the blue wing and b) the red wing of the sum of the [OI] 6300,6363Å lines from 10 December 1987 and 29 February 1988. The dispersion is 10 km/s. The blue wing contains features of higher frequency.

favours the formation of many small clumps, and the lack of evolution suggests that the clumps were formed well before the first observation – probably shortly after collapse. The features on the red wing of the profiles are systematically broader than for the blue wing as seen in the power spectra for each side of the line (Fig.4). This effect might be due to electron scattering which would have a greater effect on red–shifted than blue–shifted photons since they have a longer pathlength. The larger scale features reveal the interesting possibility that there may exist large scale global symmetry in the stability. In Fig.5 are plotted the residuals in the [OI] profile reflected about the zero of velocity for SN1987A. All the major features appear to be correlated on a scale of 300 – 500 km/sec, although the correlation disappears on a smaller scale. This may reflect a degree of anisotropy in the original explosion.

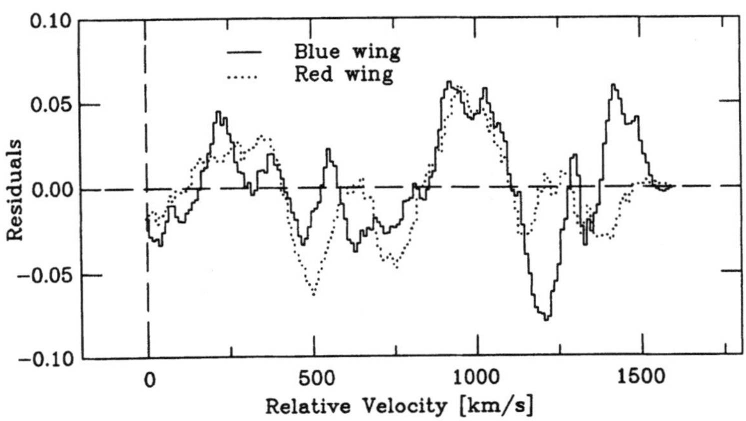

Figure 5: The blue and red wings of the [OI] data with the overall line shape removed. The blue wing has been reflected and overlaid on the red wing. Some large–scale symmetry is apparent.

4. Late-time Spectra of Other Supernovae

While our understanding of SN1987A is rapidly increasing, this knowledge is of little use if it cannot be transferred to other SNe. Late time spectra of other SNe are rare, and few objects have been observed over a sufficient time span to map the spectral evolution. With this in mind EMS and RAS began a program early this year to obtain late-time low dispersion red spectra of other SNe on the AAT. The emphasis is on Type II SNe, though in practice most SNe available with declination $< 20°$ and brighter than 17th magnitude at maximum are being observed. The objects observed to date fall into three groups - Type II, peculiar Type II and Type Ia. The Type Ia observations do not include many epochs since the spectra of these SNe at late times do not contain strong red emission lines and so the SNe are difficult to locate within their parent galaxies. Peculiar Type II SNe SN1988Z and SN1989C were observed both in the red and the blue. Their spectral lines are very narrow and have an unusual shape, and the evolution of the spectra is totally different from that seen in SN1987A and many other Type II SNe. We are collaborating with Filippenko and Shields (Berkeley) in investigating these SNe and our combined results are described by Filippenko elsewhere in this volume.

SN1988A and SN1988H were followed most successfully, to day 483 in the case of SN1988H. These two objects closely resembled SN1987A, also entering a supernebular phase. Figures 6a, b, and c compare the three SNe at approximately 190 days past core collapse. All three are dominated by Hα, [OI], CaII] and the CaII triplet. The CaII]/[OI] ratio is very similar for the three, though these lines are both weaker in SN1988A and SN1988H than for SN1987A. The major differences observed between these objects are in the behaviour of the line widths. Once entering the supernebular phase, the lines of SN1987A remained fairly constant in width with FWHM $\sim$ 2500 - 3000 km/sec. The lines of SN1988A also remained constant in width but approximately half as broad. SN1988H started with broader lines, but the line widths decreased with time, to be less than 900km/sec (unresolved) in our last observation. This decrease may be explained if the radioactive material in SN1988H was not mixed into the envelope and so the SN cooled more rapidly than for SN1987A. Alternatively, this narrowing of the lines might be due to dust absorption. If so, then dust formation was far more efficient in SN1988H. Though the time-evolution of the nebular features in SN1988A is similar to SN1987A, lines from the core became visible far more quickly. Figures 6d, e, and f compare SN1988A at 366 days past maximum to spectra of SN1987A at this epoch, and at 692 days. The [FeII] lines match at the later epoch. This is evidence that the envelope of SN1988A has also cooled more quickly than in SN1987A, so that the envelope is transparent to emission from the core at an earlier epoch. Our last observation of SN1988A at 528 days, on 29 June 1989, did not detect the SN, but only an HII region situated behind the SN. While observing conditions were difficult and blind offsets were used to position the telescope, we nevertheless suggest that the SN may have faded suddenly due to some form of 'Infrared Catastrophe' as suggested by FRANSSON and CHEVALIER [7,8]. This would be consistent with the accelerated spectral evolution observed to that date.

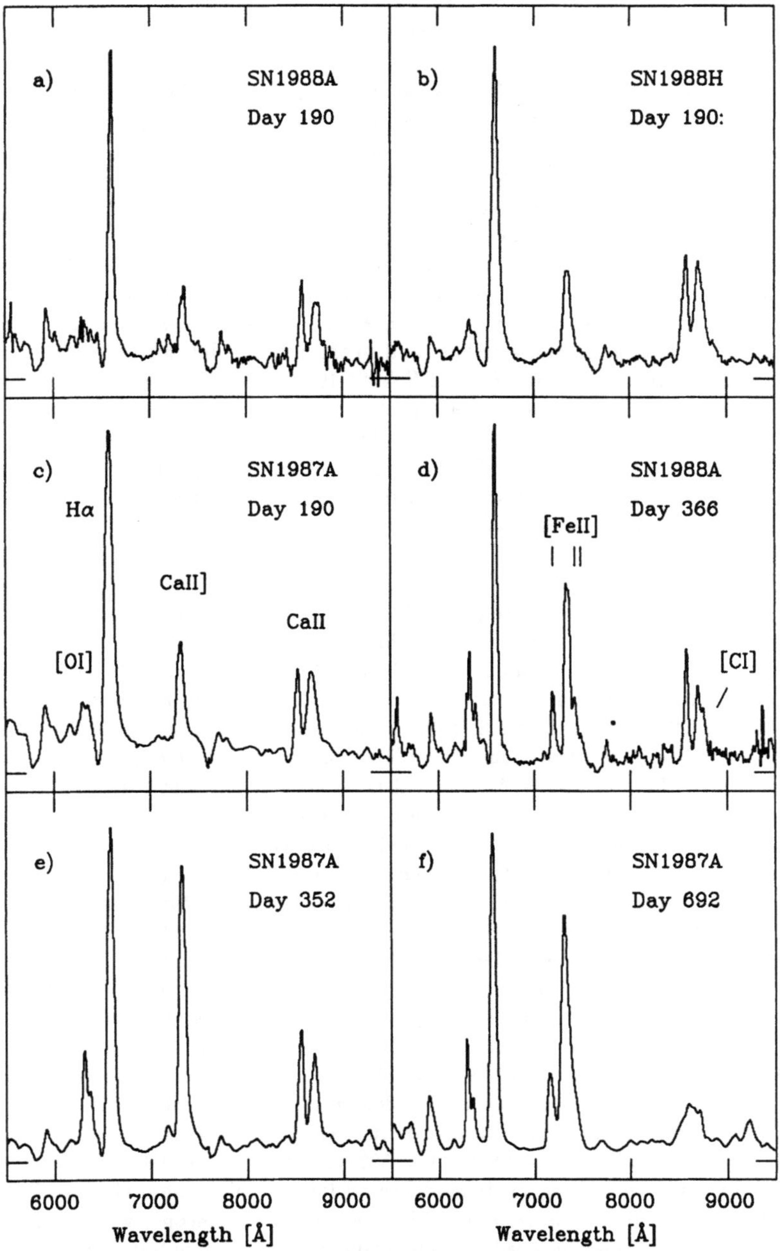

Figure 6: FORS observations of SN1988A and SN1988H compared to SN1987A at similar epochs. SN1988H was not discovered until after maximum, so its age is only known approximately. All three supernovae are very similar at day 190, but the [FeII] lines from the core appear much sooner in SN1988A than in SN1987A. On day 366 these lines in SN1988A are more similar to SN1987A at day 692. Three of the strongest [FeII] lines are identified in figure 6d.

5. Future Plans

EMS and RAS hope to continue this program in 1990 with a second observing season, with the ultimate aim of building up a sample of about 20 well-observed type II SNe which will be suitable for statistical analysis. Apart from comparison with SN1987A, we hope to investigate the relationships between SNe and their parent galaxies, and with this in mind are concurrently obtaining both direct and spectral information on these galaxies to determine their metallicity and morphology.

RAS, RDC and MAD plan to continue to follow SN1987A with the AAT for as long as it is observationally possible. RAS and RDC intend to use the medium and high dispersion data to analyse the line profiles using stellar atmosphere techniques and MAD and Steve Rawlings (MSSSO) will use the low dispersion spectra to test their model spectra. In addition we encourage interested parties to make use of these data as they becomes available, and all such requests should be forwarded to the Director at AAO.

6. References

1. R. W. Hanuschik, J. Dachs: Astr. Ap. 205, 135 (1988).
2. R. M. Catchpole *et al.*: M. N. R. A. S. 229, 15P (1987).
3. R. M. Catchpole *et al.*: M. N. R. A. S. 231, 75P (1988).
4. R. M. Catchpole *et al.*: M. N. R. A. S. 237, 55P (1989).
5. P. A. Whitelock *et al.*: M. N. R. A. S. 234, 5P (1988).
6. P. A. Whitelock *et al.*: (1989), in press.
7. C. Fransson, R. A. Chevalier: A. J. 322, L15 (1987).
8. C. Fransson: In ESO Workshop on the SN1987A, ed. by I. J. Danziger (ESO, 1987.)

Spectroscopy of SN 1987A at 1–4 Microns

W. Peter S. Meikle, David A. Allen, Jason Spyromilio, & Gian Varani

1. Introduction

Near-IR (1–4 μm) spectra of SN 1987A provide in a single wavelength range several important features, only some of which are found at other wavelengths. These include:

a). A large variety of elements are represented, typically in their neutral or singly-ionized states. These include the elements created in the explosion such as iron, cobalt and nickel. Molecules are also observed in this region. Compared with certain other wavelengths, there is less difficulty in line identification and measurement since continuum/line opacities are smaller, and line blending is less severe.

b). The high signal to noise attained, coupled with resolutions of order 200 or 300 km s^{-1} allows precise measurement of the Doppler-broadened line profiles which, in turn, can provide detailed information about the physical location of the material.

c). Acquisition of near-IR spectra will be feasible to beyond day 1000. Coupled with the relatively low optical depths which characterise the 1-4 μm range, exceptionally deep penetration of the core should be possible.

2. Observations

All the major southern observatories have been carrying out near-IR spectroscopy of SN 1987A (BOUCHET *et al.* [1], LARSON *et al.* [2], CATCHPOLE *et al.* [3, 4], OLIVA *et al.* [5], ELIAS *et al.* [6], McGREGOR [7], WHITELOCK *et al.* [8], MEIKLE *et al.* [9]). We describe here observations obtained with the Anglo-Australian Telescope. We use the common-user cooled grating spectrometer (FIGS) to obtain the spectra. The observation and calibration procedures are described in MEIKLE *et al.* [9]. Fourteen runs have been carried out, between day 18 and day 840. FIGS operates at resolutions of $\lambda/\Delta\lambda$ ~500 ('lo-res') or ~1500 ('hi-res'). However, due to falling intensity, the last complete hi-res run was day 574, although hi-res sections of interesting features have been obtained since then. Spectra obtained between days 18 and 349 have already been published (MEIKLE *et al.* [9]).

3. Early Time Spectra

As in the optical region, the near-IR spectra were initially dominated by a black-body-like continuum. Superimposed were P-Cygni lines, mostly of the hydrogen Paschen, Brackett and Pfund series – the first time these lines have been observed in a supernova (ELIAS *et al.* [6]; MEIKLE *et al.* [9]). Hu 14–6 and Hu 15–6 may also have been detected (BOUCHET *et al.* [1]).

As we move to higher series, there is a trend towards decreasing absorption depth in the P-Cygni profiles, with far more energy appearing in emission than is taken out in absorption. Also, the absorption trough blueshifts decrease. On day 18 the trough blueshifts with respect to the emission maxima were 9500 km s^{-1} in $H\alpha$ and $H\beta$ (mean value), 4300 km s^{-1} in $P\beta$ and $P\gamma$, 4150 km s^{-1} in $Br\alpha$ and $Br\gamma$ and 2600 km s^{-1} in $Pf\gamma$. The strong emission excess and the range of trough blueshifts indicate that we are not dealing with a pure resonant-scattering situation. There has been some comment in the literature that the slowest-moving hydrogen in the ejecta is indicated by the $Pf\gamma$ trough blueshift which is only 2600 km s^{-1}. While there almost certainly is such slow moving hydrogen (see below), we should not directly use the $Pf\gamma$ shift as evidence for it, since this is based on the pure resonant scattering intrepretation which is clearly not the only mechanism operating. More sophisticated modelling is necessary (e.g. HÖFLICH [10]).

4. Late Time Spectra

Spectra obtained on day 349 are shown in Fig 1 (a-d) and the subsequent evolution to day 840 is illustrated in Fig 2 (a-d). Line identification was carried out as described in MEIKLE *et al.* [9]). As SN 1987A aged, its near-IR spectrum became increasingly dominated by a rich variety of emission lines. The absorption troughs became weaker and the continuum contribution steadily declined. By day 200 emission lines constituted about 50% of the total flux and all absorption troughs, apart from those of helium (see below), had completely disappeared. Around day 200 the emission lines arose mostly from allowed transitions (H, HeI, NaI, OI, CaI, MgI, SiI), plus forbidden lines of FeII. In addition CO and CO^+ have been identified for the first time in a supernova. By day ~650 the hydrogen lines were less prominent, dominant lines being due to HeI, [FeII] and [NiI]. Some of the more prominent features are now briefly considered. (For a comprehensive discussion of the first year, see MEIKLE *et al.* [9]).

4.1 Hydrogen & Helium

The persistence of the hydrogen and helium recombination lines to the end of the second year (much longer than the natural recombination time) implies the presence of a continuous source of ionisation. γ-ray emission from ^{56}Co decay is a likely candidate (ELIAS *et al.* [6], GRAHAM [11]). By the end of year 1, Brα and Pfγ were approximately Case B relative to Pa β ($N_e = 10^6$ cm^{-3}, T = 5000 K), but Brγ lay persistently below, while Br10 and higher Brackett lines lay well above.

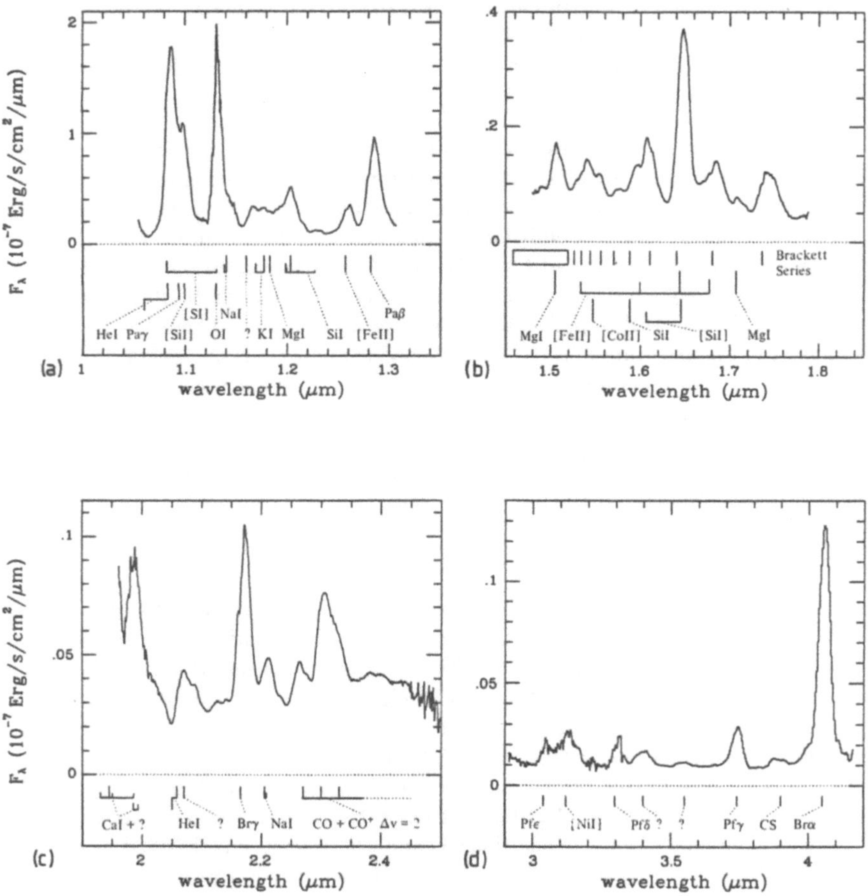

Figure 1(a–d). Spectra obtained on day 349. For identified species, the vertical markers are placed at the rest wavelengths. Within a multiplet, the lengths of these markers are proportional to the component intensities. For unidentified species, the vertical markers are placed near to the observed centres of the features.

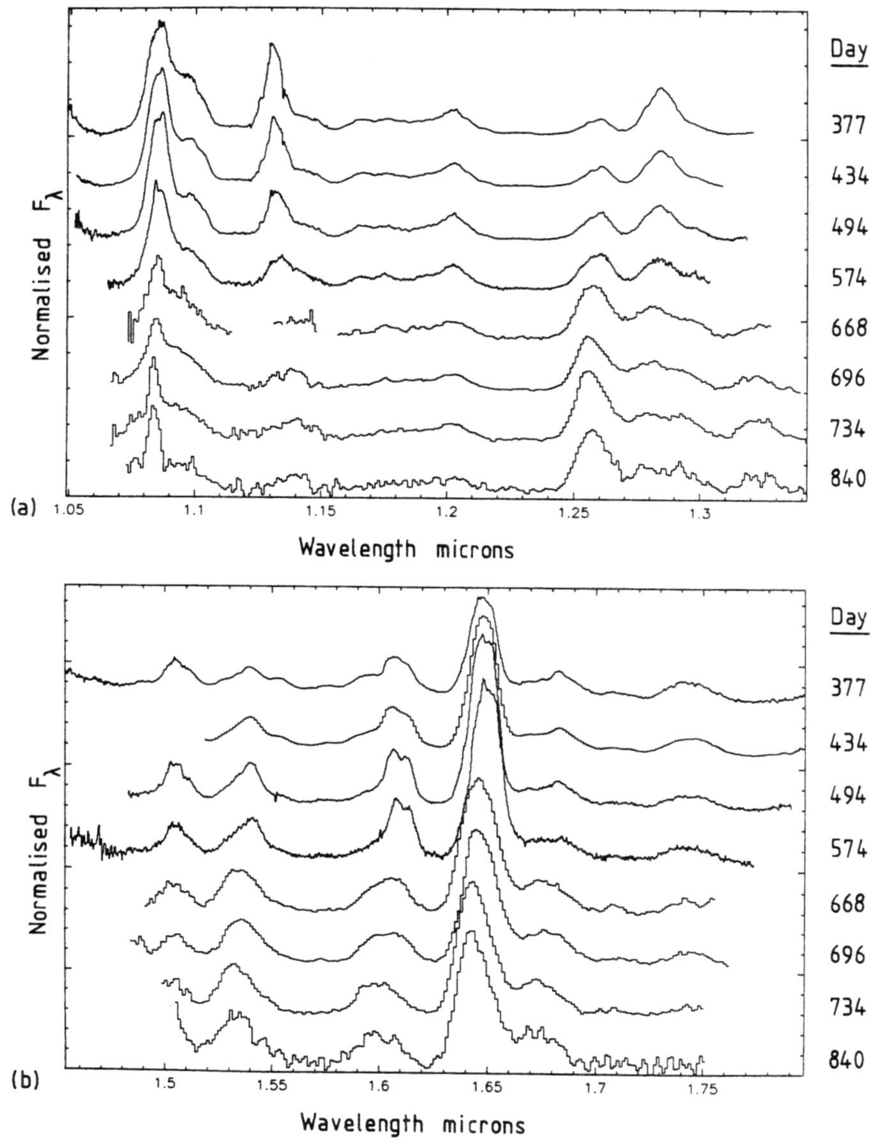

Figure 2(a & b). Spectra in the J & H windows obtained between days 377 and 840. They have been normalised to the total flux in each spectrum, and displaced vertically for clarity.

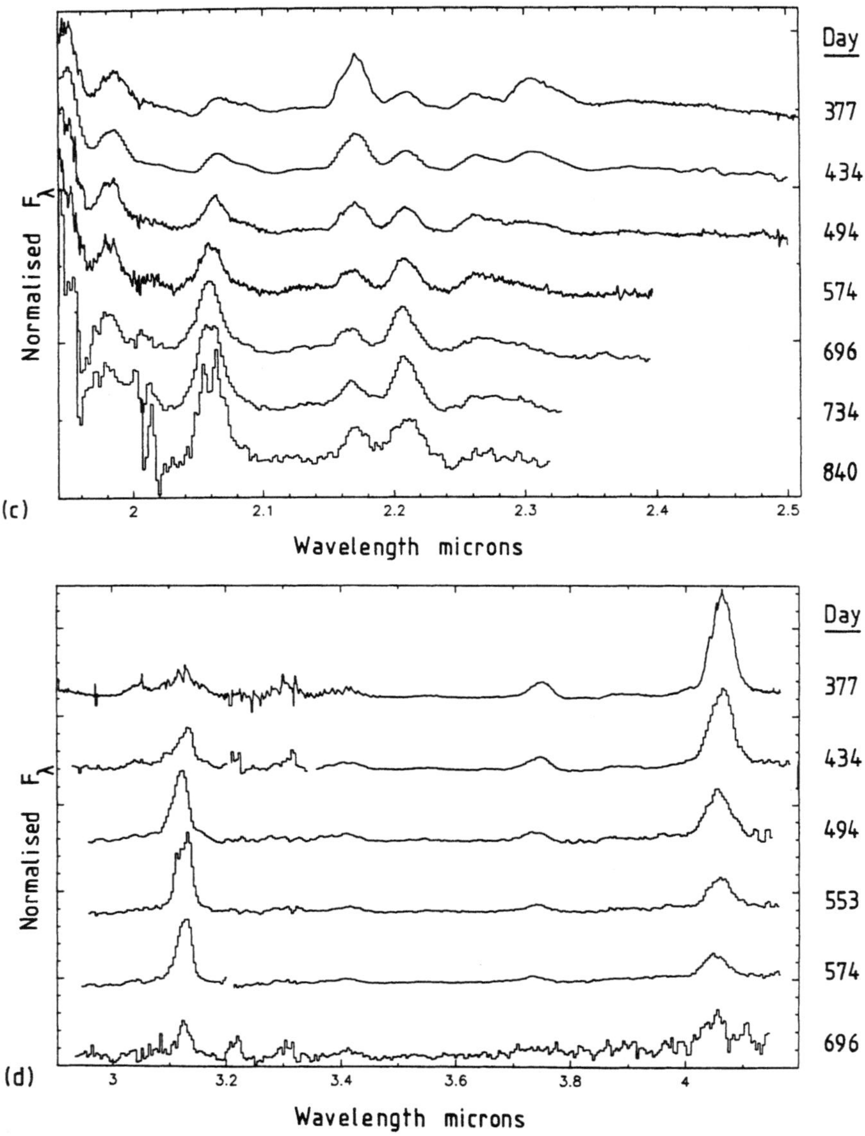

Figure 2(c & d). Spectra in the K & L windows obtained between days 377 and 840. They have been normalised to the total flux in each spectrum, and displaced vertically for clarity.

Broad P-Cygni profiles of HeI 1.083 μm $2s^3S$—$2p^3P^0$ and HeI 2.058 μm $2s^1S$—$2p^1P^0$ were present by day 112. In contrast with other species, the 1.083 μm and 2.058 μm lines of HeI continued to exhibit blueshifted absorption troughs for over a year after the explosion. We believe that these lines formed as a result of resonant scattering from enhanced populations in the metastable $2s^3S$ and $2s^1S$ levels, following recombination of HeII (MEIKLE *et al.* [9]). By day 434 a narrow component to the 1.083 μm line emerged, increasing in prominence as the broad component declined (Fig 2a). At a resolution of 300 km s^{-1}, the component was unresolved and undisplaced from the LMC rest wavelength (ALLEN *et al.* [12]). We attribute the narrow component to emission from the BSG/RSG shell ionized by the EUV flash (ALLEN *et al.* [12, 13]).

4.2 Oxygen

By day 192 a strong, relatively narrow (V_{FWHM} ~2100 km s^{-1}) emission line had appeared at ~1.13 μm, persisting into the second year. OLIVA *et al.* [5] and MEIKLE *et al.* [9] attribute this feature to the $3p^3P$—$3d^3D^0$ 1.1287 μm transition of OI, and argue that it resulted from the cascade following the pumping of the OI 102.5 nm triplet by Lyβ photons (the Bowen resonance fluorescence mechanism). The subsequent and better-known 844.6 nm $3s^3S^0$—$3p^3P$ transition was undetected in optical spectra, probably due to strong blanketing by the CaII absorption trough at about the same wavelength.

A notable aspect of the 1.13 μm feature is its exceptionally narrow width, equivalent to V_{FWHM} ~2100 km s^{-1} - considerably less than that of other lines. This characteristic continued from day 192 to ~400. For OI pumping to operate, we require a large optical depth to Lyβ photons. Such optical depths would be encountered deep in the ejecta and so the pumped OI line would arise from relatively slow-moving material, accounting for the narrow line width. Since the rest wavelengths of Lyβ and the pumped OI transition (triplet UV4) are separated by only ~15 km s^{-1}, the OI must lie close to the driving HI otherwise Doppler effects would put the OI 102.5 nm transition out of resonance. It can therefore be argued that the narrow width of this strong OI line consitutes evidence for slow moving hydrogen. In addition, the line shows no evidence of the flat top we would expect if it lay in a shell, implying the presence of hydrogen at depths below 2000 km s^{-1}.

4.3 Silicon

From day 192 to at least day 734 a complex feature lay between 1.16 and 1.22 μm. It has been identified as a blend of KI 4^2P^0—$3d^2D$, MgI $3p^1P^0$—$4s^1S$, SiI $4s^3P^0$—$4p^3D$ plus an unidentified component (OLIVA *et al.* [5], MEIKLE *et al.* [9]). However, if the feature does have this multiple origin, the persistence of its shape over such a long era seems remarkable. In addition, if much of it is due to allowed SiI emission, it is difficult to understand why the [SiI] lines had vanished by day 734 (see below).

The strongest of the [SiI] lines should be the 1.0991 μm $3p^1D$—$3p^1S$ transition, with the 1.6068/1.6455 μm $3p^3P$—$3p^1D$ multiplet being about 10-15% of this at 300 days (FRANSSON & CHEVALIER [14]). Identification of these lines is made difficult by their coincidence with hydrogen or iron lines. Simple modelling of the iron spectrum, assuming the hydrogen spectrum to be Case B leads us to conclude that on day 192 [SiI] made a negligible contribution but was significant by day 349. [SiI] emission persisted until at least 500 days, but faded by day 734. The disappearance of the [SiI] emission is interesting, and may be associated with the possible condensation of dust commencing at about 500 days (DANZIGER *et al.* [15], LUCY [16]).

4.4 Iron, Cobalt & Nickel

As the supernova evolved, [FeII] lines, especially those at 1.257 μm and 1.644 μm became increasingly prominent. [CoII] 1.547 μm was present between days 255 and 574 and [NiI] 3.119 μm dominated the L-window spectrum for much of the second year.

Masses for iron, cobalt and nickel have been inferred both from near-IR interterm spectra (OLIVA *et al.* [5], BOUCHET & DANZIGER [17], MEIKLE *et al.* [9]) and far-IR fine structure spectra (AITKEN *et al.* [18], ERICKSON *et al.* [19], RANK *et al.* [20]; MOSELEY *et al.* [21]) using simple LTE calculations. There is general agreement that the derived iron and cobalt masses are comparable to what should have been present in the ejecta from the decay of the initial $0.08M_\odot$ of ^{56}Ni. However, these mass estimates should be regarded as only approximate. While interterm or excited state fine structure lines have the advantage of arising from relatively low populations and so tend to be optically thin, they are very sensitive to adopted temperature, which different estimates place in the range ~2000 K to ~6000 K. On the other hand, ground state fine structure lines are relatively insensitive to temperature but since the populations are high, they may be optically thick allowing only lower limits for the masses to be derived.

A promising method for reducing the effects of temperature sensitivity and so allow reliable abundance estimates, is to consider the intensity *ratios* of lines from different species. Application of this technique to the ratio of [CoII] 1.547 μm and [FeII] 1.533 μm has provided direct, independent evidence for the presence of ^{56}Co in the ejecta (MEIKLE *et al.* [22]).

5. Evolution of the Spectra

Between days 200 and 700 the JHK continua exhibited approximately exponential decline with e-folding times of ~70 days. The L continuum declined more slowly at first, and then steepened around day 500 to the same rate as seen in JHK. Between days 300 and 700, the hydrogen lines declined with e-folding times of around 70 days. Lines from heavier elements tended to decline more slowly at first. For example [FeII] 1.257 μm and [SiI]+[FeII] 1.64 μm had e-folding times of ~200 days between days 300 and 450. About day 500 the rates of decline of the heavier element lines steepened, becoming comparable to those of hydrogen.

5.1 Line Widths

Up to about day 500 line widths (V_{FWHM}) of 2500 to 4000 km s^{-1} were typical (Fig 3), with hydrogen and helium exhibiting somewhat higher velocities than 'core materials' such as iron. (An exception is the OI 1.13 μm line discussed earlier). After day 500 the lines all tended towards ~3400 km s^{-1}. Simple stratified models (e.g. WOOSLEY's [23] model 10H) predict considerably lower velocities ($<$ 1000 km s^{-1}) for iron-group elements, implying that mixing into the outer layers must have occurred. There is also little evidence in the profiles of the flat-top structure characteristic of stratification. By day ~300, pronounced asymmetrical structure had become apparent in the [FeII] 1.257μm profile. Similar behaviour was seen in line profiles of nickel. SPYROMILIO *et al.* [24] interpret this behaviour in terms of fragmentation of a low density inner region (the nickel bubble).

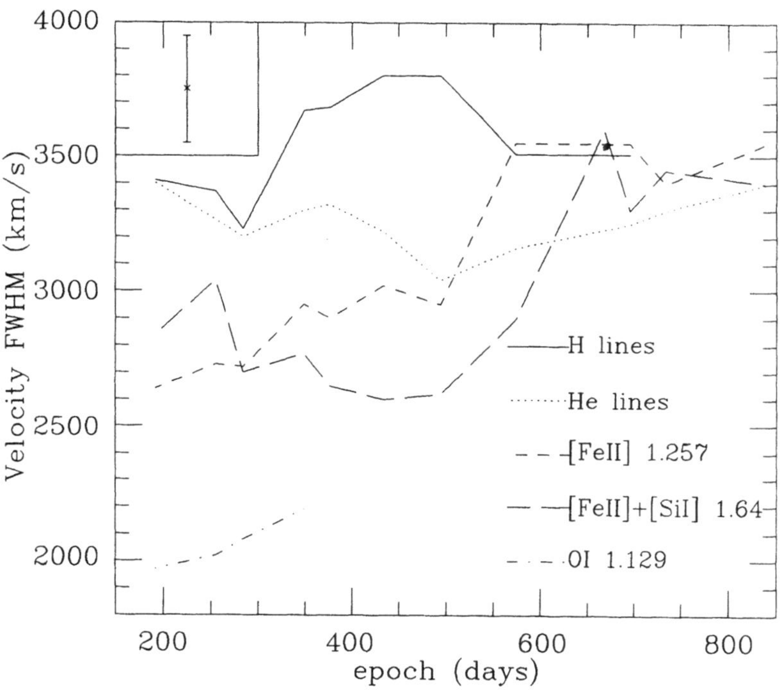

Figure 3. Evolution of line widths. A typical error bar is shown in the box.

5.2 Wavelength Shifts

During the first ~200 to ~500 days, the IR emission line peaks exhibited redshifts of typically 400 km s^{-1} with respect to the LMC rest frame (Fig 4). Similar redshifts were seen in optical lines. At least some of the shift can be attributed to electron scattering in the homologously expanding ejecta (FRANSSON & CHEVALIER [25], WITTEBORN *et al.* [26]; WOOSLEY *et al.* [27]). Between days 574 and 668 the redshifts disappeared and in most cases were replaced by blueshifts of 100 to 500 km s^{-1}. Again, a similar behaviour was observed at optical wavelengths. DANZIGER *et al.* [15]), LUCY [16] suggest that the blueshifts were due to commencement of dust condensation around day 600 which attenuated the emission from the far side of the ejecta. However, for this explanation to account for the blueshifts at near-IR wavelengths we need to invoke a very unusual grain size distribution. Intrinsic localised cooling is an alternative explanation (SPYROMILIO *et al.* [24]).

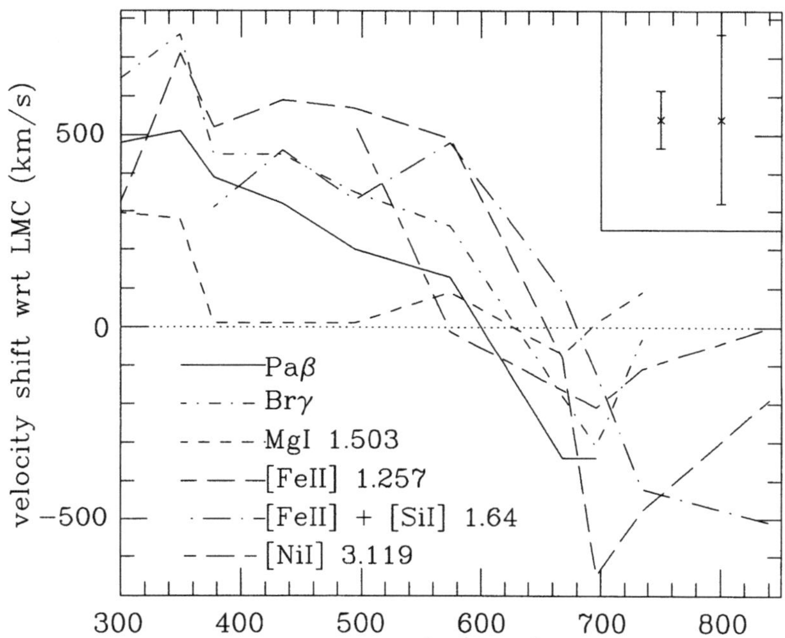

Figure 4. Evolution of line peak shifts with respect to the LMC rest frame (0 km s^{-1}). The small and large error bars shown in the box are typical for J window hi-res L window low-res measurements respectively (see text).

Acknowledgements

We are grateful to Someera Butt for assistance in preparing this review.

References

1. P. Bouchet *et al.*: In ESO workshop on the SN 1987A Conf. Workshop Proc. 26, p 159, ed. by I.J. Danziger (ESO 1987).
2. H.P. Larson, S. Drapatz, M.J. Mumma & H.A. Weaver: In ESO workshop on the SN 1987A Conf. Workshop Proc. 26, p 147, ed. by I.J. Danziger (ESO 1987).
3. R.M. Catchpole *et al.*: MNRAS, 229, 15P (1987).
4. R.M. Catchpole *et al.*: MNRAS, 231, 75P (1988).
5. E. Oliva, A.F.M. Moorwood & I.J. Danziger: The Messenger, 50, 18 (1987).
6. J.H. Elias *et al.*: Ap.J., 331, L9 (1988).
7. P.J. McGregor: Proc. Astr. Soc. Aust., 7, 450 (1988).
8. P.A. Whitelock *et al.*: MNRAS, 234, 5P (1988).
9. W.P.S. Meikle, D.A. Allen, J. Spyromilio & G.-F. Varani: MNRAS, 238, 193 (1989).
10. P. Höflich: Proc. Astr. Soc. Aust., 7, 434 (1988).
11. J.R. Graham: Ap.J., 335, L53 (1988).
12. D.A. Allen, W.P.S. Meikle & J. Spyromilio: IAU Circ. no 4747 (1989).
13. D.A. Allen, W.P.S. Meikle & J. Spyromilio: Nature (submitted).
14. C. Fransson & R.A. Chevalier: Ap.J., 322, L15 (1987).
15. I.J. Danziger, C. Gouiffes, P. Bouchet & L.B. Lucy: IAU Circ. no 4746 (1989).
16. L.B. Lucy: In this volume (1989).
17. P. Bouchet & I.J. Danziger: IAU Circ. no 4575 (1988).
18. D.K. Aitken *et al.*: MNRAS, 235, 19P (1988).
19. E.F. Erickson *et al.*: Ap.J., 330, L39 (1988).
20. D.M. Rank *et al.*: Nature, 331, 505 (1988).
21. H. Moseley *et al.*: Ap.J. (submitted).
22. W.P.S. Meikle, G.-F. Varani, J. Spyromilio & D.A. Allen: In this volume (1989).
23. S.E. Woosley: Ap.J., 330, 218, (1988).
24. J. Spyromilio, W.P.S. Meikle & D.A. Allen: MNRAS, in press (1989).
25. C. Fransson & R.A. Chevalier: Ap.J., 343, 323 (1989).
26. F.C. Witteborn *et al.*: Ap.J., 338, L9 (1989)
27. S.E. Woosley, P.A. Pinto & T.A. Weaver: Proc. Astr. Soc. Aust., 7, 355 (1988).

Direct Observation of ^{56}Co Decay in Supernova 1987

W. Peter S. Meikle, Gian –F. Varani, Jason Spyromilio, & David A. Allen

A promising method for reducing the effects of temperature sensitivity and so allowing reliable abundance estimates from spectroscopy, is to consider the intensity ratios of different species. We have examined the intensity ratio of [CoII] 1.547 μm to [FeII] 1.533 μm for the period 255 to 574 days (Fig. 1) (VARANI *et al. in preparation*). Although the two lines are somewhat blended, inspection of Fig. 1 clearly shows that the [CoII] line steadily declined relative to the [FeII] line.

To extract the intensity ratio for each epoch, it was first necessary to estimate and then subtract the Brackett series and continuum contributions to the flux. The continuum was determined by interpolation from other parts of the near-IR spectra. A case B recombination spectrum was used to model and remove the Brackett lines (HUMMER & STOREY [1]). The iron/cobalt blend was then modelled using the relatively unblended 1.257 μm [FeII] line as a template, with intensity and redshift as free parameters. The intensity ratios were converted to M_{Co^+}/M_{Fe^+} by using standard atomic data (NUSSBAUMER & STOREY [2],[3]). Although by considering the ratio, sensitivity to adopted temperature was greatly reduced, there was still some dependence. Temperatures were therefore derived from the intensity ratios of the [CoII] 1.547 μm and 10.52 μm lines, the latter being taken from AITKEN *et al.* [4].

The ionisation potentials of neutral iron and cobalt are similar, and the singly-ionised state was believed to have been dominant (MOSELEY *et al.* [5]). Thus the derived M_{Co^+}/M_{Fe^+} ratio should have been about equal to the total mass ratio, M_{Co}/M_{Fe}, at each epoch. In Fig. 2 we show the derived M_{Co}/M_{Fe} ratio versus time since explosion, and find that between days 284 and 494 the M_{Co}/M_{Fe} ratio exhibited an exponential decline with a timescale of 94±13 days. This compares with 111.3 days for the decay timescale of ^{56}Co. Thus, the IR lines provide direct, independent evidence for the presence of ^{56}Co in the ejecta. Fig. 2 also shows predicted M_{Co}/M_{Fe} ratios for scenarios in which, a) the cobalt and iron result from the decay of 0.08 $M_\odot$

^{56}Ni only, b) additional sources of iron and cobalt are included (WOOSLEY *et al.* [6], HASHIMOTO *et al.* [7]). We have evidence that the tendency of the observed ratios to lie below the model values may be due to contamination of the 1.533 μm [CoII] line by another unidentified species (VARANI *et al. in preparation*).

Figure 1. Spectra showing evolution of the [FeII] 1.533 μm and [CoII] 1.547 μm blend. The spectra have been scaled and displaced vertically for clarity. The dotted vertical lines indicate the approximate positions of the line peaks and correspond to a redshift of about 1000 km/s with respect to the LMC rest wavelength (see SPYROMILIO *et al.* [8]).

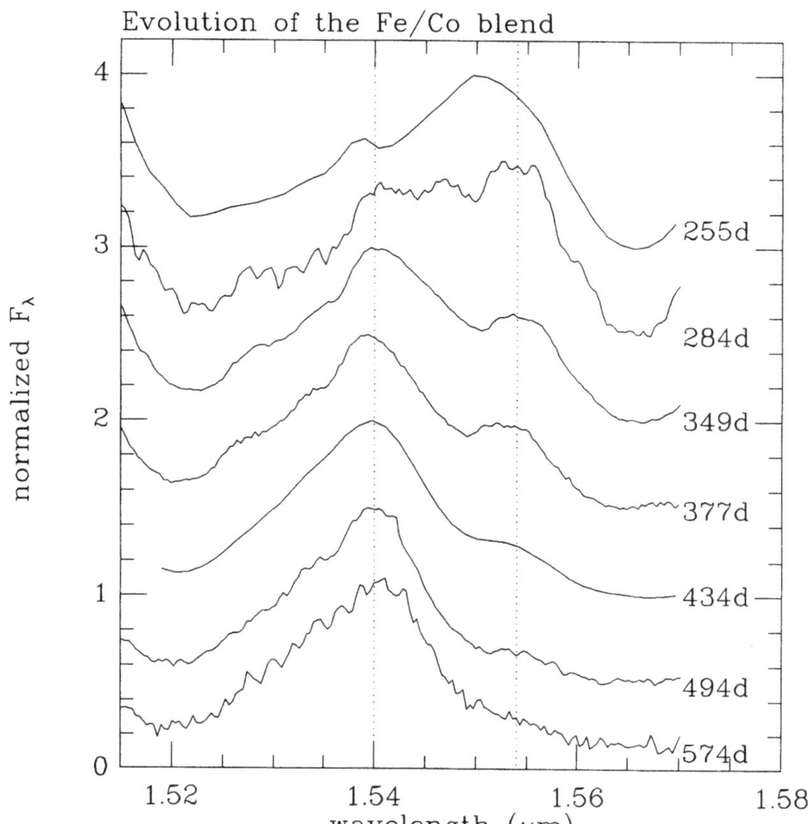

Figure 2. Mass ratios of cobalt to iron derived from the spectra shown in Fig. 1. Also shown are the predicted mass ratios for scenarios in which a) the cobalt and iron result from the decay of 0.08 $M_\odot$ ^{56}Ni only (long dashes), b) additional sources of cobalt and iron are included (dots: WOOSLEY *et al.* [6]; short dashes: HASHIMOTO *et al.* [7]).

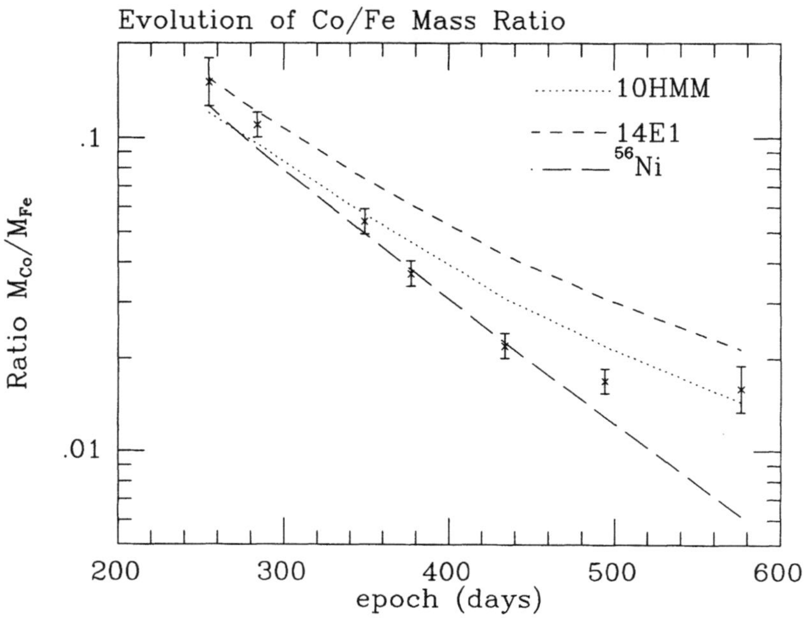

References

1. D.G. Hummer & P.J. Storey: MNRAS, **224**, 801 (1987).
2. H. Nussbaumer & P.J. Storey: Astr. Ap., **193**, 327 (1988a).
3. H. Nussbaumer & P.J. Storey: Astr. Ap., **200**, L25 (1988b).
4. D.K. Aitken *et al.*: MNRAS, **235**, 19P (1988).
5. S.H. Moseley, E. Dwek, W. Glaccum, J.R. Graham, R.F. Loewenstein & R.F. Silverberg: Ap.J. submitted (1989).
6. S.E. Woosley, P.A. Pinto & T.A. Weaver: Proc. astr. Soc. Aust., **7**, 355 (1988).
7. M. Hashimoto, K. Nomoto & T. Shigeyama: Astr. Ap., **210**, L5, (1989).
8. J. Spyromilio, W.P.S. Meikle & D.A. Allen: MNRAS (in press).

Spectral Line Profiles of Iron and Nickel in Supernova 1987A

Jason Spyromilio, W. Peter S. Meikle, & David A. Allen

High resolution ($\lambda/\Delta\lambda \sim 1500$) near-infrared spectra of SN 1987A taken at the Anglo-Australian Telescope (MEIKLE *et al.* [1], SPYROMILIO *et al.* [2], MEIKLE *et al.* in preparation) show that during the second year after explosion the iron-group line profiles were strikingly different from those of other elements, but were similar to mid-infrared iron-group line profiles (E.F. ERICKSON & M. COHEN *private communication*). The 1.257 μm [Fe II] and 3.119 μm [Ni I] profiles are compared with Pa β in Fig. 1. The iron group line profiles are characterised by an inflection near the LMC rest wavelength and excess emission redward of this. We rule out blends, optical depth effects, and electron scattering as the cause of this distinctive shape. We propose instead that the iron-group elements have a unique spatial distribution in the ejecta, and that the profiles were due to emission from a low-density inner region (the nickel bubble) which has fragmented into high-velocity bullets.

About 650 days after the explosion the profiles shifted to the blue (see Fig. 2). Similar changes were observed in the optical region and LUCY [3] has attributed the changes to the effects of dust condensation in the ejecta. However, the 1.257 μm line exhibits a shift to the blue comparable to that observed in the optical lines. While the shift of the optical lines can be explained by an increase in visual extinction of about 2 magnitudes, a figure of 5 magnitudes would be required to account for the change in the 1.257 μm line, assuming an interstellar reddening law. This seems far too high to be compatible with the optical data even if we invoke clumping. We therefore propose that the blueshift in the IR profiles arose from the combination of the effects of dust condensation and an intrinsic fading of the redder emission possibly due to accelerated cooling.

Figure 1. [Fe II] 1.257 μm, Paschen β from day 494 and the [Ni I] 3.119 μm line from day 553 in velocity space with respect to their rest wavelengths. The velocity of the rest frame of the supernova is also shown. To avoid overlap the profiles have been normalized in intensity and displaced vertically. The horizontal lines at the left indicate the zero flux levels. The peak intensities are 1.5×10^{-13}, 1.0×10^{-12} and 1.2×10^{-12} erg/s/cm^2/Å for the [Ni I], [Fe II] and Pa β lines respectively.

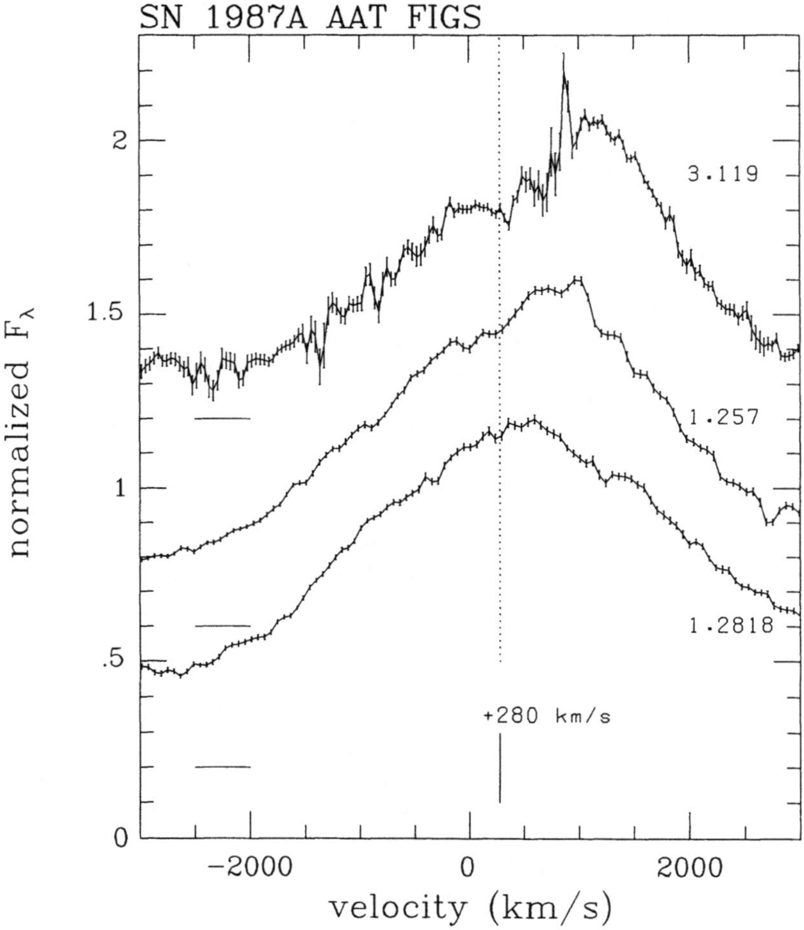

Figure 2. Evolution of the [Fe II] 1.257 μm line profile from day 192 to 735.

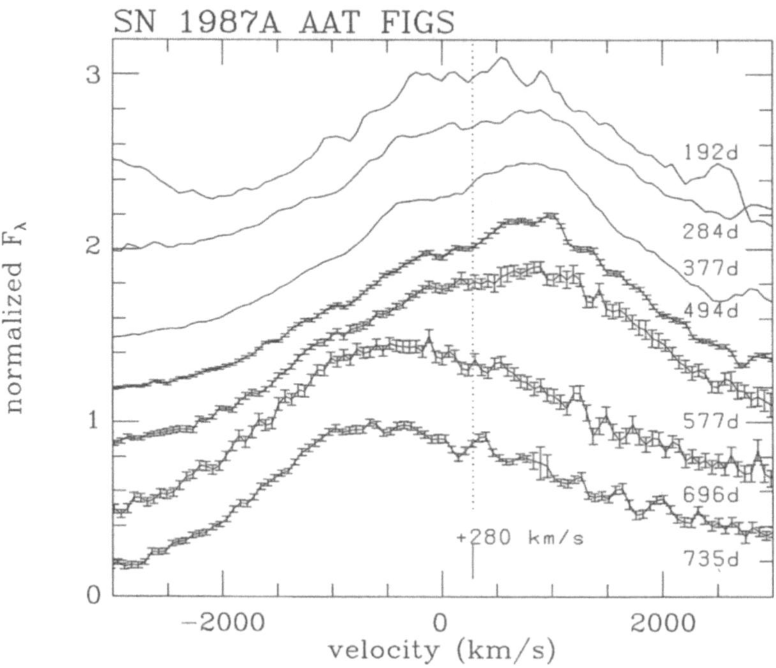

References

1. W.P.S. Meikle, D.A. Allen, J. Spyromilio & G.-F. Varani: MNRAS, **238**, 193, (1989)
2. J. Spyromilio, W.P.S. Meikle & D.A. Allen: MNRAS, (in press).
3. L.B. Lucy: In this volume (1989).

Supernova 1987A: Astron Observations of UV Absorption Spectra and Their Interpretation

L. A. Lyubimkov

The Astron UV observations of SN 1987A have been performed from 4 March 1987 to 23 March 1988, i.e. they cover the period from 9 to 394 days after the explosion. Fifteen spectra with 28 Å resolution were obtained in different dates (ref. [1,2]). All observational data are shown in Fig.1. The spectra are shifted along the wavelength axis according to radial velocity curve (Fig.2). Moreover each spectrum is shifted relatively preceding one by 0.2 dex along the vertical axis in order to avoid a confusion.

For each date of observation the shift $\Delta\lambda$ of the 3230 Å absorption blend is determined and corresponding radial velocity V_r is found. Fig.2 shows that the mean V_r values obtained from Balmer lines H_α (in absorption), H_β and H_γ (ref.[3]) are in a good agreement with UV values. The relation between V_r and the

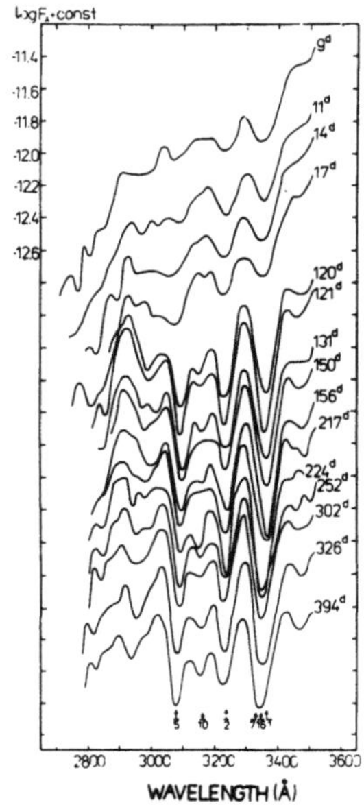

Fig.1. Observed spectra of SN 1987A shifted along both axes (see text). Location and number of strongest TiII multiplets are pointed out.

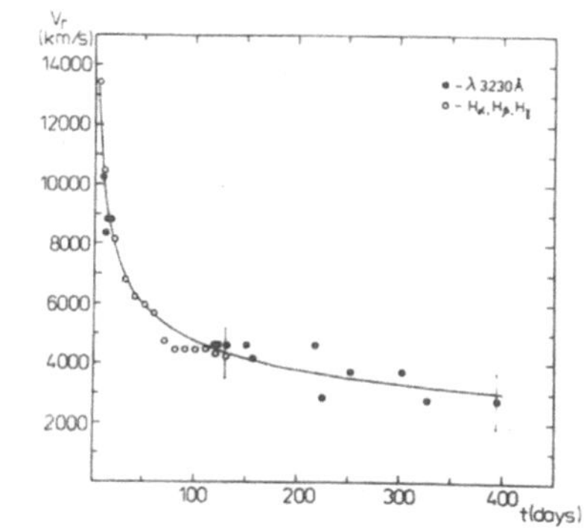

Fig.2. Radial velocity as a function of time. Solid line corresponds to the power law $V_r \propto t^{-1/3}$.

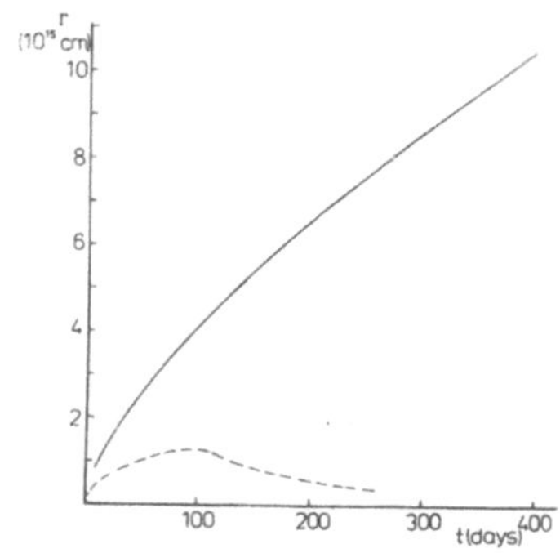

Fig.3. Distance of layers, where the 3230 Å blend is formed, as a function of time (solid line). Dashed line - radius of SN photosphere.

time t can been approximated by power law $V_r = 21200 \cdot t^{-1/3}$ (solid line in Fig.2). The V_r decreasing with t is due to shell expansion and its transparency increase.

Soon after explosion the shell enter into the stage of free inertional expansion, when distance r from the center and radial velocity V_r are connected by linear relation $r = V_r \cdot t$ (see [4]). From here and from V_r-t relation one can obtain $r = 1.83 \cdot 10^{14} \cdot t^{2/3}$ where t in days and r in centimetres. In Fig.3 this formula is presented by the solid line. For comparison the photospheric radius variations (see [5,6]) are shown by dashed line. Fig.3 shows that the layers where 3230 Å blend is formed move quickly enough away from the photosphere of Supernova.

The most pronounced details on the observed spectra are three blends with wavelengths $\lambda \approx 3350$, 3230 and 3080 Å. All the three are clearly visible also on IUE spectra of Supernova obtained in March 1987 (ref.[7]). As a first step in the interpretation the synthetic spectra for model atmosphere of supergiant with effective temperature T_{eff}= 5000 K and surface gravity log g = 0 were calculated to establish the origin of the blends (the temperature of Supernova photosphere for t≥10 d is close to 5000 K, see [5,6]). However LTE computations do not permit to reproduce main details of the observed spectrum exept absorption at 3230 Å. In LTE-computations the leading role in the spectrum belongs to numerous FeI absorption lines. But in the SN envelope we can expect considerable departures from LTE in ionization of FeI atoms. Such departures was found for yellow supergiants for instance (ref.[8]). It was shown that strong overionization of FeI atoms in the supergiant atmospheres can exist because photoionization processes,which are controlled by a "hotter" photospheric radiation,dominate strongly over recombination processes. Such situation can be expected for SN envelope too as followed from [9] .

Over-ionization must lead to considerable weakening of lines of FeI and other neutral atoms, then lines of ionized atoms are dominant. Among the ions of iron group an exeptional role may belong to TiII, because all strong resonance lines of TiII are located in the investigated spectral section. In Fig.1 positions and numbers of corresponding TiII multiplets (see [10]) are indicated by pointers. As can be seen from Fig.1, the location of the multiplets coincides with observed blends.

The synthetic spectra calculations were repeated for the condition that the spectrum is formed by lines of ionized atoms only. The results are shown in Fig.4 by thin lines for

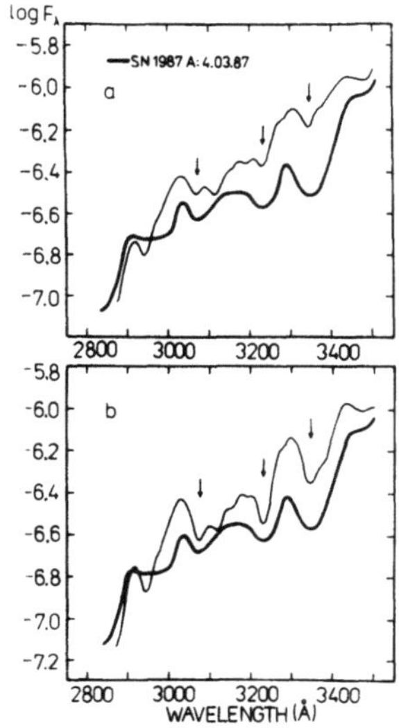

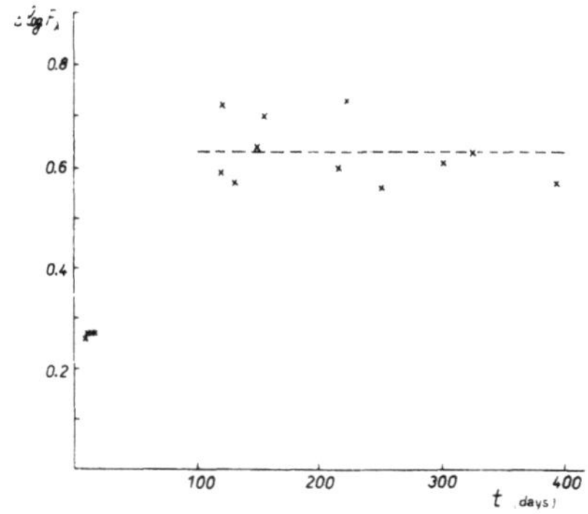

Fig.5. Evolution of the depth of the 3350 Å blend. Dashed line - mean value for t ≥ 120 d.

Fig.4. Comparison of observed and synthetic spectra. Three blends formed fully by TiII lines are pointed out on synthetic spectra.

normal (a) and enhanced (b) metal abundance. For instance the Ti and Fe abundances in Fig.4b are increased by order of magnitude. The general conclusion from Fig.4 is that the blends at 3350, 3230 and 3080 Å undoubtedly belong to TiII . It should be noticed that no quantitative estimation of Ti abundance is followed from Fig.4 because the calculations were made for stationary atmosphere of supergiant but not for expanding envelope of Supernova. The aim of the calculations was to give qualitative explanation of observed blends.

As can be seen from Fig.1, there is a strong difference between the March spectra (t = 9-17 d) and spectra obtained from 23 June 1987 (t ≥ 120 d). To evaluate these changes the relative depth $\Delta \log F_\lambda$ of the strong blend 3350 Å in its center was measured. Fig.5 shows that in the period between March and the end of June 1987 the depth $\Delta \log F_\lambda$ considerably increased (about two times) and afterwards remained at the same level. It is known that in July 1987 the X-ray flux from SN 1987A has been detected (ref. [11]). So early appearance of X-ray radiation is considered by many authors as the evidence of mixing in the SN envelope and ^{56}Co emergence (see, for example, [12]). However together with ^{56}Co other elements, produced in explosive nucleosynthesis, must be mixed into the envelope. From Fig.5 it can be followed that the products of nucleosynthesis, in particular freshly synthesized titanium, were observed in the envelope already on the 120th day after explosion. For confirmation of this conclusion an accurate determination of Ti abundance in different dates is needed.

REFERENCES

1. A.A.Boyarchuk et al.: Pis'ma v Astron. Zh., 13, 739 (1987).
2. Astrophysical investigations with the space station ASTRON, ed. by A.A.Boyarchuk (Moskow, Nauka), in press.
3. M.M.Phillips et al.: Astron.J., 95, 1087 (1988).
4. V.S.Imshennik, D.K.Nadyozhin: Uspehi fizicheskih nauk, 156, 561 (1988).
5. R.M.Catchpole et al.: MNRAS, 229, 15P (1987).
6. R.M.Catchpole et al.: MNRAS, 231, 75P (1988).
7. R.P.Kirshner et al.: Astrophys.J., 320, 602 (1987).
8. A.A.Boyarchuk, L.S.Lyubimkov, N.A.Sakhibullin: Astrofizika, 22, 339 (1985).
9. R.P.Kirshner, J.Kwan: Astrophys.J., 197, 415 (1975).
10. W.L.Wiese, J.R.Fuhr: J.Phys. and Chem. Ref. Data, 4, 263 (1975).
11. T.Dotani et al.: Nature, 330, 230 (1987).
12. R.A.Sunyaev et al.: Pis'ma v Astron. Zh., 14, 579 (1988).

SECTION III

SN 1987A — THE CIRCUMSTELLAR ENVIRONMENT

Ultraviolet Emission Lines from Circumstellar Gas Surrounding SN 1987A

George Sonneborn

Discovery in 1987 of narrow emission lines in the ultraviolet spectrum of SN 1987A (WAMSTEKER et al. [1]) provided the first direct evidence for circumstellar gas in the vicinity of the supernova. It was apparent from the outset that an emission line spectrum consisting of N V 1240, N IV] 1485, He II 1640, O III] 1664, N III] 1750, and C III] 1909 (see Figure 1) was characteristic of a low density photoionized gas, and could not plausibly arise in the ejecta of the SN 1987A explosion only a few months after the outburst. Analysis of subsequent UV spectra obtained with the *IUE* satellite have shown the emission lines to be a sensitive probe of the supernova's circumstellar environment and its progenitor's evolution. The discovery and initial analysis of these lines is presented and discussed in FRANSSON et al. [2]. In this contribution I review their results and present more recent *IUE* observations of the emission lines, which allows us to determine the physical size of the circumstellar shell.

High-dispersion *IUE* spectra (1150-2000 Å , $\lambda/\Delta\lambda \approx 10{,}000$) have recorded the C III] and N III] lines near *IUE*'s limit of detectability in 14-16 hour exposures. Both components of the C III] multiplet and three of the five components of the N III] multiplet are clearly present in the data. These lines are unresolved, hence the line widths are less than 30 km s^{-1}. An estimate of the electron density may be made from the C III] line ratio, $F(1906.7)/F(1908.7) \approx 1.0$, hence $n_e \approx 3 \times 10^4$ cm^{-3}.

Starting with the time of discovery, the line fluxes increased in an approximately linear fashion. This evolution is consistent with that expected from a fluorescent light echo by a circumstellar shell. Initial analysis (see [2]) indicated that the shell would be of order 10^{18} cm in radius.

A nebular analysis of the emission lines ([2]) based on low dispersion *IUE* spectra showed a large nitrogen overabundance in the circumstellar gas. The measured abundance ratios are N/C= 7.8 ± 4 and N/O= 1.6 ± 0.8, which correspond to factors of 37 and 12 times higher than solar, respectively. These abundance ratios indicate that the circumstellar material has undergone a significant amount of CNO processing.

These results lead to the conclusion (see FRANSSON *et al.* [2]) that the narrow emission lines arise in a circumstellar shell of gas which had its origin in a stellar wind during a red supergiant phase of the progenitor's evolution. This is the principal line of evidence that Sk -69° 202 was in a post-red giant branch stage of evolution at the time it exploded. The slow red supergiant wind ($v \approx 10$ km s^{-1}) would have been swept up by the much higher velocity wind ($v \approx 500$ km s^{-1}) from the blue supergiant progenitor to produce the shell. This picture is consistent with the early detection of the prompt radio burst from SN 1987A (TURTLE *et al.* [3]), implying a very low density of material within 10-100 stellar radii of the progenitor (CHEVALIER and FRANSSON [4]).

The most likely source of ionizing radiation to produce the observed line spectrum is the EUV burst at the time of shock breakout. Modelling of the recombination line spectrum and photoionized circumstellar shell by FRANSSON and LUNDQVIST [5] and LUNDQVIST [6] constrains the temperature of the EUV burst to be between $4-8 \times 10^5$ K.

The N V and N III] line fluxes reached a well-defined maximum in April-May 1988 (SONNEBORN *et al.* [7]). The evolution of the N V, N IV], N III], and C III] fluxes is shown in Figure 2. The emission line light curves in Figure 2 include spectra through early August 1989. The time of maximum flux for N V and N III] is $t = 405 \pm 10$ days after outburst (1987 February 23.316). Assuming a spherical geometry for the circumstellar shell, this time corresponds to a shell radius of about 200 light days, or $\approx 5 \times 10^{17}$ cm.

In a related contribution to these proceedings, Lundqvist (ref [6]) has calculated detailed photoionization models for the circumstellar interaction region around SN 1987A using this radius as input parameter. The overall agreement between the model light curves and the observations if very good. Lundqvist's models also consider the detectability of the undisturbed red supergiant wind outside the shell. The trend toward a plateau in the N V line may be an indication that we are now detecting this wind. If so, this would provide exciting new information about the presupernova history of Sk -69° 202 .

References

1. Wamsteker, W. 1987. *et al.* IAU Circular No. 4410.
2. Fransson, C., Cassatella, A., Gilmozzi, R., Kirshner, R., Panagia, N., Sonneborn, G., and Wamsteker, W. 1989. *Ap. J.,* **336**, 429.
3. Turtle, A., 1987. *Nature* **327**, 38.
4. Chevalier, R., and Fransson, C., 1987. *Nature* **328**, 44.
5. Fransson, C., and Lundqvist, P., 1988. *Astr. Ap.,* **192**, 221.
6. Lundqvist, P., 1989, (this volume)
7. Sonneborn, *et al.* 1988. IAU Circular No. 4685.

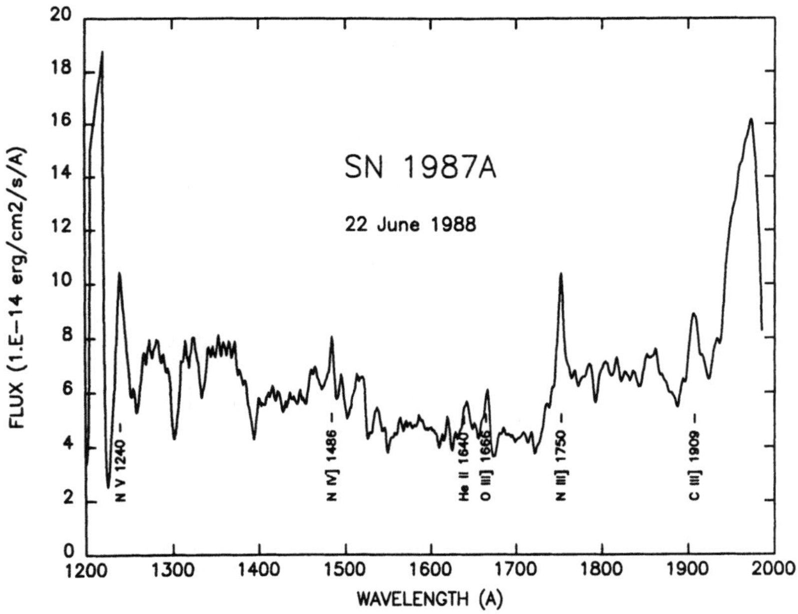

Figure 1. An *IUE* SWP spectrum of SN 1987A showing the narrow emission lines superposed on the UV continuum of Stars 2 and 3. The sharp rise in the continuum above 1915 Å is from the supernova.

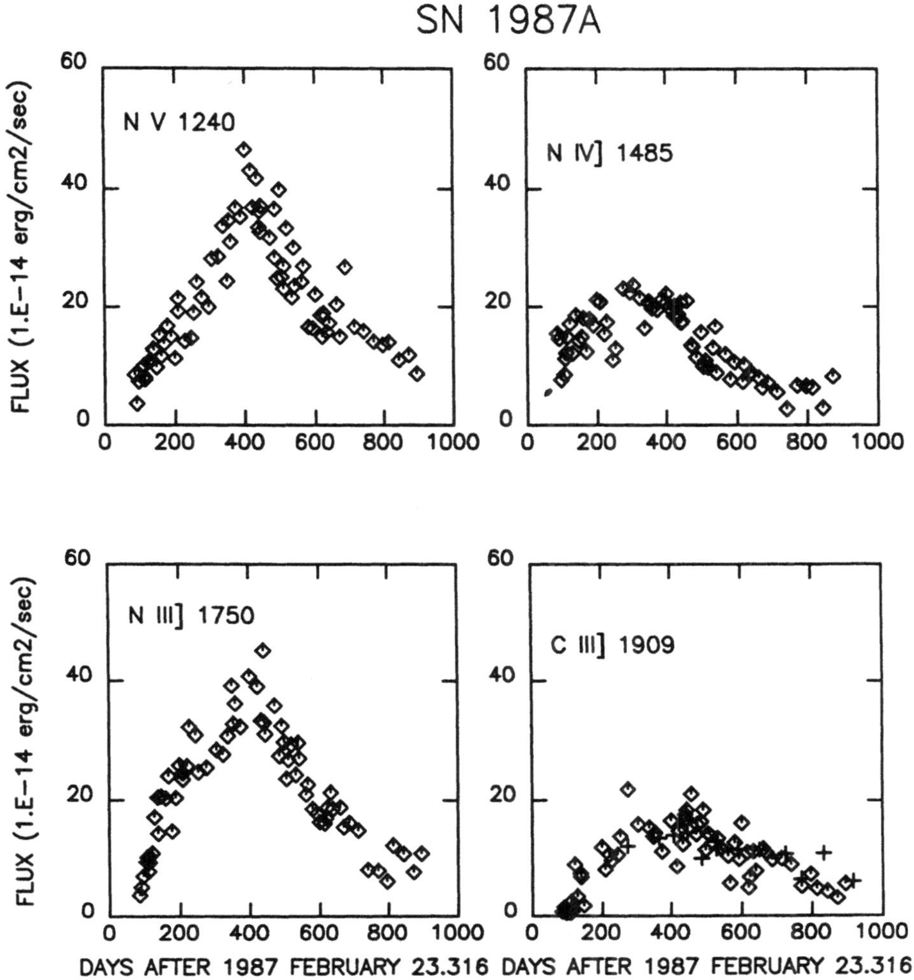

Figure 2. Light curves for the N V 1240, N IV] 1485, N III] 1750, and C III] 1090 lines. The plus signs in the C III] light curve are the C III] fluxes measured from *IUE* high-dispersion spectra, showing that the two give similar results.

Asymmetry in the Wind from the Precursor to SN 1987A

Jason Spyromilio, David A. Allen, & W. Peter S. Meikle

This work has been accepted for publication by Nature (ALLEN *et al.* [1]). The initial UV flash from the surface of SN 1987A ionized the circumstellar gas to produce narrow emission lines which were seen first in the UV (IUE) (WAMSTEKER *et al.* [2]) and later in the optical spectra (WAMPLER [3]). Since May 1988 we have recorded a narrow component to the 1083 nm He I line. Measurements were obtained on six occasions between days 434 and 737 during which time the narrow component exhibited little evidence for variation, remaining at an intensity of $15 \pm 2 \times 10^{-13}$ erg/s/cm^2 (see Fig. 1.) On day 737 we explored the spatial distribution of the narrow component (Refs [1],[4]). We set the grating so that one of the detectors in the spectrometer (FIGS) received radiation from the narrow component while several straddled the broad component. By repeated raster scanning of the telescope we thus created simultaneous maps of both the broad and narrow components. While the broad component, due to the supernova itself, originated in an unresolved point source, approximately half the total intensity of the narrow component arose in a compact source, of diameter less than $\sim$ 1.5 arcsec, displaced by 0.8$\pm$0.3 arcsec at PA 200$\pm$15 deg (see Fig 2). Geometrical and other considerations demonstrate that the source lay behind the supernova, at a distance of about one light year.

We have constructed a simple model of the compact HeI source consisting of a slowly moving cloud of helium and hydrogen only. We assume it was fully ionized by the UV flash. We have allowed for recombination, collisional effects and ionisation of neutral atoms by photons released in the recombination of He^{++}. Contemporary optical observations of circumstellar [OIII] lines (STATHAKIS private communication) indicate an electron temperature of about 20000 K. Since this might not apply to the HeI knot, we adopted a somewhat higher temperature (30000 K) as this would tend to provide a conservative estimate of the helium overabundance. In comparing the model and observations we used the result that $\sim$0.3 of the total narrow component of the Hα flux came from the direction of the HeI knot (CROTTS [5] and private communication). For T=30000 K the model satisfactorily reproduces the observed fluxes if $N_e \sim 10^5$ cm^{-3} and the helium abundance, Y, is 0.4–0.5. The cloud mass density is about $\times$10 larger than the typical value for the whole shell obtained from UV observations.

The compact He I source may be an interstellar cloud that successfully resisted the momentum flux of the RSG wind. However, it is unlikely that an interstellar cloud with so high a helium abundance would lie within 1 *ly* of the supernova. An alternative is a bipolar RSG wind (WOOD & FAULKNER [6]), possibly also interacting with the interstellar medium. A *symmetrical* bipolar form is incompatible with our data. Moreover, a frontal lobe must have been less dense than the rear if the density is to be compatible with the UV line ratios. We therefore interpret the cloud as part of an asymmetric RSG outflow.

Extensive mass-loss and mixing of the progenitor envelope have been invoked to account for its blue-red-blue evolution before explosion and to explain the high N/C and N/O ratios inferred from the ultraviolet data (MAEDER [7], SAIO *et al.* [8], WOOSLEY *et al.* [9]). With the additional constraint that the progenitor envelope was massive, SAIO *et al.* [8] find that complete mixing of the hydrogen-rich envelope together with dredge-up of about a solar mass of material from the helium core is required, giving a surface helium abundance 0.45. Our observations support this model. However, the high HeI 1083/Hα intensity ratio and the inferred helium overabundance could be partly the result of reddening by dust in the vicinity of the cloud.

Fig. 1. Spectra of SN 1987A taken in 1988 September and 1989 January. The broad component is helium and hydrogen emission from the supernova ejecta. The narrow component is helium in the circumstellar shell.

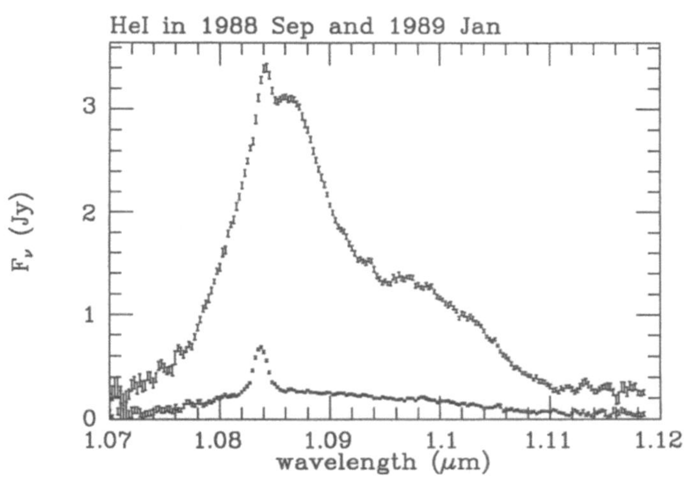

Fig. 2. Grey scale representations of maps of the broad (left) and narrow (right) components seen in Fig. 1. The pixels are 0.7 arcsec square. Both maps have been automatically scaled to cover the full tonal range. North is up and East is to the left.

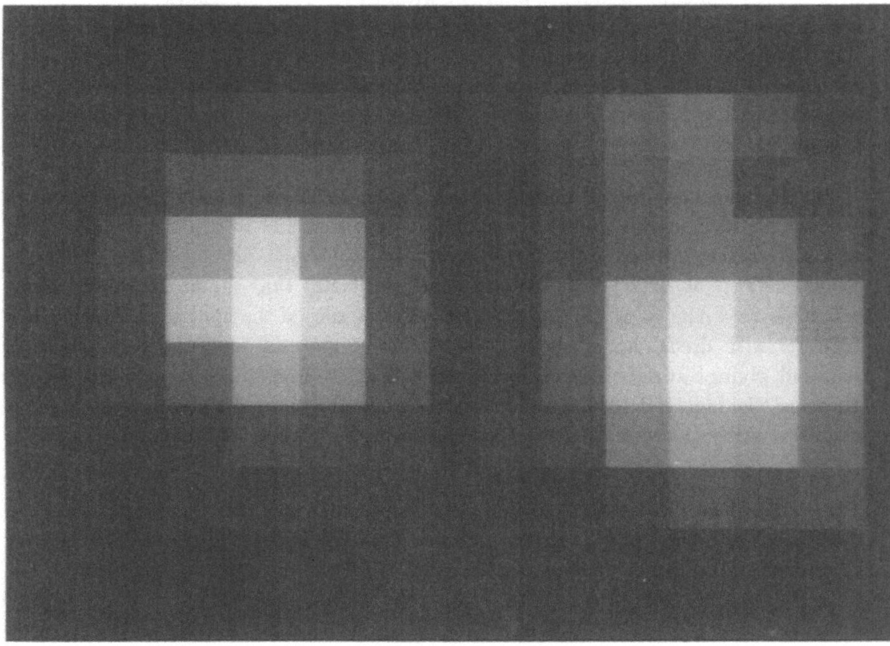

Acknowledgements

We thank Raylee Stathakis and Arlin Crotts for valuable discussions and for providing data prior to publication.

References

1. D.A. Allen, W.P.S. Meikle & J. Spyromilio: Nature, (in press).
2. W. Wamsteker, R. Gilmozzi & A. Cassatella: IAU Circ., 4410 (1987).
3. E.J. Wampler, IAU Circ., 4541 (1988).
4. D.A. Allen, W.P.S. Meikle & J.Spyromilio: IAU Circ., 4747, (1989).
5. A.P.S. Crotts, W.E. Kunkel & P.J. McCarthy: Ap.J. (in press)
6. P.R. Wood & D.J. Faulkner: IAU Circ, 4739 (1989).
7. A. Maeder: In ESO Workshop on the supernova 1987A Conf. Workshop Proc. 26, p 251, ed I.J. Danziger (ESO 1987).
8. H. Saio, K. Nomoto & M. Kato: Nature, 334, 508 (1988).
9. S.E. Woosley, P.A. Pinto & T.A. Weaver: Proc. Astr. Soc. Aust, 7, 355 (1988).

Comments on the IR Emission from the Subarcsec Vicinity of SN 1987A

A. A. Chalabaev

Our group recently published results of speckle-interferometry of SN 1987a in the near IR (2-5 μm) carried out in May, June and August of 1987a (CHALABAEV, PERRIER, MARIOTTI [1]). Here, I would like to comment on the possibility of observational artefacts in those data and to discuss the rigidity of the proposed interpretation.

Although the speckle-interferometry in its basic ides is a fairly simple technique, in practice, measuring fine details of an object through the atmosphere of a rapidly varying turbulence may prove to be a complicate affaire with sometimes unreliable results. The intrinsically small field of view (few arcsec) forces the observers to calibrate the optical modulation transfer function (MTF) sequentially in time. As a result, in the case of bright sources, the accuracy is defined by the uncompensated variations of the atmospheric and possibly telescope transfer functions. Givens of SN 1987a conspired so that its observations were a particularly challenging problem. Consider for example the rationning of atmospheric MTF's which is quite sensitive to the air mass diference and in such an unpleasant way that different spatial frequency parts of the MTF depend on the air mass in a different way (e.g. RODDIER [2]). In some cases, the efect is to produce an illusion of a companion object. At the declination of -69°, most of the SN observations from ESO (latitude -27°) were conducted at considerable air masses making this effect considerable. Obviously, a special care had to be taken in the choise of the reference star. But, stars of comparable brightness which could serve as references (mag V about 6) lie at 3°-4° from the SN while the distance of 1° is considered as the limit of safety. Bringning those details to the discussion helps to make clear that the calibration procedure for the atmospheric MTF took an important weight in the final results. Furthermore, it was realized (PERRIER [5]) that, in the case of the SN, even the telescope transfer function needed attention. The reason for that was the unusual color temperature of the SN : in August, it was about 1700 °K in the L' band (3.8 μm). Clearly, when the available reference stars have T_{col} about 5000 °K, a simple rationning of Fourier transforms will leave residuals due to unmatched telescope transfer functions which could mimic the signature of a partially resolved structure. One is therefore allowed to consider seriously the danger of artefacts in the observations of SN 1987a similar to those unmasked in observations of other sources (e.g. PERRIER AND MARIOTTI [3], MARIOTTI, PERRIER, LACOMBE [4]).

The chosen observing procedure was to use 2 or 3 reference stars above and beneath the SN and to switch the telescope every 1-2 minutes. Change of the reference star was done every 10-15 minutes. Each set of scans SN-ref. star was treated as an independent observation, producing one visibility. The final visibility was the average of all independent ones obtained during 2 or 3 hours, the statistical error reflecting the uncertainty of the MTF calibration. The data reduction included computing visibilities "reference/reference" as an additional test. Well known double stars were measured each night as a control and calibration of the equipment.

The unresolved visibilities in the K (2.2 μm) and L' (3.8 μm) band obtained on May 8-9 and June 7 and 8 (see [1]) are quite clean. They provide the figure of the achieved quality. Logically, the same credit will be given to other June visibilities obtained in the similar conditions, and in particular to those which indicate the emergence of the infrared emission from the region of about 0.1 arcsec around the SN on June 16 and its further presence on June 17-22 with the contribution of at least 10% to the total flux in the L' band. However, a lesser statistical weight should be ascribed to the results obtained in August. This is mainly due to the significantly lower amplitude of the resolved signal : about 4% of the visibility. Still, the same pattern repeats independently in three filters which favors its stellar origin.

The observational results can be summarized as folllows : 1) May 8-9, June 7-8 : no sources brighter than 5% of the supernova flux as measured in K, L' and M (4.6 μm) filters; 2) June 16-22 : a partially resolved source is well detected in the L' band; the best fit model gives a halo with FWHM about 0.1 arcsec; the visibilities in K and M remain those of the unresolved SN; 3) August 5-6 : a weak oscillation appears on the visibilities in K, L' and M. It corresponds to spots of emission at 0.3 - 0.4 arcsec from the SN. More details can be found in [1].

The interpretation we have given to this set of data is based on the following assumptions : 1) the resolved IR emission is due to the thermal echo from dust clumps heated by the initail UV-flash, and 2) the dust clumps are relicts of the "antique" wind on the red supergiant (RSg) stage of the progenitor [1]. The first assumption together with the allowed range of the angular distances of the IR sources and the dates of observations immediately gives the linear distances of the sources to the SN. It is about $1.5 \cdot 10^{15}$ cm for the June source(s) and about $3 \cdot 10^{15}$ cm for the August source(s). The distances place the clumps in the region where the fast blue supergiant (BSg) wind goes through the "antique" RSg wind material. The clumpiness of this region was indeed expected due to the Rayleigh-Jeans instability (RENZINI [6], CHEVALIER [7]). Within this framework, we proposed estimates of two parameters of the supernova phenomenon. The first one is the upper limit on t(RSg), the time elapsed since the RSg wind ceased to blow. Indeed, one may expect that "switching off" the dust supply in the envelope would create around the SN a dust free cavity, expanding with the velocity of the dust outflow. Then the distance at which the dust is placed by the IR speckle observations indicates the upper limit on the dimension of the cavity. The conservative estimate gives $R_c < 1.6 \cdot 10^{15}$ cm. Further, assuming that the cavity expands at a constant velocity, V_d, one gets $t(RSg) < 5100 \cdot (10\ km \cdot s^{-1} / V_d)$. The second estimate concerns the luminosity of the UV-flash, L(UV). It was deduced from the measured distances and T_{col} of the dust spots : $T_{col} \leq 2000$ °K on June 16 and $T_{col} = 2000 \pm 900$°K on August 5-6. Assuming that the absorption coefficient of the dust depends on the wavelegth as λ^{-1} and considering the allowed combinations of the spots, one gets for the luminosity $7 \cdot 10^{42} erg \cdot s^{-1} \leq L(UV) \leq 3 \cdot 10^{43} erg \cdot s^{-1}$.

Let me now to examine alternative explanations. Suppose that the IR speckle emission is not due to an echo. Given the apparent velocity of at least 0.18c for the June source(s) and at least 0.65c for the August source(s) , the alternative is to consider relativistic jets from the SN. The strong June-to-August acceleration makes this alternative unrealistic. Suppose further that the observed echo is not due to dust but gas. One expects such an echo to emit in narrow spectral lines. Given the strength of the resolved signal in L' band in June, not less than 10% of the SN flux, those lines would have been a prominent feature of the supernova spectrum. This was not the case. Suppose now that the dust does not originate from the RSg wind. At the distance of ~ 10^{15} cm from the SN, any dust of interstellar origin would have been blown away since long time ago. The dust should then have been formed in the BSg wind, but the conditions in this wind is much more hostile for dust condensation than in the RSg wind. Let me now to examine whether the dust echo traced the maximum of the bolometric light curve occured in May or the UV flash at the explosion as we assumed. In the former case, the shorter time delay places the June source(s) at the distance of 20 light-days (l.-d.) from the SN (cf. 60 l.-d. if the heating was due to the UV flash). Some 80 days before, the dust should have been heated by the UV-flash of $L \approx 3 \cdot 10^{43} erg \cdot s^{-1}$ (WOOSLEY [8]). As a simple calculation shows, at that time the thermal equilibrium temperature of the dust should have been as high as

2300°K, much more than the evaporation temperature of about 1300°K. The dust would have been evaporated and thus could not give rise to the second echo. The situation is more ambiguous for the August source(s). The shorter time delay places the clumps at 107 l.-d. instead of 117 l.-d. The earlier heating by the UV flash would increase the dust temperature up to 1200°K; the dust could survive and reobserved in August. However, the measured $T_{col} = 2000 \pm 900$°K (corresponding to the dust temperature of 1350 ± 500°K) requires the luminosity of the heating radiation of at least $7 \cdot 10^{42}$ erg$\cdot$ s^{-1} while the peak luminosity in May was only $8 \cdot 10^{41}$ erg$\cdot$ s^{-1} (CATCHPOLE ET AL [9]).

One may therefore conclude that the interpretation proposed in [1] appears as the most likely one. Discussion can be focused now on the range of the accepatable values of L(UV) and t(RSg).

The upper limit on L(UV) is given by the June 16 data while the lower limit follows from the August data. As it was mentioned above, the strength of the resolved June emission and a rich calibration material place those data at a high level of confidence while the August data needs caution and should be taken as indicative ones rather than strongly confirmative ones. One shall conclude that L(UV) was *certainly* less or equal to $3 \cdot 10^{43}$ erg$\cdot$ s^{-1} (assuming the dust absorption falls as λ^{-1}) and that L(UV) was *probably* more $7 \cdot 10^{42}$ erg$\cdot$ s^{-1}.

Considering the estimate of t(RSg), one notes that there are two factors which contribute in. The first one, R_c is defined by the distance from the SN to the June source(s). It can hardly be changed. The second factor, the scaling one, (10 km$\cdot$ s^{-1} / V_d), is only loosely fixed. The dynamical history of the dust clumps in the region of the fast BSg wind was certainly more complicate than the outflow at a constant velocity V_d as it was assumed in the first crude estimate. Nevertheless, in spite of the uncertainty related to the dust velocity, finding dust in such a close vicinity of the SN points to a short time scale of the progenotor evolution from the red to the explosion. This is corroborated by the time scale estimate obtained by WAMPLER AND RICHICHI [10] from the spatial extent of the fluorescence echo in narrow spectral lines. They give 7000$\cdot$ (50 km$\cdot$ s^{-1} / V_{sh}) yr for the time interval elapsed since the BSg wind began to blow. The value of V_{sh}, the velocity of the BSg wind propagation into the RSg wind material, is more open to a reasonably accurate estimate than the value of V_d. Therefore, the estimate of the evolutionary time scale obtained in [10] should be prefered. The crude estimate of t(RSg) obtained from IR speckle data should be then regarded as an independent evidence in favor of the short time scale.

References

1. A. Chalabaev, C. Perrier, J.-M. Mariotti : Astron. & Astrophys., **210**, L1 (1989)
2. F. Roddier : Progress in Optics, 19, **281** (1981)
3. C. Perrier and J.-M. Mariotti : Astrophys. J., **312**, L27 (1987)
4. J.-M. Mariotti, C. Perrier, F. Lacombe : Astron. & Astrophys., **182**, L11 (1987)
5. C. Perrier : in Proc. of NATO Advanced Study Institute *"Diffraction limited imaging with very large telescopes"*, ed. by D. Alloin and J.-M. Mariotti, Gargese (1988)
6. A. Renzini : in Proc. *ESO Workshop on SN 1987a*, ed. by I.J. Danziger, Garching, F.R.G., p. 295 (1987)
7. R. Chevalier : in Proc. *ESO Workshop on SN 1987a*, ed. by I.J. Danziger, Garching, F.R.G., p. 481 (1987)
8. S. Woosley : Astrophys. J., **330**, 218 (1988)
9. R. Catchpole et al. : M.N.R.A.S., **229**, 15p (1987)
10. E.J. Wampler and A. Richichi : Astron. & Astrophys., **217**, 31 (1989)

A Third, Inner Light Echo Ring Around SN 1987A

Nino Panagia

At the beginning of this year evidence has been found for a third, inner echo near SN 1987A (BOND *et al.* [1]). This discovery is of great importance because the detection of an echo so close to the SN 1987A reveals the existence of diffuse matter *physically* associated to SN 1987A, hence, to its progenitor Sk $-69°202$.

A careful analysis of CCD imagery observations in the B, V and R bands obtained on 1989 January 24 at the CTIO (BOND *et al.* [2]) has revealed the presence of a third, inner light-echo ring around SN 1987A, in addition to the two previously known outer rings (see Figure 1). Its mean radius of 8.3″ corresponds to a projected radius of 2.1 pc from the supernova and implies that the light scattering occurs off interstellar dust located only 4.5 pc in front of SN 1987A. The width of the ring is 2.2″, so that the the scattering material is actually distributed between radii 3.6 and 5.4 pc from the supernova. The ring is brightest on the eastern side where the average surface brightnesses in B, V and R are 23.3 ± 0.4, 22.1 ± 0.2 and 21.4 ± 0.3 mag arcsec^{-2}, respectively. These values imply an average column density of 2.2×10^{18} cm^{-2} and a local density of 1.7 cm^{-3}. The surface brightness in the western side is about 0.5 mag arcsec^{-2} fainter, indicating that the scattering material is not uniformly distributed around the SN.

Confirmation of the reality of the echo has been provided by similar observations made by COUCH and MALIN [3] and CROTTS and KUNKEL [4]. The latter investigators found that the inner echo had been detectable as early as 1988 October.

Another important piece of information has been provided by SPARKS, PARESCE and MACCHETTO's [5] polarimetric observations made on 1989 February 10 with the ESO Faint Object Spectrograph and Camera (EFOSC) operating in the coronographic mode on the 3.6-m. telescope. These observations have confirmed the existence of an arc of V-band emission with a sharp inner and outer boundary and most prominent in the Eastern quadrant between 45° and 135° position angle. More importantly, in this region the detected radiation is partially linearly polarized with a degree of polarization 0.15 ± 0.04 and an electric vector orientation of $\sim 0°$ position angle. Therefore, the observed ring is indeed a light echo produced by scattering material that is distributed in a ragged shell at a distance of about 5 pc from the supernova.

Further insight is gained considering the lack of detection at earlier times. In

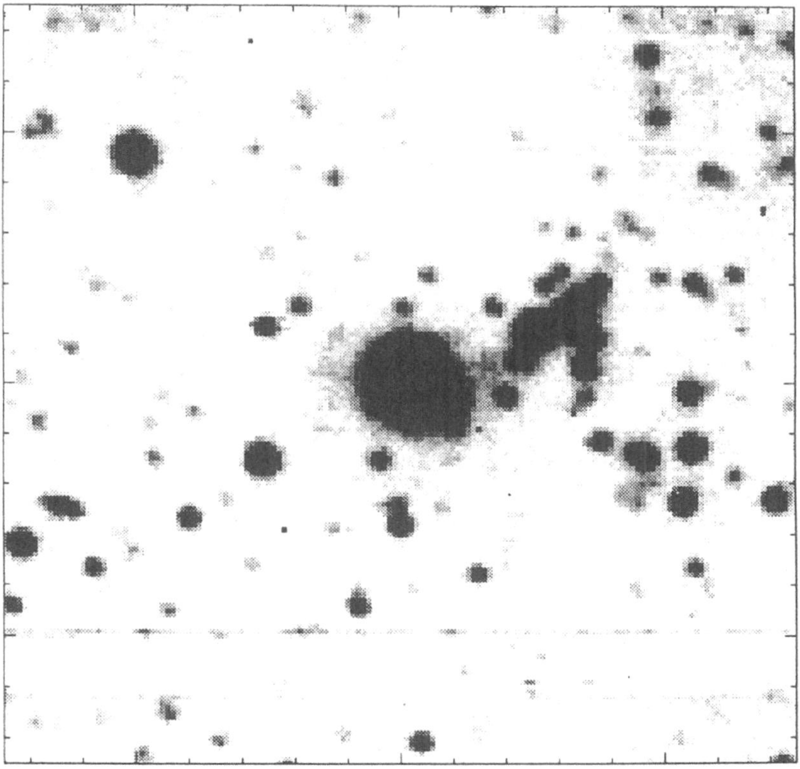

Figure 1: Image of a 75″ × 75″ field centered on SN 1987A on 1989 January 24 (top is North and left is East). The new echo is the fuzz of about 19″ diameter that surrounds the supernova.

particular. PARESCE and BURROWS [6] could not detect the inner ring in the B and R bands in 1987 December, and BURROWS [7] found only a stringent upper limit to the R surface brightness in 1988 September. These earlier null results imply that there is a cavity in the dust along the line of sight toward the Earth, such that the shell material outside the cavity only became illuminated at the end of 1988.

As suggested by CHEVALIER and EMMERING [8], such a shell may consist of wind material shed by the progenitor star when it was in the red-supergiant phase and that has been decelerated by its interaction with the surrounding medium. However, the situation must be much more complicated than that of a smooth, regular circumstellar shell because the brightness is not the same in different sides of the ring. Furthermore, the lack of detection at early times indicates that matter is irregularly distributed also radially and not only azimuthally.

In other words, the distribution of the scattering material is *neither* uniform *nor* isotropic. Judging from the fact that no detection was achieved until September 1988, the scale of the density fluctuations can be estimated to be of the order of a parsec and the filling factor to be lower than 0.5. Therefore, unless the line of sight toward the Earth is privileged, we should expect that other zones with low density be intercepted by the light paraboloids in the near future. When that will happen the echo, or parts of it, will dim or even disappear for a while. And may reappear again when more dense material will come again on the paraboloid's path.

In this context, the possible detection of scattering material as close to SN 1987A as a light year, and mostly behind it (ROCHE *et al.* [9]), may also be evidence of a patchy distribution of the pre-supernova circumstellar material and provide precious clues about the *real* evolution of massive stars.

References

1. Bond, H.E., Panagia, N., Gilmozzi, R., Meakes, M.: IAU Circ., No. 4733 (1989).

2. Bond, H.E., Panagia, N., Gilmozzi, R., Meakes, M.: Ap. J. (Letters), in press.

3. Couch, W.J., Malin, D.F.: IAU Circ., No. 4739 (1989).

4. Crotts, A.P.S., Kunkel, W.E.: IAU Circ., No. 4741 (1989).

5. Sparks, W.B., Paresce, F., Macchetto, F.D.: Ap. J. (Letters), in press.

6. Paresce, F., Burrows, C.: Ap. J. (Letters), 337, L13 (1989).

7. Burrows, C.: IAU Circ., No. 4781 (1989).

8. Chevalier, R.A., Emmering, R.T.: Ap. J. (Letters), 342, L75 (1989).

9. Roche, P.F., Aitken, D.K., Smith, C.H., James, S.D.: Nature, 337, 533 (1989).

Nimba Around Supernova 1987A: Kinematics and Evolution in Ages

A. A. Filyukov

A suggestion of possible origin and nature of light nimba around the line of observation of Super Nova 1987A was made in our paper (FILYUKOV [1]). The phenomenon was first observed by CROTTS [2] and later by others (ROSA [3], HEATHCOTE *et al.* [6]). Here some explanations of given suggestion are presented and qualitative consequences are discussed.

According to the observational data of ROSA, obtained on February 13, 1988, angular dimensions of luminous rings were 32″ and 51″ (+1″) of arc, the rings being strictly concentric. The last circumstance represents the main geometric property of discovered phenomenon. At present time it excludes the existence of possible large scale random inhomogeneities in the supernova surrounding medium. The strict concentricity of luminous nimba with respect to supernova line of observation may be only the consequence of two kinds of geometries: planar and spherical. Recently, CHEVALIER and EMMERING [4] have considered a model of planar inhomogeneity as a cause for appearance of the so called light echo. In this paper a spherical geometry is considered and the appropriate data presented gives the opportunity to establish by observations in the near future the real nimba model.

A planar inhomogeneity model of CHEVALIER and EMMERING suggests that two sheets of enhanced density intersect the observational line of supernova. The sheets are proposed to be located orthogonally to the ray of sight of the supernova. The typical feature of this model proves to be a constant area of luminous nimb in time. As the radius expands the ring width becomes proportionally narrower so that the surface brightness remains constant. Another significant feature of the plane inhomogeneity model is connected with the nimbus sizes which may grow up infinitely. Considering the above mentioned maintenance of surface brightness it causes to creation of peculiar starlike ring objects. However, accounting for the finiteness of plane inhomogeneity layer thickness removes this paradox (Fig. 1). The mutual *concentricity* of both rings will be violated if the planes of inhomogeneity

The physical nature of interaction between the light flash and the inhomogeneity medium may be of two kinds. It may be determined either by scattering of light on the dust particles - DWEK [7], CHEVALIER, EMMERING [4], SCHAEFER [8], or the reradiation of flash light due to the gas fluorescence - MORRISON, SARTORI [9]. In both cases the interaction process is determined by the first order of matter density, however in the first case the range of arbitrariness is wider due to additional information on the dust content and its spectral properties are needed. In the case of fluorescence mechanism the effect should be considered as a result of the flash short waves transformation by the inhomogeneity matter into the real nimba luminosity, rather than a mere scattering by a dust or the "light echo." Irrespective of the real luminosity mechanism one may consider geometric and kinematic properties of nimba following from the spherical geometry of continuous distributed medium.

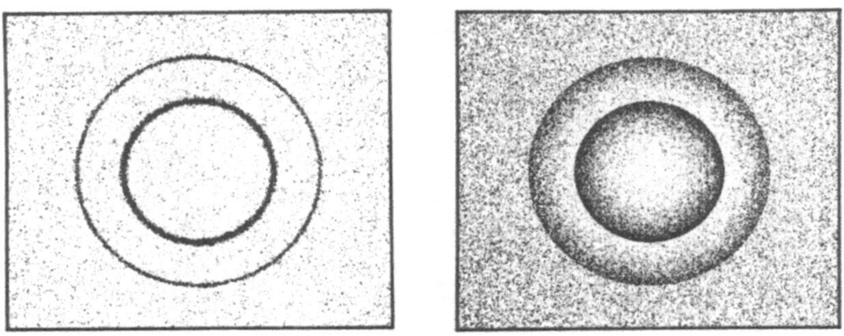

A B

Figure 3. Computer images of nimba in plane (A) and spherical (B) models.

The main geometric properties of nimba depend on simple kinematic relations connected with the finiteness of light velocity. They develop against the background of much more slower secular motion of inhomogeneities themselves, the latter may be considered as fixed. For a distant observer the light front of transformed radiation represents a paraboloid of revolution about the line of observation with a focus at supernova. Only from this surface the light signals propagating parallelly with the observation line are synchronized for the moment of arrival. Each surface point of a synchronism paraboloid, the form of which is determined by the time interval after flash moment, acts as a source of reradiated emission due to the matter of inhomogeneity. In accordance with certain phase diagram the rays parallel to arbitrary observation line exist. That is in contrast to the plane model the observational picture of the phenomenon does not depend on the observer's position due to the spherical symmetry.

are of angles of bending. This specific inevitable feature of plane model implies the condition that six geometric angular parameters determining space orientation of the planes and of line of sight should have incidentally definite values in order to secure the concentricity. Thus, one can consider the plane model only as an approximation of a more realistic spherical model at its initial stage. DWEK [7] proposed the spherical inhomogeneity model with a layer of constant density imitating the action of stellar wind on the exterior medium. This approach is the closest to the model discussed below, however it cannot give the explanation for the natural origin of binary observed nimba.

The spherical model sketched in Fig. 2 assumes the continuous distribution of inhomogeneity with two maxima at the distances R1 and R2 from the supernova due to propagation of the shock waves of unknown origin located near by supernova. Any preconception motives are completely absent in this model the unique hypothesis of which is connected with an unknown scenario of progenitor history and the near-by starts. The perturbation of surrounding medium could arise also due to the blow-off of the exterior envelope of the progenitor.This event in the history of progenitor assumed by WOOSLEY [10] or the explosion of relatively neighbour stars should lead according to our conception to formation of the binary spherical density distribution approximately concentric to the protostar. Propagation of the light front from the supernova flash through the medium perturbed in such a way, results in successive symmetric illumination of spherical inhomogeneity layers since they are taken as concentric to the star. The case of nonconcentric with the supernova inhomogeneity sphere will be discussed elsewhere.

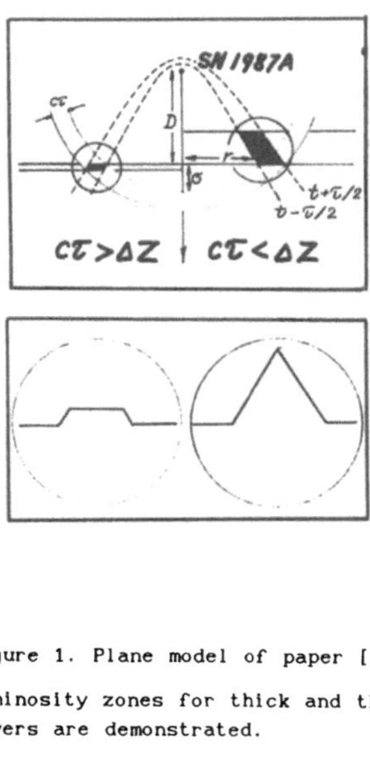

Figure 1. Plane model of paper [4].

Luminosity zones for thick and thin layers are demonstrated.

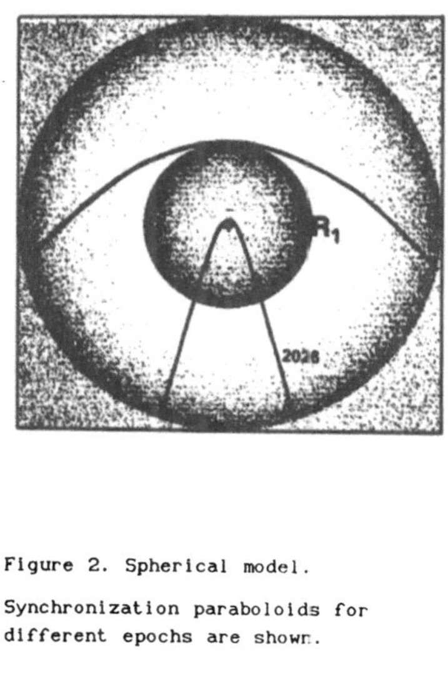

Figure 2. Spherical model.

Synchronization paraboloids for different epochs are shown.

The luminosity of the matter illuminated by the flash occurs from the region determined by the synchronism paraboloid and it arrives to the observer simultaneously with the radiation of the flash itself. However, at the initial stage of star flash observation this region adjoins directly the line of observation. The flash radiation is summed up, in fact, with the transformed radiation of the whole medium along the observational line, the latter representing a degenerated paraboloid of synchronization. That is why the nimba luminosity at the beginning of the blow-up cannot be distinguished from the luminosity of star itself. The luminosity of regions already illuminated by the flash light cannot be observed behind the surface of synchronism paraboloid because the radiation from these regions is yet on its path to the observer. After about a flash duration time the observation of formerly illuminated regions located farther from the observational line becomes possible. The observational zone expands with the synchronism paraboloid to periphery, and in the vicinity of line of sight the luminosity ceases. All these effects determine the creation of luminous nimba and the law of their evolution in ages.

Now we consider some quantitative results that follow from the above picture of phenomena at the assumptions of plane and spherical models. The distances from SN to inhomogeneities are determined by following expressions:

Plane model distance:

$$D_i = (r_i^2 - \sigma^2)/2\sigma;$$

Spherical model radius:

$$R_i = (r_i^2 + \sigma^2)/2\sigma;$$

where r_i is the i-th linear nimbus radius, $\sigma = c(t-t_*)$ with c - the light velocity and t_* - the local flash time. The nimbus size evolution in time and the difference between the plane and spherical models are shown in Fig.4. According to the observations by HEATHCOTE *et al.*[6] on March 19, 1988, the angular dimensions of nimba radii, measured with accuracy (+0″.5), were 33″.0 and 54″.0. Taking into account the growth of nimba dimensions with time the above accuracy is sufficient for the observational comparison between the plane and spherical model predictions in 1991 when the difference in models radii predictions will exceed 1%. In contrast to the plane model, where the radii of nimba unlimitedly grow with time, the nimbus radius of our model is limited by the radius of corresponding inhomogeneity at the time $t-t_* = R_i/c$. Later on the radius decreases and change cycle for each nimbus is proportional to the dimensions of inhomogeneity.

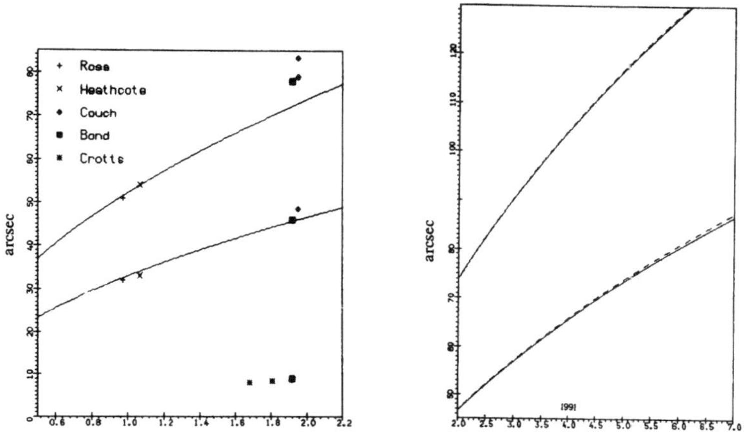

Figure 4. Nimba radii versus time in plane and spherical models.

The time is measured in years after SN 1987A explosion. Observational data of different authors are marked. Dushed line-plane, solid line-spherical model.

At the final evolution stage of internal ring, before its disappearance, the spherical model predicts emergence of a remarkable phenomenon of the *cosmic eye*. At this epoch the most convex side of the synchronization paraboloid will cross the internal sphere of the shock wave front. Since at a crossing time the minimal radius of paraboloid curvature is equal to the spherical inhomogeneity front radius the internal nimbus will convert into the bright "pupil" which is to disappear some time later.

At the time of appearance of the internal "pupil" the precise measuring of density distribution in the spherical front region along the radius is possible with the help of brightness variation depending on the local density. The most sinificant consequence of this measurement consist in the direct definition of the front radius by means of the time interval of paraboloid displacement the speed of which along the the observation line is equal to a half of the light velocity (Fig.5).

Recent emergence of the third small size nimbus discovered by CROTTS and KUNKEL [12] and observed also by COUCH and MALIN [13] and BOND, PANAGIA, GILMOZZI and HEAKES [11] one may consider in the frames of our model as a definite displacement of the center of spherical inhomogeneity relative to the supernova. On the Fig.6 the last stage of evolution of this nimbus is shown as well as two side projections of mutual location of inhomogeneity center (+) and supernova (*) are given.

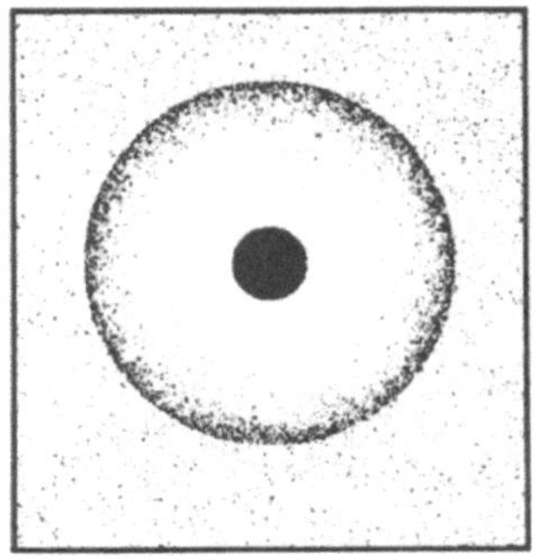

Figure 5. The "Cosmic eye" phenomenon.

On this figure the final stage of evolution of the intrinsic nimbus is shown in case of coincidence of supernova with center of inhomogeneity. The emergence time of this phenomenon gives the possibility to measure the absolute linear dimension of the intrinsic inhomogeneity sphere and by comparison with its maximal observational size to define the distance to the inhomogeneity center with the accuracy of the order of tens of parsecs.

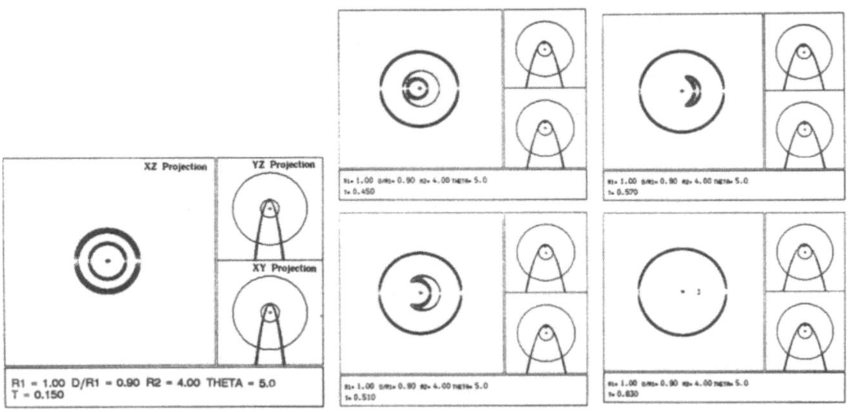

Figure 6. The "Cosmic eye" phenomenon.

On the following figures different stages of evolution of the third small nimbus are shown as well as the two side projections of mutual location of inhomogeneity center and supernova. The last stage corresponds to emergency of a "halfmoon" symbol. D is the distance from the center to supernova placed in Z=0 plane. Theta in grads is an angle between Y-axis and SN vector. The radii of spherical inhomogeneities are in arbitrary units connected with time units by equality c=1.

Since the angular radius of the intrinsic sphere is great enough to be defined with exellent accuracy the opportunity of *direct measurement* the distance of SN 1987A arises by the standard method of angular observation of this *Grand Sphere - the astroobject with potentially known large radius.*

Modern means of angular measurements allow in perspective the determination of the supernova distance $\mathfrak{D}$ with accuracy within tens parsecs provided that there are no considerable perturbations in the spherical front. The distance $\mathfrak{D}$ can be measured much earlier than the date noted above but with mach lower accuracy. To do this one should be capable to distinguish the synchronization paraboloid from the corresponding ellipsoid. The separation between ellipsoid focuses is the desired distance $\mathfrak{D}$. For these aims let us represent the radii of nimba for ellipsoidal front of synchronism in quasiparabolic approximation:

$$r_i^2 = 2\mathfrak{R}_i\Sigma_i - \Sigma_i^2; \qquad \Sigma_i \equiv \sigma\,[\mathfrak{D}-\mathfrak{R}_i + \sigma]/[\mathfrak{D}+\sigma];$$

whence it is obvious that the internal and external nimba grow up in time at somewhat different rates Σ_i. In other words, they are located on different paraboloids. Since the radii ratio R_1/R_2 does not depend on Σ and this ratio has been already measured, the two unknown parameters of the problem are $\mathfrak{D}$ and the true "elliptical" value of one radius $\mathfrak{R}_i$. For these aims it is necessary and sufficient to measure just two radii r_1 and r_2. A possible date for performing the measurements depends on the true value of $\mathfrak{D}$. One can easily see that the first opportunity opens when the discrepancy between the plane and spherical models will arise as far as the linear radius of inhomogeneity R_i can be expressed in the following form:

$$R = \sigma\,[r_{i\ plane}^2 + r_{i\ sphere}^2] \,/\, [r_{i\ plane}^2 - r_{i\ sphere}^2]$$

where the nimba radii can be measured in arcseconds.

References

1. A. Filyukov: Sov. Astron. J., 65, 1094 (1988).
2. A. Crotts: IAU Circ. No. 4561 (1988).
3. M. Rosa: IAU Circ. No. 4564 (1988).
4. R. Chevalier and R. Emmering: Ap. J. (Letters) 331, L105 (1988).
5. A. Filyukov: Sov. Doklady, 302, N5, 1082 (1988).
6. S. Heathcote and N. Suntzeff, et al.: IAU Circ., No. 4567 (1988).
7. E. Dwek: Ap. J., 274, 175 (1983).
8. B. Schaefer: Ap. J. (Letters) 323, L47 (1987).
9. P. Morrison and L. Sartori: Ap. J., 158, 541 (1969).
10. S. Woosley: SN 1987A: After the Peak, submitted to Ap. J. (1987).
11. H. Bond, N. Panagia, R. Gilmozzi, and M. Heakes: IAU Circ. No 4733 (1989).
12. A. Crotts and W. Kunkel: IAU Circ. No. 4741 (1989).
13. W. Couch and D. Malin: IAU Circ. No. 4739 (1989).

Circumstellar and Interstellar Interaction Around SN 1987A

Roger A. Chevalier

1. Introduction

A remarkable property of SN1987A is that it exploded as a blue star. Because such stars have fast stellar winds, the gas density immediately surrounding the supernova was low and the observable consequences of the initial shock interaction were limited to radio emission over the first few weeks (see [1] for a review). While the shock interaction phenomena have been disappointing, the radiation from the supernova has illuminated a variety of structures in the surrounding medium. Because light travel time effects play a crucial role in interpreting these observations, the phenomena are referred to as light echoes.

The nearest dense gas to the supernova appears to be in a shell that was initially highly ionized by the flash from the supernova at the time of shock break-out. Emission lines from the shell have been observed at ultraviolet (FRANSSON et al. [2]) and optical (WAMPLER and RICHICHI [3]) wavelengths. Modeling of the emission by FRANSSON and LUNDQVIST ([4] and this meeting) shows that the shell has a radius of about 5×10^{17} cm, a mass of about 0.03 $M_{\odot}$, and a density of $(1-3) \times 10^{4}$ cm^{-3}. These properties are consistent with expectations for a shell created by the fast wind from the blue progenitor star sweeping up the dense wind from a previous red supergiant phase [1,5]. However, the shell velocity appears to be 30 km s^{-1} cm or less [3], which is lower than expected for spherically symmetric evolution. The implication is that the coupling of the fast wind to the shell was incomplete and the fast wind material has broken through the shell leaving behind dense blobs.

The duration of the red supergiant phase of the progenitor star may be of order 6×10^{5} to 1×10^{6} years (WOOSLEY [6]). At 15 km s^{-1}, a red supergiant wind would extend to many pc from the supernova, although the extent of the outflow is probably limited by the pressure of the surrounding medium [7]. In any case, it is likely that the ultraviolet line emitting shell involves only the inner part of the red supergiant wind. A scattered light echo feature was observed at a radius of about 8" in early 1989 [8,9,10]. Assuming that the light was emitted near supernova maximum, the radial distance from the supernova is 4-5 pc. CHEVALIER and EMMERING [7] suggested that this echo is a shell created by the interaction of the red supergiant with the surrounding pressure. They assumed that $3M_{\odot}$ was lost over a red supergiant lifetime of 6×10^5 yr to yield a mass loss rate of 5×10^{-6} $M_{\odot}$ yr^{-1}; this is probably an upper limit to the average mass loss rate and the actual rate could be considerably smaller. Even with this high rate, a low surrounding pressure ($p/k \simeq 10^3$ cm^{-3} K) is needed to produce a sufficiently large shell. While a model of this type is plausible for the surroundings of SN1987A, CROTTS [11] has recently found that the diffuse emission previously attributed to the inner echo [10] is not expanding. In addition, emission has been detected between the ultraviolet line echo and a few pc from the supernova [12,13]. While some of the emission is diffuse, as expected for an undisturbed red supergiant wind, other parts show structure; in particular, there is evidence for a sheet of matter about 200 light days (0.17 pc) behind the supernova [14,11,15]. CROTTS [11] has noted that star 2 near the supernova may play a role in creating this structure; in section 2 of the present work, I examine the possible effects of a wind from star 2.

Interstellar echoes were discovered in the form of rings around SN1987A early in 1988 [16,17,18]. The 2 major features can be attributed to 2 dust clouds that are at distances of 120 and 330 pc in front of the supernova. The column density of dust required to produce the typical observed surface brightness is comparable to that required to produce the extinction along the line of sight to the supernova in the LMC [19]. This column density is comparable to that seen over the surrounding area of the LMC [1] and can be attributed to diffuse interstellar clouds. Higher density clouds can be expected and, in fact, regions of high surface brightness in the echoes have recently been detected (N. SUNTZEFF, this meeting; SCHAEFFER [20]). In section 3, I discuss the interpretation of the high surface brightness features, taking into account extinction along the echo path. The maximum surface brightness of an echo feature is noted.

2. Wind Interaction with Star 2

As discussed above, the red supergiant star wind from Sk -69 202 might extend to several pc from the supernova. Star 2 is at a distance 0.7/cos α_2 pc and star 3 at 0.3/cos α_3 pc from the supernova, where α is the angle between a line joining the star to the supernova and the plane of the sky. The star density in the group of stars 1 (Sk -69 202), 2 and 3 is sufficiently high to suggest an association in space and not merely a chance superposition [21] so that some effect on the circumstellar environment of star 1 is likely [1].

We examine the interaction between the red supergiant wind from star 1 with the fast wind from star 2. As noted in section 1, the red supergiant wind velocity is presumably $v_1 = 15$ km s^{-1}. The mass loss rate is less well determined; the estimate $\dot{M}_1 = 5 \times 10^{-6}$ $M_\odot$ yr^{-1} is an upper limit and the actual value could be an order of magnitude smaller. The nature of the wind interaction is determined by the pressure generated by the winds, which is related to the momentum flux, $\dot{M}v$, in the wind. For star 1, we have $\dot{M}_1 v_1 \leq 7.5 \times 10^{-5}$ ($M_\odot$ yr^{-1}) (km s^{-1}). The wind properties of star 2 depend on its stellar type. WEST et al. [22] and GILMOZZI et al. [23] have determined that it is a B0 V star based on optical and ultraviolet observations, respectively. A Galactic star of this type has $\dot{M} \approx 8 \times 10^{-8}$ $M_\odot$ yr^{-1} and $v \approx 2000$ km s^{-1} [24], yielding $\dot{M}_2 v_2 = 1.6 \times 10^{-4}$ ($M_\odot$ yr^{-1}) (km s^{-1}). However, WEST et al. and GILMOZZI et al. take a total $E_{B-V} = 0.2$ to the supernova and take a visual magnitude $V_2 = 15.50$ and 15.30, respectively. It has recently become possible to make optical observations of star 2 again, yielding $V_2 = 14.88$ (N. SUNTZEFF, this meeting); also, recent estimates of E_{B-V} yield 0.15 or 0.16. The net effect is that star 2 may be somewhat more luminous and somewhat redder than a B0 V star. The B0.5 III star ϵPer is estimated to have $\dot{M} = 1.3 \times 10^{-7}$ $M_\odot$ yr^{-1} and $v = 1500$ km s^{-1} [24]; the change in the momentum flux may not be large. Thus we find that the momentum flux in the wind of star 2 is likely to be greater than or equal to that of the red supergiant wind from star 1.

If the two winds have the same momentum flux, the interaction region is expected to be approximately planar midway between the two stars. This is only approximate because the red supergiant wind may cool after being shocked, while the fast wind from star 2 probably does not; the thickness of the shocked region is thus thinner for the red supergiant wind. A 15 km s^{-1} shock front requires a postshock column density of about 3 x 10^{18} cm^{-2} in order to cool. This column density is produced by the red supergiant wind for the standard parameters inside a radius of about 0.8 pc. The inner part of the shocked wind may cool, while the outer part is energy conserving.

If the interaction region is assumed to be thin, its position along the line between the two stars is given by

$$\frac{r_2}{r_1} = \left(\frac{\dot{M}_2 v_2}{\dot{M}_1 v_1}\right)^{1/2}$$

where $r_{2,1}$ is the distance between star 2, 1 and the interaction region. If $\dot{M}_2 v_2$ is considerably greater than $\dot{M}_1 v_1$, then the interaction region is close to star 1 and its shape close to star 1 is in a region where the properties of the wind from star 2 change slowly. The situation is then close to the case considered by HOUPIS and MENDIS [25] of the cavity created by the wind from a comet inside of the solar wind. On the assumptions of hypersonic flow and Newtonian thin layer theory, the position of the inner shock front is given by

$$\frac{r_{sh}}{r_{sh}\ (\Theta = 0)} = \frac{\Theta}{\sin \Theta}$$

where Θ is the angle between a line joining star 1 to a point on the shock (at r_{sh}) and the line joining the two stars. This expression is only expected to be accurate for $\Theta \lesssim 60°$ because it assumes that the shock front in the cometary wind and that in the solar wind are parallel. However, it shows that the interaction region can be curved around star 1 with a bow shock extending out away from the direction of star 2.

These considerations can be applied to the surroundings of SN1987A. There is infrared [14] and optical [11] evidence for a sheet of matter 200 light-days (0.17 pc) behind SN1987A. The optical observations extend to about 0.5 pc from the line of sight. To form a plane by the wind interaction requires that it be midway between the supernova and star 2. Considering the projected distance of star 2 from the supernova and the requirement that star 2 be behind the supernova shows that this is impossible. It may be that $\dot{M}_2 v_2 > \dot{M}_1 v_1$ and the dense matter behind SN1987A has a curved structure that has not yet been revealed by the light echo effects. In this case, the red supergiant wind would be limited by a bow shock to expand primarily in front of the supernova. There could be a conical region of dense gas in front of the supernova, which would explain the lack of expansion found by CROTTS [11] for the claimed 8" light echo.

Star 3 has a smaller projected distance from the supernova than star 2 but it is somewhat fainter and its properties are not as well determined. GILMOZZI et al. [23] estimate that it is a B1.5 V star, so the effects of a wind would be reduced. However, some effect on the red supergiant wind from star 1 is certainly possible; continued observations of the circumstellar light echoes should make possible a more detailed interpretation of the circumstellar structure. Finally, we note that the echo with the apparently small curvature is close to the ultraviolet line emitting shell. It is possible that at the time of the explosion the fast wind from the immediate supergiant progenitor (a B3 I star) was beginning to interact with the wind region created by star 2.

3. High Surface Brightness Interstellar Echoes

Continued observation of the interstellar echoes around SN1987A gives a unique opportunity to determine the three-dimensional structure of clouds in the interstellar medium of the LMC. In addition, structure within the echoes gives information on clumping within the clouds. SCHAEFER [20] has reported high surface brightness regions in the echoes. Here, we estimate the maximum possible surface brightness in an interstellar echo.

A useful starting point is a result from [1]:

$$\Sigma = 18.8 + (m_{sn}-3) - 2.5 \log \left[\left(\frac{\tau_{sc}}{0.025}\right) \left(\frac{F}{10}\right) \left(\frac{R}{50\ \mathrm{pc}}\right)^{-2} \right] + 1.086\, (\tau - \tau_{sn}) \ \mathrm{mag\ arcsec}^{-2} \quad (1)$$

where Σ is the echo surface brightness, m_{sn} is the apparent magnitude of the original supernova light, τ_{sc} is the scattering optical depth through the echo cloud, F is the scattering phase function, R is the distance of the echo cloud from the supernova, τ is the dust optical depth along the echo line of sight, and τ_{sn} is the optical depth along the supernova line of sight. The coefficient in the last term is corrected from that given in [1] (J. E. FELTEN pointed out this error). This expression assumes that the cloud is in a thin sheet perpendicular to the line of sight; the effect of broadening a cloud of a given optical depth is to decrease the peak surface brightness. Single scattering and extinction are included, but multiple scattering is not;

this approximation should be adequate for extinction optical depths less than a few [26].

Equation (1) can be rewritten by setting $\tau_{sc} = \omega\tau_c$, where ω is the dust albedo and τ_c is the dust optical depth through the echo cloud, and $\tau = \tau_c + \tau_o$, where τ_o is the dust optical depth along the echo line of sight outside of the echo cloud. If the dust properties do not vary inside an echo cloud and the lines of sight outside the cloud are similar, the only quantity to vary around an echo ring is τ_c. It is easy to show that the surface brightness is maximized for $\tau_c = 1$, yielding

$$\Sigma_{max} = 19.9 + (m_{sn}-3) - 2.5 \log \left[\left(\frac{\omega}{0.6}\right)\left(\frac{F}{10}\right)\left(\frac{R}{100\ \mathrm{pc}}\right)^{-2}\right]$$
$$+ 1.086\ (\tau_o - \tau_{sn})\ \mathrm{mag\ arcsec}^{-2}. \qquad (2)$$

This expression can be used to estimate the maximum V surface brightness from the 2 observed interstellar echoes. The maximum V magnitude of the supernova was $m_{sn} = 3.0$ [27]. We assume that the dust is characterized by $\omega = 0.6$ and $F = 10$. The two clouds are at distances $R = 120$ and 330 pc [19]. The inner echo cloud makes a small contribution to the extinction along the line of sight so we have $\tau_o \approx \tau_{sn}$. The outer cloud might contribute about half of the extinction along the line of sight which is $\tau_{sn} = 0.5$. The resulting peak surface brightnesses are 18.3 mag arcsec^{-2} for the inner echo and 20.3 mag arcsec^{-2} for the outer echo. An expected property of the high surface brightness features is reddening due to the additional extinction. For standard parameters, $\tau_c = 1$ gives a reddening of about 0.35. For larger optical depths, the surface brightness drops but the reddening increases.

SCHAEFER [20] has found three bright knots in the inner echo with an estimated FHWM of 2". The radial FHWM due to the light curve is 2".5 [19]; it is likely that the clouds can be considered thin. The core of the brightest knot has V = 18.0 and B - V = + 1.2 [20], so the surface brightness does approach the estimated maximum. However, the supernova near maximum had B - V = + 1.6 [27] and B - V = + 1.2 characterized the lower surface brightness regions originally observed in the echo; there is no evidence for a reddening effect. In the outer ring, SCHAEFER [20] found a bright section with 20.0 mag arcsec^{-2} in V and B - V = + 0.98, U - B = - 0.70. This V surface brightness is at the estimated maximum but, again, there is no evidence for increased reddening. In fact, the blue surface brightness, especially at U, is surprisingly high. We have $\Sigma_U = 20.3$ mag arcsec^{-2} observed. At maximum optical light, the U magnitude was only 7.1 [27] because of the strong line blanketing that set in after day 5. Assuming the standard grain parameters, equation 2 yields $\Sigma_{Umax} = 24.6 + 1.086\ (\tau_o - \tau_{sn})$ mag arcsec^{-2}. Along the line of sight to the supernova, we estimate $A_U = 0.7$, about half of which may be due to the outer echo cloud so that $\Sigma_{Umax} = 24.2$ mag arcsec^{-2}. The maximum predicted Σ is about 4 mag fainter than the observed Σ. While increasing the grain albedo to unity does not significantly help the discrepancy, there is a larger range available in the scattering function, F. The scattering angle is 3°.6 and grains are expected to become more strongly forward scattering if the grain size is larger than the wavelength of the light. However, it does not seem that any plausible grain model can explain such a large factor.

The alternative is that the U echo is scattered light that was emitted during the first 5 days, when the supernova was bright in the blue. It can be seen from fig. 2 of reference [19] that the U surface brightness from the initial flash of light might approach the observed value; the prediction at U from this figure for the light at visual maximum in an overestimate because the observed line blanketing is not taken into account. If this is the case, the radial peak in U should be several arcsec outside of the peak in V. Searches for the echoes farther in the ultraviolet (e.g. 1000 - 2000 Å) should take the maximum surface brightness effect into account; a region of the echo that is bright in V might have high extinction in the ultraviolet.

Finally, we comment on the gas density in the high surface brightness features. Using $N_H/E_{B-V} = 2 \times 10^{22}$ cm^{-2} mag^{-1} [28], a cloud with $\tau_c = 1$ at V has a density of 2×10^3 H atoms cm^{-3} for a cloud thickness of 1 pc. The knots in the inner echo found by SCHAEFER [20] have sizes in the plane of the sky of about 0.5 pc so that densities of several times 10^3 H atoms cm^{-3} are plausible. In a dense cloud, the ratio of total to selective extinction, A_V/E_{B-V}, can be considerably larger than in the more diffuse interstellar medium; this is presumably due to grain growth in the cloud. Large grains have greater forward scattering and give less reddening from extinction, which helps to alleviate some of the problems in the interpretation of the observed high surface brightness echo features. However, there would be less blue enhancement of the light due to scattering in the echo cloud, so that the blue color of the high surface brightness echoes remains a problem. More detailed observations of the SN1987A echoes (colors, spatial structure, and evolution) are needed to better determine the dust properties in the scattering clouds.

I am grateful to H. Bond, K. Borkowski, A. Crotts, E. Dwek, R. Emmering, J. Felten, C. Fransson, and N. Soker for helpful discussions. This research was supported in part by NSF Grant AST-8818362.

References

1. R. A. Chevalier: in ESO Workshop on SN1987A, ed. I. J. Danziger (ESO, Garching 1987), p. 483.
2. C. Fransson, A. Cassatella, R. Gilmozzi, R. P. Kirshner, N. Panagia, G. Sonneborn, W. Wamsteker: Ap. J., 336, 429 (1989).
3. E. J. Wampler, A. Richichi: Astr. Ap., 217, 31 (1989).
4. C. Fransson, P. Lundqvist: Ap. J. (Letters), 341, L59 (1989).
5. R. A. Chevalier: Nature, 332, 514 (1988).
6. S. E. Woosley: Ap. J., 330, 218 (1988).
7. R. A. Chevalier, R. T. Emmering: Ap. J. (Letters), 342, L75 (1989).
8. H. E. Bond, N. Panagia, R. Gilmozzi, M. Meakes: IAU Circ. No. 4733 (1989).
9. W. J. Couch, D. F. Malin: IAU Circ. No. 4739 (1989).
10. A. Crotts, W. E. Kunkel: IAU Circ. No. 4741 (1989).
11. A. Crotts: Invited talk at 174th A.A.S. meeting (1989).
12. S. Heathcote, N. Suntzeff, A. Walker: IAU Circ. No. 4753 (1989).
13. A. Crotts, W. E. Kunkel, P. F. McCarthy: IAU Circ. No. 4791 (1989).
14. P. F. Roche, D. K. Aitken, C. H. Smith, S. D. James: Nature, 337, 533 (1989).
15. J. E. Felten and E. Dwek: Nature, 339, 123 (1989).
16. N. Suntzeff, S. Heathcote, W. G. Weller, N. Caldwell, J. P. Huchra, R. P. Olowin, K. C. Chambers: Nature, 334, 135 (1988).
17. C. Gouiffes, M. Rosa, J. Melnick, I. J. Danziger, M. Remy, C. Santini, J. L. Sauvageot, P. Jakobsen, M. T. Ruiz: Astr. Ap. (Letters), 198, L9 (1988).
18. A. P. S. Crotts: Ap. J. (Letters), 333, L51 (1988).
19. R. A. Chevalier, R. T. Emmering: Ap. J. (Letters), 331, L105 (1988).
20. B. E. Schaefer: IAU Circ. No. 4817 (1989).
21. S. van den Bergh: Nature, 328, 768 (1987).
22. R. M. West, A. Lamberts, H. E. Joergensen, H. E. Schuster: Astr. Ap. (Letters), 177, L1 (1987).

23. R. Gilmozzi, A. Cassatella, J. Clavel, C. Fransson, R. Gonzalez, C. Gry, N. Panagia, A. Talavera, W. Wamsteker: Nature, 328, 318 (1987).
24. R. Gathier, H. J. G. L. M. Lamers, T. P. Snow: Ap. J., 247, 173 (1981).
25. H. L. F. Houpis and D. A. Mendis: Ap. J., 239, 1107 (1980).
26. R. A. Chevalier: Ap. J., 308, 225.
27. M. Hamuy, N. B. Suntzeff, R. Gonzales, G. Martin: Astr. J., 95, 63 (1988).
28. J. Koorneef: in IAU Symposium 108, Structure and Evolution of the Magellanic Clouds, eds. S. van den Bergh and K. deBoer (Reidel, Dordrecht 1984), p. 333.

Interaction of Supernova Ejecta with Circumstellar Clumps

K. Masai & K. Nomoto

It is widely believed that Sk69-202 had once evolved to a red supergiant (RSg) before the blue supergiant (BSg) identified to SN1987A. Several theoretical models have been proposed for the mechanism to work on such a peculiar evolutionary track.[1),2)] Observational study of the circumstellar matter ejected from Sk69-202 in the past can give a clue for understanding its evolution. From this point of view, Itoh et al. investigated the interaction of the expanding ejecta with the RSg wind materials and thermal X-ray emission thereby.[3)] The circumstellar structure of SN1987A has been investigated in various wavelengths since the explosion. The early detection of radio emission just after the explosion indicates that the amount of the circumstellar matter is too small to produce detectable X-rays through the interaction with the ejecta.[4)] Hence, one has thought that the BSg wind blew up to a distance of several times 10^{17} cm, outside which the rich RSg wind materials are likely located.

About 100 days after the explosion, X-rays were detected from SN1987A with Ginga[5)] and with KVANT-Mir.[6)] The X-rays below 10 keV should be attributed to any other origin but Compton degradation which is responsible for the hard component. Masai et al.[7)] demonstrated that the soft component could be produced by interaction with circumstellar matter. In January 1988, ~330 days after the explosion, Ginga observed a soft X-ray enhancement.[8)] The gross feature of this flare can be accounted for in terms of interaction with a circumstellar clump on the far side.[9)] This interpretation is favorable also for explanation of an enhancement in high energy γ-rays observed in coincidence with the X-ray flare.[10)] Recent infrared speckle observations suggest a presence of clumps on the far side of SN1987A.[11)] Such circumstellar clumps may be associated with the mass loss activity of Sk69-202 in the past. We present some characteristics of thermal X-ray emission due to interaction of the ejecta with a dense clump, which could be responsible for the X-ray flare in 1988. We also discuss wind-wind interaction in view of a possibility for the formation of dense circumstellar clumps.

Interaction with a clump and thermal X-ray emission

Collision with a circumstellar clump gives rise to a forward shock propagating outward into the clump and a reverse shock inward into the ejecta envelope. Thermal X-rays are emitted from both shock-heated regions. Unless the mass of ambient matter swept-up by the ejecta exceeds the initial ejecta mass, thermal X-rays are dominated by those from the reverse-shocked ejecta because of its higher density or larger emission measure.

For the ejecta envelope, we assume a density distribution in proportion to r^{-u} at a radius r. By the reverse-shock wave arising from the collision with a clump of density n_{c1}, the density of the outer envelope of the ejecta is raised to

$$n_R \sim [(u-3)(u-4)/3] n_{c1}. \quad (1)$$

The reverse-shock velocity relative to the preshock matter of the ejecta envelope is approximately expressed as,

$$V_R \sim 0.32[12/(u-3)(u-4)]^{1/2} V_F. \quad (2)$$

Here $V_F=(4/3)^{1/3}(5/3)^{1/2}V_E$ is the forward (blast) shock velocity, where $V_E=(2E/M)^{1/2}$ is the expansion velocity of the ejecta expressed with the mass M and the initial kinetic energy E of the ejecta. When the forward shock wave propagates out of the clump, the rarefaction occurs and the pressure of the shocked matter rapidly decreases due to adiabatic expansion. Therefore, the shock can heat up the envelope for $t_H=D_{c1}/V_F$, where D_{c1} is the thickness of the clump. These equations with u=9 give $V_R=6\times10^8$ cm s^{-1}, $n_R=3\times10^6$ cm^{-3} and $t_H=4\times10^5$ s for $V_E=2\times10^9$ cm s^{-1}, $D_{c1}=10^{15}$ cm and $n_{c1}=3\times10^5$ cm^{-3}.

At the shock front protons are predominantly heated up to $\sim(3/16)\mu m_p V_R^2$, and their kinetic energy is transferred to electrons through Coulomb collisions in the post-shock region. Then the electron temperature of the reverse-shocked envelope rises in t_H up to

$$T_e \sim 20\mu(n_R/10^6)^{2/5}(t_H/10^6)^{2/5}(V_R/10^9)^{4/5} \text{ keV}, \quad (3)$$

where μ is the mean molecular weight. The X-ray spectrum is dominated by free-free emission, and ionization nonequilibrium is of minor importance because the ionization time scale, $10^6(n_R/10^6)^{-1}$ s, is comparable to t_H. Then, the X-ray luminosity is estimated with free-free emissivity as,

$$L_X \sim 3\times10^{37}\Omega(T_e/20\text{keV})^{1/2}(n_R/10^6)^2(R_C/10^{16.5})^3(\Delta_R/0.04) \text{ erg s}^{-1}, \quad (4)$$

where Ω is the solid angle of the clump, and Δ_R is the depth of the shocked envelope normalized to R_C, the radius of the contact interface. After t_H, the X-ray luminosity turns to decline with a time scale,

$$t_L \sim (1/4)(3/5)^{1/2}(R_C/V_E) \sim 3\times10^6(R_C/10^{16.5})(V_E/10^{9.3}) \text{ s}, \quad (5)$$

which is much shorter than the cooling time scale due to radiation. Hence, the thermal X-rays rise in about 10 days to reach their maximum and decay in about 40 days, as demonstrated in numerical calculations of hydrodynamical evolution and nonequilibrium thermal emission.[9]

Such a hot plasma emits iron K-lines with significant equivalent widths. With an energy resolution of LAC aboard Ginga, these lines cannot be resolved from each other but are observed in a blend. Masai et al.[9] predict a K-line blend at 6.8 keV with an equivalent width of 250 eV for the metallicity of the envelope $Z=(1/3)Z_\odot$. The line energy predicted is in good agreement with the observation by Ginga. Also the observed equivalent width ~200 eV is consistent if the metallicity is taken to be $0.27Z_\odot$, which is a value derived from Compton-degraded X-rays.[12] If soft X-rays came from a layer under the envelope,[13),14] the iron K-line blend would be at ~6.4 keV due to fluorescence through the envelope in the same way as the case for Compton-degraded X-rays. Therefore, the energy of iron K-line blend observed by Ginga gives a constraint on the emission mechanism and the emission region.

Possible origin of circumstellar clumps

A dense clump at 3.2×10^{16} cm on the far side was proposed to explain the X-ray flare.[9] This distance is smaller by an order of magnitude than that expected for the location of RSg wind materials. An interstellar molecular cloud may be a candidate for such a material. However, the thickness $\sim10^{15}$ cm required for the flare time scale seems too small to be that of a molecular cloud. Also the clumps on infrared speckle[11] are located in the BSg wind region, i.e. at a distance shorter than that to the inner edge of the RSg wind remnant. Such clumps may be resulted from Rayleigh-Taylor instability on a thin dense shell, which is formed by a reverse compression wave arising from the collision of the matter ejected later with the matter ejected earlier. This hypothesis may be discussed on the analogy of planetary nebulae.[15]

We consider the wind-wind interaction on a spherical geometry as shown in Fig. 1. The mass loss rate and the wind velocity are denoted by $\dot{m}$ and v for the fast wind and by $\dot{M}$ and V for the slow wind, respectively, where $\dot{m}<\dot{M}$ and $v>V$ in usual cases.

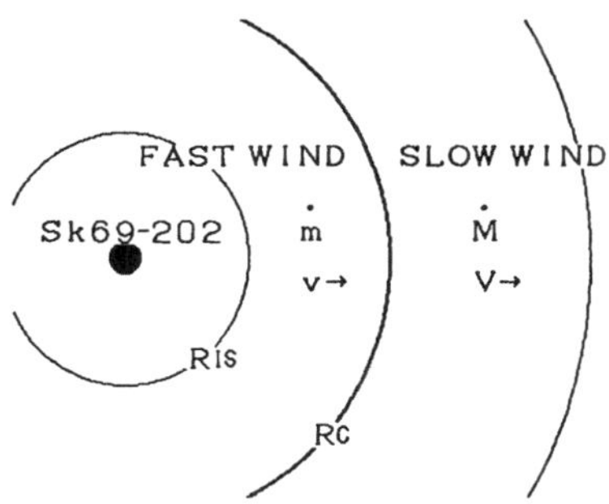

Fig. 1 Schematic view for wind-wind interaction

The matter ejected with a faster velocity catches up and collides with the matter ejected in the previous evolutionary phase. Then shocks arise to propagate toward the both sides from the interface as well as the matter ejected earlier is accelerated by the matter ejected later. The following couple of equations is considered:

$$d/dt(M_S dR_C/dt)=\dot{M}(dR_C/dt-V)+4\pi R_C^2 P, \tag{6a}$$

$$4\pi R_{IS}^2 P=2(\gamma+1)^{-1}(\dot{m}/v)(v-dR_{IS}/dt)^2, \tag{6b}$$

$$M_S=\dot{M}(R_C/V-t_2-t_1). \tag{6c}$$

Here R_C is the radius of the contact interface, M_S is the mass of the shell of the outward-shocked slow wind materials, R_{IS} is the radius of the inward shock front formed in the fast wind, P is the pressure, and t_1 and t_2 are the epoch of collision after the cease of earlier mass ejection and the time after the collision, respectively. So that,

$R_{IS}=R_C=Vt_1$ at $t_2=0$ just a moment of collision, and $R_C=Vt_1+V_Ct_2$, where $V_C=dR_C/dt$.

From eqs.(6a)-(6c), the location of the inward shock front is approximately obtained in a form of

$$R_{IS}/R_C \sim Uv/V_C+(1-Uv/V_C)[1+(V_C/V)(t_2/t_1)]^{-1/U}, \quad (7)$$

where U is a parameter related to the mass loss and is given by

$$U=(3/4)^{1/2}(\dot{m}/\dot{M})^{1/2}(v/V)^{-1/2}(1-V/V_C)^{-1}. \quad (8)$$

If the slow wind matter ejected earlier is accelerated by the fast wind ejected later to

$$V_C>2.7\text{x}10^6(v/10^8)\ \text{cm s}^{-1}, \quad (9)$$

then the inward shock front asymptotically approaches and stands at

$$R_{IS}/R_C \sim Uv/V_C, \quad (10)$$

at $t_2>10^3$ yr. For $\dot{m}/\dot{M}\sim10^{-2}$-$10^{-1}$, $v/V\sim10^2$ and $V/V_C\sim10^{-1}$, U takes the values on the order of $\sim10^{-2}$. Thus, $R_{IS}/R_C\sim10^{-1}$ is attained for the stellar wind parameters in acceptable ranges.

The above analysis implies that the front of the compression wave could be standing in the BSg wind at 10^{16}-10^{17} cm when the BSg wind is extended to 10^{17}-10^{18} cm. The presence of the standing shock front may give a possibility for the formation of a thin dense shell resulting in dense clumps at a few times 10^{16} cm. Although the time evolution of the shock front after the BSg wind is terminated is to be further investigated, it may be suggested that there left dense clumps in the vicinity of the supernova as a result of the interaction between the BSg wind in the later evolutionary phase and the RSg wind remnant.

References

1. Saio, H., Nomoto, K. & Kato, M. 1988, Nature 334, 508.
2. Woosely, S.E., Pinto,P.A. & Ensman, L., 1988, Astrophys. J. 324, 466.
3. Itoh, H., Hayakawa, S., Masai, K. & Nomoto, K., 1987, Publ. Astron. Soc. Japan 39, 529.
4. Chevalier, R.A. & Franson, C., 1987, Nature 328, 44.
5. Dotani, T. et al., 1987, Nature 330, 230.
6. Sunyaev, R. et al., 1987, Nature 330, 227.
7. Masai, K., Hayakawa, S., Itoh, H. & Nomoto, K., 1987, Nature 330, 235.
8. Tanaka, Y., 1989, in *Big Bang, Active Galactic Nuclei and Supernovae* (eds. S.Hayakawa & K.Sato), Universal Academy Press, p.481.
9. Masai, K., 1989, ibid., p.519; Masai, K. et al., 1988, Nature 335, 804.
10. Honda, M., Sato, H. & Terasawa, T., 1989, Prog. Theor. Phys. 82, 315 and references therein.
11. Chalabaev, A.A., Perrier, C. & Mariotti, J.M., 1989, Astron. Astrophys. 210, L1.
12. Mastichiadis, A., Kylafis, N. & Ventura, J., 1989, Astron. Astrophys. 208, L11.
13. Bandiera, R., Pacini, F. & Salvati, M., 1987, Nature 332, 418.
14. Fabian, A.C. & Rees, M.J., 1988, Nature 335, 50.
15. Volk, K. & Kwok, S., 1985, Astron. Astrophys. 153, 79.

SECTION IV

MODELS FOR SN 1987A AND OTHER TYPE II SUPERNOVAE

Massive Star Evolution in the LMC

Y. Tuchman & J. C. Wheeler

To completely understand the evolution of SN 1987A, one must understand the evolution of massive stars in the LMC. That is, one must determine the evolution of the average stars before one can judge whether Sk -69°202 was average or not, and in either case what the systematics of its evolution were. A key question to which we will return is whether Sk -69°202 was an average star.

A crucial new tool for the study of the of massive star evolution in the LMC is the Hertzsprung-Russell diagram of the LMC produced by FITZPATRICK and GARMANY [1]. There are a number of aspects of the "average" evolution of the stars in the LMC that can be explored with this data: the main sequence, the post-main sequence thermal contraction gap, the phase of hydrogen shell burning on a nuclear timescale, the Hertzsprung gap where stars undergo a contraction on the thermal timescale to the Hyashi track, and the Hyashi track itself. In exploring this H-R diagram one must bear in mind that the LMC is not a single cluster displaying a set of coeval stars, so it does not look like the H-R diagram of a classical open or globular cluster. Rather, for the massive stars at least, it is a sample of stars of all masses, weighted by the appropriate mass function, born at approximately a constant rate. In this case, the features tend to be more blurred, but the population in any interval of luminosity and effective temperature is still proportional to the evolutionary timescale in that phase. The potential of this H-R diagram is its ability to test the theoretically predicted distribution of stars in such a diagram. A successful theory must not merely give blue and red stars, it must distribute them in the quantitatively proper manner. Many of the theories proposed for the evolution of Sk -69°202 fail this test, if this star is supposed to have been typical of the LMC.

Examination of the H-R diagram of the LMC (FITZPATRICK and GARMANY [1], see also GARMANY and FITZPATRICK [2], HUMPHREYS and McELROY [3], BLAHA and HUMPHREYS [4]) shows that the middle portion of the diagram from T_{eff} of 5000 to 30,000 K is well populated. This means that stars which leave the main sequence do not immediately and permanently jump to the Hyashi track, but evolve on a hydrogen shell-burning nuclear timescale of about 10^6 years. The systematics of the evolution in the LMC have been explored in a series of three papers to better understand the context of the evolution of Sk -69°202 ([5],[6],[7]). In these papers envelope models are constructed assuming hydrogen shell burning on the evolution from blue to red or helium shell burning on the evolution from red to blue. The structure above the burning shell (but not necessarily that below it) is assumed to be in thermal equilibrium. An inward integration is thus performed assuming L = constant. This technique is not a substitute for full evolutionary calculations, but is a powerful complement because it allows a full survey of parameter space to delineate where stars are and are not in thermal equlibrium and hence what the distribution should be in terms of what portions of the H-R diagram are and are not densely populated.

A useful, if somewhat unorthodox, tool in this analysis has proved to be plots of T_{eff} versus the helium core mass for the blue to red evolution, or the carbon core mass for the red to blue evolution. With such plots we can understand the variation of the basic features of the H-R diagram of the LMC with such parameters as the core mass, the envelope mass, L, Z, Y(r), mass loss, and mixing length.

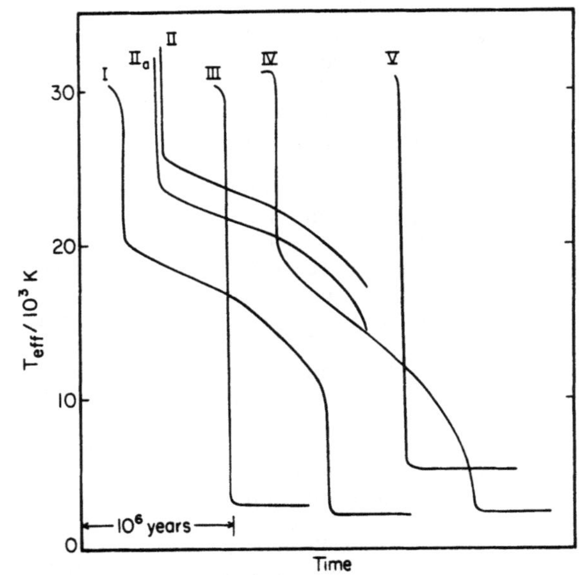

Figure 1. The effective temperature evolution with time during the hydrogen shell burning phase is plotted for different published evolutionary calculations. The vertical portions of the track represent evolution on thermal timescales, $\sim 10^4$ years, and the more horizontal portions are on the nuclear shell-burning timescale, $\sim 10^6$ years. Of these tracks, ones similar to I most closely reproduce the basic features of the LMC.

Figure 1 shows an array of schematic evolutionary tracks taken from the literature for the blue to red phase of evolution. These tracks start after core hydrogen exhaustion and hydrogen shell ignition. At first none of the models are in thermal equilibrium, but for some parametrizations, some of the models come into thermal equilibrium after a brief thermal contraction phase and then proceed to evolve on a nuclear shell-burning timescale of about 10^6 years (lines I, II, and IV). Some of the models never get to the red (lines II, IIa) which would preclude a populated Hyashi track. Others jump to the Hyashi track on a rapid thermal timescale (lines III, V) and are manifestly incorrect for the LMC because of the observed high population of stars in the intermediate effective temperature portions of the diagram. The qualitatively correct evolution for the LMC seems to be that shown by line I, with a slow shell-burning phase interrupted by a thermal-timescale jump to the Hyashi track from relatively cool temperatures.

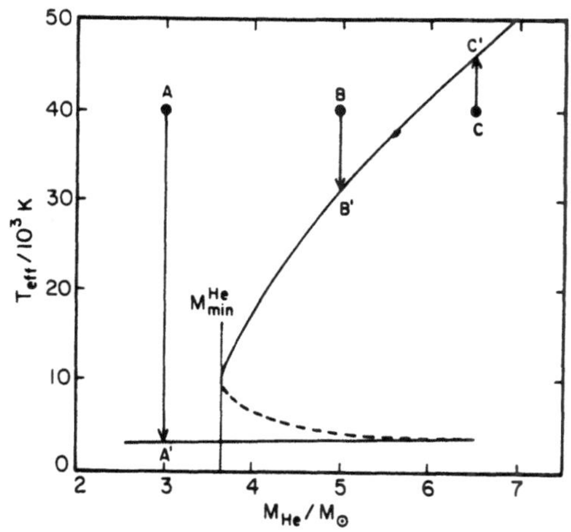

Figure 2. A schematic plot presenting the relation between the effective temperature and the mass of the helium core (interior to the hydrogen-burning shell) for envelope models in thermal equilibrium with a given total mass, composition, and luminosity. The dashed portion represents solutions that are thermally unstable. Three possible starting points for shell burning after core hydrogen exhaustion are labeled by A, B, C and the tracks AA′, BB′, CC′represent the initial thermal readjustment phase. M^{He}_{min} is the minimum helium core mass for solutions hotter than the Hyashi track.

Figure 2 shows the basis for analysis of the behavior depicted for various assumptions of the models in Figure 1. Figure 2 gives T_{eff} as a function of helium core mass for the case of a hydrogen burning shell as pertains to the evolution from the main sequence toward the Hyashi track. The curve shown in Figure 2 is the locus of models in thermal equilibrium above the hydrogen shell. The upper, high temperature branch, and the lower, Hyashi track branch, are stable. The intermediate portion of the locus, shown by the dashed line, is unstable. This configuration defines a minimum helium core mass such that for cores less massive there is no thermal equilibrium solution (line AA′) and hence stars will jump from the main sequence to the Hyashi track, in

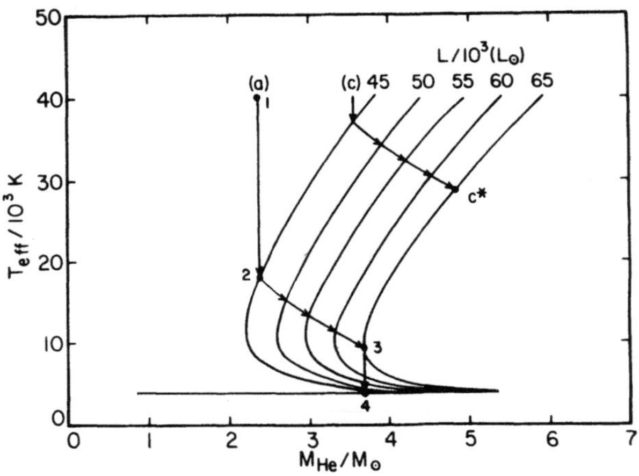

Figure 3. The relation between the effective temperature and the mass of the helium core is presented for different luminosities for models of 15 $M_\odot$ and a composition of Z = 0.001, Y = 0.3. Examples of possible hydrogen shell-burning evolution are given by tracks (a) and (c). Track cc* shows an evolution that halts in the blue, never reaching the Hyashi track. Track (a) shows the post-main sequence thermal contraction after core hydrogen exhaustion (1,2), the slower phase of nuclear shell-burning evolution with growing L(M_{He}) (2,3) and the final jump to the Hyashi track when thermal equilibrium is lost again (3,4).

contradiction to the observed behavior in the LMC. Line BB′ represents the correct evolution by joining the stable thermal equilibrium locus (line CC′ is not physical but illustrative of the stability of the upper portion of the locus). Note also that there is a maximum helium core mass for stars on the Hyashi track such that above that core mass, there is no stable solution in thermal equilibrium. If a model were to evolve to a core mass above this limit, it would be forced to return to the blue. This point will be examined below in a related context.

As the model evolves, the core grows in mass, and the model becomes somewhat brighter. Figure 3 shows the shift in the locus of the thermal equilibrium, hydrogen shell-burning loci for increasing luminosity. Track (a) 1-4 shows the evolution of a model which comes into thermal equilibrium, evolves on a nuclear shell-burning timescale, but then loses thermal equilibrium again when it "falls off the knee" as the maximum helium core mass for thermal equilibrium becomes greater than the helium core mass of the model. Line c-c* represents a model which never makes the jump to the Hyashi track. The loci in Figure 3 will shift to higher core masses if Z is increased, and to lower core masses if M or Y are increased. In this way one can determine the parameters that put the nuclear shell-burning evolution and the Hertzsprung contraction gap in the right place.

Figure 4 shows the H-R diagram from FITZPATRICK and GARMANY [1] with several relevant loci determined by this sort of analysis of the systematics of the blue to red evolution as presented in [5]. Note especially the paucity of stars beginning at about log T_{eff} ~3.8. This gap is identified as the feature corresponding to the

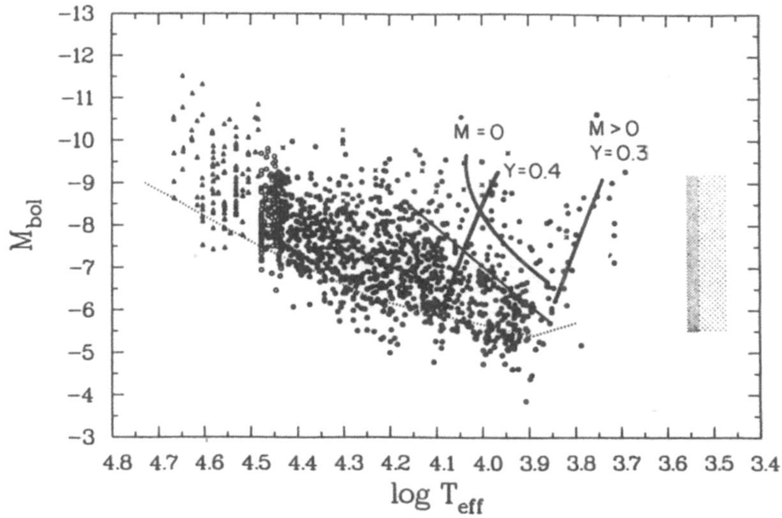

Figure 4. The Hertzsprung-Russell diagram from FITZPATRICK and GARMANY [1] is presented along with several observational and theoretical loci. The dashed line is an estimate of the completeness limit and the thin solid line represents the boundary of the "ledge." Other lines represent the blue edge of the Hertzsprung gap where models are predicted to jump to the Hyashi track for the standard case with moderate mass loss ($\dot{M} > 0$) and normal helium abundance in the envelope (Y = 0.3), with moderate mass loss and enhanced helium abundance (Y = 0.4), and normal helium abundance and no mass loss, as indicated.

Hertzsprung gap in clusters of moderate turnoff mass. Models with Y = 0.3, Z = 0.005, and no mass loss place the locus of the breakdown of thermal equilibrium too far to the blue and with the wrong slope. Models with moderate mass loss (adopted from BRUNISH and TRURAN [8] put the locus of the breakdown of the hydrogen shell-burning thermal equilibrium very close to the observed boundary of the gap. Models with mass loss but Y = 0.4 again put the locus much too far to the blue. An observational caveat should be noted here. TUCHMAN and WHEELER [5] take some satisfaction that their models with mass loss match the apparently observed blue boundary of the Hertzsprung gap in the LMC. GARMANY (private communication, 1989) cautions, however, that the yellow portion of the H-R diagram may simply not be searched thoroughly because of the problem of contamination by abundant Galactic yellow dwarfs. HUMPHREYS (private communication, 1989) responds that she has never had a problem with contaminating foreground stars and believes the gap is real. Clearly, this is an issue that must be resolved with some certainty lest the theorists overinterpret the data. For now the agreement with the observations seems satisfactory, and physically well motivated.

This analysis of the basic blue to red evolution yields several useful conclusions. One is that the boundary of the Hertzsprung gap seems well reproduced with models with Z = 0.005 and moderate mass loss with no anomolous enhancement of helium in the envelope. While the exact position of the gap may be uncertain, the data rule out a gap as blue as that predicted for models with Y as much as 0.4. This strongly suggests that there is no process such as meridional circulation that enriches the envelope in helium before or during the first passage of stars to the Hyashi track in the LMC. Another conclusion of interest is that models invoking the Ledoux criterion for convection (WOOSLEY [9], WEISS [10]) are not appropriate for massive stars in the LMC. It is this class of models, and some invoking convective overshoot, that evolve directly to the Hyashi track on a thermal timescale and completely fail to account for the number of stars populating the central portions of the H-R diagram of the LMC.

There are two striking and surprising features in the Fitzpatrick and Garmany H-R diagram of the LMC. One is that there is no sign of the post-main sequence contraction gap that is predicted by all evolutionary models, and by the analysis presented by TUCHMAN and WHEELER [5,7]. The second feature is the "ledge" which runs across the diagram from upper left to lower right. Note for now two juxtapositions that may be coincidental or significant. One is that Sk -69°202 fell near the edge of the ledge ($M_{bol} \sim -8$, $log\ T_{eff} \sim 4.2$). The second is that the locus of the boundary of the Hertzsprung gap in models with no mass loss falls near and roughly parallel to the locus of the boundary of the ledge. We will return to these points below.

TUCHMAN and WHEELER [6] address the problem of the return of stars from the Hyashi track to the blue in the LMC. Figure 5 shows the locus of a class of relevant thermal equilibrium models in which there is assumed to be a helium but no hydrogen burning shell. Figure 5 gives the effective temperature locus of thermal equilibrium models as a function of the mass of the carbon/oxygen core (i.e. the core interior to the helium burning shell). The locus of thermal equilibrium models and the systematics are qualitatively similar to Figure 2, with a high temperature stable branch, an intermediate unstable branch, and a stable Hyashi track. Note that there are again maximum and minimum (C/O) core masses. If the carbon core is below the relevant minimum for a given set of parameters ($M^{I}_{C/O}$ in Figure 5) then the only equilibrium solution is the Hyashi track. For a core mass above the relevant maximum core mass ($M^{II}_{C/O}$ in Figure 5), the only equilibrium solution is in the blue. If, during the natural evolution of the star, the C/O core mass exceeds this maximum, then there is no solution on the Hyashi track, and the star must make the transition back to the blue to a new thermal equilibrium location corresponding to the maximum critical core mass, denoted in Figure 5 by T^{*}_{eff}. The maximum carbon core mass and the corresponding blue solution are functions of the helium core mass, the envelope mass, L, Z, Y(r), and the mixing length.

Figure 5 suggests that the only legitimate way to return to the blue in a condition of having a helium burning, but no hydrogen burning, shell is to have the core grow above the maximum critical carbon core mass such that there is no equilibrium Hyashi track solution. There is another way to obtain a blue solution. Note that as the carbon core mass increases toward the maximum critical mass, the unstable intermediate temperature solution gets increasingly close to the stable Hyashi branch. This means that a stable model could be perturbed to a condition where it fell above the unstable locus (small upward arrow at $M_{C/O} = 4\ M_{\odot}$ in Figure 5). In such a case, the only equilibrium solution available is the one on the upper stable branch at high effective

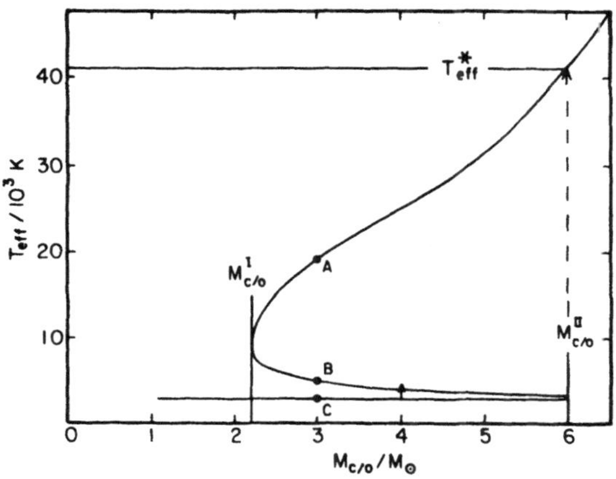

Figure 5. A schematic plot presenting the relation between the effective temperature and the C/O core mass (interior to the helium-burning shell) for envelope models in thermal equilibrium given total mass, composition, and luminosity. The two turning points, $M^{I}_{C/O}$ and $M^{II}_{C/O}$, represent the minimum core mass below which models in thermal equlibrium must be on the Hyashi track or above which they cannot be on the Hyashi track, respectively. T^{*}_{eff} represents the effective temperature to which a model must evolve if it develops a core exceeding $M^{II}_{C/O}$. The small arrow shows an "illegal" jump that might lead artificially to an excursion to the blue.

temperatures. One must be very careful not to artificially perturb a model above the unstable branch in the process of, for instance, appending new envelopes of different mass and helium abundance onto cores evolved to the Hyashi track with evolutionary calculations. One can imagine that such a perturbation might occur naturally if, for instance, the star in question were a long period variable naturally undergoing periodic changes in structure, but such a case has not been explored carefully.

Figure 6 shows a series of solutions like those in Figure 5 for models with helium cores of 6 $M_{\odot}$, an envelope mass of 10 $M_{\odot}$, Z = 0.005, and mixing length parameter equal one pressure scale height. The helium abundance in the hydrogen-rich envelope is assumed to be distributed in a step function with abundance Y = 0.5 out to a mass, M_{st}, where it steps down to a "primordial" value of Y = 0.3. This approximates the distribution found by the retreat of the convective core during central hydrogen burning. The low temperature region is expanded in Figure 6 to show clearly the effect of the variation of the helium distribution, via M_{st}, on the resulting structure of the thermal equilibrium loci and the maximum stable carbon core mass on the Hyashi track. If the helium enriched portion extends only out to 9 $M_{\odot}$ or less, the maximum carbon core mass is too large and the core mass from an evolutionary calculation could never exceed the critical value to return to the blue. On the other hand, if the helium enrichment extends nearly throughout the envelope, M_{st} = 14 and 16 $M_{\odot}$ in Figure 6, then the corresponding blue solutions are far too blue to correspond to the location of Sk -69°202. Only for intermediate cases with M_{st} ~10, that is enrichment extending about 4 $M_{\odot}$ above the helium core, can one have the

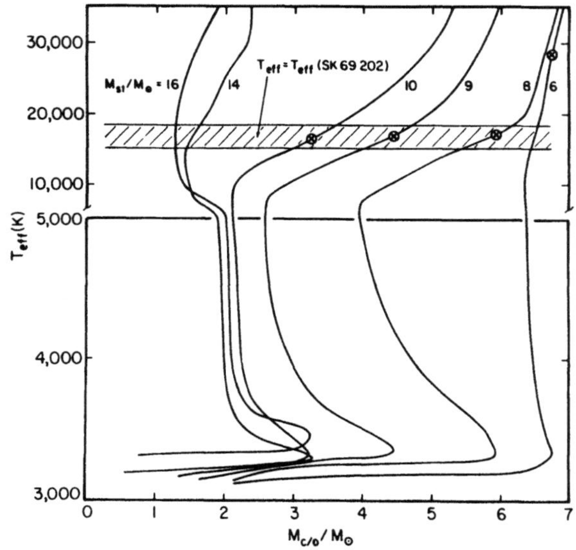

Figure 6. The relation between the effective temperature and the C/O core mass is given for the thermal equilibrium solutions for models with total mass 16 M$_\odot$, L = 100,000 L$_\odot$, helium core mass 6 M$_\odot$, Z = 0.005, and for a variety of values of the mass, M_{st}, at which there is a step in the envelope helium composition such that Y = 0.295 for $M_{st} < M$ and Y = 0.495 for 6 M$_\odot$ $< Y < M_{st}$.

core grow above the critical upper limit and have the corresponding blue solution be in the observed range of Sk -69°202.

TUCHMAN and WHEELER [6] conclude that the return of stars from the Hyashi track to the blue is a very sensitive function of the helium distribution just above the helium core. Models using the Ledoux criterion naturally return to the blue ([9],[10]) because they generate a narrow, highly enriched, helium layer during main sequence burning, but, as mentioned earlier, they are not adequate to explain the original relatively slow transition from blue to red. Models using the Schwarzschild criterion for convection are very close to making a natural transition back to the blue. Only a slight rearrangement, concentrating the helium in the models of SAIO *et al.* ([11],[12]) at smaller radii, brings such a natural transition. Homogenization of the helium in the envelope works in the wrong direction, although a net enhancement in the amount of helium in the envelope by core penetration (BARKAT and WHEELER [13]) would aid the transition to the blue.

Consider again the distribution of stars in the H-R diagram of Figure 4 and its implication for the overall systematics for the evolution of stars from the blue to the red and return. First, a comment is appropriate concerning the concentration of stars at log T_{eff} ~4.45 and the paucity of stars to the blue of that. The concentration arises because of a tendency for observers using low dispersion classification spectra to identify a star as simply "OB" or "OB0." Such stars then get plotted with the

T_{eff} of a B0 star, i.e. log T_{eff} ~4.45. This increase is thus largely artificial, although it is also affected by the increase in stellar density expected as one goes blueward of the TAMS and begins to pick up main sequence stars. The low density at log $T_{eff} \lesssim 4.5$ is also presumably due to selection and classification effects. There must be an increase in stellar density as one moves into temperatures corresponding to the main sequence, and yet this is not seen. This is surely a major challenge to the observers to find and catalog the missing main sequence population of massive stars in the LMC. There are main sequence stars in associations that are beginning to be recorded (MASSEY *et al.* [14]), but which are not yet in the study of Fitzpatrick and Garmany.

Bearing these uncertainties in mind, the distribution of stars in Figure 4 is still very interesting. FITZPATRICK and GARMANY [1] summarize a number of evolutionary calculations and, although they have different assumptions and different behavior at cooler temperatures, they have one feature of post-main sequence evolution in common. They all have a post-main sequence thermal contraction gap such that no models or combination of models predicts that there should be any concentration of stars in the range of effective temperature log $T_{eff} = 4.3 - 4.4$. For individual models the predicted gap can shift around somewhat, and is generally wider. TUCHMAN and WHEELER [7] emphasized that this gap was a generic feature of all evolutionary models, and that it was barely discernible in the preliminary H-R diagram of GARMANY and FITZPATRICK [2] which they analyzed. That is, the temperature range to the red of the main sequence which should be evacuated due to the rapid evolution of the stars is, in fact, well-populated.

Figure 7 gives the observed distribution of stars in the range $M_{bol} = -7$ to -8 versus log T_{eff} from the H-R diagram of GARMANY and FITZPATRICK [2]. Note that while there is a dip in the stellar density from log T_{eff} of 4.3 to 4.45, this temperature range is well-populated. The dashed line in Figure 7 is the theoretical density distribution derived by TUCHMAN and WHEELER [7] based on the timescale of thermal contraction in regions where thermal equilibrium must break down in the envelope, phases I and III, and the nuclear hydrogen shell-burning timescale in phase II. TUCHMAN and WHEELER [7] pointed out that the large increase in the stellar density at log T_{eff} ~4.45, while confused by the tendency to artificially assign this effective temperature, was, nevertheless, the locus of the TAMS for reasonable models. Furthermore, the boundary between phase I and phase II where thermal equilibrium is reestablished in the hydrogen shell-burning phase, corresponded to the observed increase in stellar densities at log T_{eff} ~4.3, despite the fact that the observed rise is so steep that some crudeness or artificiality in spectral type assignment is again suspected. TUCHMAN and WHEELER [7] concluded that the distribution in the shell burning phase II was in reasonable agreement with the observed distribution. They speculated that the predicted thermal gap in phase I was filled in by stars which had accreted helium-enriched matter from a companion star. They showed that such a helium enrichment in the outer layers moves the locus at which thermal equilibrium is reestablished to the blue by an amount which could move the red edge of the "gap" for helium enriched stars to the blue of the TAMS for normal stars where it would be hidden in the increased density there. They also argued that for such massive stars, mass transfer, once started, would continue until the helium-rich regions in the interior of the primary were reached so that such helium enrichment of the outer portions of the envelope of the secondary was unavoidable. Tuchman and Wheeler estimated that one star in four would need to accept mass from a suitable companion in order

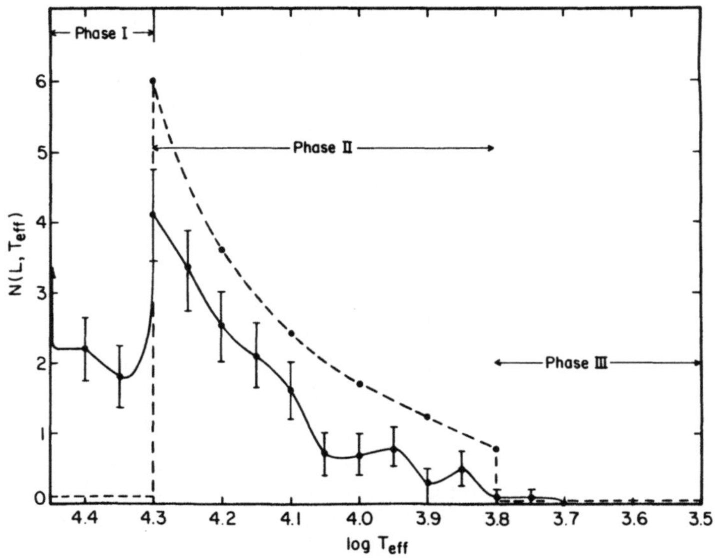

Figure 7. The observed normalized number density of stars across the H-R diagram at $M_{bol} = -7$ *to* -8 according to GARMANY and FITZPATRICK [2] is compared with the theoretical normalized duration function. The three evolutionary phases corresponding to the post-main sequence thermal contraction gap (Phase I), the nuclear shell-burning phase (Phase II), and the Hertzsprung gap thermal contraction to the Hyashi track (Phase III), are also shown.

to fill in the gap to the extent seen in Figure 7. Some support for this hypothesis is obtained from KUDRITZKI *et al.* [15] who fortuitously analyzed five stars in the "gap" region and found all five to be helium enriched.

This argument must be rethought with the presentation of the final H-R diagram of Fitzpatrick and Garmany as given in Figure 4. Between the preliminary analysis of GARMANY and FITZPATRICK [2] and the final version [1], more stars from the "OB" bin were assigned effective temperatures by means of UBV photometry rather than objective prism spectra. Many of these stars moved to cooler effective temperatures (FITZPATRICK 1989, private communication). The result is that there is now absolutely no sign of the post main sequence "gap." The conflict with all evolutionary models is thus more severe. The hypothesis of TUCHMAN and WHEELER [7] is weakened, however, since the binary hypothesis does not provide a natural way of precisely obliterating the gap. Considering the different rates of evolution of the normal and putative helium-enriched stars filling the gap, one would now need about one star in two to be suitably helium enriched, and this begins to seem extreme. The mystery of the missing "gap" thus is more severe and currently eludes solution.

The other important feature of the H-R diagram of Fitzpatrick and Garmany is the prominent "ledge" which cuts across the diagram from $M_{bol} = -9$ and log $T_{eff} = 4.15$ to $M_{bol} = -6$ and log $T_{eff} = 3.9$. Fitzpatrick and Garmany estimate that the stellar density jumps by a factor of about 6 across the ledge. The origin of this ledge, and its possible relation to Sk -69°202 which sits astride it, emerges as one of the central problems of the evolution of massive stars in the LMC. Fitzpatrick and Garmany point out that CHIOSI and SUMMA [16] computed models with the Ledoux criterion that depart from the Hyashi track, or near it, and move back to the blue to spend an appreciable time in core helium burning in the blue before moving back to the Hyashi track near the end of their lives. Similar models using a parametrized semiconvection were presented at this meeting by Langer. Fitzpatrick and Garmany suggest that such models might give a natural explanation for the ledge. The idea is that the models should evolve to the red, then return to the blue for most of core helium burning. The models of Chiosi and Summa then lose thermal equilibrium once again and return to the Hyashi track. This onset of the rapid return to the red would then represent the ledge.

FITZPATRICK and GARMANY [1] examine the models of BRUNISH and TRURAN [8] and argue that models invoking the Schwarzschild criterion undergo a thermal jump to the Hyashi track at a nearly constant effective temperature, independent of luminosity. This seems to be the case for their models with Z = 0.01, but not in general for their models with either higher or lower Z, as illustrated by TUCHMAN and WHEELER ([5] and Figure 4), where the boundary is a function of luminosity, mass loss, and Y for Z = 0.005. Fitzpatrick and Garmany argue that models with the Ledoux criterion have a blueward extent during core helium burning that depends on mass, but the blueward extent is not relevant to the ledge. The red boundary of the ledge is not related to how blue the star gets during core helium burning but is hypothesized to correspond to the breakdown in thermal evolution as the stars evolve back to the Hyashi track. Fitzpatrick and Garmany imply that the Ledoux models do seem to put the redward thermal contraction gap in the correct place, but this is not clear since many of the models they cite also seem to give the gap at log T_{eff} ~3.95 which is too red for the location of the ledge for a star of 20 $M_{\odot}$, M_{bol} ~−8, log T_{eff} ~ 4.1. Furthermore, no model with such blue loops places any stars sufficiently blue to obliterate the post main sequence gap, so such models are surely no panacea. Any theory based on blue loops during helium core burning must be able to explain the distribution of stars from log T_{eff} ~ 4.3 to log T_{eff} ~ 3.8 in Figure 7 at least as well as the hydrogen-shell burning interpretation presented there.

Returning to Figure 4, we emphasize that the boundary of the Hertzsprung gap derived for models with Z = 0.005, Y = 0.3, and no mass loss falls close to the ledge of Fitzpatrick and Garmany. Our models with mass loss fall very close to the apparent gap at log T_{eff} ~3.8. This raises the speculation as to whether there are two populations of stars in the LMC. It is not at all clear why one should have mass loss and the other not, and perhaps some other parameter is involved.

We note that our analysis of thermal equilibrium envelopes should be useful to better understand the existence and properties of blue loops during core helium burning. In this case the pertinent diagrams are those like Figures 2 and 3 rather than Figures 5 and 6, because the hydrogen shell is still active during the blue loops with core helium burning. The tentative hypothesis would be that the helium cores exceed the maximum permissible value to have thermal equilibrium solutions on the Hyashi track. This maximum core mass is again presumably a function of the parameters of the model and especially the distribution of helium in the hydrogen-rich envelope as affected by assumptions concerning convective mixing.

This analysis serves to focus on some of the important issues to be addressed in order to determine the evolution of "average" stars in the LMC. The "ledge" suggests two populations of stars. To what population did Sk -69°202 belong? Was it a single star partaking precisely of the evolution of all other stars of its mass? Or was it affected in some manner by a binary companion or other "anomolous" physical effect? The absence of the post-main sequence gap and perhaps the presence of the ledge suggest evolutionary effects in the LMC that depart in some way from current theory. If there are peculiar blue giants in the LMC, as argued by Tuchman and Wheeler, then one must be careful to ensure that Sk -69°202 did not share that peculiarity before using the "average" evolution in the LMC to constrain its evolution.

This work results in the following conclusions:

- The average evolution from blue to red in the LMC is "normal" with no anomolous envelope helium enrichment and with moderate mass loss on the main sequence.

- The Ledoux criterion gives thinner, steeper helium zones at the base of the hydrogen envelope. The resulting evolution across the H-R diagram is on a thermal timescale and tends to give no Hyashi track or one that is too blue (a function of the mixing length) and hence does not represent evolution in the LMC.

- A caveat to the previous statement is the possibility of "blue loops" that put stars back to the blue during core helium burning, but these do not seem to quantitatively account for the Hertzsprung gap, the Fitzpatrick-Garmany "ledge," or the closing of the post-main sequence contraction gap.

- The evolution from the Hyashi track back to the blue after hydrogen shell exhaustion is sensitive to the amount and distribution of helium at the base of the hydrogen envelope.

- Uniform helium enrichment is not necessary for the red to blue transition and may even be counterproductive.

- A given amount of envelope helium is more conducive to red to blue evolution if it is concentrated near the helium core.

- The Ledoux criterion does promote evolution back to the blue, but a very small perturbation to models with the Schwarzschild criterion would suffice.

- A "perturbation" of the Hyashi track may "illegally" put a model in the unstable region where it must evolve to the blue.

- The lack of a post main sequence "gap" in the H-R diagram is a major feature of the LMC which requires explanation since it is a generic feature of all theoretical models.

- The Fitzpatrick-Garmany "ledge" is also a major new feature of the H-R diagram of the LMC which so far eludes natural explanation.

- It is still unclear why Sk -69°202 fell near the "ledge" and hence whether it was a typical LMC star, or not.

We are grateful to Katy Garmany, Ed Fitzpatrick, Roberta Humphreys and Rolf Kudritzki for discussions of the H-R diagram of the LMC and for providing copies of their most recent work prior to publication. This research is supported in part by NSF Grant 8717166. The computations were performed on the Cray X-MP of the University of Texas System Center for High Performance Computing.

REFERENCES

1. Fitzpatrick, E. L., and Garmany, C. D. 1990, preprint.

2. Garmany, C. D., and Fitzpatrick, E. L. 1989, Proceeding of I.A.U. Colloq. 113, "Physics of Luminous Blue Variables," ed. K. Davidsen and H. J. G. L. M. Lamers (Dordrecht: Reidel), in press.

3. Humphreys, R. M., and McEloy, D. B. 1984, *Ap.J.*, **284**, 565.

4. Blaha, C., and Humphreys, R. M. 1990, *Ap. J.*, in press.

5. Tuchman, Y., and Wheeler, J.C. 1989a, *Ap.J.*, **344**, 835.

6. Tuchman, Y., and Wheeler, J. C. 1989b, *Ap.J.*, in press.

7. Tuchman, Y., and Wheeler, J. C. 1990, *Ap. J.*, submitted.

8. Brunish, W. M., and Truran, J. W. 1982, *Ap. J. Suppl.*, **49**, 447.

9. Woosley, S. E. 1988, *Ap. J.*, **330**, 218.

10. Weiss, A. 1989, *Ap. J.*, **339**, 365.

11. Saio, H., Kato, M., and Nomoto, K. 1988a, *Ap. J.*, **331**, 388.

12. Saio, H., Nomoto, K., and Kato, M. 1988b, *Nature*, **334**, 508.

13. Barkat, Z., and Wheeler, J. C. 1989, *Ap. J.*, **342**, 940 .

14. Massey, P., Garmany, C. D., Silkey, M., Degoia-Eastwood, K. 1989, *A. J.*, **97**, 107.

15. Kudritzki, R. P., Gabler, A., Gabler, R., Groth, H. G., Pauldrach, A.W.A. and Puls, J. 1989, Proceeding of I.A.U. Colloq. 113, "Physics of Luminous Blue Variables," ed. K. Davidson and H.J.G.L.M. Lamers (Dordrecht: Reidel), in press.

16. Chiosi, C., and Summa, C. 1970, *Ap. and Space Sci.*, **8**, 478.

Towards an Understanding of the Progenitor Evolution of SN 1987A

Norbert Langer, Mounib F. El Eid, & Isabelle Baraffe

Out of the many attempts to model the progenitor evolution of SN 1987A (see reviews by Hillebrandt and Höflich, 1989; Arnett et al., 1989), only two kinds of models are still discussed in the context of the observations: one using "restricted convection" (see e.g. contributions of S. Woosley and T. Weaver to this volume), and another assuming substantial mixing of core material into the hydrogen envelope (cf. e.g. K. Nomoto, this volume). The basic drawback of both kinds of models is a lack of theoretical justification of the applied physics. However, stellar models involving just the standard physics apparently fail in reproducing the blue–red–blue evolution of the SN 1987A progenitor.

In this contribution, we want to briefly present new stellar evolution calculations for a $20\,M_\odot$ star of LMC composition (see Langer et al., 1989, for more details), some of which closely match the observational constraints mentioned above and also agree well with the observed supergiant distribution in the HR diagram of the LMC. Our computations were performed by using an implicit hydro-code including up-to-date physics, e.g. mass loss according to Kudritzki et al. (1987,1989) and de Jager et al. (1988), nuclear reaction networks (cf. Langer et al., 1988), and detailed opacity tables from the Los Alamos Opacity Library. A further important ingredient to be mentioned here is the semiconvection model of Langer et al. (1983) together with the solution of the mixing equation in a diffusion scheme as described by Langer et al. (1985).

The basic properties of our convective model are the following. According to Kato (1966), zones where the Ledoux criterion ($\nabla > \nabla_L := \nabla_{ad} + [\beta/(4 - 3\beta)](d\ln\mu/d\ln P)$) is fulfilled are convectively unstable, while zones where the Schwarzschild criterion is fulfilled, but the Ledoux criterion is not ($\nabla_L > \nabla > \nabla_{ad}$) are only vibrationally unstable on a thermal timescale, i.e. semiconvective. Langer et al. (1983) computed the timescale of semiconvection and the corresponding diffusion coefficient D_{sc} on the basis of Kato's analysis. It involves an efficiency parameter α, whose order of magnitude has been estimated to be 0.1 (Langer et al., 1985). The diffusion coefficient depends on the molecular weight gradient through ∇_L, leading to larger mixing times for larger μ-gradients, reflecting their stabilizing nature. The divergence of D_{sc} for $\nabla \to \nabla_L$ indicates the decrease of the semiconvective mixing time for conditions becoming close to the convectively unstable situation (in our calculations we limited D_{sc} by the convective diffusion coefficient; see Langer et al., 1985), while $D_{sc} \to 0$ for vanishing superadiabaticities, i.e. radiative equilibrium. Note that such prescription recovers the Schwarzschild criterion for convection in evolutionary phases where the evolutionary timescale τ_{ev} is large compared to the semiconvective mixing time τ_{sc} — e.g. during core hydrogen burning — while the Ledoux criterion is recovered for $\tau_{ev} \ll \tau_{sc}$. Furthermore, adopting $\alpha = \infty$ will lead to the same result as adopting the Schwarzschild criterion for convection (imposing $\tau_{sc} = 0$), while $\alpha = 0$ is equivalent to adopting the Ledoux criterion for convection and no semiconvection (imposing $\tau_{sc} = \infty$).

We have computed a set of $20\,M_\odot$ sequences ($Z = Z_\odot/4$) with different values for the semiconvective efficiency parameter α, ranging from $\alpha = \infty$ to $\alpha = 0$ (see Table 1 for some key quantities of our models). We obtained a BSG SN progenitor structure for α in the range $0.05 \geq \alpha \geq 0.008$ (i.e. for $\alpha = 0.05$, 0.04, 0.03, 0.02, 0.01, and 0.008), while we obtained a RSG SN progenitor structure for larger and for smaller values (i.e. $\alpha = \infty$, 1.0, 0.1, and $\alpha = 0.005$, 0.001, 0). Note that the value of α suggested earlier (Langer et al., 1985) is $\alpha \simeq 0.1$, which is close to the upper limit of the range which yields BSG explosions. By recomputing several sequences with different mass loss rates, we found the above results to be insensitive to the amount of mass loss.

Table 1: Properties of our $20\,M_\odot$ computations for LMC composition with mass loss for various valuse of the semiconvective efficiency parameter α, i.e. He- and C/O-core mass after central helium exhaustion, the central carbon mass fraction at that time, the approximate fraction of time during central He-burning spent in the blue and red part of the HR diagram, and the final effective temperature.

$20\,M_\odot$ $Z = 0.25 Z_\odot$						
α	M_{He}	$M_{C/O}$	C_c	$\%\tau_{He}$ blue	red	$T_{eff,expl.}$
∞	6.71	4.69	0.06	99	1	4500
1.0	6.73	4.67	0.06	99	1	4500
0.1	6.79	3.03	0.05	99	1	4500
0.01	6.27	2.21	0.15	50	50	17000
0.001	6.01	2.00	0.19	0	100	4500

In all sequences which ended up blue, the stars had a much lower surface temperature (corresponding to the Hayashi line in most cases) some $10^4\,yr$ before the end of the surface temperature evolution. The final bluewards evolution started at a central temperature in the range $3 - 4 \cdot 10^8\,K$, i.e. at or shortly after core helium exhaustion. This behavior is in close agreement to the observational constraints of SN 1987A. A further feature of those tracks computed with α-values towards the upper limit of the range which yielded BSG presupernova stars is, that they perform a first blue loop from the Hayashi line during central helium burning. This loop is initiated approximately in the middle of helium burning, the first part of which has been spent as a RSG at the Hayashi line, and lasts until helium is almost exhausted in the core. Thus such tracks can account for the existence of both, red and blue supergiants in the LMC, and are also in agreement with the recent study of LMC supergiant stars by Fitzpatrick and Garmany (1989), who conclude that most LMC blue supergiants are in a post RSG stage.

A further argument in favor of the higher values of α within the range allowing terminal BSG solutions is the requirement of fitting the BSG population of the Milky Way. It has been shown by Langer et al. (1985) that the blue-to-red supergiant ratio tends to zero for decreasing α. A $20\,M_\odot$ sequence for 2% metallicity computed with $\alpha = 0.01$ had no BSG phase at all.

Note that the physical mechanism which brings the star back to the blue part of the HR diagram at central He-exhaustion — which is still far from being understood; cf. Langer et al., 1989 — may be the same in the models presented by Woosley and co-workers and in our models, since their prescription of convection/semiconvection resembles ours in several respects (see Weaver et al., 1978). It is, however, not straightforward to relate their semiconvection parameter to ours, since our diffusion coefficient involves the term $(\nabla - \nabla_{ad})/(\nabla_L - \nabla)$, while the Weaver et al. prescription involves some interpolation between radiative and convective diffusion coefficient. Therefore, the upper bound of the semiconvection parameter of 0.003 in order to allow for a BSG explosion as reported by Woosley et al. (1988) is not necessarily a contradiction to our upper boundary, which is in the range $0.1 < \alpha \leq 0.05$. However, our results clearly disagree concerning the lower bound, since we get RSG solutions for $\alpha \leq 0.005$ — especially also for $\alpha = 0$ —, while Woosley et al. find BSG solutions even for the case of the Ledoux criterion for convection and no semiconvection ($\alpha = 0$).

Despite this discrepancy of our work to that of Woosley et al., which may be related to differences in physical ingredients of the computer codes and which is currently under investigation, we want to point out that both works correspond to the same physical explanation of the blue-red-blue evolution of the SN 1987A progenitor (see above). The fact that in our case we find very good agreement with the observational constraints concerning the LMC supergiant distribution in general and SN 1987A in particular, without any artificial manipulation, may favor this type of model compared to the core-envelope mixing scenario (cf. Nomoto, this volume, and references therein). The latter model, however, agrees better with the high observed N/C and N/O ratios in the envelope of SN 1987A (Fransson et al., 1988). Our models, as they have been in the RSG stage before exploding, also produce N and O overabundances and C deficiency throughout the hydrogen envelope, but not quite the amount required by the observations. However, the amount of dredged-up material in the RSG stage varies considerably in our different sequences, and it remains to be investigated which is the maximum amount and under what conditions can it be achieved.

Finally, we want to discuss the ZAMS-mass range of LMC stars, for which a supernova explosion as a BSG may occur in the framework of our models, which may be of some importance for the estimates of SN rates. We have not yet performed computations for masses below $20\,M_\odot$, but Woosley et al. (1988) argue that blue solutions may be possible down to $M_{ZAMS} = 10\,M_\odot$. Furthermore, we got a blue solution for $M_{ZAMS} = 25\,M_\odot$ and $\alpha = 0.01$, indicating the possible extension of the BSG SN range toward higher masses. For $30\,M_\odot$ and $\alpha = 0.01$ we found a RSG terminal structure, which does, however, not exclude the possibility of blue solutions for different values of α. On the other hand, the existence of blue solutions for a (e.g.) $25\,M_\odot$ star does not exclude, that red solutions exist and are preferred by nature. Anyway, one has to face the possibility that below a certain metallicity stars within a fairly wide ZAMS mass range may explode as a BSG, like the SN 1987A progenitor. It may even be that no RSG explosions occur at all in low metallicity systems, stars below the BSG-SN range evolving into White Dwarfs, and those above into Wolf-Rayet star. Because of the huge parameter space for stellar evolution calculations related to this problem (M_{ZAMS}, Z, α, $\dot{M}$, $^{12}C(\alpha,\gamma)^{16}O$, ...) it may still take a long time until the fraction of this space, which yields BSG supernovae, can be determined, unless a fundamental physical understanding of the occurrence of the "final blue loop" can be achieved. Even the evolution towards the RSG branch after core hydrogen exhaustion is still far from being completely understood, though it is a much simpler problem.

Acknowledgment. The authors are grateful for valuable discussions with S. Woosley. This work has been supported by the Deutsche Forschungsgemeinschaft (DFG) through grants La 587/2-1, and Fr 325/28/2, and through NASA grant NAGW-1273. N.L. gratefully acknowledges support by the Astronomische Gesellschaft (AG) through the Ludwig Biermann award 1989.

References

Fitzpatrick, E.L., Garmany, C.D.: 1989, Astrophys. J. , in press

Fransson, C., et al.: 1988, Astrophys. J. **336**, 429

Hillebrandt, W., Höflich, P.: 1989, Progress in Physics, in press

de Jager, C., Nieuwenhuijzen, H., van der Hucht, K.A.: 1988, Astron. Astrophys. Suppl. **72**, 259

Kato, S.: 1966, Publ. Astron. Soc. Japan **18**, 374

Kudritzki, R.P., Pauldrach, A., Puls, J.: 1987, Astron. Astrophys. **173**, 293

Kudritzki, R.P., Pauldrach, A., Puls, J., Abbott, D.C.: 1989, Astron. Astrophys. **219**, 205

Langer, N., Sugimoto, D., Fricke, K.J.: 1983, Astron. Astrophys. **126**, 207

Langer, N., El Eid, M.F., Fricke, K.J.: 1985, Astron. Astrophys. **145**, 179

Langer, N., Kiriakidis, M., El Eid, M.F., Fricke, K.J., Weiss, A.: 1988, Astron. Astrophys. **192**, 177

Langer, N., El Eid, M.F., Baraffe, I.: 1989, Astron. Astrophys. *Letters*, in press

Sonneborn, G., Altner, B., Kirshner, R.P.: 1987, Astrophys. J. **323**, 235

Weaver, T.A., Zimmerman, G.B., Woosley, S.E.:1978, Astrophys. J. **225**, 1021

Woosley, S.E., Weaver, T.A.: 1986, Ann. Rev. Astron. Astrophys. **24**, 205

Woosley, S.E., Pinto, P.A., Weaver, T.A.: 1988, Proc. Astron. Soc. Austr. **7**, 355

Confronting Observations with the Theory of SN 1987A: Progenitor, Nucleosynthesis, and Mixing

K. Nomoto, T. Shigeyama, S. Kumagai, & H. Yamaoka

1. Introduction

Among the challenges from SN 1987A to the theory of stellar evolution and explosion are the blue supergiant progenitor and large scale mixing in the ejecta. Theories of nucleosynthesis and grain formation can also be tested with observations. These issues will be discussed in the following sections.

2. Blue-Red-Blue Evolution of the Progenitor

The progenitor of SN 1987A has been identified with a blue supergiant star Sk–69°202. The large N/C and N/O over the solar ratios and the slow expansion velocity of the circumstellar matter are the clear evidences that the progenitor had evolved once to a red supergiant stage losing mass and then contracted to the size of the blue supergiant ([32], [14]).

Several types of evolution of the low metallicity massive stars have been presented. Saio *et al.* [36] and Yamaoka *et al.* [51] adopted the Schwarzschild criterion for convection and showed that the blue - red - blue evolution in the HR diagram is well modeled by choosing appropriate parameters for mass loss and mixing as seen in Fig. 1.

2.1. Blue to Red Evolution

During early helium burning, the star is a blue supergiant and the subsequent move from the blue to red supergiant is driven by mass loss in the following way. As $M_{\rm env}$ decreases, the surface luminosity that blue supergiant can radiate gets smaller, while the core luminosity does not change appreciably because it depends on $M_{\rm core}$. As a result the envelope starts to expand to absorb the excess luminosity.

Whether the star remains blue or moves to red can be understood from the existence or non-existence of envelope solutions of blue supergiants; a certain relation between $M_{\rm env}$ and the surface helium abundance Y should be satisfied in order for the envelope to fit to the core for given L, $T_{\rm eff}$ and $M_{\rm core}$ [37, 4, 45]. As the star loses a significant

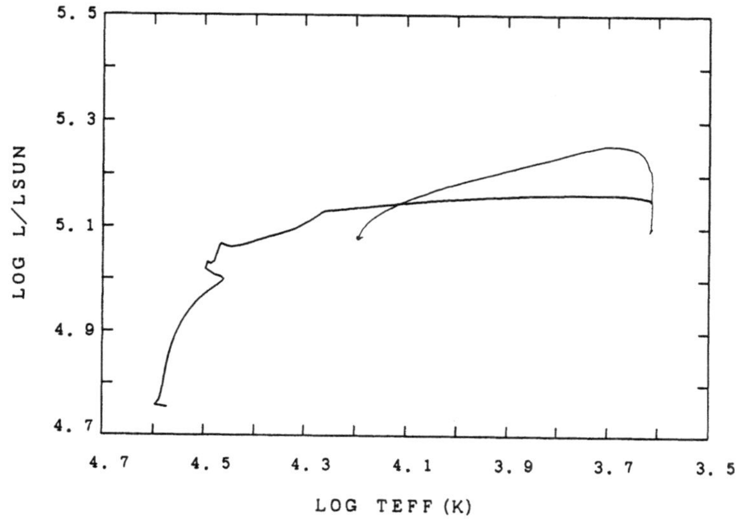

Fig. 1: Evolutionary path in the HR diagram from the main-sequence through carbon ignition for the star with the initial mass of $M_{lms} = 23\ M_{\odot}$ and metallicity $Z = 0.005$. During the evolution from the blue supergiant to the red, the stellar mass decreases from 23 $M_{\odot}$ to 16 $M_{\odot}$ and forms a helium core of $M_{core} = 6.7\ M_{\odot}$. During the red phase, the helium layer of 0.7 $M_{\odot}$ is mixed into the hydrogen-rich envelope yielding $M_{core} = 6.0\ M_{\odot}$ and $M_{env} = 10\ M_{\odot}$. The dredge up of the helium layer enhances the surface helium abundance to $Y_{surf} = 0.43$, which is large enough for the star moves from the red to the location of Sk–69°202.

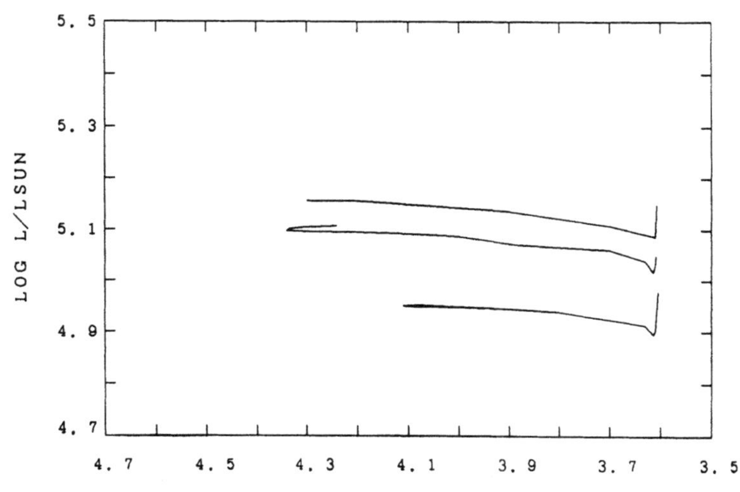

Fig. 2: Envelope solutions for the 16 $M_{\odot}$ star having a helium core of $M_{\alpha} = 6\ M_{\odot}$. At the inner boundary, the envelope (including the helium layer) is fitted to the helium burning shell at $M_r = 3.4\ M_{\odot}$. Shown are the three sequences for the surface helium abundance of $Y = 0.50$, 0.43, and 0.25, where the star is more luminous for larger Y.

fraction of its envelope mass, M_{env} becomes too small for the star to remain blue for its L and Y. In this sense, the mass loss is the driving mechanism of redward evolution.

The time scale of evolution in the H-R diagram depends mainly on the mass loss rate [51, 30] and can be consistent with the observed blue to red ratio for the LMC supergiants [21].

2.2. Red to Blue Evolution and Mixing

The evolution from the red to the blue is driven by mixing as follows: After the exhaustion of helium in the central region, the C+O core starts to contract and helium shell burning becomes active. During the core contraction, the core luminosity, L_{core}, gradually increases as the core radius decreases, while the surface luminosity, L_{surf}, depends on the surface opacity as well as the core radius. Whether the star remains red or returns to the blue depends on the balance between L_{core} and L_{surf}.

Sequences of envelope models in thermal equilibrium (i.e., $L_{surf} = L_{core}$) are integrated from the surface to the helium burning shell for several helium abundances, Y, as shown in Fig. 2 [22]. Along the line, the core radius changes. It should be noted that L_{surf} is larger for larger Y (and smaller Z) because of smaller opacity. If the hydrogen-rich envelope is metal deficient and becomes sufficiently helium-rich by mixing, the resultant decrease in the opacity causes the increase in the surface luminosity which eventually exceeds the core luminosity. To compensate this luminosity imbalance by adding the gravitational energy release, the envelope starts to contract undergoing the extensive excursion from the red back to the blue.

The mechanism of such an enhancement of helium would be convective mixing of the material near the bottom of the hydrogen-rich envelope including a part of the helium layer. Though the surface convection zone of the current model is not deep enough, the deep mixing model has an advantage that the observed large N/C and N/O ratios over the solar values can naturally be explained since the materials in the deeper layers are more nitrogen- rich [37].

This blueward evolution occurs more easily for a lower metallicity envelope ($Z < 0.01$). The envelope with larger Z has larger opacity and, to compensate, the blue supergiant envelope solution with larger Z has a larger M_{env} or larger Y for a given L. In other words, a greater enhancement of Y is required for given M_{env} if Z is larger, so evolution from red to blue is more likely to occur for smaller Z.

3. Explosive Nucleosynthesis

3.1. Explosive Nucleosynthesis and Dependence on the Progenitor Model

In the hydrodynamical calculation of the supernova explosion, an energy E is deposited instantaneously in the central region of the core to generate a strong shock wave. (For model 14E1, $M_{ej} = 14.6M_\odot$ and $E = 1 \times 10^{51}$ ergs. The initial composition structure of the ejecta is a heavy element layer of $2.2M_\odot$, a helium-rich layer of $2.2M_\odot$, and a hydrogen-rich envelope of 10.2 $M_\odot$.) As the shock wave propagates through the Si and O-rich layers, explosive nucleosynthesis takes place behind the shock. The peak temperatures T_p at $M_r \leq 1.69\ M_\odot$ exceed 5×10^9 K so that the materials, originally composed of Si, S and Ca, are burned to form elements almost in nuclear statistical equilibrium, which are mostly radioactive ^{56}Ni and some other radioactive species, ^{57}Ni and ^{44}Ti.

In the inner O-rich layers at $1.69\ M_\odot < M_r \leq 1.75\ M_\odot$, explosive oxygen burning produces ^{28}Si, ^{32}S, ^{36}Ar, ^{40}Ca and trace ^{56}Ni and ^{54}Fe. At $1.75\ M_\odot < M_r \leq 1.88\ M_\odot$, ^{20}Ne is burned to produce ^{16}O and trace ^{28}Si and ^{32}S. Explosive burning of trace carbon takes place at $1.88\ M_\odot < M_r \leq 2.06\ M_\odot$. The resulting stratified composition is given in ([18]; hereafter HNS) and [44]).

The isotopic ratios among iron-peak elements including ^{57,58}Ni, ^{61,62}Zn, and ^{44}Ti can be compared with observations. These abundances depend on the distribution of Y_e in the Si-rich region. In our presupernova model, Y_e changes from 0.4987 to 0.494 at $M_r = 1.63\ M_\odot$, which corresponds to a change in the neutron excess $\eta = 1 - 2Y_e$ from 2.6×10^{-3} to 1.2×10^{-2}.

In the presupernova evolution [28], this position marks the outer boundary of the O-burning convective shell which extends from 1.05 to 1.63 $M_\odot$ after exhaustion of oxygen in the central region. Because of larger abundances of silicon-rich products and higher densities, this layer undergoes more electron captures than the outer layers during oxygen-shell burning and the subsequent contraction of Si-rich core. This layer also contains trace neutron-rich material which was convectively mixed from the Si-burning shell. The outer layers did not experience much electron captures because of low density.

It should be noted that the outer boundary of the convective layer may not be so accurately determined because of rather flat entropy distribution there. Therefore it is quite possible to have a smaller size of the convective O-burning shell and (or) a smaller degree of mixing of Si shell-burning products into outer layers. In order to take into account this uncertainty, Hashimoto and Nomoto [19] examined nucleosynthesis where the region with $Y_e = 0.4987$ extends down to $M_r = 1.60\ M_\odot$. The resulting abundances are shown in Table I.

The isotopic ratios ^{57}Fe/^{56}Fe and ^{44}Ca/^{56}Fe reflect the ratios of ^{57}Ni/^{56}Ni and ^{44}Ti/^{56}Ni after the radioactive decays of $^{57}\text{Ni} \rightarrow {}^{57}\text{Co} \rightarrow {}^{57}\text{Fe}$ and $^{44}\text{Ti} \rightarrow {}^{44}\text{Sc} \rightarrow {}^{44}\text{Ca}$. The ^{58}Ni/^{56}Fe ratio strongly depends on η near the mass cut and is close to the solar in Table I while it is about 5 times the solar in HNS. On the other hand, the ^{57}Fe/^{56}Fe and ^{44}Ca/^{56}Fe ratios are not so different from HNS and the ^{57}Fe/^{56}Fe is still 1.7 times the solar.

3.2. Comparison to the Observed Abundances in SN 1987A

What can we learn about nucleosynthesis from the photometric and spectroscopic observations of SN 1987A? The mass of ^{56}Ni is determined as $\sim 0.07\ M_\odot$ from the optical

Table 1. Nucleosynthesis products for the model of SN 1987A

Species	Mass $(M_\odot)^a$	$< X_i/X(^{16}O) >^b$	Species	Mass $(M_\odot)^a$	$< X_i/X(^{16}O) >^b$
^{4}He	2.10	0.0580	^{48}Ca	2.41×10^{-16}	1.13×10^{-11}
^{12}C	0.114	0.243	^{45}Sc	1.04×10^{-7}	0.0173
^{13}C	1.17×10^{-10}	2.07×10^{-8}	^{46}Ti	6.81×10^{-6}	0.197
^{14}N	0.00272	0.0159	^{47}Ti	1.73×10^{-6}	0.0538
^{15}N	6.48×10^{-10}	9.60×10^{-7}	^{48}Ti	1.85×10^{-4}	0.557
^{16}O	1.48	1.0	^{49}Ti	4.89×10^{-6}	0.193
^{17}O	9.86×10^{-9}	1.64×10^{-5}	^{50}Ti	1.12×10^{-10}	4.39×10^{-6}
^{18}O	8.68×10^{-3}	2.59	^{50}V	2.15×10^{-10}	0.00150
^{19}F	7.84×10^{-11}	1.25×10^{-6}	^{51}V	6.40×10^{-6}	0.110
^{20}Ne	0.229	0.914	^{50}Cr	3.54×10^{-5}	0.308
^{21}Ne	3.03×10^{-4}	0.474	^{52}Cr	8.64×10^{-4}	0.376
^{22}Ne	0.0293	1.46	^{53}Cr	7.12×10^{-5}	0.268
^{23}Na	0.00115	0.224	^{54}Cr	6.26×10^{-9}	9.29×10^{-5}
^{24}Mg	0.147	1.85	^{55}Mn	2.27×10^{-4}	0.111
^{25}Mg	0.0185	1.77	^{54}Fe	0.00252	0.229
^{26}Mg	0.0174	1.45	^{56}Fe	0.0732	0.405
^{27}Al	0.0155	1.73	^{57}Fe	0.00307	0.696
^{28}Si	0.0850	0.842	^{58}Fe	3.70×10^{-9}	6.48×10^{-6}
^{29}Si	0.00980	1.85	^{59}Co	1.31×10^{-4}	0.253
^{30}Si	0.00719	1.98	^{58}Ni	0.00371	0.485
^{31}P	0.00105	0.833	^{60}Ni	0.00218	0.718
^{32}S	0.0229	0.374	^{61}Ni	1.59×10^{-4}	1.19
^{33}S	8.84×10^{-5}	0.177	^{62}Ni	7.26×10^{-4}	1.69
^{34}S	0.00126	0.437	^{64}Ni	2.06×10^{-15}	1.83×10^{-11}
^{36}S	4.23×10^{-7}	0.0291	^{63}Cu	3.00×10^{-6}	0.0338
^{35}Cl	6.05×10^{-5}	0.154	^{65}Cu	7.02×10^{-7}	0.0172
^{37}Cl	4.96×10^{-6}	0.0375	^{64}Zn	1.78×10^{-5}	0.116
^{36}Ar	0.00378	0.316	^{66}Zn	2.08×10^{-5}	0.229
^{38}Ar	3.25×10^{-4}	0.137	^{67}Zn	6.39×10^{-8}	0.00471
^{40}Ar	4.65×10^{-9}	1.14×10^{-3}	^{68}Zn	5.33×10^{-9}	8.48×10^{-5}
^{39}K	3.24×10^{-5}	0.0604	^{70}Zn	3.79×10^{-21}	1.77×10^{-15}
^{41}K	1.28×10^{-6}	0.0314	^{69}Ga	2.21×10^{-12}	3.60×10^{-7}
^{40}Ca	0.00325	0.351	^{71}Ga	1.54×10^{-18}	3.67×10^{-13}
^{42}Ca	9.45×10^{-6}	0.146	^{70}Ge	1.51×10^{-14}	2.26×10^{-9}
^{43}Ca	3.38×10^{-6}	0.244	^{72}Ge	1.19×10^{-20}	1.30×10^{-15}
^{44}Ca	9.15×10^{-5}	0.417	^{73}Ge	3.91×10^{-24}	7.20×10^{-19}
^{46}Ca	1.12×10^{-11}	2.59×10^{-5}	^{74}Ge	9.85×10^{-23}	7.84×10^{-18}

[a] Mass integrated at $1.60\ M_\odot \le M_r \le 6.0 M_\odot$.
[b] $< X_i/X(^{16}O) > \equiv [X_i/X(^{16}O)]/[X_i/X(^{16}O)]_\odot$

light curve. This places the mass cut between the ejecta and the remaining neutron star at $1.60 M_\odot$. Future optical light curve could make it possible to determine the masses of other longer-lived radioactive species, ^{57}Co and ^{44}Ti. (See [25] for prediction.)

The ionic masses estimated from the spectroscopic observations are given by Danziger [10]. The sum of Co and Fe is consistent with the ^{56}Ni mass derived from the light curve. The mass of stable nickel (^{58}Ni etc.) has been estimated to be 3 - 5×10^{-3} $M_\odot$ at day 400 [10, 48]. This is in better agreement with Table I than with HNS. The determination of ^{57}Co abundance would provide further test of Table I.

Observed mass estimates of other elements are approximately consistent with Table I. The crucially important is oxygen. If the oxygen mass is as large as $\sim 1.5\ M_\odot$ as in Table I, the progenitor models with $M < 20 M_\odot$ could be excluded and also the higher value of the $^{12}C(\alpha, \gamma)^{16}O$ rate is preferred.

4. Optical Light Curve and Mixing

After the shock breakout at the surface, the supernova shows its brilliant optical display. SN 1987A was identified with a Type II supernova yet its light curve has a quite unique shape, which is a useful tool to probe the supernova interior [29, 40, 49].

1) The first optical detection is only 3 hours after the neutrino burst. This imposes important constraints on the progenitor's radius and explosion energy, though detailed non-LTE atmospheric model needs to be constructed.

2) Owing to the small contribution of the initial shock heating, the light curve between day ~ 20 and ~ 100 has provided unique information on the internal energy source and the material distribution.

Shigeyama and Nomoto [38] refined the theoretical light curve for the updated hydrodynamical model of explosion, 14E1, where the ejecta mass is $M = 14.6\ M_\odot$ ($M_{\rm env} = 10.2\ M_\odot$) and $E = 1 \times 10^{51}$ erg. They found that the pre-peak smooth rise and the broad peak of the light curve are much better modeled by the *mixed* model which assumes mixing of ^{56}Ni into the hydrogen-rich envelope and of hydrogen into the core (as seen in Fig. 3) than the *unmixed* model with stratified composition.

4.1. Increase in Luminosity, Mixing of ^{56}Ni, and Bochum Event

The bolometric light curves for the mixed and unmixed cases are compared with observations [8, 16] in Fig. 4. For the unmixed case, the interior temperature is lower than in the mixed case because of the lack of radioactive heating. Hence the calculated bolometric luminosity starts to decrease at day ~ 25 (dashed line), which is clearly incompatible with the observation.

On the contrary, for the mixed model with ^{56}Ni closer to the surface, the heating of the outer layers due to radioactive decays is significant. Its effect starts to appear in the light curve from $t \sim 25$ d and forms a smooth increase in the optical light curve as observed (solid line in Fig. 4). Around this date, the photosphere reaches the layers at

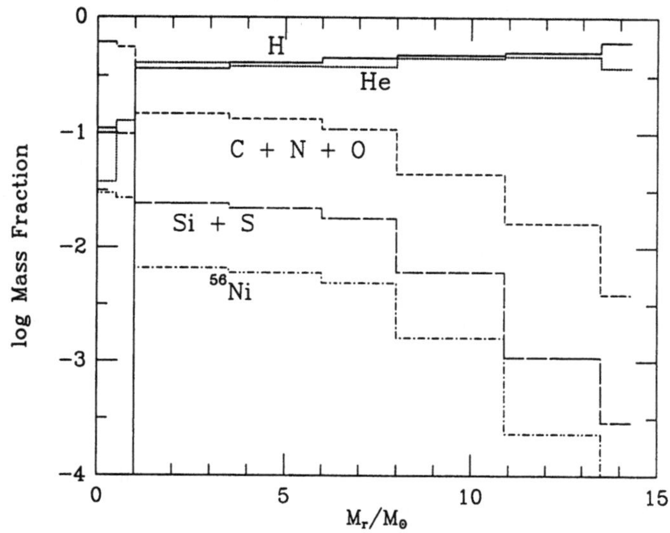

Fig. 3: The assumed abundance distribution in the ejecta after mixing for model 14E1.

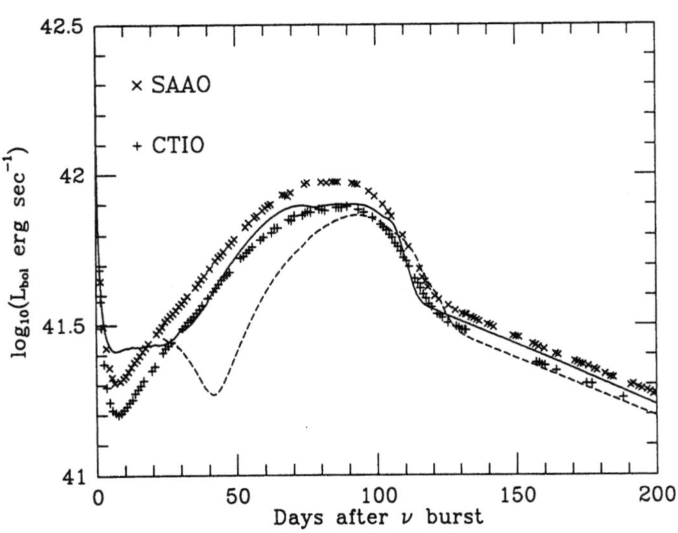

Fig. 4: The bolometric light curves for 14E1 with different distributions of ^{56}Ni and hydrogen. The solid curve assumes mixing of both ^{56}Ni and hydrogen as shown in Fig. 3. For the dashed curve, ^{56}Ni is confined in the innermost layer. For the dotted curve, hydrogen is mixed down to $M_r = 3\ M_\odot$ (in Fig. 3).

$M_r \sim 11\ M_\odot$ (Fig. 3) which contain ^{56}Ni more than 1×10^{-3} in mass fraction and whose expansion velocities are ~ 3000 km s^{-1}.

From the observational side, the satellite emission bumps appeared in the wings of H_α starting from day 25 [17, 7]. Among the proposed interpretations of this Bochum event is that X-rays and γ-rays from radioactive decays became additional source of ionization at the photosphere [27, 33]. Our model support this interpretation. Such a mixing of ^{56}Ni would occur significantly earlier than day ~ 25 and would be due more likely to the Rayleigh-Taylor instability induced by the reverse shock rather than to the ^{56}Ni bubble.

4.2. Plateau-Like Peak and Hydrogen Recombination

After the almost exponential increase up to day ~ 60, the observed bolometric light curve formed a plateau-like broad peak through day ~ 100. The calculated optical light curve for 14E1 is successful in reproducing the plateau-like peak. In forming the light curve for this phase, crucial is the existence of the hydrogen recombination front which moves inward in mass in ~ 100 days. The radius of the front is almost stationary at 8×10^{14} cm during the plateau around the peak (60 d - 100 d). Such a stationary nature of the recombination front is responsible for forming a plateau-like peak rather than simply the smoothed opacity [49, 2]; this is essentially the same as occurs in Type II supernovae.

The width of the plateau depends on the expansion velocity of the mixed hydrogen layer [40]. For the deeper mixing of hydrogen, the plateau lasts longer, and vice versa. In our favorite model shown by the solid line, hydrogen is mixed down to the shell of $M_r = 1\ M_\odot$ (Fig. 3) where the expansion velocity is as low as $\sim$800 km s^{-1}. The minimum hydrogen velocity has previously been reported to be $\sim$800 km s^{-1} according to the recent analysis by Höflich [20]).

For explosion energy, both the pre-peak light curve shape and the plateau- like peak are in good agreement with the observations for $E/M_{\rm env} = (1.1 \pm 0.3) \times 10^{51}$ erg $M_\odot{}^{-1}$. The constraint on $M_{\rm env}$ is obtained as $M_{\rm env} = 7$ - $10\ M_\odot$ [37] to be consistent with the observed enhancement of N/C and N/O in the circumstellar matter. Therefore, the explosion energy should be in the range of $E = (1.0 \pm 0.4) \times 10^{51}$ erg.

5. X-ray Light Curve: Clumpy Mixing or Pulsar?

As the column density of the ejecta decreases with expansion, the hard X-rays and γ-rays emerge from the supernova owing to the decreasing scattering and photoelectric absorption. The theoretical light curves of X-rays and γ-rays again depends on the mixing.

Figure 5 shows the calculated light curve of hard X-ray (16 - 28 keV) for the cases with *mixing and clumps* (solid curve), *mixing without clumps* (dashed), and *no mixing* (dotted) cases. It is clearly seen that only the solid curve is consistent with the Ginga

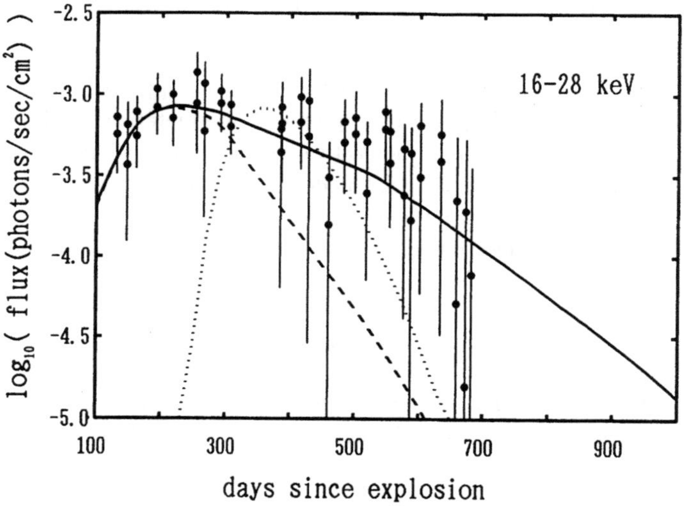

Fig. 5: The calculated light curve of hard X-rays (16 - 28 keV) for 14E1 as compared with the Ginga observations. Three curves correspond to the cases with *mixing and clumps* (solid curve), *mixing without clumps* (dashed), and *no mixing* (dotted).

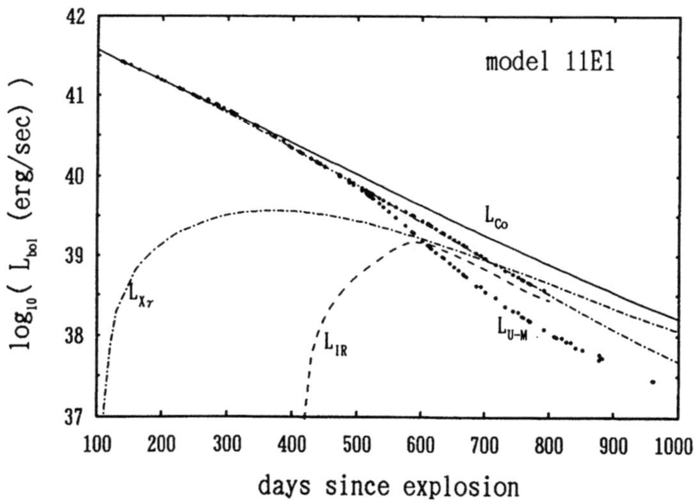

Fig. 6: Observed U to M bolometric luminosity (lower filled circles), observed infrared luminosity $L_{\rm IR}$ (dashed), energy generation rate due to ^{56}Co decay $L_{\rm Co}$ (solid curve), and the calculated X-ray and γ-ray luminosity $L_{X\gamma}$ (dash-dotted). Sum of the observed luminosity $L_{\rm U-M}$ + $L_{\rm IR}$ (upper filled circles) is almost exactly coincides with the calculated luminosity $L_{\rm Co}$ $-$ $L_{X\gamma}$ (dash-dotted) for 11E1. $L_{X\gamma}$ for 14E1 is a little lower than for 11E1.

observations [43]. To account for the early emergence of X-rays and γ-rays, ^{56}Ni needs to be mixed up to $\sim$ 3000 km s^{-1} and 4000 km s^{-1}, respectively.

At the later stages with $t > 300$ d, the observed hard X-ray flux declines much more slowly than the calculated X-ray flux which assumes the homogeneous and spherically symmetric mixing (dashed curve). The decrease in the theoretical flux is due to the photoelectric absorption by the mixed heavy elements.

The observed X-rays might originate from the buried pulsar [31]. If so, the pulsar X-ray luminosity should be $\sim 2 \times 10^{38}$ ergs s^{-1} which should have appeared in the bolometric light curve. However, such a effect was not observed.

We interpreted the X-ray light curve as the effect of clumps [25]. The ejecta must be clumpy [41] and the heavy elements would be localized in some particular clumps. Then a large fraction of X-rays could be transported through the hydrogen and helium-rich regions without suffering much photoelectric absorption; this would effectively reduce the opacity. The solid curve in Fig. 5 assumes the reduction of the photoelectric opacity by a factor of 9 in the core and shows much slower declines than the dashed curve, and looks most consistent with the Ginga observations.

6. Bolometric Light Curve and Dust Formation

With the increasing fraction of the X-ray and γ-ray luminosity, $L_{X\gamma}$ (dashed curve in Fig. 6), the calculated optical bolometric luminosity decreases faster than the energy generation rate of the radioactive decays, $L_{\rm Co}$. This prediction is consistent with the observed U to M bolometric luminosity, $L_{\rm U-M}$, from 260 d to 450 d [46, 42].

Afterwards, however, the observed $L_{\rm U-M}$ is decreasing faster than our theoretical luminosity $= L_{\rm Co} - L_{X\gamma}$ [47]. During this phase, the increase in the 10-13 μm emission was reported [35] as indicated as the IR luminosity $L_{\rm IR}$ in Fig. 6. The upper dotted curve is the sum $L_{\rm U-M} + L_{\rm IR}$, which is almost exactly coincides with $L_{\rm Co} - L_{X\gamma}$. This agreement strongly support the idea that the IR flux is due to the emission of newly formed dust in the ejecta, not from the circumstellar matter [24, 26].

Kozasa *et al.* [23] have predicted the formation of dust grains such as graphite at the temperature of 1800 K, Al_2O_3 at 1600 K, Mg_2SiO_4 at 1400 K and Fe_3O_4 at 1140 K. The grain species depend on the mixing. If mixing occurs, oxygen is more abundant than carbon for all layers so that oxidic grains such as Al_2O_3, Mg_2SiO_4, and Fe_3O_4 would condense. Kozasa *et al.* [24] have shown that the first condensate Al_2O_3 grain is a good emitter around 10 μm and its formation can account for the early rise in the observed IR light curve. Afterwards the heterogeneous nucleation of silicate (Mg_2SiO_4) on pre-condensed Al_2O_3 grains would produce the observed extinction and emission.

If the grains form uniformly, the optical depths of Mg_2SiO_4 and Fe_3O_4 grains are so large that they should black out the supernova. The observed extinction, however, shows only 10^{-4} condensation efficiency. The X-ray luminosity is far too small to destroy dust grains. Thus the low efficiency of extinction must be due to the clumpiness of the ejecta.

Anyway the formation of dust grains in SN 1987A is interesting in view of the origin of the isotopic anomalies of heavy elements in carbonaceous chondrites as well as the origin of interstellar dust grains in general.

7. Rayleigh-Taylor Instability and Mixing

7.1. Two Dimensional Hydrodynamical Calculation

As discussed in §§3-6, the light curves at all wave bands indicate the occurrence of mixing in SN 1987A. The most promising mechanism to mix the ejecta of SN 1987A is the Rayleigh-Taylor instability [9]. A linear stability analysis for 14E1 model shows the interfaces between hydrogen/helium (H/He) and helium/heavy element (He/C+O) are strongly Rayleigh-Taylor unstable and the instability can grow to nonlinear regime before the shock breakout at the stellar surface [11, 6].

Arnett *et al.* (AFM) [3] carried out 2-D axisymmetric hydrodynamical simulations and found a pronounced nonlinear growth of the Rayleigh-Taylor instability around the He/C+O interface, but not around the H/He interface.

Hachisu *et al.* [15] used an axisymmetric 2-D hydrodynamic code with more than ten times larger number of mesh points than that used by AFM and adopted a progenitor model 14E1. They have confirmed the nonlinear growth of the Rayleigh-Taylor instability but found some differences from AFM, i.e., the growth of the instability is more pronounced at the He/C+O interface than at the H/He interface. The possible source of the difference would be the difference in the initial model as discussed in refs. [12] and [15].

Just after the blast shock hits the hydrogen-rich envelope, a reverse shock forms behind the H/He interface; then the R-T instability sets in at the H/He interface. The growth time to nonlinear regime is about 1000 s and the mushroom structures become apparent at $t = 2000$ s. The number of mushroom structures is ~ 10 for random perturbations.

The Rayleigh-Taylor instability induces mixing between the core and the envelope. Figure 7 shows the positions of the marker particles at $t = 3109$ s. It is clear that the heavy elements (core material) are mixed up to the middle of the hydrogen-rich envelope. The extent of mixing can be represented by the maximum and minimum velocities of the interfaces (Fig. 8). The core materials composed of C+O and silicon-rich elements are mixed up to the layers having expansion velocities of ~ 2200 km s^{-1}. At the same time, hydrogen is mixed down to the core of expansion velocities as low as ~ 800 km s^{-1}.

7.2. Comparison with Observations

The indications of a large scale mixing is summarized as follows:

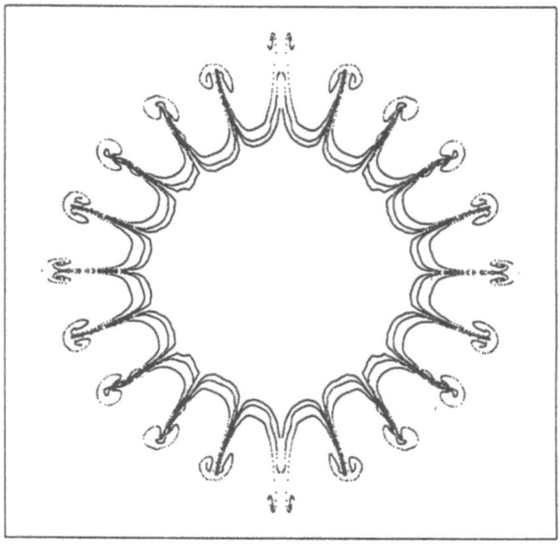

Fig. 7: Nonlinear growth of the Rayleigh-Taylor instability. Shown are the positions of the marker particles at the H/He, He/C+O, and C+O/Si interfaces. The core material finally reaches the top of the mushroom head.

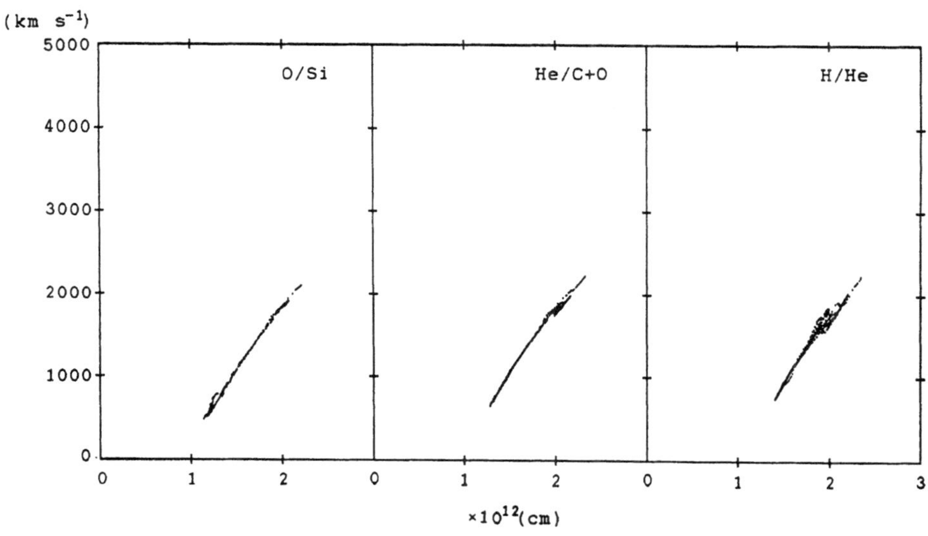

Fig. 8: Velocity vs. radial position of the marker particles at the (a) H/He, (b) He/C+O, and (c) O/Si interfaces.

a) The broad infrared lines of heavy elements and the γ-ray-lines of ^{56}Co have provided direct evidence that Ar, Ni, Co, and Fe are mixed from the low velocity cores ($v \sim 500$ km s^{-1}) to the high velocity outer envelope ($v = 2000 - 3000$ km s^{-1}) [13, 48, 5].

b) The unexpectedly early emergence of hard X-rays and the smooth increase in the pre-maximum optical luminosity have shown indirect evidence for the presence of the mixed radioactive materials (larger than 10^{-3} in mass fraction) at the velocity of ~ 3000 km s^{-1} [25, 34].

c) Larger scale mixing of radioactive material up to the expansion velocity of ~ 4000 km s^{-1} is required from the unexpectedly early emergence of γ-rays [25, 34].

d) Conversely, mixing of hydrogen into the metal-rich core down to such a low expansion velocity as 800 km s^{-1} is indicated by the plateau-like peak of the optical light curve and the minimum velocity of hydrogen found to be as low as ~ 800 km s^{-1} [20].

e) Formation of chemically inhomogeneous clumps, in particular, hydrogen/helium-rich *hole* in the core is suggested from the very slow decline in the hard X-rays observed with *Ginga* [43], which requires the effective reduction of photoelectric absorption [25]. Clumps are also indicated from the spectroscopic features [41].

The results of our 2D hydrodynamical calculations [15] clearly show that the Rayleigh-Taylor mixing can reproduce most of the observational indications of the mixing summarized above. As for the clumpy structures, heavy elements are concentrated into the high-density fingers so that the hydrogen-helium-rich fingers may effectively be *holes* for X-rays because of much smaller photoelectric absorption. Note also that dust may form preferentially in the high-density clumps, which can account for as to why the dust formed in the ejecta has not led to the blackout of the supernova [26, 24].

Regarding the early emergence of gamma-rays, the Rayleigh-Taylor instability alone may not convey ^{56}Ni to the very surface layer whose expansion velocity is as large as $\sim 3000 - 4000$ km s^{-1}. Further acceleration of the core material on a much longer time scale may be possible because of energy input by the decay of ^{56}Ni [1, 50].

K.N. and T.S. would like to thank Prof. Stan Woosley for his effort to organize this fruitful workshop and his hospitality in Santa Cruz. K.N. also would like to thank Drs. W. Hillebrandt, E. Müller, J. Truran, F.-K. Thielemann, R. Canal, and W. Benz for stimulating discussion during the stay at MPA in August 1989. We are grateful to Drs. H. Saio, M. Kato, M. Hashimoto, T. Kozasa, H. Hasegawa, T. Ebisuzaki, I. Hachisu, and T. Matsuda for collaborative work. This work has been supported in part by the grant-in-Aid for Scientific Research (63302015, 01540216, 01652503, 01790169) of the Ministry of Education, Science, and Culture in Japan, and by the Japan-U.S. Cooperative Science Program operated by the Japan Society for the Promotion of Science and the NSF.

References

1. Arnett, W.D. 1988, *Ap. J.*, **334**, 500.

2. Arnett, W.D., and Fu, A. 1989, *Ap. J.*, **340**, 396.
3. Arnett, W.D., Fryxel, B.A., and Müller, E. 1989, *Ap. J.*, **341**, 163.
4. Barkat, Z., Wheeler, J.C., 1988, *Ap. J.*, **332**, 247.
5. Barthelmy, S., Gehrels, N., Leventhal, M., MaCallum, C. J., Teegarden, B. J., and Tueller, J. 1989, *IAU Circular* 4767.
6. Benz, W., and Thielemann, F.-K. 1989, *Ap. J. (Letters)*, in press.
7. Blanco *et al.* 1987, *Ap. J.*, **320**, 589.
8. Catchpole, R., *et al.* 1989, *M. N. R. A. S.*, **237**, 55p.
9. Chevalier, R. A. 1976, *Ap. J.*, **207**, 872.
10. Danziger, I.J., 1990, in this volume.
11. Ebisuzaki, T., Shigeyama, T., and Nomoto, K. 1989, *Ap. J. (Letters)*, **344**, L65.
12. Ebisuzaki, T., Shigeyama, T., and Nomoto, K. 1989, in this volume.
13. Erickson, E. F., Haas, M. R., Colgan, S. W. J., Lord, S. D., Burton, M. G., Wolf, J., Hollenbach, D. J., and Werner, M. 1988,
14. Fransson, C. Cassatella, A., Gilmozzi, R., Kirshner, R.P., Panagia, N., Sonneborn, G., and Wamsteker, W. 1989, *Ap. J.*, in press.
15. Hachisu, I., Matsuda, T., Nomoto, K., and Shigeyama, T. 1989, *Ap. J. (Letters)*, submitted.
16. Hamuy, M., Suntzeff, N.B., Gonzalez, R., and Martin, G. 1987, *Astr. J.*, **95**, 63.
17. Hanuschik, R.W., and Dachs, J. 1988, *Astr. Ap.*, **205**, 135.
18. Hashimoto, M., Nomoto, K., and Shigeyama, T. 1988, *Astr. Ap.*, **210**, L5.
19. Hashimoto, M., Nomoto, K., 1989, in preparation.
20. Höflich, P. 1988, *Proc. Astron. Soc. Australia*, **7**, 434.
21. Humphreys, R.M., and Davidson, K. 1978, *Ap. J.*, **232**, 409.
22. Kato, M., Yamaoka, H., Saio, H, and Nomoto, K., 1990, in preparation.
23. Kozasa, T., Hasegawa, H., and Nomoto, K. 1989, *Ap. J.*, **344**, 325.
24. Kozasa, T., Hasegawa, H., and Nomoto, K. 1989, *Ap. J. (Letters)*, in press.
25. Kumagai, S., Shigeyama, T., Nomoto, K., Itoh, M., Nishimura, J., and Tsuruta, S. 1989, *Ap. J.*, **345**, 412.
26. Lucy, L. B., Danziger, I. J., Gouiffes, C., and Bouchet, P. 1989, *Structure and Dynamics of the Interstellar Medium, IAU Colloquium No. 120*, ed. G. Tenorio-Tagle, M. Moles, and J. Melnick (Berlin: Springer-Verlag), in press.
27. Lucy, L.B., 1988, in *Supernova 1987A in the LMC*, ed. M. Kafatos and A. Michalitsianos (Cambridge: Cambridge University Press), p.323.
28. Nomoto, K., and Hashimoto, M. 1988, *Physics Report*, **163**, 13.
29. Nomoto, K., Shigeyama, T., and Hashimoto, M. 1987, in *SN 1987A*, ed. I.J. Danziger (Garching: ESO)
30. Nomoto, K., Shigeyama, T., Kumagai, S., and Hashimoto, M. 1988b, in *Proc. Astron. Soc. Australia*, **7**, 490.
31. Pacini, F., 1988, in *Physics of Neutron Stars and Black Holes*, ed. Y. Tanaka (Tokyo: Universal Acad. Press), p. 461.
32. Panagia, N. *et al.* 1987, *IAU Cir.* No. 4514.
33. Phillips, M.M., and Heathcote, S.R. 1989, *Pub. Astr. Soc. Pacific*, **101**, 137.
34. Pinto, P. and Woosley, S.E., 1988, *Nature*, **333**, 534.

35. Roche, P.F., Aitken, D.K., Smith, C.H., and James, S.D., 1989, *Nature*, **337**, 553.
36. Saio, H., Kato, M., and Nomoto, K. 1988, *Ap. J.*, **331**, 388.
37. Saio, H., Nomoto, K., and Kato, M. 1988, *Nature*, **334**, 508.
38. Shigeyama, T., and Nomoto, K. 1989a, *Ap. J.*, submitted.
39. Shigeyama, T., and Nomoto, K. 1989b, in this volume.
40. Shigeyama, T., Nomoto, K., and Hashimoto, M. 1988, *Astr. Ap.*, **196**, 141.
41. Stathakis, R.A., Dopita, M.A., and Cannon, R.D., 1989, preprint.
42. Suntzef, N.B., Hamuy, M., Martin, G., Gomez, A., and Gonzalez, R., 1988, *Astr. J.*, **96**, 1864.
43. Tanaka, Y. 1988, in *Physics of Neutron Stars and Black Holes*, ed. Y. Tanaka (Tokyo: Universal Acad. Press), p. 431.
44. Thielemann, F.-K., Hashimoto, M., Nomoto, K., 1989, *Ap. J.*, **348**, No.2, in press.
45. Wheeler, J.C., 1990, in this volume.
46. Whitelock, P. *et al.* 1988, *M. N. R. A. S.*, **234**, 5P.
47. Whitelock, P. *et al.* 1989, *M. N. R. A. S.*, **240**, 7P.
48. Witteborn, F. C., Bregman, J. D., Wooden, D. H., Pinto, P. A., Rank, D. M., Woosley, S. E., and Cohen, M. 1989, *Ap. J. (Letters)*, **338**, L9.
49. Woosley, S.E. 1988, *Ap. J.*, **330**, 218.
50. Woosley, S.E., Pinto, P., and Weaver, T.A.1988, *Proc. Astron. Soc. Australia*, **7**, 335.
51. Yamaoka, H., Kato, M., Saio, H., and Nomoto, K. 1989, in preparation.

Binary Models for the Progenitor of SN 1987A

Juliana J. L. Hsu, Paul C. Joss, Philipp Podsiadlowski, & Saul Rappaport

A significant fraction of all stars are in binaries. It is thus reasonable to examine the possibility that the progenitor of SN 1987A was a component of a binary system. We here consider some of the ranges of parameter space in which binary evolution produces progenitors that are applicable to models of SN 1987A [1, 2, 3]. Other binary models have been proposed by FABIAN *et al.* [4] and CHEVALIER and SOKER [5]. Such models may help to explain several of the anomalies of this supernova, including the blue color of the progenitor, the asymmetries in the ejecta, the barium anomaly, and the mystery spot (see [1, 2, 3] for further discussion and references). Unlike many single-star models (see [1, 2, 3] for references), the fits of these models to the observations do not rely on the assumption of low metallicity, a non-standard treatment of convection, and/or the *ad hoc* stipulation of severe mass loss from the presupernova star. The binary models presented here make definite predictions that can be checked against observations; moreover, the models are sufficiently general that the results may be relevant to other supernovae as well.

1. An Asymptotic-Branch Giant Companion of Sk $-$ 69°202 [1]

First, we consider the possibility that the progenitor was an undetected red star (star 4) in orbit with Sk$-$69°202 (star 1), and that Sk$-$69°202 is merely hidden in the ejecta. After all, an asymptotic-branch giant was the preferred model for a type II supernova progenitor prior to the occurrence of SN 1987A [6, 7]. The red star must satisfy the following criteria: (a) it must have been sufficiently faint relative to the blue stars in the field to have escaped detection on photographic plates taken prior to the supernova [8]; (b) it must have been initially the more massive star; (c) the orbital parameters must have been such that the mass transfer occurred at the appropriate stage of evolution; and (d) the supernova light curve resulting from a red progenitor must be consistent with that observed.

The first three conditions determine the initial parameters of the binary system. The initial masses of stars 4 and 1 must have been in the ranges $M_4^i \approx 10 - 15 M_\odot$ and

$M_1^i \approx 4M_\odot - M_4^i$, repectively, with an initial orbital separation, a_i, of $\sim 4\,\mathrm{AU}$ and an initial orbital period $P_i \simeq 2$ yr. We have carried out evolutionary calculations using a Henyey-type code [9], for several values of M_4^i in the above range. Possible mass and angular momentum loss from the system were taken into account by the use of two free parameters, the fraction β of the mass lost by star 4 that is accreted by star 1 and the specific angular momentum, α, of any matter lost from the system, in units of $2\pi a^2/P$, where a and P are the instantaneous values of the orbital separation and period, respectively, at each evolutionary stage.

In this scenario, star 4 overflows its Roche Lobe when it is an asymptotic-branch giant (case C mass transfer) [10] and transfers much of its hydrogen-rich envelope to star 1, which then becomes the observed blue supergiant [1]. For a wide range of values of of α and β, star 4 retains $\sim 0.5 - 3\,M_\odot$ of hydrogen in the envelope prior to the supernova event; this residual hydrogen should have been sufficient to produce the Balmer lines observed during the explosion. In particular, for $M_4^i = 12M_\odot$, the mass of star 4 just before the supernova event is $\sim 3 - 6M_\odot$. After the supernova, the system remains bound, with an orbital period of $\sim 3 - 10$ yr and an eccentricity of $\sim 0.1 - 0.4$.

A particular difficulty with this model is in fitting the rise time of the light curve. Although the final masses of these models are much lower than those of models evolved without binary mass exchange, the shock breakout time is still high. If the Kamiokande [11] and IMB [12] neutrino events indicate the time of the core collapse, the initial rise time for the optical supernova must have been < 3 hours. To investigate this matter, we have carried out a series of hydrodynamic calculations; the detailed results of these calculations will be published elsewhere. For the best case of $M_4^i \simeq 12M_\odot$ and final hydrogen envelope mass $\simeq 0.5M_\odot$, we found that the shock breaks through the surface of the star in ~ 3.8 hours if the assumed input energy is 10^{52} ergs; the shock breakout time decreases with increasing energy input. Therefore, unless current estimates [13] of the energy available from core collapse are much too low, this scenario is probably excluded. Moreover, unless Sk$-69°202$ soon reappears, this scenario can be ruled out conclusively once the bolometric luminosity of the supernova falls below that of Sk$-69°202$.

2. Mass Transfer from a Companion [2]

We next consider whether the presence of a close-binary companion (star 4) might cause the progenitor itself (now assumed to be star 1) to become blue rather than red. We find that mass accreted by the progenitor from the companion can restructure the star in such a way that the progenitor is driven blueward.

There is a maximum luminosity, L_{crit}, that can be carried in the envelope of a star by radiative diffusion. (See the discussion in [2].) L_{crit} depends sensitively on the radiative opacities in the envelope and the fractional core mass $q_c = M_c/M$, where M_c is the mass of the hydrogen-exhausted core and M is the total mass of the progenitor. As q_c decreases, L_{crit} increases, the fractional amount of the envelope that is unstable against convection concomitantly decreases, and the star tends to evolve toward the blue. Hence, the progenitor could have been driven blueward by the accretion of mass from its companion during the final stages of its evolution (see also BARKAT and WHEELER [14, 15]). Figure 1 shows tracks that we have calculated for the evolution of the progenitor in the H-R diagram, for a few different amounts of transferred mass.

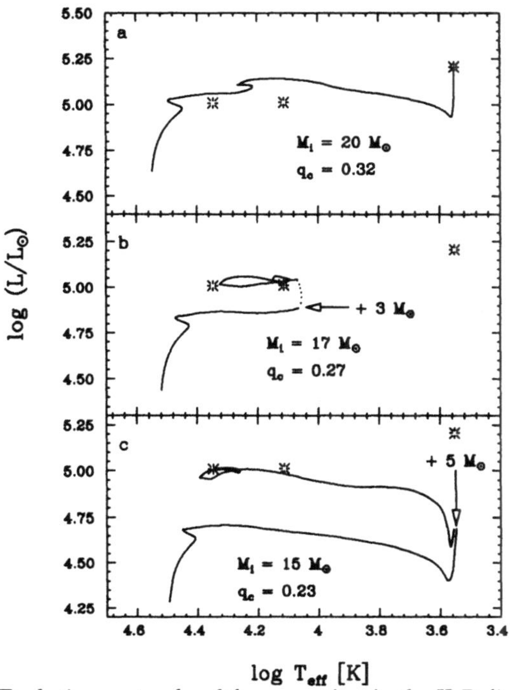

Fig. 1 Evolutionary tracks of the progenitor in the H-R diagram for several assumed amounts of accreted mass. *a*, Evolutionary track of a $20M_\odot$ star without accretion. *b*, Evolutionary track of a star with an initial mass of $17M_\odot$ which accretes $3M_\odot$ after the termination of core hydrogen burning (the dotted portion of the curve denotes the mass-transfer phase; the arrow indicates the onset of mass transfer). *c*, Evolutionary track for a star with an initial mass of $15M_\odot$ which accretes $5M_\odot$ after the core hydrogen-burning phase; here, the small break in the curve denotes the mass-transfer phase. In each case, M_i denotes the initial mass and $q_c \equiv M_c/M$ is the final fractional core mass, M_c being the mass of the hydrogen-exhausted core and $M = 20M_\odot$, the final mass of the progenitor. The starbursts indicate the parameter values of the models just before the supernova event. For comparison, the final parameter values for all three cases are shown in each panel.

In this scenario, the progenitor must have been the initially less massive star. We consider both case B and case C mass transfer [16, 10]. In case B, both star 1 and star 4 are on the first red-giant branch when mass transfer commences, but star 4 has evolved further. In order for the two stars to be red giants simultaneously, the initial masses of the two stars must have been within $\sim 0.3\%$ of each other. After the mass transfer, star 1 becomes the more massive star and is driven blueward. Star 1 evolves to become the supernova. After the supernova event, star 4 would become gravitationally unbound from star 1 and should become observable in $\sim 1 - 3$ yr as a red star within the thinning supernova remnant.

In case C mass transfer, star 4 is on the asymptotic giant branch and star 1 is in an earlier post-main-sequence phase when mass transfer commences. In this case, the two initial masses could differ by as much as $\sim 10\%$. Star 4 would have become a supernova $\sim 10^6$ yr prior to the supernova event in star 1, leaving a remnant neutron star gravitationally bound to star 1. After the supernova explosion in star 1, the older neutron star would become unbound from the remnant of SN 1987A and possibly detectable as a radio pulsar or as an X-ray source powered by the accretion of material from the current supernova remnant.

3. Merger Models [3]

In the preceding scenario, the mass transfer was assumed to be stable. In fact, this need not be the case. Here we examine a third scenario in which the mass transfer is unstable and the two stars form a common envelope [17].

If the initial masses of the two stars are sufficiently different, $M_4^i << M_1^i \simeq 13 - 18 M_\odot$, the mass-transfer time scale is much shorter than the thermal time scale of the companion. As a result, the companion will expand substantially shortly after the transfer phase begins, and the two stars will develop a common envelope [18]. Dynamical friction between the smaller star and the envelope causes the former to spiral in toward the system center-of-mass [19]. The evolution of the system depends on the frictional luminosity generated, the efficiency of energy transport in the common envelope, and the transport of orbital angular momentum within the envelope.

The mass transfer and subsequent common-envelope phase could develop as either case B or case C transfer. For case B, we modeled the merger of a $3M_\odot$ main-sequence star with a $16M_\odot$ red giant. The frictional luminosity was held constant [20] at $L_{fric} = 4 \times 10^4 L_\odot$. For this value of L_{fric}, the secondary spirals in and completely dissolves after 1200 yr. Thus, the final outcome of the common-envelope phase is a single star of $19M_\odot$ with a $4.5M_\odot$ core. This star is very similar to models in the second scenario described above, where the addition of mass from the companion changes the structure of the envelope in such a way that it appears as a single blue supergiant until the supernova explosion.

For case C, we similarly modeled the merger of a $6M_\odot$ main-sequence star with a $16M_\odot$ red supergiant. We here take L_{fric} to be $10^4 L_\odot$ for this model. The evolutionary time scale of the asymptotic-branch giant is no more than $\sim 10^4$ yr, which is comparable to the spiral-in time scale; hence, the merger may or may not have been completed at the time of core collapse of the more massive component. In our model calculations, the progenitor has a final core mass of $5.5M_\odot$ and the companion a mass of $3.4M_\odot$ at the time of the supernova explosion. The evolutionary tracks of the case B and case C mergers are shown in Fig. 2.

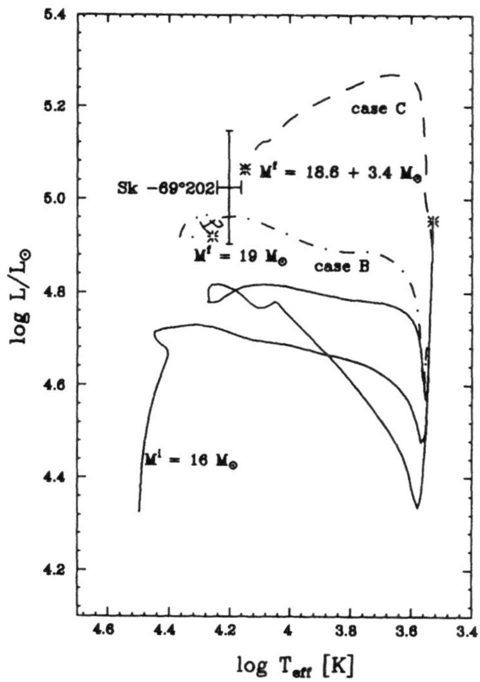

Fig. 2 Evolutionary tracks in the H-R diagram of a single $16\,M_\odot$ star (solid curve) and of two illustrative merger models (dashed and dot-dashed curves). The dashed curve represents the evolutionary track of a $16\,M_\odot$ star after it has completely merged with a $3\,M_\odot$ main-sequence star as a result of a case B common-envelope phase (the common-envelope phase itself is not shown), and the dot-dashed curve shows the evolution of a $16\,M_\odot$ asymptotic-giant branch star which is still in the process of merging with a $6\,M_\odot$ main-sequence star (case C). In the latter case, the merger has not been completed by the time of the supernova explosion. The starbursts mark the locations of the models at the time of the supernova explosion, and the error bars indicate the observationally inferred location of Sk−69°202 [21].

If a remnant companion is indeed present at the time of the supernova event, the companion is likely to be completely disrupted by the supernova blast wave. Nevertheless, we speculate that a small fractional amount of material from the companion could have remained gravitationally bound to the collapsed core. If the companion dissolves in the process of spiral-in but leaves behind a torus of enhanced density around the progenitor core, the refraction of the supernova shock wave as it passes the torus may also leave some material bound to the collapsed core. In either case, the bound matter may subsequently form a self-gravitating disk, which could be subject to instabilities that lead to its fragmentation and the formation of one or more discrete self-gravitating bodies. (However, the thermal time scale for the formation of the companion is $\sim 1 - 10$ yr. Thus, the detected companion may still be in the process of formation, and its observationally inferred properities may vary with time. If this is the case, we expect that subsequent detections of the pulsar will reveal a companion [or companions] with different but similar masses and orbital periods.) This scenario thus provides possible sources of material with the appropriate specific angular momentum to provide an explanation for the detection of a $\sim 10^{-3} M_{\odot}$ object in an ~ 8 hr orbit about the pulsar that was left behind by the supernova [22]. In contrast, material that was ejected by the collapsed core or that fell back from the supernova ejecta [23] would not have the appropriate specific angular momemtum to be in such an orbit.

Although the physics of common-envelope evolution is not well understood, a large range of binary parameters should lead to the merger of the binary components (provided that the common envelope is not ejected). For example, if a progenitor with an initial mass of $16 M_{\odot}$ had a $6 M_{\odot}$ companion, initial orbital periods between ~ 3 d and ~ 3 yr would lead to the merger of the two components in a case B event, while initial orbital periods between ~ 3 and ~ 5 yr would result in a case C common-envelope phase.

Unlike the first two scenarios described above, direct verification of the case B merger model is difficult. Confirmation of this model is more likely to be indirect and may come from a reconstruction of the chemical composition profile of the progenitor's envelope (as in the case of all single-star models). In the case C model, if the remnant companion was not completely disrupted by the supernova blast wave, a part of the companion core with mass of the order of a fraction of a solar mass (but probably much more massive than $10^{-3} M_{\odot}$) could still be detectable. The detection of this remnant, either bound or unbound from the pulsar, would verify the model conclusively.

This work was supported in part by the National Aeronautics and Space administration under grants NAGW-1320 and NAGW-1545.

References

1. P. C. Joss, Ph. Podsiadlowski, J. J. L. Hsu, and S. Rappaport: *Nature* **331**, 237 (1988).

2. Ph. Podsiadlowski and P. C. Joss: *Nature* **338**, 401 (1989).

3. Ph. Podsiadlowski, P. C. Joss, and S. Rappaport: *Astron. Astrophys. Lett.* (in press).

4. A. C. Fabian, M. J. Rees, E. P. J. van den Heuvel, and J. van Paradijs: *Nature* **328**, 323 (1987).

5. R. A. Chevalier and N. Soker: *Astrophys. J.* **341**, 867 (1989).

6. S. W. Falk and W. D. Arnett: *Astrophys. J. Suppl.* **33**, 515 (1977).

7. S. E. Woosley and T. A. Weaver: In *Proc. 5th Moriond Astrophys. Conf.: Nucleosynthesis and its Implications on Nuclear and Particle Physics*, ed. J. Audouze and N. Mathieur (Reidel, Dordrecht) 145 (1985).

8. N. Walborn: In *Supernova 1987A in the Large Magellanic Cloud*, ed. M. Kafatos and A. Michalitsianos (Cambridge University Press) 1 (1988).

9. R. Kippenhahn, A. Weigert, and E. Hofmeister: In *Methods in Computational Physics 7*, ed. B. Adler, S. Fernbach, and M. Rothenberg (Academic Press, New York) 129 (1967).

10. D. Lauterborn: *Astron. Astrophys.* **7**, 150 (1970).

11. K. Hirata et al.: *Phys. Rev. Lett.* **58**, 1490 (1987).

12. R. M. Bionta et al.: *Phys. Rev. Lett.* **58**, 1494 (1987).

13. E. Baron and J. Cooperstein, preprint (1989).

14. Z. Barkat and J. C. Wheeler: *Astrophys. J.* **332**, 247 (1988).

15. Z. Barkat and J. C. Wheeler: *Astrophys. J.* **342**, 940 (1989).

16. R. Kippenhahn and A. Weigert: *Z. Astrophys.* **65**, 251 (1967).

17. B. Paczyński: In *IAU Colloq. No. 6: Mass Loss and Evolution in Close Binaries*, ed. K. Gyldenkerne and R. M. West (Copenhagen University) 139 (1970).

18. R. Kippenhahn and E. Meyer-Hofmeister: *Astron. Astrophys.* **54**, 539 (1977).

19. H. Bondi and F. Hoyle: *Mon. Not. Roy. Astr. Soc.* **104**, 273 (1944).

20. F. Meyer and E. Meyer-Hofmeister: *Astron. Astrophys.* **78**, 167 (1979).

21. S. E. Woosley: *Astrophys. J.* **330**, 218 (1988).

22. J. Kristian et al.: *Nature* **338**, 234 (1989).

23. S. E. Woosley and R. A. Chevalier: *Nature* **338**, 321 (1989).

Light Curve Models for SN 1987A: Roles of the Recombination Front of Hydrogen

T. Shigeyama and K. Nomoto

1. Introduction

Type II supernovae have been subclassified into IIp and IIℓ according to the presence and the absence of a plateau in the light curve. The observed light curve of SN 1987A is different from these two subclasses during the first 4 months. However, Shigeyama and Nomoto [1] have shown that its broad peak from day 60 to 100 can be interpreted as a plateau. In the light of new understandings of SN 1987A, we will analize how the plateau is formed and discuss the subclassification of Type II supernovae.

2. Analytical Models for the Recombination front and SN 1987A

It has been thought that the recombination front of hydrogen (R-front) makes the plateau in the light curve of a Type IIp supernova (Grassberg, Imshennik and Nadyozhin [2], Woosley and Weaver [3]). However, it remains an open question whether the R-front necessarily makes the plateau. The observed bolometric light curve of supernova 1987A (SN 1987A) showed a unique shape during the first 4 months (SAAO [4-6], CTIO [7]). The plateau appeared between days 65 and 100. On the other hand, the observed photospheric temperature was fixed at $\sim$ 5500K from day $\sim$ 20 to 100. This means that the R-front existed during this period and did not always make the plateau. To describe such a unique thermal behavior of SN 1987A, we construct an analytic model for the R-front.

We divide the layers around the R-front into three regions (I, II, and III) as depicted in Fig. 1. The front is located at the radius r. In the region I, hydrogen is fully ionized (temperature $T_1 \geq 10^4$ K). The electron scattering opacity is so large that the radiative energy flux is negligible. The region II, where all the hydrogen is neutral (T_2 <5500 K) , is optically thin. The region III is the transition layer where hydrogen is partially ionized and the temperature and opacity suddenly decrease from the region I to II. The photosphere is located in this region. If the region III is sufficiently thin, the three conservation laws (mass, momentum, and energy) must be held between the regions I and II (see Shigeyama [8]) as follows,

$$\rho_1 u_1 = \rho_2 u_2, \tag{1}$$

$$\rho_1 u_1^2 + \frac{2k\rho_1 T_1}{m_{\rm H}} + \frac{aT_1^4}{3} = \rho_2 u_2^2 + \frac{\rho_2 T_2}{m_{\rm H}} + \frac{aT_2^4}{3}, \tag{2}$$

$$\frac{\rho_1 u_1^3}{2} + \rho_1 u_1 \left(\frac{5kT_1}{m_{\rm H}} + \frac{\epsilon_{\rm H}}{m_{\rm H}} + \frac{4aT_1^4}{3\rho_1}\right) = \frac{\rho_2 u_2^3}{2} + \rho_2 u_2 \left(\frac{5kT_2}{2m_{\rm H}} + \frac{4aT_2^4}{3\rho_2}\right) + \sigma T_{\rm eff}^4. \tag{3}$$

Here, ρ is the mass density, u the velocity of the matter in the rest frame of the front, T the temperature, $T_{\rm eff}$ the effective temperature. The suffices 1 and 2 denote the region to which the variables belong. The constants k, a, σ, $m_{\rm H}$, and $\epsilon_{\rm H}$ are the Boltzmann constant, the radiation constant, the Stephan-Boltzmann constant, the mass of a proton, and the ionization energy of hydrogen (=13.6 eV), respectively.

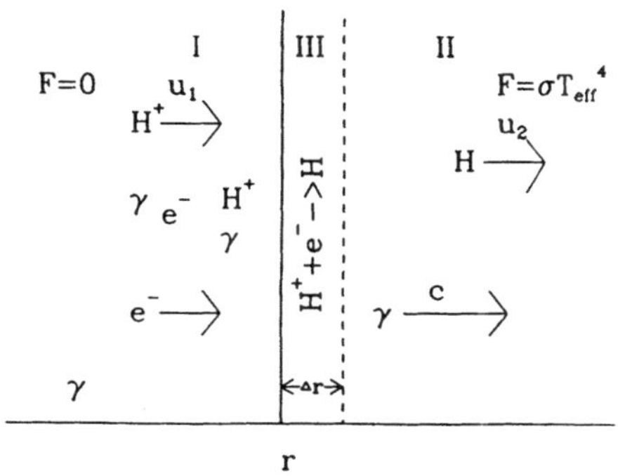

Fig 1: The schematic view around the recombination front of hydrogen.

In considering the physical conditions in the supernova ejecta, we will approximate equations (1)-(3): i) The gas pressure is negligible as compared with the radiation pressure because of the high temperature and low density. ii) The radiation pressure in the region II is much smaller than that in the region I because of the temperature difference. iii) The term containing $\epsilon_{\rm H}$ in equation (3) is negligible compared with the radiation energy. iv) The kinetic energy of the supernova ejecta is much larger than the thermal energy so that the density and velocity are hardly affected by the thermal conditions. v) The velocity of the ejecta v is proportional to the radius in the rest frame of the front. vi) Suppose the width of the region III, Δr, is smaller than r and introduce new variables u and ρ as $u_1 = u - \Delta u/2$, $u_2 = u + \Delta u/2$, $\rho_1 = \rho + \Delta\rho/2$, $\rho_2 = \rho - \Delta\rho/2$, where $\Delta\rho/\rho << 1$ and $\Delta u/u << 1$ to satisfy the steady state condition, because $\Delta r/r = n\Delta\rho/\rho$ if the density distribution is proportional to r^{-n}. Using the above variables and neglecting the small terms, the conservation laws are rewritten as,

$$\rho_1 u_1 = \rho_2 u_2 = \rho u, \tag{4}$$

$$\frac{\Delta\rho}{\rho} = \frac{\Delta u}{u} \equiv 2\xi, \tag{5}$$

$$-\xi + \frac{aT_1^4}{6\rho u^2} = 0, \tag{6}$$

$$-2\xi - \frac{\epsilon}{m_H u^2} + \frac{4aT_1^4}{3\rho u^2} = 0. \tag{7}$$

The solution gives us the propagation velocity of the front with respect to the matter,

$$\begin{aligned} u &= \frac{c}{4}\left(\frac{T_{\rm eff}}{T_1}\right)^4 \\ &= 1.35 \times 10^3 \left(\frac{T_1}{15000K}\right)^{-4}\left(\frac{T_{\rm eff}}{5500K}\right)^4 \text{ km sec}^{-1}, \end{aligned} \tag{8}$$

and the width of the front

$$\frac{\Delta r}{r} = \frac{16aT_{\rm eff}^4}{3\rho c^2 n}(\frac{T_1}{T_{\rm eff}})^{12},$$
$$\sim 7\times 10^{-3} n^{-1}(\frac{\rho}{1\times 10^{-12}\ {\rm g\ cm^{-3}}})^{-1}(\frac{T_1}{15000\ {\rm K}})^{12}(\frac{T_{\rm eff}}{5500\ {\rm K}})^{-8}. \qquad (9)$$

The values used in the above two equations are typical in the supernova ejecta model at the recombination front. The width of the front is estimated as $\Delta r \sim 7\times 10^{12}(r/10^{15}\ {\rm cm})$ cm unless $n << 1$. The time for the front to be in steady state is $\Delta r/u \sim 5.2\times 10^4$ sec and smaller than the dynamical time scale of the ejecta ($\sim$ 10d). Therefore the above solution is self-consistent.

The luminosity of the supernova is given as $L = 4\pi r_{\rm ph}^2 \sigma T_{\rm eff}^4$, where $T_{\rm eff} \sim 5500$ K when the R-front exists. In fact, the observed photospheric temperature in SN 1987A was nearly constant $\sim$ 5500 K from day 20 to 100 ([4-7]). Therefore the luminosity is determined by the radius of the photosphere (i.e., the R-front) $r_{\rm ph}(\sim r)$. If the velocity of the R-front u is smaller than the expansion velocity of the matter v, $r_{\rm ph}$ increases and correspondingly the luminosity increases. This occurs when the photosphere is in the outer region of the ejecta where v is larger. This is why the luminosity of SN 1987A increases between days 20 and 60. When the photosphere enters into the inner region of the ejecta where v is smaller, v becomes nearly equal to u. Then the inward motion of the front nearly cancells with the outward motion of the matter and $r_{\rm ph}$ remains constant. This is the mechanism to form the broad peak of SN 1987A as well as the plateau observed in the light curve of a Type IIp supernova.

Until when does the plateau last? The front is maintained by the recombination of electrons with hydrogen ions. Therefore, after the innermost hydrogen layer passes through the front, the R-front disappears. Then the thermal time scale of the ejecta becomes so small that the light curve goes into the radioactive tail. This is the end of the plateau. Thus the velocity of the innermost hydrogen layer $v_{\rm min}^{\rm H}$ can be estimated from the time at the end of the plateau, $t_{\rm pl}$, and the bolometric luminosity at $t_{\rm pl}$, $L_{\rm pl}$ as

$$v_{\rm min}^{\rm H} = R_{\rm ph}/t_{\rm pl} = (\frac{L_{\rm pl}}{4\pi\sigma T_{\rm eff}^4})^{1/2}/t_{\rm pl}$$
$$\sim 1.3\times 10^3\ {\rm km\ sec^{-1}}(\frac{L_{\rm pl}}{9\times 10^{41}\ {\rm erg\ sec^{-1}}})^{1/2}(\frac{T_{\rm eff}}{5500{\rm K}})^{-2}(\frac{t_{\rm pl}}{100\ {\rm d}})^{-1}. \qquad (10)$$

3. Light curves of other Type II supernovae

The light curves of Type IIp supernovae are well modeled for red supergiant progenitors ([2]). The larger radius leads to less adiabatic cooling than in SN 1987A, thereby keeping the ejecta temperature higher than 10^4K for longer time. When the hydrogen recombination and thus the plateau phase starts, the photospheric radius is already larger than in SN 1987A, which leads to more luminous plateau. The plateau is terminated when the inner most hydrogen passes through the recombination front, and the relation in equation (10) is applicable. For the best observed Type IIp supernova SN 1969L where $L_{\rm pl} \sim (1.5-3)\times 10^{43}$ ergs s^{-1} at $t_{\rm pl} \sim$ 100d (Kirshner *et al.* [9]), $v_{\rm min}^{\rm H}$ is estimated to be $(1.3 - 2) \times 10^3$ km s^{-1}. Further study is needed to clarify whether such a low hydrogen velocity occurres.

Without mixing of hydrogen into the core, the minimum velocity of hydrogen layer is approximately determined as

$$v_{\rm min}^{\rm H} \propto (E/M_{\rm env})^{1/2}. \tag{11}$$

Applying this relation to equation (10), we obtain

$$t_{\rm pl} \propto (L_{\rm pl} M_{\rm env}/E)^{1/2}. \tag{12}$$

Equation (12) implies that the plateau is terminated earlier for smaller $M_{\rm env}$ and larger E owing to the higher expansion velocity of the inner most hydrogen layer. The explosion with $t_{\rm pl} \sim 50$ d may be recognized as a Type IIℓ supernova. In fact, the observed light curve of SN 1970G has a plateau-like feature which terminates on day ~ 50 d (Barbon, Ciatti, and Rosino [10]). Such a case would be realized if $M_{\rm env}$ is as small as $\sim 2M_\odot$ and hence $v_{\rm min}^{\rm H}$ is as large as $\sim 2 \times 10^3$ km s^{-1}. The thermal behavior of a Type IIℓ supernova is the same as in Type IIp until the recombination starts. When the electrons start to recombine, however, the photosphere has already entered into the layer containing too small hydrogen to form a plateau because of higher expansion velocities. The light curve continues to decline and enters into the radioactive tail. (The case with low $L_{\rm pl}$ due to a small initial radius or small ^{56}Ni mass could also lead to small $t_{\rm pl}$ but might not easily be discovered.)

Possible candidates for the progenitor of Type IIℓ supernovae include: Wolf-Rayet-like massive stars that undergo iron core collapse and AGB stars whose C+O cores undergo carbon deflagration. Both of these can be distinguished from the late luminosities since they are determined by the ^{56}Ni mass which is larger for AGB stars than for massive stars (Young and Branch [11]).

Mixing of hydrogen into the low velocity core would result in larger $t_{\rm pl}$. If $M_{\rm env}$ is smaller than the core mass, however, mixing due to Rayleigh-Taylor instability might be much less efficient because there is not enough time for the instability to grow before the shock wave reaches the surface (Nomoto and Shigeyama [12]). Further 2D hydrodynamical calculation is necessary for these progenitors.

References

1. T. Shigeyama & K. Nomoto: submitted to Ap.J., (1989).
2. E.K. Grassberg, V.S. Imshennik & D.K. Nadyozhin: Ap.Space.Sci.Rev., 10, 28 (1971).
3. S.E. Woosley & T.M. Weaver: Ann.Rev.A.A., 24, 205P (1986).
4. J.W. Menzies *et al.*: MNRAS, 227, 39P (1987).
5. R.M. Catchpole *et al.*: MNRAS, 229, 15P (1987).
6. R.M. Catchpole *et al.*: MNRAS, 231, 75P (1988).
7. M. Hamuy, N.B. Suntzeff, R. Gonzalez & G. Martin: A.J., 95, 63P (1987)
8. T. Shigeyama: Ph. D. Thesis, University of Tokyo, (1989).
9. R.P. Kirshner, J.B. Oke, M.V. Penston & L. Searle: Ap.J., 185, 303 (1973).
10. R. Barbon, F. Ciatti & L. Rossino: A.Ap, 29, 57 (1973).
11 T. Young & D. Branch: Ap.J. (Letters), 342, L79 (1989).
12. K. Nomoto & T. Shigeyama: In this volume.

Supernova 1987A — Its Light Curve, Composition, and Pulsar

S. E. Woosley

Much has been written regarding the spectacular supernova of 1987. By and large, with a few exceptions to make it all the more interesting, SN 1987A was a learning experience – a chance to calibrate many previously untested models and a blueprint for how to properly study supernovae in the future – but an event not fundamentally different from what one might have expected (with considerable hindsight) from the death of a 20 $M_\odot$ blue supergiant. My own views on the evolution of the progenitor and detailed diagnostics of the explosion have been presented elsewhere [1–9] and need not be repeated. For a good overview see [10].

Here, in limited space and in keeping with the informal format of the workshop, I would like to address three issues relevant to SN 1987A. These are the light curve, the composition of the ejecta, and nature of the compact object in the middle.

1. The Light Curve of SN 1987A and Other Type II Supernovae

Most theoretical and observed Type II light curves in the literature show a luminosity that peaks as the shock breaks out (though this spike of radiation has never been observed directly) and then either declines immediately (Type II-L) or remains nearly flat (Type II-p) for ~2 months. Early on Woosley *et al.* [11] suggested that SN 1987A, which even during its first week was clearly different, might be a "Type II-b" supernova, a star that had retained a small hydrogen envelope and so spectroscopically was Type II, but which had a radioactive powered light curve at peak like a Type Ib. Later observations showed that 1987A was indeed radioactively powered from an early time, but the envelope mass turned out to be much greater than our early expectations and new models were constructed [1,2,6]. Still, just to complete the record, Figure 1 gives the bolometric light curve of a model II-b supernova. This is a theoretical model whose natural counterpart remains to be determined. It was not SN 1987A. The model was prepared by putting a 1 $M_\odot$ low density envelope of hydrogen and helium in hydrostatic equilibrium on a 6 $M_\odot$ evolved helium core. The luminosity at the base of the envelope was adjusted to give a *red* supergiant star with luminosity near that of Sk 202 -69 but with a radius of 1.7×10^{13} cm. Explosion was then simulated just as in models for SN 1987A with a comparable final kinetic energy at infinity (1.3×10^{51} erg). The light curve was then calculated including the contribution from either 0.07 $M_\odot$ or 0.20 $M_\odot$ of ^{56}Co. "Mixed" versions of the same models were also calculated. Radioactive ^{56}Ni was mixed out all the way into the hydrogen envelope (to an even greater extent than in Model 10HMM of ref. 6). Hydrogen, however, was not mixed as deeply into the helium core as in SN 1987A. Inward mixing was only into a point comparable to the envelope mass mass (*i.e.* Lagrangian coordinate 5 $M_\odot$). Figure 1 shows that this mixing did not greatly alter the light curve.

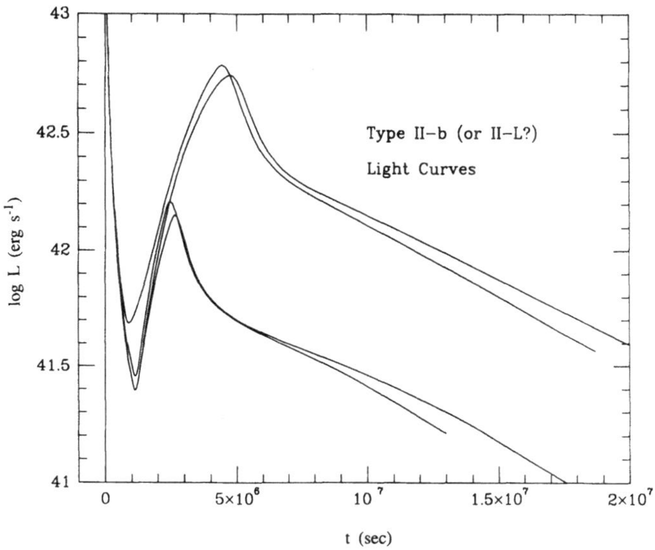

Fig. 1 – Type II-b supernovae based upon a red supergiant model of a star that has suffered extreme mass loss and has only one solar mass of hydrogen envelope at explosion. Each pair of curves shows a mixed and unmixed (brighter at late times) model for two different values of ^{56}Co synthesized (see text for details). These are *not* models for SN 1987A, but may be appropriate to observed Type II-L supernovae provided such events are discovered near peak light and not before.

All models rose quickly to a radioactive powered maximum much like SN 1987A, but the rise is quicker and the supernova brighter. For a single star, massive enough to have lost all but 1 $M_\odot$ of envelope, the core structure will favor larger production of ^{56}Ni than in SN 1987A (the mantles of massive stars have higher entropy and shallower density gradients around the collapsing iron core). Thus the upper curve (0.2 $M_\odot$ of ^{56}Ni) is preferred. One clear implication of these calculations is that one cannot search for historical analogues of SN 1987A simply based upon the existence of a "pre-maximum" halt. Here we have a local minimum in the luminosity even though the progenitor was red with a low mass envelope. The model is also interesting because many people have suggested that Type II L supernovae may arise from just such a progenitor. Mapping the bolometric luminosity into the appropriate B and V bands remains to be done, but the model doesn't look very "linear" to me, unless such supernovae are discovered at peak and only the decline observed. Perhaps this is the case.

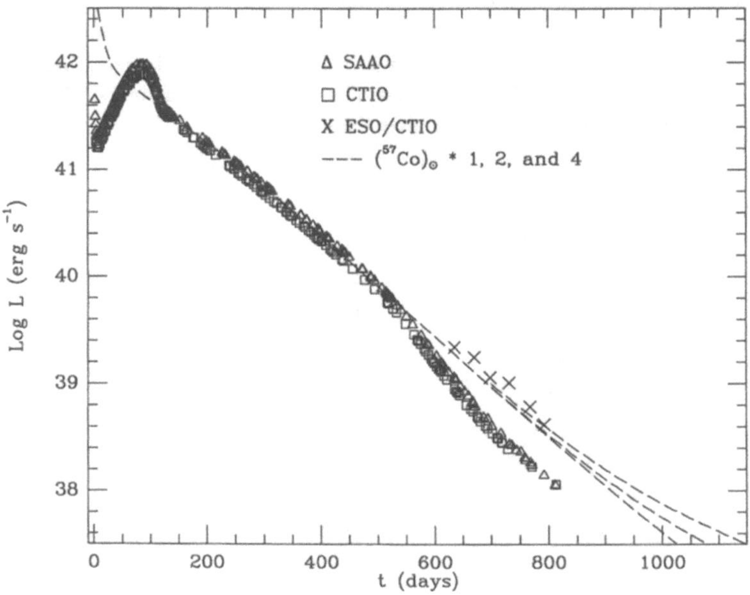

Fig. 2 – The theoretical late time light curve of SN1987A for three different values of ^{57}Co abundance (see text) is compared to the U through M measurements from SAAO and CTIO (see papers by Whitelock and by Suntzeff, this volume) and the combined IR data from ESO and U through M from CTIO (Suntzeff and Bouchet, private communication). Through day 800 all measurements were compatible with the radioactive model. Measurements at ESO and CTIO after day 800 are still being analyzed, as this book goes to press but tend to show a leveling out of the light curve after day 800 that is inconsistent with the radioactive model and would require a new power source.

While on the subject of light curves and unfinished business, Figure 2 gives the projected long term history of the light curve for 1987A based solely upon a radioactive power source. This subject has been treated in detail elsewhere [8], but here I take the opportunity to show what occurs when one varies the ^{57}Co mass. The three lines indicate the light curve when the final ^{57}Fe/^{56}Fe ratio (after all radioactive decay is complete) is equal to 1, 2, and 4 times the solar value. Values much above 2, and certainly above 4 for this ratio are disallowed, not only upon the basis of nucleosynthetic constraints (one should not greatly overproduce ^{57}Fe), but also by reasonable constraints on the neutron excess that can exist in the silicon and oxygen shells of the presupernova star. The ratio of ^{57}Ni/^{56}Ni (ultimately ^{57}Fe/^{56}Fe) made in explosive nucleosynthesis is proportional to the square root of the neutron excess parameter η. By the time ^{57}Ni/^{56}Ni rises to a value corresponding to more than 4 times solar, the dominant iron group nucleus is no longer ^{56}Ni but ^{54}Fe.

My best guess is the middle line. Values below the lowest line would indicate that observers are not seeing all the radiation. Values above would indicate the operation of an additional energy source, probably the central compact object.

2. Details of Nucleosynthesis

One of the major benefits of a nearby supernova is the possibility to study in detail the composition produced by a representative example of these element factories. Except for radioactivities, observations are unable to distinguish isotopic composition. Thus the following table summarizes the elements produced in two representative explosions of 18 and 20 $M_\odot$ supernovae [2]. Conditions were selected to span a range of probable parameters, not only in stellar mass but in the rate adopted for the $^{12}C(\alpha,\gamma)^{16}O$ rate and the efficiency of semiconvective mixing. Both explosions had kinetic energy at infinity of 1.3×10^{51} ergs as required by observations [1]. Not given in the table but of some interest are the ranges of $^{57}Fe/^{56}Fe$, which gives the ^{57}Co abundance at early times, and the mass fractions of ^{22}Na and ^{44}Ti. These were in the range 1.2 to 1.4 times solar for $^{57}Fe/^{56}Fe$, 6×10^{-5} $M_\odot$ for ^{44}Ti, and 2×10^{-6} $M_\odot$ for ^{22}Na. Note the production of all iron group nuclei (Ti thru Ni) in the 20 $M_\odot$ model needs to be reduced by a factor of 2 to agree with the known amount of ^{56}Fe produced in SN 1987A.

TABLE 1: NUCLEOSYNTHESIS SUMMARY[a]
18 and 20 $M_\odot$ Stars; 1.3×10^{51} erg

Element	Predicted Mass Ejected	Element	Predicted Mass Ejected
Fe Core	1.3 – 1.7	S	5.9(-2) – 5.3(-2)
H	8.2 – 9.0	Cl[b]	2.6(-4) – 1.2(-4)
He	6.8 – 7.0	Ar	9.3(-3) – 1.1(-2)
C	0.26 – 0.18	K	2.1(-4) – 9.4(-5)
N	0.015 - 0.015	Ca	4.3(-3) – 9.6(-3)
O	0.24 – 1.6	Sc	4.2(-7) – 1.1(-6)
F[b]	1.0(-6) – 1.1(-6)	Ti	1.4(-4) – 3.3(-4)
Ne	0.15 – 0.18	V	8.6(-6) – 1.2(-5)
Na	2.7(-3) – 1.4(-3)	Cr	8.8(-4) – 2.2(-3)
Mg	0.044 – 0.10	Mn	3.1(-4) – 4.8(-4)
Al	2.3(-3) – 7.8(-3)	Fe	7.5(-2) – 1.4(-1)
Si	0.14 – 0.11	Co	1.3(-4) – 2.8(-4)
P	3.1(-4) – 6.0(-4)	Ni	4.8(-3) – 5.9(-3)

[a] 18 $M_\odot$ results are the first number; 20 $M_\odot$ is second.

[b] Greatly increased by ν-nucleosynthesis not included here. See Table 2.

The value for hydrogen and helium here are in agreement with the properties of the presupernova star (namely its helium core mass) and the requirement that roughly 10 $M_\odot$ of low density envelope be in place at the time the star exploded. The iron core is consistent with estimates of the total energy of the neutrino burst and the ejected iron mass is in agreement with the requirements of the bolometric light curve (though an artificial adjustment is required for the 20 $M_\odot$ model). Experimental determinations of the other elements are less precise (see papers by Danziger and by Phillips, this volume). Oxygen in particular has been estimated to have a mass in the range 0.2 to 4 $M_\odot$. A more precise determination of the oxygen abundance, or as Kirshner suggested at the meeting, the ratio of oxygen to some other element whose nucleosynthesis is less model sensitive, would be very useful in limiting the parameters of the model.

These results do not include the modifications in composition induced by the neutrino burst that accompanies collapse of the iron core to a neutron star. The enormous flux of neutrinos can excite many nuclei to particle unbound levels by neutral current scattering interactions. The μ- and τ-neutrinos, which have a substantially higher energy than the electron neutrinos, are chiefly responsible. Consequent modification of the abundances of roughly a dozen isotopes occurs as is discussed by [12] and by Hartmann elsewhere in this volume.

Since the meeting, however, I have carried the calculations further, by dropping the restrictions imposed by the parameterized sampling of only a few supernova zones in the stellar explosion. Instead the neutrino physics (several hundred reactions) has actually been incorporated into an realistic supernova model of several hundred zones. After following the evolution of a 25 $M_\odot$ star through all its burning stages using a network of some 150 isotopes, and then artificially exploding the star with a piston at the base of the silicon shell (final kinetic energy equals 7×10^{50} erg) and including all the neutrino interactions, the results below are obtained [13].

TABLE 2: NEUTRINO NUCLEOSYNTHESIS
25 $M_\odot$– 7.0×10^{50} erg

Species	No ν	T_μ=4	T_μ=6	T_μ=8	E_{53}=6, T_μ=8
^{2}H	8.1(-6)	8.8(-6)	8.8(-6)	8.8(-6)	9.1(-6)
^{7}Li	2.7(-8)	2.0(-7)	4.7(-7)	1.2(-6)	2.0(-6)
	0.01	0.08	0.19	0.5	0.8
^{11}B	5.6(-9)	9.6(-7)	1.8(-6)	4.0(-6)	6.3(-6)
	0.004	0.8	1.4	3.2	5.1
^{19}F	2.3(-5)	7.7(-5)	1.4(-4)	2.3(-4)	3.7(-4)
	0.2	0.7	1.3	2.2	3.5
^{35}Cl	2.8(-4)	3.2(-4)	3.7(-4)	4.4(-4)	5.4(-4)
	0.4	0.5	0.6	0.7	0.8
^{22}Na	5.3(-6)	6.3(-6)	7.1(-6)	8.4(-6)	1.1(-5)
^{26}Al	4.8(-5)	5.5(-5)	6.0(-5)	6.8(-5)	8.1(-5)

For each isotope, the table gives the abundance ejected in solar masses for several values of assumed μ- and τ-neutrino temperature. The last column employs twice the total neutrino flux (a neutron star binding energy of 6×10^{53} ergs rather than 3×10^{53} ergs used in the other studies). For ^{7}Li, ^{19}F, ^{11}B and ^{35}Cl a second line in the Table gives the ratio of the abundance produced to that of ^{16}O, a definite Type II supernova product, compared to the same ratio in the sun [14]. A value of unity would indicate that the indicated species is produced in a solar ratio compared to ^{16}O. From this we see that all four isotopes can be produced in nearly solar proportions (though the Galactic abundance of lithium remains uncertain) if the μ- and τ-neutrino temperature is near 6 MeV, a value favored by the supernova model builders.

Though 25 $M_\odot$ is too heavy for SN 1987A and the explosion energy adopted too small by about a factor of two, similar results should obtain for SN 1987A.

3. What Blinks in SN 1987A?

The detection of the neutrino burst simultaneously at IMB, Kamiokande, and Baksan and, especially its 10 second duration showed conclusively that a neutron star had formed in the supernova. As others have noted, the statistics are inadequate to say much about the explosion mechanism, but it is an unavoidable conclusion that an iron core collapsed to supranuclear density. Further, as Burrows has emphasized [15], the continued implosion of the neutron star into a black hole did not occur during the first 10 seconds. During such a collapse the matter flows faster into the event horizon than the neutrinos can diffuse. The neutrino signal is abruptly truncated. Thus at age 10 seconds a neutron star surely existed inside the supernova, but of its magnetic field strength, rotation rate, and mass accretion rate at that time, we know nothing.

Did it stay a neutron star? We really don't know. Two hours after the core collapsed, just as first light was appearing from the surface, the "reverse shock", the reflection of the expanding helium core from the star's own hydrogen envelope, arrived back at the center of the star. This slowed the heavy element ejecta considerably and some matter certainly fell back. How much depends upon several things, especially details of the unknown explosion mechanism and the exact energy of the explosion. Empirically I have found [1], when artificially exploding presupernova models with pistons, that once the total kinetic energy at infinity of all the ejecta falls below about 3×10^{50} ergs, solar mass quantities of helium and heavier elements fall back onto the neutron star and that this fall back occurs when the reverse shock reaches the center. Thus the assumption that we still have a neutron star in SN 1987A rests largely upon estimates of the explosion energy, roughly 1.5×10^{51} ergs. It is also worth noting, as Chevalier points out [16], that blue supergiants may be particularly prone to mass reimplosion. This is because in a compact star the expanding helium core encounters its own mass in a much shorter time than in a red supergiant. The reverse shock thus arrives back at the core earlier when the density is still high. Chevalier estimated that $\sim$0.1 $M_\odot$ may have fallen back in SN 1987A, but this is a very uncertain number.

Another very important constraint, however, is that 0.07 $M_\odot$ of ^{56}Ni needs to have been ejected by the explosion in order to power the light curve. During the explosion some of this radioactivity was mixed to layers farther out, but most of its stayed localized near the center. In all past models of supernova explosion and nucleosynthesis it has proved very difficult to raise more than a few tenths of a solar mass to temperatures in excess of 5 billion degrees for any realistic presupernova model. I conclude that no more than 0.2 $M_\odot$ of matter fell back onto the neutron star, quite possibly a lot less. Whether this turns the neutron star into a black hole depends upon how close it already was. The best calculations, which ignore rotation and make uncertain assumptions about the explosion mechanism suggest a value near 1.5 $M_\odot$ (baryon), though 1.7 $M_\odot$ is not excluded. Thus a black hole in SN 1987A would be likely only if the critical mass is substantially below 2 $M_\odot$. Such models are common [17–19].

That's where our knowledge of the neutron star stood on day one (reconstructed later of course), and pretty much the way it stayed until 1989. During those two years we were able to place increasingly tight constraints on the brightness of any central source. Since a model based solely upon radioactivity fits the data so well through day 800 (Fig. 2), there can be no additional source brighter than about 10^{38} erg s^{-1}. Any energy deposited below about 20 keV is quickly thermalized and contributes to the bolometric (non-γ and non-X-ray) light curve. Using simple scaling laws for pulsar emission various people, myself included, placed limits on the combination of magnetic field and rotation rate to stay within that boundary. In retrospect it was probably naive to use such relations in an environment where accretion could be continuing and when a large fraction of the pulsar energy may come out in the form of energetic particles rather than pulsed electromagnetic emission. By late 1989 there were some indications that the light curve might be bottoming out near the Eddington limit, $\sim 2 \times 10^{38}$ erg s^{-1}. This would be a natural occurrence if as little as 10^{-8} solar masses of material fell back each year onto the neutron star (or black hole). A larger fall back rate might also give emission near the Eddington limit.

On January 19, 1989 a group from the California and Los Alamos detected pulsed optical emission from SN 1987A with a frequency of 1969 Hz. These measurements and the compelling case for their association with a pulsar in the LMC are summarized elsewhere in these proceedings. The dilemma posed by these observations is illustrated by the following list of "peculiarities". In many cases major revisions would have to be made to current theory to accommodate the data.

1) The early appearance of the pulsar. At optical wavelengths the supernova was still calculated to be thick in the optical bands in January, 1989. The early detection came as a surprise, but might be understood in terms of the near infrared efficiency of the detector. But a source several per cent as bright as the supernova in a near IR band of width less than 2000 A leads to some concern about the optical efficiency.

2) The optical efficiency. In the band observed the pulsar was from one to several hundred times brighter than the Crab pulsar Yet it is hard to make optical emission from a millisecond pulsar (a search for optical emission from other millisecond pulsars is planned and should be carried out). Yet, if anything, one would expect the optical efficiency of a millisecond pulsar to be *less* than the Crab. Wang *et al.* [20] have pointed out that the small radius of the light cylinder requires a large electron energy to avoid the blackbody limit. But then hard radiation from the synchrotron process would dominate the bolometric light curve unless the magnet field were extremely weak ($\lesssim 10^3$ gauss). Their alternative invokes a pulsating neutron star instead of a rotating one, but the the emission mechanism is very contrived – sound waves steepening into shocks that accelerate iron nuclei that then emit ionic cyclotron radiation specifically in the optical.

3) The subsequent invisibility of the pulsar. All though 1989 attempts continued to reacquire the pulsar. The attempts were unsuccessful, but placed a limit at least ten times fainter than the night of January 19. Where could the pulsar have gone? It is extremely unlikely it accreted enough material during that interval to become a black hole. Limits from the bolometric light curve show that the accretion luminosity is not super-Eddington, suggesting an accretion rate less than 10^{-8} $M_{\odot}y^{-1}$. Larger accretion rates could, in principal, be accommodated if the excess energy were lost to neutrinos, but could not be so large as to substantially alter the mass of ^{56}Co inferred from the light curve ($\sim 10^{-2}$ $M_\odot$ y^{-1}). A solid upper limit is given by the fact that the pulsar frequency remained so constant during the 8 hours of observation. This means that the moment of inertia did not change and limits the accretion rate to less than about 10^{-3} $M_\odot$ y^{-1}. It also means that a rotationally stabilized neutron star is very unlikely to have slowed and contracted into the event horizon.

The observers suggest that the supernova is cloudy and that the observations on January 19 occurred at a time of unrepeated clarity. The variation of the magnitude observed over an 8 hour interval is suggestive of of a variable optical depth. However, this explanation becomes increasingly contrived as the months pass and the pulsar does not reappear. Rather than clouds one would need to have peered through a small hole in an otherwise opaque spherical shell, a very unlikely event.

If the signal is real, then, in my opinion, more reasonable explanations for its disappearance must be sought, probably a transient emission mechanism. In the case of a rotating model this might involve the fall back of matter and short circuiting of the pulsar mechanism. For a vibrating model, one might invoke a large neutron star quake just prior January 19 and a subsequent damping of the oscillations.

4) The mysterious companion(s). These had to be formed after the explosion, yet in order to fit inside of their Roche radii the objects must have a high density characteristic of solid matter ($\gtrsim 1$ g cm^{-3}). On the other hand, leaving aside for now the issue of "strange" matter, the density could also not be much *greater* than a few g cm^{-3}. Gravity does not compress Jupiter-mass objects to extremely high degeneracy. Take an average density of 5 g cm^{-3} and a mass of 0.0006 $M_\odot$ as representative. For the orbit given, 2 $R_\odot$, the planet would presumably be formed of heavy elements, some of which are radioactive. From its density and mass, the gravitational binding energy per nucleon of the planet would be $\sim$ 10 eV. A typical weak decay, *e.g.* of ^{56}Co, releases several MeV, or $\sim$50 keV per nucleon. The time required for radiation to diffuse out of a Jupiter-mass solid object is much greater than two years. Thus in order that the planet not be disrupted by its own radioactive decay, the mass fraction of ^{56}Co (or other radioactivities having similar lifetimes) must be less than a few times 10^{-4}. Yet we have evidence that the mass fraction of ^{56}Co was $\gtrsim$ 1%, even at the base of the hydrogen shell [5]. I conclude that a planet of the type reported is unlikely to have formed from well-mixed supernova ejecta. It's not neutron star ejecta either. Pieces of neutron star of mass less than $\sim$0.05 $M_\odot$ would explosively decompress and mix with the rest of the supernova. Moreover, such matter would be full of neutron-rich radioactivities. Strange matter is a possibility, but one I'd just as soon not count on. Even with strange matter it is hard to see how a planet having 0.1% of the mass could be ejected with an angular momentum comparable to that of the entire neutron star. One way out of this dilemma might be to form planets from clumps outside of the explosive silicon burning region where the concentration of radioactive nuclei is small, but this only helps if the smallest scale of the Rayleigh-Taylor instability (responsible for the outward mixing of ^{56}Co) is larger than 0.001 $M_\odot$. Present 2D-calculations (*eg.* Fryxell and Arnett, this volume) lack sufficient resolution to comment on this.

5) The rapid rotation rate and the critical mass for neutron stars. So far, this has received the greatest attention in the literature. Those few models that use baryonic equations of state and allow (or nearly allow) the existence of 0.5 ms periods require that the neutron star have a mass very nearly equal, within a few hundredths of a solar mass, of the maximum value allowed (ref. 18,19; see also Lattimer this volume). A slight increase in the mass or decrease in the rotation rate pushes the star over the limit and forces collapse to a black hole. Compelling as I find the observations of the Berkeley group, I also find it difficult to believe that the core of the star would be clever enough, in the one case where we have watched a neutron star being born, to evolve to, explode with, and accrete just enough matter to end up within 0.05 $M_{\odot}$ of its maximum mass. Perhaps this constraint can be relieved by appealing to strange or mesonic matter [21, but see also Lattimer this volume), but one must take care not only to allow the fast rotation rate of the putative 87A pulsar and the existence of a stable slowly rotating pulsar of mass 1.442 $M_{\odot}$ in PSR1913+16, but also to utilize only (protoneutron star, baryonic) masses for the 87A pulsar greater than 1.3 $M_{\odot}$. This is because, with obvious prejudice, I believe that the stellar evolution calculations have shown it extremely unlikely that the iron core which collapsed in SN 1987A was any smaller. Subsequent processes – delayed explosion, mass fall back, *etc.* – can only increase the baryonic mass above 1.3 $M_{\odot}$. Thus solutions, such as those of Glendenning [21] that invoke strange matter, must use a "bag constant", $B^{1/4}$, that is less than 200 MeV and again require the final pulsar mass to be narrowly tuned. Of course, hand in hand with the rapid rotation rate goes a much smaller magnetic field than was expected from observation of other young pulsars.

6) The birthrate of millisecond pulsars. The slow spin down time, at least 1000 years, reported for the new pulsar, implies that it is not a short lived phenomenon. No other young supernova remnant contains a millisecond pulsar and the birthrate of millisecond pulsars is constrained, by their long life times *e.g.* ref. 22) and relative infrequency in sensitive surveys [23], to a small fraction of that of longer period pulsars. Why in this one case, the first neutron star birth ever witnessed, are we presented with such an unusual phenomenon? One possibility is that the high angular momentum of the pulsar is related to the blue supergiant nature of the progenitor star [24], but then supernovae like SN 1987A would still need to be very rare, probably considerably less than 1% of Type II's.

If one assigns a value significantly less than unity to each of the above six probabilities (though 5 and 6 are obviously related), one is left with fairly convincing evidence (Perlmutter, this volume) for a phenomenon that seemingly has no right to exist. What can be done about this paradoxical situation?

The nice thing is that we should ultimately know whether a pulsar was or is there or not. Bets can be made and collected. The best window for making observations is in the x-ray at about 30 keV. Below that, the photoelectric optical depth remains very large for years. Above it, any reasonable emission spectrum falls off. Figure 3 (adapted from ref. 8) shows the effect on the hard x-ray spectrum at day 1000 of a central source having a spectrum like that of an accreting magnetic neutron star (Her X-1) and a luminosity of 1.6×10^{38} erg s^{-1}. A source having the spectrum of the Crab pulsar and a similar luminosity would show a similar rise around 20 keV [8]. On this day $\sim$10% of the 30 keV emission will be getting out unattenuated and could be studied for pulsations. Unfortunately no experiment, US or foreign, now or planned for the near future, has the requisite sensitivity. Perhaps one can be constructed.

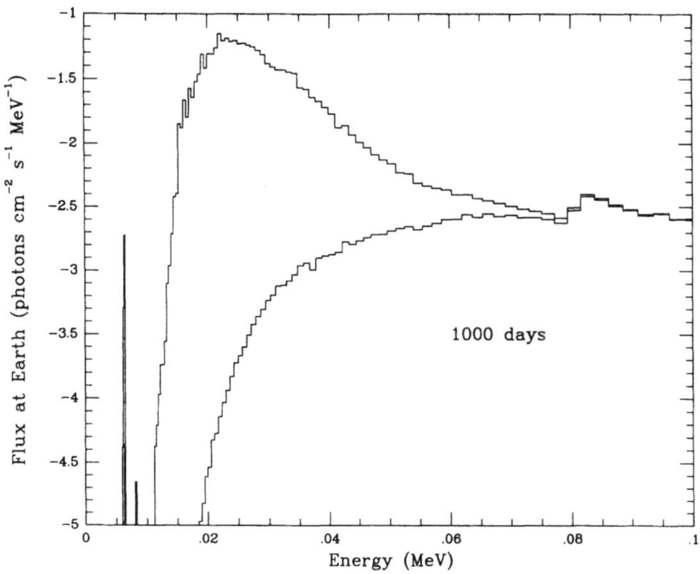

Fig. 3 – The hard x-ray spectrum of SN 1987A at day 1000 if an accreting compact object having the spectrum of Her X-1 and a luminosity of 1.6×10^{38} erg s^{-1} is embedded in the supernova[8]. Two lines show the spectrum with radioactivity only and radioactivity plus the pulsar. Similar curves would be obtained if the spectrum were that of the Crab pulsar. Note the enhancement of the emission lines when the pulsar is present. An energy band near 20–30 keV would be the best place to look for the compact object and its pulsations.

Evidence for hard emission might also be sought in the infrared by looking for highly ionized atoms [25] and the energy from a central compact object may ultimately cause a leveling out of the bolometric light curve. Indeed a leveling out of the light curve was being reported in all wavebands as 1989 came to an end. Unfortunately the light curve is simply a bolometer and says noting about whether the source is pulsed or not. Nor does it distinguish an accretion powered source from one powered by rotation. Only time-resolved measurements of unscattered radiation can do that. Perhaps in a year or so – meanwhile my bet is that the supernova still has a few surprises up her sleeve.

This research has been supported by the National Science Foundation (88-13649) and by NASA (NAGW-1273).

REFERENCES

1. Woosley, S. E. *Astrophys. J.*, **330**: 218-253, (1988).
2. Woosley, S. E., Pinto, P. A., and Weaver, T. A. *Proc. Astron. Soc. Australia*, **7**, No. 4, 355, (1989).
3. Woosley, S. E. *Ann. N. Y. Acad. Sci.*, Proc. 14th Texas Symposium on Relativistic Astrophysics, in press.
4. Pinto. P. A., Woosley, S. E., and Ensman, L. M. *Ap. J. Lettr.*, **31**, L101, (1988).
5. Pinto, P. A., and Woosley, S. E. *Ap. J.*, **329**, 820, (1988).
6. Pinto, P. A., and Woosley, S. E. *Nature*, **333**, 534, (1988).
7. Woosley, S. E. *Proc. of the National Academy of Sciences*, in press.
8. Woosley, S. E., Pinto, P. A., & Hartmann, D. *Astrophys. J.*, in press, November 1, (1989).
9. Woosley, S. E., and Weaver, T. A. *Scientific American*, August, 1989.
10. Arnett, W. D., Bahcall, J. N., Kirshner, R. P., and Woosley, S. E. *Annual Reviews of Astronomy and Astrophysics*, **27**, 629, (1989).
11. Woosley, S. E., Pinto, P. A., Martin, P. G., and Weaver, T. A. *Ap. J. Lettr.*, **318**, 664, (1987).
12. Woosley, S. E., Hartmann, D., Hoffman, R., and Haxton, W. *Ap. J.*, in press, (1990).
13. Woosley, S. E., Haxton, W., and Weaver, T. A., *Bull. Am. Astron. Soc.*, **21**, No. 4, 1095 (1989).
14. Cameron, A. G. W. in *Essays in Nuclear Astrophysics*, ed. D. D. Clayton, C. A. Barnes, and D. N. Schramm, (Cambridge: Camb. Univ. Press), p. 23, (1982).
15. Burrows, A. *Astrophys. J.*, **334**, 891-908, (1988).
16. Chevalier, R. A. preprint submitted to *Astrophys. J.*, (1989).
17. Arnett, W. D., and Bowers, R. L. *Ap. J.*, **33**, 415 (1977).
18. Friedman, J. L., Ipser, J. R., and Parker, L. *Astrophys. J.*, **304**: 115-139, (1986).

19 Friedman, J. L., Ipser, J. R., and Parker, L., *Phys. Rev. Lettr.*, **62**, 3015, (1989).

20. Wang, Q., Chen, K., Hamilton, T. T., Ruderman, M., and Shaham, J. *Nature*, **338**, 319, (1989).

21 Glendenning, N. K., preprint submitted to *Journal of Physics G: Nuclear and Particle Physics*, July, 1989.

22. Kulkarni, S. R., and Naryan, R. *Ap. J.*, **335**, 755, (1988). 23. Segelstein, D. J., Stinebring, D. R., Fruchter, A. S., and Taylor, J., *Nature*, **322**, 714, (1986). 24. Woosley, S. E., and Chevalier, R. A., *Nature*, **338**, 321, (1989).

25. Colgan, S. W. J., and Hollenbach, D. J. *Astrophys. J. Lettr.*, **329**, L25, (1988).

Modelling of the Early Light Curve of SN 1987A with the Multi-Group Time-Dependent Radiative Transfer

S. I. Blinnikov & D. K. Nadyozhin

ABSTRACT

The fully implicit high-order scheme has been developed for time-dependent multi-group radiative transfer coupled with implicit hydrodynamics. The application of this scheme to SN 1987A explosion shows that shortly after the shock breakout there forms a dense shell.

After the SN 1987A outburst numerous papers have described its early light curve and the behavior of its effective temperature and photospheric velocity (see, for example, Woosley et al., 1988; Grasberg et al., 1987; Arnett 1987; Shigeyama et al., 1987). Then the researchers have concentrated their efforts on the modelling of a broad maximum and radioactive tail of SN 1987A light curve. All these numerical models were based on approximation of equilibrium diffusion of radiation in accordance with one or another flux-limited scheme.

Here we report some preliminary results of our calculations of SN 1987A outburst with a new radiation gas-dynamic code describing the radiative transport in multi-group approximation with variable Eddington factors.

The supernova gasdynamics with nonequilibrium radiative transport has been used earlier by Falk and Arnett (1977) and by Chevalier and Klein (1979). In the theory of gravitational collapse, the multi-group neutrino transport has been employed by Bowers and Wilson (1982), Bruenn (1985), and Myra et al. (1987).

Let us describe now main points that differ our code from those employed by above mentioned authors. They used only a constant Eddington factor $f = 1/3$. We calculate with a space variable factor $f(r)$ evaluated from equation of transfer by the Feautrier (1964) method for instantaneously static atmosphere to which our code calls after certain number of steps. Optionally, the user can take into account the retardation effect more precisely by calculating factors $f(r,t)$ from time-dependent equation of transfer (Mihalas and Mihalas, 1984) at every time step.

Falk and Arnett neglected the ionization and recombination of matter, whereas our code take ionization equilibrium into account in the Saha approximation. Moreover, in the outermost rarefied layers, the option for the kinetic (NLTE) treatment of the hydrogen ionization is provided for. Chevalier and Klein (1979) used one-group diffusion approximation to incorporate radiative transport in gasdynamics while multi-group spectral calculations were performed in a static model. In our code, fully time-dependent radiative transport accounting for all the effects of the order of v/c is combined with gasdynamics in a common fully implicit difference scheme. Instead of the static expression for radiation flux used in most of previous works, we calculate the flux in every energy group from the time-dependent equation as described by Falk and Arnett (1977) and Mihalas and Mihalas (1984).

Thus, we have realized the numerical scheme which is able to calculate the radiation transport together with gasdynamic flow. This scheme involves all the necessary effects and is close to one described by Mihalas and Mihalas (1984). Due to a low capacity of our computers at present we treat the Eddington factor as being the same for all the energy groups. However, our gasdynamic scheme (as well as sections of the code that control the radiation transport with related atomic kinetics) is implicit and thereby can be successfully applied to quasi-static phases of stellar evolution.

Gasdynamic part of our computer code was thoroughly and successfully tested in investigations of gravitational collapse with kinetics of beta-processes (Blinnikov and Rudzsky, 1984) and of white dwarfs cooling (Blinnikov, 1988).

Moreover, there was undertaken a special testing of the gasdynamic code for the problem of strong explosion in degenerate stellar matter with allowing for kinetics of ^{12}C nuclear burning. These calculations were performed also with the use of previous (not fully implicit) code (Imshennik and Nadyozhin, 1983, 1988) that proved to be a highly effective tool during a long-term investigation of supernova gasdynamics. The comparison of results showed an exceptionally close agreement - in the range better than 1 percent accuracy for all physical quantities, even at the stage of free inertial expansion of supernova envelope.

For SN 1987A, were used the models of mass 16 $M_{\odot}$, radius 30-47 $R_{\odot}$ and explosion energy $(1-3)x10^{51}$ erg. The initial hydrostatic model was constructed with the use of special code for initial models (Nadyozhin and Razinkova, 1986) and was close to polytrope of index $n = 3.5$.

We present now the main results for the simplest case: LTE-ionization, opacity independent of frequency (but all the effects v/c are taken into account in transport of radiation), Compton scattering is treated as pure absorption. The run of the calculated light curve proves to be very close to results of Grasberg et al. (1987) and Utrobin (1989). In particular, the effective temperature reached a maximum value of $\sim 5 \cdot 10^5$ K for the moment of the shock outbreak at stellar surface.

The most important qualitative difference of these new calculations from the results obtained in equilibrium diffusion approximation is the formation of a dense peak (with density contrast of 30-100 times) in the outermost layers of a compact star. Such a peak has been discovered earlier also by Falk and Arnett (1978), Chevalier and Klein (1979) and for more extended models by Grasberg and Nadyozhin (1969). Contrary to Chevalier and Klein, the radiative acceleration of matter outside the peak proves in our calculations to be fairly high and a new hightemperature shock is therefore absent.

Fig.1 shows the structure of the outermost layer of SN 1987A envelope at the moment $\Delta t \approx 250$ s since the shock outbreak when the density peak looks most prominent (it takes ~ 4700 s for the shock wave to propagate from centre of the star to its surface in the model $R_\sigma = 47\ R_\odot$) . The layer containing the density peak has a mass of $\sim 2 \cdot 10^{-6}\ M_\odot$ and optical thickness of $\tau \approx 10$. This is in excellent agreement with an analytical estimate of parameters of the outermost layer, where the shock cumulation described by the self-similar solution has to be cut off (Imshennik and Nadyozhin, 1988 a,b). Thus, this calculation gives an example for a physically correct description of the structure of region where the shock wave cumulation is saturated.

Further calculations with other parameters of presupernova models and with more accurate treatment of the comptonization effect are expected to show how the density peak properties can vary.

The described method is now in extensive use for calculation of supernova outbursts and of gravitational collapse hydrodynamics.

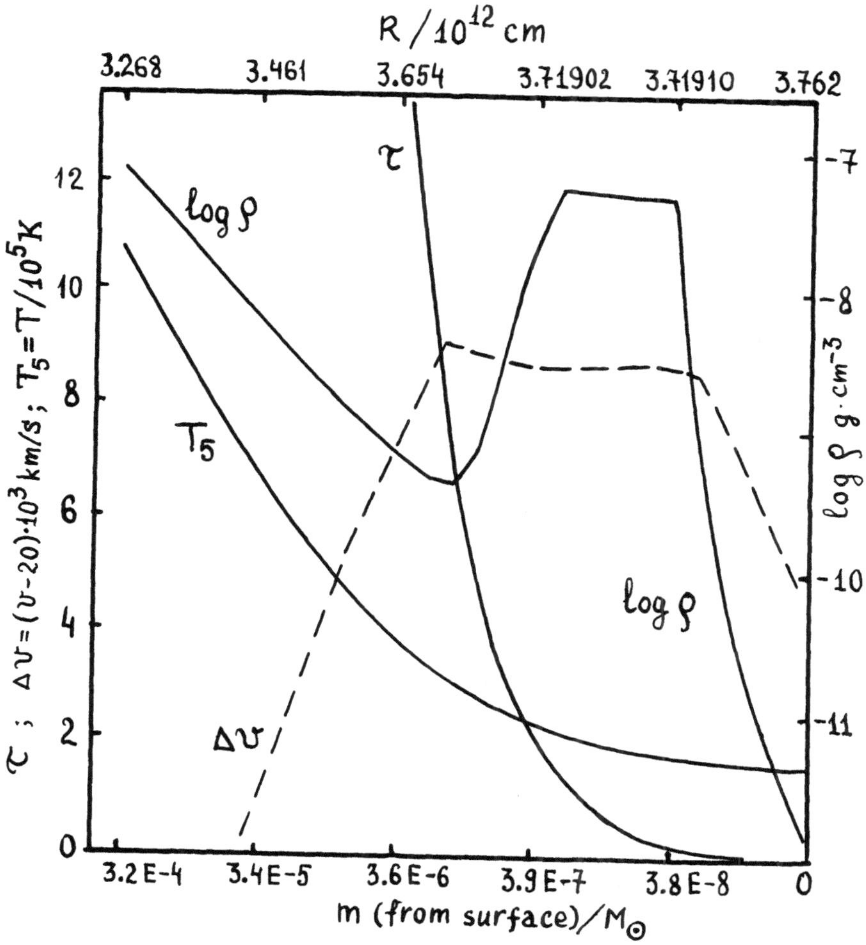

Fig.1. Distributions of density ρ , temperature T , velocity v and optical depth τ near the edge of SN 1987A envelope for the model $M = 16\ M_\odot$, $R_o = 47\ R_\odot$, $E_{expl} = 2 \cdot 10^{51}$ erg at time t = 4919 s (N_{step} = 2600) after explosive energy release in stellar centre.

We express sincere gratitude to O.S.Bartunov for his constant help in calculations.

REFERENCES

Arnett W.D. 1987, Ap.J. 319, 136.

Blinnikov S.I. 1988, Preprint ITEP No 19.

Blinnikov S.I., Rudzskiy M.A. 1984, Pis'ma Astron.Zh. 10, 363 (Translated in Sov.Astronomy Letters).

Bowers R.L., Wilson J.R. 1982, Ap.J.Suppl. 50, 115.

Bruenn S.W. 1985, Ap.J.Suppl. 58, 771.

Chevalier R.A., Klein R.I. 1979, Ap.J. 234, 597.

Falk S.W., Arnett W.D. 1977, Ap.J.Suppl. 33, 515.

Feautrier P. 1964, CR, 252, 3189.

Grasberg E.K., Nadyozhin D.K. 1969, Astron.Zh. 46, 745 (Translated in Sov.Astronomy).

Grasberg E.K., Imshennik V.S., Nadyozhin D.K., Utrobin V.P. 1987, Pis'ma Astron.Zh. 13, 547 (Translated in Sov.Astronomy Letters).

Imshennik V.S., Nadyozhin D.K. 1983, Soviet Science Reviews, Ser.E: Astrophys. and Space Phys. 2, 75.

Imshennik V.S., Nadyozhin D.K. 1988a, Uspekhi Fiz.Nauk 156, 561 (Translated in Sov.Sci.Rev. E).

Imshennik V.S., Nadyozhin D.K. 1988b, Pis'ma Astron.Zh. 14, 1059 (Translated in Sov.Astronomy Letters).

Mihalas D., Mihalas B.W. 1984, Foundations of Radiation Hydrodynamics. Oxford University Press.

Myra E.S., Bludman S.A., Hoffman Y., Lichtenstadt I., Sack N., Van Riper K.A. 1987, Ap.J. 318, 744.

Nadyozhin D.K., Razinkova T.L. 1986, Nauchnye Inform. 61, 29.

Shigeyama T., Nomoto K., Hashimoto M., Sugimoto D. 1987, Nature, 328, 320.

Utrobin V.P. 1989, Pis'ma Astron.Zh. 15, 97 (Translated in Sov.Astronomy Letters).

Woosley S.E., Pinto P.A., Ensman L.1988,Ap.J. 324,466.

Explosion of a Supernova with a Red Giant Companion

E. Livne, Y. Tuchman, & J. C. Wheeler

The problem of the explosion of a supernova in a binary system and the effect on the binary companion have been explored semi-analytically (COLGATE [1], CHENG [2], SUTANTYO [3], WHEELER, LECAR, and McKEE [4], BURROWS and LATTIMER [5]) and numerically (FRYXELL and ARNETT [6], TAMM and FRYXELL [7]). Main sequence stars are found to be very resiliant because of their large central mass concentration. They are subject to some mass stripping by the effects of direct momentum transfer and ablation, but the effect is relatively small, of order 0.01 M. The momentum transfer to the companion is even smaller than that given by simple estimates because the blast wave tends to wrap around the star and provide a back pressure neglected in the semi-analytic estimates but computed in two-dimensional simulations [6,7]. Wheeler *et al.* included estimates of the impact of a supernova on a red giant star as well, and concluded that the loosely bound envelope would be stripped from the core. Fryxell and Arnett and Taam and Fryxell only computed main sequence companions in their 2-D simulations. Because the numerical simulations showed features that the simpler estimates did not, a two-dimensional computation of a supernova exploding next to a red giant companion was computed with a 2-D interface-tracking code developed by Livne.

The calculation consisted of impacting a plane wave of ejecta corresponding to the explosion of a "Type Ia" supernova on a red giant star of total mass 2.1 $M_\odot$, of which 1.4 $M_\odot$ is in the envelope, and 0.7 $M_\odot$ is in the core with an outer radius of 3.4×10^{13} cm. The supernova ejecta are modeled with 0.25 $M_\odot$ moving at 9×10^{8} cm/s. The red giant envelope is strongly compacted by the blast wave, but rather than being merely stripped as the simple estimates give, the envelope rebounds strongly due to the compressional effects. This gives a strong backwards and tangential flow in addition to the flow induced by brute momentum conservation in the forward direction.

Figure 1 shows the pattern of the subsequent outflow 2.48×10^5 seconds after the impact. Note that while there is the strong flow expected in the direction of the original velocity of the ejecta (left to right) there is also a considerable flow in the tangential and backward directions. The latter is partially "ablated" mass but is

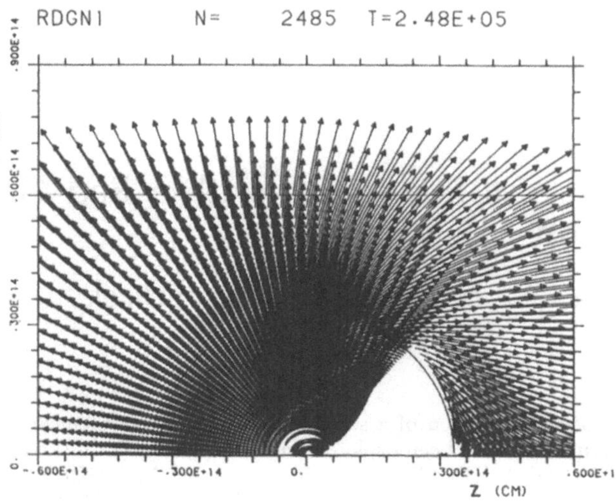

Figure 1. The velocity field 2.48×10^5 seconds after the impact of the ejecta of the supernova on the red giant envelope is shown. The length of the arrows is proportional to the velocity of the matter. The original red giant envelope was a hemisphere on this plot with a radius of 3.4×10^{13} centimeters containing a mass of 1.4 $M_\odot$ surrounding a core of 0.7 $M_\odot$.

mostly the result of the compression and rebound, as is the tangential flow that is a feature that is not treated at all in the simple estimates.

Figure 2 shows the density contours at the same epoch. The heavy line demarks the dividing line between the ejecta of the supernova (beyond the boundary) and the matter which has rebounded from the giant's envelope (within the boundary). Some of the hashed matter at 1, 2, and 3 o'clock is a mixture of supernova ejecta and envelope matter which can not be portrayed easily in this single color plot.

Type Ia supernovae (SN Ia) do not display any evidence for hydrogen in their spectra. It is very important to establish the upper limit to the abundance of hydrogen implied by this non-detection. The current limit is not very stringent. Models in which 0.1 $M_\odot$ of hydrogen were added to successful models for SN Ia showed no evidence for hydrogen in the theoretical spectra corresponding to the epoch of maximum light (HARKNESS and WHEELER [8], WHEELER and HARKNESS [9]). In principle, evolutionary models for SN Ia containing as much as 0.1 $M_\odot$ of hydrogen could be rather different than standard models in which one strives to have "no hydrogen" in the explosion. The models that have been computed to set limits on the hydrogen were based on notions that if there were any hydrogen in the system it would be surrounding the exploding white dwarf, or perhaps surrounding the putative binary system thought to be involved in the evolution of SN Ia events. Thus in the calculations reported in Harkness and Wheeler and Wheeler and Harkness, the hydrogen was added to the outer portion of the ejecta, or mixed with the outermost

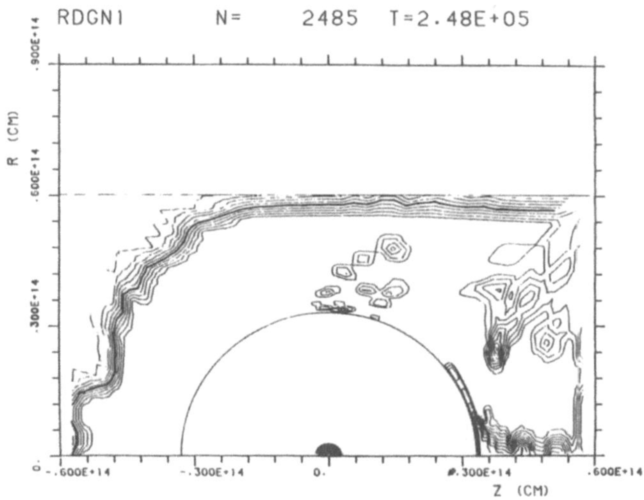

Figure 2. The density profile corresponding to the explosion of Figure 1 is given. The radius of the original red giant envelope is shown by the semicircle. The heavy line through the ejecta on the left and top is the dividing line between the supernova ejecta beyond this line, and the hydrogen-rich envelope within the line. Note that the hydrogen lags behind the heavy element rich supernova ejecta.

heavy element-rich ejecta. Other speculations concerning the possible role of hydrogen ejected from a companion have been based on the notion that the stripped matter would be ejected predominantly in the direction of the blast wave and probably in a rather narrow solid angle [5] raising the unpleasant likelihood of aspect effects and difficult, angular dependent, radiative transfer problems to address the issue.

The importance of the current calculation in this regard is the predicted location and kinematics of the hydrogen ejected from the companion's envelope. The current calculation suggests that it may be that the hydrogen is in a rather unexpected place, not in the outer, high velocity portions of the ejecta, but lagging behind in the inner, low velocity ejecta. The reason for this is clear from Figures 1 and 2. In the forward direction, the blast wave moves rapidly past the star, wrapping around it and tending to precede the envelope ejecta that follows in its wake. In the backward direction, the supernova ejecta compress the envelope and then rebound, again preceding the later rebounding, more slowly moving hydrogen layers. Moreover, the ejected hydrogen comes off more or less uniformly filling 4 π steradians, so that a spherically symmetric radiative transfer calculation of such configuration might not be such a bad approximation. Clearly appropriate calculations should be attempted to explore the degree to which SN Ia could be contaminated with hydrogen contained in the innermost, more slowly moving portions of the ejecta, with $v \sim 2 \times 10^8$ cm/s. One suspects there could be a great deal.

This work was supported in part by NSF Grant 8717166. The calculations were done on the Cray X-MP of the University of Texas System Center for High Performance Computing.

REFERENCES

1. Colgate, S. A. 1970, *Nature*, **225**, 247.

2. Cheng, A. 1974, *Ap. Space Sci.*, **31**, 49.

3. Sutantyo, W. 1974, *Astr. Ap.*, **31**, 339.

4. Wheeler, J. C., Lecar, M., and McKee, C. F. 1975, *Ap. J.*, **200**, 145.

5. Applegate, J. H., and Terman, J. L. 1989, *Ap. J.*, **340**, 380.

6. Fryxell, B. A., and Arnett, W. D. 1981, *Ap. J.*, **243**, 994.

7. Taam, R. E., and Fryxell, B. A. 1984, *Ap. J.*, **279**, 166.

8. Harkness, R. P., and Wheeler, J. C. 1989, "Classification of Supernovae," in *Supernovae*, ed. A. Petschek (New York: Springer-Verlag), in press.

9. Wheeler, J. C., and Harkness, R. P. 1990, *Reports on Prog. in Phys.*, in press.

The Stability of Point Explosion Models

M. Den, H. Hanami, Y. Yamada, & T. Nakamura

Introduction

Observations of SN1987A strongly suggest that mixing has occured in the star and that the shock propagating system within inhomogeneous medium is very unstable. Several authors are working on this probelm and whether the instability occurs or not [1] ~ [3] and where is the most unstable are clarified[4], the exact growth rate of such instability is not still known. Then using the class of Sedov solutions which are analytical solutions of point explosion model, we study the problem of a strong shock in a medium witha power-law density profile and obtain the growth rate, the characteristic mode and the behavior of the perturbation. Although real star has not power-law density distribution, it is expected that the analysis here can be applized to local region of the star.

As for the previous works, Vishniac et. al.[5] presented the dispersion relation of linear perturbations to one of the Sedov solutions but their initial density profile is limitted to homogeneous. Chevalier[2] and Bandiera[3] specified the unstable solutions in the Sedov solutions with the initial power-law density distribution, but the exact growth rate was not given. So it is worthwhile obtaining the dispersion relation, and it is expected that this result will help to understand the mixing problem of SN1987A and other problems such as star formation theory around OB association.

Sedov Solution and Formulation of Perturbation

As was denoted in Refs.[6] and [7], there are some variations in topology of the solutions, which is classified by an initial density power index ω ,i.e., $\rho_i \propto r^{-\omega}$. Figures 1a and 1b are density, pressure and velocity profile for several ω respectively. There are some critical values of ω. In the case of $3/\gamma < \omega < (7-\gamma)/(\gamma+1)$ and $\omega > 6/(\gamma+1)(\equiv \omega_c)$, the density has negative gradient. Noting that the pressure gradient is always posituve, these solutions are unstable against Rayleigh-Taylor instability. Especially we pay attention to the case of $\omega > \omega_c(\sim 2.57$ when $\gamma = 4/3)$, because this case is globally unstable and the real star is classified in it. Figure 2 shows the density profile of real star model and polytrope type $N = 3$. Nomoto's model is accidentaly fitted to $r^{-2.6}$ very well. This means that real star is globally unstable although Ebiszaki et. al. [4] suggest that the boundary of the chemical component is the most unstable. On the other hand, polytrope type is almost stable except the outer region. Then the interesting case is the larger $\omega(> \omega_c)$, however, at present we can only show the result for the smaller ω because of some problems concerned with regularity of solutions.

Now let us define the normalized perturbations. The perturbations are expanded by the spherical harmonics. Their time dependence is assumed to be fatorlized out

Figure Captions

1a. The density profile of Sedov solution for several values of ω.

1b. The pressure and velocity profile of Sedov solution for several values of ω.

2 . The density distribution of the real star model and polytrope type $N = 3$.

3 . The dispersion relation in the case of $\omega = 1$ and 2.1.

Fig. 1a

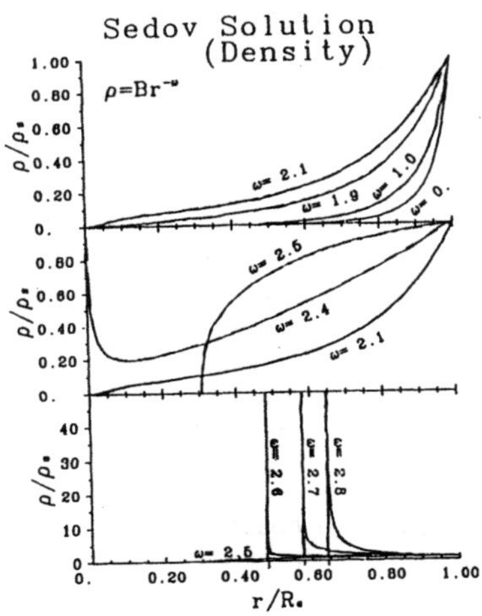

Fig. 1b

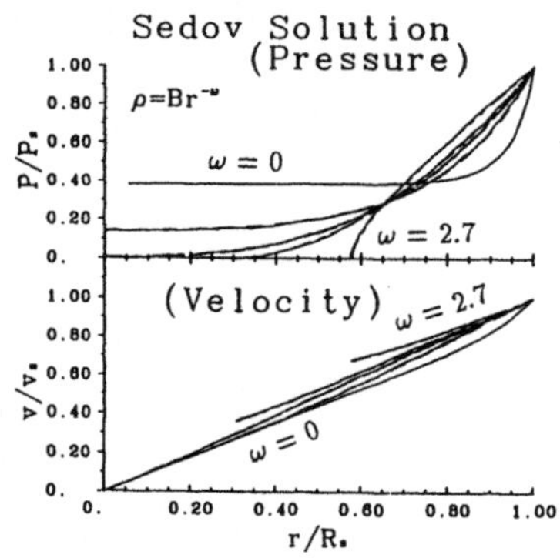

Fig. 2

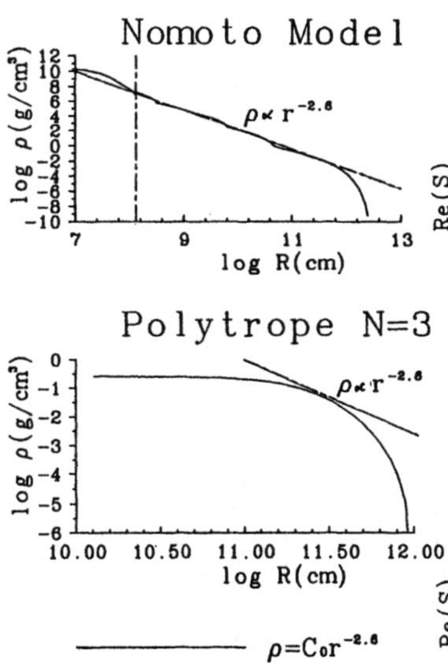

Fig. 3

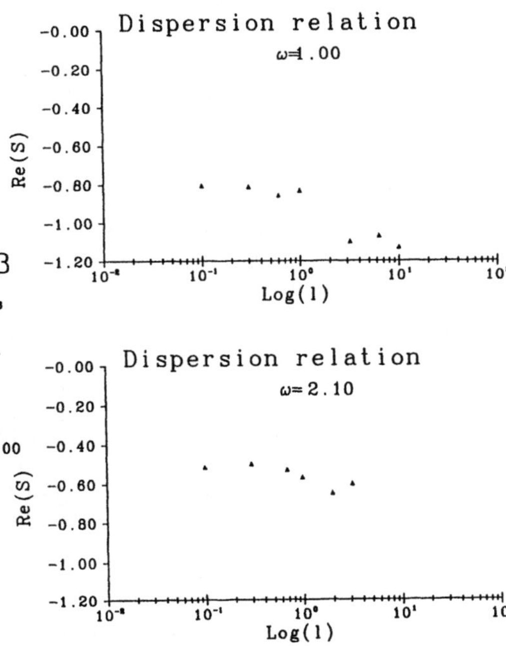

and to be power-law because of no characteristic time scale in this model. That is to say,

$$\delta Q(\mathbf{x},t) = \delta\tilde{Q}(\xi)t^s Y_{lm}(\theta,\phi)$$

where $\delta Q(\mathbf{x},t)$ means perturbed quantities and ξ is radial coordinate normalized by the distanse from the origin to the shock front. So we will obtain power index s as eigen values, by solving the linearlized perturbation equations. The perturbed continuity equation is

$$(\tilde{v} - \frac{\gamma+1}{2}\xi)\frac{d}{d\xi}\delta\tilde{\rho} + [\frac{\gamma+1}{4}(5-\omega)s + \frac{d\tilde{v}}{d\xi} + (2-\omega)\frac{\tilde{v}}{\xi}]\delta\tilde{\rho}$$
$$+\tilde{\rho}\frac{d}{d\xi}\delta\tilde{v}_r + [\frac{d\tilde{\rho}}{d\xi} + \frac{1}{\xi}(2-\omega)\tilde{\rho}]\delta\tilde{v}_r - \frac{l(l+1)}{\xi}\tilde{\rho}\delta\tilde{v}_T = 0$$

where $\tilde{v}$ and $\tilde{\rho}$ are normalized velocity and density by the quatities on the shock front respectively, $\delta\tilde{v}$ and $\delta\tilde{\rho}$ are their linear perturbations and subscripts r and T are ξ - and its transverse direction respectively (in detail, see Ref.[5]). We do not describe the equation of motions and the adiabatic condition here because of saving space. We solve this eigen value problem by adopting shooting method. Namely we impose the appropriate boundary conditions on the origin and the shock front and solve the perturbation equations from one side numerically. Figure 3 shows the results. We considerd two cases, $\omega = 1$ and 2.1. For both cases, eigen value s is negative for all wave numbers l, so these solutions are stable. These results are consistent with Rayleigh-Taylor stability criterion. It is noticed that there are four independent modes for s because the system is composed of four the 1st differential equations. We presented here only one of them, so it can be possible that these solutions are unstable when other modes are given. But the result obtained here is reasonable in view of the criterion.

Summary

We investigated the stability of the point explosion model and obtained the dispersion relation in the case of $\omega = 1$ and 2.1. It is found that the both solutions are stable. This result is consistent with Rayleigh-Taylor stability criterion. So it can be found that the shock propagating system in a medium with a power-law density distribution whose power is smaller than 2.1 is stable. From this analysis it can be explained that polytrope type $N = 3$ is almost stable and the real star model is unstable.

For future work, we will study the stability of the solution with a larger power index, $\omega > \omega_c$. It is expected that this work will help to understand the mixing problem of SN1987A.

References

1. M. Nagasawa, T. Nakamura and S. Miyama: *Publ. Astron. Soc. Japan* **40**, 691 (1988).
2. E. Müller, W. Hillebrandt, M. Orio, P. Höflich and R. Mönchmeyer: submitted to *Astron. Astrophys.* (1989).
3. Y. Yamada and T. Nakamura: private communication.
4. T. Ebiszaki, T. Shigeyama and K. Nomoto: *Astrophys. J. Lett.*, in press.
5. D. Ryu and E. Vishniac: *Astrophys. J.* **313**, 820 (1987).
6. R. A. Chevalier: *Astrophys. J.* **207**, 872 (1976).
7. R. Bandiera: *Astron. Astrophys.* **139**, 368 (1984).

On the Nature of Hydrodynamic Instabilities in Supernovae

Rino Bandiera

1. Introduction

Can supernova explosions be adequately described by spherically symmetric models? Before SN 1987A the usual answer to this question was "Yes". However it was known that, in supernova models, there are regions in which the Rayleigh-Taylor instability criterion, based on a density inversion with respect to the direction of the effective gravity $\vec{g} = \nabla p/\rho$, is satisfied (CHEVALIER [1]). Such instability was expected to produce either clumps (CHEVALIER and KLEIN [2]) or turbulence (FALK and ARNETT [3]), and therefore it was argued that its onset could change to some extent the conditions in the ejecta, compared to theoretical predictions based on one-dimensional models.

But, strictly speaking, Rayleigh-Taylor instability postulates fluid incompressibility, while the supernova ejecta consist of compressible gas. Expressions for the dispersion relation of "compressible Rayleigh-Taylor" have been presented (e.g. BLAKE [4]): they are usually derived from the general dispersion relation given by CHANDRASEKHAR [5], by using a steady solution appropriate for a compressible gas. However even in the general case incompressible perturbations, namely with $\nabla \cdot v = 0$, are assumed: therefore those dispersion relations do not properly describe the behaviour of a compressible fluid.

Alternatively, one may wonder what are the conditions in which a normal gas behaves as an incompressible fluid. This may happen when the instability time scale $(1/\nu)$ is smaller than the sound crossing time $(1/kc)$: here c is the sound speed, while ν and k are the instability growth rate and wavenumber, respectively. For standard Rayleigh-Taylor $\nu \sim \sqrt{gk} \sim kc\sqrt{|\nabla p/p|/k}$: the condition for incompressibility then becomes $k \ll |\nabla p/p|$. But, being ν larger at larger k, for the most linearly unstable modes the assumption of incompressibility is not justified.

Furthermore, a fluid looks incompressible even when there is nothing compressing it: this happens if the pressure is almost uniform (anyway $\nabla p \neq 0$ is needed in order to trigger the instability). Since Rayleigh-Taylor instability appears at wavelengths larger than the density scale height, this regime of incompressibility requires $|\nabla \rho/\rho| \gg k \gg |\nabla p/p|$, and therefore $|\nabla \rho/\rho| \gg |\nabla p/p|$.

2. Convective instabilities

A linear study of hydrodynamic instabilities with scales smaller than density and pressure scale heights has been already presented (BANDIERA [6]), in which the effects of these instabilities on the evolution of supernovae are also investigated. By taking the hydrodynamic equations for a perfect gas with adiabatic index γ, and assuming a static zeroth order solution in the comoving reference frame, one can derive the following dispersion relation:

$$\nu^4/c^4 + k^2\nu^2/c^2 - \mathcal{P}^2 k^2 \sin^2\theta\,(1-\gamma\mathcal{R}/\mathcal{P})\,/\gamma^2 = 0, \tag{1}$$

where $\mathcal{R} = \nabla\rho/\rho$, $\mathcal{P} = \nabla p/p$, $c^2 = \gamma p/\rho$, while θ is the angle between k and $\mathcal{P}$. The dispersion relation can be split in two modes; one corresponding to normal sound waves, while the other with the following dispersion relation:

$$\nu = \pm\mathcal{P}c\sin\theta\sqrt{1-\gamma\mathcal{R}/\mathcal{P}}/\gamma. \tag{2}$$

The latter mode can grow with time only when $(1-\gamma\mathcal{R}/\mathcal{P}) > 0$ (Schwarzschild Criterion), and is most unstable when $\theta \simeq \pi/2$ (vertical shear motions): for this reason this instability can be classified as "linear convection". From a linear analysis, convection looks more unstable than Rayleigh-Taylor. In fact the former instability criterion ($\mathcal{R}/\mathcal{P} < 1/\gamma$, or entropy inversion), is more easily fulfilled than the latter ($\mathcal{R}/\mathcal{P} < 0$, or density inversion). Moreover the former growth rate ($\nu \sim \mathcal{P}c\sqrt{1-\gamma\mathcal{R}/\mathcal{P}}$) is larger than the latter ($\nu \sim \mathcal{P}c\sqrt{k/\mathcal{P}}$, with $k \ll |\mathcal{R}|$).

3. Convection in supernovae

Let me review now the basic results of BANDIERA [6], in which the effects of instabilities during an explosion in a power-law density profile $\rho \propto r^{-\omega}$ (SEDOV [7]) have been considered. As a first result, the Sedov solutions turn to be unstable for $2.25 < \omega < 3.18$: for comparison, according to the Rayleigh-Taylor criterion the unstable solutions are restricted to $2.53 < \omega < 3.10$ (CHEVALIER [1]). A saturation factor has been computed as $F = \int \nu\,dt$, where the integral is evaluated by following a given fluid element; in this way it has been found that for $\omega \sim 3$ a large mass fraction reaches high saturation factors. From an inspection of standard models of SN progenitors (WEAVER *et al.* [8]) it turns out that, while in the inner region a profile with $\omega \simeq 3$ is well reproduced, a flatter profile is expected in the outer regions.

An estimate of the effectiveness of convection has also been obtained by defining a convective flux as $F_c = Q\gamma Pc/(\gamma-1)$, where Q is an unknown factor. By using diffusive approximation, one can write:

$$\frac{P}{\gamma-1}\left(\frac{1}{p}\frac{dp}{dt} - \frac{\gamma}{\rho}\frac{d\rho}{dt}\right) + \nabla\cdot F_c = 0. \tag{3}$$

A convective factor Q of order unity is usually required in order to have in (3) a diffusive term of the same magnitudine as the others; even smaller Q are required, for $\omega \sim 3$, in the inner regions.

A formal approach to highly saturated, non-stationary convection represents a very difficult task; for this reason I used instead a very crude approach, namely that of mixing length. If l is the mixing length, the convective velocity and flux turn out to be, respectively:

$$V_c = \frac{c}{2}\frac{l\mathcal{P}}{\gamma}\sqrt{1-\gamma\mathcal{P}/\mathcal{R}};\quad F_c = \frac{1}{2}\frac{\gamma pc}{\gamma-1}\left(\frac{l\mathcal{P}}{\gamma}\right)^2(1-\gamma\mathcal{R}/\mathcal{P})^{3/2}. \tag{4}$$

The convection factor is therefore:

$$Q = \frac{1}{2}\left(\frac{l\mathcal{P}}{\gamma}\right)^2(1-\gamma\mathcal{R}/\mathcal{P})^{3/2} = 2\left(\frac{V_c}{c}\right)^2\sqrt{1-\gamma\mathcal{R}/\mathcal{P}}. \tag{5}$$

If $V_c \simeq c$, a convection factor of order unity can be attained if $|\mathcal{R}| \sim |\mathcal{P}|$, and even larger if $|\mathcal{R}| \gg |\mathcal{P}|$; while the mixing length is $l \sim 1/\mathcal{P}$ if $|\mathcal{R}| \sim |\mathcal{P}|$, or $l \sim 1/\sqrt{-\mathcal{R}\mathcal{P}}$ if $|\mathcal{R}| \gg |\mathcal{P}|$. If the convection is dynamically important, its effect on the global structure, apart from the well known mixing, is that of levelling entropy gradients in the ejecta.

4. Conclusions

The original purpose of my work was to convince modellists that, due to the onset of instabilities, one-dimensional spherically symmetric models can be inadequate for describing the details of the dynamics of supernovae. But after SN 1987A the bulk of observational evidence led to the common belief that strong dynamical instabilities occurred in that supernova.

For this reason my purpose has changed, and it is now to show that those instabilities are basically convection, rather than Rayleigh-Taylor. This point is of little relevance for multidimensional numerical codes (e.g. Arnett, Fryxell, Nagasawa, Thielemann, Yamada in this workshop), whose results come out independently of our belief. It is instead useful when one tries to understand the results of those numerical codes, as well as for the study of instabilities by using semianalytical models. This kind of models (e.g. Ebisuzaki, Nomoto, Shigeyama, Thielemann in this workshop), consists in applying an instability criterion to the solution of a one-dimensional code, in order to determine, in the linear regime, to what extent original perturbations can grow.

Speaking of "convection" instead of "Rayleigh-Taylor" is not just a fact of terminology. The important fact is that these instabilities are compressible and follow the Schwarzschild instability criterion. And the interesting fact is that the instability conditions are wider, and the growth rate is faster than expected for normal Rayleigh-Taylor instabilities.

5. References

1. R. A. Chevalier: *Astrophys. J.* **207**, 872 (1976).

2. R. A. Chevalier, R. I. Klein: *Astrophys. J.* **219**, 994 (1978).

3. S. W. Falk, W. D. Arnett: *Astrophys. J. Suppl.* **33**, 515 (1977).

4. G. M. Blake: *M. N. R. A. S.* **156**, 67 (1972).

5. S. Chandrasekhar: *Hydrodynamic and Hydromagnetic Stability* (Clarendon, Oxford, 1961).

6. R. Bandiera: *Astron. Astrophys.* **139**, 368.

7. L. I. Sedov: *Similarity and Dimensional Methods in Mechanics* (Academic, New York, 1959).

8. T. A. Weaver, G. B. Zimmerman, S. E. Woosley: *Astrophys. J.* **225**, 1021.

Three-Dimensional Simulation of Supernova Explosion and Growth of Instabilities

Y. Yamada & T. Nakamura

Introduction

From the observational results of supernova 1987A and its interpretation, mixing in the ejecta becomes more and more important. Some group have already performed two dimensional and three dimensional numerical simulation of supernova explosion and analyze the instability of the ejecta. Two dimensional mesh calculation is done by Dr. D. Arnett and Dr. B. Fryxell. They showed that instability occurs at near the density gap and mushroom like pattern is formed. Three dimensional mesh calculation is done by Dr. E. Müller et al. and our group. Dr. Müller et al. show that there is not enough time for evolving the instability to sufficiently large in the case of N=3 polytropic star. In our simulation, number of meshes is twice in each direction as large as Dr. Müller's. The results of N=3 star is essentially the same as them. SPH approach is done by Dr. M. Nagasawa et al. They showed that in the N=3 model Rayleigh-Taylor instability grows up to nonlinear stage and expanding shell fragments. Dr. W. Benz and Dr. F. Thielemann also performed three dimensional computation by SPH code.

2D Mesh	3D Mesh	SPH
Arnett and Fryxell	Müller et al. (74^3) mesh	Nagasawa et al.
	Ours (140^3 mesh)	Benz Thielemann

Table 1. Related works.

From the analytical study by Dr. T. Ebisuzaki et al. , density gap plays very important role in this problem. Two dimensional simulation by Dr. Arnett et al., also suggests that density gap is important. So three dimensional numerical simulations must be performed using the realistic star model. In the realistic star, density contrast is very large (about 10^{10}). So, I think mesh calculation is more suitable for this simulation than SPH.

It is very difficult problem that weather fine two-dimensional or coarse three dimensional calculation is better. But in the problem of instability, we have a chance to find different unstable mode in the three dimensional calculation that cannot show in the two dimensional simulation. So we choose coarse three dimensional simulation.

From our experiences of the simulation physics, 100 meshes in each direction is marginally sufficient. Today we show the results of $(140)^3$ meshes calculation (200 MB memory). We can use $(190)^3$ meshes using HITAC S820-80 and we are now trying to use $(220)^3 \sim 10^7$ meshes (effectively 700 MB memory).

Method of Computation and Results

We solve next formulae.

$$\frac{\partial \rho}{\partial t} + \nabla(\rho \mathbf{u}) = 0$$

$$\frac{\partial \rho \mathbf{u}}{\partial t} + \nabla(p + \rho u^2) = 0$$

$$\frac{\partial E}{\partial t} + \nabla(\mathbf{u}(E + p)) = 0$$

We use equidistance Cartesian coordinate and Eulerian mesh calculation. We use explicit finite difference scheme with Lebranc's advection treatment as transport term which is second order accurate monotonicity preserving scheme. Our computational domain is cubic encompasses an octant of the initial star, applying symmetric condition on the respective boundaries.

We use two initial model, N=3 Polytropic star and realistic star model computed by Dr. K. Nomoto. We add the explosion energy as thermal energy and its value is 10^{51}ergs.

Our code passes many check. Energy and mass conservation is exact *i.e.*the same as numerical accuracy of machine. Spherical symmetric point explosion in the homogeneous medium can reproduce the Sedov's analytical solution. Other tests we have checked is shock tube problem, movement of box in each direction, and some analytical solutions.

The results of Polytropic Star is as follows. If we do not impose initial fluctuation, there is no evidence for growth of instability. At the stage when the shock reaches outer boundary of the star, magnitude of the density fluctuation is about the same as the initial seed.

Next, we turn to the model of the realistic star. Star evolution person always uses $\rho - M_r$ graph for displaying the density distribution. But we are now treating with this mode using equidistance cartesian coordinate. So we must treat such a steep density profile. As a first step, we compute not whole of the star but a part of the star near the gap of chemical composition. Our computational boundary is twice as large as density gap and we stopped our computation, when shock reached the numerical boundary. In such a computation, true effects of reflected shock cannot be included. But we can see the growth of instability in the contact of chemical composition. We show the example of the H-He contact.

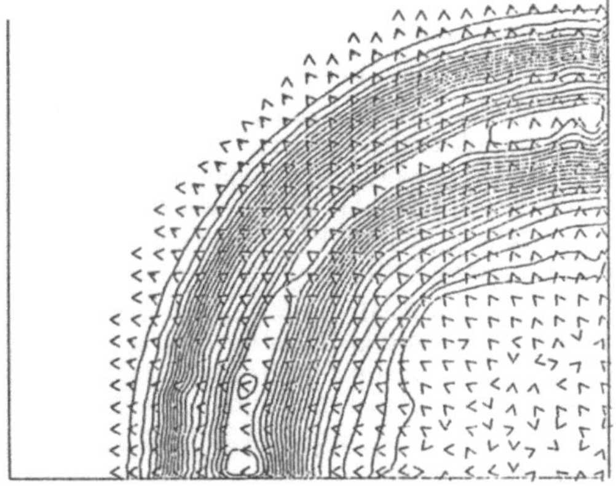

Figure 1. Density Contour and Velocity of N=3 model. Each contour is 5 % step, initial seed is 10 % in density at 2221 sec after the explosion.

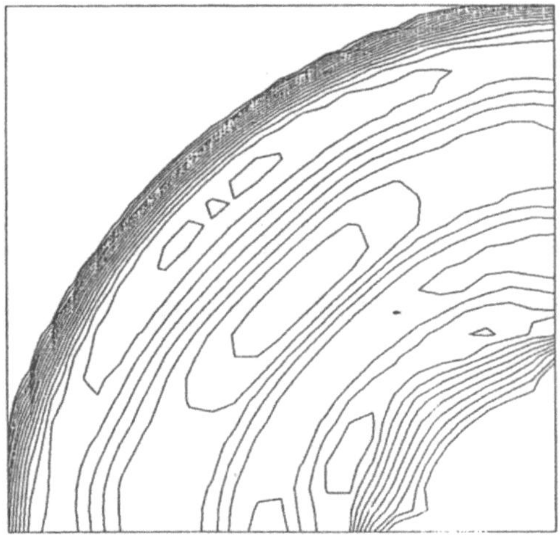

Figure 2. Density Contour of Nomoto model at 12.32 sec after the explosion. Each contour is 12% step. Each sides of square is 10^{11} cm and initial radius of density gap of H-He layer is 5×10^{10} cm.

Discussion and Future Problem

Our results can be summarized as follows.

1. In the case of N=3 Polytropic star, the magnitude of density fluctuation is about the same as the initial seed. There is no sufficient time for growing instability up to nonlinear.
2. In the case of realistic star, instability occurs clearly. We have not been checked what kind of the instability is dominate. But density gap or gradient of density would play the important role of this instability.

As our future problem we must clarify the method of the treatment of unstable problem. Dr. Miyama show the perturbation of analytical solution which can be solved by ordinary differential equation. This is a good check problem and now we are analyzing this problem in our code. Another problem is the treatment of realistic star. Realistic star has high density contrast and particle method cannot be realize such a high contrast. But in such condition, almost all of the mass concentrates in the central small region and whole of the distribution cannot be realize by equidistance Eulerian coordinate. So it is important to construct moving mesh code for this problem.

Instabilities and Mixing in SN 1987A

David Arnett, Bruce Fryxell, & Ewald Muller

1. Introduction

There appears to be increasing evidence that the observations of Supernova 1987A can not be explained unless substantial nonradial motion and mixing occurred during the explosion. The earlier than expected detection of X-rays (eg. [1-3]) and γ-rays [4] can most easily be explained if the radioactive ^{56}Co had been mixed from the interior regions of the star into the envelope where the optical depth was much smaller. Moreover, the width of infrared spectral lines of Fe II [5] and Ni II, Ar II, and Co II [6-8] indicate that these elements were mixed from the slower moving inner regions of the supernova into the faster moving outer layers. The smoothness of the light curve is also indirect evidence of the need for mixing to occur [9-11]. For additional observations evidence of instabilities and mixing, see elsewhere in this volume.

The idea that nonradial motion would occur in Type II supernovae was first discussed by FALK and ARNETT [12]. CHEVALIER [13] used the stability analysis of CHANDRASEKHAR [14] to show that in the idealized case of a blast wave propagating down a power law density gradient, Rayleigh-Taylor instabilities could develop for a range of power law indices. The criterion for the gas behind the shock front to become unstable was that the density and pressure gradients should be in opposite directions.

The recent observations of mixing in SN1987A has caused renewed theoretical interest in this problem. A number of numerical simulations have been performed recently in order to attempt to determine if Rayleigh-Taylor instabilities do occur in supernovae and, if they do, how much mixing would result. The first of these calculations was performed by NAGASAWA, NAKAMURA, and MIYAMA [15] using a three dimensional smooth particle hydrodynamics (SPH) code for the idealized case of a shock propagating through a polytropic density distribution. They found that a small amount of mixing occurred, but probably not enough to explain the observations. However, BENZ and THIELEMANN [16] repeated their calculations also using SPH and found that the development of the instability depended very sensitively on how the explosion was initiated. They concluded that the instability found by NAGASAWA, NAKAMURA, and MIYAMA [15] was probably a numerical artifact. The same conclusion was reached by MÜLLER et al. [17] who repeated the same calculations using several different finite difference codes in both two and three dimensions. The results for a polytropic density distribution showed no instability. However, a power law density distribution with a power law index in the range which was predicted to be unstable by CHEVALIER [13], did show signs of a Rayleigh-Taylor instability. Higher resolution two dimensional calculations of ARNETT, FRYXELL, and MÜLLER [18] using a realistic density distribution for a Type II supernova progenitor found a very strong Rayleigh-Taylor instability, which causes substantial mixing on a time scale of a few hours after the explosion.

We report here on an even higher resolution calculation which shows a well resolved Rayleigh-Taylor instability using a realistic supernova progenitor. In the following section, the numerical techniques are described briefly and the method used for initiating the explosion is discussed. In Section 3, the dependence of the nature of the instability on the resolution of the numerical grid and the seed perturbation is illustrated. The amount of mixing produced by the instability is shown in Section 4. The final section discusses additional effects which must be included before the results can be compared with observations and plans for future calculations.

2. NumericalTechniques

The calculations described below were carried out using the PROMETHEUS computer code [19]. The code solves Euler's equations of hydrodynamics using the piecewise-parabolic method (PPM) of COLELLA and WOODWARD [20], which is ideally suited to problems of this type. The method was extended to include an arbitrary number of separate fluids which are used to keep track of the amount of mixing of various nuclear species. Ten different elements were included in the calculations presented here. In addition, a nuclear reaction network was included in the code (which is irrelevant on the timescales considered here), and modifications were made to handle a non gamma-law equation of state [21]. The code uses sophisticated techniques to obtain very high resolution and resolve fine structure. Artificial viscosity is not required to stabilize shocks, and as a result, shocks and composition discontinuities are spread over only one or two zones. Resolving composition discontinuities is a special problem for Eulerian methods. Shocks have a self-steepening mechanism which keeps them from spreading more than a few zones, even with very diffusive methods. On the other hand, composition discontinuities continue to spread diffusively without limit in most Eulerian codes. PPM solves this problem by including a special algorithm which detects regions in the flow containing composition (and density) discontinuities and keeps them as sharp as possible.

Figure 1 shows a comparison of the amount of diffusive spreading of such discontinuities for PPM and several other widely used hydrodynamic methods. The curves show the width (in zones) of the discontinuity versus the number of zones it has moved across the grid. The results for PPM (the bottom curve) show that the discontinuity spreads quickly over two zones, but no additional spreading takes place. For the most diffusive methods, the discontinuity can be spread over more than 60 zones on the size grids needed to resolve the instability. This difference in resolution (more than a factor of 30 in each dimension) is very important for these calculations, since the amount of numerical mixing in the most diffusive methods will probably dominate the real physical mixing. We should also note that the amount of computer time needed to calculate a two dimensional flow is proportional to the cube of the number of zones needed in each dimension, while for a three dimensional calculation, the time required is proportional to the fourth power. Since these calculations (especially in three dimensions) are near the limit of what can be done on current supercomputers, it is clear that using a method which is even a factor of two lower in resolution can have disastrous effects.

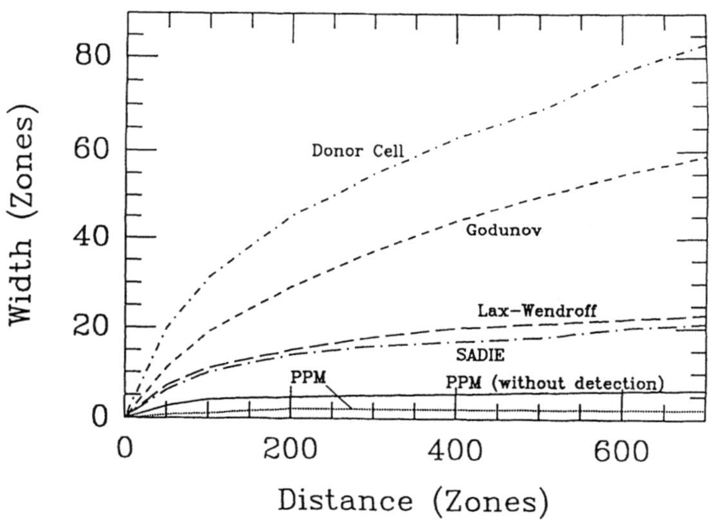

Figure 1. Width of a composition discontinuity vs. the number of zones it has moved for several different numerical methods.

The calculations presented here were carried out on a two dimensional rectangular grid with cylindrical symmetry about the vertical axis and equatorial symmetry about the horizontal axis. Grid sizes ranging from 50x50 to 500x500 were tried to show the effects of increasing the resolution. The initial model was taken to be a 15 $M_{\odot}$ star which was evolved using a one dimensional Lagrangian stellar evolution code.

This model was then mapped onto a uniformly zoned Eulerian grid using 10,000 zones. The explosion was initiated by instantly depositing a mixture of internal and kinetic energy into the inner few zones. The propagation of the shock through the inner portion of the model was then followed using a one dimensional version of PROMETHEUS. This enabled us to obtain a very accurate representation of the flow behind the shock. In particular, a large spike in density is produced at the trailing edge of the mass shell. In order to get the peak density in this spike correct (even to within a factor of two), more than a thousand zones were required. Following the initial propagation with a one dimensional code has another important advantage. After the initial explosion energy is put into the calculation, the central temperature is extremely high. As a result, the timestep required for stability is very small. Evolving this stage with a multidimensional code would require enormous amounts of computer time, most of which would be wasted, since it takes some time before any multidimensional effects become important. Finally, a few hundred seconds after the explosion, the flow begins to become unstable and the one dimensional results are mapped onto the multidimensional grid. At this point, a random perturbation is added to the flow in order to get the instability started. The results are sensitive to what this perturbation is, as will be discussed in the next section. The further propagation of the shock and the development of the instability is then followed on the two dimensional grid using PROMETHEUS. The shock is allowed to propagate off the grid rather than expanding the grid to follow the expansion of the outer layers of the star. This allows us to keep more zones and therefore higher resolution in the region of the instability. This is one advantage of using an Eulerian code for this problem rather than a Lagrangian code such as SPH.

3. Results

In this section we discuss the how the instability is affected by grid resolution and the size of the seed perturbation. Grid resolutions ranging from 50x50 to 500x500 were tried. Figure 2 shows the density contours obtained on the 50x50 grid at approximately 2 hours after the explosion. The axes are labeled in units of the radius of the pre-explosion star (ie. 3×10^{12} cm). There is a hint of fingers developing in the flow due to a Rayleigh-Taylor instability but it is difficult to distinguish between physical and numerical effects. For example, the clump near the top left of the figure is a result of the boundary condition along the symmetry axis. The corresponding result on a grid of 100x100 zones is plotted in Fig. 3. There is now a clear indication of fingers, although they are clearly not well resolved. Substantial improvement is obtained by increasing the resolution to 200x200 (Fig. 4). The fingers which developed near the horizontal axis are beginning to show the characteristic shape of the nonlinear development of a Rayleigh-Taylor instability, namely a narrow finger topped

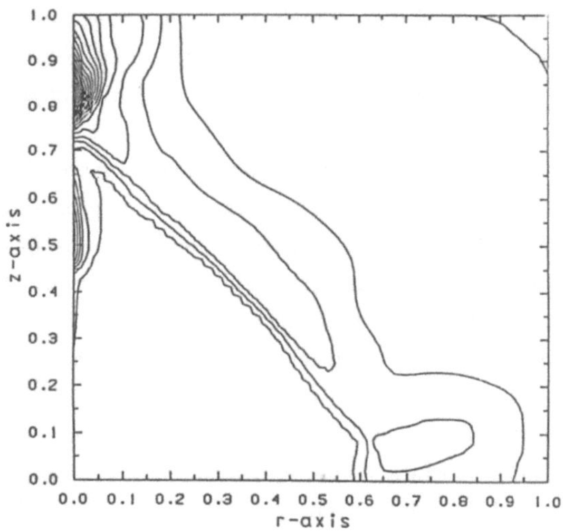

Figure 2. Density contours for the Rayleigh-Taylor instability after approximately 2.5 hr calculated on a 50x50 grid. The plot shows only one quadrant of the star. There is rotational symmetry about the vertical axis and equatorial symmetry about the horizontal axis. The numbers on each axis are in units of the initial stellar radius ($3. \times 10^{12}$ cm). The instability was started using a random seed perturbation with a ten per cent amplitude.

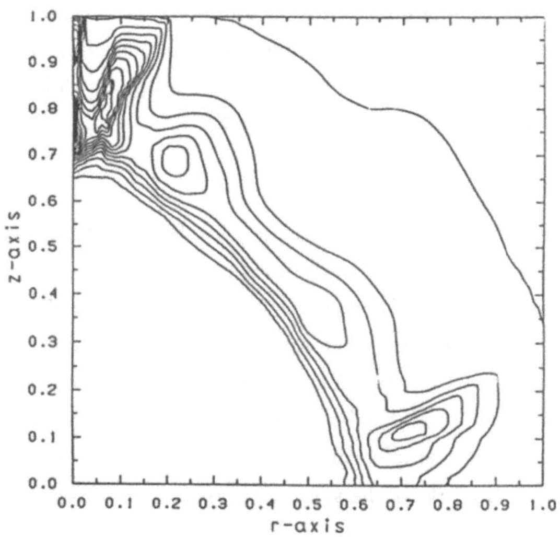

Figure 3. Same as Fig. 2 expect the results was obtained on a 100x100 grid.

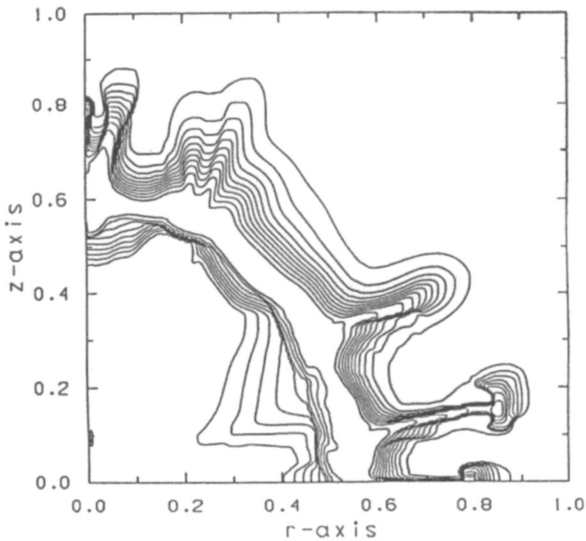

Figure 4. Same as Fig. 2 except the result was calculated on a 200x200 grid. At this resolution, the fingers near the horizontal axis are almost resolved.

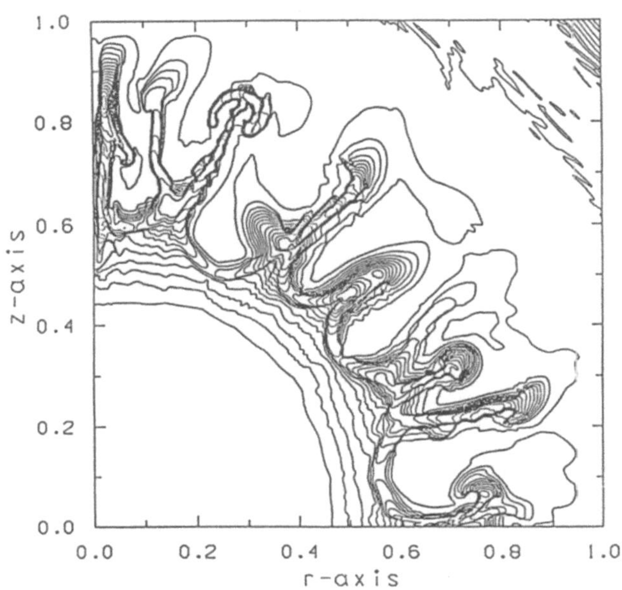

Figure 5. Same as Fig. 2 except the result was calculated on a 380x380 grid. At this resolution, all of the fingers, except for those near the diagonal, are beginning to be resolved.

by a "mushroom cap" which actually develops as a result of the Kelvin-Helmholtz instability. Although the fingers are beginning to be resolved in some parts of the flow, most of the fingers are still completely unresolved, especially along the diagonal where the effective resolution is the lowest. Increasing the resolution to 380x380 (Fig. 5) produces a nice set of fingers which are roughly uniformly spaced in angle. The nonlinear structure of the fingers along the diagonal is still not quite resolved although the flow appears to be approaching a converged solution with the spacing between the fingers remaining roughly constant as the resoltuion is increased. Finally, Fig. 6 shows the results obtained on a 500x500 grid. All of the fingers are now well resolved, most of them showing the characteristic "mushroom cap" shape.

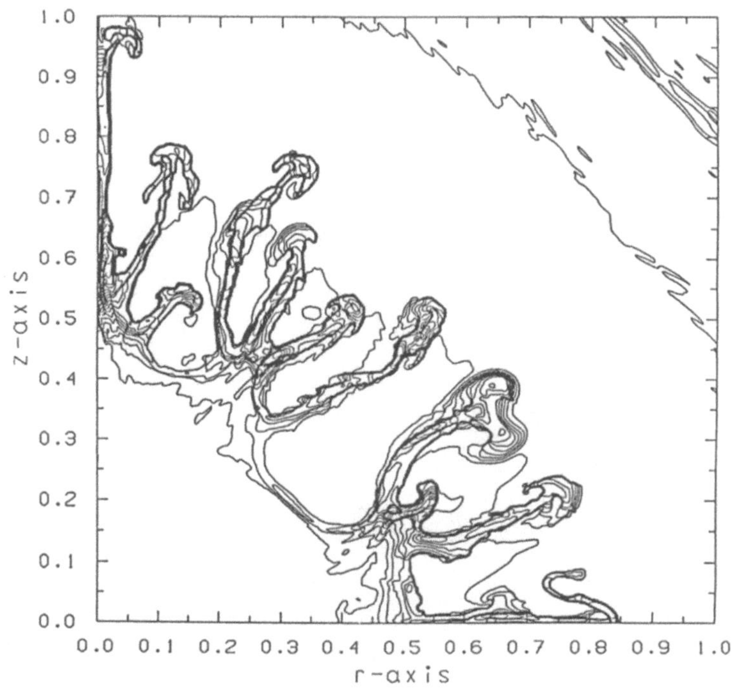

Figure 6. Same as Fig. 2 except the result was calculated on a 500x500 grid. All of the fingers are now fairly well resolved and show the typical mushroom cap shape typical of a Rayleigh-Taylor instability.

It appears that some of these caps are about to break off which would form dense knots of about $10^{-3} M_{\odot}$. One would expect that the fingers would also break up into smaller knots not long after this time. The density contrast between the densest parts of the fingers and the surrounding material can be up to a factor of 10 in some places. The column density integrated along radial lines from the surface of the star to the center as a function of angle is plotted in Fig. 7. The locations of the fingers are clearly visible. The variation in column density is roughly a factor of two to three for this model. Density variations (clumping) of this order would decrease the diffusion time for escaping gammas and thermal photons. Mass estimates from diffusion times would be less than the actual masses. Such an error could be significant for SNIb's.

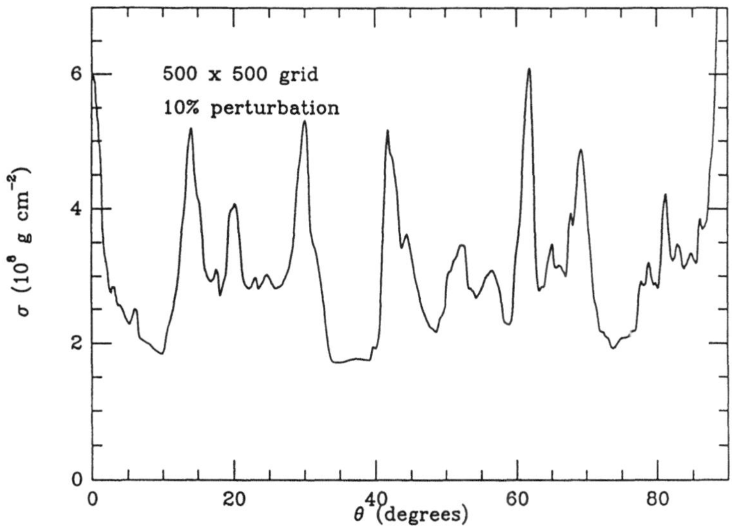

Figure 7. Column density as a function of angle for the results shown in Fig. 6.

The instabilities shown in the previous plots were initiated by inserting a 10 per cent random perturbation in the density, both components of velocity, and energy throughout the entire grid. When the size of the perturbation was reduced by a factor of 10, the results shown in Fig. 8 were obtained. As expected, the fingers are smaller in this case and not quite as well developed. For comparison, the column density for this model is plotted in Fig. 9, which shows a much smaller variation than Fig 7.

One can also produce an instability with any number of fingers by inserting a sinusoidal seed perturbation with the appropriate wavelength. Not knowing what the seed perturbation in SN 1987A was makes it difficult to calculate the correct instability to match the observations. However, it is hoped that be trying several different perturbations and comparing each with the observations, it may be possible to reverse the question and determine what type of perturbations actually existed at the time of the explosion. This may even be able to provide information on the explosion mechanism. In addition, SN1987A is becoming resolvable, so that direct evidence of the instability can be obtained.

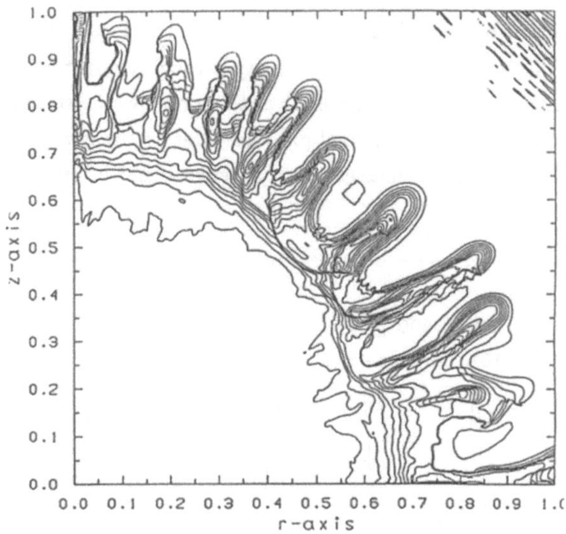

Figure 8. Evolution of the Rayleigh-Taylor instability started with a random perturbation of one per cent amplitude on a 500x500 grid. The qualitative nature of the results is the same, although the instability is somewhat less developed than the one obtained using a larger perturbation.

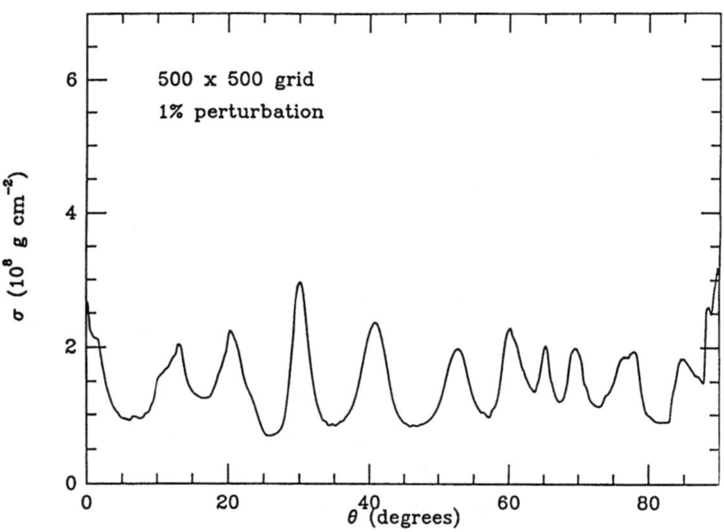

Figure 9. Column density as a function of angle for the results shown in Fig. 8.

4. Mixing

By computing the advection of several separate fluids in the calculation, each of which representing a different nuclear species, it is possible to compute the amount of mixing which occurs in each model. For example, contours of constant helium mass fraction are shown in Fig. 10. This figure corresponds to the density contours in Fig. 6. It is immediately obvious that the originally spherical helium shell has become extremely distorted. In fact, although it is difficult to tell from the contour plot, the shell is beginning to break up into various knots and fragments. The material outside of this shell, which is mostly hydrogen, has been pushed downward between the fingers into the interior of the star. Likewise, the heavy elements which were originally located interior to the helium shell, are moving outward within the fingers themselves. This is a macroscopic mixing process in which large fragments of material of different composition are mixed together rather than a microscopic mixing process in which composition is mixed together to form a a region of homogeneous composition.

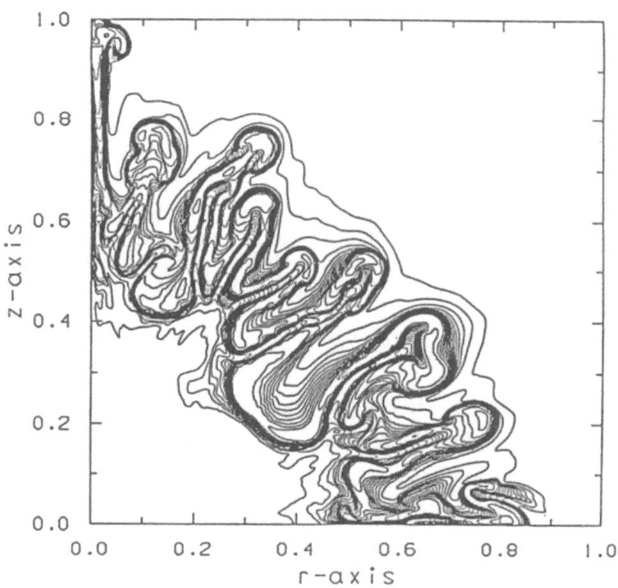

Figure 10. Contours of helium mass fraction corresponding to the density profile shown in Fig. 6. At this time, the helium shell is beginning to fragment.

Mixing in velocity space, which has been inferred from spectral line observations is illustrated in Figs. 11 and 12. The dashed line indicates the distribution of the element for a spherically symmetric model, while the solid line shows the distribution which results after the development of a Rayleigh- Taylor instability. The effect of mixing on the helium distribution is shown in Fig. 11. Most of the helium was originally moving at about 1500 km s^{-1}. When the effect of the instability is included, however, the distribution becomes almost uniform between 1500 km s^{-1} and 2500 km s^{-1}. The carbon distribution (Fig. 12), which is typical of the heavy elements, shows a similar effect.

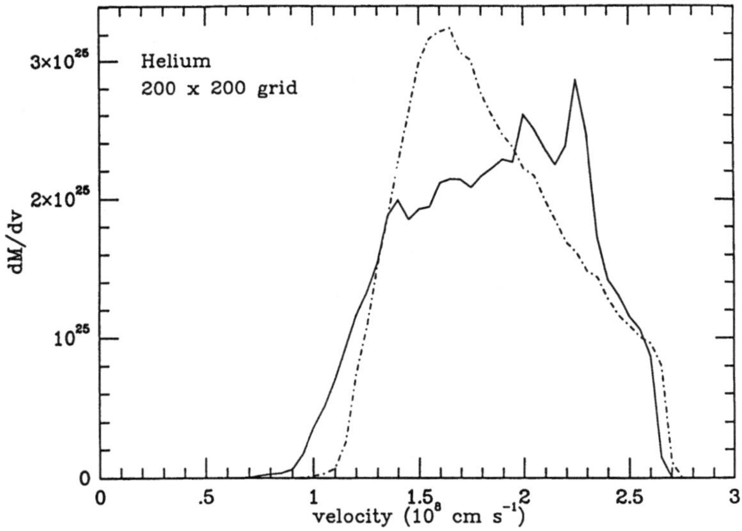

Figure 11. Mass distribution of helium vs velocity. The dashed line shows the distribution in a spherically symmetric model, while the solid line indicates the results obtained when the effects of the Rayleigh-Taylor instability are included. Much of the helium has been accelerated to a higher velocity by the instability.

5. Discussion

The amount of mixing discussed in the previous section may or may not be sufficient to explain the observations, but the results presented here only represent the amount of mixing one could expect at a very early time (a few hours after the explosion). The ^{56}Ni which is formed in the explosion undergoes radioactive decay on a timescale of about a week. This will have two significant effects on the mixing. First, a significant amount of energy will be released, some of which will go into kinetic energy. It does not seem unlikely that the material which was mixed out to a velocity of 2500 km s^{-1} could be accelerated to velocities of 3000 km s^{-1} or higher. Second, the decay will lead to a second Rayleigh-Taylor instability. Unfortunately the effects of the interaction of these two instabilities is impossible to predict. However, it is clear that this problem must be calculated before the results can be compared to the observations. One must also worry about whether the nature of the instability will be very different in three dimensions.

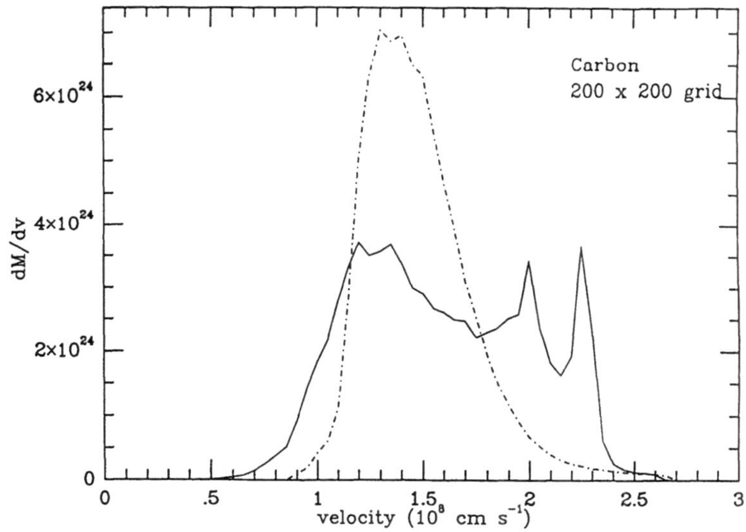

Figure 12. Same as Fig. 11 for carbon. This plot is also typical of the other heavy elements.

It is our intention to proceed as follows. First we intend to do a more thorough study of the effects of different seed perturbations, changing the wavelength, amplitude, and location of the perturbation. After finishing this, three dimensional calculations will be performed to compare the amount of mixing which results with the two dimensional calculations. The calculations will then be carried to much later times, first to investigate the continued nonlinear evolution of the original Rayleigh-Taylor instability, and finally to include the radioactive decay of ^{56}Ni .

References

1. Dotani, T. et al.: Nature 330, 230 (1987).

2. Sunyaev, R. et al.: Nature 330, 227 (1987).

3. Wilson, R. B. et al.: in Nucelar Spectroscopy of Astrophysical Sources, ed. by N. Gehrels and G. Share (AIP, New York 1988), p. 66.

4. Matz, S.M., Share, G.H., Leising, M.D., Chupp, E.L., Vestrand, W.T., Purcell, W.R., Strickman, M.S., Reppin, C.: Nature 331, 416 (1988).

5. Erickson, E.F., Haas, M.R., Colgan, S.W.J., Lord, S.D., Burton, M.G., Wolf, J., Hollebach, D.J., and Werner, M.: Ap. J. (Letters), 330, L39 (1988).

6. Rank, D.M. et al.: Nature 331, 505 (1988).

7. Witteborn, F., Bregman, J.D., Wooden, D.H., Pinto, P.A., Rank, D.M., Woosley, S.E., and Cohen, M.: Ap J. (Letters) 338, L9 (1989).

8. Barthelmy, S., Gehrels, N., Leventhal, M., MacCallum, C.J., Teegarden, B.J., Tueller, J.: IAU Circular 4764 (1989).

9. Arnett, W.D.: in Supernova 1987A in the Large Magellanic Cloud, ed. by M. Kafatos and A. Michalitsianos (Cambridge University Press, Cambridge 1988), p. 61.

10. Woosley, S.E.: in Supernova 1987A in the Large Magellanic Cloud, ed. by M. Kafatos and A. Michalitsianos (Cambridge University Press, Cambridge 1988), p. 289.

11. Arnett, W.D. and Fu, A.: Ap. J. 340, 396 (1989).

12. Falk, S.W. and Arnett, W.D.: Ap. J. (Letters), 180, L65 (1973).

13. Chevalier, R.: Ap. J. 207, 872 (1976).

14. Chandrasekhar, S.: Hydrodynamic and Hydromagnetic Stability (Clarendon Press, Oxford 1961).

15. Nagasawa, M., Nakamura, T., and Miyama, S.: Pub. Astr. Soc. Japan, 40, 691 (1989).

16. Benz, W. and Thielemann, F.-K.: in preperation (1989).

17. Müller, E., Hillebrandt, W., Orio, M., Höflich, P., Mönchmeyer, R., and Fryxell, B.A.: Astr. Ap., 220, 167 (1989).

18. Arnett, D., Fryxell, B., and Müller, E.: Ap. J. (Letters), 341, L63 (1989).

19. Fryxell, B., Müller, E., and Arnett, D.: in Numerical Methods in Astrophysics, ed. by P.R. Woodward (Academic, New York 1989), in press.

20. Colella, P., and Woodward, P.R.: J. Comp. Phys. 54, 174 (1984).

21. Colella, P., and Glaz, H. M.: J. Comp. Phys. 59, 264 (1985).

Mixing in Convective Supernova —Bubble Solution vs Blast Wave Solution

Mikio Nagasawa

I SEARCH FOR 3-D SOLUTION

In hydrodynamical studies of supernova explosions, there are several suggestions that expanding spherical shells are unstable against non-radial perturbations. The criterion for the convection in the stellar interior is called Brunt-Väisälä frequency,

$$\nu^2 = \frac{1}{\rho}\frac{dp}{dr}\left(\frac{1}{\rho}\frac{d\rho}{dr} - \frac{1}{\gamma p}\frac{dp}{dr}\right) = -gA \quad . \tag{1}$$

Generally speaking, this frequency depends on the direction between effective acceleration and the entropy gradient. The condition for Rayleigh–Taylor instability is included in the first term of Eq.(1) in the limit of incompressible fluid.

We can understand the hydrodynamical instability of type II supernovae as follows. First, the positive entropy gradient is expected in the shock passed region because the Mach number increases during the shock propagation in stellar envelope. Then the entropy gradient satisfy the convective condition behind the shock and the mixing motion of hot particles is triggered. In the nonlinear stage, convection of hot bubble causes the fragmentation of shocked shell.

Under the assumptions of spatial symmetry, some important non-axisymmetric freedom may have been omitted from the beginning. We try to study supernova explosions including non-axisymmetric modes in both initial and later evolutions. Adiabatic supernova explosions of polytropic stars are investigated by a three dimensional Smoothed Particle Hydrodynamics (Nagasawa et al. 1988). They are found to be unstable against convection. As a result, we find a porous density structure on the expanding shell.

Observationally, it seems true that non-spherical mixing motions of the clumpy ejecta in SN1987A exist. The broad line width of infrared heavy metal suggests the mixing of the ejected iron with lighter elements (Erickson et al. 1988). The interpretation of large linear polarizations suggests that the supernova ejecta itself is not spherical but prolate or oblate (Barret 1987). X-ray observation from SN1987A requires that the ^{56}Ni is mixed toward the surface due to convection or Rayleigh–Taylor instability (Itoh et al. 1987).

However, 2-D simulations of Type II supernovae in Euler scheme do not get the clumpy structure of polytropic explosions (Müller et al. 1989; Arnett et al. 1989). Is there an inconsistency of numerical simulations ? The apparent discrepancy arises from the set up of problem. The 3-D simulations have two aspects. One is studying the stability of 1-D solutions and the other is searching the 3-D solutions. The evolution of point explosions which has a singularity at the origin shows two types of solutions, even if we are restricted to search the almost spherical expansions in a global sense.

If we consider the infinite envelope with the power-law density structure $\rho \propto r^{-\omega}$, the Sedov solutions are the good standards of shock structure. The spherical shock condition decides the topology of shocked gas. For the steep gradient envelope such as $\omega > \omega_{crt} = (7-\gamma)/(\gamma+1) = 2.43$ for $\gamma=4/3$, the Sedov solution which we call *"bubble solution"* shows the shell structure and has a real singularity at inside discontinuity. While in the relatively uniform envelope, Sedov solution is the *"blast wave solution"* which has only one discontinuity, that is, shock front. For the polytrope $n = 3$ progenitor, the density slope changes and we can not decide apriori which types of shock will be triggered. The *"bubble solution"* can not satisfy the regularity condition at the origin so that the special treatment is required to simulate it with Euler scheme. On the other hand, SPH can easily handle this by setting the off-central energy deposition at the initial.

II DYNAMIC RANGE OF SPH

In order to check the validity of 3-D simulations in detail, we have to calculate the explosion of realistic progenitors besides the polytrope. However, there are some difficulties to overcome. In principle, SPH can represent infinite range of density contrast of ρ_{max}/ρ_{min} when we use the varying smoothing kernel $h(r)$. However, there is a reliable range of calculation for a given particle number N.

First, the low density region are represented by the overlap of large size particles. In 3-D representations of spherical structure, the particle size h have to be small as to represent the curvature radius R of the gas distribution. This condition tells us the required angular resolution on the polar coordinates, $f_1 \approx 4\pi R^2/\pi h_{max}^2$, where $h_{max} = R(N\rho_0/\rho_{min})^{1/3}$. In the stellar structure, for example, the outer diffuse envelope of ρ_{min} is difficult to be reproduced as an symmetric envelope unless $f_1 \gtrsim 400$.

Secondly, to minimize the numerical correction due to the spatially varying $h(r)$, the difference of adjacent particle should be small as $|\nabla h| \lesssim 1/3$. This condition has the physical meaning of $|\nabla\rho/\rho| \lesssim 1/h$. For example, to resolve the gradient of $(\rho_{max} - \rho_0)/\rho_0 R \sim \rho_{max}/\rho_0 R \lesssim 1/h_{min}$, the above condition determines the required particle numbers, that is, $\rho_{max}/\rho_0 \lesssim \sqrt{N}$.

Combining the above two conditions, we get the dynamic range of stratified SPH, $\rho_{max}/\rho_{min} \lesssim N^{3/2}/1000$. The particles of N=4000 can simulate the polytropic star of $\rho_{max}/\rho_{min} \sim 250$, while $N = 10^6$ particle is required to simulate the realistic stellar envelope of $\rho_{max}/\rho_{min} \sim 10^6$. We have developed a new fast code which can treat the 3-D supernovae of realistic progenitors with $N = 10^6$.

III POLYTROPE REVISITED

There is a criticism about the small particle number and the energy conservation of our former simulations (Benz et al. 1989). To make these effects clear, we have developed our SPH code to be able to guarantee the energy conservation and the momentum conservation. We solved the adiabatic explosion of n=3 polytrope with $N = 10^6$ particles.

The point explosions induce two types of dynamical solutions. Thermal point explosions induce spherical *"blast wave"* expansions similar to that of 1-D calculations (see Fig 1.). Off-center kinetic energy deposition triggers *"bubble solution"* as in Fig 2. The *"blast wave solution"* get only a few factor of enhancement of the initial perturbations (see Fig 3.). These result of our SPH simulation are consistent with mesh calculations of Müller et al. (1989) . While in the *"bubble solution"*, the convective instability raises the density fluctuations in that shell due to the displacement of hot blobs. In the linear stage, the small fluctuations appear on the contact discontinuity. The convection between shock front and contact discontinuity

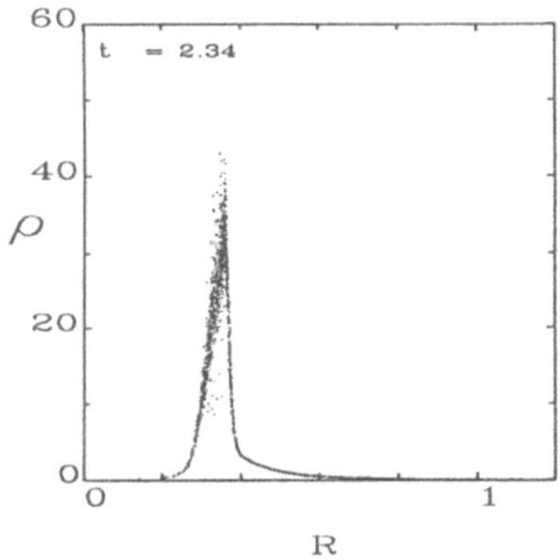

Fig.1. The projected radial density profile of *"blast wave solution"*. As in 1-D calculation, the topology of the expansion is not a shell but a sphere with the limb enhanced by shock front.

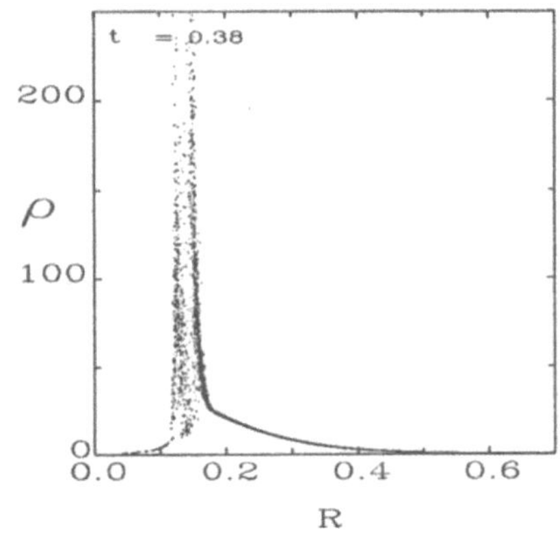

Fig.2. The radial density profile shows two discontinuity in 3-D explosion of *"bubble solution"*. The shock front and the contact discontinuity are the characteristic of Sedov solution in steep density gradient.

causes the effective mixing in the shell. The typical unstable wavelength, in the nonlinear stage, is determined by the thickness of the spherical shell as in Fig 4. The mixing motion in the nonlinear stage is resolved with $N = 10^6$ particles. The ejected gas, plotted with a certain density level, seems very clumpy. While, due to the phase cancellation of complicated fragments, there is no dominant contribution to the degree of polarization. The initial random fluctuations are introduced as a numerical noise ($\lesssim 1\%$) when we construct the 3-D equilibrium

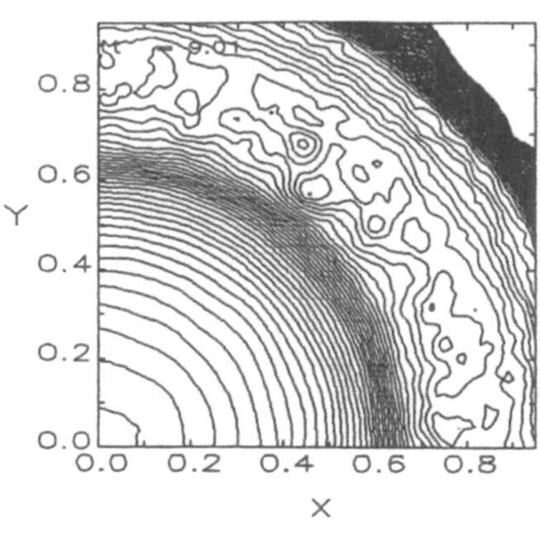

Fig.3. The density contour of *"blast wave solution"*. Only the narrow region of shocked gas is unstable as indicated by linear stability analysis.

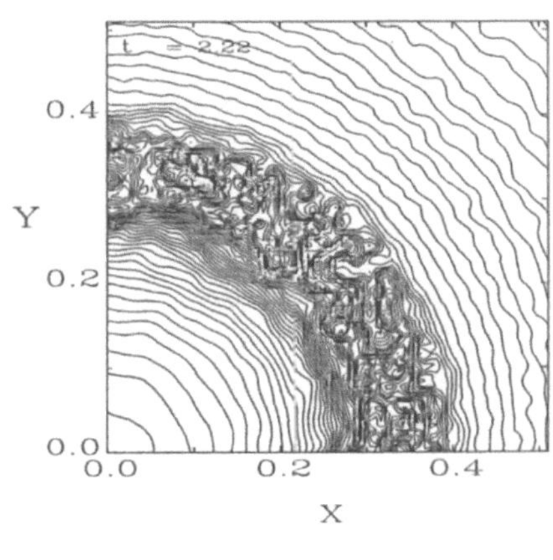

Fig.4. The density contour of *"bubble solution"* in contour level of 10% spacing. The convective motion in the whole shell makes the clumpy structure.

spheres by relaxation. However, we found that the fluctuation of the hot bubble at the initial is more important than the fluctuation in the outer envelope. In this way, the fragmentation and the mixing of the ejecta is reconfirmed.

We have to be careful to discuss Rayleigh–Taylor instability of compressible gas. It is different from the ordinary growth of density fluctuations. Since the boundary of density jump is disturbed by instability, focusing on the density at a certain position, the fluctuation appears as if there is no stage of linear growth. The buoyancy of hot blob makes this nonlinear growth possible, which is inhibited in 1-D simulations as a shell crossing of fluid element. In 3-D spaces, it means the existence of interior motion of hot blobs in the shocked shell. This means the heavy elements which locate at the center of progenitor can appear at the surface of supernova most effectively. The mixing motion of gas element is most effective in the wholly convective case. We can interpret this interchange motion as a result of convection.

Our 3-D SPH makes the fluid motion clear: the mixing is not the result of fragmentation but the fragmentation is the result of wholly convective motion. The evolutions of off-central point explosions are almost spherically symmetric as a whole, but they are found to be convective and to grow non-spherical structures on the shell. The inner hot blobs creep out the outer mass shell. Our former results should be distinguished from the relatively stable *"blast wave solutions"*. To investigate the mixing degree, 3-D Lagrange method is preferable and the Euler methods need much efforts to treat multi-fluids system and difficulty to get void structure. The advantage of our SPH is that most of particles are used to resolve this interesting discontinuous region.

REFERENCES

Arnett, D., Fryxell, B.A., and Müller, E., 1989, *Asrtrophys. J. Lett.*, in press.

Barret, P., 1987, *ESO Workshop on SN1987A (ed. I.J.Danziger)*, 174.

Benz, W., Thielemnn, F.K., 1989, *Asrtrophys. J. Lett.*, in press.

Erickson, E.F., Haas, M.R., Colgan, S.W.J., Lord, S.D., Burton, M.G., Wolf, J., Hollenbach, D.J., and Werner, M., 1988, *Asrtrophys. J. Lett.*, **330**, L39.

Itoh, M., Kumagai, S., Shigeyama, T., Nomoto, K., and Nishimura, J., 1987, *Nature*, **330**, 233.

Müller, E., Hillebrandt, W., Orio, M., Höflich, P., Mönchmeyer, R., and Fryxell, B.A., 1989, *Astr. Astrophys.*, in press.

Nagasawa, M., Nakamura, T., and Miyama, S.M., 1988, *Publ. Astron. Soc. Japan*, **40**, 691.

Mixing in SN 1987A

Willy Benz & F. C. Thielemann

1. Introduction

Many supernova remnants (like e.g. Cas A, Kirshner and Chevalier 1977) show evidence that some form of mixing has to take place in the ejecta. While there existed some theoretical indications that this mixing is due to instabilities associated with the propagation of a shock wave (e.g. Falk and Arnett 1973, Gull 1973, Chevalier and Klein 1978) it has usually been related to the expansion into the interstellar medium. SN1987A now showed clearly that mixing has to take place already during the supernova explosion itself (or possibly very shortly thereafter). Several independent observations lead to such a conclusion: the early appearance of x-rays and gamma-rays, the spread of expansion velocities seen in line widths of infrared observations and in the gamma ray lines of ^{56}Co, and the details of the rise of the light curve after the initial adiabatic decline and a flattened rather than a sharp maximum before the exponential decline of ^{56}Co.

This extensive mixing must have resulted from hydrodynamic instabilities taking place within the first two or three weeks after the explosion. Two major driving mechanisms were suggested: 1. Rayleigh-Taylor instabilities associated with the additional expansion of the central nickel bubble triggered by the energy release from the radioactive decay of ^{56}Ni and ^{56}Co (Arnett 1988, Woosley 1988). 2. Instabilities associated with the propagation of the shock through the star itself, dependent on the detailed internal structure of the progenitor (Chevalier 1976) or associated with the reverse shock originating at the inner edge of the hydrogen envelope (Shigeyama, Nomoto, Hashimoto 1988). Both driving mechanism are discussed.

2. The Ni-Bubble

We followed the homologous expansion phase of SN1987A from 1.1 days to 160 days after the explosion by modelling the inner $6M_\odot$ numerically with a 3D Smooth Particle Hydrodynamics code. The initial conditions were taken from Woosley (1988) on which small (2-4%) random density perturbations were superimposed. 0.07 $M_\odot$ of ^{56}Ni were deposited in the center and a nuclear network for the Ni-decay chain added to the code as in Benz, Hills, and Thielemann (1989). Various numbers of particles were used, ranging from 10 000 for the whole sphere to 40 000 in one octant.

After 160 days only small density fluctuations (clumping) of less than a factor 2 (peak to peak) emerged. Fig. 1 shows density contours in a 2D slice which contains the center of the explosion. Contrary to prior suggestions the Ni-Co-decay energy, when superimposed onto a strong spherical expansion, is not sufficient to result in large

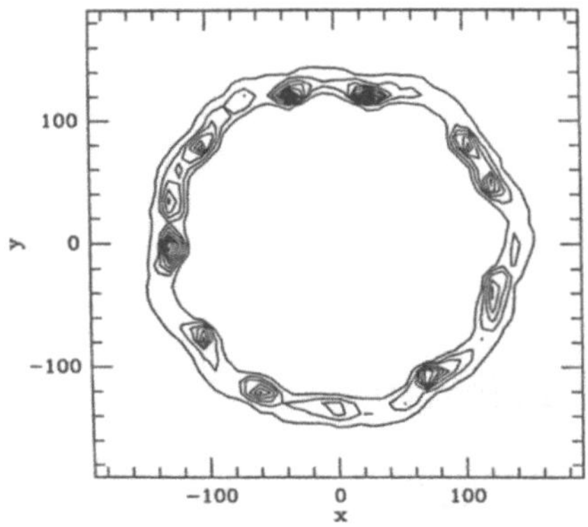

Fig. 1: Isodensity contours after 160 days. The density contrast is of about a factor 2 peak to peak.

enough density inversions which can cause Rayleigh-Taylor instabilities and mixing within the first three weeks.

This picture might change if, as it turns out to be the case, strong instabilities are already associated with the propagation of the blast wave ***inside*** the star itself resulting in a strongly inhomogeneous abundance and density distribution, even before Ni-decay. The additional energy release by the radioactive decay of the nickel can in this case quite conceivably have more dramatic consequences. We address this question in the following sections.

3. 3D Calculations of Blast Wave Propagation

Nagasawa ***et al.*** (1988) followed the propagation of a supernova blast wave by depositing the appropriate amount of energy in the center of a n=3 polytrope of $10M_{\odot}$. Their simulations, done with a 3D SPH code, resulted in rapid and severe clumping (density fluctuations of 400%) regardless of the explosion energy. Since the same calculations done by Müller ***et al.*** (1989) and by Clancy and Bowers (this volume), but using classical Eulerian finite-difference methods, did not result in any major clumping or instabilities (density fluctuations less than 15-20%), it became clear that the stability of blast waves propagating in even simple density distribution is by far not definitively established.

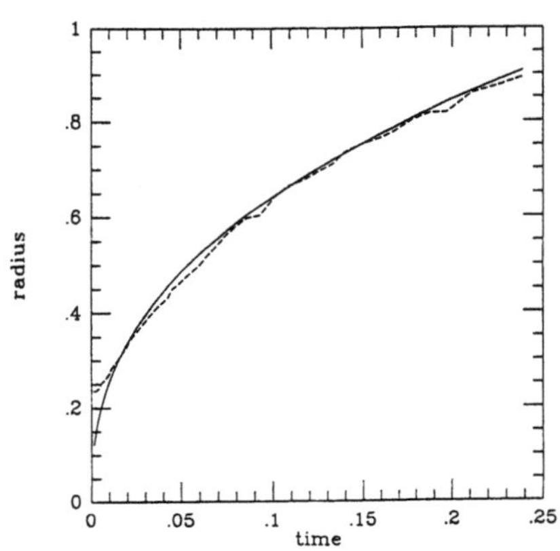

Fig. 2: Location of the shock in a blast wave propagating in a constant density medium.
solid line : exact
dashed line : numerical

In order to establish that SPH is able to accurately model the propagation of spherical blast waves, we modelled a central explosion in a constant density medium.

Constant, as well as simple power law, density distributions are ideal testing grounds for numerical calculations as the analytical solutions are available (Sedov 1959). In Fig. 2 we plot the location of the shock as a function of time for both the analytical and numerical solution in the case of a strong explosion in a constant density medium. Notice that for this case, the analytical solution is completely determined, no scaling is therefore necessary. Clearly, the agreement between both (once the initial conditions are forgotten) is very good.

As a second test we have run the same polytrope explosion simulation as Nagasawa *et al.* (1988), using from 10000 particles for the whole star up to 40000 particles in one octant only. These simulations did not show the extensive clumping and mixing claimed by Nagasawa *et al.* (1988) but rather confirm the results obtained by the Eulerian finite-difference methods. The 2% density perturbations that were introduced in the initial configuration resulted indeed in 15-20% density fluctuations at the end of our simulation (see Fig. 3). Extensive clumping might, however, be obtained when using highly perturbed initial conditions or unrealistic explosion energy deposition schemes.

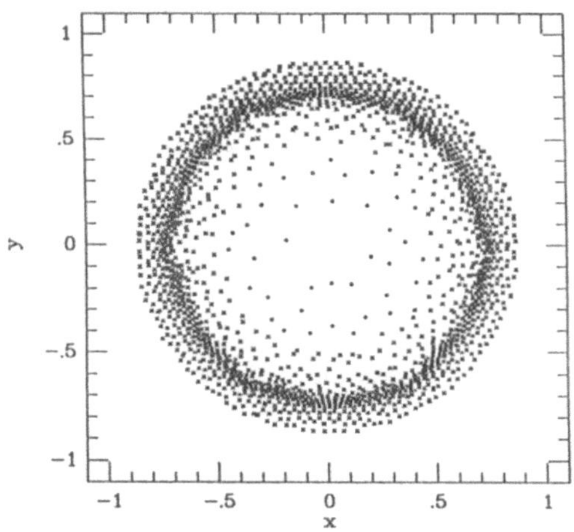

Fig. 3: Blast wave in a n=3 polytrope. Shown are the particles in a small slice through the central plane.

The search for instabilities in SN1987A has to be performed with a realistic supernova progenitor model, rather than a polytrope. Arnett, Fryxell, and Müller (1989) could indeed show in 2 and 3D numerical simulations that instabilities were growing in the high density shell left behind the shock. They identified the finger-like structures obtained with a classical Rayleigh-Taylor instability. Ebisuzaki, Shigeyama, and Nomoto (1989) performed a linear stability analysis of a spherical explosion using the usual Rayleigh-Taylor criterion (see section 4) and were able to show that essentially the mass zones between the metal-He and He-H interfaces are unstable.

4. Stability Analysis

Chandrasekhar (1961) discussed the stability of an initially static, incompressible fluid in a gravitational field. In the case of supernova explosions, gravity is negligible and pressure gradients provide the required relative acceleration. The criterion for a Rayleigh-Taylor instability to develop in this case can be written as

$$\frac{\mathcal{R}}{\mathcal{P}} < 0, \tag{1}$$

where $\mathcal{P} = \frac{1}{P}\frac{\partial P}{\partial z}$ and $\mathcal{R} = \frac{1}{\rho}\frac{\partial \rho}{\partial z}$ are the reciprocals of the pressure and density scale heights, respectively. As emphasized by Bandiera (1984) and Benz and Thielemann (1989) a local linear stability analysis show that for a compressible fluid instabilities will develop and grow exponentially provided the Scharzschild criterion is met

$$\frac{\mathcal{R}}{\mathcal{P}} < \frac{1}{\gamma}, \tag{2}$$

where γ is the adiabatic index of the gas. Comparing equations (1) and (2), it appears that the incompressible Rayleigh-Taylor criterion is more stringent than the actual stability criterion for a compressible fluid. Consequently for a compressible fluid, instabilities, i.e. convective motions, will set in even if condition (1) is not satisfied. Taking the limit $\gamma \longrightarrow \infty$ in (2), that is assuming incompressibility, we indeed recover the classical criterion (1). We conclude that applying the Rayleigh-Taylor criterion (1) to determine the stability of compressible fluid flows can be misleading since instabilities will occur for situations predicted to be stable. Stability of compressible fluids should be checked using (2). Whether these instabilities are called Rayleigh-Taylor or convection, is only of a semantic interest, but to keep with astronomical traditions, we probably should call them convective instabilities.

We have performed explosion simulations with a 1D Lagrangian hydrodynamics code, using the stellar model for the progenitor of SN1987A with a $10M_\odot$ H-envelope by Nomoto *et al.* (1988). We deposited 10^{51} erg in the center and followed the blast wave propagation through the entire star. For simplicity we adopted a polytropic equation of state with $\gamma = \frac{4}{3}$ and neglected radiation transport over the time scale considered. When zones were found unstable, according to (1) or (2) an estimated time integrated growth rate was computed (Benz and Thielemann 1989). The results are presented in Fig. 4 against the Lagrangian mass coordinate of ejected mass (excluding the $1.6M_\odot$ neutron star - Thielemann, Hashimoto, Nomoto 1990) for a time corresponding to one hour after the explosion. The dotted curve gives the integrated Rayleigh-Taylor growth rate, whereas the solid line corresponds to the integrated convective instability growth rate. Notice the large spikes at the metal-He and He-H interfaces. When using instability criterion (2) rather than (1), the growth of the perturbations in the center is substantially larger although still not as large as at the metal-He interface, and the barrier at $3M_\odot$ is partially removed. Therefore convective motions will lead to substantial mixing inside the inner core, and even into the hydrogen envelope but with a significantly smaller growth rate. This makes it possible for some material (most probably only a small fraction) originating from the inner core to be convected into the envelope and conversely, for some hydrogen from the envelope to be transported into the core.

Further improvements in understanding the quantitative extend of the mixing and the nonlinear behavior of the instabilities have to come from a global stability analysis including boundaries (Goodman 1989) and multi-dimensional, high-resolution numerical simulations of realistic stellar explosions similar to the ones by Müller *et al.* (1989) and Arnett, Fryxell and Müller (1989). We are currently investigating these questions with a 3D Smooth Particle (SPH) code.

This research was supported in part by NASA grant NGR 22-007-272, NSF grant 86-12647 and the National Center for Supercomputer applications at the University of Illinois (AST 890009N). One of us (W.B.) also acknowledges partial support from

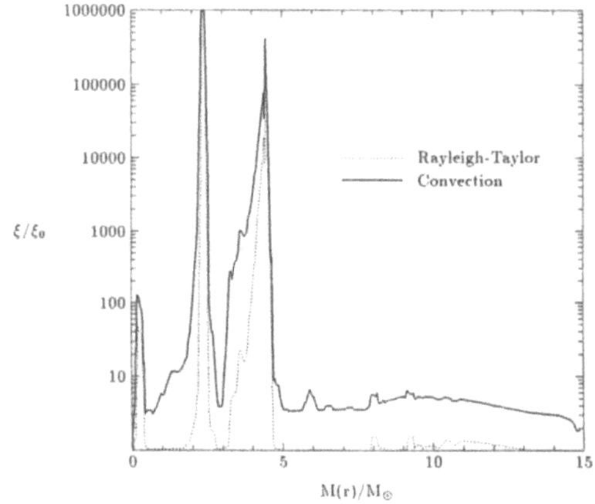

Fig. 4: Integrated growth rate for the convective (solid line) and Rayleigh-Taylor (dashed line) instabilities

the Swiss National Science Foundation and from the Milton Fund.

References

Arnett, W.D. 1988, ***Ap. J.*** **331**, 377

Arnett, W.D., Fryxell, B., Müller, E. 1989, ***Ap. J. Lett.*** **341**, L63

Bandiera, R. 1984 ***Astron. Astrophys.*** **139**, 368

Benz, W., Hills, J. G., and Thielemann, F.-K., 1989, ***Ap. J.*** **342**, 986

Chandrasekhar, S. 1961, ***Hydrodynamic and Hydromagnetic Stability***, (Dover, New York)

Chevalier, R.A. 1976, ***Ap. J.***, **207**, 872

Chevalier, R.A., Klein, R.I. 1978, ***Ap. J.***, **219**, 994

Ebisuzaki, T., Shigeyama, T., Nomoto, K. 1989, ***Ap. J. Lett.***, in press

Falk, S.W., Arnett, W.D. 1973, ***Ap. J. Lett*** **180**, L65

Goodman, J. 1989, preprint

Gull, S.F. 1973, *M.N.R.A.S.* **161**, 47

Kirshner, R.P., Chevalier 1977, ***Ap. J.*** **218**, 142

Müller, E., Hillebrandt, W., Orio, M., Höflich, P., Mönchmeyer, R. 1989, ***Astron. Astrophys.***, **220**, 167

Nagasawa, M., Nakamura, T., Miyama, S. 1988, ***Publ. Astron. Soc. Japan*** **40**, 691

Nomoto, K., Hashimoto, M., Shigeyama, T., Kumagai, S. 1988, ***Proc. Astron. Soc. Australia*** **7**, 490

Sedov, L.I. 1959, ***Similarity and Dimensional Methods in Mechanics*** (Academic Press, New York)

Shigeyama, T., Nomoto, K., Hashimoto, M. 1988, ***Astron. Astrophys.*** **196**, 141

Thielemann, F.-K., Hashimoto, M., Nomoto, K. 1989, ***Ap. J.*** **348**, in press

Woosley, S.E. 1988, ***Ap. J.*** **330**, 218

Rayleigh-Taylor Instability in Supernova 1987A: Dependence on the Presupernova Model

T. Ebisuzaki, T. Shigeyama, & K. Nomoto

1. Rayleigh-Taylor Instability

The Rayleigh - Taylor instability is the most probable mechanism to mix the ejecta of SN 1987A as suggested from observations. Although the gravity is negligible in the explosion, acceleration of the matter acts as an effective gravity and can be responsible for the Rayleigh - Taylor instability. The growth rate, $G_{\rm RT}$,of this instability is estimated as

$$G_{\rm RT} = \sqrt{-\frac{\rho_+ - \rho_-}{\rho_+ + \rho_-}\frac{1}{\rho}\frac{dP}{dr}k}, \tag{1}$$

where k is the wave number of the perturbation and ρ_+ and ρ_- are the densities in the upper and lower layers. Equation (1) shows that the layer is Rayleigh - Taylor unstable when $(dP/dr)(d\rho/dr) < 0$.

Ebisuzaki, Shigeyama, and Nomoto [1; hereafter ESN] performed a linear stability analysis of the explosion using the realistic hydrodynamical model 14E1 [2, 3], where the ejected masses of the hydrogen-depleted core and the hydrogen-rich envelope are 4.4 $M_\odot$ and 10.2 $M_\odot$, respectively, and the kinetic energy of explosion is 1×10^{44} J. The model 14E1 with mixing of hydrogen and ^{56}Co well reproduces the light curves in the optical/infrared, X-ray, and gamma-ray regions [4, 5].

Figure 1 shows the evolution of the pressure profile of the exploding star. The dashed lines indicate the H/He interface and He/Metal interface. Before the explosion, pressure gradient is negative being balanced with gravitational attraction (stage 0). After the blast shock passed, the layer expands to decrease the pressure rapidly. When the blast shock is propagating through the hydrogen-rich envelope with a much less steep pressure gradient, an inwardly propagating reverse shock (R) forms (stages 2 and 3). The layer between the blast wave (B) and the reverse shock (R) is decelerated and thus has a positive pressure gradient. When the blast shock reaches the surface layer with a steep pressure gradient, an inward-moving rarefaction wave forms and produces a negative pressure gradient that accelerates the matter outward (stage 5). As a result of this complicated pressure evolution, the H/He interface is first accelerated, then decelerated, and finally accelerated again; i.e., the pressure gradient changes its sign twice. The He/Metal interface also experiences similar acceleration and deceleration.

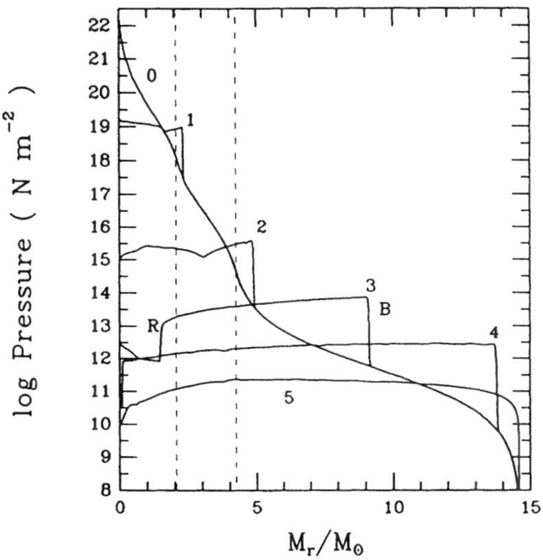

Figure 1: Change in the pressure profile in the ejecta of model 14E1. Stage numbers correspond to (0) $t = 0$, (1) 9.0 s, (2) 166 s, (3) 1060 s, (4) 3330 s, and (5) 6710 s after the explosion. The two dashed lines indicate the H/He and He/Metal interfaces. The letters B and R indicate the positions of the blast shock and the reverse shock, respectively. The pressure gradient is positive between the blast and reverse shocks.

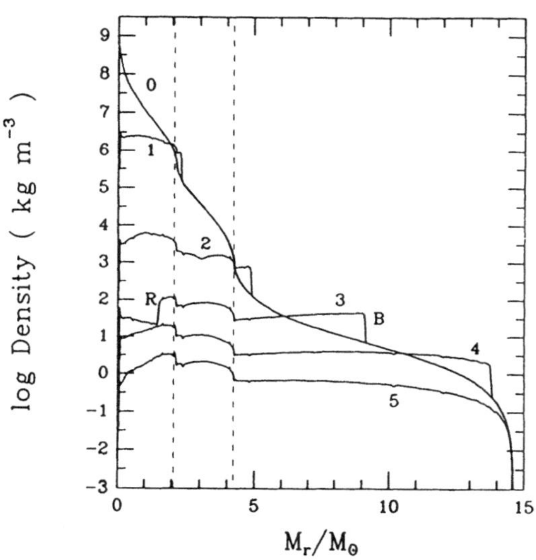

Figure 2: The same as Figure 1 but for the change in the density profiles. The density steeply decreases outward at the H/He and He/Metal interfaces (dashed lines).

Figure 2 shows the time evolution of the density profile. Near the H/He and the He/Metal interfaces (dashed lines), the density steeply decreases with radius because of the changes in the mean molecular weight, μ, and the specific entropy, s. Such distributions of μ and s originate from the presupernova model. During the hydrodynamical stages of explosion, the sign of the gradients of μ and s and, hence, of density near the composition interfaces do not change.

Figures 1 and 2 clearly show that pressure gradient is positive during the deceleration phase (stages 2 - 4 for the H/He interface), while the density gradient remain negative near the H/He and He/Metal interfaces. These layers are Rayleigh - Taylor unstable.

The distribution of the amplification factor ζ/ζ_0 at t = 18,000 s is plotted against M_r in Figure 3 for $l = 20$ (solid curve). Here the amplification factor is given as

$$\zeta/\zeta_0 = \exp(\int_0^t \mathrm{Re}(G_{\mathrm{RT}})dt') \tag{2}$$

and well exceeds 100 near the H/He interface.

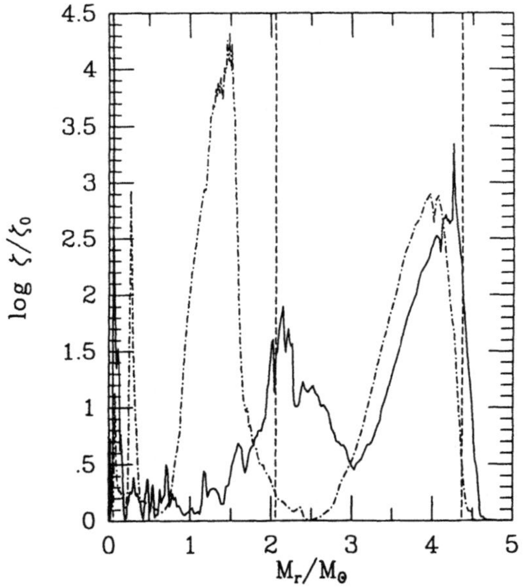

Figure 3: Amplification factor (ζ/ζ_0) at t = 18,000 s is plotted against mass (M_r). The H/He and He/Metal interfaces are Rayleigh - Taylor unstable and the amplification factors are well above 100 for both $l = 10$ and 20.

2. Dependence on the Presupernova Structure

Arnett, Fryxell, and Müller [6; hereafter AFM] performed two-dimensional simulations of the explosion for the 15 $M_\odot$ progenitor model of SN 1987A (Arnett [7]). They found a pronounced nonlinear growth of the Raleigh - Taylor instability around the He/Metal interface, which is basically in agreement with the linear stability analysis. However, the instability around the H/He interface is *not* significant, which is *not* consistent with the above linear analysis.

Hachisu *et al.* [8] have also carried out a two-dimensional simulation for the model 14E1 and found that the instability develops more rapidly at the H/He interface than at the He/Metal interface. Their results are consistent with ESN's linear analysis.

The difference between the two models can be seen from linear stability analysis for the Arnett's [7] progenitor model with the mass cut at $M_r = 0$. The amplification of the pertur bation (Eq. 2) is larger at the He/Metal interface than at the H/He interface as shown by the dash-dotted curve in Fig. 3.

This difference may stem from the difference in i) the initial models and ii) the assumed mass cut that divides the neutron star and the ejecta:

i) Compared with model 14E1 which has a 6 $M_\odot$ helium core [9], Arnett's (1987) 15 $M_\odot$ model has a smaller helium core (4 $M_\odot$) that is more centrally concentrated; the density gradient is much steeper around $M_r = 1.5M_\odot$, i.e., the density changes by a factor of $\sim$ 1000 over the narrow region from the He/Metal interface to the outer part of the central core.

ii) AFM [6] deposited energy at the center of the progenitor model, i.e., the mass cut is $M_r \sim 0$ neglecting the neutron star residue. After the shock passage, only the central region suffers from rarefaction and the density gradient is much steeper around $M_r = 1.5M_\odot$ than in the model that has a mass cut around $M_r \sim 1.2M_\odot$ (see Fig. 6 of Arnett [7]). (The mass cut for 14E1 is 1.6 $M_\odot$.)

These differences yield much larger density contrast at the He/Metal interface in Arnett's model than 14E1. Equation (1) gives a higher growth rate for a larger density contrast, if k and the pressure gradient is fixed.

As can be seen in the above example, the development of the Rayleigh - Taylor instability in the supernova explosion is quite sensitive to the density structure of the progenitor. Since ^{56}Ni is synthesized near the mass cut, the above differences could yield a significant difference in the mixing process of ^{56}Ni. Further systematic studies are necessary to figure out how the Rayleigh - Taylor instability depends on the structure of the progenitor model.

REFERENCES

(1) Ebisuzaki, T., Shigeyama, T., and Nomoto, K. 1989, *Ap. J. (Letters)* , in press (ESN).
(2) Shigeyama, T. 1989, Ph.D. Thesis, University of Tokyo.
(3) Shigeyama, T., and Nomoto, K. 1989 *Ap. J.,* submitted.
(4) Nomoto, K., Shigeyama, T., and Kumagai, S. 1989, in *Particle Astrophysics* , ed. E.B. Norman (Singapore: World Scientific), in press.
(5) Nomoto, K., Shigeyama, T., and Kumagai, S. 1989 in this volume.
(6) Arnett, D., Fryxell, B. and Müller, E. 1989, *Ap. J. (Letters)* , **341** , L63 (AFM).
(7) Arnett, D., 1987, *Ap. J.* , **319** , 136.
(8) Hachisu, I, Matsuda, T., Nomoto, K. and Shigeyama, T. 1989, *Ap. J. (Letters)* , submitted.
(9) Nomoto, K., Hashimoto, M. 1988, Physics Report, **163,** 13.

SECTION V

X-RAYS AND γ-RAYS FROM SN 1987A

SMM γ–Ray Observations of SN 1987A

Mark D. Leising & Gerald H. Share

Supernovae have long been considered important sites of nucleosynthesis of many nuclides found in nature, including some thought to be ejected in the form of radioactive parents. These unstable progenitors can reveal themselves through their power input to the ejecta [1], or more directly through their emission of signature γ-ray line photons [2]. Supernova 1987A has yielded an awesome quantity of data which has for the most part confirmed theories of Type II supernovae, as is demonstrated by many articles in these Proceedings. Among the important observations was the the precise agreement between the decline of the "bolometric" light curve and the 113 day exponential decay of ^{56}Co [3,4], implying the production of its radioactive parent ^{56}Ni by the explosion. The ^{56}Co nucleus was further implicated by observations of large abundances of the element cobalt in infrared lines [5,6]. The direct observation of γ-ray lines from ^{56}Co decay (e.g., [7]) laid to rest any doubts about its production in the explosion. It is the detection of these γ-ray lines by the γ-ray spectrometer (GRS) on the Solar Maximum Mission (*SMM*) satellite we discuss here. We also describe how the time evolution of the fluxes in these lines can be used to probe the distribution of the ^{56}Co within the ejecta. These γ-ray light curves show that the radioactivity is found over a large range of optical depths in the ejected matter, a conclusion consistent with many other observations discussed in this volume.

I. Data Analysis *SMM* was launched in February 1980 to observe the Sun and has operated for most of the time since then. The GRS [8] has been used successfully to detect cosmic γ-ray emission [9]. Here we use data accumulated by the GRS since 1984. This analysis is described in detail elsewhere [10]. Basically, each one-minute spectrum accumulated with the Large Magellanic Cloud (LMC) visible to the detector is corrected by subtracting spectra taken under similar background conditions before 1987, and by also subtracting similarly corrected spectra accumulated during the same orbit but with the LMC occulted by the Earth. These difference spectra are then summed over periods of about 36 days. (For about 17 days out of each 53 day orbital precession period, there is little or no time when the Earth blocks the LMC.) Each such summed spectrum is then fit over two energy intervals, 0.7 to 1.5 MeV, and 2.3 to 3.6 MeV, with models of smooth continua and gaussian peaks. The peak positions were chosen at energies of instrument calibration source lines and at the energies of four lines of ^{56}Co decay, namely 0.847, 1.238, 2.599, and 3.250 MeV (the last is a composite of three lines).

The resulting best-fit intensities of these four lines are shown versus time in Fig. 1. For each line the intensities before the outburst of SN 1987A (day 2611) are consistent with zero counts. All four show positive excesses after July 1987. Figure 2 shows the sum of all spectra from August 1987 through May 1988, in which at least four lines from ^{56}Co

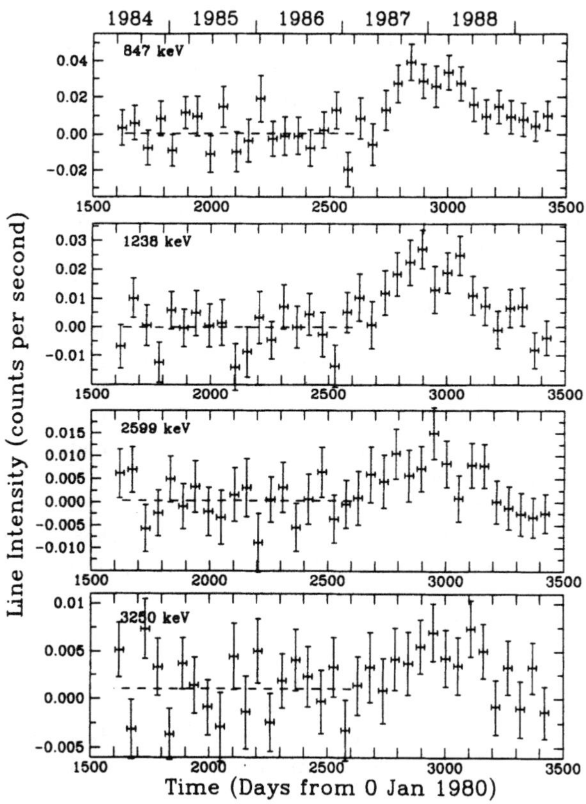

Figure 1. The count rates in each of four ^{56}Co lines versus time. The dashed lines are the mean counts in each line before February 1987. Each mean is consistent with zero. Errors are statistical only.

decay are apparent. Other features of this spectrum are from known background or celestial sources. Figure 3 shows the measured fluxes in the four ^{56}Co decay lines for the time after the explosion of SN 1987A. Generally the 0.847 and 1.238 MeV fluxes are consistent with those measured by balloon experiments [11-15]. The results presented here are also consistent with earlier analyses of some of the same data [7,16].

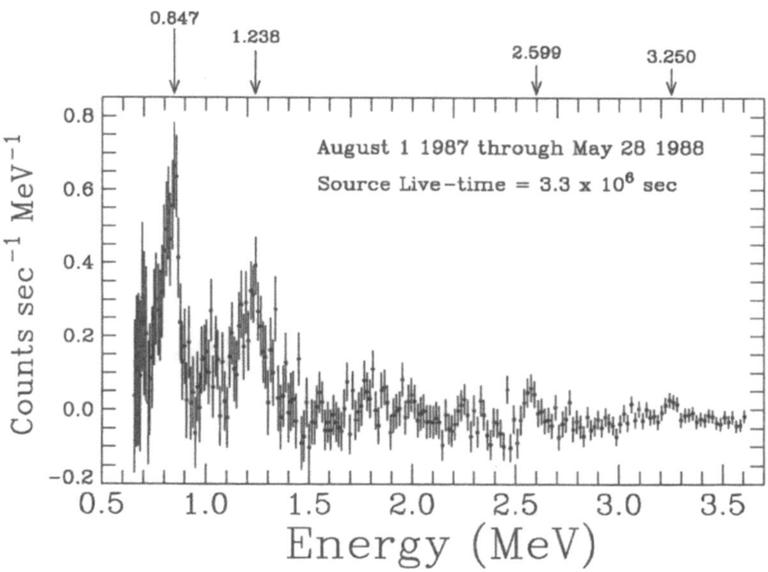

Figure 2. The mean of all SN 1987A spectra from August 1987 through May 1988. This is the time over which significant fluxes are found in the lines.

<u>II. Discussion</u> The γ-ray light curves can be understood in terms of a simple description of a supernova envelope. The escaping fluxes depend only on the amount of ^{56}Ni initially produced and the column depth of scatterers between the ^{56}Co and the observer at a given time. Conventional descriptions of supernova ejecta, with the ^{56}Co located entirely at the inner edge, cannot fit the γ-ray fluxes, or the shape of the peak of the optical light curve [17-19]. The γ-ray light curves can be satisfactorily described only by models where the ^{56}Co is distributed over a large range of optical depths (e.g., [10,20]), but given the precision of the γ-ray measurements,

many different such models can fit the data. For example, it was shown that two different distributions of ^{56}Co mass versus optical depth gave equally good fits to the γ-line fluxes [10]. One model had two "spikes" in the ^{56}Co mass distribution, with ~5% of the ^{56}Co at very low optical depth and ~95% located at large depth as in the pre-SN 1987A picture. Another successful model utilized a continuous distribution of ^{56}Co in optical depth.

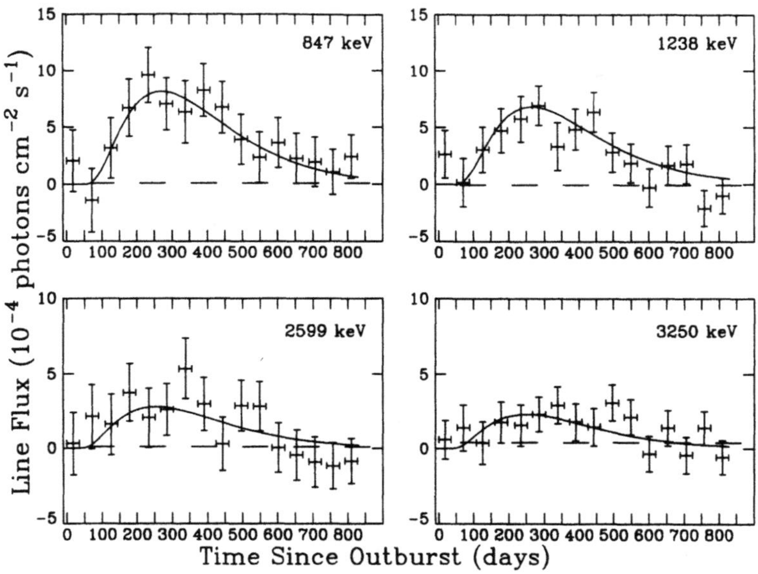

Figure 3. The measured fluxes in four ^{56}Co lines versus time since the explosion (crosses) and the flux in each line from the best fit of equation (1) (solid lines). For reference, the mean measured fluxes before the explosion are also shown (dashed lines).

Here we describe the latter model in some detail. Consider the case where the emitting nuclei are distributed over a region of the inner ejecta of finite thickness (e.g., homogeneous "mixing" of the ^{56}Co). The outer envelope, overlying the mixed region, will exponentially attenuate the escaping γ-rays, and its thickness will decrease as 1/time^2. Because of the finite optical thickness of the inner zone, the outer, unmixed, zone will effectively "see" only the outermost unit optical depth of the mixed region. The fraction of the mass (and therefore number of ^{56}Co nuclei) within this depth will increase with time, approximately proportional to time^2 (this approximation becomes less good at later times). Obviously the fraction visible will increase only until all of the inner zone is thin. Thus the emerging luminosity in a line of energy E is

$$L_E(t) = b_E\, L_o\, (e^{-t/\tau_{Co}} - e^{-t/\tau_{Ni}})\; e^{-\frac{\sigma_E}{\sigma_{847}} (t^2_{o,847} / t^2)} \times \frac{\sigma_{847}}{\sigma_E} (t^2 / t^2_{i,847}) \qquad \text{for } t < t_{i,847}\,, \tag{1}$$

where b_E is the branching ratio of the line, L_o is the total (input) 847 keV line intensity at $t=0$, $\tau_{Co}=113$ days, $\tau_{Ni}=8.8$ days, the σ's are the total Klein-Nishina scattering cross-sections, $t_{o,847}$ is the time at which the escape of 847 keV photons through the outer shell is 1/e of the input luminosity (i.e., when the effective optical depth of the unmixed shell at 847 keV is unity) and $t_{i,847}$ is the time when the inner, mixed, region becomes thin to 847 keV γ-rays. Assuming we know the initial ^{56}Ni mass (which we fix at 0.07 $M_\odot$) and distance (we assume 50 kpc), this function requires only two parameters to define the γ-ray light curves at all energies.

We fit the four light curves of Fig. 3 simultaneously with this function and find an acceptable fit (reduced $\chi^2 = 0.7$), shown as the solid curves in Fig. 3. The single function seems to fit all four lines equally well, and so is an acceptable description of the average depth of the γ-emitters. The best-fit values of the parameters are $t_{o,847}=124\pm31$ days and $t_{i,847}=1860\pm111$ days. That is, nearly the entire ejecta must be mixed to explain the γ-ray data. This outer zone is only as thick as 1/3 $M_\odot$ of hydrogen expanding at 3000 km s^{-1}! Clearly the data require some of the ^{56}Co at very low optical depth at very early times. This model also requires a very thick, either massive or slow moving, mixed region to keep the fluxes low after the initial peak. Because it is encouraging for future work, we note that with the mass of ^{56}Ni as a free parameter in the fit, we find a value in reasonable agreement with that determined from the optical light.

A significant component of the bolometric luminosity, especially at later times is the hard emission, at X-ray and γ-ray energies. So far, efforts to account for all the expected ^{56}Co decay power in the bolometric luminosity have utilized only theoretical estimates of the X-ray and γ-ray luminosities (e.g., [3]). We have measured only a few of the ^{56}Co γ-ray lines, but we can infer from models, such as the above one, the escape of all lines. Also, we can obtain the escape fraction of γ-rays from any species (e.g., ^{57}Co) at a given time from (1). With the aid of Monte Carlo simulations of Compton scattering and photoelectric absorption we can also calculate the emerging continuum luminosity. Shown in Fig. 4 are the γ-ray line and continuum luminosities, based on this model, and the total ^{56}Co decay power.

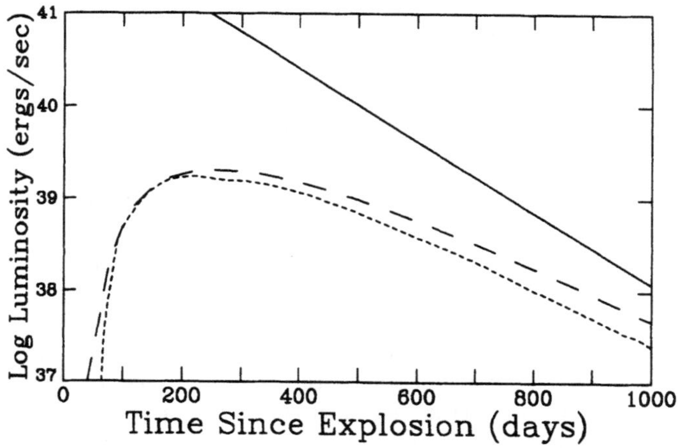

Figure 4. The total γ-ray line luminosity from (1) versus time (dashed line); the > 10 keV continuum luminosity, from Monte Carlo simulation (dotted line); and the ^{56}Co decay power (solid line).

Shown in Fig. 5 is the flux of ^{57}Co 122 keV γ-rays predicted by equation 1 assuming that 1.7×10^{-3} $M_\odot$ of ^{57}Ni was ejected by the explosion (i.e., that the solar ratio of $^{56}Fe/^{57}Fe$ holds for the ejection of the unstable progenitors). Theoretical estimates of the mass of ^{57}Ni ejected range from this value to 2.5 times greater (e.g., [19,21]). The ejected mass of ^{57}Ni is important because it reflects both the neutron richness of the burning region and the division between the ejected and accreted portions of the star. It is possible that the ^{57}Co mass will be measured from its influence on the bolometric light curve, as was that of ^{56}Co. However, it is becoming increasingly difficult to measure the true bolometric light curve. The determination of this mass from γ-rays requires a measurement of the line flux and an estimate of its attenuation by the ejecta. The model discussed here has the greatest attenuation at late times of those with which we have successfully fit the earlier ^{56}Co γ-line fluxes. Even so, the 122 keV line and its Compton scattered continuum fluxes are marginally detectable by future experiments [22,23] if they are launched on current schedules. The attenuation can be derived from measurements of the line to continuum ratio or from multiple observations of the line fluxes, considered in the context of simple models of the ejecta.

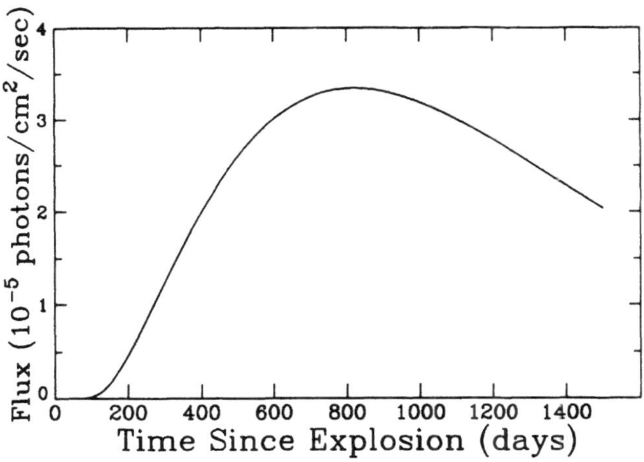

Figure 5. The 122 kev line flux versus time, assuming 1.7×10^{-3} $M_\odot$ of ^{57}Ni ejected, from (1). This is about the minimum flux expected at late times, as this model has larger attenuation than others and most calculations suggest more ^{57}Ni is ejected.

<u>III. Conclusions</u> With the *SMM* GRS, we have detected significant γ-ray line fluxes from the decay of radioactive ^{56}Co in the ejecta of SN 1987A, and monitored their evolution over two years. This has allowed us to infer the thickness of the envelope and, to a degree, the distribution of ^{56}Co within it. It is clear that the ^{56}Co was distributed over a large range of depths in mass in the ejecta, although the statistical precision of the data do not allow us to probe that distribution in detail. We have also derived the luminosity in all the >10 keV emission from ^{56}Co in a semi-empirical manner. Finally we note that although γ-ray spectroscopy has added insight into this supernova, type I supernovae emit relatively more information regarding their nature in γ-rays. At distances of 10 to 20 Mpc, where they are likely to be found, type I supernovae will be detectable with near-future γ-ray experiments at levels of significance comparable to those in this work. We are therefore hopeful that we will learn much about these objects, such as the mass of radioactivity ejected and something of their structure, from γ-ray measurements made in the next decade.

<u>References</u>

1. S.A. Colgate, C. McKee: *Ap. J.* **157**, 623 (1969).
2. D.D. Clayton, S.A. Colgate, G.J. Fishman: *Ap. J.* **155**, 75 (1969).
3. P.A. Whitelock *et al.*: These proceedings and references therein.
4. N.B. Suntzeff *et al.*: These proceedings and references therein.

5. W.P.S. Meikle: These proceedings and references therein.
6. D.M. Rank: These proceedings and references therein.
7. S.M. Matz *et al.*: *Nature* **331**, 416 (1988).
8. D.J. Forrest *et al.*: *Solar Physics* **65**, 15 (1980).
9. G.H. Share *et al.*: *Ap. J.* **326**, 717 (1988).
10. M.D. Leising *et al.*: in preparation (1989).
11. W.R. Cook *et al.*: *Ap. J. Lett.* **334**, L87 (1988).
12. W.R. Mahoney *et al.*: *Ap. J. Lett.* **334** L81 (1988).
13. W.G. Sandie *et al.*: *Ap. J. Lett.* **334**, L91 (1988).
14. A.C. Rester *et al.*: *Ap. J. Lett.* **342**, L71 (1989).
15. J. Tueller *et al.*: These proceedings.
16. S.M. Matz, G.H. Share, E.L. Chupp: *AIP Proceedings* **170**, 51 (1988).
17. W.D. Arnett: These proceedings.
18. K. Nomoto: These proceedings.
19. S.E. Woosley: These proceedings.
20. P.A. Pinto: These proceedings.
21. F.-K. Thielemann *et al.*: submitted to *Ap. J.* (1989).
22. P. Durouchoux: These proceedings.
23. J.D. Kurfess: *AIP Proceedings* **170**, 368 (1988).

X-Ray Observation of SN 1987A from GINGA

Yasuo Tanaka

1 Introduction

X-rays from SN1987A (hereafter abbreviated as SN) were detected first time in July 1987 from Ginga. The first result was published by DOTANI et al. [1], hereafter refered to as Paper I. In Paper I, it was shown that X-rays from the SN appeared to consist of two separate components; a hard component which was essentially flat in the range 10 - 40 keV and a soft component which rose towards low energies. The observed intensity of the hard component was in general agreement with the result of an independent observation from the Kvant experiments by SUNYAEV et al. [2], which measured the spectrum up to about 1 MeV. However, the Kvant observation did not confirm the soft component.

In this paper, we report on the X-ray intensity history of the SN obtained from the continued Ginga observations over 900 days after the outburst. After the error box was determined, we employed the pointing mode only (see Paper I). The X-ray light curves are obtained in two energy bands, 6 - 16 keV and 16 - 28 keV. These two energy bands are so chosen that the separation of the hard and soft comoponents is optimum, since the observed energy spectrum indicates that the hard X-ray component turns down below 20 keV. The results on the X-ray energy spectrum will be reported separately.

In obtaining the X-ray intensity of the SN, there are three different sources of systematic errors in addition to the statistical errors. These are (1) the cosmic-ray induced background, (2) the absolute pointing direction, and (3) the contributions from neighbouring sources in the field of view. The last one will be discussed in the next section. The results published in Paper I were also re-examined for these systematic errors.

A background measurement was performed immediately before or after each SN observation. In addition, the background rate during each SN observation can be estimated independently by an empirical method established by HAYASHIDA et al. [3]. This method also provides an estimate of the systematic error in the background estimation, which is about 1% of the background counts. The directly measured background and that estimated from this empirical method are in agreement within this systematic error. The systematic error is usually much larger than the statistical error, and is incorporated in the present errror estimation.

The error in the pointing direction influences not only the aspect determination for the SN but also the corrections for the nearby source contaminations. The error in the pointing direction estimated from the comaprison of the gyro data with the star tracker data is found to be at most 1.5 arcminutes. Although this error is not of statistical nature, we include the effect corresponding to an offset of ± 1.5 arcmin. in the error estimation.

2 Corrections for Nearby Sources

The region of the SN is populated with many X-ray sources as shown in Fig. 1 (LONG, HELFAND and GRABELSKY [4]). Among these sources, LMC X-1 and four supernova remnants SNR0540-69, N132D, N157B and SNR0519-69 (source No. 26 from the *Einstein* survey by LONG, HELFAND and GRABELSKY [4]) are relatively intense, and their contributions are corrected for. LMC X-1 is by far the brightest among them and also time variable. We therefore attempt to keep LMC X-1 outside the field of view. The exposure to LMC X-1 has been less than 2 %, with only one exception for the observation of July 4, 1987 in which the exposure was about 9 %. In this way, the contribution of LMC X-1 turns out to be much smaller than that of the SNR0540-69.3 described below. The contamination by LMC X-1 through reflection on the collimator walls is absolutely negligible in the range above 6 keV. Since the intensity of LMC X-1 at the time of each observation is unknown, we employ a large enough systematic error in order to cover the maximum variation based on the results from the frequent LMC X-1 observations from Ginga.

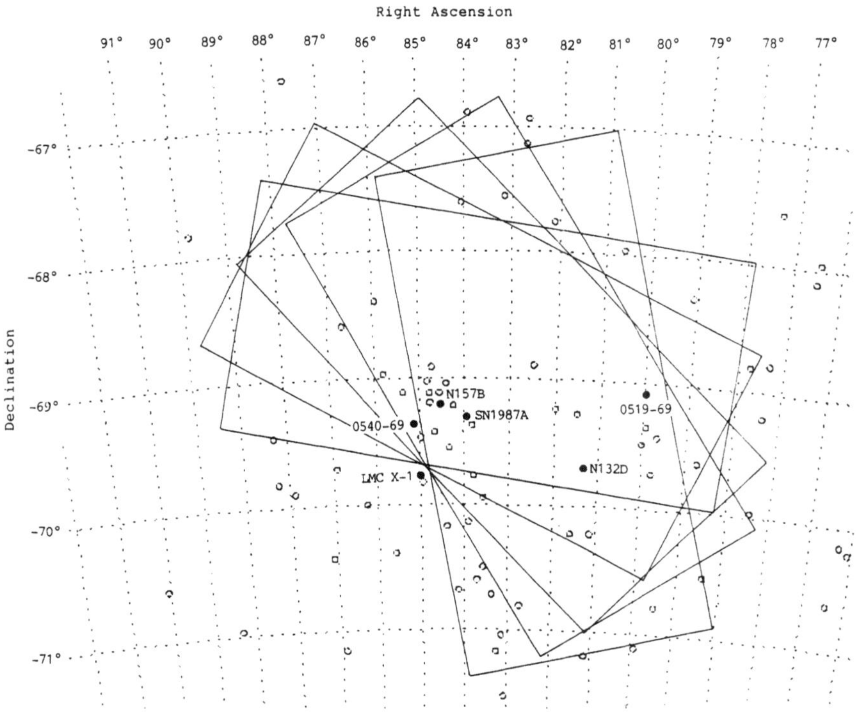

Fig. 1. The sky region of the Large Magellanic Cloud including SN1987A. X-ray sources detected from the Einstein survey (LONG et al. [4]) are indicated, and various orientations of the Ginga field of view employed for the SN observations are also illustrated.

Contributions of the supernova remnants SNR0540-69.3, N132D, N157B and SNR0519-69 were individually estimated for every SN observation. The spectrum of each of these SNR's was determined from separate Ginga observations with various pointings, putting the SN outside the field of view. However, the statistical uncertainties at high energies remain fairly large, since these sources are all weak for the Ginga sensitivity. We therefore determined the acceptable range of intensity for each source using two different model spectra, power-law model and thin thermal model. Power-law model gives higher flux estimates at high energies than the thin thermal model.

Among these sources, SNR0540-69.3 requires the largest correction, because this source, located only half a degree away from the SN, is the strongest in the energy range concerned and known to possess a hard spectrum (CLARK et al. [5]). This SNR includes a 50 ms. pulsar and is similar to the Crab Nebula in several respects (SEWARD et al. [6]). This pulsar component is measured in every SN observation, and an accurate intensity was obtained. The power-law index determined over the range 1 - 24 keV was 1.77 ± 0.09 in good agreement with that obtained in the Einstein energy range [6]. In the overlapping energy range, the measured fluxes from Ginga and the Einstein Observatory also agree with each other within statistical errors. The pulsed fraction in the range 1.7 - 4 keV is 27 ± 4% according to SEWARD et al. [6]. For obtaining a high estimation, we assume that the pulsed fraction is constant over the entire energy range. However, the pulsed fraction may increase with energy as in the case of the Crab Nebula. For a low estimation, we assume a thin thermal spectrum with kT = 10 keV for the remnant. In this case, the flux in the range 16 - 28 keV is essentially due to the pulsed component alone.

N157B is a SNR whose spectrum is similarly hard to SNR0540-69.3 in the Einstein energy range [5]. We assume the intensity of N157B to be approximately one quarter of SNR0540-69.3, the ratio in the Einstein range [5], over the entire energy range of Ginga.

N132D is a bright, oxygen rich SNR similar to Cas A. According to the result of the Einstein observation by CLARK et al. [5], the energy spectrum is composed of a soft (kT = 0.57 keV) component dominating in the low energy range and a hard (kT = 4.0 keV) component. However, an extrapolation of their best-fit model gives a much smaller flux than measured from Ginga, although the Ginga field of view includes another weaker source SNR0519-69. Since the Einstein spectrum of N132D [5] is very steep, a straight extrapolation of this spectrum from the Einstein range to the Ginga range is subject to a large ambiguity. We estimated the intensities of these two SNR's from our own Ginga observations assuming the same spectral shape for both and employing the intensity ratio of eight to one obtained from the Einstein observation [4]. A high estimation is obtained from the best-fit power law with a slope of 2.2, and a low estimation is given by an acceptable thin thermal spectrum with kT = 10 keV.

Thus obtained high and low estimates of the source intensities are given in Table 1. Contributions from these sources quickly diminishes towards higher energies. The other sources in the field of view are much fainter and estimated to be negligible. The diffuse background intensity in the SN region above 6 keV was determined for a field directly north of the SN, which excluded all major sources listed above.

Table 1. Estimated Fluxes of the Nearby Sources

Source Name	6 - 16 keV (counts/s)		16 - 28 keV (counts/s)	
	High Estimate	Low Estimate	High Estimate	Low Estimate
LMC X-1	20.30 ± 12.2		0.65 ± 0.48	
N132D	1.27	0.54	0.06	0.005
SNR0540-63	5.40	4.30	0.45	0.10
SNR0519-69	0.16	0.07	0.008	0.00
N157B	1.24	0.99	0.10	0.00
Diffuse Bkgd.	0.16		0.02	

3 X-Ray Light Curves of the SN

The light curves of the SN so far obtained are shown in Fig. 2(a) and 2(b) for the soft X-ray band in 6 - 16 keV and the hard X-ray band in 16 - 28 keV, respectively. The flux from the SN is plotted for two cases in which the high estimates (open circles) and the low estimates (crosses) of the nearby source intensities are employed. (See Table 1.) The 1σ errors (including all systematic errors) are shown on only one side of the individual data points. For an approximate conversion of a count rate (counts/sec) to an energy flux in units of 10^{-12} ergs/cm^{2}sec, multiply 4.6 for the soft X-ray band and 32 for the hard X-ray band.

The first positive detection of hard X-rays from the SN was on July 4, 1987 (131 days after outburst). The difference between the open circle and the cross in each observation is mostly due to the different intensity values of SNR0540-69.3 employed. The observed light curve indicates that the intensity in the hard X-ray band increased gradually by roughly a factor of two from July through December, 1987. In January 1988, hard X-ray intensity increased by a factor of two, which was however much less pronounced than the increase in the soft X-ray band. From the study of the energy spectrum, this increase in the hard X-ray band is explained as due to a spill over from the soft X-ray band. Excluding this period, the maximum of the hard X-ray intensity seems to have occurred near the end of 1987. Since then, the hard X-ray intensity has been steadily decreasing with time. If one assumes an exponential decline, the decay constant lies

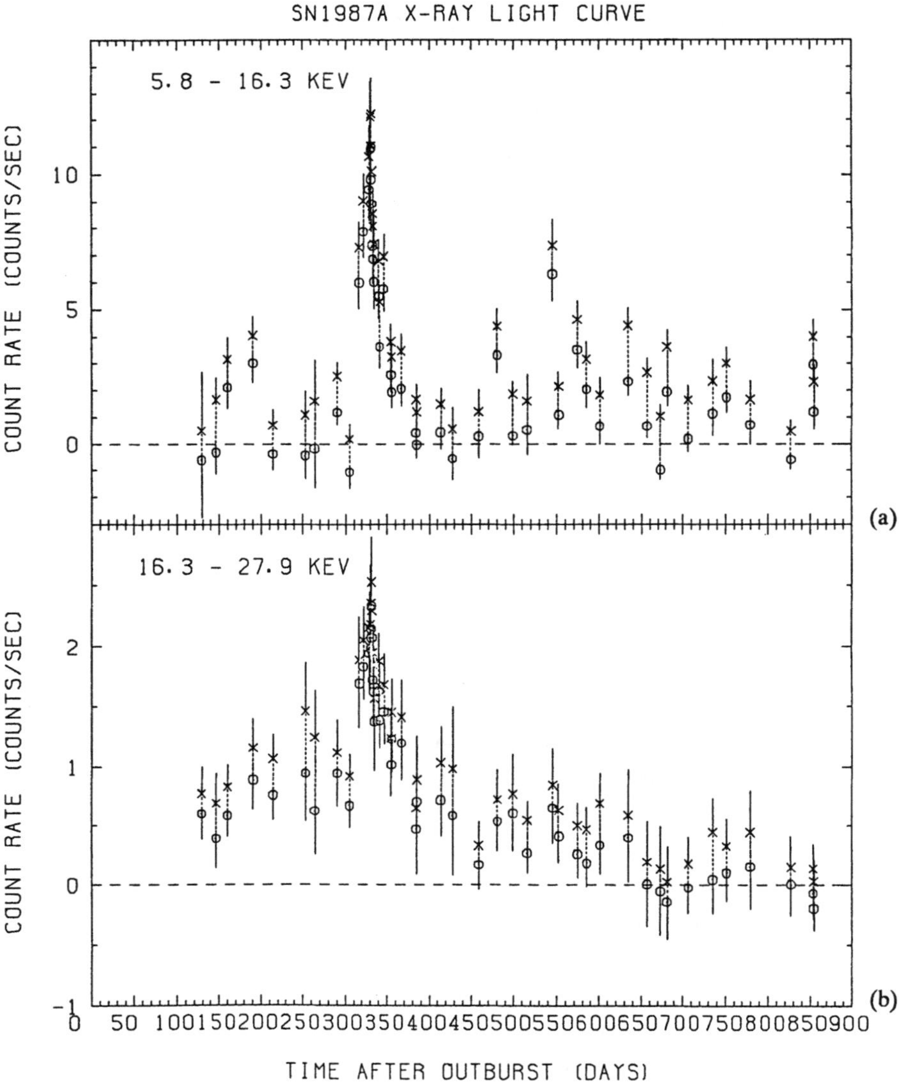

Fig. 2. The X-ray light curves of SN1987A in two energy ranges; (a) the soft X-ray band in 6 - 16 keV, and (b) the hard X-ray band in 16 - 28 keV. Open circles and crosses correspond to the cases in which the high estimates and the low estimates of the nearby source intensities are employed for the corrections, respectively. The error bars shown on only one side include all systematic errors.

in the range between 200 and 350 days. The flux approached to the detection limit around Januray 1989. Later, the average hard X-ray flux over the last two hundred days (650 - 850 days) is lower than the detection limit.

In the soft X-ray band, the difference between the open circle and the cross is due mostly to that between the high and low intensity values of SNR0540-69.3 and/or N132D employed, depending on the orientaion of the field of view. On January 7, 1988 (318 days), a dramatic intensity increase was detected. The intensity increased further and stayed at the maximum level from January 19 (330 days) through 22 (333 day). The intensity started to decrease from January 23 (334 days) steadily through March 14 (357 days) and came back to the general level of 1987. We shall hereafter call this event the "January flare". The light curve in the soft X-ray band for the part of the January flare is shown in Fig. 3. For this flare, we were able to determine the line position of the source by means of an aspect switching method, as drawn in Fig. 4. In addition, the spectrum during the flare was very hard, much harder than that of any known class of galactic X-ray sources. From these results, we conclude that the source of the flare is SN1987A.

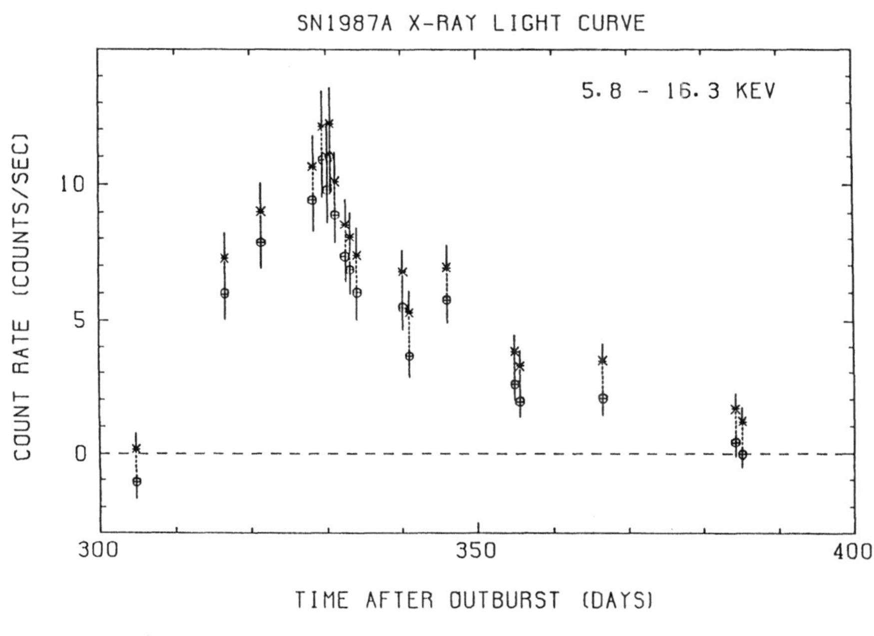

Fig. 3. The X-ray light curves during the January flare. See the caption for Fig. 2.

Besides the January flare, a significant soft X-ray flux was observed occasionally. For the crosses in which the high estimations of the nearby source intensities are used, about two thirds of the points are consistent to be null. On the other hand, there are several points for which the

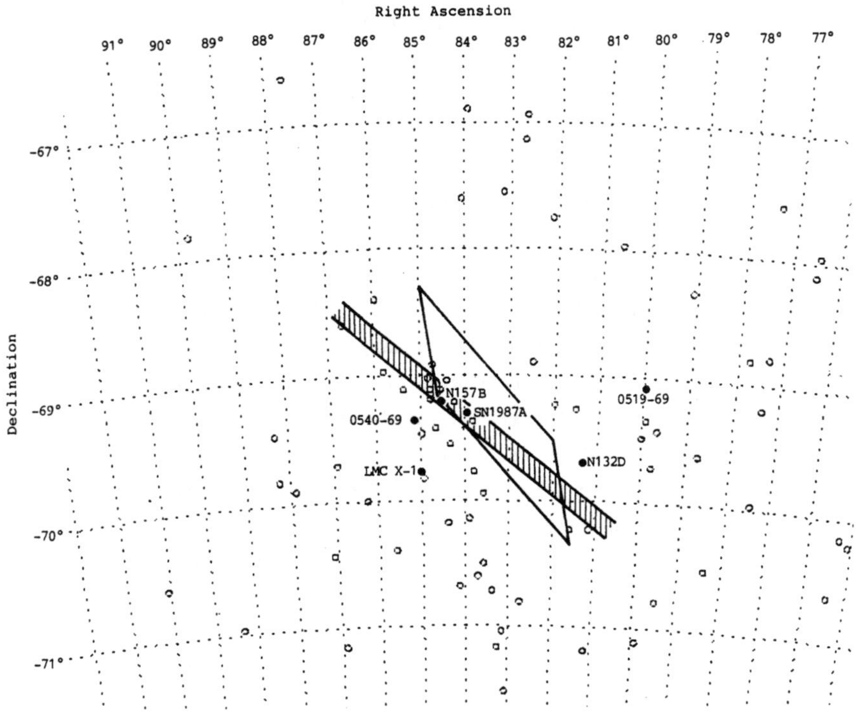

Fig. 4. The line position of the January flare (hatched) and the error box of the variable soft X-ray source.

observed soft X-rays are quite intense. The presence of a variable soft X-ray source in the region of SN1987A is beyond doubt. The soft X-ray intensity is highly variable on a time scale of the order of a day.

The accurate source position has not been determined yet. In order to obtain the source position, we are currently performing aspect switching observations. In this mode, the pointing direction is switched back and forth by 0.2° . The line position of the source can be determined from the amount of change in the count rate, when a significant soft X-ray flux is present. A two-dimensional error box will be obtained from the crossing of line positions determined at different orientaions of the field of view. At present, thus localized source region shown in Fig. 4 includes the SN. Yet, the error box is still fiarly large, and we cannot firmly exclude the possiblity that some other highly variable source which was not detected by LONG et al. [4] exists near the SN.

Search for pulsar with a high time resolution mode has been continued.

The result so far is negative. It is to be noted, however, that, if the pulsation period were shorter than 2 milliseconds, it would not be possible for Ginga to detect the pulsar due to the limitation of the time resolution.

Summary of the Observational Results

Hard X-rays from the SN is consistent to be of Co^{56} origin from the shape of the energy spectrum and the long, smooth decay. However, they emerged much earlier and lasted much longer than originally predicted. In addition, a soft X-ray flare occurred in the SN in January, 1988. Variable soft X-ray flux is also observed occasionally in the SN region, although we still reserve conclusion of the SN origin. No X-ray pulsation is detected as yet.

Acknowledgment

The author is indebted to the Ginga Team for making frequent observations of SN1987A posible in spite of extremely tight schedule of Ginga. This analysis was performed by K. Hayashida, H. Inoue, M. Itoh and the author.

References

1. T.Dotani et al.: Nature **330**, 230 (1987).

2. R.Sunyaev et al.: Nature **330**, 227 (1987).

3. K.Hayashida et al.: Publ. Astron. Soc. Japan **41**, 373 (1989).

4. K.S.Long, D.J.Helfand and D.A.Grabelsky: Astrophys. J. **248**, 925 (1981).

5. D.H.Clark et al.: Astrophys. J. **255**, 440 (1982).

6. F.D.Seward, F.R.Harnden Jr. and D.J.Helfand,D.J.: Astrophys. J. Letters **287**, L19 (1984).

High Resolution Observations of Gamma-Ray Line Profiles from SN 1987A

J. Tueller, S. Barthelmy, N. Gehrels, M. Leventhal, C. J. MacCallum, & B. J. Teegarden

Supernova 1987A was a unique opportunity for gamma-ray astronomers to observe freshly synthesized radioactive material from a type II supernova. Gamma-ray lines were first detected by the spectrometer on the SMM satellite (MATZ *et al.* [1,2]), which was the only instrument to provide a continuous monitor of the gamma-ray line light curves. Unfortunately, no high resolution satellite instrument was in space to make the critical continuous measurements of the shape of the gamma-ray lines, which are necessary to extract direct information about the distribution of ^{56}Co in the expanding supernova shell. Balloon-borne instruments were able to partially fill this gap with a few snapshots of the line profile (see Table 1). The Gamma-Ray Imaging Spectrometer (GRIS) was the most sensitive of a new generation of high resolution spectrometers which were rushed to completion in time to observe SN1987A (see TUELLER *et al.* [3] for a description of the instrument).

The gamma-ray measurements have forced revisions of the relatively simple pre-existing models of the supernova, which predicted a delay of about a year before the hydrogen envelope of SN1987A would be thin enough to allow the escape of gamma rays. Figure 1 summarizes the gamma-ray line flux measurements and illustrates the discrepancy with the initial models (5L). To allow the early escape of gamma-rays from the supernova, mixing was introduced to get the radioactive elements up near the surface where the optical depths are low. The 10HMM model is representative of these spherically-symmetric homogeneous mixed models and Fig. 1 illustrates that it can produce an acceptable fit to the gamma-ray line light curves. Gamma rays in this energy range interact by Compton scattering, thus the scattering cross-sections are not dependent on ionization state of the scattering medium and photons are usually scattered far away from the energy of the line. As a result, the line profiles can be calculated by a straightforward integration of the distribution of ^{56}Co multiplied by the transparency, which is determined by the mass distribution. The gamma-ray line profiles yield a direct measure of the distribution of radioactive material in the supernova without the complications of ionization state and resonant scattering effects that confuse the interpretation of line profiles from atomic transitions. While different in detail, all of these mixed models make qualitatively similar predictions of the gamma-ray line shape. Because the ^{56}Co is mixed far into the fast moving hydrogen envelope, large Doppler shifts, up to 3000 to 4000 km s^{-1}, will be observed at the edges of the lines. Because the source is optically thick and homogeneous, only the blueshifted photons from the approaching surface can reach the observer and, therefore, the line centroid will be blue shifted.

On day 433 (after the core collapse) the 847 and 1238 keV lines from ^{56}Co were observed by GRIS at 2.3 and 4.3σ significance. On day 613, the lines at 847, 1238, and 2599 keV were observed at 4.6, 3.4, and 1.9σ respectively. The combined significance for the three line complex in both flights is 7.8σ. Gaussian profiles yield acceptable least-squares fits to the lines. Figures 2, 3, and 4 are plots of photon spectra for the spring 1238 keV line and the fall

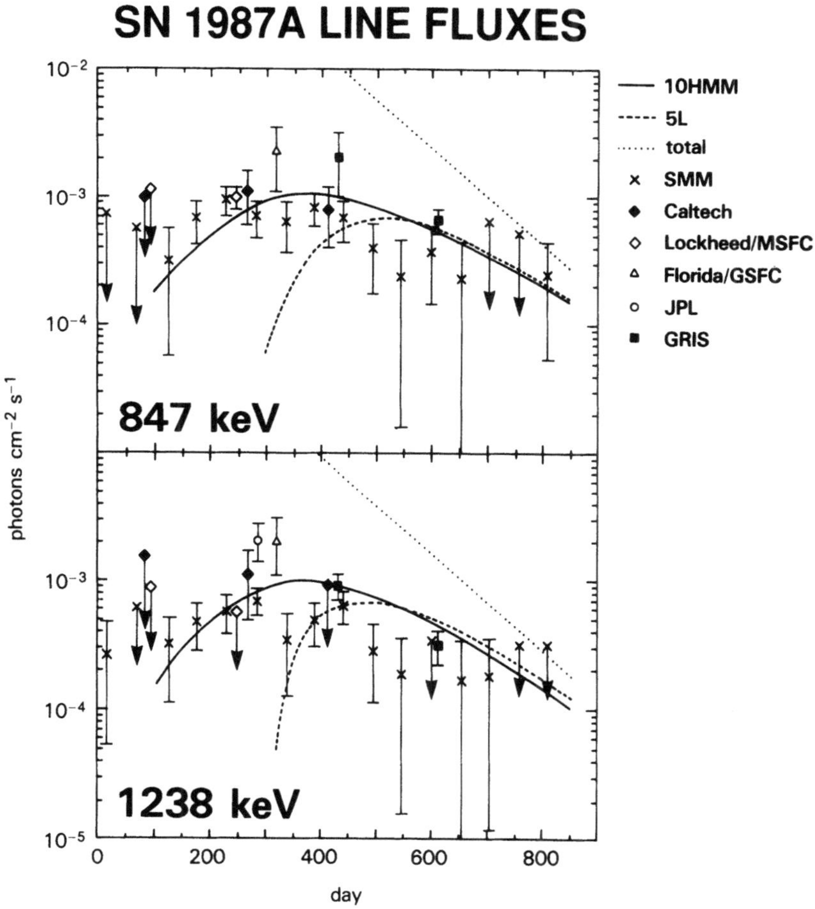

Figure 1. The GRIS line flux measurements are shown with theoretical predictions and the measurements of other instruments. The GRIS fluxes on day 433 are 21.+12.,-9. and 9.1±2.1 x10^{-4} photons cm^{-2} s^{-1} for the 847 and 1238 keV lines. On day 613 the fluxes are 6.5±1.4 and 3.1±0.9 x10^{-4} photons cm^{-2} s^{-1} for the 847 and 1238 keV lines. All of the measurements are in reasonably good agreement with the 10HMM model. (solid line: 10HMM model; dashed line: 5L model without mixing; dotted line: unattenuated line flux; all models from PINTO and WOOSLEY [6,7]; SMM: LEISING [8]; Caltech: COOK *et al.* [9,10,11]; Lockheed/MSFC: SANDIE *et al.* [12,13]; Florida/GSFC: RESTER *et al.* [14]; JPL: MAHONEY *et al.* [15]) This figure is from TUELLER *et al.* [5].

Table 1. SN1987A High Resolution Measurements

Group	Day	847 keV Line		1238 keV Line	
		Centroid (keV)	FWHM (keV)	Centroid (keV)	FWHM (keV)
Lockheed, MSFC	248-250	~844	~12	—	—
JPL	287	—	—	1240.8 ± 1.7	8.2 ± 3.4
Florida, GSFC	319-322	844.1 ± 1.0	6.0 ± 2.6	1239.0 ± 1.7	10.4 ± 3.4
GSFC, Bell Labs, and Sandia Labs	433	844.6 ± 1.2	8.7 +4.9,-3.3	1235.0 ± 2.2	14.8 ± 5.2
	613	844.2 ± 1.3	10.8 +2.9,-2.4	1233.5 ± 1.8	11.0 +3.7,-3.1

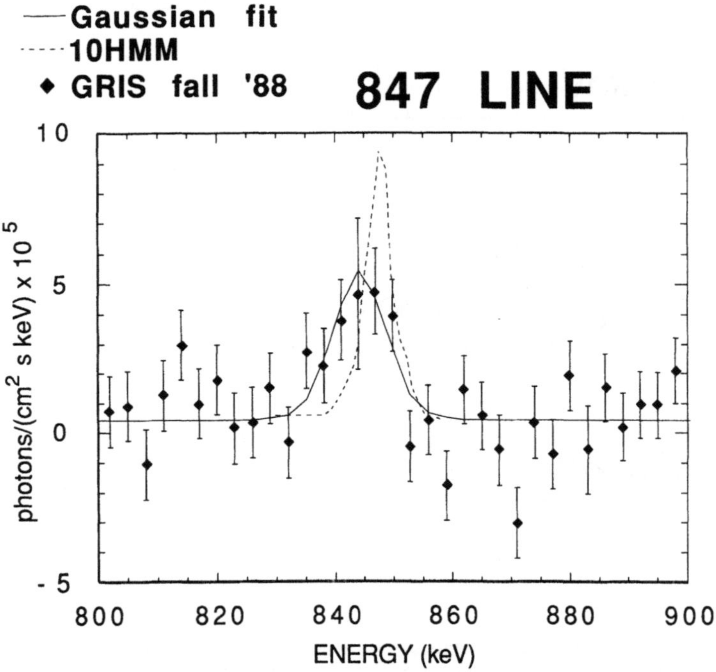

Figure 2. GRIS data for the 847 keV line from the decay of ^{56}Co in SN87A is shown for day 613. The top-of-atmosphere photon spectrum is shown in 3 keV bins which are larger than the FWHM resolution of the instrument. The solid line is a best fit to a Gaussian line profile. The dashed line is the predicted line profile for the 10HMM model. Although this model and similar spherically symmetric mixed models predict the line flux correctly there is a clear discrepancy with the measured line profile.

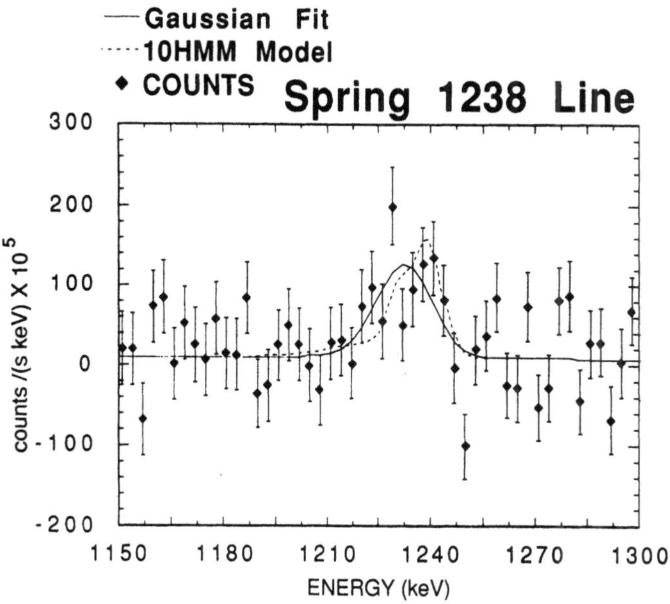

Figure 3. GRIS data for the 1238 keV line on day 433 is shown. The 10HMM model has been folded with the instrument response function to account for the spring flight resolution problem.

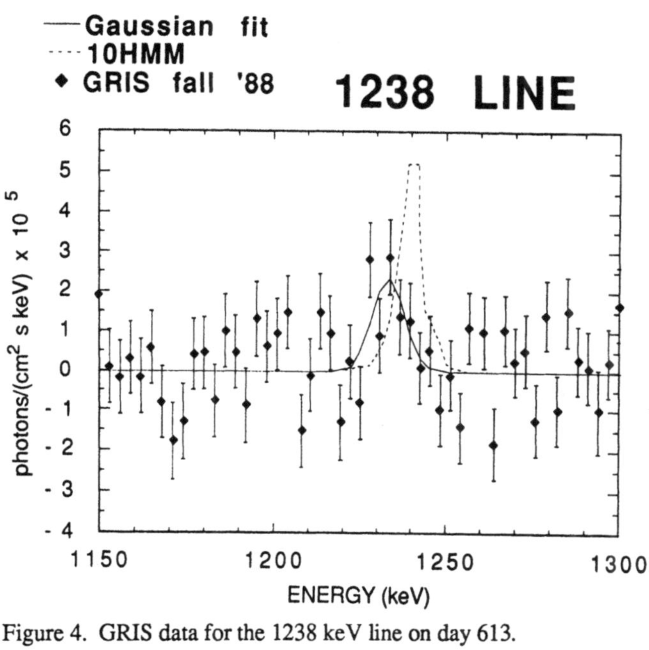

Figure 4. GRIS data for the 1238 keV line on day 613.

847 and 1238 keV lines. On each plot the the best fit Gaussian line profile (solid line) and the 10HMM line profile (dashed line) are shown. The observed lines are not blueshifted and are broader than the 10HMM prediction. The combined significance of the differences between our data and the 10HMM model is 4.9σ for the line centroids and 3.0σ for the widths. (The spring 847 keV line was not included because potentially significant systematic uncertainties have not been evaluated.) For the details of the data analysis see TEEGARDEN *et al.* [4] and TUELLER *et al.* [5]. Table 1 is a summary of the line profiles from all the high resolution measurements of SN1987A gamma-ray lines. At 847 keV, all of the instruments are in good agreement: the centroid of the line is redshifted from the rest energy of 846.0 keV (corrected for LMC redshift) and the width is consistently broader than the value of <5.5 keV predicted by the 10HMM model. While the situation at 1238 keV is less consistent, the GRIS results are the most statistically significant and are in excellent agreement with the 847 keV line profiles. The line centroids are redshifted from the rest energy of 1237.2 keV and the widths are consistently broader than predicted by 10HMM (<7.3 keV). The JPL results for the 1238 keV line are in better agreement with the 10HMM model, but they are only different from the GRIS values at the ~2σ level of significance. Although the results of the Florida/GSFC collaboration at 1238 keV do not appear to be consistent with their own 847 keV line profile, the uncertainties assigned by the authors to the line centroids are significantly smaller than the conventional value of FWHM/(2.36 x significance). This suggests that a more conservative analysis might find that these results are consistent. We conclude that the consensus of the high resolution gamma-ray line measurements contradicts predictions of the line profiles by 10HMM and similar models.

The observed gamma-ray lines have the profile of an optically thin source, but the fluxes are $\lesssim$30% of the emission of 0.075 $M_\odot$ of ^{56}Co determined from the bolometric light curve. The line widths imply velocity dispersions of ~3500 km s^{-1} in the distribution of radioactive material, which is much larger than predicted by unmixed models. To be consistent with the gamma-ray results, a theory must explain optically thin emission for the lines at times when most of the gamma rays are still absorbed in the source and when a uniform hydrogen envelope would still be optically thick. It must also explain the acceleration of a significant fraction of the ^{56}Co to velocities characteristic of the outer regions of the expanding supernova shell. To accommodate the apparent contradictions and explain the shape of the gamma-ray lines will require new and more complex models for SN1987A, such as a theory combining mixing and fragmentation.

References

1. S. M. Matz *et al.*: Nature 331, 416 (1988).
2. S. M.Matz, G. H.Share, and E. L. Chupp in Nuclear Spectroscopy of Astrophysical Sources (AIP Conf. Proc. 170), ed. N. Gehrels and G. H. Share (New York,AIP 1988), p. 51.
3. J.Tueller *et al.* in Nuclear Spectroscopy of Astrophysical Sources (AIP Conf. Proc. 170), ed. N. Gehrels and G. H. Share (New York,AIP 1988), p. 439.
4. B. J. Teegarden *et al.*: Nature, 339, 122 (1989).
5. J. Tueller *et al.*: submitted to Ap. J. Let. (1989).
6. P.Pinto and S. E.Woosley: Ap. J., 329, 820 (1988).
7. P.Pinto and S. E.Woosley: Nature, 333, 534 (1988).
8. M. Leising: private communication (1989).
9. W. R. Cook *et al.*: IAU Circ., 4400 (1987).
10. W. R. Cook *et al.*: IAU Circ., 4584 (1988).
11. W. R. Cook *et al.*: Ap. J. Let., 334, L87 (1988).
12. W. G. Sandie *et al*:: IAU Circ., 4463 (1987).
13. W. G.Sandie *et al.*: Ap. J. Let., 334, L91 (1988).
14. A. C. Rester *et al.*: Ap. J. Let., 342, L71 (1988).
15. W. A.Mahoney *et al.*: Ap. J. Let., 334, L81 (1988).

An Observation of SN1987A with a New High Resolution Gamma-Ray Spectrometer

J. Matteson, M. Pelling, B. Bowman, M. Briggs, R. Lingenfelter, L. Peterson, R. Lin, D. Smith, K. Hurley, C. Cork, D. Landis, P. Luke, N. Madden, D. Malone, R. Pehl, M. Pollard, P. von Ballmoos, M. Neil, & P. Durouchoux

1. Introduction

The discovery (Matz et al. [1]) of gamma-ray line emission at 847 and 1238 keV from radioactive ^{56}Co in the recent supernova SN1987A proved that explosive nucleosynthesis occurred in this supernova. Gamma-ray light curves derived from these and subsequent observations (Cook et al. [2], Mahoney et al. [3], Sandie et al. [4], Rester et al. [5], Teegarden et al. [6]) have a broad plateau from August 1987 to October 1988, with an 847 keV flux of $\sim 7\times10^{-4}$ ph/cm^2-sec (Tueller et al. [7]). The early detection of gamma rays required the inclusion of mixing or clumping in the models, (e.g. Pinto and Woosley [8] and Chan and Lingenfelter [9]). The gamma-ray fluxes are predicted, e.g. Bussard et al. [10], to peak at about day 400 and then decrease, by a factor of $\sim$ 6 at day 800, as the effect of increasing transparency becomes dominated by radioactive decay. Then they should depend primarily on the amount of ^{56}Co produced and little on the degree on mixing since most of the ^{56}Co should be exposed. Thus measurements of the ^{56}Co gamma-ray line fluxes and profiles will continue to be important during the decline of SN1987A.

Radioactive ^{57}Co is also expected to have been produced in SN1987A initially as ^{57}Ni by neutron rich nucleosynthesis in the deepest layers of the ejecta (Clayton [11]). Its production ratio, ^{57}Ni/^{56}Ni has been estimated (Woosley and Pinto [12], Kumagi et al. [13]) to lie between the solar value of the decay products, ^{57}Fe/^{56}Fe = 0.024 (Cameron [14]), and about twice this value. Depending on the mixing, velocity and mass of the overlying ejecta, the ^{57}Co's 122 keV gamma -rays are expected to have a broad maximum from about day 700 to 1600, with a peak flux from 5 to 15×10^{-5} ph/cm^2-s.

Observations of these phenomena require high resolution gamma-ray spectroscopy with a sensitivity to narrow lines of $\sim 10^{-4}$ ph/cm^2-s. Below we describe a new instrument with these capabilities and its 22 May 1989 observation of SN1987A.

2. The 12-Detector Germanium Spectrometer

This new balloon-borne, high resolution gamma-ray spectrometer was developed (Matteson et al. [15]) by a collaboration of US and French scientists in order to observe astrophysical sources of gamma-ray lines, with a sensitivity of $\sim 10^{-4}$ ph/cm^2-s, and to prove new instrumental techniques that may be applied to future space missions. Gamma rays are detected in high purity germanium detectors which measure $\sim$ 55 mm in length and $\sim$ 55 mm in diameter. They are cooled to 85 K by liquid nitrogen and have an energy resolution of 1 keV below 100 keV and 2 keV at 1 MeV. The detectors' total volume is 1568 cm^3 and total area is 285 cm^2. The instrument's new techniques include detector β-decay background rejection by detector segmentation and pulse shape discrimination (Roth et al. [17], Smith et al. [18]), a large array of 12 detectors and the use of 5 cm thick bismuth germanate (BGO) for the anticoincidence shield. The detectors' inner electrodes are segmented, providing a 1.2 cm thick front segment and a 4.3 cm thick rear segment. Segmentation provides information on the energy losses' axial position and dispersion. Pulse shape discrimination is used to determine the radial dispersion of energy losses. The position and dispersion data are then used to discriminate against β-decays (single site events) while accepting gamma rays (multiple site and front segment events). Data for accepted energy losses are telemetered in an event-by-event format at 40 kbps.

The sides and rear of the anticoincidence shield containing 51 BGO bars, with a maximum size of $5\times7\times21$ cm, each with its own 2 inch photomultiplier tube (PMT). The excellent light collection geometry results in an anticoincidence threshold of 30 keV even though BGO's light output is only $\sim$ 10 percent of NaI. Conventional CsI(Na), 10 cm thick, is used for the front of the shield. It has apertures over each detector which define a 20° FWHM field of view and nine 2 inch PMTs, which also give a 30 keV threshold. The BGO and CsI weigh 240 and 55 kg, respectively, and the entire instrument weighs 540 kg. Alt-azimuth pointing control is provided by the balloon gondola, whose complete weight, including the instrument, is 1134 kg.

3. Balloon Flight and Instrument Performance

The 12-Detector Germanium Spectrometer had its first balloon flight from Alice Springs, Australia on 22 May 1989, day 819 of SN1987A. Four of its detectors had fully operational segmentation and pulse shape discrimination. Carried on a Raven 28 million cubic foot balloon, it floated at 3.5–4.7 g/cm^2 for 17.4 hours and observed SN1987A for 9.9 hours, the Galactic Center for 6.3 hours, and the Crab Nebula and the transient x-ray source 0535+26 for 1.3 hours. Observations were performed by a series of target and background pointings, each for 20 minutes. Pointing aspect was verified by a star camera.

The instrument and gondola performed extremely well. The only problems were the failure of two detectors and the presence of spurious peaks below 200 keV in the background spectra. The causes for these are under investigation, and the latter is certain to be due to recovery from large energy losses. Neither problem had a significant effect on the scientific results. The background levels at 122, 847 and 1238 keV were 4×10^{-4}, 4×10^{-5} and 2×10^{-5} c/cm^2-s-keV, respectively. The background reductions from segmentation and pulse shape discrimination have not yet been exploited.

4. Preliminary Scientific Results

Preliminary analysis of the SN1987A data has been performed. We expect the final results to be up to a factor of 2 more sensitive due to finer energy resolution, which will result from more accurate gain corrections, the use of all the data, lower background through the use of segmentation and pulse shape discrimination and detector-to-detector anticoincidence, which will reduce the background by $\sim$ 20 percent.

The 2σ limits on broad (FWHM/E = 1 percent) gamma-ray line emission from SN1987A are 2×10^{-4} ph/cm^2-s at 122 keV from ^{57}Co, and 3.4 and 2.4×10^{-4} ph/cm^2-s at 847 and 1238 keV, respectively, from ^{56}Co. The ^{57}Co flux limit is a factor of 1.5 above the prediction of Kumagi et al. [13]. It can be interpreted to limit the ^{57}Co/^{56}Co ratio to < 0.086, or 3.6 times the solar value for ^{57}Fe/^{56}Fe. Here we assume the same optical depth at 122 and 847 keV. The 2σ limits on the two ^{56}Co gamma-rays have been combined using the branching ratios for their production and assuming that lines have about the same optical depth. This yields an equivalent limit at 847 keV of 2.4×10^{-4} ph/cm^2-s, which is a factor of $\sim$ 3 below the plateau of 847 keV flux measurements from August 1987 to October 1988, showing that the flux decreased after October 1988. This flux limit is only 0.75 of the flux that would be expected if the supernova ejecta were completely transparent, exposing the entire 0.075 $M_\odot$ of ^{56}Co, as determined from the bolometric light curve. Thus the optical depth for the 847 keV line must still have been > 0.3 on day 819, indicating that the decrease in the optical emission at about that time did not result from greatly increased gamma ray escape. This limit is nonetheless still a factor of 1.7 above the 10HMM model prediction (Pinto and Woosley [8]).

Acknowledgements

The successful development and balloon flight of the 12-Detector Germanium Spectrometer was the result of the work of many people. The authors express their appreciation for the efforts of F. Duttweiler, C. James, G. Huszar, D. Gruber, D. Philips, G. Beriones and G. Allen at UCSD, H. Primbsch, S. McBride, J. Penegor, and B. Campbell at UCB, G. Vedrenne, F. Cotin, J. Couteret and J. Coutelier at CESR and J. Poulalion at CEN–Saclay. The balloon flight and supporting logistics were ably handled by R. Kubara, D. Gage and D. Ball of NSBF and R. Sood of UNSW. This work was supported under NASA Grant NAGW-449.

References

1. Matz, S. M., et al. 1988, *Nature*, **331**, 416.
2. Cook, W. R., et al. 1988, *Ap. J. (Letters)*, **334**, L87.
3. Mahoney, W. A., et al. 1988, *Ap. J. (Letters)*, **334**, L81.
4. Sandie, W.G., et al. 1988, *Ap. J. (Letters)*, **334**, L91.
5. Rester, A.C., et al. 1989, *Ap. J. (Letters)*, in press.
6. Teegarden, B.J., et al. 1989, *Nature*, published.
7. Tueller, J., et al. 1989, in *Proceedings of the Gamma-Ray Observatory Workshop*, in press.
8. Pinto, P.A., and Woosley, S.E. 1988, *Nature*, **333**, 534.
9. Chan, K. W., and Lingenfelter, R. E. 1988, in *Nuclear Spectroscopy of Astrophysical Sources*, ed. N. Gehrels and G. Share (New York: Am. Inst. Phys.), p. 110.
10. Bussard, R. W., Burrows, A., and The, L.S. 1989, *Ap. J.*, **341**, 40.
11. Clayton, D. D. 1974, *Ap. J.*, **188**, 155.
12. Woosley, S.E., and Pinto, P.A., 1988 in *Nuclear Spectroscopy of Astrophysical Sources*, ed. N. Gehrels and G. Share (New York: Am. Inst. Phys.), p. 98.
13. Kumagi, S., et al. 1989, *Ap. J.*, in press.
14. Cameron, A. G. W. 1982, in *Essays in Nuclear Astrophysics*, ed. C. A. Barnes and D. N. Schramm (Cambridge: Cambridge Univ. Press) p. 23.
15. Matteson, J. L., et al. 1985, in *19th Internat. Cosmic Ray Conf. Papers*, **3**, 326.
16. Matteson, et al. 1989 in *21st Internat. Cosmic Ray Conf. Papers*, in press.
17. Roth, J., Primbsch, J.H., and Lin, R.P. 1984, *IEEE Trans. Nuc. Sci.*, **31**, 367.
18. Smith, D.M., et al. 1988, in *Nuclear Spectroscopy of Astrophysical Sources*, ed. N. Gehrels and G. Share (New York: Am. Inst. Phys.), p. 484.

Pre-Discovery Hard X- and Gamma-Ray Luminosity of SN 1987A from Optical Spectra

N. N. Chugai

Modelling of the emergent hard X- and gamma-ray emission in SN1987a is an important diagnostic tool for ^{56}Co distribution in the envelope as well as for the envelope structure (SUNYAEV *et al.* [1]; KUMAGAI *et al.* [2]; GREBENEV and SUNYAEV [3], and ref. there). The time of apearence of escaping gamma-rays is crucial point for the model. GRAHAM [4] proposed to estimate pre-discovery gamma-ray flux from HeI 10830 A scattering line, which had been recognised as a probe of ^{56}Co decay in SNII just before detection of this line in SN1987A (CHUGAI [5]). In the present paper I use Hα absorption to probe escaping gamma-rays. New estimates of gamma-ray flux dramatically exceed those by GRAHAM [4]. This controversy stems from previously missed processes, namely, Penning ionization and radiative transition $2^3S-2^3P-1^1S$, which strongly depopulate 2^3S level of He.

1. Non-Thermal Ionization and Excitation of H and He

In outer layers presumably transparent to gamma-rays ($\tau_\gamma<1$) the energy deposition rate [ergs cm^{-3} s^{-1}] due to Compton scattering is

$$\epsilon=(4\pi)^{-1}(vt)^{-2}\eta\langle\sigma_\gamma\rangle(n_H+2n_{He})L_\gamma, \quad (1)$$

where $4\pi\eta$ is a ratio of the average intensity of gamma-rays to the flux (we adopt $\eta=1.4$), $\langle\sigma_\gamma\rangle=0.13\sigma_T=0.86\times10^{-25}$ cm^2 is the cross-section $\sigma_C(\Delta E/E)$ weighted over the gamma-ray spectrum. Primary (Compton) electrons loose their energy "on the spot" on ionization and excitation of H and He, and heating of thermal electrons. The characteristic rate of non-thermal ionization is $\zeta^0=\epsilon_H W_H n_H^{-1}$ [s^{-1}], where W_H=36 eV is the average energy spent on creation of H^+ ion (cf. DALGARNO and Mc CRAY [6]). Taking into account main processes of the degradation of the deposited energy and adopting ionization degree x_H $10^{-3}-10^{-2}$, and hydrogen abundance X=0.6 one obtains rates of non-thermal ionization, ζ, and excitation, ξ, of hydrogen and helium:

$$\zeta_H = 1.14\zeta^o, \quad \zeta_{He} = 0.14\zeta^o,$$
$$\epsilon_H = 0.77\zeta^o, \quad \epsilon_{He}(2^3S) = 0.026\zeta^o. \tag{2}$$

In the "two level plus continuum" approximation values n_e and n_2 are defined by equations (we assume $n_e = n_H$ and $v = r/t$)

$$dn_e/dt = -3n_e/t + \zeta_H n_H + P_2 n_2 - \alpha_B n_e^2,$$

$$n_2(A_{2q} + P_2 + A_{21}\beta_{12}) = \epsilon_H n_H + \alpha_B n_e^2, \tag{3}$$

where $A_{2q} = 2$ [s^{-1}] is the Einstein coefficient for the two-photon transition 2-1 (2s and 2p states are farely mixed), P_2 is the photoionization rate in a given point, β_{12} is Lα escape probability. In steady-state regime (i.e for $n_e/\zeta_H n_H \ll t$), equations (3) can be reduced to

$$n_2(A_{2q} + A_{21}\beta_{12}) = (\zeta_H + \epsilon_H)n_H. \tag{4}$$

The equation for He^+ concentration is

$$dn_{He^+}/dt = -3n_{He^+}/t + \zeta_{He}n_{He} - \alpha_B^{He}n_e n_{He^+}, \tag{5}$$

while for the population of 2^3S level of HeI the equation is

$$n_{He}(2^3S)(C + C_P + R) = \epsilon_{He}n_{He} + \alpha(^3L)n_e n_{He^+}, \tag{6}$$

where $C = qn_e$ is the rate of collisional transition 2^3S-2^1S, $q = 1.9\times10^{-8}$ cm^3s^{-1} (cf. OSTERBROCK [7]), $C_P = \langle\sigma_P u_{th}\rangle n_H$ is the rate of Penning process $He(2^3S) + H - He + H^+ + e$, equal to 3×10^{-9} cm^3s^{-1} for $T = 5000$ K (OLSON [8]; BELL [9]; SHAW *et al.* [10]). R is the rate of two-stage radiative transition 2^3S - 2^3P,

$$R = B_{23}f_\nu^c(4\pi)^{-1}(D/vt)^2 e^{0.92A(\lambda)}A_{32}^{-1}A_{31}\beta_{13}, \tag{7}$$

where f_ν^c is the continuum flux at $\lambda = 10830$ A, determined from IR spectra, see: BOUCHET *et al.* [11]; ELIAS *et al.* [12] (distance D = 50 kpc and $A_V = 0.6$ mag are adopted here). Both, Penning process and two-stage radiative transition, dominate depopultion of 2^3S due to electron collisions in the outer layers (v>4000 km/s) of SN1987A at the relevant epoch.

2. Density in the Outer Layers of SN1987A

Data on hard X- and gamma-rays from SN1987A (cf. COOK *et al.* [13]; MATZ *et al.* [14]; GREBENEV and SUNYAEV [3]) imply that in mid-October 1987 gamma-ray luminosity of SN1987 was $L_\gamma = 4\times10^{39}$ ergs s^{-1} (cf. GREBENEV and SUNYAEV [3]). With the known value of L_γ one can find from (1) and (2) concentration of hydrogen in the outer layers for a given n_2. On the other hand, n_2 can be obtained from the blue wing of Hα or Hβ absorption. On 13 Oct. 1987 according to Hβ profile in spectrum of SN1987A (cf. CATCHPOLE *et al.* [15]) n_2 in the range $4000<v<7000$ km/s can be described by $n_2 = 0.26(v/5000\ \mathrm{km/s})^{-10}$ (CHUGAI [16]). This result and the analysis of hydrogen excitation lead to the density in the range of velocities 4000-7000 km s^{-1}

$$n_H = 0.8\times10^{8}(v/5000\ \mathrm{km\ s^{-1}})^{-m}(t/232\ \mathrm{d})^{-3},$$

where $5<m<6$. For $X=0.6$, $Y=0.4$ one obtains at $v=5000$ km s^{-1} on 232 day the density $\rho=2.2\times10^{-16}$ g cm^{-3}, which is only two times lower than the density in the model by SHIGEYAMA *et al.* [17].

3. Hα Absorption as an Indicator of Gamma-Ray Escape

The relative intensity in the blue wing of Hα absorption ($v_z = -7000$ km s^{-1}) at the epoch 20 - 150 days after explosion in accordance with the spectra of SN1987A (MENZIES *et al.* [19]; CATCHPOLE *et al.* [20]; CATCHPOLE *et al.* [15]), as well as the model curves (with and without nonthermal excitation) are shown in Fig 1. Equations (3) were solved for the ionization degree $x_H=0.00124$ on 25 day. $P_2 = 2.9(t/100\ \mathrm{d})^{-2}(v/7000\ \mathrm{km/s})^{-2}$ s^{-1} was adopted in accordance with the UV data in the range 2500-3400 A from IUE (KIRSHNER [21]) and ASTRON (BOYARCHUK *et al.* [22]). The dependence of ζ_H on time was parametrized as

$$\zeta_H = 2\zeta_H^{o}(t/50\ \mathrm{d})^{k}[1+(t/50\ \mathrm{d})^{k-1}]^{-1},$$

where parameters k and ζ_H^{o} were determined from the best fit of Hα absorption evolution (see Fig. 1).

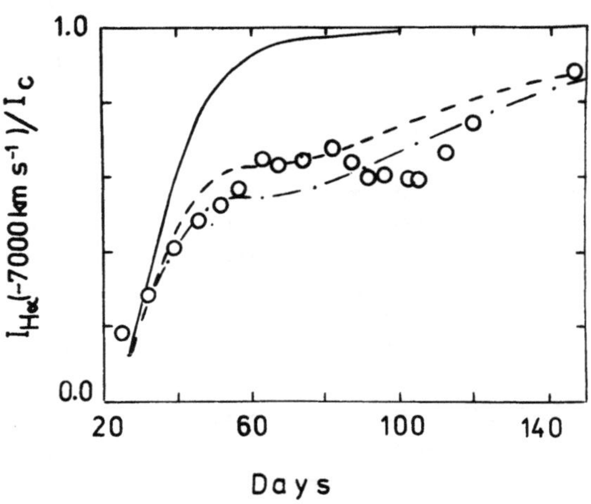

Figure 1. Evolution of intensity of Hα absorption at $v_z=-7000$ km/s. Circules: data from SAAO (see text); solid line: model without non-thermal excitation; dashed line: model with non-thermal excitation, $\zeta_H^{o}=4.65\times10^{-10}$ s^{-1}; and dash-dotted line: with non-thermal excitation, $\zeta_H^{o}=5.8\times10^{-10}$ s^{-1}

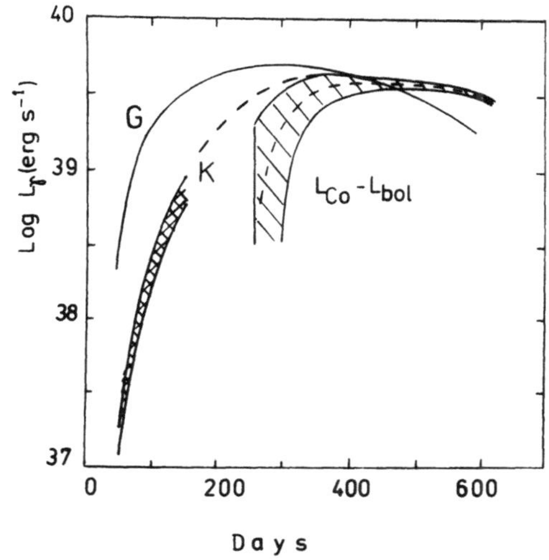

Figure 2. Evolution of gamma-ray luminosity of SN1987A. Cross-hatched field: this work; line-hatched field: from WHITELOCK *et al.* [23]; curve G: model by GREBENEV and SUNYAEV [3]; K: model by KUMAGAI *et al.* [2]

Figure 1 shows that model without non-thermal excitation fails for t>40 days. Models with non-zero $\zeta_H{}^0$, 5.8×10^{-10} and 4.65×10^{-10} s^{-1}, (in Fig. 1 the case m=6, k=4 is shown) fit better (especially latter case). For m=5 best fit values are k=4 and $\zeta_H{}^0 = 3\times10^{-10}$ s^{-1}.

With known n_H and ζ_H the hard X- and gamma-ray luminosity of SN1987A was found from (1). Two extreme set of parameters (a) m=5, $\zeta_H{}^0=3\times10^{-10}$ s^{-1}, and (b) m=6, $\zeta_H{}^0=4.65\times10^{-10}$ s^{-1} give lower and upper curves $L_\gamma(t)$ in Fig. 2. For t>250 days evolution of $L_\gamma(t)$ is given according to WHITELOCK *et al.* [23]. Gamma-ray luminosity calculated by KUMAGAI *et al.* [2] (K) and by GREBENEV and SUNYAEV [3] (G) are presented at this plot too.

Pre-discovery (untill Jul-Aug 1987) gamma-ray luminosity is substantially higher than that suggested by early behavior of L_γ curve given by WHITELOCK *et al.* [23]. Yet, the model calculations (both suggest mixing of ^{56}Co, up to 4200 km s^{-1} in K, and up to 5000 km s^{-1} in G model) generally agree with the high early gamma-ray luminosity implied by hydrogen lines.

4. HeI 10830 A Line and Escaping Gamma-Rays

On 76, 105, and 135 day gamma-ray luminosity found from Hα, respectively, 10^3, 10^2, and 10 times exceeds those found by GRAHAM [4] from HeI 10830 A line. This discrepancy was removed by taking into account Penning process (He^*+H) and radiative mechanism of the 2^3S depopulation. For two sets of parameters m, k, and $\zeta_H{}^0$, corresponding to upper and lower limit for

Table 1. Theretical and "observed" values of $n_{He}(2^3S)$

t days	L_y, 10^{38} ergs/s m=5	m=6	$n(2^3S)$, cm^{-3} m=5	m=6	obs.
76	0.63	1.0	0.012	0.018	0.008
105	2.0	2.9	0.018	0.025	0.02
135	4.5	7.1	0.026	0.039	0.03

$L_y(t)$ implied by Hα (see Fig. 2 and Table 1), equations (3,5) and (7) were solved to find n_e, n_{He^+}, and $n(2^3S)$ on 76, 105 and 135 days at the level v=7000 km s^{-1}. Results are presented in the fourth and fifth column of Table 1. Values of $n_{He}(2^3S)$, found from profiles of HeI 10830 A in the spectra published by ELIAS *et al.* [12] for the same level v=7000 km s^{-1}, are given in the last column. Agreement between theoretical and "observed" values of $n(2^3S)$ is fairly good, thus supporting reliability of pre-discovery gamma-ray luminosity found from hydrogen lines.

References

1. R.A. Sunyaev et al.: Nature 330, 227 (1987).
2. S. Kumagai et al.: Astr. Ap. 197, L7 (1988).
3. S.A. Grebenev, R.A. Sunyaev: Preprint IKI No. 1560 (1989).
4. J.R. Graham: Ap. J. (Letters) 335, L53 (1988).
5. N.N. Chugai: Pisma Astr. Zh. 13, 671 (1987).
6. A. Dalgarno, R.A. Mc Cray: Ann. Rev. Astr. Ap. 10, 375 (1972).
7. D. Osterbrock: In Astrophysics of gaseous nebulae (Freeman, San Francisco 1974).
8. R.E. Olson: Phys. Rev. 6A, 1031 (1972).
9. K.L. Bell: J. Phys. B. 3, 1308 (1970).
10. M.J. Shaw et al.: Chem. Phys. Lett. 8, 148 (1971).
11. P. Bouchet et al.: ESO Preprint No. 592 (1988).
12. J.H. Elias et al.: In CTIO 25th Aniversary Symposium (San Francisco 1988).
13. W.R. Cook et al.: Ap. J. (Letters) 334, L87 (1988).
14. S.M. Matz et al.: Nature 331, 416 (1988).
15. R.M. Catchpole et al. Mon. Not. R. astr. Soc. 231, 75P (1988).
16. N.N. Chugai: Pisma Astr. Zh. 14, 1079 (1988).
17. T. Shigeyama et al.: Astr. Ap. 196, 141 (1988).
18. W.D. Arnett: Ap. J. 331, 377 (1988).
19. J.W. Menzies et al.: Mon. Not. R. astr. Soc. 227, 39P (1987).
20. R.M. Catchpole et al.: Mon. Not. R. astr. Soc. 229, 15P (1987)
21. R.P. Kirshner: Preprint of Harvard-Smithsonian Center for Astrophysics No. 2888 (1989).
22. A.A. Boyarchuk et al.: Pisma Astr. Zh. 13, 739 (1987).
23. P.A. Whitelock et al.: Preprint SAAO No. 631 (1989).

Future Missions for Gamma-Ray Astronomy — Sigma and the Nuclear Astrophysics Explorer

P. Durouchoux & J. Matteson

1. Introduction

Continuing progress in gamma-ray astronomy requires new instruments which provide significant improvements in sensitivity, angular resolution and energy resolution. The Gamma-Ray Observatory, scheduled for launch in June 1990, will obtain a substantial improvement in sensitivity over earlier space missions. However, in the MeV energy range its angular resolution, $\sim 5°$, will not be adequate to resolve dense source regions and locate sources with high accuracy, and its energy resolution, $E/\Delta E \sim 10$, will not be adequate to resolve gamma-ray lines and measure their profiles. In this paper we describe two future instruments which will obtain much better angular and energy resolution while maintaining high sensitivity. The SIGMA was launched on the USSR's GRANAT spacecraft on 1 December 1989. It will obtain 13 arc minute resolution imaging in the 30 keV to 2 MeV range. The Nuclear Astrophysics Explorer is a possible NASA Explorer mission which could be launched in the late 1990's. It would perform high resolution spectroscopy in the 15 keV to 10 MeV range, with an energy resolution of 2 keV at 1 MeV, i.e. $E/\Delta E = 500$.

2. SIGMA

The SIGMA telescope is the first space instrument that obtains high quality imaging in the low energy gamma-ray range. Developed in France, it is expected to operate for several years and has the following features.

- a wide operational bandwidth, 30 keV – 2 MeV, which will link high energy x-ray and gamma-ray astronomy.
- good angular resolution, typically 13 arc minutes, within a field of view measuring 4.7° x 4.3°, fully coded, or 11.5° x 10.6°, partially coded. Stronger sources will be positioned to ~1 arc minute.
- moderate energy resolution, 30% to 6%, depending on energy.
- high sensitivity for both continuum, of the order of a few times 10^{-7} to few times 10^{-6} ph/cm^2-s-keV, and line measurements, $\sim 3 \times 10^{-5}$ ph/cm^2-s.

2.1 Description of SIGMA

A cross section of the SIGMA telescope is shown in Fig. 1. Its four key subsystems, the coded mask, position sensitive detector, detector electronics and anticoincidence shield are described below.

Coded Mask – This consists of 49 x 53 elements, based on a 29 x 31 Hadamard matrix. The elements are made of tungsten, measuring 9.4 x 9.4 x 15 mm thick. The thickness has been choosen in order to give good absorption of photons in the overall energy range. These elements are attached to a carbon-Nomex honeycomb plate which provides the necessary stiffness and transparency to gamma-rays. The mask is located 250 cm above the detector plane,

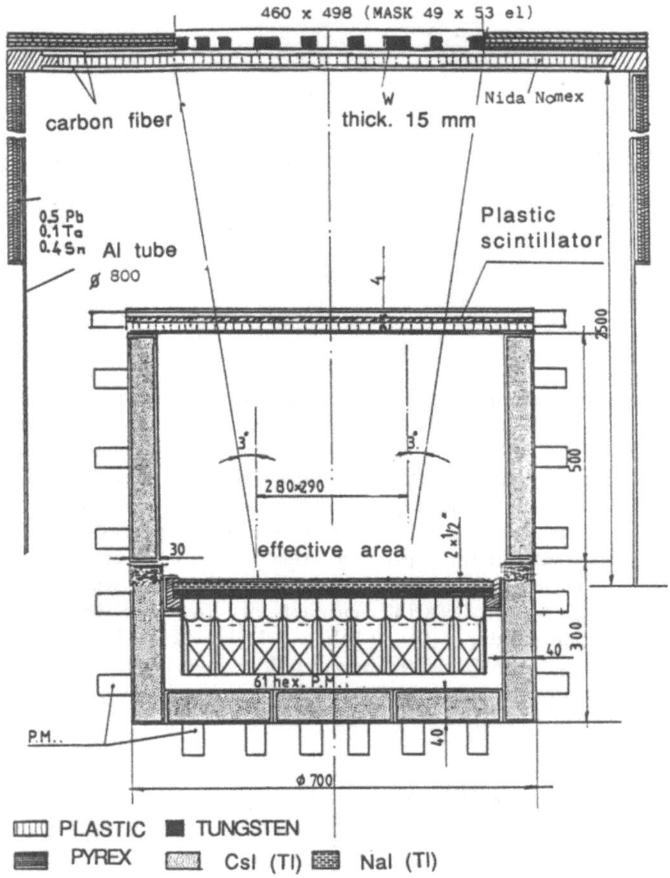

Figure 1. A cross section of the SIGMA instrument.

attached to a tube which provides the requisite stiffness to achieve the required alignment accuracy.

Detector – This is directly derived from a medical camera, and consists of a NaI crystal, 57 cm diameter x 1.25 cm thick, which is viewed by 61 hexagonal photomultiplier tubes (PMTs).

Detector Electronics – These process the signals from the PMTs to determine the coordinates of the sites of the photon interactions and their energy losses. The coordinate data are collected in a 232 x 248 element matrix.

Anticoincidence Shield – This shields the detector in order to reduce the background from gamma-rays and charged particles. It contains CsI crystals, with a thickness of 3 to 4 cm on the sides and 4 cm on the bottom. A thin plastic scintillator is located at the top of the shield.

2.2 Performance

The sensitivity of SIGMA depends on the background, which is directly dependent on the orbit. SIGMA is in a high altitude, highly eccentric orbit with a 2000 km perigee, 200000 km apogee and 4 day period. This has higher background than the typical low altitude orbit, but allows a continuous observations during the 3 days of each orbit above 60000 km. In the range 20–100 keV, the sensitivity is of the order of few milli-Crab. Table 1 gives the continuum sensitivity and Table 2 gives the line sensitivity. The continuum sensitivity and spectra of selected sources are shown in Fig. 2.

Table 1. 3σ Continuum Sensitivity for 1 and 10 day Observations

E (keV)	Sensitivity (ph/cm^2-s-keV) (1 day)	(10 days)
20–30	1.7x10^{-5}	5.4x10^{-6}
30–50	7.5x10^{-6}	2.4x10^{-6}
50–100	3.7x10^{-6}	1.2x10^{-6}
100–300	1.6x10^{-6}	5.0x10^{-7}
300–500	2.0x10^{-6}	6.5x10^{-7}
500–1000	1.3x10^{-6}	4.2x10^{-7}
1000–2000	1.0x10^{-6}	3.2x10^{-7}

Table 2. 3σ Line Sensitivity (10 day Observation)

E (keV)	Sensitivity (ph/cm^2-s)
30	3.3x10^{-5}
70	2.5x10^{-5}
200	3.9x10^{-5}
511	1.6x10^{-4}

The sensitivity to the 122 keV gamma-ray from ^{57}Co gamma-ray is of the order of 3 x 10^{-5} ph/cm^2-s, which is below the theoretical predictions for SN1987A. At 511 keV the sensitivity is a factor of $\sim$ 10 below the maximum flux reported from the Galactic Center region. This should allow maps to be obtained which will provide $<$ 2 arc minute localization of the variable point source of 511 keV gamma-rays.

The angular resolution results from the distance between the mask and the detector, 250 cm, and the size of the mask elements, 9.4 mm. The nominal resolution is 13 arc minutes. Strong sources can be localized to about 10 times this accuracy, i.e. $\sim$ 1 arc minute.

Gamma-ray bursts will be detectable with SIGMA, both with the detector and the shield. For the detector, with an area of 400 cm^2 in the energy range 20–100 keV, a sensitivity of 1.6 $\times$ 10^{-8} erg/cm^2 is expected for time bins of 0.25 s.

2.3 Observation Program

SIGMA will perform observations of a wide range of galactic and extragalactic objects including active galaxies and quasars, clusters of galaxies, selected regions of the galactic plane, interstellar clouds, supernova remmants, the COS-B sources, radio pulsars, known x-ray sources and targets of opportunity such as supernovae and novae. Most observations will require from 1 to 10 days, depending on the sensitivity desired.

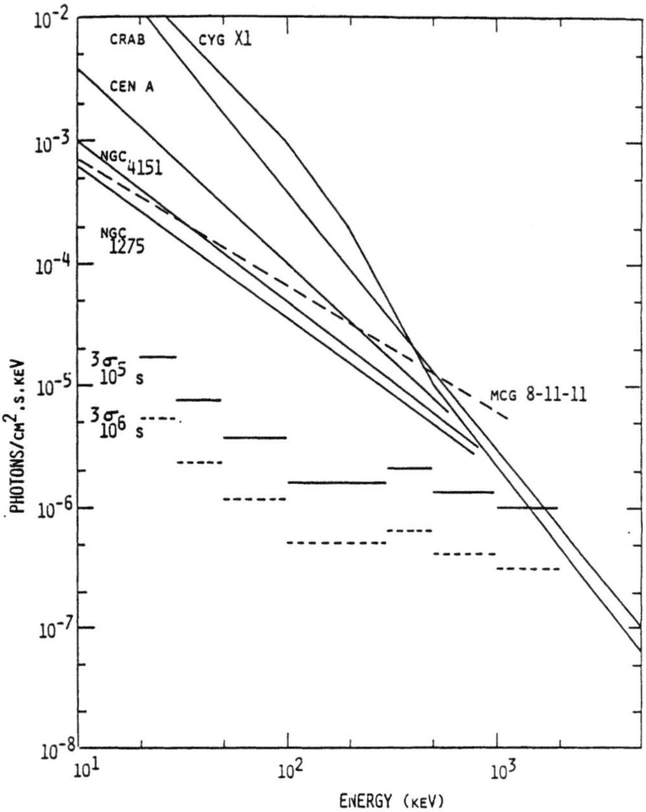

Figure 2. SIGMA continuum sensitivity for 1 and 10 day observations and the spectra of selected sources.

The study of active galaxies and quasars is one of the major objectives of the mission. The unambiguous identification of a large number of active galaxies will lead directly to the compilation of a gamma-ray luminosity function for these objects in the region of the spectrum where their luminosity is maximum. SIGMA will also study their variability both in hard x-rays and gamma-rays, as well as the correlation of these variations.

A survey of the central region of the Galaxy is scheduled, in particular at the annihilation energy of 511 keV. Due to the expected point source flux at this energy, $\sim 1\text{x}10^{-3}$ ph/cm^2-s, SIGMA should be able to localize the point source with an accuracy of ~2 arc minutes, which is equivalent to 5 pc at the center of the Galaxy. This goal is very important in terms of understanding the object, which might be at the galactic nucleus itself.

2.4 Status and Plans

The SIGMA was launched on the USSR's GRANAT satellite on 1 December 1989 and is operating nominally, with a slightly lower than expected background. The planned operational life is 18 months, but the onboard consumables permit the mission to be extended by a few years. No Guest Observer program is officially set up, but people interested in collaboration should contact one of the two French PI's:

J. Paul	P. Mandrou
CEN/SACLAY	CESR
SAP/GERES	9, avenue du Colonel Roche
91191 Gif Sur Yvette	31400 Toulouse
Cedex, France	France

3. The Nuclear Astrophysics Explorer

The Nuclear Astrophysics Explorer (NAE) is a concept for a future NASA Explorer mission which would obtain high resolution, $E/\Delta E \sim 500$, observations of gamma-ray lines, at a sensitivity of $\sim 3x10^{-6}$ ph/cm^2-s, or 100 times below the presently known fluxes, in order to study many fundamental problems in astrophysics. In July 1989 NASA's Explorer Concept Study Program completed a 1-year feasibility study of the NAE and three other missions. Two of these were to be selected for development for missions in the mid 1990's. Although the NAE was not selected, there is strong interest within NASA and the US and international scientific communities in performing the NAE mission. The late 1990's would be the earliest date for this. The scientific motivation, instrument concept, mission concept and expected results, and status and plans are discussed below.

3.1 Scientific Motivation

Gamma-ray lines are the most direct probe of cosmic nuclear processes. High resolution spectroscopy of these lines can provide basic information on fundamental astrophysical problems such as nucleosynthesis, supernovae dynamics, neutron star and black hole physics, and particle acceleration and interactions. Gamma-ray line observations have, in fact, already made major contributions to our understanding of a variety of astrophysical objects and phenomena. Table 3 contains a summary of these observations, which were made with instruments that had relatively low sensitivity, greater than $\sim 2x10^{-4}$ ph/cm^2-s, and, in many cases, low energy resolution, $E/\Delta E \sim 10$.

Gamma-ray lines are known to be produced in astrophysical objects by electron-positron annihilation, radioactive decay, nuclear excitation, neutron capture and cyclotron processes in strong magnetic fields, i.e. transitions between quantized electron energy levels. The observed lines' energies and production processes are listed in Table 3. Lines can be produced when the temperature exceeds $\sim 10^8$ K, energy exceeds $\sim$1MeV/nucleon or magnetic field exceeds $\sim 10^{12}$ gauss. Although these are extreme conditions by terrestrial standards, they occur frequently in the cosmos. Unique astrophysical information is encoded in gamma-ray lines; not only do their energies indicate the presence of specific nuclei and excitation processes, electron-positron pairs or the magnetic field strength, but the line parameters, i.e. intensities, energy shifts from rest values, widths and profiles, carry information on abundances, bulk velocities, gravitational potentials, densities, temperatures, the spectra of the exciting particles and the magnetic field geometry. The literature contains many reviews of gamma-ray line astrophysics [1–9].

The observations summarized in Table 3 have lead to many important conclusions and promise a wealth of new information when new instruments are flown in space. Some of these are discussed further below.

Electron-positron pair plasmas are expected to form at the extreme densities and temperatures of matter accreting onto black holes. Electron-positron annihilation radiation is the signature of this phenomenon, appearing as a broad line shifted above 511 keV. This feature has been observed in the spectra of Cyg X-1 and the Galactic Center, both of which also contain a narrow 511 keV annihilation line as well. These sources have been observed to vary on 6-month time scales, but the variability has not been resolved and its connection to underlying mechanisms is unknown. Electron-positron annihilation and the decay of radioactive ^{26}Al, which has a 740,000 year half-life, have been observed to occur in the central 5 kpc of the Galaxy, producing gamma-rays at 511 and 1809 keV. The latter proves that nucleosynthesis is an ongoing process in the Galaxy and the former is thought to be primarily due to the decay of nucleosynthetic ^{56}Co produced in supernovae [54]. This produces positrons, some of which escape the remnant and have lifetimes of 10^5 to 10^6 years in the interstellar medium.

These gamma-rays provide tracers of the sites and nature of galactic nucleosynthesis in the present epoch and the physical conditions of the interstellar medium [55]. The latter results from the dependence of the profile of the 511 keV line and associated positronium continuum on the physical conditions of the annihilation medium. Gamma-rays from radioactive ^{56}Co synthesized in the recent Type II supernova SN1987A proved that explosive nucleosynthesis of heavy elements occurred in this supernova and the line profile requires that the ejecta is fragmented [37,38]. Gamma-rays from radioactive material synthesized in supernovae and novae can be used as diagnostics of the yield, expansion and energetics of these events. ^{44}Ti is expected to be produced in significant quantities in supernovae [56] and its 54 year half life allows its gamma-rays to be used as tracers of the undiscovered $\sim$ 10 galactic supernovae thought to have occurred in the past 500 years. About 5 of these should be discovered by the NAE. Assuming half a solar mass of ^{56}Ni is produced in a Type I supernova, the NAE should be able to observe the ^{56}Co gamma-rays from these events out to distances of 20 Mpc, i.e. the Virgo cluster, with 1 week time resolution for several months after the optical outburst.

Gamma-ray lines due to nuclear excitation, neutron capture and electron-positron annihilation are diagnostics of particle acceleration to high energies, 10-100's of MeV, and their subsequent interactions. Solar flares are copious producers of these lines, which have been interpreted to study the spectrum of the accelerated particles and the abundances in the solar atmosphere. With future high resolution measurements many more lines will be observed and they will be used to determine the geometry of the accelerated particles confinement, transport and interactions. Nuclear excitation must also occur throughout the Galaxy due to cosmic ray interactions with the interstellar medium. The predicted fluxes [57] are below present detection capabilities, but the NAE should detect the stronger lines and use them to obtain new information on the abundances of the interstellar gas and dust, and the flux and spectrum of the low energy, <100 MeV, cosmic rays, which cannot penetrate the solar system.

Absorption and emission lines in the 20 to 70 keV range, interpreted as due to cyclotron absorption and emission in intense, $\sim 10^{12}$ gauss, magnetic fields have been observed in the spectra of many objects thought to be neutron stars, e.g. gamma-ray bursters, X-ray pulsators and the pulsar in the Crab Nebula. Since the transport of radiation is strongly dependent on photon energy, propagation direction and electron temperature [58,59], the lines are sensitive diagnostics of the geometry of the magnetic field and the conditions in it.

Generally only the intensities and energies of gamma-ray lines have been determined to a useful accuracy and the information carried by the other line parameters has not yet been exploited. However, observations with the NAE would obtain detailed measurements of the parameters of known lines and many weaker lines as well. It would simultaneously obtain 1) a hundred fold improvement in sensitivity to point sources of narrow lines, i.e. $\sim 3 \times 10^{-6}$ ph/cm^2-s, and a diffuse flux sensitivity of $\sim 2 \times 10^{-5}$ ph/cm^2-s-rad, 2) an energy resolution that is less than most lines' predicted widths in order to determine energy shifts, widths and profiles, i.e. $E/\Delta E \sim 500$, and 3) an angular resolution of a few degrees in order to map diffuse emission, resolve source complexes and locate sources. With these capabilities the NAE could effectively pursue all the presently defined objectives of high resolution gamma-ray spectroscopy.

3.2 Instrument Concept

The NAE instrument concept is shown in Figure 3 and its parameters are given in Table 4. It contains 9 large, $\sim$ 300 cm^3 each, Ge detectors in a heavily shielded 3 x 3 array. These are cooled by a Stirling Cycle mechanical refrigerator. The detectors have very low background, because of 4 essential features. 1) They use position sensitivity, obtained through axial segmentation and pulse shape discrimination, to discriminate against induced β-decay radioactivity in the detectors themselves. Here the multiple-site signature of Compton scattered gamma-rays is distinguished from the single site signature of β-decays, which are a major background component in heavily shielded instruments. 2) A 10 cm thick anticoincidence shield made of bismuth germanate (BGO), which greatly attenuates the background

Table 3. Astronomical Gamma Ray Line Observations

Process	Observed Energy	Source	Flux, ph/cm^2-s	Ref
$e^{\pm}$ Annihilation Radiation	511	Galactic Center	0.6-1.8x10^{-3}	10-15
	511	Interstellar Gas	~2x10^{-3}/rad	15-17
	511	Solar Flares	up to ~0.1	18-19
	400-460	Gamma Ray Bursters	up to 70	21-23
(Redshifted)	~400	CrabPulsar Transient	2-7x10^{-3}	24,25
	~413	10June74 Transient	7x10^{-3}	26,27
	500-2000	Cygnus X-1	up to 2x10^{-2}	28-31
Radioactive Decay				
^{56}Co($\epsilon\gamma,\beta^+\gamma$)^{56}Fe	847	Supernova 1987A	~10^{-3}	32-39
	1238	" "	~10^{-3}	32,33-39
	2598	" "	~10^{-3}	36,39
^{26}Al($\beta^+\gamma$)^{26}Mg	1809	Interstellar Gas	4.8x10^{-4}/rad	40,41
Nuclear Excitation				
^{56}Fe (p,p$'\gamma$)	847	Solar Flares	up to ~0.05	18-20
^{24}Mg (p,p$'\gamma$)	1369	" "	up to ~0.08	18-20
^{20}Ne (p,p$'\gamma$)	1634	" "	up to ~0.1	18-20
^{28}Si (p,p$'\gamma$)	1779	" "	up to ~0.08	18-20
^{12}C (p,p$'\gamma$)	4438	" "	up to ~0.1	18-20,42
^{16}O (p,p$'\gamma$)	6129	" "	up to ~0.1	18-20
Neutron Capture				
^{1}H (n,γ)^{2}H	2223	Solar Flares	up to ~1	18-20,42,43
^{1}H (n,γ)^{2}H	2223	10June74 Transient	1.5x10^{-2}	26,27
(Redshifted)	1790	" "	3x10^{-2}	26,27
^{56}Fe (n,γ)^{57}Fe	5947	" "	1.5x10^{-2}	26,27
(Redshifted)				
Cyclotron Emission & Absorption in ~10^{12} gauss fields	20-70	Gamma Ray Bursters	up to 3	22,44-46
	20-58	X-Ray Pulsators	1-3x10^{-3}	47-51
	73-79	Crab Pulsar Transient	4x10^{-3}	25,52,53

due to ambient gamma- rays. Its transmission is 1 percent at 1 MeV. 3) A 10° FWHM field of view, defined by apertures in the BGO shield, reduces the 1 MeV background due to aperture gamma-rays to the level of the residual, non-rejected detector radioactivity. 4) The low inclination, low altitude orbit, e.g., 10° inclination and 500 km altitude, has a small exposure to the background producing fluxes of trapped protons, cosmic rays and their secondaries.

Imaging and aperture chopping, for suppression of background systematics, are simultaneously obtained through the use of a 25-element coded mask/antimask system. The mask and antimask codes, or patterns, form complementary 5 x 5 arrays, which are produced by open elements and 7 cm thick BGO elements that are nearly opaque to gamma-rays. They pro-

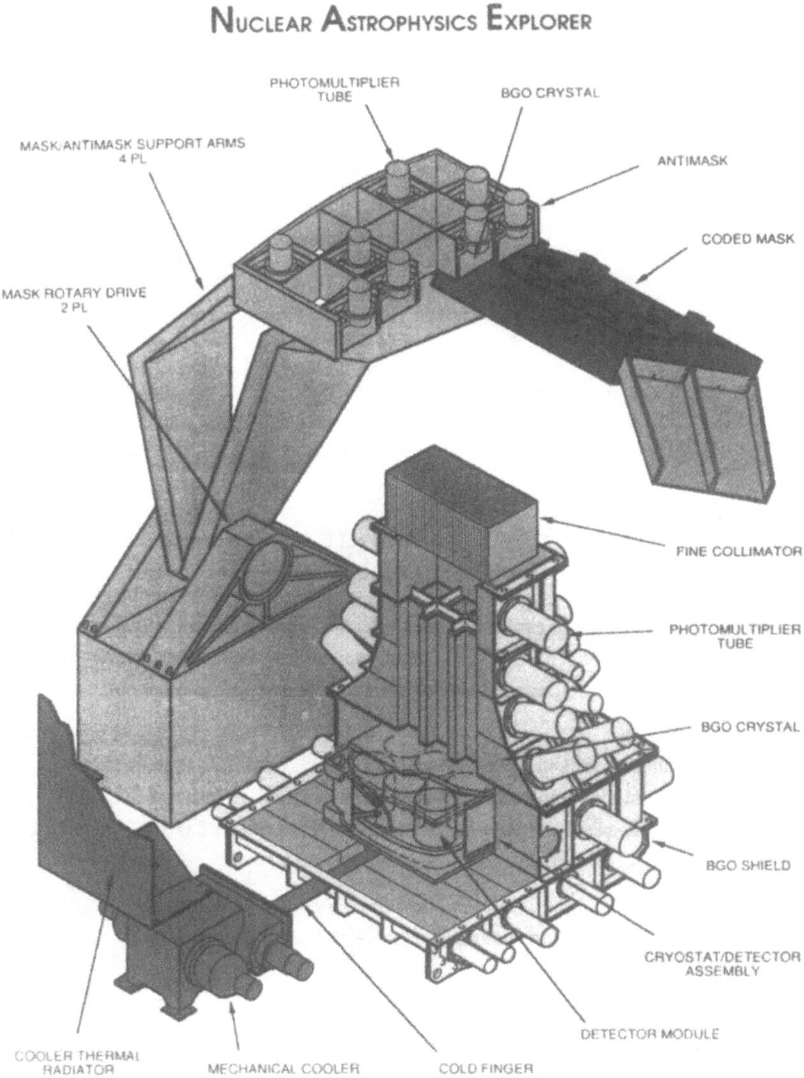

Figure 3. The Nuclear Astrophysics Explorer instrument concept. The BGO anticoincidence shield and cryostat are shown in a cutaway view to allow the germanium detector modules to be seen. The mask/antimask system is also cutaway. Its support arms extend to a second rotary drive (not shown) at the right side of the instrument.

duce shadowgrams on the detector array which are deconvolved by matrix multiplication to produce a sidelobe-free 2-D image of point and point-like sources with 4° angular resolution.

Table 4. NAE Instrument Concept Parameters

Energy Range	15 keV to 10 MeV
Energy Resolution	2 keV at 1 MeV
Sensitivity, 3σ	3 x 10^{-6} ph/cm^2-s in 10^6 sec
Detector System	9 cooled Ge detectors, 300 cm^3 each
Anticoincidence Shielding	10 cm thick bismuth germanate (BGO)
Imaging and Aperture Modulation System	25 element mask/anti-mask system, 7 cm thick bismuth germanate
Field of View	10° FWHM (Image Mode, Knife-Edge Mode) 20° FWOM (Image Mode, Knife-Edge Mode) 1° FWHM (Fine Collimator Mode)
Angular Resolution	4° FWHM (Image Mode) 2° FWHM (Knife-Edge Mode)
Instrument Size	1m(W) x 1.3m(L) x 1.8m(H)
Cryostat Thermal Load	0.5 W at 85 K
Cryostat Cooling System	Mechanical refrigerator, thermoelectric cooler
Power	242 W
Mass	1500 kg
Bit Rate	8 kbps (av),25 kbps (peak)
Pointing Requirement	0.05°, unrestricted viewing in any direction
Orbit Requirement	low exposure to trapped radiation and cosmic rays, e.g. <10° incl. x 500 km alt.

The mask and antimask are alternately placed in the aperture during imaging observations. This is performed with a few minute cycle in order to suppress the systematic effects of varying background caused by changing cosmic ray cutoff energies around the orbit. The mask/antimask combination can also be moved together in and out of the aperture to obtain either 1) a "knife-edge" which is smoothly moved over the aperture to produce a 1-D image with 2° angular resolution, in order to study complex source regions and better locate sources or 2) a totally blocked or totally open aperture, in order to modulate and detect diffuse flux with high efficiency. A 1° FWHM collimator that is effective below 150 keV can be placed in the aperture to improve the sensitivity and angular resolution at low energies.

As a consequence of its very low background, the NAE would become background limited at very low flux levels, $\sim 1\text{x}10^{-5}$ ph/cm^2-s at 1 MeV. It would obtain sensitivity to larger fluxes very rapidly. Only 1/2 hour would be required to reach $\sim 3\text{x}10^{-4}$ ph/cm^2-s, the limiting flux for previous instruments, and 1 day to reach $\sim 1\text{x}10^{-5}$ ph/cm^2-s. With 10^6 sec of good data, a sensitivity of $\sim 3\text{x}10^{-6}$ ph/cm^2-s would be reached. Even better sensitivity could be obtained with longer observations, for example, of the galactic plane. The NAE's sensitivity versus energy is shown in Figure 4 along with those of the Gamma-Ray Spectrometer on the HEAO-3, which used high resolution Ge detectors, $E/\Delta E \sim 300$, and the Oriented Scintillation Spectrometer Experiment (OSSE) on the GRO, which uses lower resolution NaI detectors, $E/\Delta E \sim 15$.

At 1 MeV the NAE would have a sensitivity and energy resolution that are 10 and 30 times better than the OSSE. Its sensitivity would be 100 times better than the HEAO-3. The very good sensitivity predicted for the NAE is the result of careful consideration of the many factors which affect sensitivity. The most significant of these, in comparison with the HEAO-3, are: 6 times more detector volume, 6 times more observing time (due to pointed observations), 2 times more useful data and $\sim$ 100 times lower background per unit detector

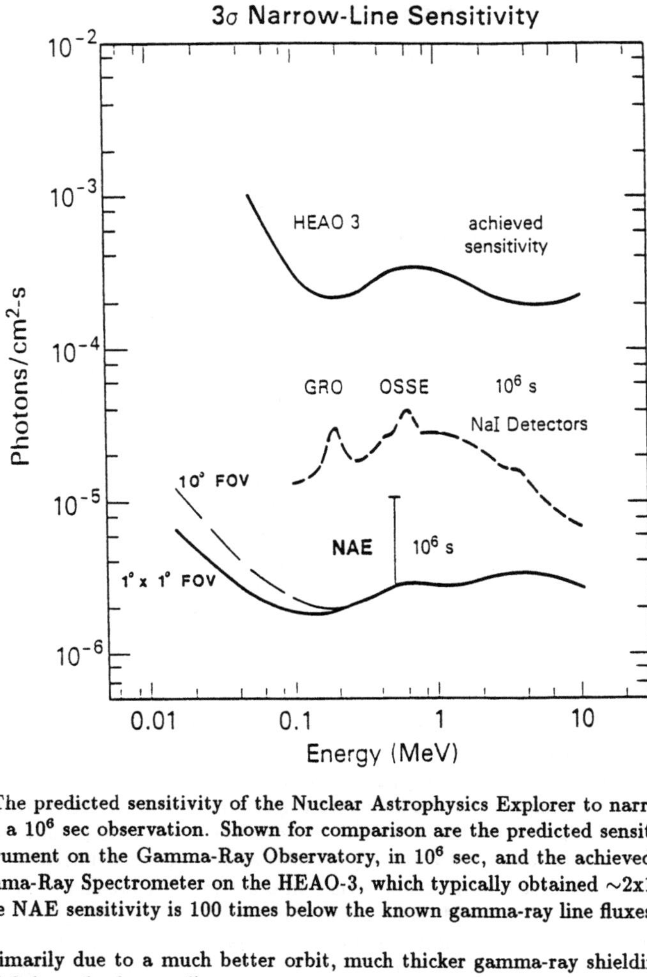

Figure 4. The predicted sensitivity of the Nuclear Astrophysics Explorer to narrow gamma-ray lines in a 10^6 sec observation. Shown for comparison are the predicted sensitivity of the OSSE instrument on the Gamma-Ray Observatory, in 10^6 sec, and the achieved sensitivity of the Gamma-Ray Spectrometer on the HEAO-3, which typically obtained $\sim 2 \times 10^5$ sec on a source. The NAE sensitivity is 100 times below the known gamma-ray line fluxes.

volume (primarily due to a much better orbit, much thicker gamma-ray shielding, and the rejection of β-decay background).

3.3 Mission Concept and Expected Results

The NAE's sensitivity and versatility will allow it to be used to pursue many astrophysical problems with relatively brief observations, e.g. a few hours to a few days. Therefore, the observing program would extend beyond the mission development team and involve a large number of scientists who would use the NAE as a facility. In a 2-year mission scenario the following observational program could be accomplished.

3.3.1 Galactic Plane Survey and Mapping (6 months)

A complete survey of the galactic plane would be performed with 5 days of observations at each 10° step in longitude. Mask/antimask and blocked/open modes would be used to detect and map the diffuse emission with 10° and 4° resolution, allowing the total galactic flux to be determined and separated into its diffuse and point-like, < 4°, components. The galactic 511 and 1809 keV gamma-rays would be detected at $\sim 50\sigma$ and $\sim 20\sigma$ significance in each 10°

step. A sensitive, high-contrast map would result which could be used to determine (1) the sites and rates of nucleosynthesis; longitude from the map, distance from the lines' Doppler velocities, determined to < 30 km/sec, and the galactic rotation model known from 21 cm observations, (2) the nature of the mixing of nucleosynthetic material into the Galaxy and temperature of the electron-positron annihilation regions from the line widths and profiles, (3) the sites and nature of ~ 5 undiscovered galactic supernova which have occurred in the past ~ 500 years from their ^{44}Ti gamma-ray emission, (4) the scale height and velocity dispersion of the postulated high-velocity plasmas in the galactic disk, bulge and corona.

3.3.2 Detailed Mapping of Selected Regions (2 months)

The knife-edge mode would be used to obtain 2° resolution images at multiple position angles to resolve source complexes and better locate point sources. The vicinity of the Galactic Center is already known to be a region where this will be required. The knife-edge mode would also be used to map the latitude distribution of galactic gamma-rays with high precision, in order to determine their scale height and study their possible association with various galactic components.

3.3.3 Extragalactic Supernovae (6 months)

About 4 Type I supernova/year are expected at distances out to the Virgo cluster, ~ 20 Mpc, and at this distance the 847 keV line flux from ^{56}Co decay should be detected for several months with ~ 1 week time resolution. Measurements of line widths will test models of supernova explosions and measurements of their intensities, profiles and time evolution will give information on the nucleosynthetic yield, and the energetics and mass distribution of the ejecta.

3.3.4 Galactic Novae (2 months)

Several galactic novae within 2 kpc are expected to be discovered during a 2 year mission. These will be close enough for sensitive tests of the predicted nucelosynthesis of ^{22}Na (2.6 year half-life) by searching for its 1275 keV gamma-ray. The predicted ^{22}Na yield is sensitive to the thermal history and dynamics of a nova outburst, so it is expected to vary greatly from one nova to the next. The NAE sensitivity corresponds to a yield of $\sim 6\text{x}10^{-9}\ M_\odot$ for a nova at 2 kpc, which is 100 times below present limits.

3.3.5 Observations of Known Point Sources (8 months)

Many known galactic gamma-ray sources will require regular monitoring to observe unpredictable temporal changes that are known to occur but have not been resolved. The Galactic Center region and Cygnus X-1 are two of the best known in this class. Other objects, such as X-ray binaries, have their unique, predictable variability that requires observations at specific times. External galaxies which have energetic nuclei characterized by high energy, nonthermal radiation, i.e. the active galactic nuclei, are prime candidates for observations. Models predict a massive, accreting black hole at the nucleus with an associated electron-positron plasma or relativistic particle jets which transport energies over vast distances. In either scenario gamma-ray lines are expected [60] and their discovery would place the theoretical ideas on a much firmer footing. The sun is known to be a source of intense gamma-ray line fluxes during solar flares which often last for tens of minutes. The NAE would perform dedicated solar observations when a large flare is deemed likely.

3.3.6 Benefits of an Extended Mission

An extended mission lasting much longer than 2 years is very desirable scientifically and technically feasible. The latter follows from the lack of consumables in the instrument and spacecraft concepts and the detectors' > 10 year resistance to radiation damage. In a 5 to 10 year scenario much more extensive observations could be performed of objects where unpredictable variability will carry key information. We already know that this is true of active galaxies, and the Galactic Center and Cyg X-1. An extended mission will allow the galactic plane survey to be repeated several times in order to obtain information on source variability, with variable sources being selected for follow up observations. A large number, ~ 20, of extragalactic Type I supernovae would be observed, providing definitive information on their range of nucleosynthetic yield and explosion mechanism(s). The likelihood of the mission overlapping solar maximum, the only time when gamma-ray producing flares are frequent, would be greatly increased and long periods, many months to years, of dedicated solar viewing could be obtained. In this scenario ~ 30 large flares would be expected, with hundreds to thousands of photons counted in the 15 strongest lines in each flare.

3.4 Status and Plans

In October 1989 NASA announced that the NAE was not selected for development. However, strong interest was expressed in the NAE and the Astrophysics Division of NASA is planning to continue the support of technology and instrument development as well as observational programs in high resolution gamma-ray spectroscopy. The next round of future Explorer selections is planned to begin with a Spring 1991 request for proposals for 1 year feasibility studies. The NAE will be proposed at that time and the earliest it could be launched is the late 1990's. In November 1989 a European-US collaboration submitted a proposal to ESA for a "Blue Box" mission in the late 1990's. This would combine ESA and NASA Explorer resources in a mission named INTEGRAL. This joins the NAE instrument with the Gamma-Ray Imager from the ESA/GRASP mission concept, which was studied in 1987-88.

Acknowledgements

Many people contributed to the development of the SIGMA and the NAE concept. We wish to acknowledge the contributions to the NAE concept by B. Teegarden, W. Mahoney, N. Gehrels, R. Lingenfelter, R. Ramaty, J. Higdon, M. Leventhal and R. Muller. This work was supported in part by NASA contract NAS5-30338 and NASA grant NAGW-449.

References

1. D. D. Clayton: in Gamma-Ray Astrophysics, eds. F. W. Stecker and J. I. Trombka (NASA SP-339, Washington, DC, 1973) p. 263.
2. R. Ramaty, R. E. Lingenfelter: Ann. Rev. Nucl. Part. Sci. 32, 235 (1982).
3. J. L. Matteson: Adv. Space Res. 3, No. 4, 135 (1983).
4. R. E. Lingenfelter and R. Ramaty, in Conference Papers of 19th International Cosmic Ray Conference, Vol. 5, 1985, p. 19.
5. L. E. Peterson: in Nuclear Spectroscopy of Astrophysical Sources, eds. N. Gehrels and G. H. Share (AIP, New York, 1988) p.1.
6. R. E. Lingenfelter: in Nuclear Spectroscopy of Astrophysical Sources, eds. N. Gehrels and G. H. Share (AIP, New York, 1988) p. 17.
7. E.L. Chupp: in Nuclear Spectroscopy of Astrophysical Sources, eds. N. Gehrels and G. H. Share (AIP, New York, 1988) p. 24.
8. R. Ramaty and R.E. Lingenfelter: Nature, 278, 127 (1979).
9. R. E. Lingenfelter and R. Ramaty: Physics Today, 31, No. 3, 40 (1978).
10. R. C. Haymes et al.: Ap. J. 201, 593 (1975).
11. M. Leventhal et al.: Ap. J. (Letters) 225, L11 (1978).
12. G. R. Riegler et al.: Ap. J. (Letters) 248, L13 (1981).
13. G. R. Riegler et al.: Ap. J. (Letters) 294, L13 (1985).

14. M. Leventhal et al.: Nature, 339, 36 (1989).
15. J. L. Matteson et al.: IAU Circular No. 4889 (1989).
16. G. H. Share et al.: Ap. J. 326, 717 (1988).
17. W. A. Mahoney: in Nuclear Spectroscopy of Astrophysical Sources, eds. N. Gehrels and G.H. Share (AIP, New York, 1988) p. 149.
18. E. L. Chupp et al.: Nature 241, 333 (1973).
19. M. Yoshimori et al.: Solar Physics 86, 375 (1983).
20. E. L. Chupp et al.: Ann. Rev. Astr. Ap. 22, 359 (1984).
21. E. P. Mazets et al.: Nature 282, 587 (1979).
22. E. P. Mazets et al.: Nature 290, 378 (1981).
23. B. J. Teegarden and T. L. Cline: Ap. J. (Letters) 236, L67 (1980).
24. M. Leventhal et al.: Ap. J. 216, 491 (1977).
25. C. A. Ayre et al.: Mon. Not. R. Astr. Soc. 205, 285 (1983).
26. A. S. Jacobson et al.: in Gamma-Ray Spectroscopy in Astrophysics, eds. T. L. Cline and R. Ramaty (NASA TM 79619, Greenbelt, MD, 1978) p. 228.
27. J. C. Ling: in Gamma-Ray Transients and Related Astrophysical Phenomena, eds. R. E. Lingenfelter, H. S. Hudson and D. M. Worrall (AIP, New York, 1982) p. 143.
28. P. L. Nolan and J. L. Matteson: Ap. J. 265, 389 (1983).
29. J. C. Ling et al.: Ap. J. (Letters) 321, L117 (1987).
30. J. C. Ling: in Nuclear Spectroscopy of Astrophysical Sources, eds. N. Gehrels and G.H. Share, (AIP, New York, 1988) p. 315.
31. J. C. Ling, W.A. Wheaton: Ap. J. 343 (1989).
32. M. Matz et al.: IAU Circular No. 4568 (1988).
33. W. Sandie et al.: IAU Circular No. 4526 (1988).
34. W. R. Cook et al.: IAU Circular No. 4527 (1988).
35. W. A. Mahoney et al.: IAU Circular No. 4584 (1988).
36. A. C. Rester et al.: IAU Circular No. 4535 (1988).
37. S. Barthelmy et al.: IAU Circular No. 4593 (1988).
38. B. J. Teegarden, et al.: Nature 339, 122 (1989).
39. J. Tueller, et al.: in Proceedings of the Gamma-Ray Observatory Workshop, ed. N. Johnson (1989) p. 4–258.
40. W. A. Mahoney et al.: Ap. J. 286, 578 (1984).
41. G. H. Share et al.: Ap. J. (Letters) 292, L61 (1985).
42. H. S. Hudson et al.: Ap. J. (Letters) 236, L91 (1980).
43. T. Prince et al.: Ap. J. (Letters) 255, L81 (1982).
44. B. R. Dennis et al.: in Gamma-Ray Transients and Related Astrophysical Phenomena, eds. R. E. Lingenfelter, H. S. Hudson and D. M. Worrall (AIP, New York, 1982) p. 153.
45. G. J. Hueter in High Energy Transients in Astrophysics, ed. S. E. Woosley (AIP, New York 1984) p. 373.
46. T. Murakami et al.: Nature, 355, 234 (1988).
47. J. Trümper et al.: Ap. J. (Letters) 219, L105 (1978).
48. W. A. Wheaton et al.: Nature 282, 240 (1979).
49. D. E. Gruber et al.: Ap. J. (Letters) 240, L127 (1980).
50. J. Tueller et al.: Ap. J. 279, 177 (1984).
51. G. S. Maurer et al.: Ap. J. 254, 271 (1982).
52. J. C. Ling et al.: Ap. J. 231, 896 (1979).
53. M. S. Strickman et al.: Ap. J. (Letters) 253, L23 (1982).
54. R. E. Lingenfelter, R. Ramaty: in High Resolution Gamma-Ray Cosmology, eds. D.B. Cline and E. Fenyves, Nuclear Physics B Proc. Suppl. 10B (North Holland, Amsterdam, 1989) p. 67.

55. R. E. Lingenfelter, R. Ramaty: in Proceedings of the 21st International Cosmic Ray Conference, paper OG7.2–6, in press.
56. S. E. Woosley, et al.: Ap.J. 301,601 (1986).
57. R. Ramaty, B. Kozlovsky, and R.E. Lingenfelter: Ap.J. Supp., 40, 487 (1979).
58. R. W. Bussard: Ap.J. 237 970 (1980).
59. E. Nagel: Ap.J. 251 288 (1981).
60. J.L. Matteson: in Electron-Positron Pairs in Astrophysics, eds. M.L. Burns, A.K. Harding and R. Ramaty (AIP, New York, 1983) p. 292.

SECTION VI

THE NEUTRON STAR IN SN 1987A AND OTHER SUPERNOVAE

The Sub-Millisecond Pulsar in SN 1987A: A Review of the Discovery Data After Six Months

Saul Perlmutter, Richard A. Muller, Carlton R. Pennypacker, & Timothy P. Sasseen

1. THE PULSAR DISCOVERY

1.1. Experiment Design

Two years ago, when Supernova 1987A was discovered, we were presented with an unusual opportunity to catch a pulsar at its birth. It was of course not known how long it would take before the expanding supernova photosphere would become transparent enough to see into a pulsating center. In collaboration with researchers from the United States, Canada, Chile, and Australia, we began an optical search for pulsations in SN 1987A about a month after the supernova discovery, and this search set upper limits on pulsed light until January, 1989 (PENNYPACKER *et al.* [1]).

The experiment itself is very simple: A silicon photodiode detector collects light from 2-, 3- or 4-meter telescopes in Chile and Australia (PENNYPACKER *et al.* [1]). The voltage from the photodiode is digitized at 5 or 10 kHz, based on a stable clock, and the resulting data are written to magnetic tape. The analysis of these data tapes is performed at Los Alamos on the Cray XMP computers. This is necessary because we search many 10^8-point modified Fourier transforms, each with a different pulsar slow-down rate (dP/dt). The Cray CPU time for the analysis is comparable to the time for the observation at the telescope, and in the two years of observations since the discovery of SN 1987A , we have used hundreds of hours of Cray time.

1.2. The Observed Signal

On January 18, 1989 we observed the supernova for seven hours at the Cerro Tololo 4-meter telescope, with a sampling rate of 5 kHz.(MIDDLEDITCH *et al.* [2], KRISTIAN *et al.* [3]). The Fourier transforms of these data show a very strong peak at 1968.63 Hz that reaches as high as 600 times the background noise level in this frequency region. As shown in Figure 1, the signal amplitude stays relatively constant for the first half of the observation and then increases by a factor of three in the last half. This means that the fraction of the light from the supernova that is pulsed varies from one part in 300 to one part in 100. (Note that these numbers have been re-calibrated to account for instrument inefficiencies since the first reports. See MIDDLEDITCH *et al.* [5]) This corresponds to about 100 times the optical luminosity of the Crab pulsar at the same distance.

Twelve minutes after the seven hours of observation of the supernova we observed Globular Cluster NGC 3201; this serves as an instrument check. No pulsations were seen near 1968 Hz in a half-hour observation of this object, at a limit 12 times less than the weakest SN 1987A pulsations (the arrow in Figure 1), indicating that the SN 1987A pulsations were not instrumental artifacts.

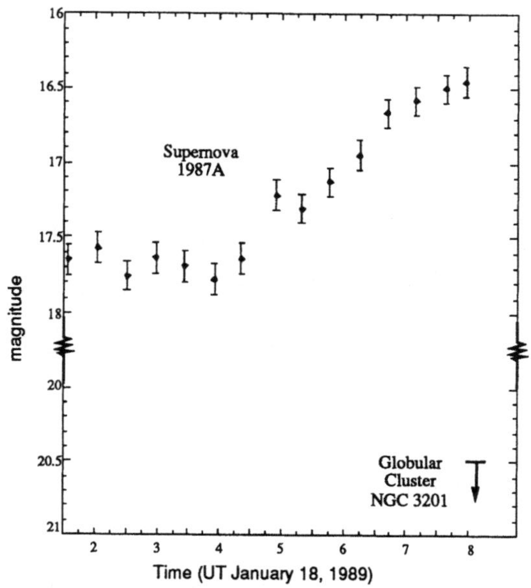

Figure 1. The magnitude of pulsed light from Supernova 1987A versus time for seven-hour observation on January 18, 1989. The arrow shows (on a "broken" magnitude axis) the upper limit on pulsed light from an observation on globular cluster NGC 3201. (These magnitudes are newly calibrated for instrument inefficiencies. See MIDDLEDITCH *et al.* [5].)

The SN 1987A signal at 1968.63 Hz was also visible in higher harmonics. Since the Nyquist frequency of 2.5 kHz is below the second harmonic, it is necessary to look at the "folded" frequency where the aliased harmonics should appear. The second harmonic of the 1968.63 Hz signal is very strong and tracks the fundamental in amplitude, while the third harmonic is visible towards the end of the observation when the signal was stronger. After correcting for the effects of aliasing and the rolloff of the photodiode amplifier, the ratios of the power in the first three harmonics, 1 : 1.8 : 1.6, is similar to those of the Crab pulsar in its optical light, 1 : 1.92 : 1.36. Figure 2 shows the SN 1987A pulse profile reconstructed from these first three harmonics, with the analogous curve for the Crab pulsar drawn for comparison.

We conclude from this data that we are seeing pulsations from SN 1987A. In the 36 previous observations in the two years since the supernova was discovered, no comparable signal was seen. The largest Fourier-transform peaks in these two years were 6σ events (PENNYPACKER *et al.* [1]), while the 1968.63 Hz signal increased from 11σ to 37σ during the January 18 observation.

1.3. A Pulsar at 1968 Hz

If we interpret this signal as a classical rotating-neutron-star pulsar, we find it to be a surprising object. At this rotation frequency, the star's self-gravitation is barely stronger than its centrifugal force: it is almost flying apart (more about this later). The velocity at the equator is about 40% of the speed of light—this is a highly relativistic object. The kinetic energy of rotation is on the order of 10^{53} ergs, comparable to the total amount of energy released in the supernova collapse/explosion.

A pulsar rotating this fast inside SN 1987A must have a small magnetic field, or else the supernova would appear brighter than it does ($L_{1987A} \approx 3 \times 10^{38}$ erg/sec). The least luminosity that could be emitted from a pulsar with a dipole magnetic field B would be:

$$L_{\text{dipole}} \propto B^2 \omega^4$$

where ω is the pulsar rotation frequency. For comparison, the Crab pulsar, which is believed to power the Crab nebula with a luminosity of 5×10^{38} erg/sec, has a rotation frequency of 30.2 Hz and a magnetic field strength of 4×10^{12} Gauss. Scaling from the Crab pulsar, we find that the SN 1987A pulsar must have a magnetic field a factor of (1968 Hz / 30 Hz)^2 smaller, or about 10^9 Gauss. This is of course an upper limit.

1.4. Follow-up Observations and Other Group's Observations

The natural question at this point is, have we seen the pulsar again? In the six months since the January 18 observation, we have observed SN 1987A on about 16 separate nights, for a total integrated observation time of approximately 20 hours. None of the resulting data sets have shown significant pulsations at frequencies near 1968 Hz. The limits set by these observations are all about 20.5 magnitude, a factor of 13 below the weakest January 18 pulsations.

Ours was not the only group searching for a pulsar in SN 1987A. A European group, observing in Chile, and an Australian group also had searches underway. In fact, Peterson and Manchester, in Australia, observed the supernova on the same night that we found the pulsar. Their system sampled the signal at 100 kHz, although they originally analyzed their data in 1 kHz bins. (Our experiment was apparently the only one that was sensitive to a pulsar faster than previously known pulsars.) Peterson and Manchester later reanalyzed their data with shorter time bins to look for the 1968 Hz signal, but found no signal. The significant difference between their experiment and ours is probably their choice of a blue-pass filter, blocking light redwards of 4800 Å. According to the SN 1987A atmosphere models of PINTO, AXELROD, and WOOSLEY [4], the atmosphere becomes much more transparent for wavelengths longer than 8000–9000 Å where our silicon photodiode still had sensitivity. It appears likely that a detector with sensitivity at even longer wavelengths would be appropriate for this search.

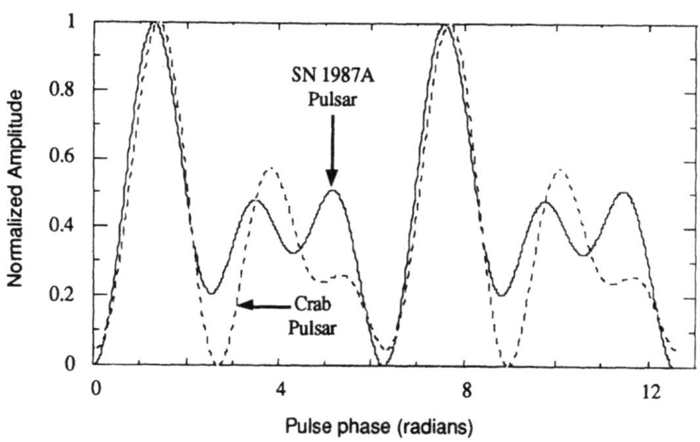

Figure 2. Pulse profiles synthesized from the first three harmonics. The solid line is the pulse profile of the signal in SN 1987A and the dotted line is the profile of the Crab Pulsar.

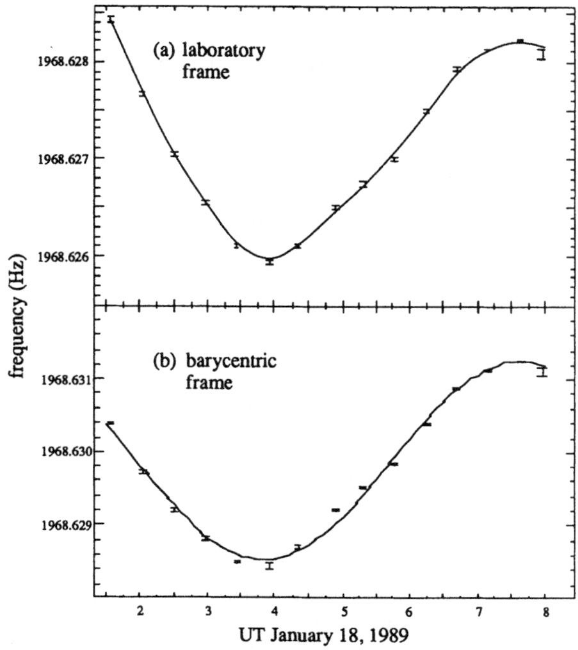

Figure 3. a) Observed frequency versus time for seven-hour run on January 18, 1989. The solid line is to guide the eye. b) Same as a), but corrected for the Doppler shift due to the rotation of the Earth about its axis and about the Sun. The solid line is the best fit sine function. (Based on KRISTIAN *et al.* [3].)

1.5. Frequency Modulation

The signal at 1968.63 Hz is very narrow in frequency; it is stable to better than a part in 10^6. There is, however, a very small frequency modulation observable over the 7 hours of data, at the millihertz level. Figure 3a shows this modulation. The timescale of this frequency drift was difficult to understand until we corrected the frequencies for the doppler shifts due to the Earth's rotations about its axis and about the Sun. As Figure 3b shows, this corrected frequency modulation is close to a sine wave, which would be the doppler modulation for an orbiting companion in a circular orbit.

The 7 hours of data contain a strong enough signal that we can do a more detailed analysis by breaking the data into 56 seven-minute segments (MIDDLEDITCH *et al.* [5]). For each segment we can find the frequency and phase of the 1968.63 Hz signal so precisely that the exact number of cycles in each seven minute segment can be found and linked to the next segment. We can then calculate the arrival time of the pulsar pulse that occurred closest to the middle of each segment, and use this as an independent "pulsar clock." Since this clock is very stable, a plot of "pulsar time" versus lab-clock time is essentially a straight line. The deviations from this straight-line can be magnified as in Figure 4 to show the pulses arriving a few milliseconds early and then a few milliseconds late as the pulsar moves forward and back, perturbed (apparently) by the 7-hour orbit of a companion.

If the deviation is due to an orbiting companion, what are its properties? Using Kepler's Law and the 7-hour period, we can easily find the semi-major axis, R, of the companion's orbit. Scaling from the Earth's orbit about the Sun, we find

$$R \approx \left(\frac{7\ \text{hours}}{365\ \text{days}}\right)^{2/3} [\text{AU}] \approx 10^{-2}\ \text{AU}$$

in Astronomical Units, the distance from the Earth to the Sun. This is approximately $R \approx 10^{11}$ cm. The modulation amplitude of $\sim 1.4 \times 10^{-3}$ Hz corresponds to a doppler velocity of ($\sim$0.2 km/sec) / sin i, where i is the inclination angle of the orbit to our line of sight. Taking sin i to be of order unity, this gives a semi-major axis for the pulsar about the center-of-mass, $r \approx 10^8$ cm. The ratio of these semi-major axes is of course the ratio of the masses of the pulsar and its companion, about 1 to 1000. For a neutron star with a typical mass of 1.4 solar masses, this makes the companion slightly more massive than Jupiter.

Pulsars with a short-period companion are not completely unknown. Recently, FRUCHTER, STINEBRING, and TAYLOR [6] discovered pulsar PSR 1957 +20, with a 9-hour orbit. The companion of PSR 1957 +20 (or its surrounding ionized region) appears to eclipse the pulsar as it passes in front. So far the data for the SN 1987A pulsar shows no significant evidence of a similar eclipse by its companion, so we are probably not looking at an orbit edge on.

2. IMPLICATIONS

2.1. Problems for Current Pulsar Models

If the signal described in the preceding section is a pulsar, it presents a number of puzzles for current pulsar theories. First, there is the surprising period—shorter than any other known pulsar by a factor of three. This short period suggests a low magnetic field, a second puzzle. Third is the problem of the companion orbiting so close to the pulsar.

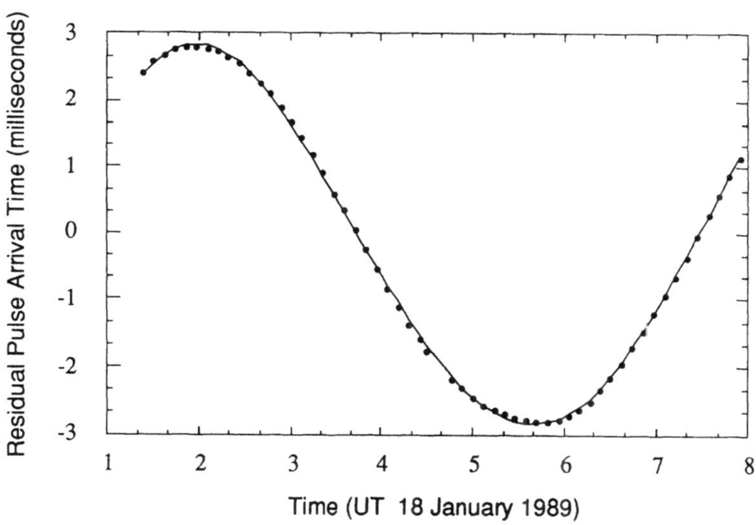

Figure 4. Residuals from fit of pulse arrival times to straight line ($f_0t + \varphi_0$). The error bars are smaller than the data points. The solid line is the best fit Keplerian orbit with a 7.1 hour period and eccentricity $e = 0.093$. (Based on MIDDLEDITCH *et al.* [5].)

Fourth, and perhaps most perplexing, why were the pulsations so bright, and why are they not visible now?

Most fast pulsars are currently believed to have been "spun up" over millions of years by accreting matter from an orbiting companion. The SN 1987A pulsar, however, has had only two years to reach its current period. WOOSLEY and CHEVALIER [7] have suggested that an accretion-driven spin-up did occur within the first few hours after the explosion. When the expanding shock wave hits the hydrogen envelope of the progenitor star, a reverse shock forms, which can throw material back down onto the neutron star. Woosley and Chevalier point out that the progenitor star for SN 1987A is thought to be much smaller than is typical before a supernova explosion, and therefore there is much denser debris to be thrown back by the reverse shock wave. This scenario might also explain the weak magnetic field, since material accreting on the neutron star with a disordered magnetic moment can bury its magnetic field (WOOSLEY and CHEVALIER [7]).

WANG *et al.* [8] have proposed a fundamentally different model to explain the short pulse period, and to avoid the problem of a weak magnetic field. They point out that the half-millisecond period is very close to the natural period of radial oscillation of a neutron star. If an optical pulse could be generated by some mechanism every time the star "breathed" out or in, this would avoid the necessity of considering high spin rates or weak magnetic fields. One constraint on this model is that the optical-pulse mechanism must generate the interpulse structure that is visible in Figure 2, or, equivalently, the dominant power at the second harmonic.

The radial oscillation model will be easily tested when and if the pulsar is reobserved, since a rotating neutron star should slow down as it radiates energy, while an oscillating neutron star should remain at the same (or higher) frequency with a reduced amplitude of oscillation. It is possible that the oscillation observed on January 18 has already damped out, and that we are now waiting for another star-quake to "ring the bell" of the neutron star and thus produce more optical pulsations. If the neutron star is oscillating at 1968 Hz *and* rotating (at a slower rate) then we should also be able to see frequency-modulation sidebands about the main Fourier peak.

With either a rotating or oscillating pulsar model it is still difficult to understand how a companion can exist at a distance of only 10^{11} cm from the pulsar. This is well within the progenitor star's radius and the companion is unlikely to have survived the supernova explosion. Aside from the energy released in the explosion, enough mass is lost from the region inside of 10^{11} cm that a bound object should become unbound. Alternatively, we can look for mechanisms that could create such an object after the explosion. (The other pulsar with a similar orbit, PSR 1957 +20, has been explained by postulating a companion star spiraling its way in towards the pulsar over many millions of years, ablating mass as it goes. Clearly this is not a viable solution in the case of SN 1987A.)

Why have we--and other groups--been unable to reobserve the 1968 Hz pulsations? For the reasons given above and in MIDDLEDITCH *et al.* [5], it is difficult to attribute this signal to a spurious local source, and we have been forced to ask what is going on at SN 1987A that could be hiding or suppressing the signal source. It is possible that we saw the signal through a momentary clearing in the opaque debris surrounding the pulsar, and that dust has been forming to add to this shroud. If this is the case, a detector more sensitive in the IR may help. Given that the pulsations were intrinsically much brighter than the Crab's, and that the amplitude changed by a factor of three in three hours, it is also plausible that we observed a flare-up in the pulse emission mechanism. If this is the case the observing task is a little more daunting, since we are then spot checking the pulsar with the hope of catching another flare-up.

2.2. Nuclear Equations of State

A pulsar rotating at 1968 Hz puts strong constraints on the nuclear equation of state, the equation that describes the compressibility of matter at nuclear densities. This is because

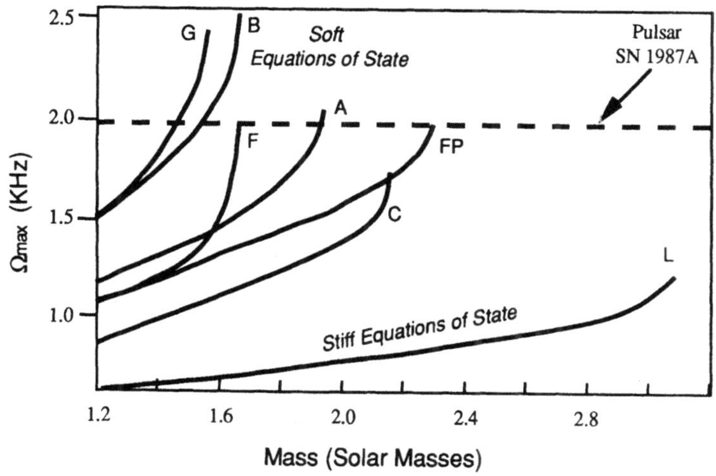

Figure 5. The maximum rotation frequency versus neutron star mass, for a sampling of equations of state. (Based on figure from FRIEDMAN, IPSER, and PARKER[9].)

the compressibility of a neutron star sets a limit on the maximum rotation frequency of the star: A "stiff," incompressible equation of state gives a larger star for a given mass, while a "soft," compressible equation of state allows a star of the same mass to "pull itself in" to a smaller radius. The smaller star can rotate faster before its surface becomes unbound or unstable.

The neutron star mass is the other factor in the calculation of a rotational "speed limit." For a given equation of state, a more massive star will be smaller and thus able to spin faster. One cannot however add mass indefinitely in a stellar model to permit higher and higher angular velocities, because above a certain mass the star collapses into a black hole. Figure 5, based on a figure from FRIEDMAN, IPSER, and PARKER [9], shows these competing factors for a few different equations of state. The stiff equations of state give a lower maximum velocity, Ω_{max}, than the soft equations of state, and for each equation of state increasing the mass increases Ω_{max}. The endpoint for each curve is at the maximum mass that the equation of state could support against gravitational collapse, even with the help of centrifugal force from a 1968 Hz rotation.

If the pulsar in SN 1987A is rotating at 1968 Hz, then the only viable nuclear equations of state are those that appear above the horizontal line in Figure 5. This rules out all of the stiff equations of state. Of the equations shown, only models B, G, F, and A of FRIEDMAN, IPSER, and PARKER [9] allow such a rapidly rotating neutron star. However, as their paper points out, models B and G are not stiff enough to support the mass of the binary pulsar, PSR 1913 +16, which rotates much more slowly. Equations F and A are probably also unacceptable, because 1968 Hz is so near their endpoint that nonaxisymmetric instabilities may set in. These are, however, only a sampling of possible phenomenological equations of state, and there could exist other equations between curves A and B that do satisfy all the constraints (see LATTIMER *et al.* in this volume for more discussion on this point).

It is of course possible to avoid these tight limits on nuclear equations of state if the pulsar turns out to be a radially oscillating neutron star, as described above.

2.3. Gravitational Radiation

The power emitted in gravitational radiation from a rotating neutron star with a finite quadrupole moment scales as $P_{grav} \sim \omega^6$. The SN 1987A pulsar should thus be the most efficient gravitational radiator of any known pulsar—unless it is symmetric about its spin axis. (Compared to the much nearer Crab pulsar, it would be radiating 10^8 times more power for the same quadrupole moment.) In fact, a 1968 Hz pulsar would be so efficient that any asymmetries in its mass distribution would be damped out in a time-scale of seconds. To be emitting gravitational radiation now, asymmetries would therefore have to be continuously regenerated through some dynamical mechanism. KLUZNIAK *et al.* [10] have suggested that Chandrasekhar-Friedman-Schutz gravitational instabilities would be such a mechanism for this pulsar. So far, however, one can only set limits on the total amount of gravitational radiation based on the lack of significant slowing in the pulsar period ($dP/dt < 3 \times 10^{-14}$ s s^{-1}).

The next generation of gravitational wave antennas might be sensitive enough to detect these waves, particularly if they can be tuned to a specific pulse period. Unfortunately, the Chandrasekhar-Friedman-Schutz mechanism does not generate gravitational radiation at the pulsar rotation period, so it is more difficult to look for this particular source of gravity waves.

3. CURRENT PROGRESS

3.1. Further Analysis of Pulse Arrival-time Residuals

Figure 4 showed the residuals from a two-parameter straight-line fit to the pulse arrival times. If we fit five more parameters to account for the orbit (MIDDLEDITCH *et al.* [5]), as shown by the solid line in Figure 4, we are left with the residuals of Figure 6. The plotting scale has once again been magnified to show the very small (~30 microsec RMS) deviations. At this scale the error bars are finally larger than the plotting points, and it is clear that the points are not statistically scattered about the zero line. We are currently examining possible explanations for the systematic drifting of these residuals, including systematic error in the laboratory clock, random perturbations at the pulsar, and periodic events at the pulsar.

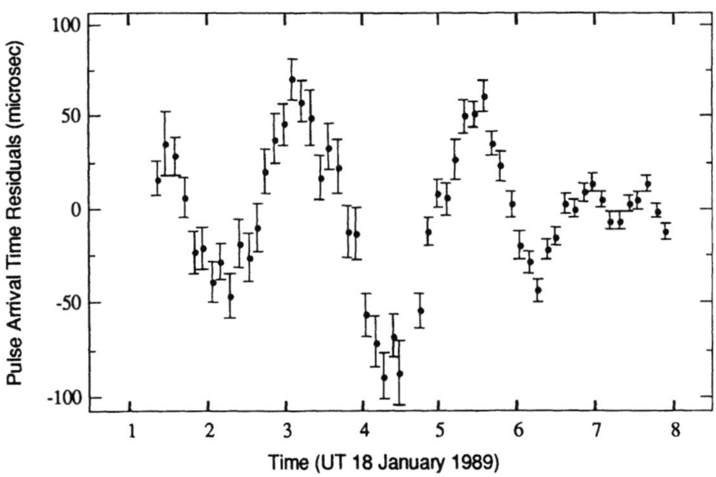

Figure 6. Residuals from the fit of a Keplerian orbit to pulse arrival times versus Universal Time. (Based on MIDDLEDITCH *et al.* [5].)

Clock Error. The clock used to record the data was a Datum model 9110 quartz crystal with a factory-specified limit on its frequency drift of 5 parts in 10^9 per day. Its stability was determined by comparison with a rubidium standard clock to be within 5 parts in 10^{10} and 3 parts in 10^9 for time intervals of minutes and several days, respectively. The drift in Figure 6 is at the part in 10^8 level, at least an order of magnitude larger than the clock's stability; thus we conclude that the clock could not have been the cause.

Random Perturbations at the Pulsar. Similar systematic drifting of pulse-arrival-time residuals have been found in the analysis of the Crab Pulsar's optical pulsations (GROTH [11], BOYNTON *et al.* [12]). In the case of the Crab, the time scale of the drifts turned out to depend on the length of the data set. This led GROTH [11] to observe that since their analysis fit the data to a second-degree polynomial ($1/2\, [df/dt]_0 t^2 + f_0 t + \varphi_0$), the residuals can only be characterized by a third-degree (or higher) polynomial. Thus any random perturbations in the frequency of the pulsar (due to material accreting on the pulsar, for example) would result in residuals with three or more zero crossings.

In the case of the data from SN 1987A, the seven parameter fit to an orbit is more complicated than a sixth-degree polynomial, so it is not obvious how the residuals should look if there are random frequency perturbations at the pulsar. The number of zero crossings is, however, comparable to the number of parameters fit. We have therefore performed Monte Carlo simulations by adding the effect of a random walk in frequency to a perfect Keplerian orbit's pulse arrival times. After fitting a new orbit (seven parameters) to these simulated data sets, we found that the residuals did show systematic drifts similar to the true data, for certain values of the random walk parameters. Figure 7 shows the residuals from one of the first five random-walk data sets that was generated this way. Although these Monte Carlo residuals look similar, they do not appear to the eye to be as regular in their zero crossings, so we have also considered the alternative possibility that the perturbing events are periodic.

Periodic Events at the Pulsar. The residuals in Figure 6 appear to vary with a period of approximately 2 hours. Fitting an additional five parameters for a second companion's orbit gives the excellent fit shown in Figure 8, but is probably ruled out because of the immense densities necessary for such a companion to survive the tidal forces from the pulsar. A precession of the neutron star's spin axis about its symmetry axis could also produce the 2 hour period (NELSON, FINN, and WASSERMAN [13]). The fit to precession, with four additional parameters, appears.almost the same as Figure 8, reminding us that is difficult to

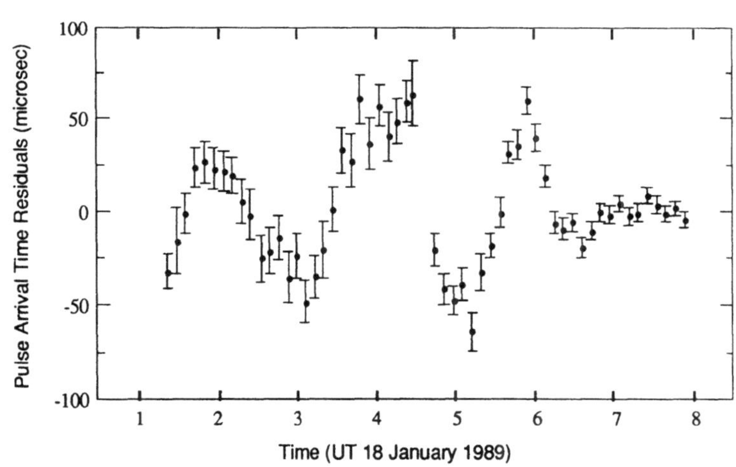

Figure 7. Same as Figure 6, but for Monte Carlo simulated data.

Implications of a Fast Pulsar for the Equation of State

James M. Lattimer

The recent observation of a sub-millisecond pulsar [1] in the remnant of SN1987A with period $P = 0.508$ ms is very exciting because the possibility exists of pinning down the equation of state (EOS) of supra-nuclear density matter, *if the pulsations are due to rotation.* The neutrinos detected by Kamioka [2] and IMB, [3], although basically confirming the standard theoretical models of neutron star birth, did not give an accurate enough estimate of the neutron star's binding energy or mass to constrain the EOS. The maximum rotation rate of a star must be less than or equal to the Keplerian rate, Ω_K, at which the equatorial surface velocity equals the orbital velocity of a particle at the equator. Recent general relativistic instability analyses [4] demonstrate that the maximum rotation rate cannot be less than about $0.9\Omega_K$. If there exist physical mechanisms for increasing the neutrino fluxes of young neutron stars, such as quarks or meson condensates, the maximum rotation rate becomes nearly equal to Ω_K.

In general, Ω_K is larger for more compact neutron stars with higher central densities. Thus, the maximum rotation rate is an increasing function of the "softness" of the EOS. The observed period of 0.508 ms, if due to rotation, sets one limit to the EOS. In addition, the masses (1.44 $M_\odot$and 1.38 $M_\odot$) of the components of the binary pulsar PSR 1913+16 [5] set a lower limit to the maximum mass of a neutron star. The maximum mass is a decreasing function of the "softness" of the EOS, and therefore establishes a second constraint. These two constraints will restrict the EOS to lie in a narrow range.

The calculations of the maximum rotation rate and the maximum mass permitted by a given EOS has to be done with full general relativity. It has been shown, however, that an approximate formula [6] for the Keplerian rotation rate in terms of the maximum mass properties of the non-rotating star is extremely accurate [7]. It is

$$\Omega_K = 7.7 \times 10^3 \left(\frac{M_{max}}{M_\odot}\right)^{1/2} \left(\frac{R_{max}}{10\text{km}}\right)^{-3/2} \text{s}^{-1}. \tag{1}$$

The subscripts "max" refer to the maximum mass non-rotating star of a given equation of state. This formula is, in fact, accurate to within 4% for all equations of state we have tested, except that of a completely incompressible fluid (for which the coefficient 7.7 is 9.6). In addition, the general relativistic calculations of Friedman, Ipser and Parker [8] follow this formula to within the same accuracy. The challenge for an EOS is therefore to compress at least 1.44 $M_\odot$within a radius limited by $\Omega_K = 2\pi/P \geq 1.237 \times 10^4 \text{s}^{-1}$, which implies $R_{max} \leq 8.2\sqrt{M_{max}/1.44\ M_\odot}$ km. By employing the schematic equation of state developed by Prakash, Ainsworth and Lattimer [9], we have shown that this constrains the EOS to be ***both*** relatively soft around nuclear densities ***and*** quite stiff at higher densities.

The softness around nuclear densities may have two possible sources: a low value for the nuclear compression modulus $K \leq 160 \pm 20$ MeV, or a phase transition in the range 1–3 times nuclear saturation density ($n_s = 0.16$ fm^{-3}). The high density (above 5–6 times n_s) equation of state, on the other hand, must be very nearly at the causal limit $\partial P/\partial \epsilon = 1$, where P is the pressure and ϵ is the energy density. Even with a phase transition, the nuclear compression modulus cannot be much larger than 200 MeV if both rotation and mass constraints are to be satisfied by the EOS. Interestingly, the phase transition apparently cannot be to a quark phase of matter, because in this case the high density EOS is very subcausal ($\partial P/\partial \epsilon = 1/3$).

Our results for the case without any phase transitions are shown in Figure 1. The graphs on the left display the case that the high density equation of state is exactly causal above the transition density n_t. The only other significant parameter of the EOS is the compression modulus K. The upper left graph shows the Keplerian rotation rate Ω_K in units of 10^4s^{-1} as a function of the two EOS parameters. At best, Ω_K must be less than about 160 MeV in order to satisfy the SN1987A rotation rate (1.237×10^4s^{-1}). The lower left graph shows the maximum mass of non-rotating stars as a function of the EOS parameters. The small region to the right of the curve 1.44 M$_\odot$is excluded by the PSR 1913+16 constraint. The graphs on the right display contours of $\Omega_K = 1.237$ and $M_{max} = 1.44$ M$_\odot$ for various values of the parameter $s = \partial P/\partial \epsilon$, which defines the EOS above n_t. The case $s = 2$ is included only for reference, since in reality the EOS cannot violate causality. It is clear that for s less than unity the allowable phase space in K and n_t rapidly shrinks: the rotation constraint implies that the maximum allowable value of K drops, and the binary pulsar mass constraint increases the size of the excluded region. By $s = 1/2$, there are no more acceptable solutions, ruling out quarks ($s = 1/3$).

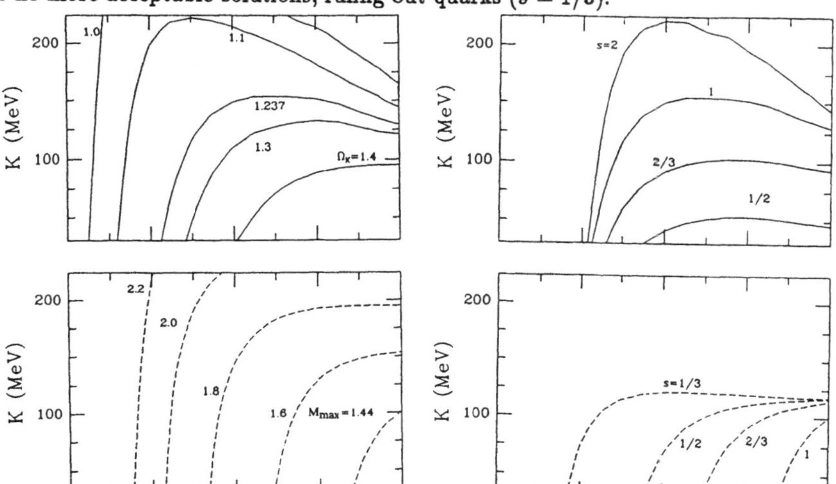

Fig. 1: Left: Contours of $\Omega_K/10^4$s^{-1} (upper) and $M_{max}/$ M$_\odot$ (lower) as functions of K and n_t for the case $s = \partial P/\partial \epsilon = 1$. Right: Contours of $\Omega_K/10^4$s^{-1} $= 1.237$ (upper) and $M_{max}/$ M$_\odot$ $= 1.44$ (lower) as functions of s.

Friedman, Ipser and Parker [8] and we [7] have examined the rotational properties of a number of published equations of state. Only equations of state with very small values of K and $s \simeq 1$ are successful. Therefore, we feel that although a schematic equation of state was used to derive Figure 1, the results are, in fact, rather general.

The effect of phase transitions is, generally speaking, to lower the effective value of the nuclear compression modulus. It is difficult to express this quantitatively, however, because the compression modulus is defined at n_s and phase transitions, if they occur, set in at higher densities. Nevertheless, from our own experience with specific models for phase transitions (pion condensates [10]; kaon condensates [11]; Chiral parity doubling transitions [12]), they are only effective in permitting rapid rotation if they occur below 3– 4 n_s, if K is not too large, and if the high density equation of state is nearly causal. Therefore, qualitatively, the results expressed in Figure 1 remain valid even if phase transitions occur.

The constraint that the high density equation of state must be nearly causal seems to rule out a phase transition to quark matter at moderate densities. The EOS models containing quark cores that have been published [13] all fail to attain sufficiently rapid rotation. But what if the entire star is made of quark matter?

Witten [14] explored the structure of neutron stars using the self-bound equation of state $P = (1/3)(\epsilon - \epsilon_o)$ that is applicable in the case of massless quarks of both 2 and 3 flavors. The constant $\epsilon_o = 4B$, where B is the MIT bag constant. The linear dependence of pressure on the energy density has the interesting consequence that non-rotating configurations with maximum mass scale with ϵ_o according to $M_{max} \propto \epsilon_o^{-1/2}$ and $R_{max} \propto \epsilon_o^{-1/2}$. For $\epsilon_o = 224$ MeV fm^{-3}, $M_{max} = 2.033$ M$_\odot$ and $R_{max} =$ 11.09 km. The success or failure of these models depends crucially on the value of ϵ_o. In the bag model, ϵ_o is bounded from above by the condition that quark matter be bound at low densities: $\epsilon_o \leq 366$ MeV fm^{-3}. This implies that $M_{max} \geq$ 2.6 M$_\odot$, $R_{max} \geq 14.2$ km, and, therefore, $\Omega_K \leq 1.20 \times 10^4 \mathrm{s}^{-1}$. Therefore, even a quark star, subject to the matter being self-bound, cannot satisfy the rotation constraint.

This work was supported in part by the DOE under grant DE–FG02–87ER40317.

References

1. C. Kristian *et al.*, Nature 338, 234 (1989).
2. K. Hirata, *et al.*, Phys. Rev. Lett. 58, 1490 (1987).
3. R. M. Bionta, *et al.*, Phys. Rev. Lett. 58, 1494 (1987).
4. J. R. Ipser and L. Lindblom, Phys. Rev. Lett.62, 2777 (1989).
5. J. M. Weisberg and J. H. Taylor, Phys. Rev. Lett.52, 1348 (1984).
6. P. Haensel and J. L.Zdunik, Nature, submitted (1989).
7. J. M. Lattimer, M. Prakash, D. Masak and A. Yahil, Ap. J., submitted (1989).
8. J. L. Friedman, J. R. Ipser and L. Parker, Astrophys. J. 304, 115 (1986); J. L. Friedman, J. R. Ipser and L. Parker, Phys. Rev. Lett. 62, 3015 (1989).
9. M. Prakash, T. L. Ainsworth and J. M. Lattimer, Phys. Rev. Lett. 61, 2518 (1988).
10. W. Weise and B. E. Brown, Phys. Lett. 58B, 300 (1975).
11. G. E. Brown, K. Kubodera, M. Prakash and M. Rho, Nucl. Phys. A479, 175c (1988).
12. T. Hatsuda and M. Prakash, Phys. Lett. 224B, 11 (1989).
13. W. B. Fechner and P. C. Joss, Nature 274, 347 (1978); P. Haensel, J. L. Zdunik and R. Schaeffer, A. and Ap. 160, 121 (1986).
14. E. Witten, Phys. Rev. D30, 272 (1984).

Surface Structure of Neutron Stars and Nuclear Reactions with High Magnetic Fields

Ikko Fushiki

There is considerable evidence for the existence of strong magnetic fields at the surfaces of some neutron stars. The strengths of magnetic fields at neutron star surfaces have been estimated for over three hundred known pulsars and lie between $10^{10.36}$ G to $10^{13.33}$ G (MANCHESTER and TAYLOR [1]). In a strong magnetic field, the electron motion perpendicular to the field lines is quantized to discrete Landau orbitals (LANDAU and LIFSHITZ [2]) and the electrons behaves as a one–dimensional gas rather than a three–dimensional gas. Many calculations of the properties of bulk matter in such fields have been performed, beginning with the work of RUDERMAN [3] and KADOMTSEV and KUDRYAVTSEV [4]. Subsequently a variety of techniques has been applied to calculate the energy of bulk matter.

An important question is whether or not matter in a high magnetic field can have a zero pressure state with lower energy per atom than the energy of an isolated atom, since some pulsar emission mechanisms (RUDERMAN and SUTHERLAND [5]) work only if such a bound state exists. The question is a difficult one because the energy differences are so small, but the most recent Hartree–Fock calculations for chains of atoms (NEUHAUSER, KOONIN, and LANGANKE [6]) suggest that for magnetic fields B in excess of 1×10^{12} G for $Z > 2$ and for $B > 5 \times 10^{12}$ G for $Z > 4$ no such bound state exists. The purpose of this work is not to discuss in detail the possibility of the existence of such a bound state, but rather to show that the magnetic field has a large effect on the equation of state and the surface structure of neutron stars, irrespective of whether or not there is a bound state of condensed matter.

In the surface of the neutron star the Tolman–Oppenheimer–Volkov equation of hydrostatic equilibrium becomes simply (GUDMUNDSSON, PETHICK, and EPSTEIN [7])

$$\frac{dP}{dz} = \rho g_s , \tag{1}$$

where z is the proper distance below the surface of the star and g_s is the surface gravity including the general relativistic corrections. The Gibbs-Duhem relation then has the form

$$dP = n_{\mathbf{atom}} d\mu_{\mathbf{atom}} , \tag{2}$$

where n_{atom} is the number density of atoms and μ_{atom} is the chemical potential of an atom. From Eqs. (1) and (2) one finds (FUSHIKI, GUDMUNDSSON, and PETHICK [8])

$$\mu_{atom}(z) = \mu_{atom}(0) + (Am_p g_s)z, \tag{3}$$

where m_p is the proton mass. The knowledge of the chemical potential as a function of density thus gives immediately the density profile of the star.

We calculate the equation of state in the Thomas-Fermi (TF) approximation, and in the Thomas-Fermi approximation with exchange, the so-called Thomas-Fermi-Dirac (TFD) approximation. Our work differs from earlier work in that we derive new results for the exchange energy of the uniform electron gas, which we use in the TFD calculations. In Fig. 1 we show a plot of the density versus zg_{14}, where g_{14} is the surface gravity in units of 10^{14} cm s^{-2} (A neutron star with a mass of 1 $M_\odot$ and a radius of 10 km has a surface gravity of 1.56×10^{14} cm s^{-2}). The results for the TFD (thick continuous lines) and TF (thick dashed lines) approximations are given for $A = 56$, $Z = 26$ with $B = 10^{11}$, 10^{12} and 10^{13} G in Fig. 1. In addition to these results we show those for the non-magnetic free gas (line with thin long and short dashes) and a non-interacting electron gas in a magnetic field (thin dashed lines) in Fig. 1. We notice that the surface structure of the neutron star is insensitive to exactly which approximation one uses except at the very lowest densities.

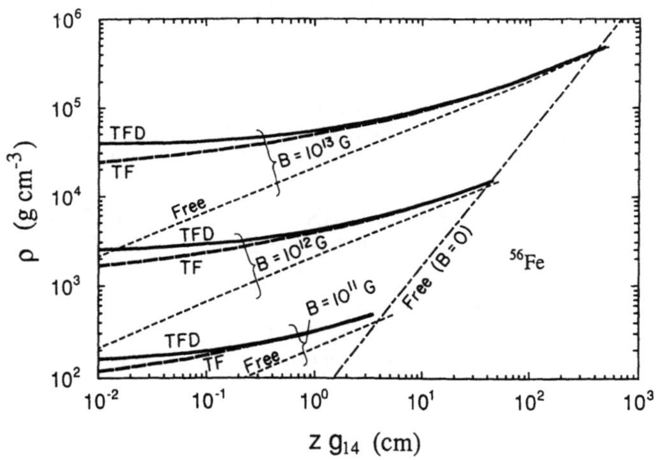

Fig. 1. Density of a neutron star for pure ^{56}Fe as a function of zg_{14} where z is the depth below the surface and g_{14} is the acceleration in units of 10^{14} cm s^{-2} (This is taken from Fig. 6. of Ref. [8]).

In the presence of a magnetic field, density of matter at a given depth is higher than in the absence of a field. For example, the density rises to $\sim 4 \times 10^3$ g cm^{-3} within 1 cm of the surface of a neutron star with $B = 10^{12}$ G while for the non-magnetic free gas the density is less than 10^2 g cm^{-3} at $z = 1$ cm. Since the surface density of a neutron star which is governed by a strong magnetic field ranges from $\sim 10^3$ to $\sim 10^6$ g cm^{-3}, the surface structure is going to influence the relation between the surface and interior temperature of the star (see [7]). This relationship is important for model calculations of neutron star cooling when one wants to compare theoretical results with observations of thermal X-ray emission from the stellar surface.

The rate of nuclear reactions in high-density matter is affected by the fact that the clouds of electrons surrounding nuclei alter the interactions among nuclei. As a consequence of the electron clouds, the reaction rate is increased by a factor which is conventionally written as $e^{-U_{sc}/k_B T}$, where U_{sc}, a negative quantity, is the so-called screening potential and T is the temperature. If the electron distribution is rigid, the screening potential is simply given by the difference of the lattice energies and it is independent of the magnetic field. However, the electrons are not rigid and they respond to the ionic potential. Their compressibility depends on the magnetic field, and this leads in turn to a field dependence of the screening potential. The change in the screening potential due to the compressibility of the electrons for a high magnetic field is (see [8])

$$\delta U_{sc} \doteq -254 \left[(Z_1 + Z_2)^{7/3} - Z_1^{7/3} - Z_2^{7/3}\right] \overline{\left(\frac{A}{Z}\right)}^{4/3} \rho^{-4/3} B_{12}^2 \,\mathrm{keV}, \qquad (4)$$

where $\overline{(A/Z)}$ is the average A/Z ratio which corresponds to the mean molecular weight per electron. The densities at which hydrogen burning by the CNO cycle occurs in X-ray burst sources are of order 10^5 g cm^{-3}, and if the magnetic field were as high as 5×10^{13} G, the value at which relativistic effects begin to play a role, the screening correction for proton capture on ^{14}N would be 4.5 keV, while in the absence of a magnetic field it is only 0.23 keV. Since temperatures of hydrogen induced X-ray bursts are around 2×10^7 K $\simeq$ 2 keV (FUSHIKI and LAMB [9]), screening corrections can be significant in the case of very high magnetic fields, even though they are negligible in the zero field case.

REFERENCES

1. R. N. Manchester, J. H. Taylor: Astron. J., 86, 1953 (1981).
2. L. D. LANDAU, E. M. LIFSHITZ: In Quantum Mechanics, 3rd ed. (Pergamon Press, Oxford), p. 457 (1977).
3. M. A. Ruderman: Phys. Rev. Lett., 27, 1306 (1971).
4. B. B. Kadomtsev, V. S. Kudryavtsev: Soviet Phys., JETP, 35,76 (1972).
5. M. A. Ruderman, P. G. Sutherland: Ap. J., 196, 51 (1975).
6. D. Neuhauser, S. E. Koonin, K. Langanke: Phys. Rev., A36, 4163 (1987).
7. E. H. Gudmundsson, C. J. Pethick, R. I. Epstein: Ap. J., 272, 286 (1983).
8. I. Fushiki, E. H. Gudmundsson, C. J. Pethick: Ap. J. 342, 958 (1989).
9. I. Fushiki, D. Q. Lamb: 1987, Ap. J. Lett., 323, L55 (1987).

Multigroup Simulation of Protoneutron Star Cooling

Hideyuki Suzuki & Katsuhiko Sato

As well known, the data of the neutrino burst from SN1987A observed by KAMIOKANDE-II and IMB show good agreement with the theoretical model of protoneutron star cooling developed by mainly BURROWS and LATTIMER[1]. Although they used the energy-integrated scheme for neutrino transfer, it is not very insufficient because of the small statistics of the data as for SN1987A. But situations would be changed when the neutrino burst from the next Galactic supernova is detected by the future experiment such as SUPERKAMIOKANDE. Since several thousands of events will be detected, we can analyze the neutrino spectrum. That means multigroup simulation of the protoneutron star cooling is required to get useful information about equation of state of high density matter, supernova mechanism, neutrino interaction and so on. This is one of the main reasons why we are doing multigroup simulation of the protoneutron star cooling.

Our code is a spherical symmetric Lagrange mesh code with implicit method. We calculate neutrino transfer with the multigroup flux limited diffusion scheme and calculate the protoneutron star structure by solving Oppenheimer-Volkoff equation with Henyey method. That is, we assume the hydrostatic structure of the protoneutron star and our simulation is corresponding to the cooling stage which starts about 1 second after the bounce. Three types of neutrinos (ν_e, $\overline{\nu}_e$, 'ν_μ' $= 1/4$ (ν_μ, $\overline{\nu}_\mu$, ν_τ, $\overline{\nu}_\tau$)) are included and general relativistic effects for neutrino transfer such as red shift are also included. The special feature of our code is flexibility. Our code consists of many module and it is very easy to change the EOS, flux limiter and so on. As for the EOS, however, in the present simulations we use the table calculated with Hartree-Fock method by R. Wolff. Matter is composed of free neutrons, free protons, α particles, a representing species of nuclei, electrons, positrons and photons.

The following neutrino interactions are included in the source term of the spectral change with the assumption of spherical symmetry.

$$e^- p \longleftrightarrow \nu_e n \ , \ e^+ n \longleftrightarrow \overline{\nu}_e p \ , \ e^- A \longleftrightarrow \nu_e A'$$
$$e^+ e^- \longleftrightarrow \nu \overline{\nu} \ , \ \nu e^\pm \longleftrightarrow \nu e^\pm$$

We mainly use the energy-dependent interaction rate summarized in BRUENN's paper[2]. We would stress that we treat the neutrino-electron scattering and pair process in a full manner and not use the Fokker-Planck approximation. Isoenergetic neutrino scattering off n, p, α, A are included only as the opacity source.

Up to now, we calculated two models. Model C40B is started from an artificial initial model with the baryon mass of $1.7M_\odot$ which is somewhat cold model. We use the flux limiter almost same as BRUENN's[2]. 40 radial grids and 12 energy grids are used and we terminate the simulation at 15.6sec. The initial model for another model, MW88, is constructed from the data given by WILSON[3]. From his data of 0.4sec after bounce, the entropy distribution and the electron fraction distribution within the baryon mass of $1.62M_\odot$ are used for the construction of initial model. In the case of model MW88, we adopt the almost same flux limiter as MAYLE & WILSON's[4]. Furthermore the radial(84) and energy(16) grids resemble theirs. Note that since we do not have the same EOS as Wilson's, we use Wolff's table and the constructed structure is different from theirs.

Model C40B shows good agreement with the results of BURROWS[5] as for the $\overline{\nu}_e$ luminosity curve and the $\overline{\nu}_e$ mean energy. But the evolution of the inner structure of the protoneutron star is somewhat different from their early work[1]. We also calculate the neutrino energy spectrum and get the information about the evolution of the spectral shape such as $< E^2 > / < E >^2$. We will investigate the evolution of the structure and spectra systematically since now.

Figure 1 shows the comparison between the neutrino mean energy of our model MW88 and MAYLE & WILSON's[3]. Our simulation results in lower mean energy than theirs. Specially our 'ν_μ' is much lower. As for the luminosity curve, while 'ν_μ' has the smallest luminosity among the three in our results, 'ν_μ' is highest in their result. (The difference between our luminosity and theirs for ν_e and $\overline{\nu}_e$ is small.) What is the origin of these differences? Although we do not know the code dependence and it may be effects of matter accretion in Mayle & Wilson's simulation, we think one of the main reasons is the difference of EOS.

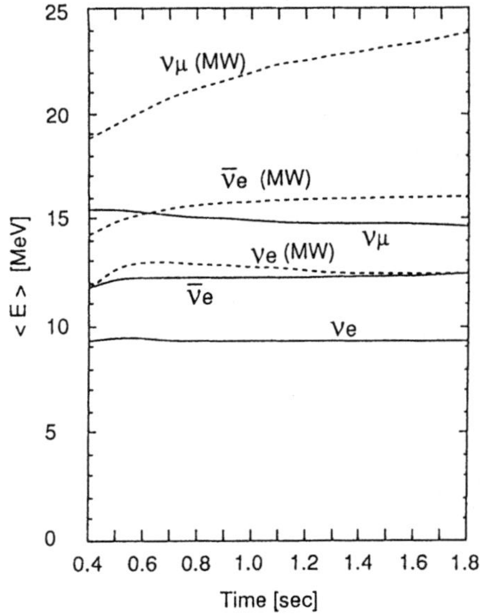

Figure 1. Mean energy of the neutrino flux. Solid lines are our model MW88 and dashed curves are Mayle & Wilson's.

In Fig.2 we plot the matter pressure of the two EOSs along the same array of density, electron fraction and entropy, P(ρ_B, $Y_e(\rho_B)$, $S(\rho_B)$). Wolff's EOS is stiffer than Wilson's. For example it has three times higher pressure than Wilson's at $\rho_B = 8.407 \cdot 10^{14}$g/cm^3, $Y_e = 0.2921$, $S = 1.19$. Stiffness of the EOS results in the low temperature in the protoneutron star because of less contraction. In our initial model maximum temperature is less than 30MeV but in the data of Wilson's it is nearly 50MeV. Since the production rate of 'ν_μ' is very sensitive to the temperature, we can interpret the low mean energy of 'ν_μ' as the effect of stiff EOS.

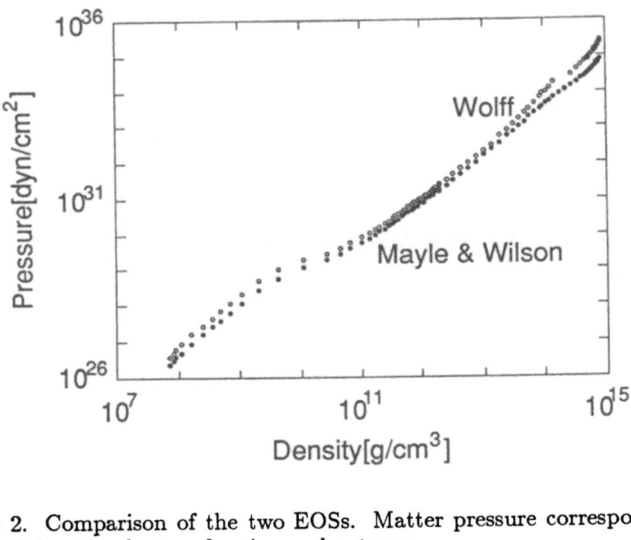

Figure 2. Comparison of the two EOSs. Matter pressure corresponding to the same sets of density, electron fraction and entropy.

The authors thank M. Fukugita, R. Wolff, R. Mayle and J. Wilson for offering us their data, H.-T. Janka for many discussions and KEK for supporting our computation. This research is supported in part by the Grant-in Aid for Scientific Research Fund from the Ministry of Education, Science and Culture(01629504,01790167) and the Japan-U.S. Cooperative Science Program(MPCR-185).

References

1. A. Burrows and J. M. Lattimer, Ap. J. **307**, 178 (1986).

2. S. W. Bruenn, Ap. J. Suppl. **58**, 771 (1985).

3. J. R. Wilson and R. W. Mayle, preprint (1988), and private communications.

4. R. W. Mayle, J. R. Wilson and D. N. Schramm, Ap. J. **318**, 288 (1987).

5. A. Burrows, Ap. J. **334**, 891 (1988).

SECTION VII

EXPLOSION MECHANISMS AND NEUTRINO BURSTS

Neutrino Astrophysics: New Adjunct of Astronomy and Elementary Particle Physics

Alfred K. Mann

Neutrinos, the neutral, elementary fermions with very small, perhaps zero mass, are the only known particles to interact only weakly (and gravitationally) with matter. Nevertheless, they carry energy and linear and angular momentum, as well as a quantum number denoting their lepton nature. Due to these properties neutrinos constitute a unique probe of stellar interiors because, unlike photons, they can emerge promptly from the central regions of stars and transmit direct information relating to, e.g., relaxation times or processes in those regions. Moreover, traversal of the stellar interiors with their extreme values of temperature and density provides a means of searching for as yet undetected intrinsic properties of neutrinos. Accordingly, neutrino astrophysics is an adjunct of both astronomy and elementary particle physics.

The data of neutrino astrophysics are in part neutrino flux limits acquired from searches for energetic neutrinos from distant point sources, e.g., SN1987A; from a search for relic antineutrinos (with energies from a few to 20 Mev) from all part supernovae; and from searches for neutrinos as or from dark matter candidates. More interesting are the data on atmospheric (cosmic ray induced) neutrinos and the positive observations of low energy anti-neutrinos from SN1987A and their implications; and, of comparable interest, measurement of the magnitude and direction of the flux of neutrinos from the Sun. Observation of neutrinos from the Sun in turn permits tests for possible correlations of the solar neutrino flux with intense solar flares and with the solar magnetic cycle.

Each of the subjects above is material for an extended seminar and consequently only a representative sample could be discussed in a short talk. For the purpose of a talk at the Santa Cruz Workshop on Supernovae it seemed appropriate to treat briefly the less familiar material, for example, the limiting flux values from the searches for neutrinos and implication of neutrinos from the Sun. This was intended as an overview of the progress of neutrino astrophysics during the last few years. Much of this material has been published recently, and rather than reproduce it in this proceedings of the Supernovae Workshop, it appears reasonable simply to give in addition to the summary in this abstract a few references for the reader who may wish to pursue the subject further.

References

Search for distant point sources: Y. Oyama *et al.*, Phys. Rev. Lett., **59,** 2604 (1987); R. Svoboda *et al.*, Astrophys. J., **315,** 420 (1987); Y. Oyama *et al.* Phys. Rev., **D39,** 1481 (1989).

Search for relic anti-neutrinos from past supernovae: W. Zhang *et al.* Phys. Rev. Lett., **61,** 385 (1988); A. K. Mann and W. Zhang, to be published in Proc. of the NASA Workshop on Physics and Astrophysics from a Lunar Base, Stanford University, May, 1989 (in press).

Solar neutrinos: R. Davies, Jr. in Proc. Conf. on Neutrino Physics and Astrophysics, Boston, 1988. Ed. by J. Schneps *et al.* World Scientific, 1989 (p. 518); K. S. Hirata *et al.* Phys. Rev. Lett., **63,** 16 (1989); R. Davis, Jr., A. K. Mann, and L. Wolfenstein, Ann. Rev. Nucl. and Particle Science, **38,** 467 (1989).

Calculations of Neutrino Heating Supernovae

Ronald W. Mayle & James R. Wilson

I. Physical and Numerical Model

The numerical model is based on a fully general relativistic treatment of hydrodynamics and neutrino flow. The neutrino time evolution is approximated by a flux limited diffusion equation. The neutrinos are described by three functions, one representing electron neutrinos, another electron antineutrinos and a third to represent muon and tauon neutrinos and their antiparticles; these functions depend on time, space and neutrino energy. All neutrino-matter interactions thought to be important in the collapse and explosion process are included. The equation of state for matter below nuclear density is represented by the species: photons, electrons, positrons, neutrons, protons, helium nuclei and heavy nuclei. The heavy nuclei have properties dependent on the chemical potentials of the other constituents; Saha equations are solved to determine the abundance of all particles. A simple nuclear burn model consisting of He, C, O, Ne, Si, and Ni is used to carry matter up "iron" (nuclear statistical equilibrium); after matter has burned to "iron" it is assumed to be in statistical equilibrium thereafter. When the star becomes convectively unstable by either the Le Doux or the salt finger criteria, convection is treated in the mixing length approximation.

II. Collapse and Bounce

After the inner part (1.2 - 1.5 solar masses) of a massive star burns to "iron", the core cools by neutrino emission and consequently increases in density due to slow contraction. When densities reach about 10^9 [gm/cc] electron capture on heavy nuclei and the accompanying energy loss becomes fast enough that the core begins to collapse dynamically. The central part of the iron core contracts homologously with the in fall velocity proportional to the distance from the center of the star. The size of this homologous core is about the

Chandrasekhar mass (M_{ch}) = 5.8 Y_e^2 solar masses, where Y_e is the number of electrons per baryon. In what follows we will give results of a collapse calculation using as initial data the evolved core of a 20 solar mass stellar model of SN1987a supplied by T.A. Weaver and S. Woosley. This model had an initial iron core mass of about 1.45 solar masses.

At the start of the collapse $Y_e \approx .43$ giving M_{ch} = 1.12 solar masses; by the time nuclear density is reached at the center of the core $Y_e \approx .30$ and M_{ch} = 0.53 solar masses. The entropy per baryon starts at about 1.0 and rises to 1.5 as the density passes from 10^{11} to 10^{12} [gm/cc]. The star passes from being transparent to neutrinos for $\rho < 10^{11}$ [gm/cc] to being diffusive for $\rho > 10^{12}$ [gm/cc]. In the range $10^{11} < \rho < 10^{12}$ [gm/cc] the neutrinos are very interactive with the matter but not able to stay in equilibrium; this leads to the entropy increase. After the density rises above 10^{14} [gm/cc] the entropy slowly falls due to neutrino cooling. It is found that $Ye \approx Y_l$ (Y_l is the lepton number per baryon) for $\rho < 10^{11}$ [gm/cc] and (Y_l - Y_e) is constant for $\rho > 3 \times 10^{12}$ [gm/cc] reflecting again the transition of the state of the neutrinos from complete non interaction to complete equilibrium with the matter. The fraction of baryon matter contained in heavy nuclei is .98 at the start of collapse. Due to the rise in temperature and density during the core contraction, this fraction drops to about .75, keeping the pressure lower than it would have been without the break up of the heavy nuclei and allowing the collapse to accelerate.

Just before bounce, the core has attained the maximum in fall kinetic energy of 1.1×10^{52} [ergs]; below the sonic point (the position in the star where the sound speed equals the in fall velocity) the matter is collapsing homologously. The position of the sonic point is close in mass to $M_{ch} \approx .50$ solar masses. Outside the sonic point the matter is falling in supersonically with more than half the free fall velocity. After the central density exceeds nuclear matter density, $\approx 2.4 \times 10^{14}$ [gm/cc] at $Y_l = .37$, the pressure rises very rapidly and the collapse decelerates; the peak central density reached is 5.6×10^{14} [gm/cc]. The inner sonic core rebounds and starts a pressure wave moving out into the supersonic in falling material. At a mass of about 0.60 solar masses the pressure wave turns into a shock wave. Some fraction of the maximum in fall kinetic energy of 1.1×10^{52} [ergs] should be available for energizing the outward moving shock wave. However, the shock has to proceed through nearly 0.9 solar masses of in falling iron; to dissociate one solar mass of iron requires about 1.65×10^{52} [ergs] of energy. In addition to the energy expended dissociating iron the neutrino luminosity is very high, peaking at close to 10^{54} [ergs/sec]. The integral over the first 10 [ms] after bounce of the neutrino luminosity just outside the shock front minus the luminosity at the neutrinosphere (the point in the star inside of which the neutrino mean free path is less than the neutrinosphere radius) is 3.2×10^{51} [ergs]. The combination of energy losses to iron dissociation and neutrino emission weakens the shock so that after a few tens of milliseconds it turns into an almost stationary accretion shock at a radius of about 4×10^{7} [cm].

III. Late Time Neutrino Heating

After a few tenths of seconds the first phase of neutrino heating occurs. Well outside the neutrinosphere we can approximate the neutrino energy exchange with the matter in the following manner. The heating rate is given by

$$\dot{E}_{+} = K(T_p)\, L / 4\pi R^2 \tag{1}$$

where L is the total luminosity in electron neutrinos and electron antineutrinos. The opacity, $K(T_p)$, is evaluated at the neutrinosphere temperature T_p and is proportional to the square of the temperature. Since the major source of opacity in this phase of neutrino heating is the emission and absorption of electron neutrinos and antineutrinos on free baryons, the opacity is also proportional to the free baryon mass fraction (this is not explicitly shown in the notation). The cooling rate is given by

$$\dot{E}_{-} = - K(T_m)\, a'c\ T_m^4 \tag{2}$$

where T_m is the matter temperature at the point of interest, a' is the radiation constant for Fermi particles (7/8 the photon radiation constant 'a'), and c is the speed of light. If we let $L = 4\pi R_p^2\, a'\, c\, T_p^2 / 4$ then the net heating rate becomes

$$\dot{E}_{net} = a'\, c\, K(T_p)\, T_p^4 \left[\ (R_p / 2R_m)^2 - (T_m / T_p)^6\ \right]. \tag{3}$$

Thus if the matter temperature is low enough compared to the neutrinosphere temperature a positive net heating can occur.

At the times under consideration the matter falling through the shock wave is iron; the kinetic energy dissipated in the shock is approximately the gravitational energy since the shock position is nearly constant. So we may write

$$\frac{GM}{R} \approx \left(\frac{3}{2} f + 3\, Y_e \right) kT + f\ I \tag{4}$$

where the first two terms on the left are the thermal energy of free baryons and electrons; in the last term I is the dissociation energy and f is the degree of dissociation. For our example calculation M ≈ 1.5 solar masses, $Y_e \approx 0.5$ so we have with R_7 equal to the radius in units of 10^7 [cm] and temperatures and energies measured in [MeV]

$$\frac{20}{R_7} \approx 1.5\,(\,1.+f\,)\,T + 8.4\,f\ . \tag{5}$$

At the densities just below the shock (10^8 to 10^9 [gm/cc]) the decomposition temperature is about 1 to 2 [MeV]. Above a radius of about 3 x 10^7 [cm] the material is predominately undecomposed iron and helium which have smaller opacities for neutrino interactions than free baryons; heating will not occur in this region. The heating is a maximum at about 2 x 10^7 [cm].

This phase of neutrino heating does not occur immediately after bounce. As time progresses the density of in falling matter decreases which lowers the temperature at which decomposition occurs. Also the proto-neutron star contracts with time and the neutrinosphere temperature increases. These two effects lead to the heating phase being delayed for several tenths of a second after bounce. When neutrino heating occurs the heated matter expands; the amount of material in the heating region decreases due to the expansion thus resulting in a decrease of the heating rate. The heating rate is high from about 0.3 to 0.5 seconds after core bounce . The accretion shock wave begins moving outward rapidly again after 0.5 seconds passing through the iron and into the predominately silicon-oxygen region; in this region the shock raises the temperature sufficiently to burn some of the Si-O to nickel. After the shock passes a radius of about 2 x 10^8 [cm] the density becomes too low and the shock too weak to induce nuclear burn. The amount of Ni^{56}, which is an observable, produced depends directly on the energy deposited in this first phase of neutrino heating. The energy contained in the hot bubble between the neutrinosphere and the shock is still much less than the gravitational binding energy of the matter above the shock. Further energy deposition is needed.

After the hot bubble has expanded a different kind of neutrino heating can occur, neutrino-antineutrino annihilation. This was discussed as an explosion mechanism by GOODMAN, DAR and NUSSINOV [1]; COOPERSTEIN, VAN DEN HORN and BARON [2] dismissed neutrino-antineutrino annihilation as being a small effect since the annihilation can only occur near the neutrinosphere and if the density in this region is too high the deposited energy is immediately re-radiated by electron-positron capture on protons and neutrons. However, after the first phase of neutrino heating the density gradient outside the neutrinosphere becomes progressively steeper and eventually becomes sufficiently so that matter on the surface of the proto-neutron star is heated and blown off.

The cross section for neutrino antineutrino annihilation depends on the neutrino energy, ε_ν, and the angle , θ_ν, between the neutrino and antineutrino in the combination $\varepsilon_\nu^2 (1 - \cos \theta_\nu)^2$. The cross section is thus proportional to the square of the collision energy in the center of mass frame of the colliding neutrino and antineutrino. In the numerical computer model we carry only the energy density of neutrinos, F_0, as a function of the neutrino energy. We must infer the angular distribution indirectly. In the diffuse limit we approximate the angular distribution by

$$F(R,\varepsilon,\mu) = F_0(R,\varepsilon) + \frac{3D\mu}{c}\frac{\partial F_0(R,\varepsilon)}{\partial R} \tag{6}$$

where D, the flux limited diffusion coefficient, depends on $\lambda \left|\frac{\partial \log F_0}{\partial R}\right|$ with λ being the neutrino mean free path (see BOWERS and WILSON [3]) and μ the cosine of the angle of the neutrinos with respect to the radial direction. Integration of the cross section over μ gives the angular factor

$$Q_1 = 1 - \frac{3}{2}\frac{D\bar{D}}{c^2}\left|\frac{\partial \log F_0}{\partial R}\right|\left|\frac{\partial \log \bar{F}_0}{\partial R}\right| \tag{7}$$

where the bars over D and F_0 denote the flux limited diffusion coefficient and the distribution function for the antineutrinos. In the limit of an infinitely sharp neutron star boundary the angular integration gives a factor

$$Q_2 = (1 - x)^2(5 + 4x + x^2) / 8 \tag{8}$$

with $x = \sqrt{1 - (R^*/R)^2}$ where R^* is the radius of the neutron star which we will take as the neutrinosphere radius. We use the larger of Q_1 and Q_2 for our angular factor; in the limit where Q_2 dominates the net heating per unit volume goes as $(1/R)^8$ for large R.

A a time of 0.9 [sec] after bounce the density above the neutrinosphere falls with radius as $(1/R)^{27}$. The second phase of neutrino heating is just getting underway. At a density a little under 10^{10} [gm/cc] the neutrino-antineutrino annihilation energy deposition becomes greater than the electron-positron annihilation energy loss to neutrino production. The energy deposition rate is not high, only a few times 10^{50} [ergs/sec], but it will continue as long as the proto-neutron star is emitting energy in neutrinos.

An additional neutrino process also occurs at late times when matter in the bubble region has heated sufficiently that most of its internal energy is in photons and electron and positron pairs. As soon as this happens we may write

$$E = \frac{11}{4} \frac{a T^4}{\rho} \tag{9}$$

$$n_p = \frac{7}{4} \frac{a T^3}{3k} \tag{10}$$

$$\sigma_e = \frac{5}{4} \sigma_0 \frac{\varepsilon_\nu T}{(m_e c^2)^2} \tag{11}$$

$$\dot{E} = \sigma_e n_p L / 4\pi R^2 = \frac{11}{4} a \dot{T}^4 \tag{12}$$

$$\frac{\dot{T}^4}{T^4} = \frac{5}{12} \frac{\sigma_0}{k} \frac{\varepsilon_\nu}{(m_e c^2)^2} \frac{L}{4\pi R^2} \equiv \frac{1}{\tau} \tag{13}$$

where a is the photon radiation constant, k is Boltzmanns constant, c is the speed of light, m_e is the mass of an electron, $\sigma_0 = 1.7 \times 10^{-44}$ $[(\mathrm{cm})^2]$ is a fundamental weak interaction cross section (see TUBBS and SCHRAMM [4]), ε_ν is the neutrino energy, n_p is the number of electron-positron pairs, L is the neutrino luminosity, E is the internal energy of matter per gram and R is the radius. At a time of 2.5 sec after bounce at a radius of 10^7 [cm] $t = 0.5$ [sec]. This pair heating does not produce much total energy deposition but it takes matter heated by the other two neutrino heating processes and raises the entropy per baryon to a few thousand. The calculation was carried out to a time of 3.6 [sec] after bounce. At that time the density near 10^7 [cm] had fallen to about 100 [gm/cc].

The calculation becomes very expensive in computer time so it was not completed. We extrapolate the energy production rate to estimate the final explosion energy. Not until 2.5 sec after bounce is there a net explosion energy since the initial binding energy of the hydrogen helium envelope is negative. At 3.6 sec the net explosion energy is 0.3×10^{51} [ergs] and about two thirds of the final binding energy (3×10^{53} [ergs]) of the neutron star has been emitted in neutrinos that have escaped the star . At this time we find the ratio of the rate of energy deposition from neutrino-antineutrino annihilation reactions to the total neutrino luminosity to be 0.02. If we assume that the energy deposition efficiency for the remaining energy to be emitted in neutrinos is the average (0.01) found in the interval from 2.5 to 3.6 seconds then the final explosion energy should be about 1.4 x 10^{51}[ergs].

IV. Comparison of Calculations and Observations

As stated earlier the calculation described above was based on a stellar evolution calculation made by Weaver and Woosley for a 20 solar mass star that is thought to be similar to the progenitor of SN1987a. We estimate the explosion energy to be about 1.4 x 10^{51} [ergs]. From the immediate post explosion light curve an estimate of the explosion energy of 0.6 to 2.0 x 10^{51} [ergs] has been derived (see for example SHIGEYAMA, NOMOTO and HASHIMOTO [5] and WOOSLEY [6]). From the late time photon luminosity of the supernovae the Ni^{56} production is estimated to be about .075 $\pm$ .020 solar masses. Our calculation gave .065 solar masses of Ni^{56}. One half of the Ni^{56} is produced during in fall by adiabatic compressional heating and the rest is produced by shock heating after the shock wave passes the burn front. The density of the in fall burn is about 3 x 10^6 [gm/cc] while the preshock burn density is around 2 x 10^6 [gm/cc]. At the end of our calculation the baryonic mass of the neutron star is 1.63 solar masses. To calculate the neutrino signal we took the star shortly after bounce and removed all the matter outside 1.63 solar masses and then calculated the neutrino emission assuming the star is in quasi static equilibrium. To check the mean energy of the electron antineutrinos we folded our energy spectrum with the detector efficiencies of the IMB detector (see BIONTA et al. [7]) and the Kamiokande-II detector (see HIRATA et al. [8]). Our mean neutrino energy weighted by the detector efficiency for the Kamiodande-II detector is 20.2 [MeV] while the average energy of the antineutrinos detected is 15.4 [MeV]. The numbers for the IMB detector are 27.9 [MeV] for the predicted average energy and 32.5 [MeV] for the average detected particle energy. To examine the temporal emission of the neutrinos we combined the events of the two detectors with equal weight (we counted an IMB event as 11/8 of a Kamiokande-II event since IMB saw 8 events and Kamiokande-II saw 11) to produce a cumulative count curve versus time. We find good agreement between the calculated emission folded with an average detector efficiency (to take into account the two different detector efficiencies) and the combined observational emission (cumulative count curve). Thus we have good overall agreement of observation and calculation. The final neutron star gravitational mass we estimate to be 1.45 solar masses; however, this is not observable.

V. Equation of State Considerations

In Section I of this paper the sub nuclear density equation of state we use is described. Above nuclear density we use a zero temperature equation of state plus a thermal component. The zero temperature nuclear EOS was suggested by H.A. Bethe who used results described in MUTHER, PRAKASH and AINSWORTH [9]. The internal energy per baryon, E_0 (in units of [MeV]), of the zero temperature component of our EOS is taken to be

$$E_0 = -16 + \frac{1}{9} K_0 (\eta^\gamma - 1 + \gamma (\eta - 1)) / (\eta \gamma (\gamma - 1)) + E_{SYM} \tag{14}$$

$$E_{SYM} = 16 (1 - 2 Y_e)^2 \eta (1 + 72 / (1 + 4 \eta)) \tag{15}$$

where $K_0 = 200$ [MeV], $\gamma = 2.75$, and $\eta = \rho/\rho_N$ with $\rho_N = 2.656 \times 10^{14}$ [gm/cc]. This equation of state produces a cold neutron star with a gravitational mass of 1.40 solar masses and a baryon mass of 1.64 solar masses ; the binding energy is 3.04×10^{53}[ergs].

The presence of pions, which is ignored in the present calculations, will make an appreciable change in the EOS. Estimates using a model similar to that described in the work of FRIEDMAN, PANDHARIPANDE and USMANI [10] show that a lowering of the peak temperature seen in our model calculations by 10 [MeV] could be expected (inclusion of an effective nucleon mass would raise the temperature). The neutrino opacity would increase by as much as a factor of two if pions were included in the EOS. The biggest effect expected is from the reduction of the electron density by the conversion of electrons to negative pions. The effect of this latter process on the cooling of the proto-neutron star is hard to estimate at present.

The present calculation appears to account for SN1987a quite adequately. One concern we have at present if whether we have treated the prompt shock propagation well. It is necessary that the prompt shock go out to a distance of about 4×10^7 [cm] and remain there long enough (around 0.5 [sec]) for the first heating to occur. Another concern is whether our algorithm for neutrino-antineutrino annihilation is sufficiently accurate outside the neutrinosphere. We are working with H.-T. Janke on this latter problem.

This work was performed under the auspices of the USDOE at Lawrence Livermore National Laboratory under contract no. W-7405-ENG-48.

References

1. Goodman, Dar and Nussinov: Ap. J., 314, L10 (1987).
2. Cooperstein, van den Horn and Baron: Ap. J., 309, 653 (1987).
3. Bowers and Wilson: A. J. Suppl. 50, 115 (1982).
4. Tubbs and Schramm: Ap. J., 201,467 (1975).
5. Shigeyama, Nomoto and Hashimoto: Astron. Astrophys, 196, 141 (1988).
6. Woosley: Ap. J., 330, 218 (1988).
7. Bionta et al.: Phys. Rev. Lettr. 58, 1494 (1987).
8. Hirata et al.: Phys. Rev. Lettr. 60,1999 (1987).
9. Muther, Prakash and Ainsworth: Phys. Lettr. B, 199, 469 (1987).
10. Friedman, Pandharipande and Usmani : Nuc. Phys. A372, 483 (1981).

Initial Models and the Prompt Mechanism of SN II

E. Baron & J. Cooperstein

1 Introduction

Massive stars ($M \gtrsim 10 - 12\ M_\odot$) become catastrophically unstable when the fuel in their central regions is exhausted. The innermost $1 - 2\ M_\odot$ of fusion ashes (iron peak elements burnt to nuclear statistical equilibrium (NSE)) is the *iron core*. It resembles a white dwarf star dominated by the pressure of relativistic electrons, albeit a hot one. An isolated white dwarf can not support more than the appropriate Chandrasekhar mass for its composition, but the compact core of the massive star must support the overlying burning shells. Before it evolves completely to the low–temperature white dwarf configuration it loses this ability, is overwhelmed by gravitation and collapses. It is generally agreed that a Type II supernova explosion ensues but there is no general agreement about the details of the process. In fact at present there exist no self–consistent calculations of the explosive stage which adequately explain observed explosion features.

There are two well developed theories of the central moments of the explosion. In the first, the prompt mechanism, which has been studied by many authors, the star implodes to supranuclear densities and then resists further compression. The inner core rebounds and launches a shock wave which then propagates through the star and eventually erupts through the photosphere producing the supernova display. In the second theory, the delayed mechanism[1,2,3,4] there is insufficient energy in the initial shock wave to propagate all the way to the edge of the star. Instead it falters and becomes an accretion shock. However, about one second after the central bounce, it is revived by deposition of a small percentage of the energy leaving in neutrinos, and resumes its outward motion. The result is a weaker explosion than in the prompt mechanism. In this talk, we will focus on the prompt mechanism.

It is generally agreed[5,6,7,8] that using the best input physics available and initial models that result from current presupernova evolution calculations the prompt shock mechanism does not work. The prompt shock wave stalls and dies either to be revived by the delayed mechanism or to be engulfed when enough matter rains through it to drive the proto-neutron star to a black hole.

Stellar evolution calculations are not, however, without uncertainties. The treatment of convection is notoriously uncertain, as well as the complicated silicon burning stage (cf. refs.[9,10]). Recently[11], the electron capture and beta decay rates used in the presupernova calculations have been re-examined and it has been suggested that the models may be considerably cooler than current models. In this paper we examine the effect of varying the initial iron core on the viability of the prompt shock wave. Lower mass iron cores require the shock to traverse, and hence, dissociate less iron significantly reducing the energy losses of the shock. Colder cores have fewer free protons. Free protons copiously capture electrons on infall, reducing the final trapped lepton fraction and hence, the mass point where the shock forms. Thus, we consider smaller, colder cores in hopes of delineating the range of successful prompt shock waves. We find that cold initial cores of mass $M \sim 1.12\ M_\odot$ can produce reasonably strong explosions.

2 Construction of Models

Ideally, to construct a series of initial models we would tell the stellar evolutionist that we would like a core of such and such a description and that if she would change her assumptions about this piece of physics she could produce for us an iron core meeting our specifications. This method is, however, not feasible. To construct a run of models would take an enormous amount of computer time as well as frustration. Instead, we construct our own "fake" models taking guidance from the results of pre-supernova evolution calculations.

In order to construct an initial model we must specify both the composition and profile of the thermodynamic variables and the density structure and velocity profile. Now, iron cores are essentially destabilized hydrostatic stars, since the pressure of the hydrostatically stable core of the star has just been reduced by a combination of photodisintegration and electron capture. Therefore we build models that are in hydrostatic equilibrium and destabilize them. Examining the results from stellar evolution calculations we find that the models are destabilized rather uniformly. In order to accomplish this uniform destabiliztion we have chosen to build hydrostatic models using a somewhat reduced gravity. Thus, when gravity is restored in the hydrodynamical calculation the model collapses with a very uniform pressure deficit. In order to accomplish this we build models using a Tolman, Oppenheimer, Volkov code, but with Newton's constant somewhat reduced; i.e., we take

$$G = g_{\text{eff}}\ G_0, \tag{1}$$

where G_0 is the actual value and $g_{\text{eff}} < 1$.

To complete the specification of the models, we must choose a central density, a composition profile, and a temperature or entropy profile. We choose to work with the entropy since that stays roughly constant during the collapse. Figure 1 displays the entropy and Y_e profiles of the results of pre-supernova evolution calculations of 13 $M_\odot$ model of NOMOTO and HASHIMOTO[9], the 15 $M_\odot$ model of WOOSLEY and WEAVER[10], and the 18 $M_\odot$ model of WEAVER and WOOSLEY[12]. Certain general features of the entropy profile are immediately apparent. There is a roughly linear rise in the entropy in the center up to mass point M_1 which corresponds to the convective silicon burning core. In the two models of WOOSLEY and WEAVER which had a silicon burning shell ignite after core silicon exhaustion there is a steeper rise up to mass point M_2, followed by a flat region up to mass point M_3, then there is another steep rise up to mass point M_4 after which the entropy is roughly constant; this is the silicon burning shell at the onset of instability. It is important to note that

Model	102 103 106	104	105	107	109 110
M_1	0.72	0.72	0.72	0.72	0.72
M_2	1.00	1.00	0.90	1.00	1.00
M_3	1.10	1.10	0.975	1.05	1.10
M_4	1.15	1.15	1.00	1.10	1.15
$M_{\text{iron core}}$	1.125	1.125	0.988	1.075	1.125
S_c	0.60	0.50	0.60	0.60	0.50
S_1	0.86	0.70	0.86	0.86	0.63
S_2	1.60	1.60	1.30	1.60	1.60
S_4	4.00	4.00	4.00	4.00	4.00
$\langle S \rangle_{M \leq M_3}$	0.94	0.83	0.84	0.91	0.80
Y_{ec}	0.41	0.42	0.40	0.41	0.415
Y_{e3}	0.46	0.46	0.46	0.46	0.46
Y_{e4}	0.50	0.50	0.50	0.50	0.50
$\langle Y_e \rangle_{M \leq M_3}$	0.435	0.440	0.430	0.435	0.438

Table 1: Initial model compositions

The masses M_{1-4} are the points where the entropy and Y_e profile change as discussed in the text. $M_{\text{iron core}}$ is the mass of the iron core. S_c and Y_{ec} are the central values of S and Y_e. The other values of S and Y_e refer to those values at the mass points with the same subscripts. Also listed are the average values of S and Y_e where averages are taken for mass points inside of M_3. The masses are given in units of $M_\odot$, and the entropy is in k_B^{-1} per nucleon.

Model	K_0	γ	W_s	NES	$\rho_{10}^{(c)}$	$g_{\rm eff}$
102	180	3.0	31.5	off	0.538	0.975
103	180	3.0	31.5	on	0.538	0.975
104	180	3.0	31.5	on	0.500	0.975
105	180	3.0	31.5	on	0.150	0.950
106	180	2.5	31.5	on	0.538	0.975
107	180	3.0	31.5	on	0.250	0.975
109	180	3.0	31.5	on	2.000	0.975
110	180	2.5	31.5	on	2.000	0.975

Table 2: Model parameters

The parameters used in each calculation. K_0 is the incompressibility of symmetric matter at saturation in MeV and γ is the high density adiabatic index. W_s, the symmetry energy is measured in MeV, NES refers to whether the effects of neutrino electron scattering are included, and $\rho_{10}^{(c)}$ is the initial central density in units of 10^{10} g cm^{-3}. $g_{\rm eff}$ is discussed in the text.

Model	102	103	104	105	106	107	109	110
$R_{\rm dead}$	–	566.1	–	318.3	1117.3	263.2	–	–
$M_{\rm dead}$	–	1.167	–	1.125	1.208	1.125	–	–
E_{ν_e}	7.08	3.39	5.92	1.16	6.18	0.87	4.69	4.61
$E_{\bar{\nu}_e}$	5.21	1.36	4.05	0.03	4.24	0.01	2.95	2.90
$E_{\nu_{\rm pair}}$	7.13	1.98	6.20	0.20	6.17	0.13	3.96	4.64
$E_{\rm kin}$	0.01	–	0.01	–	–	–	0.14	0.26
$u_{\rm max}$	0.49	–	0.66	–	–	–	1.59	2.05
R_s	2104	–	1754	–	–	–	2682	3159
M_s	1.267	–	1.208	–	–	–	1.192	1.200

Table 3: Results of hydrodynamical simulations

The results at the end of the calculation are described for each model. The radius (in km) and mass point (in $M_\odot$) at which the wave stalls are listed as $R_{\rm dead}$ and $M_{\rm dead}$ respectively. The total amount of energy lost in electron type neutrinos, electron type anti-neutrinos, and μ and τ pair neutrinos is listed as E_{ν_e}, $E_{\bar{\nu}_e}$, $E_{\nu_{\rm pair}}$. The energies are given in units of 10^{51} ergs. If the shock wave is still propagating at about 2000 km the maximum outward velocity at that time, (in units of 10^9 cm s^{-1}) is listed as $u_{\rm max}$, the total out going kinetic energy is $E_{\rm kin}$ (in units of 10^{51} ergs), and the radius (in km) and mass point (in $M_\odot$) of the shock wave are listed as R_s and M_s, respectively.

model 102 produces a significantly more powerful shock wave than 103, the shock wave still propagating at 2100 km, albeit weakening significantly. The effects of NES reduce the final Y_l by about 0.02 which is quite significant. BRUENN[7] finds that NES reduces the final Y_l by about 0.04. However, with NES included both our calculations and those of BRUENN find a total final Y_l of about 0.365, i.e. BRUENN finds significantly less capture on infall with no NES than we do. The difference is due to differing treatments of neutrino transport, but it is important to note that we agree when NES is included.

3.3 Model 106

Model 106 is another variation on our "benchmark" model, 103. In this calculation we study the effects on the shock propagation of softening the high density equation of state. Model 106 had the value of the parameter γ in BCK eos reduced from $\gamma = 3$ to $\gamma = 2.5$, producing a soft but not excessively soft equation of state. It is well known (BCK) that softening the high density equation of state strengthens the shock wave. We see that although the shock wave is indeed strengthened, propagating out to over 1100 km, versus 550 in the stiffer case (103) the shock is still not strong enough to produce an explosion. In fact, comparing 102, 103 and 106 we see that the harmful effects of NES are not completely overcome by the somewhat softer equation of state.

3.4 Model 104

Model 104 is significantly colder in the central regions than our "benchmark" model. This extra cooling could occur, for example, if beta decays on heavy nuclei play an more important role than has been assumed in current stellar evolution calculations[11]. The effect on the collapse of having such a low initial entropy is to reduce the fraction of free protons. This reduction in the free proton abundance substantially quenches the amount of electron capture on free protons and thus leads to a higher final trapped lepton fraction at bounce. Whereas the previous models that we have considered are both hot enough and have a low enough initial Y_e that the effects of electron capture on heavy nuclei would probably not give a large contribution to the total reduction in the final trapped lepton fraction. In this model, the initial Y_e is high. The available nuclei have states for which allowed transitions can proceed, and because there are very few free protons, electron capture on nuclei could become a more important source of lepton number reduction. In these calculations we have not included the effects of such captures on heavy nuclei and so we must temper our conclusions until this process is properly included in such cold models. Another possibly harmful effect which has yet to be included is neutrino downscattering due to the neutral current process $A(\nu, \nu')A^*$ where A is a heavy nucleus[17]. This has effects similar to those of NES, however, the size should be smaller and in combination with NES it may not be much of an enhancement. In any event, it is clear that this model produces a weak explosion; while not strong enough to actually be a supernova, it shows that such a cold core is helpful.

3.5 Models 105 and 107

In models 105 and 107 we examine the effect of making the iron core smaller simply by altering the positions of the silicon burning shells, that is by moving them in. The results are interesting and perhaps somewhat paradoxical. The trapped lepton fraction at bounce is significantly lower in these models than for model 103. Both the shock waves make it out of the putative iron core and die in fact at the same mass point $1.125 M_\odot$. Specifically how changing the location of the burning shells makes a difference on the final trapped lepton fraction is not obvious. This reduction in the lepton fraction does not occur until the center reaches about 10^{13} g cm^{-3} so it must alter the flow of neutrinos, but exactly how this occurs is not clear, and we have not yet arrived at a satisfactory explanation of this complicated behavior. Comparing the initial density profiles the models appear more favorable than 103, yet they fare worse.

3.6 Models 109 and 110

These models represent more extreme versions of Model 104. The mass of the iron core is kept fixed, as is the central entropy. It is the slope of the entropy in the initially convective core that is reduced. The average Y_l is substantially higher in model 109. This is reflected in the strength of the shock wave. Models 109 and 110 have reached the end of the computational grid with substantial velocities. An estimate of the total energy in the shock wave for model 109 is about 0.5×10^{51} ergs and 1.3×10^{51} ergs for model 110. Model 110 is stronger simply because the high density equation of state was softer for that model. In these models the same caveat about electron capture on heavy nuclei that we noted in our discussion of model 104 applies. It is still gratifying to see that the prompt mechanism can work on some models with conservative input physics.

4 Discussion and Conclusions

The above hydrodynamical results seem to show that the single most important factor determining the sucess or failure of the shock wave is the value of the final trapped lepton fraction Y_l in the homologous core of the star. The models that are cold initially and hence are subject to less electron capture on infall produce viable shock waves. These models are uncertain, however, as to the effects of capture on heavy nuclei as discussed above. A variety of agencies may affect the final Y_l, including the outer regions of the core, in a complicated fashion. Regardless of the actual cause of the variation, small changes in Y_l lead to large changes in the shock's strength. Primarily this is because the shock formation point moves inward with decreasing Y_l[18,19] as has been well studied. Our results agree with the conclusions of BRUENN[8] who also studied the effects of reducing the entropy and mass of the initial model.

Our results clearly show that the mass of the iron core, at least as we have defined it, does not allow one to easily predict the effect on the shock wave when compared with some other core. In fact, our calculations seem to show that there is a minimum mass below which the effects of reducing the mass still further become detrimental. This can be seen in table 3 by comparing our "benchmark" model 103 with models 105 and 107 which all have smaller iron cores, but the shock is weaker in these models. We have no unambiguous explanation for this effect and the possibility remains that it is the result of our assumptions, although the same effect occurs with models that come from the results of stellar evolution calculations. It would seem that it may be possible that iron cores with masses below about 1.1 $M_\odot$ could produce strong shock waves, but such models would have to look very different in their entropy and composition structure than those we have considered.

Acknowledgement

We wish to thank Morry Aufderheide, for allowing us to use the results of our joint unpublished work on neutrino electron scattering, Paul Schinder for providing us with his results on plasmon emission rates, Ken Nomoto and Stan Woosley for providing us with the results of their calculations and useful discussions, and Hans Bethe for helpful discussions. The calculations reported herein were performed at the National MFE computer center under auspices of the Nuclear Physics division, US Department of Energy.

This work has been supported in part by the U.S. Department of Energy under contract no. DE-AC02-76CH00016 and grant DE-FG02-88ER40388.

References

[1] H. A. Bethe and J. R. Wilson, Astrophys. J. **295**, 14 (1985).

[2] R. Mayle, PhD thesis, University of California, Berkeley, 1985, issued as Livermore Report UCRL-53713.

[3] R. Mayle and J. R. Wilson, Astrophys. J. **334**, 909 (1988).

[4] J. R. Wilson, in *Relativistic Astrophysics*, edited by J. Centrella, J. LeBlanc, and R. Bowers (Jones and Bartlett, Boston, 1985).

[5] J. Cooperstein and E. Baron, in *Supernovae*, edited by A. Petschek (Springer-Verlag, New York, 1989) [in press].

[6] E. S. Myra and S. A. Bludman, Astrophys. J. **340**, 384 (1989).

[7] S. Bruenn, Astrophys. J. **340**, 955 (1989).

[8] S. Bruenn, Astrophys. J. **341**, 385 (1989).

[9] K. Nomoto and M. Hashimoto, Phys. Repts. **163**, 13 (1988).

[10] S. E. Woosley and T. A. Weaver, Phys. Repts. **163**, 79 (1988).

[11] M. B. Aufderheide, G. E. Brown, D. B. Stout, T. T. S. Kuo, and P. Vogel, (1989) [preprint].

[12] T. A. Weaver and S. E. Woosley, (1988) [private communication].

[13] E. Baron and J. Cooperstein, Astrophys. J. (1989) [submitted].

[14] E. Baron, J. Cooperstein, and S. Kahana, Nucl. Phys. **A440**, 744 (1985).

[15] E. Baron, J. Cooperstein, and S. Kahana, Phys. Rev. Lett. **55**, 126 (1985).

[16] S. Bruenn, Astrophys. J. Suppl. **58**, 771 (1985).

[17] W. Haxton, Phys. Rev. Lett. **60**, 1999 (1988).

[18] A. Burrows and J. M. Lattimer, Astrophys. J. **270**, 735 (1983).

[19] J. Cooperstein, H. A. Bethe, and G. E. Brown, Nucl. Phys. **A429**, 527 (1984).

Supernova Calculations and the Hot Bubble

Stirling A. Colgate

The recent calculations of James Wilson and Ronald Mayle (1989) showed that the mechanism of supernova explosions caused by collapse to a neutron star now appears to be both understood conceptually and modeled convincingly up to 3.4 s following collapse. In particular they show the formation of a hot, high entropy bubble that continues to push on the shocked matter for a time long enough that the subsequent history is not in significant doubt. The hot bubble is formed primarily due to mu, tau neutrino antineutrino annihilation as first proposed by Goodman, Dar, and Nussinov (1987). The hot bubble that separates the neutron star from the ejected matter has high entropy, 10^2 to 10^4, measured in units of the Boltzmann constant, k per free nucleon. This high entropy means that for every nucleon there are many photons and electrons (pairs) and so the molecular weight is small and the scale height is large even at modest temperatures, $\leq 1MeV$. Such a photon gas can "push" simultaneously on both the neutron star surface as well as on the expanding matter. It extends to a radius of $10^9 cm$ so that no fallback or reimplosion of any significant fraction of the ejected matter will take place. The kinetic and internal energy minus the gravitational energy of matter whose total energy is positive is $0.35 \times 10^{51} ergs$ at 3.4 s. They expect this to increase to 1 to $1.5 \times 10^{51} ergs$ by the end of the calculation, typical of Type II supernova.

It has long been my major concern (Colgate 1971) that despite a very strong shock wave a significant fraction of the ejected matter would subsequently fall back onto the neutron star. This fraction, several solar masses or more, would fall back in a time of about a half an hour despite an initial large (greater than escape) velocity leading to a black hole.

The reason for the large fallback mass is two-fold. First, normal matter falling onto a neutron star will be cooled by neutrino emission in a few tens of seconds from the high temperature, 1 to 2 MeV, that is created when this matter is shocked and compressed at the neutron star surface. Hence, the pressure that might ordinarily extend from the neutron star surface to the shocked matter, forming a piston to maintain the shock, disappears. The neutron star then acts like a black hole. Second, the internal energy in shocked material is always equal to the kinetic energy of motion of the same matter (behind a strong shock). Hence, despite the radially outward velocity that exists behind a strong ejection shock there is always enough heat to allow for an expansion inwards or backwards (in one dimension) at a velocity equal to the shock created outward velocity. The radially inward velocity of the rarefaction wave overcomes the original outward radial velocity of the shocked matter. When one adds to this backwards expansion the strong gravity existing in the central regions, the rarefaction velocity is quite sufficient to insure reimplosion even for the most energetic shocks whose total energy greatly exceeds that initially needed to eject matter from the star. Third, if the pressure deficit at the neutron star surface, due to neutrino cooling, occurs significantly later than the formation of the strong outward going shock, then the resulting rarefaction wave moving in the co-moving matter will always catch up to the shock wave and weaken it. (The flow behind a shock wave is always subsonic.) Thus we require a positive pressure extending from the neutron star surface to the shock until the shock reaches the outer low density layers of the star where the entropy generated by the shock alone is very large.

With spherical convergence the problem is more complicated due to what is known as Bondi accretion. Here if the shocked matter has a very high entropy or sound speed, the accretion or rarefaction flow tends to "choke" at a smaller radius corresponding to where the gravitational potential is equal to the specific internal energy (sound speed)^2 of the shocked matter. For example if the shock wave were four times stronger, i.e. has four times more energy per mass than is required to unbind the matter, then the radius at which the accretion would be choked by Bondi accretion (Bondi 1952) is no smaller than half the radius of the shock wave, and hence, would only reduce the flow by a factor of between two to four. Thus, there has always been the need to prevent the fallback of matter after being shock ejected from the neutron star.

The reason that measuring entropy in photons per nucleon is conceptually convenient is the following: the gravitational binding of a nucleon to the neutron star is approximately 1/10 its rest mass or 100 MeV per nucleon. (The neutron star cools in a second such that its radius is $\simeq 10^6 cm.$) Consequently, at temperatures typical of the neutron star surface, namely an MeV or 10^{10} degrees, the scale height of normal matter, i.e. just nucleons, would be roughly 1/100 of the neutron star radius. And so, despite the high temperature, nucleonic matter would be tightly bound to the neutron star, and the "push" 100 scale heights away at the shock contact surface would be trivially small (e^{-100}). Hence, one requires an extremely lightweight low atomic number gas to push equally against both the neutron star surface and the contact surface, i.e. the piston, of the ejection shock. Such a lightweight gas is the photon-pair dominated gas that forms at high temperature provided it is nucleon deficient. A measure of this deficiency is that there must be at least 100 photons with their associated electron pairs per nucleon such that the energy density per nucleon is greater than the nucleon binding energy to the neutron star or hence, entropies of the order of 100 to 1000 at $T \simeq 1MeV(\rho_{nucleon} \leq 10^6\ g\ cm^{-3})$. This is just the entropy range formed in the hot bubble in Wilson's and Mayle's recent calculations.

Older calculations of Wilson (1985) and in (Bethe and Wilson 1985) showed a weaker hot bubble, $s \leq 100$ whose existence was dependent sensitively upon input parameters. Instead the present calculation which utilize a more exact treatment of neutrino-neutrino annihilation shows a much higher entropy, larger hot bubble.

The hot bubble is aided by the bounce shock, which temporarily reduces the flow of matter onto the neutron star. At the same time the neutrino emission that carries the heat of the binding energy of the neutron star is emitted in neutrinos in all flavors: electrons, muons and taus and their antiparticles. The density of nucleons ($\rho_{nucleon} \leq 10^9\ g\ cm^{-3}$) following the bounce shock is not in itself enough to absorb these neutrinos and cause it to heat into a hot bubble. Instead, the neutrino number density itself from the heat flux is far greater than ($\times 100$) the nucleon component, and as was first shown by Goodman et al. (1987), these neutrinos and particularly the very hottest neutrinos, the mu's and tau's ($T \simeq 10 MeV$) that carry the most energy, annihilate on each other into electrons that create the hot bubble ($\nu_{\mu,\tau} + \bar{\nu}_{\mu,\tau} \rightarrow e^+ + e^-$). The energetic electron pairs immediately thermalize resulting in a photon plus pairs plus nucleon gas. As Wilson and Mayle have discussed, the extreme energy dependence of this neutrino-neutrino annihilation process requires that the neutrinos preferentially annihilate in nearly head-on collisions or at least with extreme angular dependency of the order of the eighth power of the angle. Hence, only when the neutron star surface has become extremely sharp, that is a very small scale height at the mu tau neutrino photosphere, will the neutrino-neutrino annihilation start heating the external low density matter to form the bubble. This happens some few tenths of a second following the bounce and thus forms a hot cavity that expands rapidly driving the ejection shock. Once the hot bubble starts to form, then the pairs, associated with the high temperature gas, become the dominant particle for neutrino interaction and so once started, the hot bubble is further heated by the entire neutrino flux only a small fraction of which has annihilated to form the initial hot bubble. This further heating produces at the end of 3.6 seconds a hot bubble whose entropy is as high as 10^4 in places and whose temperature adjacent to the contact surface i.e. the piston driving the shock is now low enough, a few hundred keV, that its subsequent cooling by neutrino emission is negligible.

Mixing Behind the Shock

The consequence of the hot bubble is that it will mix at the contact surface of the expanding shocked matter. This shocked matter has considerably lower entropy than the hot bubble. To investigate the degree of this mixing there are several surprising simplifications to the problem. These are (1) the expected mixing is close to the thickness of matter behind the shock and (2) the shock entropy increases due to the decreasing density of the envelope and the two entropies: that of the expanding hot bubble and that of the material behind the shock will become equal somewhere near the boundary between helium and hydrogen or when $\rho \simeq 1\ g\ cm^{-3}$. Thus the outer mass fraction of ejected hydrogen will be unmixed, and that inside everything will be mixed.

Turbulent Mixing

Read and Youngs (1984) developed measurements and theory that shows the Rayleigh Taylor unstable mixing leads to a turbulent boundary layer separating the two fluids. This is the nonlinear limit of growth from a thermal noise spectrum of an initial interface with no other initial perturbations and infinite Reynolds numbers. They found that the thickness grew as a function of acceleration and time as

$$\Delta x = 0.07at^2 \tag{1}$$

where a is the acceleration. Since the displacement of the interface increases as

$$x = 1/2at^2, \tag{2}$$

the thickness of the layer becomes 1/7 of the displacement. As an aside, the mechanism of growth is that the spikes and bubbles of the nonlinear limit of the dominant growing wavelength excite eddies of equal scale, some of which coalesce by inverse turbulent cascade and thereby excite the next larger wavelength perturbation, etc. In the case of the stellar shock, the slowing down of the contact surface between the hot bubble material at higher entropy and the shocked matter at lower entropy is what drives the instability. This slowing down is due to the increasing mass of the matter external to the shock. In general the effect of gravity is less than the deceleration due to increasing mass behind the shock, but contributes at the early phase of the expansion.

Acceleration of Contact Surface

We assume that the hot bubble has been formed and expands adiabatically. Both the hot bubble as well as the material behind the shock is radiation domination and so $\gamma = 4/3$. Hence, in a homologous expansion no pressure gradient is formed across the contact surface due to different adiabats.

The acceleration of the contact surface then is due to a decreasing shock velocity and gravity. The contact surface is always close to the shock, $\Delta R \simeq R/(3\eta)$ because the compression ratio, $\eta = \frac{\gamma+1}{\gamma-1} = 7$ is large and the entropy gradient is weak.

The density distribution is close to polytropic external to the core and so to good approximation

$$\rho = \rho_o(R/R_o)^{-3} \tag{3}$$

$\rho_o = 4 \times 10^4\ g\ cm^{-3}$ at $R_o = 2.5 \times 10^9\ cm$. This gives a logarithmically increasing mass as $M = 1.25M_\odot$ at a surface $R = 2 \times 10^{12}\ cm$ including the neutron star. With this distribution there also exists $1.25M_\odot$ of matter inside R_o of the hot bubble that has collapsed to the neutron star. The Wilson, Mayle calculations put this mass at $1.63M_\odot$ which is within the accuracy of the approximation. If we take $10M_\odot$ ejected with 1.5×10^{51} ergs - typical of the models (Woosley 1988, Arnett 1989, Nomoto 1989) then the surface velocity of the equivalent uniform density sphere is

$$(u_{surf}^2/2)(5/3)M_{ej} = 1.5 \times 10^{51} ergs \tag{4}$$

or $u_{surf} = 3 \times 10^8\ cm\ s^{-1}$. The fluid velocity behind the ejection shock that gives rise to this free surface velocity is closely one half of the surface velocity, contrary to simple energy conservation (Colgate and White 1966). Hence, the fluid velocity behind the shock near the surface is $\simeq 1.5 \times 10^8$ $cm\ s^{-1}$ neglecting the speed-up of the shock in the final density gradient. The corresponding velocity at the radius of $R_o = 2.5 \times 10^9$ cm will be larger by the square root of the internal mass ratio or by $(\ell n R_s/R_o)^{\frac{1}{2}} = 2.6$ in order to conserve energy. Thus $u_o = 4 \times 10^8$ $cm\ s^{-1}$. The corresponding internal energy of the hot bubble is then $3P_o \times Vol$ where $P_o = \rho_o u_o^2 \times 7/6 = 7.4 \times 10^{21} dynes\ cm^{-2}$, or $W_{bubble} = 1.4 \times 10^{51} ergs$. The sum of internal and kinetic energy behind the shock at $u_o = 4 \times 10^8$ $cm\ s^{-1}$ and $M_o = 1.6 M_\odot = 0.5 \times 10^{51} ergs$ or a total of $1.8 \times 10^{51} ergs$. Roughly $0.8 \times 10^{51} ergs$ is retained as binding energy giving the self-consistent value of final kinetic energy of $1 \times 10^{51} ergs$.

Entropy and Mixing

Using this model we can calculate the entropy. If we use the variable $S = P/\rho^{4/3}$, then for a bubble entropy of 10^2 the minimum value necessary to retain a pressure against the neutron star in units of k, the Boltzmann constant, then $s_b = 1.4 \times 10^{17}$. The entropy behind the shock in the same units

$$s_o = \frac{\rho_o u_0^2(\eta/\eta - 1)}{(\rho_o \eta)^{4/3}} = 8.7 \times 10^{-2} u_o^2/\rho_o^{1/3} = 4 \times 10^{14}. \quad (5)$$

Thus the shock entropy and the bubble entropy become equal when

$$s_b = s_{shock} = s_o(\rho_s/\rho_o)^{1/3} \ell n(R_s/R_o)$$

Since $(\rho_s/\rho_o)^{1/3} = R_s/R_o$, we obtain a radius $R_s = 80 R_o = 2 \times 10^{11}$, or where the stellar envelope density is $\rho_s = 0.05\ g\ cm^{-3}$.

The integral of the acceleration determines the degree of mixing up to this point in the envelope. If the velocity had decreased to zero in this distance and neglecting gravity, the mixed layer would be 1/7 the radius. However, the velocity decrease is only half and so the acceleration is reduced in half so that roughly the mixed zone is only 1/2 or $R/14$. On the other hand, for a uniform density sphere of matter compressed by η will appear as a layer $\Delta R/R = (1/3\eta)$ thick or $R/21$. Hence, a strong shock will be compressed to a thin layer in a uniform medium and somewhat thicker with our density distribution $\rho \propto R^{-3}$. Thus the mixing will penetrate close to the shock. A decreased mixing can also be expected due to the decreasing entropy of the bubble itself due to mixing with the lower entropy shocked matter. Thus more detailed calculations are warranted, but it appears that one can expect the hot bubble to mix out to roughly the helium-hydrogen zone in the star, $\rho \simeq 1\ g\ cm^{-3}$ and possibly somewhat further. This is just what is required to give the gamma ray transparency from $^{56}Ni \rightarrow ^{56}Co \rightarrow ^{56}Fe$ decay.

It is noteworthy that the astrophysical attempt to understand the explosion of supernovae using the caculational tools developed for nuclear weapons is now a tradition of 30 years at Lawrence Livermore Laboratory (Colgate and White 1966). The existence of the Z particle which permits neutrino-neutrino annihilation was not known at that time. Demonstrating the solution to the problem at LLL is a tribute to this long dedication.

References

1. J. R. Wilson and R. Mayle: Proceedings of the NATO Conf. "The Nuclear Equation of State," (Springer-Verlag, Berlin, 1984).
2. S. A. Goodman, A. Dar, and S. Nussinov: Ap. J. 314, L7 (1987).
3. S. A. Colgate: Ap. J. 163, 221 (1971).
4. H. Bondi: M.N.R.A.S (1952).
5. J. R. Wilson: In Numerical Astrophysics, ed. J. Centrella, J. LeBlanc, and R. Bowers (Jones and Bartlett, Boston 1985) p. 422.
6. H. Bethe and J. Wilson: Ap. J. 295, 14 (1985).
7. K. I. Read: Physica 12D, (North-Holland, Amsterdam) p. 45-58 (1984).
8. D. L. Youngs: Physica 12D (North-Holland, Amsterdam) p. 32-44 (1984).
9. S. A. Colgate and R. H. White: Ap. J. 143, 626 (1966).
10. S. E. Woosley: Astrophys. J. 330, 218 (1988).
11. W. D. Arnett: In Supernova 1987A in the Large Magellanic Cloud, ed. M. Kafatos and A. Michalitsianos (Cambridge: Cambridge University Press) p. 301 (1988).
12. K. Nomoto, T. Shigeyama, M. Hashimoto: In SN 1987A, ed. I. J. Danziger, ESO, Garching, p. 325 (1987).

Effects of Rotation on Collapsing Stellar Iron Cores

Ralph Mönchmeyer

O- and B- main sequence stars are found to rotate in general and it cannot be excluded that rotation may significantly influence the collapse of iron cores of Type II Supernova progenitor stars. Mönchmeyer and Müller have, therefore, numerically investigated the effects of angular momentum conservation on the axisymmetric collapse of a model iron core with a mass of 1.36 $M_{\odot}$. Several cases with parameterized initial angular momentum distributions have been considered. For details of the input physics, of the numerical scheme and the results see MÖNCHMEYER and MÜLLER [1], [2], MÖNCHMEYER [3], HILLEBRANDT et al. [4]. In the following some major results of these calculations will be summarized.

I. General Effects due to Rotation

During collapse the iron core splits up into a subsonically contracting "inner core" [IC] and a supersonically falling outer core. The mass of the IC depends both on its average electron concentration and on its angular momentum and may be up to 10% larger than in a nonrotating model. Without rotation the collapse of the IC is stopped in a "core bounce", only, when nuclear densities are reached at the core center. In rotating models, instead, even a small amount of initial rotational energy (about 1% or more of the initial potential energy of the core) may cause a bounce at subnuclear central densities, if the adiabatic index γ given by the equation of state [EOS] of iron core matter is close to the value 4/3 for densities $\rho \lesssim 10^{14}\,\mathrm{gcm}^{-3}$.

Further effects due to centrifugal forces during collapse are the following : i) The equatorial diameter of the IC may become a factor of 2 larger than the polar diameter. ii) Centrifugal deceleration may reduce the kinetic infall energy (i.e. the kinetic energy minus the rotation energy) of the IC at bounce significantly in comparison to nonrotating models. (see ref. [3]). iii) In contrast to the oblate deformation of the isopycnic surfaces the surfaces of constant angular velocity Ω have at and after bounce an overall cylindrical shape inside the IC. There the Ω–profile eventually exhibits an exponential decline with the distance from the axis. At the polar edges of the IC the evolution of local Ω-maxima causes the generation of dips in the density stratification. iv) The profile of the infall velocity becomes very asymmetric. At bounce the maximum infall velocity in the equatorial plane roughly reaches only half the value of the maximum infall velocity on the axis.

A deformed shock front is generated at the surface of the IC at bounce. The shape of the outward propagating shock surface is determined by the density and velocity stratification outside the IC and details of the EOS. The large infall velocities of the supersonic matter near the axis lead to the generation of polar entropy blobs behind the shock front, which rise and expand due to buoyancy forces. The resulting inclination of isopycnic and isobaric surfaces may drive circulation flows in the shock heated matter. The shock is mainly weakened by energy losses due to the photo–disintegration of nuclei. At larger radii the shock dissipation of kinetic energy decreases because of the decreasing infall velocities in the supersonic flow. In one of the calculated models the resulting unstable entropy gradient triggered the evolution of a Rayleigh - Taylor - instability on timescales $\Delta t < 5$ ms. The corresponding mixing of high with low entropy material leads to a contraction of the shock heated matter in the gravitational field and weakens the shock. Nevertheless the shock front can reach a larger mass coordinate in rotating than in nonrotating models (see II.).

Post bounce oscillations of the IC with amplitudes larger than in the nonrotating case were found in all rotating models. The frequencies and amplitudes of the oscillation modes depend strongly on the stiffness of the EOS at maximum central density and on the kinetic infall energy of the IC at bounce. In a specific case, in which the collapse was stopped at a central density of $1.6 \cdot 10^{14}\,\mathrm{gcm}^{-3}$ nonlinear, large scale volume oscillations occured : The shock was driven outward by the vehemently expanding IC until a minimum central density of only $7 \cdot 10^{12}\,\mathrm{gcm}^{-3}$ was reached. In all cases the oscillations are damped by the generation of non-radial pressure waves, which are emitted from the surface of the IC and eventually catch up with the shock. The IC thus transfers energy to the shock until the IC approaches a state of rotational equilibrium after a model dependent damping time between 5 ms and 50 ms. The gravitational waves of the collapsed cores reflect the bounce and post bounce dynamics in a model specific, characteristic way.

Iron cores that bounce at central densities around (or below) $10^{13}\,\mathrm{gcm}^{-3}$ produce only weak shocks and evolve into matter accreting configurations in rotational equilibrium. How their stability is influenced by neutrino transport processes on timescales of several hundred milliseconds remains to be investigated. In any case it can be shown that the deformation of the neutrino emitting region ("neutrinosphere") in the shock heated matter of collapsed rotating cores leads to a significant dependence of the observable neutrino flux on the inclination angle between the observer's line of sight and the rotational axis of the core (JANKA and MÖNCHMEYER [5]).

II. Consequences for the Shock Propagation

The shock can propagate to the surface of the iron core with large velocities, only, if the energy transferred to the shock heated matter is sufficient both to compensate energy losses (e.g. due to disintegration of nuclei) and to supply at least the minimum internal energy necessary to establish an equilibrium stratification of the shocked matter against the gravitational forces and the ram pressure at the shock surface. Otherwise the hot matter behind the shock will stop to expand. Therefore, rotation seems to be fatal for the shock propagation, because centrifugal forces reduce the kinetic infall energy (i.e. a major source of the initial shock energy and the energy dissipated by the shock) throughout the core in comparison to nonrotating models. The quantity that measures the energy transferred from the IC to its surroundings is the binding energy of the IC, i.e. the sum of its gravitational, internal and rotational energy. In all rotating models the binding energy of the IC in the final rotational equilibrium state was found to be smaller than in the nonrotating case but it was at least larger than the kinetic infall energy of the rotating IC at bounce. The latter holds especially in models, in which the post bounce oscillations of the IC have large amplitudes.

Some effects of rotation can, however, strengthen the shock : The virial theorem shows that the stabilizing effects of rotation on a gas in hydrostatic equilibrium are comparable to those of a gas with $\gamma = 5/3$. Centrifugal forces reduce the average ram pressure of the supersonic matter at the shock surface. In addition the gravitational energy of the matter outside the IC is smaller in rotating than in nonrotating models due to the larger radial extensions of rotating cores at and after bounce. According to the virial theorem the thermal energy needed to establish hydrostatic equilibrium in the shock heated matter (with $\gamma > 4/3$) outside the IC is, therefore, significantly reduced in rotating cores. It follows that the thermal energy deposited in the shock heated matter can much more efficiently be transformed into expansion work than in nonrotating cores. The shocks, therefore, reached in all calculated rotating models at least a mass coordinate of $1.3\,M_\odot$, slightly larger than in the nonrotating case.

In a specific model characterized by a maximum central density of $1.6 \cdot 10^{14}\,\mathrm{gcm}^{-3}$ at bounce and extreme large scale oscillations after bounce the shock had even penetrated the silicon shell and reached a mass coordinate of $1.42\,M_\odot$, when the calculation was stopped. Although even in this case no strong explosion is to be expected, further calculations are necessary to clarify the role of rotation for the final evolution of massive stars and supernova explosions.

<u>References</u>

1. R.Mönchmeyer, E.Müller: NATO ASI C <u>262</u>, eds. H.Ögelman, van den Heuvel, E., Kluwer, Dordrecht, 549 (1989)

2. R.Mönchmeyer, E. Müller: Astron. Astrophys. <u>217</u>, 351 (1989)

3. R.Mönchmeyer: Proc. MPA <u>P1</u>, 92 (1989)

4. W.Hillebrandt, E. Müller, R. Mönchmeyer: NATO ASI, "The Nuclear Equation of State", ed. W.Greiner, Plenum, NY (1989) in press

5. H.T.Janka, R.Mönchmeyer: Astron. Astrophys. <u>209</u>, L5 (1989)

Neutrino Inelastic Scattering and Prompt Shocks in Core Bounce Supernovae

Alak Ray

Attempts to numerically simulate "prompt explosions" in core bounce supernovae -believed to be due to a shock delivering sufficient energy at the base of the envelope, have been largely unsuccessful (COOPERSTEIN and BARON [1]). The primary reasons for the shock's failure in these models are two-fold: (1) energy loss from the shock radiated in neutrinos and (2) energy loss due to dissociation of iron group nuclei into helium and that of the latter into nucleons as the shock moves into the outer core and heats the matter to high temperatures. However, a significant fraction of the energy of the electron type neutrinos radiated in a few milliseconds burst may not be lost and can possibly go into preheating iron nuclei outside the shock front. If sufficient heat is deposited by the burst neutrinos, the material ahead of the shock will pre-dissociate before the shock actually arrives. Thus, "prompt" preheating by neutrinos would be a net energy saver for the shock. The energy of complete dissociation of $M_{Fe}/M_{\odot}$ of iron group nuclei into nucleons is $E_D = 1.7 \times 10^{52}$ erg $(M_{Fe}/M_{\odot})$. As the typical energy of explosion in a type II SN is a few times 10^{51} erg, predissociation of approximately one tenth of a solar mass of heavy nuclei is significant.

When the shock breaks out of the neutrinosphere, a burst of neutrinos is radiated much in the same way as the ultraviolet flash that occured in SN1987A when the the shock reached the photosphere of the progenitor star. Since the burst duration is so short, the radiation of $\approx 3 \times 10^{51}$ erg leads to an extraordinary peak luminosity (e.g. $L_{\nu_e} \approx 5 \times 10^{53}$ erg/s as in MAYLE et al [2]). At this early stage the mu tau neutrino luminosity is still small enough so that the preheating is primarily due to electron type neutrinos. To evaluate the preheating due to the ν_e we use the recently reprted rates by HAXTON [3] who includes both charged and neutral current cross-sections on light and heavy nuclei with their giant resonances in first - forbidden transitions (see also DOMOGATSKII and NADYOZHIN [4]). The hydrodynamic model calculation used here is from MAYLE et al [2] for neutrinos from a 25 $M_{\odot}$ star (model 25 C). The shock break out time is 4 ms when the neutrinosphere radius is at 70 km and the peak luminosity occurs at 10.5 ms after bounce.

Very soon after the shock breakout, there are two components of the ν_e emission: (1) the direct blackbody emission from the neutrinosphere and (2) the emission from a thin shell behind the shock due to the rapid electron capture on protons just behind the shock in the entering material. These two components have been investigated analytically by BETHE, APPLEGATE and BROWN (BAB) [5] and by BURROWS and MAZUREK [6]. After shock breakout and neutrinospheric decay, the electron capture neutrinos take over and continue to dominate for 20 to 30 ms after which the long term core deleptonization flux takes over. The main part of the electron captures take place between densities $10^{11} - 10^{10}$gmcm^{-3} in the timescale of interest ($\tau_{cap}\ \alpha\ (X_e\rho)^{-5/3}$). Fast shocks generate stronger shell sources. The initial neutrinospheric emission is argued to be blackbody in nature in these analyses [5]. A luminosity, half of the peak luminosity of 5×10^{53}erg/s from a neutrinosphere of radius

70 km implies a neutrino temperature close to that for BAB's x parameter of 0.5 (i.e. 6.4 MeV). In one set of our calculations we take a neutrinosphere radius and temperature of 70 km and 6.5 MeV respectively. Even though the shell emission due to electron capture from a particular density zone may be non-thermal (and are richer in high energy particles compared to a Fermi Dirac distribution of temperature T_ν and zero η_ν), taken together, the energy spectrum of neutrinos, a jumble of such emission spectra coming from different density and temperature zones can be described by an effective temperature. We also take the neutrinosphere stationary at 70 km since it moves far less than the shock during the ν_e burst.

Any mass zone overlying the position and mass coordinate of the shock when the ν_e burst is over is heated purely by the burst neutrinos and can be termed as a region of pure preheating, i.e. unaffected by shock heating. By the time the neutrino burst is clearly over the shock is about 180 km from the center [2]. The net heating rate at a distance R from the center is (BETHE and WILSON [7]):

$$\dot{E} = K(T_\nu)\left[\frac{fL_\nu}{4\pi R^2} - \left(\frac{T}{T_\nu}\right)^2 acT^4\right] \mathrm{erggm^{-1}s^{-1}} \tag{1}$$

The preheating causes a change in composition in any given mass zone. Thermodynamic evolution of matter in these zones is tracked as in our earlier work (RAY and KAR [8]). The equilibrium fractions of the four components (Fe, He, n, p) at temperature T are found by solving the Saha equations for iron and alpha dissociation. Given a net heat deposition $\dot{E}\,\Delta t$ in a time step, the change in entropy is:

$$T\Delta S = \dot{E}\Delta t - \frac{Q_\alpha}{4}\Delta X_\alpha - \frac{Q_{Fe}}{56}\Delta X_{Fe} \tag{2}$$

where Q_α (=28.3 MeV) and Q_{Fe} (=124.4 MeV). (Neutrino - nucleus inelastic scattering cross sections are used in calculating the factor $K(T_\nu)$ in eq.(1)). The temperature change is calculated along with the two independent baryon fractions self-consistently. The total entropy of the system at a given mass zone contains the contribution from the four types of baryonic matter (see e.g.,FULLER [9]), their nuclear excitation entropy and radiation and (relativistic) electron entropies. Included in this temperature change calculation is the density change of a given zone after core bounce. This density dependance outside the shocked zone is: $\rho_1 = A_1 t^{-1} r^{-3/2}$, with $A_1 = 1.2 \times 10^{18}$cgs units (BETHE and WILSON [7]). The calculation of heating is continued till $\approx$ 18 ms after bounce in which roughly 3.5×10^{51}ergs of ν_e energy is radiated away. The initial core entropy ahead of the shock is taken as a constant to a first approximation and the initial composition profile calculated. Since the initial core entropy depends on the previous stages of evolution, (i.e. presupernova and collapse phases) we take it as a variable parameter in our calculations.

The optical depth at the neutrinosphere does not abruptly drop to zero at $r = R_\nu$; in reality there is an extended neutrinosphere. As in the case of stellar atmospheres this renders the neutrino distribution more isotropic through partial reprocessing of the inner flux and increases the neutrino number density over what it would have been otherwise. We take into account the isotropy of the distribution by multiplying the heating rate in eq (2) by a factor $f = (1 + 3\,(R_\nu/r)^2)$ to test the effects of the fuzzy neutrinosphere. Detailed computations of the reprocessed flux cofirms this to be a good approximation [10]. The heating rate at a radius of e.g. 240 km at the shock breakout time for $T_\nu = 6.5$ MeV is dominated by the heating on iron nuclei. This situation changes dramatically during the peak of the ν_e burst when almost the entire heating is due to the nucleons. The results of the preheating calculations are summarized in Table I. In order to see the effects of neutrino energy spectrum on the predissociation of nuclei, we have used three temperatures for the neutrinos: 6.5 MeV, 5 MeV and 4 MeV. *The radius and mass upto which matter is completely dissociated into nucleons are quite sensitive to the difference of neutrino temperatures and energies.* The isotropy factor introduced by the factor f in eq (2) produces only minor changes in the masses of iron and helium dissociated and the dissociation energy. In addition, in the case of a higher initial entropy in the core, although the initial heating rates are somewhat higher due to the presence of a higher fraction of free nucleons, the initial entropy conditions are quickly forgotten and the dissociation energy remains close to the case for a lower entropy.

An energy of the order of 10^{51}ergs is substantial in the energy budget of the shock which is expected to produce prompt explosions. Even in the case of a neutrino temperature of 5 MeV this is in excess of 10^{51} ergs and is a reasonable fraction of the energy in the ν_e pulse. However, lower neutrino temperatures make a less effective coupling of the neutrino energy to matter ahead of the shock, although the energy of dissociation for $T_\nu = 4$MeV is not negligible and since it is delivered at the appropriate place in the star, it can possibly improve situations for shock propagation. To conclude, preheating of unshocked matter by neutrinos can contribute substantially towards shock survival and propagation in a prompt explosion *depending on the neutrino energy spectrum.* Detailed hydrodynamical simulations need to include this effect and determine the ν_e burst energy spectrum carefuly.

A detailed version of this work done with Kamales Kar will be published elsewhere. I was supported in part by a grant from the National Science Foundation.

TABLE 1

NEUTRINO PREHEATING AHEAD OF SHOCK

$\frac{T_{\nu_e}}{MeV}$	$\left(\frac{T}{10}\right) r_7^{3/4}$	$\frac{E^{th}_{in}}{foe}$	$\frac{R_{He}}{km}$	$\frac{R_{Fe}}{km}$	$\frac{\Delta M_{He}}{M_\odot}$	$\frac{\Delta M_{Fe}}{M_\odot}$	$\frac{E_D}{foe}$
$S_i = 1.2 k_B$/nucleon							
6.5	0.5	4.7	270	380	0.08	0.2	2.5
6.5 (f=1)	0.5	4.7	250	360	0.06	0.175	2.2
5.0	0.38	2.75	230	300	0.04	0.11	1.4
4.0	0.31	2.0	200	240	0.02	0.05	0.7
$S_i = 2.5 k_B$/nucleon							
6.5	0.5	4.7	275	385	0.085	0.205	2.6

REFERENCES

1. J. Cooperstein and E. Baron, to appear in this Proceedings.
2. R. Mayle, J.R. Wilson and D.N. Schramm, Astrophys. J. 318, 288 (1987).
3. W.C. Haxton, Phys. Rev. Lett. 60, 1999 (1988).
4. G.V. Domogatskii and D.K. Nadyozhin, Soviet Astronomy 22, 297 (1978).
5. H.A. Bethe, J.H. Applegate and G.E. Brown, Astrophys. J. 241, 343 (1980).
6. A. Burrows and T.L. Mazurek, Nature 301, 315 (1983).
7. H.A. Bethe and J.R. Wilson, Astrophys. J. 295, 14 (1985).
8. A. Ray and K. Kar, Astrophys. J. 319, 143 (1987).
9. G.M. Fuller, Astrophys. J. 252, 741 (1982).
10. H.T. Janka, Private Communication (1989).

A Look at Dissipation in Stellar Collapse

Edward A. Baron

1 Introduction

The possible discovery[1,2] of an optical pulsar in SN 1987A (PSR 1987A) having a period of $\sim$ 0.5 ms raises many exciting issues. Not the least of these is the effect of the rotation on the actual mechanism of the explosion. The distribution of angular momentum in the iron core prior to collapse is completely unknown.

One natural possibility is that the iron core is rotating rigidly with an angular velocity

$$\Omega_{ic} = (\frac{R_{ns}}{R_{ic}})^2 \Omega_{ns} \quad (1)$$

where *ic* refers to quantities of the iron core and *ns* refers to quantities of the neutron star. Taking $\Omega_{ns} = 1.2 \times 10^4$ s^{-1}, the value appropriate to PSR 1987A, this gives $\Omega_{ic} = 1$ s^{-1}, an enormously fast rotation rate. While this model seems quite näive, the iron core is very similar to a white dwarf and white dwarfs are known to rotate rigidly although somewhat more slowly.

Another possibility[3] is that the iron core is rotating slowly and the mantle is rotating rapidly. In this scenario, the core collapses to nuclear matter density and the neutron star is "spun up" by the accretion of $\sim$ 0.2 $M_\odot$ of material at the Keplerian velocity. This structure could arise, for example, by the transfer of angular momentum out of the core due to convection. The efficiency of this process is unknown.

A third possibility is a combination of the previous two. Core silicon burning does not occur throughout what is to become the final iron core, but stops at between 0.8 and 1.0 $M_\odot$. The rest of the iron core is built up later through shell burning events. Now it is possible that the convective core silicon burning could serve to sweep the angular momentum from the inner 0.8–1.0 $M_\odot$ of the iron core. In this scenario the inner part of the core is rotating slowly, while the outer part of the iron core is rotating rapidly.

Current one-dimensional calculations of iron core collapse fail to provide a convincing mechanism for turning the collapse into a successful explosion. Given the state of one dimensional calculations and the impetus supplied by the observation of PSR 1987A, it is natural to investigate the effects a large rotational energy might have upon the collapse of the iron core. Two dimensional calculations have been performed by several groups[4,5,6,7,8,9]. These calculations have shown that with rapid rotation nuclear matter densities may not be reached. General relativistic effects remain beyond the reach of current 2-d codes and they have also had to treat neutrino transport in a simplified manner.

In this talk I present a summary of some schematic calculations I have done with Stan Woosley that will be published in detail elsewhere[10]. We study the effect of various angular momentum distributions as well as the effect of dissipation of the kinetic energy of differential rotation. These calculations are carried out in a one dimensional framework and are only schematic. They serve to illustrate the possible size of effects and are useful bases upon which more detailed and correct calculations can be built.

2 Calculations

In the case that convection during silicon core burning is very effective at transferring angular momentum it might be thought that one can produce a strong shock wave by bouncing at supernuclear density, while at the same time strengthen the shock by keeping the outer parts of the iron core at low density due to centrifugal forces. It turns out that this is not the case. As an example we took the WEAVER and WOOSLEY[11] 18 $M_\odot$ model as an initial model. We gave it 3×10^{49} erg s of angular momentum, distributed in the core for $M \geq 0.8$. While, this is marginally helpful (the shock goes 30 km further than the non-rotating case the shock still dies at around 220 km, inside the iron core. Basically, the outer part of the core doesn't collapse enough for the centripetal acceleration to be important.

Now, let us turn our attention to models where differential rotation leads to dissipative heating in the iron core. We study the effect of dissipation in a phenomenological manner. Since dissipation will heat the matter, and will be proportional to the difference in the velocity across a surface, we simply add a term to the pressure, analogous to the pseudoviscosity used to model shock waves. The form we take for the dissipation term is

$$q_{vis} = q < R >^2 |\Omega_j^2 - \Omega_{j-1}^2|/V, \tag{2}$$

where $< R >$ is the radius at the center of the zone, V is the specific volume and Ω_j is the angular velocity of the zone, Ω_{j-1} is the angular velocity of the adjacent zone, q is simply a scale factor that we will take to be in the range 0–10. To gain some feeling for the size of q, we can examine the total work done by the dissipative term in a calculation for which $q = 1$. We find that the total work done in a zone is less than 10^{49} ergs, or between 1 and 100 percent of the maximum rotational kinetic energy of a zone. Another measure is the maximum value of the ratio q_{vis}/p, where p is the pressure in a zone. We find for $q = 1$ that $q_{vis}/p < 0.5$, corresponding to a weak shock wave. Thus, it seems reasonable to take q in the range 0–10.

Table 1 describes the models we have calculated and Table 2 displays the results. We see that the two models are reasonably vigorous explosions. The difference between them is only that neutrino electron scattering (NES) is included in model 111 and not in model 101. NES does not hurt the shock wave, in fact, since it was run a bit longer model 111 is a stronger explosion.

We have shown in our schematic calculations that dissipation of the kinetic energy could provide a potent energy source for the shock wave. In fact, dissipation is a continuous source of energy that could act as a pre-heater for the delayed mechanism as well as an energy booster over long times.

Model	Description	NES	Ω_0 s^{-1}	$L_{tot}/10^{49}$ erg s	q	ρ_{dis}
101	Rigid Rotation	off	1.0	1.0	8	5(10^{11})
111	Rigid Rotation	on	1.0	1.0	8	5(10^{11})

Table 1: Summary of models run with dissipation
NES refers to whether or not the effects of neutrino electron scattering were included. The value of q refers to the coefficient of the dissipative term, and the value of ρ_{dis} refers to the value of the central density when the dissipation term was turned on since we expect models to be stable against turbulence initially.

Model	101	111
ρ_c^{max}	4.51	4.41
R(1.1)	169.1	172.7
u(1.1)	-1.3	-1.3
ρ(1.1)	8.3(9)	8.0 (9)
T(1.1)	1.08	1.07
R_s	12.4	12.9
M_s	0.65	0.55
R(2.7)	10.8	12.9
M(2.7)	0.55	0.55
ρ_c^{equil}	2.5	2.5
E_{kin}	0.2	0.27

Table 2: Summary of results for dissipative models
The table lists various quantities of the models, ρ_c^{max} is the maximum central density at bounce in units of 10^{14} g cm^{-3}, R(1.1), u(1.1), ρ(1.1), and T(1.1) are the radius (in km), velocity (in 10^9 cm s^{-1}), density (g cm^{-3}), and temperature (in MeV), at the mass point $M = 1.1\ M_\odot$, listed for each model at bounce. R(2.7) and M(2.7) are the radius and mass point where the shock forms, R_s and M_s are the radius and mass point of the shock wave at bounce, ρ_c^{equil} is the value of the central density after rebound, and E_{kin} is the kinetic energy (in units of 10^{51} ergs) in the shock wave if the shock wave makes it to the end of the numerical grid.

Acknowledgement

I would like to thank Stan Woosley for letting me present our joint work and Jerry Cooperstein for the use of the hydrodynamical code that we have developed together as well as many helpful discussions. These calculations were performed at the National MFE Computer Center and I thank the US DOE for a generous allocation of computer time.

This work has been supported in part by the U.S. Department of Energy under grant DE-FG02-88ER40388.

References

[1] J. Middleditch et al., IAU Circ. No. 4735 , (1989).

[2] J. Kristian et al., Nature **338**, 234 (1989).

[3] S. E. Woosley and R. A. Chevalier, Nature **338**, 321 (1989).

[4] P. Bodenheimer and S. E. Woosley, Astrophys. J. **269**, 381 (1983).

[5] J. LeBlanc and J. R. Wilson, Astrophys. J. **161**, 541 (1970).

[6] R. Mönchmeyer and E. Müller, in *Timing in Neutron Stars* (Reidel, Dordrecht, 1988).

[7] E. Müller and W. Hillebrandt, Astr. Astrophys. **103**, 358 (1981).

[8] E. Müller, M. Różyczka, and W. Hillebrandt, Astr. Astrophys. **81**, 288 (1980).

[9] E. M. D. Symbalisty, Astrophys. J. **285**, 729 (1984).

[10] E. Baron and S. E. Woosley, (1989) [preprint].

[11] T. A. Weaver and S. E. Woosley, (1988) [private communication].

Shock Simulation in Type II Supernovae: A Collection of Recipes

Maurice B. Aufderheide

It is safe to say that the central riddle of type II supernovae, how a collapsing iron core manages to form a proto-neutron star and a strong shock wave blowing apart the star, has not yet been solved. Yet it is clear from SN 1987A that such events do occur. Although this central problem has not yet been solved, it is possible to study the physics outside the core by generating a shock wave by hand. There are several ways of doing this and the question is whether they are equivalent and when do these artificial methods approach what might really be seen in supernovae? In this study these questions will be discussed. They will be explored in greater detail in a forthcoming paper[1].

Before discussing different generation methods, it is necessary to examine how one can characterize the shocks generated. One method which is commonly used is to measure the "energy of the shock". This quantity is not uniquely defined and each group has its own definition. One possible definition is to quote the initial energy deposited in generating the shock. Another definition is to quote the kinetic energy of the shocked material after the shock has reached hydrogen envelope. Another measure is to compute the total energy in escaping zones, less the total energy which was present in the initial model. All of these are useful measures, but they will not yield the same values. However, one does expect a one-to-one mapping between them.

Another method which can be used for characterization is to compare temperature versus Lagrangian mass coordinate generated in various shocks. One can compare the peak temperatures reached as a function of enclosed mass in the shocked material. One can also examine temperature versus time profiles at particular points within the star. These diagnostics allow one to examine the strength and shape of the generated shock wave and its effect upon nucleosynthesis. In Figure 1 (see the last page for all plots), peak temperatures are plotted as a function of enclosed mass for two shocks of differing energy. All simulations reported here were performed upon the NOMOTO SN 1987A model[2]. The mass range plotted is the prime nucleosynthetic region. The solid line corresponds to a shock energy of 1 foe (10^{51} ergs), while the dashed line is a shock with energy of 2.9 foes. These energy assignments are obtained using the third method discussed above. Note that tripling the shock energy results in peak temperature increases of $.5 \times 10^9$ K to 1×10^9 K.

Having discussed how to characterize shocks, it is now possible to compare methods of generation. One method is to deposit a lot of kinetic energy somewhere in the outer iron core and let the shock be generated as a result of the collision with surrounding material. A second way is to deposit the energy as a surplus of internal energy in a layer of the core[3]. This layer will then expand, generating a shock wave. The third way to make a shock is to specify a violent trajectory, R(t), for a particular mass point (a piston) which forces it to smash into the surrounding material. If one knew the trajectory of such a point in the actual, self-consistent case, this is all that would be necessary to generate the actual shock wave. Such a trajectory is not known at present, so some prescription must be chosen. WOOSLEY and WEAVER have such a prescription[4] and this method will be used for comparison here.

One must be careful in the area just outside the shock initiation region because nucleosynthetic results may be unphysical. This is an artifact of the original artificiality of each method. A physically self-consistent shock wave carries its energy partitioned between kinetic and internal energy. Each initiation starts with this partition skewed unphysically. As the shock propagates, smashing into new material, the energy is gradually partitioned properly. One can think of the shock wave "maturing". The region in which the shock has sorted itself out will be called the mature region and the volume between the shock initiation point and the mature region will be called the adolescent region. Within the adolescent region the temperature profiles will not be consistent with the energy attributed to the shock wave, since either too much or too little of the shock's energy will be in internal energy. Because of the dubious temperature profiles, the abundances determined in the adolescent region will also be suspect.

This effect can be seen in Figures 2 and 3. In both graphs, peak temperature is plotted versus enclosed mass. In Figure 2, two kinetic energy bombs with an energy of 1 foe are compared. The solid line is a shock started at $.6M_{\odot}$, while the dashed line is a shock started at $1.32M_{\odot}$. Note that this shock exhibits low temperatures until roughly $1.55M_{\odot}$, after which the profiles agree. $1.32M_{\odot}$ to $1.55M_{\odot}$ is the adolescent region for the second shock wave. Notice that this implies that the first shock had matured by the time it reached roughly $.83M_{\odot}$. In Figure 3, two internal energy bombs with an energy of 1 foe are compared. The solid line is a shock started at the center of the star, while the dashed line is a shock started at $1.43M_{\odot}$. Note here that the second shock is immature until roughly $1.66M_{\odot}$, and that here the immaturity is exhibited by higher temperatures. It is interesting that in both cases, the adolescent region extends $.23M_{\odot}$. One should not trust nucleosynthetic results in these regions for such shock waves. If one does desire accurate results in such areas, one must move the shock initiation point back far enough that the adolescent region does not intersect with the region of interest. Immaturity in piston generated shock waves has not yet been examined.

A comparison of the three initiation methods is also interesting. In Figure 4, peak temperatures are again plotted versus mass for shocks with energies of 1 foe. Maturity has been verified for each shock wave. The solid line is a kinetic energy bomb, the dashed line is a WOOSLEY-WEAVER piston, and the dotted line is an internal energy bomb. Note the agreement between the first two methods. Note also that the internal energy bomb exhibits lower temperatures until roughly $2.2M_{\odot}$. This last shock wave would produce $.04M_{\odot}$ less of ^{56}Ni than the first two shocks using the THIELEMANN *et. al.* rule of thumb for the synthesis of ^{56}Ni[5]. This is over half of what was seen in SN 1987A and is not trivial matter. It is worrisome that, given the same amount of energy, significantly different temperature profiles can produced.

This may indicate that in the silicon and oxygen shells there is not a unique shock solution corresponding to a given energy. If this is true, reliable simulation of the nucleosynthesis of iron group elements will have to wait for the solution of the core collapse-explosion problem.

Acknowledgement: I would like to thank my collaborators, E. Baron, F.-K. Thielemann, and J. Cooperstein for allowing me to discuss results of our work.

This work has been supported by the U.S. Department of Energy under contract no. DE-AC02-76CH00016, and grant no. DE-FG02-88ER40388

References

[1] M. B. Aufderheide, E. Baron, F. K. Thielemann, and J. Cooperstein, in preparation, 1989.

[2] K. Nomoto and M. Hashimoto, Phys. Rept. **163**, 13 (1988).

[3] T. Shigeyama, K. Nomoto, and M. Hashimoto, Astron. and Astrophys. **196**, 141 (1988).

[4] S. E. Woosley and T. A. Weaver, in *Essays in Nuclear Astrophysics*, edited by C. A. Barnes, D. D. Clayton, and D. N. Schramm, p. 377 (Cambridge Univ. Press, 1982).

[5] F. K. Thielemann, M. Hashimoto, and K. Nomoto, preprint, 1989.

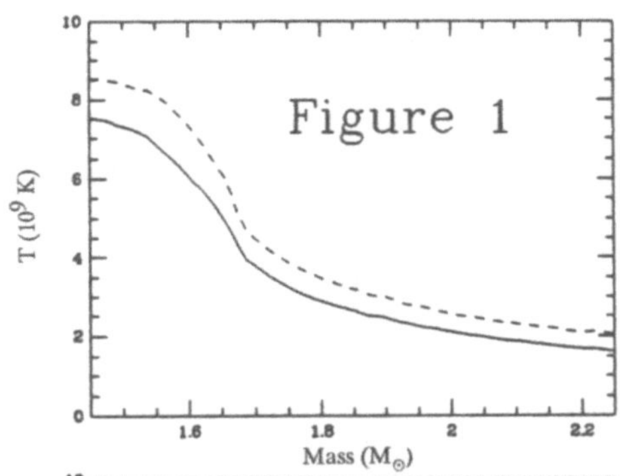

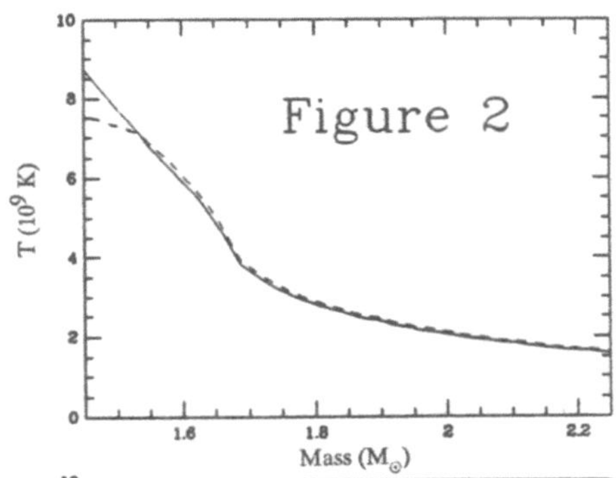

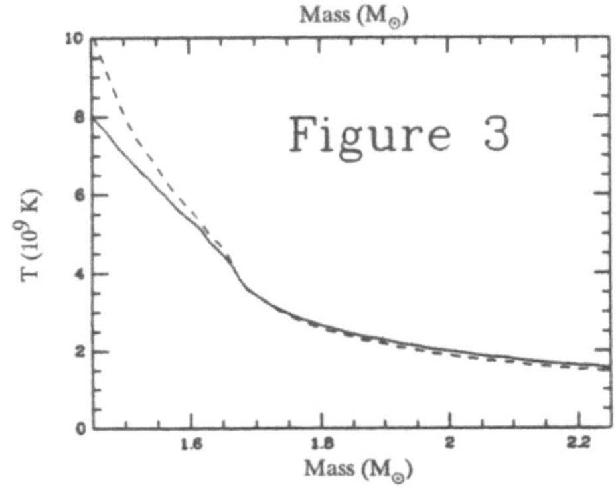

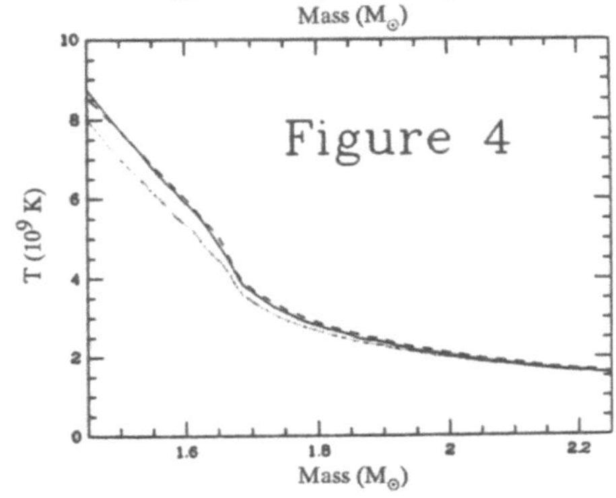

Is There Life After Fuller, Fowler & Newman?

Maurice B. Aufderheide

Weak interactions have been an important part of nuclear astrophysics ever since hydrogen burning in the sun was elucidated. In type II supernovae, $\beta^{\pm}$ decay and electron capture are extremely important because they determine the electron fraction per nucleon (Y_e) of the iron core. The mass of the iron core and the Lagrangian position of shock formation both vary as Y_e^2[1],[2]. Also, these reactions create neutrinos which, before core collapse, are able to carry core energy off to infinity. These rates are thus crucial ingredients of the stellar evolution calculation which is the initial model for a type II supernova simulation.

The inclusion of these rates is difficult to achieve because one must compute them for a large number of potentially important nuclei, over a large range of temperature, density, and Y_e. This monumental work has been undertaken several times in the past. The most recent effort was made by FULLER, FOWLER, and NEWMAN in the early 1980's[3],[4], [5],[6]. They tabulated $\beta^{\pm}$ decay and electron capture rates for nuclei with mass number up to 60 over a wide range of temperature, density, and Y_e. In their calculations, they included the effects of the Gamow-Teller and the isobaric analog resonances. These rate tables are the ones currently employed in the stellar evolution codes of NOMOTO[7] and WOOSLEY and WEAVER[8].

The study reported here is an investigation whether nuclei with masses heavier than 60 could have a significant effect upon the presupernova evolution of the iron core[9]. It is a sensitivity study for nuclei with $60 < A \leq 70$, during the hydrostatic silicon burning stages of the core's evolution. This is the most active part of the core's life as far as the pre-collapse history is concerned. It is characterized by densities of 10^6 g/cm^3 to 10^9 g/cm^3, temperatures of 2 billion K to 5 billion K and electron fractions ranging from .5 down to .42. In this study, only the strongest allowed Gamow-Teller transition was included for each nucleus. Because a lot of the Gamow-Teller strength has been neglected, these rates will be underestimates; however, if large effects are uncovered here, they will certainly be important after a full calculation is performed. This study is therefore a prelude to an exhaustive calculation.

The electron capture rate on a nucleus A due to a transition i can be written as:

$$\lambda_i^A(e.c.) = Factors \times B(GT) \int_{\mathcal{L}}^{\infty} \frac{p_e^2(\Delta E + \epsilon_e)^2 F(Z, \epsilon_e) dp_e}{1 + exp[(\epsilon_e - \mu_e)/k_B T]} \qquad (1)$$

where B(GT) is the Gamow-Teller matrix element for the transition, ΔE is E_{mother}

$-$ $E_{daughter}$, μ_e is the electron chemical potential, and T is the temperature. F(Z,ϵ_e) is the Fermi function, which includes the effect of the nuclear Coulomb field on the electron wave function. $\mathcal{L}$ is the lower limit of the electron momemtum. For ΔE $\geq -m_e$, $\mathcal{L}$ is zero, while for $\Delta E < -m_e$, $\mathcal{L}$ is $\sqrt{\Delta E^2 - m_e^2}$. The β^- decay rate can similarly be written as:

$$\lambda_i^A(\beta^-) = Factors \times B(GT) \int_0^{\sqrt{\Delta E^2 - m_e^2}} \frac{p_e^2(\epsilon_e - \Delta E)^2 F(Z,\epsilon_e) dp_e}{1 + exp[(\mu_e - \epsilon_e)/k_B T]}. \tag{2}$$

where all factors are defined as above, except that Z is now the charge of the daughter nucleus. Such transitions affect the evolution of the electron fraction in the following way:

$$\frac{dY_e}{dt}\Big|_i^{(Z,A)} = \pm \lambda_i^A Y_{(Z,A)} \tag{3}$$

$Y_{(Z,A)}$ is the number fraction of the mother nucleus. The plus sign applies for β^- decay, while the minus is used for electron capture or β^+ decay. To estimate the effect a transition will have upon the electron fraction, form $\Delta Y_e = \pm\ \lambda_i^A\ Y_{(Z,A)}\ \Delta t$ where Δt is the characteristic time of the silicon burning stages, 10^5 to 10^6 seconds.

This last expression shows that, in order for a reaction to really affect the core, the product of the abundance and the rate must be greater than 10^{-8}sec^{-1}. Computing the reaction rate is not enough; the species could still be abysmally rare. On the other hand, a mother nucleus could become abundant at a time when the rate is small. In this study the abundance of each nucleus was estimated by determining its abundance in nuclear statistical equilibrium (NSE) for a given temperature, density, and electron fraction. Of course this might yield abundances which are somewhat large during early stages of silicon burning, because NSE has not yet been achieved. But it provides an order of magnitude estimate.

Using these tools, the strongest transitions on nuclei with A between 60 and 70 were sought. Electron captures on nuclei which were on the neutron-poor side of the valley of β stability were ignored for now because they would only be abundant for values of Y_e near .5. This corresponds to the early stages of hydrostatic Si burning, during which the material is not yet in NSE and is mostly silicon and sulfur. The strongest electron capture transition which was found was ^{66}Cu + e^- $\longrightarrow$ ^{66}Ni + ν_e. This is a transition between the 1^+ ground state of ^{66}Cu and the 0^+ ground state of ^{66}Ni. The value of ΔE is -.747 MeV[10] and the β^- decay in the reverse direction has been measured, providing the desired matrix element by detailed balance. At a temperature of 4×10^9K, a density of 10^8 g/cm^3, and and electron fraction of .44, the change in Y_e is:

$$\Delta Y_e = 3.5(-3)\frac{1.3(-5)}{66}10^6 = 7(-4). \tag{4}$$

This is not very large. This was the largest effect found for electron capture reactions. Inclusion of the full Gamow-Teller strength could make these rates larger.

When one considers the possible β^- decays, one sees a vastly different picture. The contributions of these rates is largely determined by the size of ΔE in the reaction. In the region of atomic mass studied here, the ^{64}Co $\longrightarrow$ ^{64}Ni + e^- + $\bar{\nu}_e$ reaction is the largest because of its huge mass difference of 7.818 MeV[10]. The ratio of Z to A for ^{64}Co is low (.422), which ensures that it will only contribute at $Y_e \leq .44$. This corresponds to the last stages of silicon burning. If one estimates the change in electron fraction due to this single reaction, one obtains

$$\Delta Y_e = 2.09 \frac{6.3(-5)}{64} 10^5 = .2. \tag{5}$$

This would be an huge effect on the electron fraction. If one tries to estimate the thermal effects of these decays, one obtains a similarly large effect. These results are so large that it is clear that the nonlinearity of the iron core's response will regulate the effect of these decays. Although it is difficult to predict what the inclusion of such reactions will do in detail, it is clear that they will alter the evolution of the iron core. With the inclusion of the full Gamow-Teller strength in the electron capture rates, even these marginal rates will begin to make significant contributions.

In conclusion, even with deliberate underestimates of the rates for nuclei with A between 60 and 70, it has been shown that some electron captures may make a small contribution to core evolution and that several of the β^- decays will have a large effect on the core's evolution. This shows that there is a need for a full calculation of these rates and inclusion of these rates into present stellar evolution codes. There is life after FULLER, FOWLER, and NEWMAN and it is time to get to work.

Acknowledgement: I would like to thank my collaborators, G. E. Brown, T. T. S. Kuo, D. B. Stout, and P. Vogel for allowing me to discuss results of our work.

This work has been supported by the U.S. Department of Energy under contract no. DE-AC02-76CH00016.

References

[1] A. Burrows and J. Lattimer, Ap. J. **270**, 735 (1983).

[2] J. Cooperstein, H. A. Bethe, and G. E. Brown, Nuclear Physics **A429**, 527 (1984).

[3] G. M. Fuller, W. A. Fowler, and M. J. Newman, Ap. J. Suppl. **42**, 447 (1980).

[4] G. M. Fuller, W. A. Fowler, and M. J. Newman, Ap. J. **252**, 715 (1982).

[5] G. M. Fuller, W. A. Fowler, and M. J. Newman, Ap. J. Suppl. **48**, 279 (1982).

[6] G. M. Fuller, W. A. Fowler, and M. J. Newman, Ap. J. **293**, 1 (1985).

[7] K. Nomoto, Ap. J. **277**, 791 (1984).

[8] T. A. Weaver, S. E. Woosley, and G. M. Fuller, in *Numerical Astrophysics*, edited by J. Centralla, J. LeBlanc, and R. Bowers, p. 374 (Jones and Bartlett, 1985).

[9] M. B. Aufderheide, G. E. Brown, T. T. S. Kuo, D. B. Stout, and P. Vogel, in preparation.

[10] C. M. Lederer and V. S. Shirley, editors, *Table of Isotopes* (John Wiley, 1978) 7th edition.

Nucleosynthesis in Non-Spherical Supernova Explosion and Observations

Valeri M. Chechetkin, Andrei A. Denissov, & Yuri P. Popov

During the last two decades many investigators had taken attempts to construct Supernova models, that will be able to describe observed physical pattern of this phenomena. Most of them considered models of collapsing iron core of massive star, that were supposed to be a mechanism for Type II Supernova (see HILLEBRANDT [1] and references there). But despite the fact that some of them were very sophisticated one (includes rotation, relativity effects, neutrino transport and some other physical processes), models of this class have failed to explain some aspects of Type II Supernova and Type I Supernova. That was the reason why thermonuclear model of Supernova explosion was suggested by FOWLER and HOYLE [2], ARNETT [3] and others for Type I Supernova.

But the main unresolved question in this model is the question about mode of burning that should realize in supernova explosion - deflagrational or detonational one. The mode depends on many factors, including degree of degeneration of electron gas, initial temperature distribution and chemical composition. WOOSLEY [4] and BLINNIKOV and KHOKHLOV [5] have studied formation of detonational wave depending on instantly burning mass. IVANOVA et al [6] have studied dependence of the mode on instantly burning mass and found that not only detonational mode may take place, but also deflagrational burning can appear. But all these models were spherically symmetric, so rotation was not included in them. Including rotation into thermonuclear model is a further step towards real physical situation, since we know, that most of compact Supernova remnants rotates quite rapidly.

In our study two-dimensional calculations of burning of a rotating CO core were carried out. As an initial configuration a degenerate CO core with a mass of 1.4 $M_{\odot}$ and a central density of $2\cdot10^{9}$ g cm^{-3} was assumed. We have considered solid-state rotating core with several angular velocities. At the initial moment roughly 20% of the total core mass was instantly burned off and 10^{50} erg of nuclear energy was released, generating the detonation wave.

Figure 1 shows propagation of the detonation front through the stellar core for the variant with rapid rotation. Angular velocity equal roughly 0.5 of critical value $(G\cdot M/r^{3})^{1/3}$, where G is gravitational constant, M – stellar mass, r – equatorial radius of the core. The detonation burning in external layers of the core occurs in over-compressed mode. Such a detonation wave has a tendency to be dumped in case when supporting pressure profile behind the burning front is decreased. One can see from Fig 1 that burning front first reaches the surface of core in polar regions. It leads to appearance of a rarefaction wave going from poles to equatorial region, so the detonation wave may damp here under certain conditions and unburned remnant will appear.

So the main difficulty in our simulation arise from impossibility to calculate detonation and deflagration wave using the same code. Now we couldn't determine in simulation the mode of burning really exist in the model described earlier, it is possible only to postulate existence one of the modes and calçulate process of explosion. So, there appear necessity for special computational algorithms, that should give us possibility to simulate all types of modes of burning.

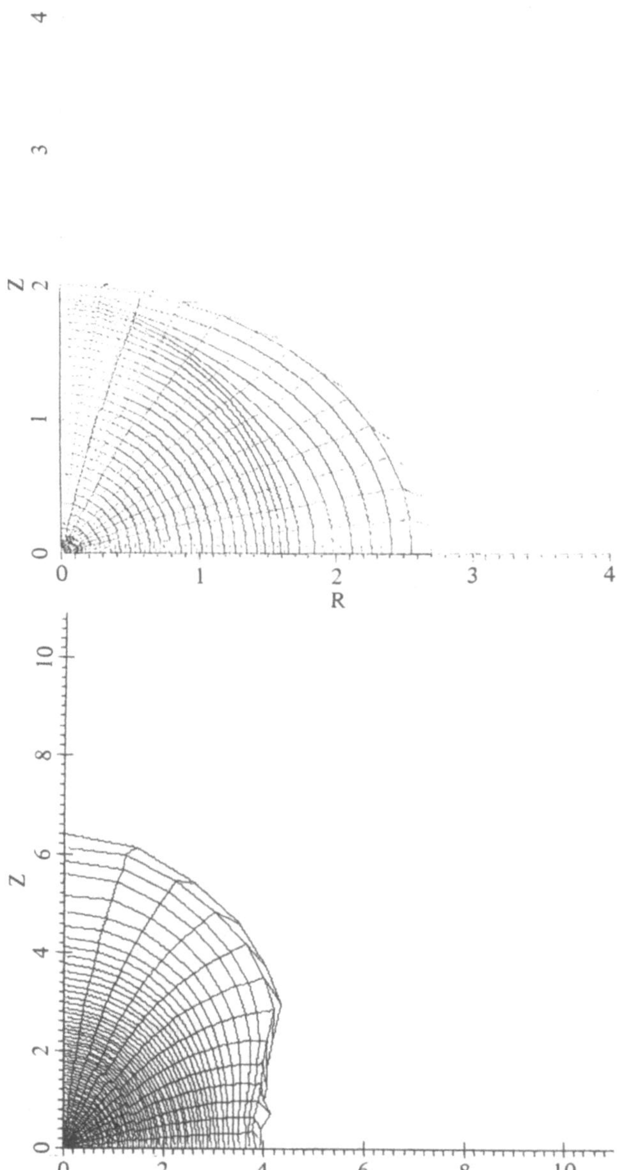

Figure 1. Detonation front at the moment 0.110 sec from the begining of explosion; matter expandes along the polar axis at the moment 0.190 sec. Z is axis of rotation, R – axis is located in equatorial plane. All distances are shown in 10^8 cm units.

What astrophysical consequence follow from the results which include a possibility of formation of unburned carbon-oxygen remnant? With further evolution such a remnant will resemble a colder region against a hotter region with increased content of primary matter from degenerate stellar core. Similar structures have been observed in young supernova remnants (LOSINSKAYA [7], CHUGAI [8].

Some interesting conclusions from the results of our simulation may be done for the resent Supernova 1987A in Large Magellanic Cloud. It is known from observations that presupernova SK-69202 has rotate quite rapidly (WEISS, HILLEBRANDT, TRURAN [9]). Explosion along polar directions may lead to mixing of core and envelope matter, which leads to appearance of large amounts of radioactive Ni^{56} in envelope. Such a process should "trigger" hard X-ray emission from supernova remnant.

References

1. Hillebrandt W., Preprint MPI-PAE/Astro 279, (1981).
2. Fowler W.A., Hoyle F., Astrophys. J. Suppl. 9, 201, (1964)
3. Arnett W.D., Astrophys. Space Sci., 5, 180, 1969.
4. Woosley S.E., Ann. Rev. Astron. Astrophys., 24, 205, (1986).
5. Blinnikov S.I., Khokhlov A.M., Pis'ma Astr. Zh., 12, 318, (1986).
6. Ivanova L.N., Imshennik V.S., Chechetkin V.M., Astrophys. Space Sci., 31, 477, (1974).
7. Losinskaya T.A., Stellar winds and supernova remnants, Nauka, Moscow, (1986).
8. Chugai N.N. Astr. Circular No. 1469, (1986).
9. Weiss A., Hillebrandt W., Truran J.W. Astron. Astrophys. 197, L11 (1988).

Relativistic Neutrino Transport in Stellar Collapse

Sidney A. Bludman & Paul J. Schinder

1. POLAR-SLICED GENERAL RELATIVISTIC HYDRODYNAMICS

In stellar core collapse, the initial bounce shock energy derives from core compression almost high enough to form an event horizon and black hole. We are interested in maximizing this compression in order to make a strong supernova explosion and also in collapses that, after emitting an observable neutrino signal, accrete enough matter to form a black hole rather than remnant neutron star. We have therefore developed a neutrino radiation hydrodynamics [1,2,4 that gives complete coverage of the space-time domain of outer communication as seen by a distant observer.

1.1 Reasons for an Improved General Relativity

Previous relativistic stellar collapse calculations have used comoving (Lagrangian) Gaussian normal coordinates in which the metric

$$ds^2 = -\alpha^2 dt^2 + \Gamma^{-2} dR^2 + R^2 d\Omega$$

is diagonal and R is the areal radius. These coordinates are "synchronous" in the sense that as soon as any trapped surface begins to form, they become singular everywhere in a finite time. Because we consider only spherical collapse, the gravitational field is always non-dynamic.

1.2 Advantages of Polar Slicing

In order to allow evolution to proceed until each mass zone approaches the horizon, we have performed the coordinate shift

$$dR = \dot{R}dt + R'dr = R'(\beta^r dr + dr) \qquad (1)$$

on each time slice, so that the metric becomes

$$ds^2 = -(\alpha^2 - \beta^2)dt^2 + 2\beta_r dr dt + (R'/\Gamma)^2 dr^2 + R^2(r,t)d\Omega,$$

in which the shift $\beta^r = \dot{R}/R', \beta_r = (R'/\Gamma)^2\beta^r$, dots and primes represent partial derivatives with respect to t and r respectively. In these comoving polar-sliced coordinates, the lapse function is determined by

$$\frac{\alpha'}{\alpha} = R' \frac{M + 4\pi R^3(P + \tilde{P})}{R^2(1 - 2M/R)} \tag{2}$$

where $\tilde{P} \equiv (P + \rho)U^2/(1 - 2M/R - U^2)$ is the contribution to the pressure by fluid flow with velocity $U = \partial R/\alpha\partial t$. Because $d\alpha/\alpha = dP/(P + \rho)$, when the fluid is stationary $(\beta^r = 0, R = r)$ Eq. (2) reduces to the Tolman-Oppenheimer-Volkov equation. The advantage of these non-normal coordinates is that they are non-singular everywhere outside the apparent horizon (outermost trapped surface): because these coordinates are non-synchronous, each mass zone carries its own clock, so that evolution continues in outer mass zones even as it slows down in inner mass zones, as they approach the apparent horizon, which is never reached in finite time.

1.3 Implicit Hydrodynamics

In order to economically follow the core evolution through quasistatic stages (as in shock revival and later neutrino cooling), we have also converted our code from explicit to implicit hydrodynamics. An implicit code escapes the Courant time step restriction and is always numerically stable. We solve the large set of non-linear equations self-consistently at each time step, by not only inverting numerically but also by calculating the many differential coefficients numerically. Our code is therefore (a) comoving, to give maximum coverage of material zones, (b) polar-sliced, to give maximum coverage of space-time, and (c) implicit, to allow efficient calculation over long times.

1.4 Adiabatic Hydrodynamic Tests

Our implicit polar-sliced code tested well in three adiabatic hydrodynamic calculations: relaxation to equilibrium of a stable polytrope, collapse to a black hole of an unstable polytrope, collapse of an initially homogeneous dust cloud from rest to a black hole.

2. GENERAL RELATIVISTIC RADIATION TRANSPORT

2.1 Exact Solutions of the Transport Equation

Almost all previous calculations of radiation transport during stellar collapse assumed some form of diffusion approximation. Because this reduces a hyperbolic wave equation to a parabolic equation, any diffusion approximation is inherently non-relativistic. Although the flux can be limited so as to maintain causality, any flux limited approximation is ad hoc and of uncertain accuracy in regions intermediate between the diffusion and free-streaming regions. This semi-transparent intermediate region is crucial in all radiation transport problems and especially so in supernovae in this region, the angular distribution of the emergent radiation is determined and the weak coupling between neutrinos and the outer collapsing core will or will not explode matter.

For these reasons, we believe an exact treatment of neutrino transport to be necessary for an ultimate understanding of the supernova explosion mechanism. In spherical symmetry, the relativistic phase-space distribution function f(t,r,E/μ) depends on two coordinates (t,r) and two momentum variables (E,μ), $\mu \equiv \cos\vartheta$ being the direction cosine of the neutrino momentum with respect to the radius vector and the neutrino energy E depending upon the gravitational field. The transport equation is a partial differential equation which can be solved at each time step: (1) by finite difference methods in the spatial elements r, μ, provided sufficient computing power exists; (2) By the method of characteristics, provided there is sufficient symmetry in the problem. The first method [3] is being pursued with A. Mezzacappa and will not be described here. Instead, we describe the second method, a relativistic generalization of the tangent ray method, which reduces the partial differential equation to a system of ordinary equations [4].

The spherical symmetry can always be used to eliminate the variable μ from the transport equation, which now reads for rays characterized by the constant angular momentum $L = RE\sin\vartheta$,

$$\gamma^{-1}\{[1+\beta\mu]\frac{\partial}{\partial r} + \frac{\Gamma\gamma^2}{R'\alpha}\mu\frac{\partial}{\partial R} - [\mu A + \mu^2 B + \frac{\dot{R}}{R}(1-\mu^2)]E\frac{\partial}{\partial E}\}f$$

$$= \eta/E^3 - \chi f \equiv \chi(S-f), \tag{3}$$

where the source function $S \equiv \eta/E^3\chi, \eta$ is the emissivity, χ the total absorption coefficient and

$$A \equiv \frac{\Gamma\gamma}{R'\alpha}[\gamma' + \frac{\partial}{\partial t}(\frac{\beta r}{\gamma})]$$

$$B \equiv \frac{1}{\gamma^2}[\beta_r \alpha \frac{\partial}{\partial t}(\frac{\beta r}{\alpha}) + \alpha^2 \frac{\partial}{\partial t}(\frac{R'}{\Gamma})]$$

$$\gamma^2 \equiv \alpha^2 - \beta^2, \quad \beta^2 \equiv \beta_r \beta^r = (\dot{R}/\Gamma)^2 \tag{4}$$

depend on the gravitational field.

2.2 Stationary Transport in Static Gravitational Fields

We now restrict ourselves to stationary transport in a static gravitational field described by time-independent metric coefficients $\Gamma, \alpha = \gamma$ so that $\beta_r = 0$, r = R. (Although a dynamic phenomenon when radiation transport in a collapsing star is treated by operator splitting, the gravitational field is held fixed during each transport step.) The stationary transport equation is now

$$\Gamma\{\mu\frac{\partial}{\partial R} + (1-\mu^2)(\frac{1}{R} - \frac{\alpha'}{\alpha})\frac{\partial}{\partial \mu} - \mu(\frac{\alpha'}{\alpha})E\frac{\partial}{\partial E}\}f = \chi(S-f). \tag{5}$$

The terms in α'/α describe gravitational red shift and angular abberation.The characteristic equations of Eq. (5) are

$$\frac{dR}{\Gamma\mu} = \frac{d\mu}{\Gamma(1-\mu^2)(\frac{1}{R} - \frac{\alpha'}{\alpha})} = \frac{-dE}{\Gamma\mu E(\alpha'/\alpha)} = \frac{df}{\chi(S-f)} \equiv d\lambda, \tag{6}$$

where λ is the arc length along a characteristic ray, so that the transport equation is

$$\frac{df}{d\lambda} = \Gamma\mu\frac{df(R, E_\infty b)}{dR} = \chi(S-f), \quad \mu \equiv \sqrt{1-(\alpha p/R)^2}. \tag{7}$$

Each ray is now parametrized by two constants obtained by integrating the first two equations in Eq. (4):

$$\alpha E = E_\infty, \qquad \frac{R\sin\vartheta}{\alpha} = p, \tag{8}$$

the neutrino energy at infinity and the impact parameter p = angular momentum/E_∞. In the non-relativistic regime (weak gravitational fields $\alpha, \Gamma \to 1$), the neutrino energy E would be constant along characteristic trajectories which are straight lines $R\sin\vartheta = p$. Strong gravitational fields increase the neutrino energy as they propagate blue shift inwards on trajectories that bend inwards toward the radial direction as $\alpha(R)$ increases inwards.

3. VARIABLE EDDINGTON FACTOR METHOD FOR STATIONARY TRANSPORT

3.1 Radiation Moment Equations in Stars

We are generally most interested in the mean intensity and flux $J \equiv 2\pi \int E^3 f d\mu \quad H = 2\pi \int E^3 F \mu d\mu$. These could be obtained directly from the zeroth and first angular moments of the transport equation, which for stationary transport in a static background geometry with isotropic absorptivity χ and source function S, are

$$\Gamma \frac{d}{dR}(\alpha R^2 H) = R^2 \alpha (4\pi E^3 S - J) \tag{9}$$

$$\Gamma[\frac{dK}{dR} + \frac{3K - J}{R}] \equiv \frac{\Gamma}{q}\frac{d}{dR}(q f_K J) = -\chi H, \tag{10}$$

provided we knew the second moment $K \equiv 2\pi \int E^3 f \mu^2 d\mu$ or Eddington factor $f_K \equiv K/J = \langle \mu^2 \rangle$. In these equations, the relativistic effects are carried by the metric coefficients Γ, α and the sphericity effects are contained in the sphericity factor q(R) defined by

$$\frac{d \ln q}{d \ln R} \equiv \frac{3f_K - 1 + (R\alpha'/\alpha)}{f_K}. \tag{11}$$

The radiation angular distribution can only be obtained by solving the original transport equation subject to appropriate boundary conditions on the flux H(R). Even if the Eddington factor $f_K(R)$ could somehow be prescribed, the flux boundary values $H(0)$ and $H(R_s) \equiv f_H(R_s)J(R_s)$ are needed before Eqns. (9) and (10) can be (trivially) integrated.

In the variable Eddington factor method, some approximate $f_K(R)$ is used to calculate $J(R)$ from Eqns. (9) and (10) which is then inserted in the transport equation source term S[J(R)] to approximately solve for the distribution function and an improved value for $f_K(R)$. This procedure of using the moment equations to iteratively solve the transparent equation accelerates the numerical convergence of the solution which is otherwise poor.

3.2 Conservative Scattering and Power Law Operators

We now consider flux-conservative scattering in which $S = J/4\pi E^3$ so that $\alpha R^2 H$ = constant. This is the case for pure isotropic coherent scattering in a gray structure in radiative equilibrium. In terms of $\mathcal{J} \equiv 4\pi r^2 J, \kappa \equiv 4\pi R^2 \mathcal{H}$ = luminosity / 4π, the moment equations (9) and (10) now are

$$\alpha \mathcal{H} = \text{constant} \tag{12}$$

$$\frac{(\Gamma\alpha)R^2}{q}\frac{d}{dR}(\frac{q}{R^2}f_K\mathcal{J}) = -\chi(\alpha\mathcal{H}). \qquad (13)$$

In the diffusive region $f_K \approx 1/3, q \propto \alpha^3$,

$$\Gamma\frac{d}{dR}(\alpha^3 J) = 3\chi(\alpha^3 H) \qquad (14),$$

this reduces to planar scattering. In the free-streaming region $(f_K \approx 1, q \propto \alpha R^2$,

$$\Gamma\frac{d}{dR}(\alpha\mathcal{J}) = -\chi(\alpha\mathcal{H}). \qquad (15)$$

We now consider the (Kosirov) power law opacity $\chi(R) = R_o^{n-1}/R^n$ with scale height $h(R) \equiv (-d\ln\chi/dR)^{-1} = R/n$. For $n > 1$, the optical thickness to the center is infinite. For large n, the opacity decreases from a large value just inside R_o to a small value just outside R_o. The neutrinosphere is therefore located at radius R_o and of thickness R_o/n.

We distinguish three different kinds of stellar atmospheres by the ratio of stellar radius R_S to neutrinosphere radius R_o: (1) $R_s/R_o >> 1$, extended atmospheres; (2) $R_s/R_o \approx 1$, semiplanar atmosphere; (3) $R_s/R_o << 1$, planar atmosphere. Except in spherical shells, planar atmospheres (3) do not occur astrophysically. This case is nevertheless interesting formally because, in the limit $R_s/R_o \to 0$, it reduces the spherical problem to the classical Milne problem of conservative scattering in a semi-infinite planar medium. The Eddington approximation $f_K = 1/3$ is then a good approximation nearly up to the boundary.

Case (2), with large n, is realized in the very thin atmosphere surrounding a neutron star (X-ray bursters, Goodman-Dar-Nussinov heat deposition in a stalled shock, cooling neutron stars). Because absorption is very non-gray in these situations, surface cooling and back-warming are important, so that the neutrinosphere does not radiate as a black-body.

3.3 Sphericity Effects in an Extended Atmosphere

Although our ultimate interest is in the semiplanar atmospheres surrounding compact objects, our methods will be most severely tested in (1), extended atmospheres where the sphericity effects become important, so that f_K, f_H approach each other and approach unity as R_s/R_o or n increases. Because non-relativistic numerical solutions are well known [4,5], we have used this case (1) to test our relativistic tangent-ray method.

Choosing starting values $f_K = 1/3, f_H = 1/2$, we iterated between the moment equations (9), (10) and the transport equation (7). Using variable spatial zoning to give adequate coverage in both the interior and the atmosphere, after less than 5 iterations, we obtained good agreement with the Hummer-Rybicki solutions. The Eddington factor $f_K(R) \equiv K/J$ and streaming function $f_H(R) \equiv H/J$ are plotted in Fig. 7, 13, 14, 15 of ref. 3 and will not be repeated here. They show (f_K, f_H) changing over from interior values (1/3, 0) to surface values $(f_K(R_s), f_H(R_s))$ in the outermost mean free path. In a weak gravitational field, these surface values are [4,6]:

(1) For the planar atmosphere $R_s/R_o = 0.1, (f_K(R_s), f_H(R_s)) =$ (0.414, 0.581), (0.448, 0.616), (0.529, 0.685) for n = 3, 2, 3/2 respectively.

(2) For the semiplanar atmosphere $R_s/R_o = 1$, $(f_K(R_s), f_H(R_s))$ = (0.63, 0.78), (0.67, 0.80), (0.68, 0.81) for n = 3, 2, 3/2.

(3) For the extended atmospheres $(R_s/R_o, n)$ = (10,2), (100, 3/2), (30, 2), (10, 3), $(f_K(R_s), f_H(R_s)$ = (0.930, 0.959), (0.943, 0.966), (0.974, 0.985), (0.982, 0.990) respectively.

The emergent Eddington factors all fall within a few percent of the linear trajectory $f_K(R_s) = -\frac{1}{3} + \frac{4}{3} f_H(R_s)$, i.e. although the Kosirov opacities depend on two parameters (R_o, n), the surface Eddington factors both depend on one parameter. In the semiplanar atmosphere outside the stalled shock, we expect $f_H(R_s) \sim 0.75, f_K(R_s) \sim 0.5$, a radiation pattern that is sufficiently forward-peaked to disfavor appreciable neutrino momentum deposition by the Goodman-Dar-Nussinov $\nu + \bar{\nu} \to e^- + e^+$ mechanism.

We now digress to discuss the origin of "universal" Eddington trajectories and their limited applicability.

3.4 Maximum Entropy Methods

Maximum entropy methods have been used [6, 7] to derive the most probable gray atmosphere angular distribution consistent with two specified angular moments J, H. For non-degenerate neutrinos or photons, these methods lead to a Boltzman distribution of direction cosines, $f(\mu) = e^{\eta} e^{a\mu}$, that depends on only two parameters η and a. The Eddington ratios depend only on a, which ranges from zero to infinity as the angular distribution ranges from isotropic to free-streaming. Eliminating a, a universal Eddington trajectory $f_K = f_K(f_H)$ is obtained that is expected to be most probable for each depth in all radiation fields.

Because boundary conditions are ignored, allowing μ to range over the full interval $-1 \leq \mu \leq 1$ everywhere, the full-range moments of *MINERBO* [7] give only a fair approximation: $f_K(f_H)$ is underestimated in an extended atmosphere, especially at depth, and overestimated in a planar atmosphere, especially near the surface. By allowing only the half-range $0 \leq \mu \leq 1$, Rybicki (1984) [9] obtained half-range moments

$$J, H, K = e^{\eta} \int_{o}^{1} \mu^{l} e^{a\mu} d\mu \quad l = 0, 1, 2$$

and Eddington factors

$$f_H(R_s) = \frac{1}{J}\frac{dJ}{da} = \frac{1}{1 - e^{-a}} - \frac{1}{a} \tag{16}$$

$$f_K(R_s) = \frac{1}{J}\frac{d^2J}{da^2} = \frac{1}{1 - e^{-a}} - \frac{2}{a} f_H, \tag{17}$$

that well represent the emergent radiation field for all the Kosirov models considered. Graphically eliminating a $=$ a(R_o, n) [8], we find the trajectory is approximately linear $f_K(R_s) \approx -\frac{1}{3} + \frac{4}{3} f_H(R_s)$, to within a few percent.

What use is the Eddington trajectory of the emergent radiation? The streaming function $f_H(R_s)$ still must be obtained by solving the transport equation subject to appropriate boundary conditions. Nevertheless, a good initial guess for $f_K(R)$ in the moment equations (9) and (10) allows a good starting approximation for the source function in the transport equation whose iterative solution then converges rapidly.

3.5 Stationary Transport in Strong Gravitational Fields

After testing the tangent-ray method in the non-relativistic Kosirov problem, we studied [4,10,11] the stationary radiation transport out of two 1.65 $M_{\odot}$ polytropic stars with $P = K\rho^2$ and power-law opacity $\chi(R) = R_o/R^2$ in an extended atmosphere $R_s = 10R_o$. (Although compact objects have semiplanar atmospheres,we chose an extended atmosphere to test our treatment of sphericity effects.)

The two 1.65 $M_{\odot}$ stars chosen were (1) A "white dwarf" of central density $\rho_c = 10^9 \text{gcm}^{-3}$ and stellar radius $R_s = 1370$ km; as indicated by the central values for the relativity index $(P/\rho c^2)_c = 8.9 \times 10^{-4}$ and lapse $\alpha_c = 0.996$ and by the surface red-shift $(GM/RC^2)_s = 0.0018$, this is a non-relativistic or weak gravity case. (2) A "neutron star" of $\rho_c = 4 \times 10^{15} \text{gcm}^{-3}$, $R_s = 9.08$ km which, because $(P/\rho c^2)_c = 0.45$, $\alpha_c = 0.326$, $(GM/Rc^2)_s = 0.269$, is practically maximal for a stable polytrope with causal equation of state, this is a strong-gravity case.

We found that strong gravity enhanced the sphericity q(r) in Eq. (11) by about a factor two and reduced the mean intensity J and the luminosity $4\pi\mathcal{H}$ by the red-shift factor $\alpha(R)$ and the surface Eddington factor $f_K(R_s)$ by about 20%.

4. CONCLUDING SUMMARY

We have (1) prepared an implicit hydrodynamics code to simulate stellar core collapse and bounce shock for long times; (2) gone to asynchronous coordinates that are polar-sliced and comoving, so as to allow maximum coverage of space-time and of material zones. This adiabatic hydrodynamics code has been tested in the collapse of a polytrope and of a dust cloud to a black hole; (3) prepared the formalism for treating general relativistic radiation hydrodynamics by variable Eddington factor methods in order to accurately describe neutrino momentum deposition in semi-transparent material regions. We have generalized the tangent ray method (method of characteristics) to radiation transport in curved space-time. This code has been tested in stationary transport in a static geometry with power-law (Kosirov) opacity.

We intend to go on to (1) compare exact solutions of the transport equations by the method of characteristics and by finite differences; (2) compare the neutrino fluxes emerging from slow collapse into a black hole with that from collapse to a remnant neutrino star; (3) study the late time evolution of the bounce shock and the cooling of the nascent hot neutron star by the radiation of neutrinos of all flavors.

References

1. P. J. Schinder, S. A. Bludman and T. Piran, Phys. Rev. D **37**, 2722, (1988).

2. P. J. Schinder, Phys. Rev. D **38**, 1673, (1988).

3. A. Mezzacappa and R. A. Matzner, Astrophys. J **343**,853, (1989).

4. P. J. Schinder and S. A. Bludman, Astrophys. J **356**, 350, (1989).

5. D. G. Hummer and G. B. Rybicki, M.N.R.A.S. **152**, (1971).

6. P. G. Martin and C. Rogers, Astrophys. J. **284**, 317, (1984).

7. G. N. Minerbo, J. Quant. Spectros. Rad. Transf. **20**, 541, (1978).

8. A. Fu, Astrophys. J. **323**, 227, (1987).

9. Referred to in ref. 6, 8.

10. S. A. Bludman and P. J. Schinder in Proc. XIV Texas Symposium on Relativistic Astrophysics, Proc. N.Y. Academy of Sciences (1989).

11. S. A. Bludman in Proc. Workshop on Particle Astrophysics: Forefront Experimental Issues, edited by E. B. Norman (World Scientific, 1989).

Neutrino Transport in Supernovae and Protoneutron Stars by Monte Carlo Methods

H.-Thomas Janka

The still unsolved puzzle of how to make a type II supernova explosion from the collapse of a massive star's iron core, the detection of 19 neutrino events associated with SN 1987A in the IMB and Kamiokande 2 experiments (BIONTA *et al.* [1], HIRATA *et al.* [2]) and, most currently, the idea of making nucleosynthesis by non-conservative scattering of neutrinos off heavy nuclei (EPSTEIN *et al.* [3]; WOOSLEY *et al.* [4]) keeps interest in neutrino emission from newly formed neutron stars very vivid.

Although considerable effort has been payed to neutrino transport during stellar core collapse and neutron star formation, still a number of questions are not answered yet. E.g. it is not yet clear how the spectra of the various neutrino kinds can be described, how the released gravitational binding energy is shared between the different neutrino types and how efficiently neutrinos can deposit energy during the shock propagation phase. Transport methods employed in hydrodynamical codes are constrained in accuracy by the need of computational fastness. So most of the workers in the field employ flux-limiters of some kind. However, it is clear that such an approach will lead to uncertainties and errors which have to be checked and, if possible, reduced. In order to do this and to try to investigate the points mentioned above we employed a Monte Carlo (MC) method for simulating neutrino transport. Some of the results will be shortly described here.

1. Spectra

Spectra of all types of neutrinos were found to be different from a black body distribution. Although they are actually non-thermal it is possible to approximate them very closely by Fermi-Dirac functions employing 'effective degeneracy' parameters $\eta_\nu^{\rm eff}$. Typical values for these were found to be $\eta_{\nu_e}^{\rm eff} \approx 3 - 3.5$ for electron-type neutrinos, $\eta_{\bar\nu_e}^{\rm eff} \approx \eta_{\nu_x}^{\rm eff} \approx 2 - 2.5$ for electron antineutrinos and neutrinos of other flavors ($\nu_x \hat{=} \nu_\mu, \bar\nu_\mu, \nu_\tau, \bar\nu_\tau$). These numbers seem to be quite general and turned out to hold roughly for all investigated cases, i.e. for core collapse models of the 1.36 $M_\odot$ iron core of a 20 $M_\odot$ star (stiff equation of state) at stages 12 and 315 milliseconds after core bounce (HILLEBRANDT [5]; JANKA and HILLEBRANDT [6], [7]) and for the Kelvin-Helmholtz cooling phase of a 1.64 $M_\odot$ protoneutron star (soft equation of state) at times 3.32, 5.77 and 7.81 seconds after bounce (WILSON [8]). So we conclude that they reflect general properties of neutrino-matter interaction and do not sensitively depend on specific features of the background model and its physical (and numerical) description (JANKA and HILLEBRANDT [7]). The inferred narrowing of the energy distributions leads to a strong suppression of the high energy tails. This is important for the evaluation of neutrino detections (JANKA and HILLEBRANDT [7]) as well as for the strongly energy dependent reactions between neutrinos and heavy nuclei that are supposed to be responsible for nucleosynthesis during a supernova explosion.

2. Flow Pattern

Using flux-limiting causes severe uncertainties in the absolute values of the luminosities and the local 'flow pattern' of the radiation field. By the latter we mean the distribution of the outgoing particle flow in angle. As a rough measure for its degree of anisotropy one can use the first and second Eddington moments, $\langle z^1\rangle \equiv j/(n\cdot c)$ and $\langle z^2\rangle$, respectively (j being the flux density, n the local energy or particle density and c the speed of light). Let us estimate the errors due to a 'vacuum description' of the radiation field outside the neutrino sphere (COOPERSTEIN *et al.* [9]), when a number for the local energy deposition by neutrino-antineutrino annihilation is to be derived. We will adopt the notation employed in COOPERSTEIN *et al.* [10]. Assuming vanishing phase space blocking, separability in energy and angle and coincidence of neutrino and antineutrino spheres they get for the local energy deposition rate:

$$Q_{e^+e^-} = \frac{4\,G^2}{9\,\pi}\,D\,\varepsilon_\nu\,\varepsilon_{\bar\nu}\,(\omega_\nu+\omega_{\bar\nu})\cdot\chi \quad , \tag{1}$$

where ε_ν, $\varepsilon_{\bar\nu}$ are the energy densities, $\omega_\nu = T_\nu\cdot\frac{F_4(\eta_\nu)}{F_3(\eta_\nu)}$ are energy moments of the neutrinos, and χ is given by

$$\chi = \frac{3}{4}\cdot\left[1-2\langle z^1\rangle\langle\bar z^1\rangle+\langle z^2\rangle\langle\bar z^2\rangle+\frac{1}{2}(1-\langle z^2\rangle)(1-\langle\bar z^2\rangle)\right] \quad . \tag{2}$$

Assuming vacuum outside an isotropically emitting neutrino sphere of radius R_ν implies that $\varepsilon_\nu^{\mathrm{vac}}(z) = 2\cdot(1-z)\cdot L_\nu/(4\pi\,R_\nu^2 c)$ (L_ν = neutrino luminosity) and $\langle z^1\rangle_{\mathrm{vac}} = \frac{1}{2}(1+z)$ and $\langle z^2\rangle_{\mathrm{vac}} = \frac{1}{3}(1+z+z^2)$ for $z\equiv\sqrt{1-(R_\nu/r)^2}$, so $\chi_{\mathrm{vac}} = \frac{1}{8}(1-z)^2(5+4z+z^2)$. Most of the annihilation heating will occur at radii between R_ν and $1.1R_\nu$, i.e. for z-parameters between 0 and 0.417. Therefore deviations from the vacuum description due to non-vanishing neutrino-matter interactions outside the 'neutrino sphere' (which is an artifact by definition rather than a physical sphere, anyway) will be very important. MC simulations (JANKA and HILLEBRANDT [6]) suggest that they are significant, indeed. It was found that having a density profile according to $\rho\propto r^{-n}$ a very good description is given by $\varepsilon_\nu(z)^{\mathrm{MC}} = 2(1-z)\left(1+(1-z^2)^{(n-1)/2}\right)\cdot L_\nu/(4\pi R_\nu^2 c)$, i.e. $\langle z^1\rangle_{\mathrm{MC}} = \frac{1}{2}\frac{1+z}{1+(1-z^2)^{(n-1)/2}}$. In addition $\langle z^2\rangle_{\mathrm{MC}}$ increases much slower than in the vacuum case. So we end up with a local ratio of

$$\frac{(Q_{e^+e^-})_{\mathrm{MC}}}{(Q_{e^+e^-})_{\mathrm{vac}}} \cong \frac{\chi_{\mathrm{MC}}}{\chi_{\mathrm{vac}}}\cdot\left[1+(1-z^2)^{(n-1)/2}\right]^2 \quad , \tag{3}$$

where we used equality of the quantities L_ν, $L_{\bar\nu}$ and ω_ν, $\omega_{\bar\nu}$ in the nominator and denominator, an assumption which turns out to be actually not true, when e.g. flux-limited diffusion results are compared to MC studies (see 3.). Inserting typical numbers for the region of interest ($z\lesssim 0.5$), $\langle z^1\rangle\approx\langle z^2\rangle\approx z\approx\frac{1}{3}$ we find for density profiles of $n = 4, 10, 20$:

$$\frac{(Q_{e^+e^-})_{\rm MC}}{(Q_{e^+e^-})_{\rm vac}} \approx \frac{0.8333}{0.3580} \cdot (3.378\,,\ 2.524\,,\ 1.760) = 7.864\,,\ 5.874\,,\ 4.096\ . \qquad (4)$$

Although these results were derived on grounds of a vacuum approximation outside an artificially defined 'neutrino sphere', it must be expected that also flux-limited diffusion cannot do any better due to its inability of describing the angular distribution of the radiation field especially in situations of rapid variations in space and time.

3. Luminosities

The possibility of a direct conclusion from the MC transport results on influences due to flux-limiting and discretization in energy space is most easily possible in the case of muon neutrinos, because their number flux is conserved at densities below 10^{13} g cm^{-3}, where (in the 'late' protoneutron star models discussed here) their emission and annihilation processes are essentially unimportant. At the inner boundary of the stellar window where the transport is followed explicitly by the MC code we impose an isotropic ν_μ-distribution according to the given local neutrino density resulting from WILSON's [6] data. Without including general relativistic effects the multigroup flux-limited diffusion scheme yields particle fluxes of $J_{\nu_\mu}^{\rm MFL} \approx 1.03 \cdot 10^{56}, 3.26 \cdot 10^{55}, 2.78 \cdot 10^{55}$ s^{-1} and average energies of $\langle \epsilon_{\nu_\mu} \rangle^{\rm MFL} \approx 23.6, 32.7, 30.8$ MeV for times $t = 3.32, 5.77, 7.81$ seconds after core bounce, whereas with the MC treatment we find $J_{\nu_\mu}^{\rm MC} = 1.27 \cdot 10^{56}, 1.19 \cdot 10^{56}, 5.90 \cdot 10^{55}$ s^{-1} and $\langle \epsilon_{\nu_\mu} \rangle^{\rm MC} = 24.5, 27.0, 25.5$ MeV, respectively. Roughly, the differences increase as the neutron star becomes more compact, i.e. the density decline near its surface gets steeper.

4. General Relativistic Effects

Gravitational redshift and light bending effects were included into the MC transport code. Ray bending and time dilatation yield a reduction of the local Eddington moment $\langle z^1 \rangle$ by about 15–30% in the diffusion regime (where $\langle z^1 \rangle \lesssim 0.1$), decreasing to 1–2% when free streaming is achieved ($\langle z^1 \rangle \gtrsim 0.5$). Accordingly, the local neutrino concentrations in the free streaming region are smaller by 15–30%, too. Gravitationally redshifted energies are lower by 5–10% at the protoneutron star 'surface'. Altogether the emitted fluxes turn out to be reduced by 20–40%. For an observer at infinity an additional reduction by 20–35% occurs, so that including general relativity causes the observable fluxes in particle number to be only 60–80% and in energy only 40–60% of those obtained in the Newtonian case.

It is the aim of this short contribution to point out that there are significant uncertainties in the presently used approximate treatments of neutrino transport in hydrodynamical core collapse simulations. They are of a size which makes them important when neutrino detections are to be evaluated and when dynamical consequences of neutrinos for the supernova are to be investigated.

References

1. R. M. Bionta, G. Blewitt, C. B. Bratton, D. Casper, A. Ciocio, R. Claus, B. Cortez, M. Crouch, S. T. Dye, S. Errede, G. W. Foster, W. Gajewsky, K. S. Ganezer, M. Goldhaber, T. J. Haines, T. W. Jones, D. Kielczewska, W. R. Kropp, J. G. Learned, J. M. LoSecco, J. Matthews, R. Miller, M. S. Mudan, H. S. Park, L. R. Price, F. Reines, J. Schultz, S. Seidel, E. Shumard, D. Sinclair, H. W. Sobel, J. L. Stone, L. R. Sulak, R. Svoboda, G. Thornton, J. C. van der Velde, C. Wuest: Phys. Rev. Lett. 58, 1494 (1987).

2. K. Hirata, T. Kajita, M. Koshiba, M. Nakahata, Y. Oyama, N. Sato, A. Suzuki, M. Takita, Y. Totsuka, T. Kifune, T. Suda, K. Takahashi, T. Tanimori, K. Miyano, M. Yamada, E. W. Beier, L. R. Feldscher, S. B. Kim, A. K. Mann, F. M. Newcomer, R. Van Berg, W. Zhang, B. G. Cortez: Phys. Rev. Lett. 58, 1490 (1987).

3. R. I. Epstein, S. A. Colgate, W. C. Haxton: Phys. Rev. Lett. 61, 2038 (1988).

4. S. E. Woosley, D. H. Hartmann, R. D. Hoffman, W. C. Haxton: The ν-Process, preprint (1989).

5. W. Hillebrandt: in High Energy Phenomena around Collapsed Stars, ed. by F. Pacini, (Dodrecht, Reidel 1987) (p. 73).

6. H.-T. Janka, W. Hillebrandt: Astron. Astrophys. Suppl. 78, 375 (1989).

7. H.-T. Janka, W. Hillebrandt: Astron. Astrophys. (1989), in press.

8. J. Wilson: private communication (1988).

9. J. Cooperstein, L. J. van den Horn, E. A. Baron: Astrophys. J., 309, 653 (1986).

10. J. Cooperstein, L. J. van den Horn, E. A. Baron: Astrophys. J. (Letters), 321, L129 (1987).

The SN 1987A Neutrino Signal and the Future

Adam Burrows

1. Introduction

As brilliant as SN1987A was and is in light, on February 23, 1987 at $7^h35^m41^s$ universal time, startling proof was obtained that photons were not its most spectacular radiation. It was then that two massive underground detectors, the Irvine-Michigan-Brookhaven (IMB) detector in the U.S. [1] and the Kamiokande II (KII) detector in Japan [2], registered supernova neutrinos for the first time. Within only about ten seconds, the general theory developed by a generation of astrophysicists over the past thirty years connecting supernova explosions and the death of massive stars with prodigious neutrino bursts was transformed into a concrete fact, and extragalactic neutrino astronomy was born. Neutrinos interact very weakly with matter and can penetrate with ease the earth, the sun, and even the envelope of the 15 to 20 solar mass ($M_\odot$) blue star that exploded as SN1987A. The transparency of the progenitor star to the neutrinos we now know are generated in its heart as it dies makes a supernova's neutrino signal the only good diagnostic of the violent internal convulsions that attend stellar death.

The supernova trigger is otherwise shrouded in mystery by the profound opacity of the star to photons. In theory, the $\sim$1.5 $M_\odot$ core of a massive star (>8 $M_\odot$) whose thermonuclear life may have lasted $\sim 10^7$ years collapses within one second to a "protoneutron-star" (PNS), whose subsequent transformation into a dense neutron star proper (later, perhaps a pulsar) takes seconds. It is during the birth of the neutron star that the neutrinos are emitted. The shock wave generated during these seconds travels from the core to the stellar surface in between an hour and a day and disassembles the star in an explosion that lasts months to years. This explosion is the supernova. Though 10^{49} ergs ($\sim 2.5\times10^{26}$ megatons of TNT equivalent) is eventually radiated in supernova light and the kinetic energy of the supernova debris is $\sim 10^{51}$ ergs, more than one hundred times again as much energy ($>10^{53}$ ergs) is liberated in the brief, but massive burst of neutrinos. During such a burst, the supernova core is as bright in neutrinos as the entire observable universe is in optical light. It was such a neutrino burst from SN1987A that was detected simultaneously in North America and Asia on February 23, 1987, and it is such bursts, their detection, and what we can learn from them that is the subject of this paper.

2. Neutrinos from Supernovae: Theory

Neutrinos dominate the emissions from the high-energy density plasma in the core of a massive star before and after collapse almost by default. Since the electrodynamic interaction is much stronger than the weak interaction, the photon production rate in this hot, dense "pit" far exceeds that of neutrinos. However, once created, a photon can travel no more than a few angstroms before being absorbed. It is completely trapped in the "matter," which is actually a soup of heavy nuclei, alpha particles, electrons, positrons, free neutrons and protons, and photons in thermal equilibrium. The neutrinos, on the other hand, once created can much more easily stream unimpeded out of the core. By surveying all the particles in the particle physicist's zoo, one arrives at the conclusion that, of the known particles, only neutrinos are both copiously produced and weakly coupled in the hot core of a massive star. Hence, neutrinos dominate the luminosity of a massive star in the advanced stages of its life. Even though the core neutrino luminosity of a massive star just before core collapse can be ten orders of magnitude greater than its surface photon luminosity, this neutrino luminosity and the average per neutrino energy during pre-supernova stages are still not adequate to trigger extant or projected terrestrial neutrino detectors. However, once the core has grown to the "critical" Chandrasekhar mass of $\sim 1.5\ M_\odot$ and can no longer support itself against gravity, it implodes to even higher densities and temperatures. It is under these extreme conditions that the neutrino luminosities and energies are finally adequate to be detected on Earth from anywhere in our galaxy, the LMC, or the Small Magellanic Cloud (SMC).

Before we address this neutrino burst, we must first describe collapse dynamics. After some ten or twenty million years, the core of a massive star (Mass $> 8.0\ M_\odot$) has exhausted its thermonuclear fuel. It has a mass of approximately $1.5\ M_\odot$ (solar masses), a radius of approximately three thousand kilometers, a central density near 10^{10} gm/cm^3, and a central temperature near ~ 0.5 MeV (approximately equivalent to 6×10^9 K). Embedded inside the massive star (envelope mass $> 6.5\ M_\odot$), which has an outer radius (photosphere) of no less than 30 million kilometers, the core would seem insignificant. However, as the exhausted core reaches the critical "Chandrasekhar" mass of ~1.5 $M_\odot$, it can no longer support itself against persistent gravity. After eons of stability, the core becomes dynamic and collapses "under its own weight." Note that the rest of the star does not collapse with the core. It is still unaware of its impending death, and its photosphere is quiet. In less than one second, parts of the imploding core reach speeds of 70,000 kilometers per second, one quarter the speed of light. So strong is the gravitational pull of this compact imploding sphere, that collapse is halted only after nuclear densities (2.7×10^{14} gm/cm^3) and beyond are achieved. It is only then that the strong repulsive nuclear force can stiffen the core sufficiently to finally halt and reverse collapse. If the matter did not stiffen, the entire core would plunge into a black hole.

During collapse, the core marches though almost two decades in radius, five decades in central density, and more than one decade in central temperature to the most extreme thermodynamic conditions since the big bang. In bouncing, the inner core (~0.8 $M_\odot$) "pulls" as many as one hundred billion (10^{11}) "g's." It collides violently with the still-collapsing outer core of 0.7 $M_\odot$ and thereby generates a strong off-center, through spherically symmetric, shock wave. This shock is launched into the stellar envelope with a speed near ten thousand kilometers per second and is the supernova in its infancy. Though core collapse takes almost one second, rebound and shock-wave formation take only a few milliseconds. After launching the shock wave, the spent core does not reexpand, but quickly settles into hydrostatic equilibrium. This dense, hot (temperature ~ 10^{11} K) and extended (radius ~ 100 kilometers) "protoneutron-star" is not the cold, compact (radius ~ 10 kilometers) neutron star of the textbooks, but an intermediate state through which the core must go in birthing a neutron star.

Having described core dynamics, we can now return to a general theoretical description of the neutrino signature. Compression during the early stages of collapse increases the electron energies above the threshholds (a few MeV) for electron capture on the heavy nuclei in the core. This weak interaction process, e^- (electron) + p (proton) $\rightarrow$ n (neutron) + ν_e (electron-neutrino), leads to the further loss of electron pressure and the conversion of protons into neutrons in the nuclei. Core nuclei are thereby made more neutron-rich and electron-neutrinos are liberated in substantial numbers for the first time. The above process assumes lepton conservation, which is now an important fixture of the standard theory of collapse, but is not etched in stone and may yield to future experiments. Nevertheless, since lepton conservation is assumed in current supernova calculations, it is the results of such calculations that we report here.

The electron capture accelerates as higher core densities are achieved. However, when the central density reaches and exceeds 10^{11} gm/cm^3, the product of the nuclear scattering cross section for the progressively more energetic capture neutrino and the matter density becomes so large that the electron-neutrino mean-free-path, its "interaction" length, reaches, then exceeds, the size of the core. Electron-neutrinos, that only hundreds of milliseconds earlier streamed freely out of the imploding core, are now trapped in the flow. Though no more than a few hundred kilometers in radius, the core quickly achieves the neutrino opacity of a "light-year of lead." Beyond 10^{11} gm/cm^3, decades in density before bounce, electron-neutrino capture on neutrons, the inverse of electron capture, stabilizes electron loss. If it were not for trapping, much of the supernova neutrino burst would be in electron-capture neutrinos emitted during the few hundred milliseconds of collapse. Because of trapping, "only" about 1% ($\sim 10^{51}$ ergs) of the neutrino burst is in such "infall" neutrinos. The lion's share of the energy has yet to be radiated.

The lepton-rich core continues to collapse beyond trapping, achieves nuclear densities ($>2.7\times10^{14}$ gm/cm^3), bounces, drives the shock wave into the outer core, and settles into the non-dynamical, hydrostatic PNS state. The shock is formed in a neutrino-opaque region near densities of 10^{13} gm/cm^3. However, within milliseconds, it plows to lower densities where the neutrino mean-free-paths are longer and the matter is transparent. The boundary between neutrino opacity and transparency is called the "neutrinosphere," in analogy with the more common photosphere. As the shock traverses the neutrinosphere, there is a brilliant burst of electron-neutrinos [3]. These electron-neutrinos are liberated by electron capture behind the matter-compressing shock, and are dammed-up until the shock "breaks out" of the neutrinosphere. However, this mini-burst lasts no more than a few milliseconds and also involves only a few times 10^{51} ergs. Though it is a distinctive signature of shock dynamics, the break-out burst is difficult to detect.

Most of the neutrino burst of a supernova is actually radiated during the non-dynamical transformation of the bloated PNS into the compact neutron star [4]. The PNS (again, "protoneutron-star") is energy- and lepton-rich and only marginally bound. A cold neutron star is so compact and dense and of such a large mass ($\sim 1.5\ M_\odot$) that it is tightly bound, with a large binding energy near 3×10^{53} ergs. Therefore, energy conservation alone demands that to form a neutron star this binding energy must be radiated, and theory concludes that it must be in neutrinos. Since the PNS is opaque to neutrinos (indeed, at this stage there are $\sim 10^4$ mean-free-paths from the center to the neutrinosphere!), the energy cannot stream out quickly, but must diffuse out slowly with the neutrinos. The PNS is like a hot ember that cools and shrinks via neutrino emission. Since the neutrino mean-free-paths are so short (meters to kilometers, inside to outside), diffusion takes not milliseconds, but seconds. Net electron capture and final neutronization (to form the "neutron" star) are paced by the slow diffusion and loss of capture ν_e's.[5)] In addition, so hot is the PNS (10-50 MeV) that neutrino pairs of all species (electron (ν_e) and anti-electron ($\bar{\nu}_e$) types, muon (ν_μ) and anti-muon ($\bar{\nu}_\mu$) types, and tauon (ν_τ) and anti-tauon ($\bar{\nu}_\tau$) types), are created. Therefore, neutrinos of all species, not just electron-type, carry away by slow diffusion the prodigious binding energy of the neutron star. The average energies of the electron-types are roughly 10-20 MeV, while those of the mu- and tau-types are roughly 25 MeV. However, each of the six neutrino species carries away roughly the same fraction (~one-sixth) of the binding energy. Curiously, theory has about two-thirds of the neutrino burst in the difficult-to-detect mu and tau channels. Importantly, theory also predicts that $\bar{\nu}_e$'s are radiated in respectable amounts. As stated previously, $\bar{\nu}_e$'s have the largest interaction cross sections in water and are expected to dominate the detected signals, despite the fact that they do not dominate the emissions. The different phases of neutrino emission are summarized in Table 1. Calculations show that more than 90% of the signal in water should be due to the $\bar{\nu}_e$ Cowan-Reines reaction, even though only 15% of the energy is in $\bar{\nu}_e$'s. The total number of events in one kilotonne of water at one kiloparsec (~ 3×10^{21} centimeters) is predicted to be between 10,000 and 20,000 [3]. This number can easily be scaled to any distance and to any detector size.

A few additional facts can illustrate the almost incredible magnitude of the ten-second PNS phase neutrino emission. The $\sim 3\times10^{53}$ ergs radiated is equivalent to ($E=mc^2$) a mass of 0.15 solar masses, 50,000 times the mass of the Earth. Over the age of our galaxy, approximately ten million solar masses in neutrinos have issued from neutron star births. This is equivalent to the mass of ~10 giant globular clusters. Indeed, at any one time, due to the finite speed of light (and neutrinos), there are 10-100 solar masses of burst neutrinos in very thin expanding shells within our galaxy. Though neutrinos are notoriously weakly interacting, so brilliant is a neutrino burst that it is lethal to humans within a distance equal to the radius of Pluto's orbit around our sun (~40 astronomical units $\equiv 6\times10^9$ kilometer). (However, given the supernova's photon radiation and blast wave, neutrino lethality is rather academic for civilizations near dying stars.)

Table 1. Neutrino Signature Sketch

	E (10^{51} ergs)	Δt (seconds)	Species
Collapse (infall)	~1.0	0.01→0.1	ν_e
Prompt burst	1-3	~0.003	ν_e
*Cooling and neutronization	250	~1.0→10.0	$\bar{\nu}_e$, ν_e, "ν_μ"

Protoneutron-Star → Neutron Star
Shock Break-out → Prompt Burst

*Inside, Deg. 200 MeV ν_e ---> Outside, many, 10-30 MeV (All species)

$$<\epsilon_\mu> \sim 25 \text{ MeV}$$
$$<\epsilon_{\bar{\nu}_e}> \sim 15 \text{ MeV}$$
$$<\epsilon_{\nu_e}> \sim 10\text{-}15 \text{ MeV}$$

$$\underset{*(\nu)}{2.5 \times 10^{53}} >> \underset{(SN)}{10^{51}} >> \underset{(Light)}{10^{49}} \text{ ergs}$$

3. The IMB and KII Data

The epochal neutrino data gathered by the IMB [1] and KII [2] collaborations are shown in Table 2. Quite a bit of information has been derived about not only "collapse," but also neutrino and exotic particle properties. This paper will not recapitulate the detailed analyses of the author [5] or others that can be found now in many journals. Rather, we sketch the important conclusions concerning protoneutron-stars and "supernovae" and make some general remarks about the new status of the field.

Table 2 demonstrates that the problems of small-number statistics must play a central role in any but the most general conclusions. However, even those conclusions are quite useful to a field emerging from three decades of data starvation. The IMB and KII detections demonstrate above all else that neutrinos, or particles just like neutrinos, are indeed radiated from Type II supernovae.

The large angles of the events with respect to the LMC (shown in Table 2) imply that most were not ν-e^- scatterings, which are quite forward-peaked, but that most or all were indeed $\bar{\nu}_e$ absorptions. Positron emission following $\bar{\nu}_e$ absorption is almost

Table 2. Data from the Kamiokande II and IMB detectors

Event #	Time (sec)	Electron energy (MeV)	Angle with respect to Large Magellanic Cloud (degrees)
Kamiokande II			
1	0.000	20.0 ± 2.9	18 ± 18
2	0.107	13.5 ± 3.2	40 ± 27
3	0.302	7.5 ± 2.0	108 ± 32
4	0.323	9.2 ± 2.7	70 ± 30
5	0.507	12.8 ± 2.9	135 ± 23
6[a]	0.685	6.3 ± 1.7	68 ± 77
7	1.540	35.4 ± 8.0	32 ± 16
8	1.728	21.0 ± 4.2	30 ± 18
9	1.915	19.8 ± 3.2	38 ± 22
10	9.219	8.6 ± 2.7	122 ± 30
11	10.432	13.0 ± 2.6	49 ± 26
12	12.439	8.9 ± 1.9	91 ± 39
IMB			
1	0.000	38 ± 7	80 ± 10
2	0.411	37 ± 7	44 ± 15
3	0.650	28 ± 6	56 ± 20
4	1.141	39 ± 7	65 ± 20
5	1.562	36 ± 9	33 ± 15
6	2.683	36 ± 6	52 ± 10
7	5.010	19 ± 5	42 ± 20
8	5.581	22 ± 5	104 ± 20

isotropic. However, the IMB data show a marked preference for the forward hemisphere that may well be a statistical fluke, but is as yet unexplained [6].

The best inferred average source temperature is 3.0-5.0 MeV. There is a slight indication from the data that the source is actually cooling, as one would expect. The late-time events are all lower in energy. The shorter duration of the IMB signal (5.58 seconds) with respect to the KII signal (12.44 seconds) also suggests that the source is cooling, since a given decay in temperature implies an even more rapid decay in the high energy tail of a thermal, or near-thermal, spectrum.

The evolution of the signal can be fit reasonably well, not by a single exponential, but by two exponentials, one after the other, with τ's of $\lesssim 1.0$ seconds and ~ 4.0

seconds, respectively. In other words, two phases, an early short one and a later long one, seem indicated. No neutrino mass, source pulsing, or neutrino oscillations are necessary to explain the data, though none of these exotica are absolutely eliminated. The total $\bar{\nu}_e$ energy radiated is ~3-5x10^{52} ergs, assuming a distance of 50 kpc. If we multiply by 6 to approximately account for the other five neutrino species, a total neutron star binding energy of 2-3x10^{53} ergs is derived. This is our first "direct" measurement of a neutron star's binding energy and it is surprisingly close to what was expected.

Both the KII and IMB data must be fit simultaneously. The two detectors must have sampled the same underlying source. The two major uncertainties in determining which model best represents the LMC remnant are our imprecise knowledge of D (±10%) and the inevitable $\sqrt{\mathcal{N}}$ statistical fluctuations in the sampled event number. The latter are, for 8 and 11 events, 35% and 30%, respectively. The resultant uncertainty in the fit of the data sets when taken individually is ~40% and when taken together is ~30%. In addition, though the KII signal is only weakly dependent on $\langle T_{\bar{\nu}_e} \rangle$ ($\propto \langle T_{\bar{\nu}_e} \rangle^{1.35}$ at T = 4.0 MeV), the IMB signal, sampling as it does the high energy tail, depends steeply on $\langle T_{\bar{\nu}_e} \rangle$ ($\propto \langle T_{\bar{\nu}_e} \rangle^4$ at T = 4.0 MeV). A 10% error in the calculation of $T_{\bar{\nu}_e}$ will result in a ~50% error in the predicted IMB signal. From this fact alone, we can conclude that $\langle T_{\bar{\nu}_e} \rangle$ is now known to better than ~20% and cannot be less than 3.0 MeV or greater than 5.0 MeV. Final baryon masses smaller than 1.2 $M_\odot$ or larger than 1.7 $M_\odot$ do not fit the data. From these considerations, we can conclude that M_G of the LMC neutron star can be 1.3-1.5 $M_\odot$ with little difficulty. It would be difficult to square with the neutrino data the accretion of more than 0.3 $M_\odot$ onto a protoneutron star whose initial mass is ~1.3 $M_\odot$ (baryon). The masses we here derive for the LMC neutron star are in the standard neutron star mass range. For these best-fit models, E_T(20 seconds) is 2.0-2.5x10^{53} (D/50 kpc)^2 ergs and $E_{\bar{\nu}_e}$(20 seconds) is 3.0-4.0x10^{52} (D/50 kpc)^2 ergs. Independently, the observed SN1987A ^{56}Ni mass of 0.075 $M_\odot$ and recent likely models for SN1987A progenitors have led NOMOTO et al. [7] to conclude that the residue mass is indeed ~1.6 $M_\odot$ (baryon).

4. The Fallout and the Future

The epochal detections by IMB and KII of a neutrino burst from a Type II supernova serve to verify the general, not the specific, features of stellar collapse. We now know that neutrinos (specifically, anti-electron neutrinos) are indeed important in such supernovae. This in itself is important since it forcefully suggests that our theories have not been devoid of content. From the detected events and the detector characteristics, we can infer a neutrinosphere temperature near 4.0 MeV (4.6×10^{10} K). This is ten million times hotter than the surface of our sun and the hottest emission temperature since the big bang (or the last stellar collapse). That the IMB and KII signal lasted seconds, not milliseconds, fits nicely into our current models which demand that the PNS is opaque to neutrinos, as bizarre as this seems. Opacity considerably (by factors of hundreds) lengthens neutrino escape times and, hence, the duration of the neutrino burst [4]. In fact, since a set amount of binding energy must be radiated to form a neutron star and the radius of the emitting surface is set by basic neutron star physics to be near ten kilometers, or only a few times ten kilometers, the duration of the cooling burst and the temperature of the neutrinosphere must be complimentary: If the signal duration is shorter, then the emission temperature is higher, the more quickly to radiate that set binding energy. A duration of ten milliseconds, a number from the 1960's, implies a temperature of a least 20 MeV, far in excess of what was observed. However, and gratifyingly, modern theoretical durations of seconds imply temperatures near 3-5 MeV, in precisely the temperature range observed. Hence, the detected neutrino energies and signal durations are mutually reinforcing observational facts.

Nevertheless, the capture of but nineteen events between the two multi-kilotonne detectors should humble even the most exuberant astrophysicist. The statistics of small numbers will hobble any attempt to write more than the first line on the final granite tablets of "stellar collapse." Little astrophysics beyond the important fundamentals cited above has been learned. What in fact is the mechanism of Type II supernovae? Is there a break-out burst? What is the neutrino spectrum and how does it evolve? Is lepton number trapped on infall? Are neutrinos of all species in fact radiated? What is the mass of the residue? What is the pressure-density relation for neutron star matter? What is the role of convection in PNS evolution? None of these important questions was directly answered by SN1987A.

However, there is a great deal of information in these sparse data about the properties of neutrinos themselves. The fact that a neutrino (strictly a $\bar{\nu}_e$) traveled for approximately 160,000 years before reaching Earth gives us our best limit on its lifetime and eliminates neutrino decay as a reason for the outstanding solar neutrino problem [8]. This is new fundamental particle physics information, garnered on the cheap. If $\bar{\nu}_e$s had a nonzero rest mass, they would not all travel at the speed of light, but at speeds that would depend on their energies. A spread in $\bar{\nu}_e$ energy would imply, therefore, a spread in speeds and, hence, an increased spread in arrival times to our northern hemisphere. Since the neutrino signal was dispersed over only a few seconds after traveling for 160,000 years, a good upper limit on the $\bar{\nu}_e$ mass of ~20 eV/c^2 can be derived, assuming little about burst theory [9]. This is 25,000 times smaller than the mass of the electron.

Furthermore, the properties (and existence) of a zoo of hypothetical particles, from majorons [10] to axions [11], each of which has been evoked to solve some outsanding problem in particle physics, can be constrained by the fact that, though these particles would have been created and emitted at the high temperatures in the PNS, they would also have competed with the neutrinos for the finite neutron star binding energy available. Since the standard neutrino burst alone can explain the signals and registered in full in IMB and KII, little room is left for the exotic particles to have carried away some of the energy. Assuming that exotica could be buried in the noise or statistics gives us usable limits on the coupling and properties of these particles, still inaccessible by standard laboratory experiments.

The above list of non-supernova conclusions from the SN1987A neutrino data illustrates the manifold and unexpected uses to which good data can be put. There is always much more than 19 bits of information in 19 events. However, a collapse not fifty, but five kilosparsecs away, in our own galaxy, promises to incite a revolution. A factor of ten decrease in distance leads to a factor of one hundred increase in integrated signal. Table 3 depicts predicted event totals from a collapse at the galactic center in proposed or extant neutrino telescopes (see this volume for details). What could not be learned about supernova physics and fundamental physics with one thousand events?

The first detection of neutrinos from a supernova has been an exhilarating experience for all those involved. Nevertheless, we were lucky. Both IMB and KII had undergone crucial upgrades only six months to one year before the epiphany of SN1987A. Had these upgrades not been successfully completed when they were, neither detector would have seen the neutrinos, even if they had been on-line. It is curious to note that the detection of neutrinos emitted long before the end of the last ice age, when Neanderthals still roamed Europe, should have depended on such recent serendipity. Furthermore, these detectors were not built to catch supernova neutrinos, but to observe proton decay. When proton decay was not seen, these huge tanks of water searched for monopoles or solar neutrinos. Supernova neutrino detection was rarely considered and, when considered, was not considered feasible by the

Table 3. Sample Future Detector Event Totals (from the Galactic Center (8.5 kpc))

	Total #	Prompt ν_e's ($\dot{E}_{\nu_e}/10^{51}$ ergs)	Infall ($E_{\nu_e}/10^{51}$ ergs)	"ν_μ"'s
KII (H_2O)	548	0.5	0.5	~20
SNO (D_2O+H_2O)	1179	~4.1	~4.1	~500
ICARUS (^{40}Ar)	169	~3.6	~3.6	~25
Homestake (^{37}Cl)	5.7	~0.15	~0.15	--

experimenters. To date, SN1987A has been each detector's greatest triumph.

The future looks good for supernova neutrino detection. Around the world, many large mass detectors are being built or planned, not just for burst detection, but for a variety of astrophysical and particle physics missions. The scientific community has been educated by SN1987A in the next simple steps: upgrade phototubes, expand data buffers, reduce dead times, maintain accurate clocks, and, importantly, coordinate and link the international network of detectors to ensure that at least one detector (preferably more) is on-line at all times. The mutually confirming data from the U.S. and Japan emphasize the importance and usefulness of such an international approach.

Acknowledgments

This work was supported by an NSF grant (No. AST87-14176) and the Alfred P. Sloan Foundation.

References

1. Bionta, R. M. et al., Phys. Rev. Lett., 58, 1494 (IMB Collaboration) (1987).

2. Hirata, K. et al., Phys. Rev. Lett., 58, 1490 (KII collaboration) (1987).

3. Burrows, A., Mazurek T. J., Nature, 301, 315 (1983).

4. Burrows, A. and Lattimer, J. M., Ap. J., 307, 178 (1986).

5. Burrows, A., Ap. J., 334, 891 (1988).

6. Matthews, J., in proceedings of the 4th George Mason Workshop in Astrophysics, ed. M. Kafatos and A. G. Michalitsianos (Cambridge University Press, Oct. 12-14, 1987).

7. Nomoto, K., Shigeyama, T. and Hashimoto, M., to appear in the proceedings of the IAU Colloquium 108 on Atmospheric Diagnostics of Stellar Evolution: Chemical Peculiarity, Mass Loss and Explosion, Tokyo, Japan, ed. K. Nomoto (Springer-Verlag 1987 (p. 319)).

8. Frieman, J. et al. (1988), preprint.

9. Burrows, A., Ap. J. (Letters), 328, L51 (1988).

10. Aharnov, Y., Avignone, F. T., and Nussinov, S., Phys. Rev. D, 37, 1360 (1988).

11. Burrows, A., Turner, M. S., Brinkemann, R. P., Phys. Rev. D, 39, 1020 (1989).

Implications of the SN 1987A Neutrinos for Supernova Theory and the Mass of ν_e

Thomas J. Loredo & Don Q. Lamb

The detection of neutrinos from supernova SN 1987A in the Large Magellanic Cloud by the Kamiokande II (KII) and Irvine-Michigan-Brookhaven (IMB) detectors was a landmark event in astrophysics [1-4]. Although only about two dozen of the $\sim 10^{28}$ neutrinos which passed through the Earth were detected, they provide us with the first glimpse of the collapse of a dying star, and hence deserve careful scrutiny.

There is already an extensive literature analyzing these epochal detections. Our work improves on previous studies in three important respects. First, our comparison of the data with parametrized models of the neutrino emission uses a consistent and straightforward statistical methodology. Second, our analysis uses an improved detector model which explicitly includes the empirically measured detector background spectra. Third, we compare the data with a much wider variety of neutrino emission models than was explored previously.

Our improved methodology allows a complete and rigorous comparison of the implications of the data with the expectations of the theory of neutron star formation, taking into account the effects of the strong correlations between the inferred neutrino emission model parameters. Further, it provides correct confidence regions for the electron antineutrino mass $m_{\bar{\nu}_e}$, unlike previously used methods that produce regions that do not enclose the unknown true value of $m_{\bar{\nu}_e}$ with the stated probability. Our inclusion of the background spectra in our analysis allows us to correctly account for the probability that some of the detected events are background events, and significantly weakens the constraint placed on $m_{\bar{\nu}_e}$ by the data. Our exploration of a wide variety of emission models ensures that our inferences regarding $m_{\bar{\nu}_e}$ are robust. The importance of an analysis exploring such a broad class of models has been emphasized by KOLB, STEBBINS, and TURNER [5].

A detailed discussion of our work is presented elsewhere [14-16], including a full comparison with earlier work; here we present our principal conclusions.

The class of neutrino emission models we have explored spans a wide variety of temporal and spectral behavior. Our calculations show that a simple exponential cooling model adequately explains the KII and IMB data when background is included, though more complicated models cannot be ruled out. In the exponential cooling model, the neutrinosphere has a fixed observed radius, $R_{\rm obs}$, and an exponentially decreasing $\bar{\nu}_e$ temperature, $T(t) = T_0 \exp(-t/4\tau)$, with initial temperature T_0 and luminosity decay timescale τ. The best-fit parameter values are $(R_{\rm obs}, T_0, \tau) = (22.5$ km, 4.17 MeV, 4.15 s$)$. These parameter values are in remarkable agreement with the basic predictions of supernova theory. The implied number of $\bar{\nu}_e$ is comparable to that expected from degenerate ν_e diffusing out of the inner core and heating the outer core by neutral current scattering and absorption. The $\bar{\nu}_e$ energy $3.15\ T_0 \approx 15$ MeV is typical of that expected for neutral current diffusion of degenerate $\bar{\nu}_e$ out of the hot outer core. The cooling time scale $\tau \approx 4$ sec is of order that expected for the neutral current diffusion of ν_e out of the inner core.

Assuming the neutron star emits its binding energy equally into 6 species (3 flavors) of neutrino, the best-fit parameters imply a binding energy of $E_b = 2.85 \times 10^{53}$ erg. The best-fit observed radius and binding energy are unusually large, and it has been suggested that the observations therefore rule out soft equations of state [7,8]. However, the inferred parameter values have a large statistical uncertainty. More importantly, the parameter values are highly correlated; confidence volumes in the three dimension parameter space are very asymmetrical, as can be seen by examining their two dimensional projections [14,15]. To allow full comparison of the inferred parameter values with neutron star models, Fig. 1 shows the 95% confidence region projected onto the $(R_{\rm obs}, E_b)$-plane. Also shown are $(R_{\rm obs}, E_b)$ curves for a representative set of equations of state from the compendium of neutron star models compiled by ARNETT and BOWERS [17]. Although the best-fit values of the radius and energy lie outside of the range permitted by these neutron star models, the projected confidence region comfortably overlaps all of the curves.

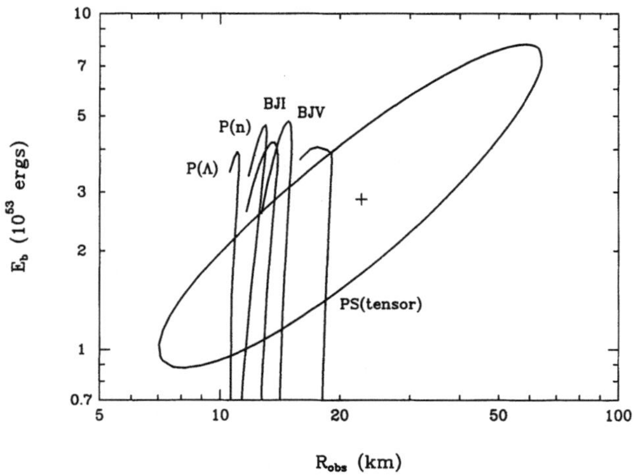

Figure 1. Comparison of the best-fit values and the 95% confidence volumes projected onto the $(R_{\rm obs}, E_b)$-plane, and the $(R_{\rm obs}, E_b)$-curves for neutron star models based on a representative set of equations of state.

The intersection of the projected confidence region with the $(R_{\rm obs}, E_b)$ curves defines the ranges of neutron star gravitational mass M_G that are permitted for each model. For each neutron star model, Fig. 2 shows the range of M_G bounded by the projected 95% confidence contour for the exponential cooling model including background. The allowed values of the neutron star mass are in remarkable agreement with the expected gravitational core mass $M \approx 1.4\ M_\odot$ expected from gravitational collapse and measured for neutron stars in binary systems.

To assess the implications of the data for $m_{\bar{\nu}_e}$, we include it as an additional parameter in our model. It enters the predicted rate function in a manner completely analogous to that of other model parameters, thus inferences regarding $m_{\bar{\nu}_e}$ are made simply by calculating best-fit parameter values and confidence regions in a larger parameter space.

All models considered give a best-fit mass at or near 0 eV. The exponential cooling model, with best-fit mass $m_{\bar{\nu}_e} = 0$ eV, gives the most conservative (*i.e.*, highest) upper limit on $m_{\bar{\nu}_e}$, indicating that $m_{\bar{\nu}_e} < 25$ eV with 95% confidence. If this limit is calculated incorrectly by omitting the background spectra, it decreases to 18 eV. Thus proper inclusion of the background is crucial for obtaining the proper upper mass limit. Our 95% upper limit is 1.5 - 5 times larger than found previously using incorrect statistical methods and ignoring the background [10-13]. We note that our limit is not significantly better than current laboratory limits.

In conclusion, our calculations show that a simple exponential cooling model

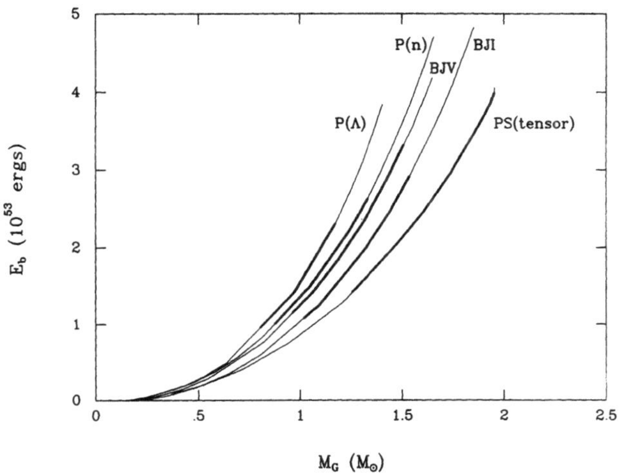

Figure 2. (M_G, E_b) curves for the neutron star models shown in Fig. 2, with the portion of each curve that lies within the projected 95% confidence volume drawn more thickly, indicating the allowed range of neutron star gravitational mass for each model.

adequately explains the KII and IMB data when background is included. The inferred characteristics of this model are in spectacular agreement with the salient features of the theory of gravitional collapse and neutron star formation that had developed over several decades in the absence of direct observational data. We have also shown that the data do not require a nonzero $\bar{\nu}_e$ rest mass, and that they place an upper limit on this mass of $\approx$ 25 eV at the 95% confidence level. Use of a correct and rigorous statistical methodology and inclusion of the detector background are crucial for obtaining complete and correct inferences, and our consideration of a wide variety of neutrino emission models ensures that our neutrino mass limit is robust.

This work was supported in part by NASA grants NAGW-830, NAGW-1284, and NGT-50189.

REFERENCES

1. K. Hirata, *et al.*: Phys. Rev. Lett. **58**, 1490 (1987).
2. R. M. Bionta, *et al.*: Phys. Rev. Lett. **58**, 1494 (1987).
3. C. B. Bratton, *et al.*: Phys. Rev. **D37**, 3361 (1988).
4. K. Hirata, *et al.*. Phys. Rev. **D38**, 448 (1988).
5. E. W. Kolb, A. J. Stebbins, and M. S. Turner: Phys. Rev. **D35**, 3598; **D36**, 3820 (1987).
6. L. M. Krauss: Nature. **329**, 689 (1987).
7. D. N. Spergel, T. Piran, A. Loeb, J. Goodman, and J. N. Bahcall: Science. **237**: 1471 (1987).
8. S. A. Bludman and P. J. Schinder: Ap. J. **326**, 265 (1988).
9. D. Q. Lamb, F. Melia, and T. J. Loredo: 1988. In Supernova 1987A in the Large Magellanic Cloud, Proceedings of the George Mason Workshop, ed. by M. Kafatos (Cambridge U. Press, Cambridge, England 1988), p. 204.
10. E. N. Adams: Phys. Rev. **D37**, 2047 (1987).
11. L. F. Abbott, A. DeRújula, and T. P. Walker: Nuc. Phys. **B299**, 734 (1988).
12. D. N. Spergel and J. N. Bahcall: Phys. Lett. **B200**, 366 (1987).
13. A. Burrows: Ap. J. (Letters)**328**, L51 (1988).
14. T. J. Loredo and D. Q. Lamb: In Proceedings of the Fourteenth Texas Symposium on Relativistic Astrophysics (New York Academy of Sciences, in press).
15. T. J. Loredo and D. Q. Lamb: Submitted to Phys. Rev. (1989).
16. T. J. Loredo and D. Q. Lamb: Submitted to Phys. Rev. (1989).
17. W. D. Arnett and R. L. Bowers: Ap. J. Suppl. **33**, 415 (1977).

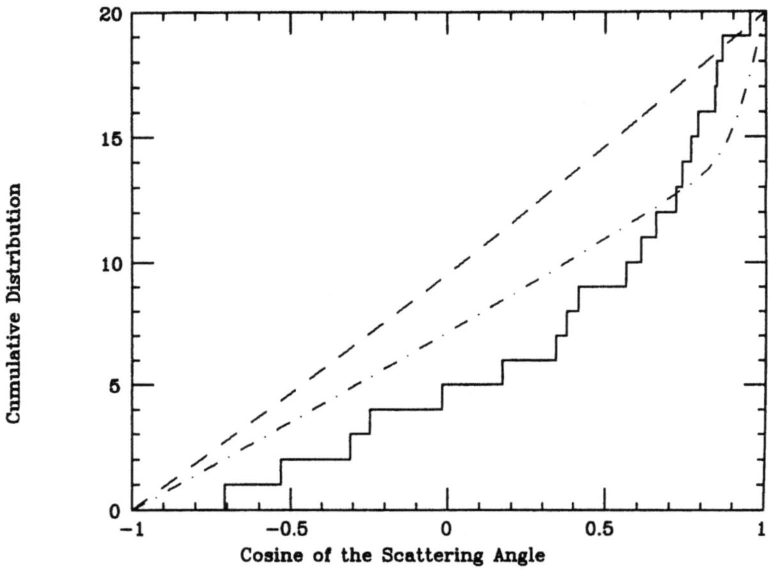

Figure 1: Comparison of Data with Expectations

Figure 1 compares the observed angular distribution with a distribution expected for purely electron antineutrino interactions (dashed curve) and with one expected if 25% of the interactions are due to neutrino electron scattering

$$\nu e \rightarrow \nu e$$

(dot dashed curve). The dot dashed curve is favored. The full argument and the likelihood is discussed in reference 5.

The presence of these additional reactions is not surprising. Since the pulse contains all neutrino types this is to be expected. But it is very difficult to draw quantitative results on these other components due to the very low data rate and the various possible reactions.

Implications. Idealy one would like to have a unique reaction for each of the possible components of the neutrino flux. But since the muon and tau neutrinos are below charged current reaction threshold they can only be observed by neutral current interactions. We do not know how to identify the type of neutrino involved in a neutral current reaction (except possibly by its cross section). The electron antineutrino is already well identified by its isotropic angular distribution and by the large fraction of the incident neutrino energy that can be measured in the recoiling positron.

It would be good to find a nuclear reaction in which only electron neutrinos can participate.

$$\nu_e N \rightarrow e^- X$$

The reaction needs to have a low energy threshold, so that interactions will occur for a substantial fraction of the flux. One can also hope for a large cross section since the detector may not be feasible for simple economic reasons if the cross section is too small.

The neutrino electron scattering reaction will record all neutrinos present in the pulse. But measurements with it are very difficult. In particular the emerging neutrino carries off some of the initial energy. Scattering and measuring errors make it very difficult to reconstruct the recoiling electron direction. Without an accurate direction the energy can not be measured.

With three measurements, the electron antineutrino spectrum through the proton reaction, the electron neutrino spectrum through the nuclear reaction and all neutrinos through the neutrino electron scattering reaction a good picture of the neutrino pulse can be reconstructed with minimal assumptions. The electron neutrino and antineutrino contribution to the neutrino electron scattering rate can be calculated and subtracted to yield a net signal from the other neutrino types.

Conclusions and Recomendations. It is important to have a well understood set of neutrino interactions to achieve the maximum astrophysics results from the next supernova neutrino observation. It would be prudent to use reactions that have been experimentally tested, and found to conform to expectations. We are fortunate that the muon decay neutrino spectrum is comparable in energy to the observed supernova neutrino pulse and can provide electron neutrinos and muon antineutrinos for a possible beam test.

An independent measurement of the electron neutrinos may confirm speculation[6] that this component carries as much as twice the energy of each of the others.

Acknowledgements. I would like to thank S. Woosley for organizing a great workshop. I would like to thank my colleagues on the IMB experiment for helpful advice and encouragement. This work was supported in part by the US DOE under Contract No. DE-AC02-87ER40366.A002.

References

1. R.M. Bionta, *et al.*, Phys. Rev. Lett. **58**, 1494 (1987)
   C.B. Bratton, *et al.*, Phys. Rev. **D37**, 3361 (1988).
2. K. Hirata, *et al.*, Phys. Rev. Lett. **58**, 1490 (1987)
   K. Hirata, *et al.*, Phys. Rev. **D38**, 448 (1988)
3. T.D. Lee and C.N. Yang, Phys. Rev. **126**, 2239 (1962)
   S. Bonetti, *et al.*, Nuov. Cim. **38**, 260 (1977)
   C. Llewellyn-Smith, Phys. Rep. **3C**, 261 (1972).
4. S.A. Bludman and P.J. Schinder, Ap. J. **326**, 265 (1988)
   I. Goldman, *et al.*, Phys. Rev. Lett. **60**, 1789 (1988).
5. J.M. LoSecco, Phys. Rev. **D39**, 1013, (1989).
6. H.-T. Janka, this conference.

SECTION VIII

SYNTHETIC SPECTRA OF SUPERNOVA 1987A, TYPE I'S, & OTHERS

Spectral Diagnostics of Type II Supernova

Peter Höflich

The investigation of the spectra of Type II supernovae (SN II) is important for different fields in astronomy and astrophysics. Firstly, SN II are one of the most luminous single objects and may reach the same brightness at maximum as a whole galaxy. Therefore, they can be used as candles (KIRSHNER and KWAN, [1]) to determine the distances of galaxies, in particular, to fix the cosmologically important Hubble constant H_o. Secondly, the observed spectrum gives direct information on the physical and chemical conditions of the supernova photosphere at a given time. Deeper layers of the expanding envelope are observable at later times. Therefore, a detailed spectral analysis of the time evolution is helpful in order to answer questions concerning the the overall structure of the envelope. In principle, this allows for an investigation of the explosion mechanism of SN II, for the test of hydrodynamic models and of the final stages of the evolution of massive stars.

Supernova 1987A in the Large Magellanic Cloud (LMC) has triggered a rapid development of tools for atmospheric diagnostics of SN II, because it has provide a lot of information due to the complete sample of observations. It has been shown that the spectra of SN1987A can be well understood by atmospheric models which take extention and NLTE effects into account. The total mass of the H-rich envelope was estimated to be about 10 $M_\odot$ both from the light curve (LC) (e.g. WOOSLEY[2]) and from the spectral analysis. Strong mixing processes of Co and Fe up to the hydrogen rich layers and of the inner few solar masses of hydrogen with the layers below are strongly indicated by the IR-lines of (RANK et al.[3]) and by the chemical analysis of the optical wavelength region, respectively. As a result of mixing hydrogen was observed at velocities of about 800 km/sec at the beginning of October 1987. In addition, the chemical abundances could be determined etc.(HÖFLICH[4,5,6,7]). The distance of SN1987A was derived by several groups (see paragraph III). A more detailed discussion of the atmospheric models for SN II and the diagnostics of SN1987A can be found in HILLEBRANDT and HÖFLICH[8]and the references therein.

In order to investigate scattering dominated atmospheres of SN II, we have extended our investigation of SN1987A to perform a more systematic study but by taking SN1987A as an example for spectral analysis. We want to show different aspects of the application of these models.

Type II SN show a wide range of individual variations of the light curves (LC) and details of the spectra (BRANCH, this volume). Therefore, detailed atmospheric models are needed for the determination of the properties of the expanding atmospheres, and the sensitivity of the observable features on the model parameters has to be investigated (paragraph II). The consequences are discussed for the use of SN II as distance indicators, and the practical usefulness of the models is demonstrated by the example of SN1987A (paragraph III). The LC of SN1987A is well understood and a lot of details concerning the progenitor and explosion of SN1987A could be derived by the LC using the grey flux limited diffusion approximation for the radiation transport. However, the explosion energy needed for the interpretation of the LC is only certain up to a factor of 2 (1...2 $10^{51} erg$). In paragraph IV the validity of the limited diffusion approximation is investigated and we discuss whether additional information can be derived by detailed atmospheric models from very early LC of SN1987A. Finally, consequences of the limb darkening effects are studied as an example in order to investigate the speckled data which are used for the determination of radii and of the geometrical structure of SN1987A (PAPALIOLIOS et al.[9]). We will show what additional information is needed in this specific case to allow for a detailed analysis, to prevent the investigation of the limb darkening from being a "marginally useful, somewhat academic investigation. ... in the absence of application".

I. The Model Construction

Stationarity and radiative equilibrium are generally assumed because the radiative timescales are much shorter than the hydrodynamic ones shortly after the expansion of the envelope begins. Spherical symmetry is adopted. The expansion of the envelopes are homologous soon after the initial acceleration. Therefore, the velocity is assumed to be a linear function of r. Power law density profiles as a function of the distance r (i.e. $\rho \propto r^{-n}$) are taken for the general investigation (paragraph II) according to WEAVER and WOOSLEY [10]. Homologous expansion of an initial stellar structure and power laws are assumed for the analysis of SN1987A for the first few months and at the later stages, respectively. Note, that the density, velocity and temperature profiles are taken from the hydrodynamic model in paragraph IV because the above assumptions are not valid for the very early stages.

Detailed atomic models are used for the three most abundant ionization stages of a number of elements (H, He, C, N, O, Ne, Mg, K, Ca, Ba). The statistical equations are consistently solved with radiation transport for the bound-bound and bound-free transitions. The radiation field is calculated by a comoving frame method very similar to that of MIHALAS et al.[11,12], but it includes line blending of the NLTE elements. Here, we presume complete redistribution for the individual line source functions but handle the blending effect between different lines as a partial redistribution of the total source function. The redistribution function is iteratively determined. Line blanketing of 5000 to 50 000 additional lines of heavy elements are taken into account by using either a two level or the pure scattering approximation for weak lines. The atomic data of the line transitions of these elements are taken from the compilation of KURUCZ [13]. See HÖFLICH [4-7]for more details of the atmospheric models.

II. Study of the Synthetic Spectra

The relation between the free model parameters and the synthetic spectra are discussed in some examples. Our models are characterized by the following free parameters (Table 1): a) the photospheric radius R_{ph}, b) the effective temperature T_{eff}, c) the exponent n of the density slope ($\rho \propto r^{-n}$), the expansion velocity v_{ex} and d) a statistical velocity field which is assumed to be 10% of v_{ex}.

Two types of models can be distinguished (Table 1). They are characterized by the recombination front R_{HII} which is located in or above the line forming region for *ionization bounded models* (IBM) and *density bounded models* (DBM), respectively. The location mainly depends on T_{eff}, and to a smaller amount on v_{ex} and on R_{ph} because their influence of the photon escape probability in lines and the scattering domination in the continua. Although hydrogen recombines inside the distance grid of the models with $T_{eff} = 7000K$, they should not be regarded as IBMs because R_{HII} is located above the line forming region.

Table 1. The distance R_{ph} at which the optical depth is 1 for true absorption at 5000 Å, the effective temperature T_{eff}, the particle density N_o, the value of n of the density profile ($N(r) \propto r^{-n}$), the expansion velocity v_{ex} and the Thomson optical depth τ_{sc} are given at R_{ph} in cgs. In addition, we show the the relative extention E of the continuum forming region (i.e. $r(\tau_{sc} = 1.)/R_{ph}$) and the distance R_{HII}, up to which hydrogen is mainly ionized. Solar composition is assumed.

No.	R_{ph}	T_{eff}	N_o	n	v_{ex}	E	τ_{sc}	R_{HII}
I	5.E14	5500.	4.6E11	3	1000.	1.10	25.	5.5E14
II	3.E15	5500.	1.3E11	3	1000.	1.13	64.	3.3E15
III	5.E14	5500.	5.0E11	5	1000.	1.07	21.	5.3E15
IV	1.E15	5500.	3.7E11	5	1000.	1.08	31.	1.08E15
V	3.E15	5500.	1.8E11	5	1000.	1.12	57.	3.3E15
VI	5.E14	5500.	7.0E11	8	1000.	1.06	19.	5.3E14
VII	3.E15	5500.	2.3E11	8	1000.	1.07	34.	3.2E15
VIII	5.E14	7000.	3.7E11	5	3500.	2.1	34.	1.1E15
IX	1.E15	7000.	2.5E11	5	3500.	2.3	46.	2.9E15
X	1.E15	7000.	2.5E11	5	7000.	2.3	45.	2.7E15
XI	3.E15	7000.	1.3E11	5	3500.	2.6	80.	1.1E16
XII	5.E14	10000.	5.3E11	5	3500.	2.3	50.	(2.7E15)
XIII	1.E15	10000.	3.5E11	5	3500.	2.5	69.	...
XIV	3.E15	10000.	1.6E11	5	3500.	2.9	99.	...
XV	1.E15	10000.	5.4E11	8	3500.	1.7	58.	...

The following qualitative consequences for the processes can be drawn which govern the spectra: a) scattering domination increases with flattening of the density (i.e. -n), with R_{ph} and T_{eff}. b) The extention effects as represented by E (Table 1) become substantial for the continuum forming region for all DBMs, but, in general, they are much smaller for IBMs and can be neglected for steep density slopes (n $\leq$-5...-6).

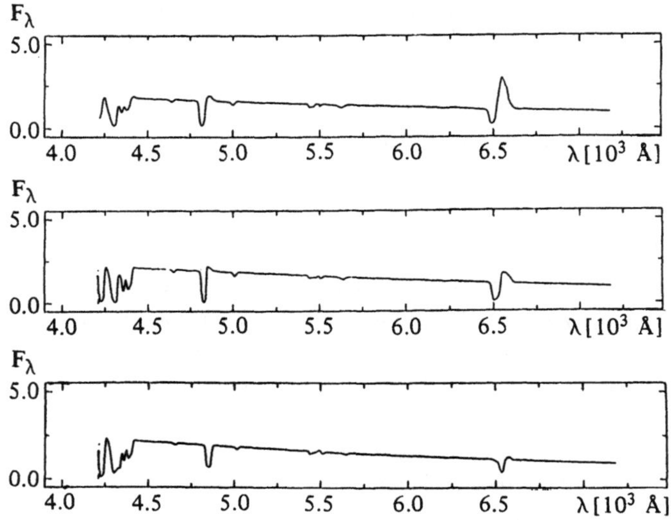

Figure 1. Synthetic optical spectra as calculated during the recombination phase (model III-V, Table 1).

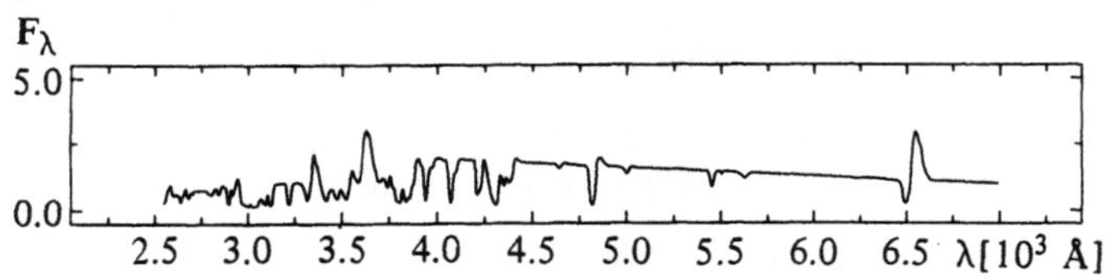

Figure 2. Synthetic spectrum of the UV to optical wavelength range as calculated by model V (Table 1).

The DBMs are already studied (HÖFLICH [14]). They show the following tendencies: The H line profiles are sensitive to v_{ex}, T_{eff} and the exponent n of the density profile. Generally, the line shift of the absorption feature decreases inside the Balmer series of H. The absorption minima of H_β and of weak lines of heavier elements occur at R $(\tau_{sc}) \approx 1$. Therefore, v_{ex} can be determined first in an analysis, then T_{eff} is derived by the energy slope of the continuum (but $T_{color} \neq T_{eff}$, see below) and n is fixed by the emission of H_α. However, the optical spectra are less sensitive to R_{ph}. The problem of a radius determination can be overcome for DBMs, if the UV flux including the Balmer jump (BJ) is measured. In particular, the BJ changes from absorption to emission with increasing radius (HÖFLICH et al.[15]).

The transition from the density bounded regime to the ionization bounded case is clearly marked by the occurance of strong optical lines due to heavy elements.

The hydrogen lines are very sensitive to T_{eff}, n and R_{ph} for IBMs. In particular, the emission components increase with T_{eff} and -n because of the larger excitation and higher cross section, respectively. In contrast to DBMs the emission of H_α is sensitive to R_{ph} (Fig.1). However, the v_{ex} by the absorption minima of the H Balmer series differs significantly from $v_{ex}(R_{ph})$ and they sensitivly depend on R_{ph}. E.g. $v_{ex}(R_{ph})$ equals $1400 km/sec$ in the models IV and V but the absorption minima of H_α & H_β correspond to 3200 & 2600 and 2600 & 2000 km/sec, respectively. In addition, the H lines of the higher Balmer series are strongly blanketed by transitions of heavier elements, causing serve problems for the determination of v_{ex}. Therefore, the expansion of the photosphere can hardly be determined by the absorption minima to a good accuracy. The UV flux increases with R_{ph}, T_{eff} and n whereas the IR flux of the continua and the lines increase with lower n. Line blanketing becomes most prominent in the UV (Fig.2). The opacity is mostly given by a large number of weak scattering lines which exceed the Thomson scattering opacity by orders of magnitude and which form a more or less smooth quasi continuum. Conclusively, observations during the recombination phase of hydrogen provides more information than the earlier phase and, in principle, allow for a more accurate analysis. Furthermore, the relation between v_{ex} and R_{ph} at a specific time can be used even if the time of the initial event is not well known, because of the increasing time base. However, the coupled influence of the different free model parameters on the synthetic spectra causes complications for the analysis. A detailed discussion of the subject of this paragraph will be given in a forthcoming paper.

Most observations of SN II are measurements of the brightness in the UBV-filter system, and the black body approximation is used for the interpretation. However, the color temperature bases on the color index (B-V) differs significantly from T_{eff} (Table 2). This is even true if the U-B and B-V color excess follows the relation as given by a black body. The physical reason is due to extention effects and the domination of the (grey) Thomson scattering which results in a dilution of the emitted fluxes but hardly redistribute the photons. The black body approximation results in a systematicly overestimation of the V flux by about 35 to 45 % during the DBM phase, with a small dependence on the model parameters. Note that the V flux may be underestimated during the recombination phase because the line blocking in the B filter is much more pronounced than in the visual wavelength range. This should not be regarded as a systematic effect because its dependency on the density structure, the metalicity, the velocity field etc. .

III. Type II SN as Distance Indicators

SN1987A has been taken as a examination of accuracy of the calculated absolute fluxes. We have used those models which allow for the best representation of the observed spectra of SN1987A. A distance of 48 ± 4 Kpc has been derived on the basis of the monochromatic optical fluxes (Table 3, HÖFLICH, [4,5]). This value is in good agreement with those infered from other distance indicators for the LMC (RR-Lyrae:

Table 2. Comparison of the calculated fluxes (see Table 1) δF in the V-filter band (at 5500 Å) with those as infered by a black body for the same color index B-V.

No.	R_{ph}	T_{eff}	n	B-V	T_{BB}	δF
IV	1.E15	5500.	5	0.58	5900	1.26
IX	1.E15	7000.	5	0.50	6500	0.85
XII	5.E14	10000.	5	0.09	10750	0.65
XIII	1.E15	10000.	5	0.08	11000	0.6
XIV	3.E15	10000.	5	0.06	11250	0.55
XV	1.E15	10000.	8	0.12	10250	0.6

48.7 ±1.1 kpc, 48.3±1.3 kpc, WALKER and MACK [16]; δ Cep.: 49.4 ±3.3 kpc; WALKER [17]; 54.9 kpc, MARTIN et al., [18]; 51.8±1.2 kpc, WELCH et al.[19]) but it clearly tends to the lower values. Our distance of SN1987A compares well with those deduced from detailed atmospheric models by other groups which used the UV flux (43 ± 4 kpc, CHILUKURI and WAGONER [20]) and which, recently, used the optical wavelength range (46 kpc,SCHMUTZ et al.[21]; 49 ± 5 kpc EASTMAN, this volume).

Table 3: The distance in Kpc of SN1987A as derived by comparison between the calculated (HÖFLICH [5]) and observed monochromatic absolute fluxes (mean value ±30 Å). A reddening of $E_{B-V} = 0.15^m$ is assumed.

No.	Date	5000	5300	5500	5700	6000	6300	at H_α
Mar. 23	53.	44	49	46	52	52	46.5	
May 14	48.	50.	48.	49.	51.	50.	45.	
June 02	48.	52.	46.	46.5	45.	50.	52.	

In the last few years several distances of SN II have been determined using the Baade-Wesselink method, black body fits and taking the minima of the absorption features as v_{ex} at the photosphere (e.g. SN 1969L: $12 \pm 4kpc$ and SN 1970G: $7 \pm 2kpc$ KIRSHNER and KWAN [1]; SN1979C: $23 \pm 3kpc$, BRANCH et al. [22]). A Hubble constant H_o of $57 \pm 4kms^{-1}Mpc^{-1}$ is implied (BRANCH[23]). Our calculations have shown that the line shift of the absorption features of weak line transitions (see above) are good first order indicators for $v_{ex}(\tau_{sc} = 1)$ before the recombination phase and, therefore, allow for the determination of the velocity at a certain distance in the envelope if the exact time of the initial explosion is well known. However, the absolute visual flux is overestimated by about 15 to 45 % , causing a systematic error of the distance in the order of 10 - 20 % and indicating a correction of H_o (65 ± 10 $km\ s^{-1}Mpc^{-1}$). The problem with the inaccuracy due to the initial event can be significantly reduced if late stages are considered. However, the difficulties of the parameter deduction for IBMs should be noted. Therefore, detailed analyses of different stages are really needed to minimize the error for the use of SN II as distance indicators. Besides the uncertainties due to the spectral analyses in respect to H_o, note that up to now only very few spectra of SN II have been measured which are in the Hubble flow (see WAGONER, this volume).

IV. Analysis of the Early Light Curve of SN1987A

In order to investigate the fluxes as emitted from the expanding envelope of SN1987A during the first day, we have used the hydrodynamic model 14E1 of NOMOTO et al.[24](Table 4). The atmospheric models correspond to the times when the shock breaks through the photosphere (A), when the first photons where detected in the V filter band (B) and when the first UBV magnitudes where available (D).

Table 4. The absolute visual brightness M_V, the color indices U-B and B-V in Johnson's filter system and the bolometric luminosity L_{bol} are given for model 14E1 of NOMOTO et al.[24](He core mass: 6 $M_\odot$, envelope mass: 10 $M_\odot$; E_{kin} : $10^{51} erg$) for different times as calculated by the atmospheric models. The correction factors C_V and C_{bol} give the ratio between the visual and bolometric fluxes, respectively, as calculated by atmospheric models in comparison with the extended grey atmosphere (EGA). In addition, the electron temperature T, the distance R and the particle density N_o are given at the distance where the Thomson optical depth equals 2/3. All quantities are given in CGS except for the brightness.

Model	time [s]	$T[K]$	$R[cm]$	$N_o[cm^{-3}]$	M_V	U-B	B-V	L_{bol}	C_V	C_{bol}
A	6916.	820000.	3.34E12	1.2E15	-11.26	-1.12	-0.28	8.9E44	0.60	0.25
B	10798.	69000.	1.60E13	6.3E11	-10.51	-1.08	-0.27	2.0E42	0.18	0.12
C	43254.	20800.	7.80E13	1.1E11	-12.73	-0.99	-0.23	2.8E41	0.23	0.26
D	129587.	10300.	1.85E14	4.4E10	-13.31	-1.02	-0.16	1.6E41	0.38	0.43

The LC of SN1987A is well understood by hydrodynamic models which use a flux limited diffusion approximation or a grey LTE solution for the radiation transport equation. However, there are several major differences between the detailed NLTE calculations and the diffusion approximations which have strong implications on the interpretation of the early observed data and the use of the measurements as tests for the hydrodynamical calculations: i) L_{bol} is smaller by a factor C_{bol} of between 0.12 and 0.43. ii) L_V is reduced by a factor C_V of between 0.18 and 0.6. iii) L_V shows a strong local minimum, i.e. in the very first stages not only the bolometric luminosity decreases but also the visual one. iv) The correction factors C_V and C_{bol} show a strong time dependence which can be well understood as a consequence of the change of the densities and temperatures at the photosphere. The reduction of the luminosities in respect to the EGA solution is a consequence of line blanketing effect and of the Thomson scattering domination at the photospheres.

Although the general observed trend is reproduced by the calculations (Fig.3), two major differences between the observed and calculated visual brightness m_V are obvious: i) The calculated m_V is just at its local minimum (i.e. 1.7^m fainter, model B) than the observation of McNaught. ii) The calculated luminosity is somewhat too low. iii) At day 1.5 (model D) the reddening corrected observed color index B-V (-0.235^m) is significantly lower than the calculated value of -0.16^m, indicating a too low local temperature in the hydrodynamic models. This interpretation is also supported by the absence of the optical HeI triplet line at Feb.24, 1987,in the synthetic spectra.

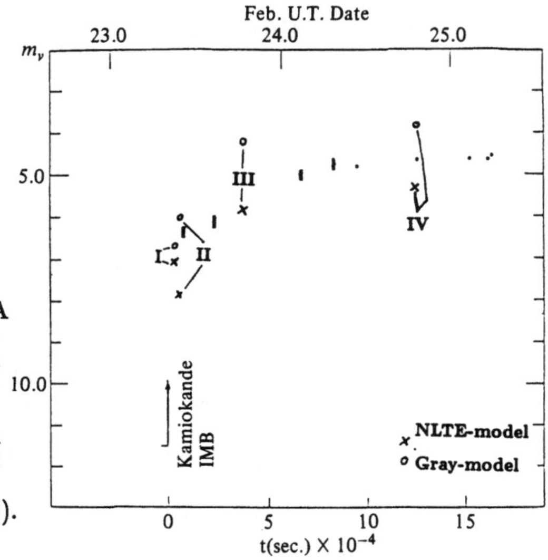

Figure 3. Observed m_V of SN1987A in comparison with those as predicted by Nomoto's hydrodynamic model 14E1 (NOMOTO et al.[24]) by using atmospheric models (x) and the grey LTE approximation (o). The distance is taken to be 48 kpc. A reddening correction E_{B-V} of $.15^m$ is applied (WAMPLER [25]).

All this strongly indicate a larger kinetic energy than $10^{51} erg$ as used for this hydrodynamic model. A quantitative estimate can be derived by the difference of the visual luminosities of model B showing that the minimum of the calculated L_V occur too late. The time of the increase of the very early light curve mainly depends on the shock travelling time through the initial stellar structure and it is approximately proportional to the square root of E_{kin} (SHIGEYAMA and NOMOTO [26]). Therefore, E_{kin} can be estimated to be 20 to 40 % larger than 10^{51} erg.

Note, that detailed atmospheric diagnostics of hydrodynamic models are needed to allow for a more accurate determination of the total kinetic energy. Furthermore, such analyses should be done for different initial stellar models. E.g. the general slopes and effects turned out to be qualitatively the same for the Arnett's stellar model but quantitative differences occur. These problems and the interpretation of the initial UV light flash are under investigation.

V. Limb Variation Effects

We want to discuss the effects of the limb variation at the example of model V (Table 1) which has typical parameter for type II SN during the recombination stage. This phase occurs some time after the initial expansion. Therefore, the envelope most likely can be resolved by speckle observations.

The normalized intensity distribution functions (IDF) differ significantly from those of a disk of constant intensity (i.e. a step function) at all frequencies (Fig.4; HÕFLICH [27]). At 6700 Å the IDF shows a relatively sharp outer boundary, because the R_{HII} is just 12 % (Table 1) above the Thomson scattering dominated photosphere. Note, that a sharp outer boundary does not occur at times earlier than the recombination phase of hydrogen. The opacity in the UV is mainly dominated by a large number of weak lines which form a quasi continuum. This results in a similar global behaviour of the IDFs as in the case of the Thomson scattering but at larger distances (HÖFLICH[27]). However, this global structure of the IDF profile cannot be generalized. In particular, even ring structures in the IDFs occur if the opacities are strongly frequency dependent in the comoving frame.

Several speckle measurements are performed in the H_α-line of SN1987A. Therefore, we want to discuss the wavelength dependence over this line in some detail (Fig. 4). Near the rest wavelength, secondary maxima of the IDFs appear which strongly depend on the wavelength of both the size and radial location. The maxima occur if a specific projected Doppler shift in the spectrum corresponds to a specific radial distance. Note that even a difference of 15 Å is sufficient to shift the position of the secondary maximum by about 20 % in distance, and a shift of 140 Å would result a radius change by a factor of 80 % . This implies that the transmission function of the filter have to be known quite well to use the correct IDF.

However, the frequency variation of the IDFs have some severe consequences not only for the determination of the radii but also for the reconstruction of images on the basis of two dimensional speckle observations. Even a small deviation from sphericity of the velocity, the density or chemical structure may imply very different radii for two orthogonal directions of a slit which is layed over the disk, because two different IDFs are seen.

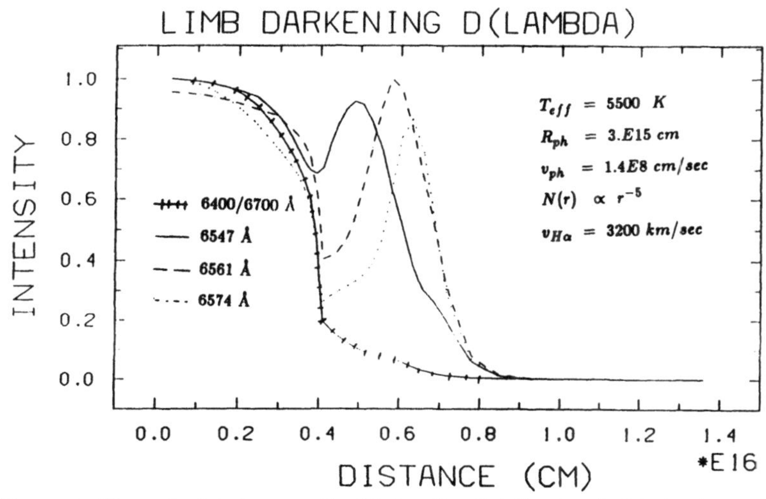

Figure 4. The calculated normalized intensity distribution (IDF) in the region of the H_α line as calculated by model V (Table 1).

The observed large frequency and time dependent axis ratio for SN1987A as proposed by PAPALIOLIOS et al.[9]may be explained as a consequence of the frequency dependent IDFs. See HÖFLICH [27] for more details.

References

1. R. Kirshner, J. Kwan: Astrophys. J. 193, 27 (1974)

2. S.E.Woosley: Astrophys.J. 330, 218 (1988)

3. D.M.Rank, J.Gregman, F.C.Witteborn, M.Cohen, D.K.Lynch, R.W.Russell: Astrophys.J. 325, L1 (1988)

4. P.Höflich: Lect.Notes in Phys. 287, ed. W. Hillebrandt, Springer, p.307, (1987)

5. P.Höflich: Proc.Astron.Soc.Austral. 7, 434 (1988)

6. P.Höflich: IAU Symp.108, ed.K.Nomoto, Springer, p.388 (1988)

7. P.Höflich: Proc.MPA P1, 111 (1989)

8. W.Hillebrandt, P.Höflich: Rep.on Prog. in Phys. (1989) in press

9. C.Papaliolios, M.Karovska, L.Koechlin, P.Nisenson, C.Standley, S.Heathcote: Nature 338, 565 (1989)

10. T.A.Weaver, S.E.Woosley: Astrophys.J. 289, 198 (1984)

11. D.Mihalas, R.B.Kunasz, D.G.Hummer: Astrophys.J. 202, 4 (1975)

12. D.Mihalas, R.B.Kunasz, D.G.Hummer: Astrophys.J. 206, 5 (1976)

13. R.L.Kurucz: Proceedings of the IAU conference, Baltimore (1989) in press

14. P.Höflich: *Particle Astrophysics Workshop*, ed. C. Pennypacker, (1989) in press

15. P.Höflich, R.Wehrse, G.Shaviv: Astron. Astrophys. 163, 105 (1986)

16. A.R.Walker, P.Mack: Astron.J. 96, 872 (1988)

17. A.R.Walker: M.N.R.A.S. 225, 627 (1987)

18. W.L.Martin, R.P.Warren, M.W.Feast: M.N.R.A.S 188, 139 (1979)

19. D.H.Welch, R.A.McLaren, B.F.Madore, C.W. McAlary: Astrophys.J., 321, 162 (1984)

20. M.Chilukuri,R.V.Wagoner: IAU Symp.108, ed.K.Nomoto, Springer, p.295 (1988)

21. W.Schmutz, C.Russell, D.C.Abbot, U.Wessolowsky: preprint (1989)

22. D.Branch, S.Falk, M.McCall, P.Rybski, A.K.Uomoto, B.J.Will: Astrophys.J. 244, 780 (1981)

23. D.Branch: Proc. of the Symp.on Extragalactic Distance Scales, ed. van den Bergh, (1988) in press

24. K.Nomoto et al.: ed. Hayakawa& Sato, Universal Academy Press (1988) in press

25. E.J.Wampler: SN1987A: One Year later, ed. M.Greco, Ed. Frontiers, p.17 (1988)

26. T.Shigeyama; K.Nomoto; M.Hashimoto: Astron.Astrophys. 196 141 (1988)

27. P.Höflich: Astron. Astrophys. (1989) submitted

Synthetic Spectrum Calculations of the Type II Supernova SN 87A

Ronald G. Eastman

I. Introduction

Hydrodynamic calculations of Type II supernova (SNII) explosions have been highly successful at reproducting the major observational properties of these events, particularly in the case of SN87A (e.g. [1], [2], [3]) Basic properties which can be determined from the data, such as the gas velocity and the bolometric luminosity are reasonably well fit by the hydrodynamic models. What makes these calculations feasible however is a compromise treatment of details concerning the radiation field. Most hydro codes treat the radiation transport by a simple flux limited radiative diffusion scheme and use a Lagrangian mass grid which is too course to yield up more than crude information on details of the emergent spectrum. Ultraviolet, optical and infrared spectrophotometry of supernovae are pregnant with information about their source, but in order to decipher most of this information it is first necessary to perform more refined calculations of the gas excitation/ionization and the radiation field structure in the photospheric and overlying layers. These calculations, which are becoming increasingly more elaborate and sophisticated, are themselves compromises in that most often the explicit time dependent behavior of the atmosphere and radiation field are ignored. The properties of the gas and radiation field are treated as being in a *quasistatic steady state.* Such an approximation may introduce errors in the transfer solution of 10 − 30%, which may or may not be acceptable, depending on the intended application.

The intrinsic properties which can be learned about supernovae through detailed modeling of their atmosphere and spectra are myriad, and include the mass, chemistry and energetics of the ejecta. One very important application of model supernova atmospheres concerns the use of SNIIs as distance indicators to external galaxies. The spectra of SNII's contain enough information, subject to simple interpretation, to deduce their intrinsic luminosities and thus their distances. Supernovae are the brightest stars in the Universe, and are observable well into the Hubble flow at distances greater than 30 Mpc, making them the ideal candidate for use in determining the scale and age of the Universe.

In this contribution I would like to present some results of model atmosphere calculations for SN87A during the first few days of its existence, and show how model atmospheres of Type II supernovae may be combined with SNII observations to derive highly accurate distance estimates.

II. Model Atmospheres of SN87A

Robert P. Kirshner and I have modeled the atmosphere of SN87A for times $t \leq 10$ days. The density structure through the atmosphere was approximated as $\rho(v,t) = \rho_0(v/v_0)^{-\gamma}(t/t_0)^{-3}$, which is an excellent description of the outer, hydrogen rich layers. Homologous expansion was assumed, so that $dv/dr = v/r = 1/t$, where t is the time since the onset of expansion.

The code we have developed performs a simultaneous solution of the transfer equation and equations of gas excitation/ionization. The transfer equation solved is given by

$$\mu\frac{\partial I_\nu}{\partial r} + \frac{1-\mu^2}{r}\frac{\partial I_\nu}{\partial \mu} + \frac{1}{ct}\left(3 - \frac{\partial}{\partial \ln \nu}\right) I_\nu = -\chi_\nu I_\nu + \sigma_\nu J_\nu + \eta_\nu, \tag{1}$$

where I_ν is the specific intensity at frequency ν, σ_ν is the monochromatic (and isotropic) scattering opacity, $\chi_\nu = \sigma_\nu + \kappa_\nu$ is the total monochromatic opacity, including both scattering and true absorption, $J_\nu \equiv 1/2\int_{-1}^{1} I_\nu d_\mu$ is the angle averaged mean intensity, and η_ν is the emission term. All quantities in equation (1) are measured in the local gas rest frame. If equation (1) was explicitly time dependent, there would be a term $(1/c)(DI_\nu/Dt)$, which is the Lagrangian time derivative, on the right hand side. In these calculations all explicit time dependence has been suppressed, and so this term was set equal to zero.

The atmosphere calculations were performed in two steps. In the first step, the gas equation of state for all elements was assumed to be given by the Saha-Boltzmann equation. Thermal balance between gas and radiation field was assumed. The temperature distribution through the atmosphere was computed by linearizing the thermal balance constraint and radiative transfer equations at each frequency. Once convergence was obtained, the temperature distribution obtained in the first step was then used in the next step where the detailed equations of excitation/ionization were solved for hydrogen and helium. Besides H and He, a complete contingent of heavy elements was included at an abundance relative to solar of 1/3. The effects of line blanketing in the UV were substantial in SN87A, and were accounted for in the calculations by the inclusion of 63,000 lines of heavy elements. The excitation and ionization of all elements heavier than helium was calculated with the Saha-Boltzmann equation.

The lower boundary condition to the transfer equation was given by specifying the radiative diffusion flux. For models of SN87A at a particular time, the value chosen was such that the bolometric luminosity at the photosphere would equal the observed value for SN87A. For more details, see Eastman and Kirshner [4].

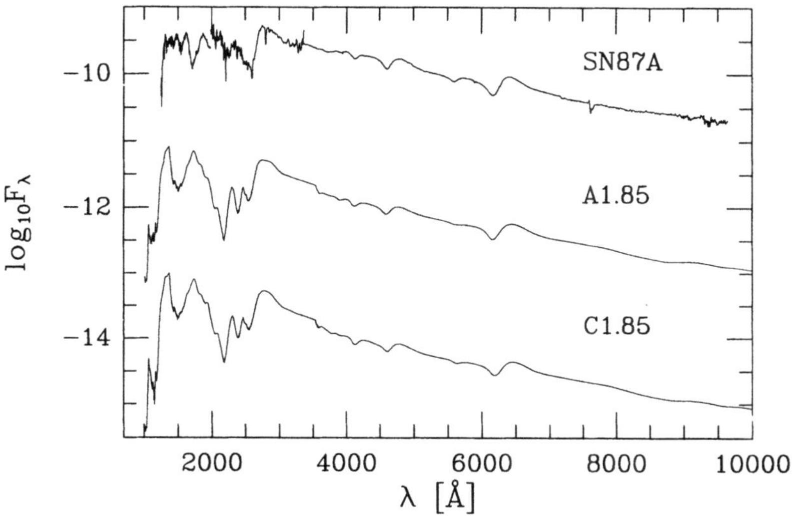

Figure 1 - Comparison of optical and ultraviolet observations of SN87A at t = 1.85 days with synthetic spectra of two model atmospheres.

Figure 1 compares combined CTIO optical (BLANCO *et al.* [5]) and IUE ultraviolet (KIRSHNER *et al.* [6]; WAMSTEKER *et al.* [7]) spectra of SN87A at $t = 1.85$ days with synthetic spectra computed for two models. The two models shown, A1.85 and C1.85, differ only in that the former has $\gamma = 9$ while the latter has $\gamma = 7$. The overall agreement between the two models and the data is not bad, with the exception of several strong features in the UV. Almost all of the strong features in the ultraviolet come from transitions in heavy elements (Mg, Si, Al, Ti, Fe, Co) which we have calculated using the Saha-Boltzmann equation, and I believe that this explains much of the discrepancy. Such an approximation is more suited for some elements than for others. Lucy [8] has obtained much better agreement between calculated and observed UV spectra by making a the nebular approximation and explicitly calculating the ionization balance for each species. Figures 2 and 3 compare synthetic spectra of models with observations of SN87A at 7.7 days and 10 days, respectively. In general, the spectra of these models agree well with the data, especially the stronger Balmer

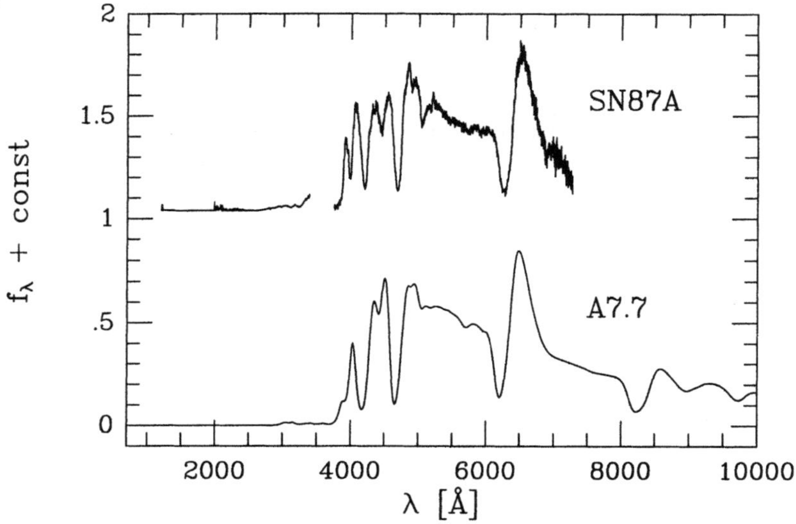

Figure 2 - Comparison of model atmosphere spectrum with optical and ultraviolet observations of SN87A at $t = 7.7$ days.

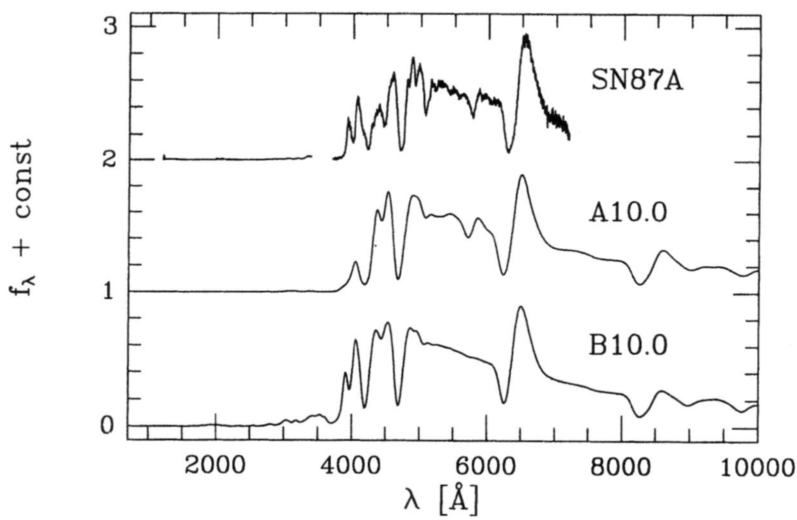

Figure 3 - Comparison of optical and ultraviolet observations of SN87A at $t =$ 10 days with synthetic spectra of two model atmospheres.

lines and the overall continuum distribution. Again, the strength of some features calculated with the Saha-Boltzmann equation are not quite right. One troublesome disagreement between models A7.7 and A10.0 with the data is that there is not enough flux at around $\lambda 4000$ Å. These models have about the same luminosity as SN87A, but it turns out that their effective temperatures are slighly lower because their photometric radii are larger. Consequently, in model B10.0 the density has everywhere been decreased by a factor of 4. This has the effect of moving the photosphere inward, and for constant luminosity gives a higher effective temperature. Both the hydrogen Balmer lines and continuum of model B10.0 agree better with the observed spectrum than does model A10.0. However, the Na I D line at $\lambda 5892$, which is well matched in model A10.0, is too weak in model B10.0. This is an indication that the Saha-Boltmann equation used to calculate the Na ionization overestimates the ionization of species above the photosphere.

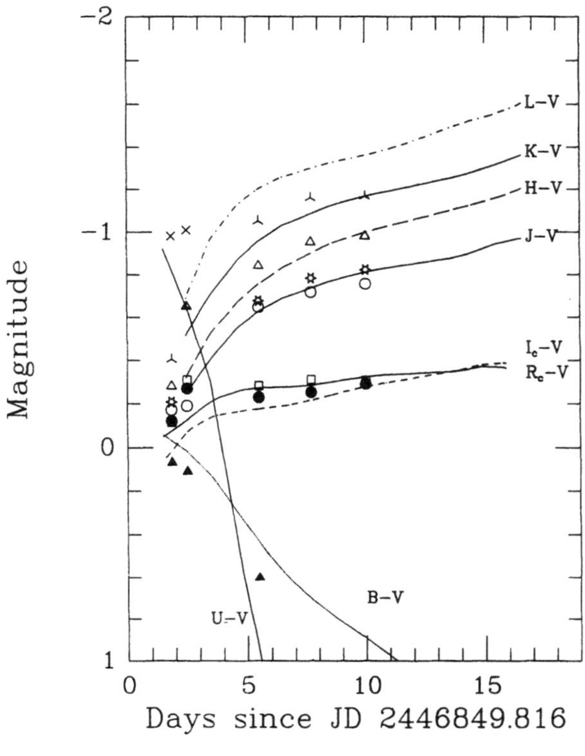

Figure 4 - Comparison of observed and computed photometric colors, $M_\lambda - V$, versus time. Labeled curves are observed colors dereddened by $E(B-V) = 0.18$. Symbols correspond to colors computed from synthetic spectra of model atmosphere. They are $L-V$ (⅄), $K-V$ ($\triangle$), $H-V$ ($\star$), $J-V$ ($\circ$), $R_c - V$ ($\square$), $I_c - V$ ($\bullet$), $B-V$ ($\blacktriangle$), and $U-V$ ($\times$).

Figure 4 shows a comparison between optical and infrared broad band photometry of SN87A, with photometry computed for a sequence of model atmospheres. The data are displayed as colors, and have been dereddened by an amount $E(B-V) = 0.2$. The agreement is adequate, with the largest deviations of ~ 0.3 magnitudes occuring between the computed and observed $U - V$. Although the computed infrared to V flux ratios do appear to be systematically shifted downward relative to the observations, the infrared-infrared colors are in excellent agreement with the observations. This is explained by the fact that, because the free-free opacity in the infrared is high, the infrared continuum most nearly matches that of a blackbody at the effective photospheric temperature.

III. The Distance to SN87A

An important application of supernovae observations and models is in the use of supernovae as distance indicators via the Expanding Photosphere Method (KIRSHNER and KWAN [9]; KIRSHNER [10]; BRANCH [11]; WAGONER [12]; EASTMAN and KIRSHNER [4]). The basic idea is that if the size of the photosphere, R_{ph}, is known, and the flux emergent from the photosphere, $\mathcal{F}_\nu$, is known, then observations of the observed flux f_ν^{obs} can be used to solve for the distance. The efficacy of this method derives from the fact that one can determine the photospheric velocity and the age of the supernovae by observing the change in apparent angular diameter with time, and from the fact that the emergent flux is observed to be very nearly Planckian.

The flux emergent from the photosphere may be written as

$$\mathcal{F}_\nu = \xi_\nu \pi B(T_{c,\nu}) \tag{2}$$

where B is the Planck function and $T_{c,\nu}$ is the continuum color temperature at frequency ν, which can be determined from spectrophotometric observations. The correction factor, ξ_ν, accounts for the fact that the emergent flux, although Planckian in its distribution, is perhaps dilute. In the analysis which follows, we will assume that $T_{c,\nu}$ is constant with frequency, and just write T_c for the color temperature and ξ for the correction factor.

Broad band photometry is often easier to obtain that spectrophotometry. Good results can be obtained using BVR_cI_c photometry, although we have also applied the method using just V and I_c. In terms of the observed, dereddened photometric magnitude, m_λ, the approximation becomes

$$m_\lambda \approx -2.5 \log \Pi - 2.5 \log \pi B_\lambda(T_c) + C_\lambda, \tag{3}$$

where $\Pi \equiv \xi\theta^2$ and $\theta \equiv R_{ph}/D$ is the angular size of the photosphere, with D being the distance from observer to supernova. We have computed an evolutionary

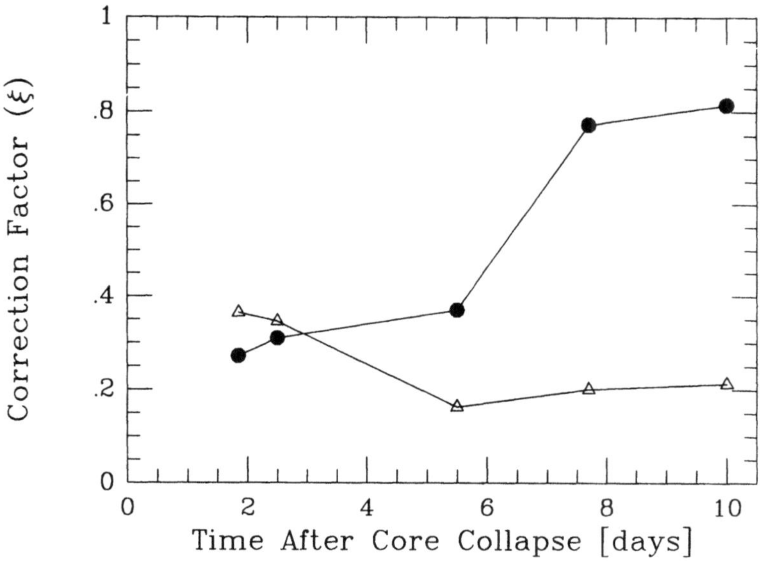

Figure 5 - Time variation of correction factors derived by fitting photometry from model atmospheres to equation (3). The filled circles correspond to correction factors derived from the BVR_cI_c magnitudes, while the triangles correspond to correction factors defined for the $(V - I_c)$ color temperature (see text).

sequence of models based on a single power law density distribution and possessing surface bolometric luminosity equal to that of SN87A. For each of these model atmospheres, we then derived a value of ξ from the computed spectrum. Figure 5 shows how the value of these correction factors change as the atmosphere expands, and both the density and the effective temperature decrease. The two curves in this figure correspond to two different sets of photometric magnitudes employed in the fit. In one case, BVR_cI_c colors were used to determine ξ (circles). At $t \lesssim 2 - 3$ days when the effective temperature is $\sim 10^4$ K, the emergent spectrum is highly diluted. As the photosphere cools the correction for dilution becomes less important, and the BVR_cI_c temperature fit approaches the true effective temperature of the supernova. If only the V and I_c magnitudes are used, a somewhat different behavior is found. This is shown in fig. 5 by the triangles. The explanation for this is that both the B and R bands are heavily absorbed at $t \gtrsim 5$ days, and their inclusion in the temperature fit leads to a lower value of T_c than the value obtained from just V and I_c. The lower value obtained in the former case is closer to the true effective temperature of the supernova, and so the corresponding Planck function is less dilute.

We have fit equation 3 to photometric observations of SN87A which were dereddened by an amount $E(B_V) = 0.2$. The fit gives a value for Π and T_c. The size of the photosphere was set equal to $v_{ph}t$, where t is the time since the onset of

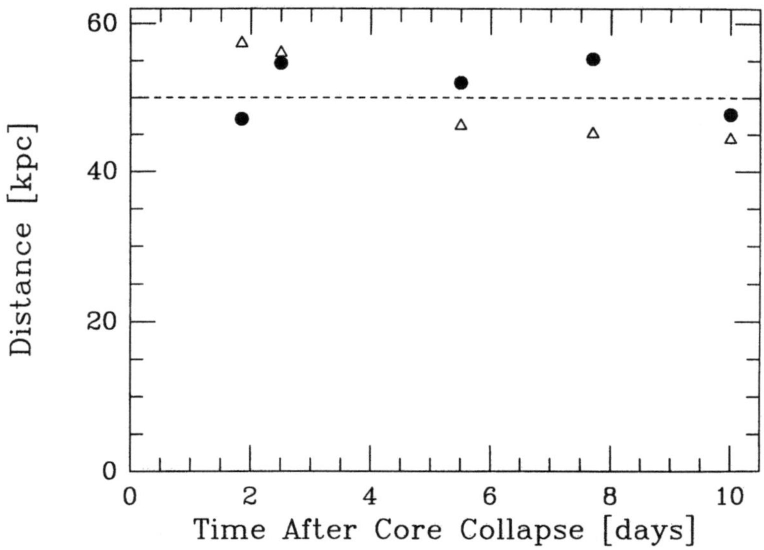

Figure 6 - Time variation of the distance derived to SN87A. The filled circles correspond to distances derived from observed BVR_cI_c magnitudes, while the triangles correspond to distances derived using only the V and I_c magnitudes. The dashed line represents the mean distance of 50 kpc obtained by the first method.

expansion and v_{ph} was determined from spectroscopic observations. The distance is given by

$$D = \sqrt{\frac{\xi}{\Pi}} v_{ph} t. \tag{4}$$

The results are shown in fig. 6, which displays the derived distance to SN87A from BVR_cI_c photometry (circles) and from just V and I_c (triangles). In the former case, the mean distance determination is 50 ± 5 kpc, while in the latter case it is 49 ± 6 kpc. These values are in good accord with the recent results of WALKER and MACK [13] who observed RR Lyrae stars in NGC 1786 and obtained an LMC distance of 49.8 kpc, and of WALKER [14], who found the distance to LMC Cepheids to be 49.4 kpc. The fact that there appears to be no strong systematic trend in the calculated distance is encouraging, especially since over the period considered, T_c changed from $\sim$ 13000 K to $\sim$ 6000 K, the age of the supernova increased by a factor of 5, the density dropped by a factor of 1000, and ξ varied by a factor of 4.

References

[1] Woosley, S. E., Pinto, P., and Ensman, L. 1988, *Ap.J.*, **324**, 466.

[2] Arnett, W. D. 1987, *Ap.J.*, **319**, 136.

[3] Nomoto, K., Shigeyama, T., and Hashimoto, M. 1987, in *SN 1987A*, ed. I.J. Danziger, (Garching bei Munchen: ESO), p325.

[4] Eastman, R. G. and Kirshner, R. P. 1989, *Ap.J.*, **347**, in press.

[5] Blanco, V. M. *et al.* 1987, *Ap.J.*, **320**, 589.

[6] Kirshner, R. P., Sonneborn, G., Crenshaw, D. M., and Nassiopoulos, G. E. 1987, *Ap.J.*, **320**, 602.

[7] Wamsteker, W., Panagia, N., Barylak, M., Cassatella, A., Clavel, J., Gilmozzi, R., Gry, C., Lloyd, C., van Santvoort, J., Talavera, A. 1987, *Astr. Ap.*, **177**, L21-24.

[8] Lucy, L. 1987, in *SN 1987A*, ed. I.J. Danziger, (Garching bei Munchen: ESO), p417.

[9] Kirshner, R. P. and Kwan, J. 1974, *Ap.J.*, **193**, 27.

[10] Kirshner, R. P. 1985, in *Supernovae as Distance Indicators*, ed. N. Bartel, (Berlin:Springer-Verlag), p171.

[11] Branch D. 1987, *Ap.J. (Letters)*, **320**, L23.

[10] Wagoner, R. V. 1988, in *Proceedings of the Berkeley Workshop on Particle Astrophysics*, ed. C. Pennypacker (in press).

[13] Walker, A. R. and Mack, P. 1988, preprint.

[14] Walker, A. R. 1987, *MNRAS*, **225**, 627.

Modelling the Late-Time Optical Spectra of Supernova 1987A

Douglas Swartz

Supernovae consist of up to tens of solar masses of material freely expanding into nearly empty space. This gas will rapidly cool and quickly fade from view. However, if even a fraction of a percent of this material happens to be radioactive ^{56}Ni, then the situation changes dramatically. The cold gas is now permeated with high energy photons. These γ-rays travel nearly unhindered through the gas but eventually they may strike one or several electrons and accelerate them to relativistic energies. The electrons, in turn, rip through the gas, stripping more electrons from atoms and heating the surrounding material. Eventually, when these hot electrons have lost most of their energy, they become indistinguishable in the sea of free electrons they have created. The cold gas, in the mean time, attempts to maintain its low energy state. Electrons recombine with ions, excited atoms radiate their excess energy in returning to the ground state, and thermal energy is equally partitioned among all the nearby particles. The heating by the high energy decay products is distributed among all the particles in the gas and eventually escapes through thermal processes. The spectrum we observe, therefore, is the signature of this thermal gas and contains a wealth of information about the physical state of the atmosphere.

If the equations governing these and other relevant atomic processes are solved for a model of such an atmosphere, then the physical state can be determined and the resulting spectrum computed. Spectra are then compared to observations. By implication, models which are best able to reproduce the observations represent most correctly the conditions within the supernova atmosphere. Observations of supernovae at the late-time stage of evolution are particularly useful in this regard. At late times the supernova ejecta is highly tenuous and even the deepest layers should be directly observable, at least in principle.

As a specific example, such a method is applied to SN 1987A in an effort to determine the physical characteristics of this supernova and as an independent means of testing conclusions drawn by others concerning this event.

Initial models based on hydrodynamic models discussed in the literature have been used here. These provide the density, velocity, and composition structure of the atmosphere. They represent the best currently accepted representations of SN 1987A

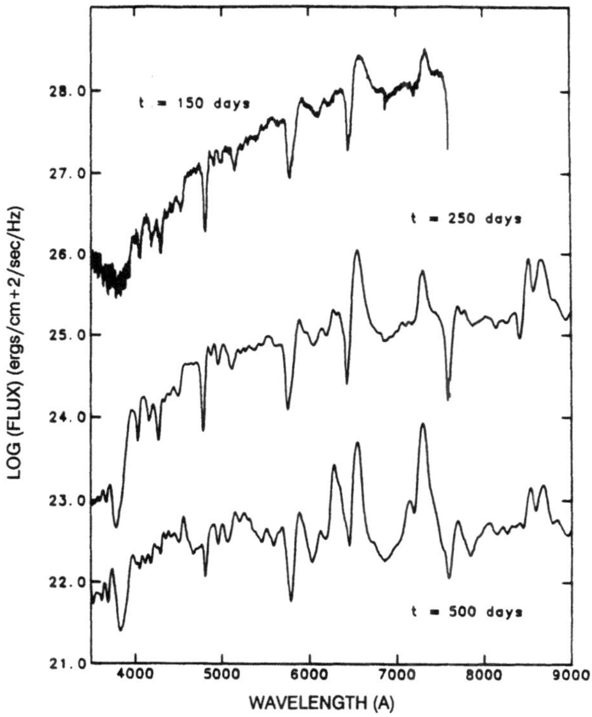

Figure 1: Observed optical spectra of SN 1987A. Spectra courtesy N. Suntzeff, CTIO.

through their ability to reproduce the known characteristics of the progenitor, the observed bolometric light curve and other properties. In the present work, these models will be shown to quantitatively reproduce the optical spectrum from 200 to 500 days. Other possible input models, such as models with different envelope masses or density distributions, have not been examined here. The effects of composition mixing and of the distribution of ^{56}Co on the emergent spectra have, however, been considered.

Figure 1 shows the observed time evolution of SN 1987A from 200 to 500 days. The spectra do not change appreciably during this time frame even though the total luminosity decreases by a factor of ≈ 20. P-Cygni type absorption features are evident indicating the presence of an underlying continuum component existing up to t>500 days. Many of the features remain at the same strength relative to the continuum throughout this phase while some strengthen ([OI] 6300-6365) and others weaken (CaII infrared triplet). All the identifiable lines are due to neutral and singly ionized species and most are associated with low lying energy levels. The electron temperature of the gas is therefore less than about 7000 to 10,000K.

Three quantities are allowed to vary in the models. These are the distribution of the elements in velocity (radial) coordinate, the distribution of radioactive ^{56}Co, and the position of a continuum emitting lower boundary. The mass distribution in velocity space is taken from the models in the literature.

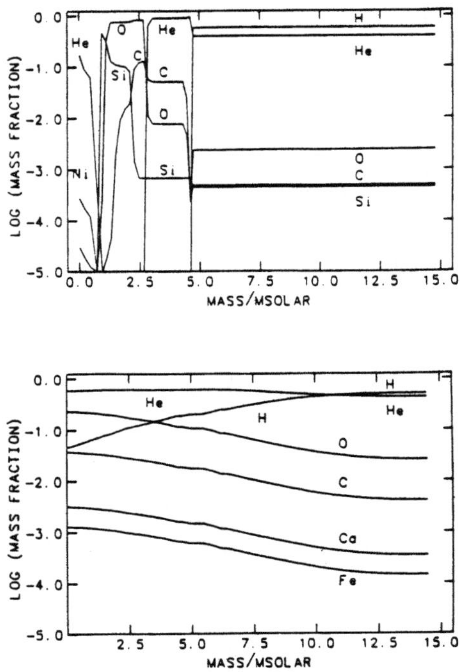

Figure 2: Composition (in mass fraction) for two models of SN 1987A.

Figure 2 illustrates the composition plotted logarithmically against the mass coordinate for a stratified (WOOSLEY [8]) and mixed (PINTO and WOOSLEY [3]) version of an ≈ 16 $M_\odot$ star containing a 6 $M_\odot$ core. The neutron star has been removed from this plot and the mass coordinate appropriately rescaled. The total mass of each element is identical in the two models.

The radioactive cobalt distribution is chosen independently of the distribution of the remaining elements as another parameter. The mass of radioactive substance is constrained to be 0.075 $M_\odot$ based on the bolometric light curve and known distance to the LMC. Radioactive cobalt (initially ^{56}Ni) is the only source of energy.

The placement of a lower, opaque, continuum-emitting boundary is also a free parameter. The use of this boundary helps to determine where the emission lines are formed by replacing line emission in the region below this boundary with a continuum flux (under the constraint of energy conservation). Implicitly, the lower boundary represents the effect one expects of many weak lines which are undoubtedly present in the optical (but which are not modelled in detail in this work) and the effect of electron scattering of radiation in broadening the lines in the observer's frame (WITTEBORN, ET AL. [7]) The shape of the continuum photon energy distribution is represented in the models as a blackbody spectrum at 5000K. The strength of the continuum is determined by the local heating rate and hence by the cobalt distribution (thus maintaining energy conservation).

Details of the models computed for comparison to observations made at 200 days, emphasizing the results for different cobalt distributions and different photosphere positions, have been discussed elsewhere (SWARTZ, HARKNESS, WHEELER [5]). Here we reiterate the basic conclusions and extend the work to later times. The best fit to the observed spectrum at 200 days occurs when the cobalt is distributed out to a Lagrangian mass point greater than 10 $M_\odot$ (corresponding to material velocities $v > 2.5 \times 10^3$ km/sec) and when the lower boundary is placed at the core/envelope interface, thus effectively obscuring the core from direct view. The cobalt is always assumed uniformly distributed below some parameterized mass point in the models. A more precise analysis using non-uniform cobalt distributions has not been undertaken.

Figure 3 illustrates the spectrum at 200 days for the mixed and unmixed models compared to the observed spectrum. There has been no scaling of the data nor the theoretical spectra, however, note the logarithmic scale. The distance to the LMC is assumed to be 50 kpc. Both models shown assume the ^{56}Co is uniformly distributed below mass point 14 $M_\odot$ and the lower boundary is placed at the core/envelope interface. These parameter values provide the best fit to the observations in both the mixed and unmixed models (SWARTZ, HARKNESS, WHEELER [5]).

The differences in the two model spectra arise from the different envelope compositions. There is a larger mass fraction of intermediate and high mass elements in the mixed model than in the stratified model. (The difference in hydrogen and helium abundance is not very significant, see Fig. 2). Metals are more efficient coolants than either hydrogen or helium and the mixed model tends to be 1500 to 2000 K (about 20 %) cooler than the unmixed model in the envelope. This shifts the bulk of the emission to lower lying energy states. This factor, combined with the increased metal abundance, is reflected in the theoretical spectrum. For example, the hydrogen H_α line is weaker in the mixed model due to the lower temperature. The [OI] 6300 Å line, in contrast, is stronger in this model because of the increased abundance and the cooler temperature which favors emission in this line. Other examples include the [CaII feature at 7300 Å which is strengthened in the cooler, higher metallicity, mixed model.

There are some features which are poorly fit in both models. Most notable is the feature at 4571 Å a semi-forbidden line of neutral magnesium. This feature is very strong in both models but is not found in emission in the observations at 200 days. The entire spectrum blueward of this feature is also poorly reproduced by the models. This is partially an artifact of the continuum shape (5000K blackbody) which was chosen principally on the basis of the observed continuum redward of 5000 Å.

The spectrum of both models are again compared to observations in Figs. 4 through 6 at 300, 400, and 500 days, respectively. The important changes occurring in the models are the decrease in density with time and the increased transparency to γ-rays. These factors act in opposition in their effect on the temperature structure. A decreased density with the same γ-ray heating rate per unit volume tends to drive the equilibrium temperature upward. The escape of γ-rays reduces the local heating rate and hence the equilibrium temperature. Prior to 500 days, neither of these factors are particularly troublesome and the spectrum evolves rather smoothly instead of shifting to the extremely low energy fine structure transitions in the far IR (AXELROD [2]). As has been described elsewhere in these proceedings, the supernova fades rapidly in the optical after 550 days and an increasing fraction of the total energy is observed at long wavelengths due in part to this effect and partially to dust.

As early as 300 days, the mixed model severely underestimates the strength in the H_α feature. The unmixed model produces a better representation of all features with the possible exception of the CaII infrared triplet at 300 days, compare particularly the FeII

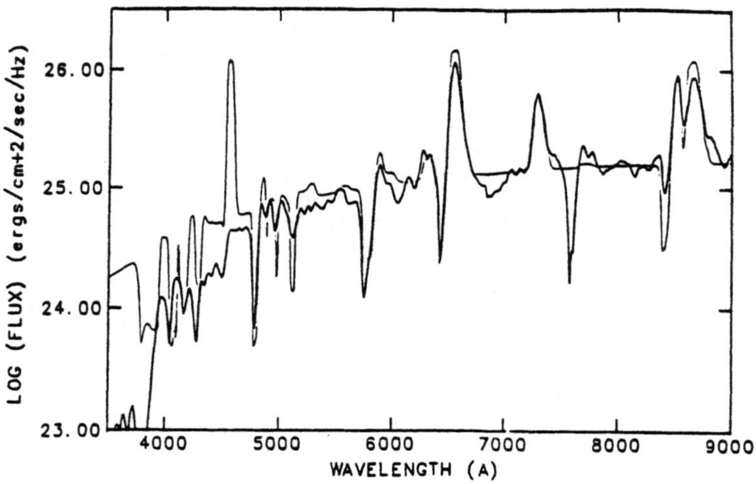

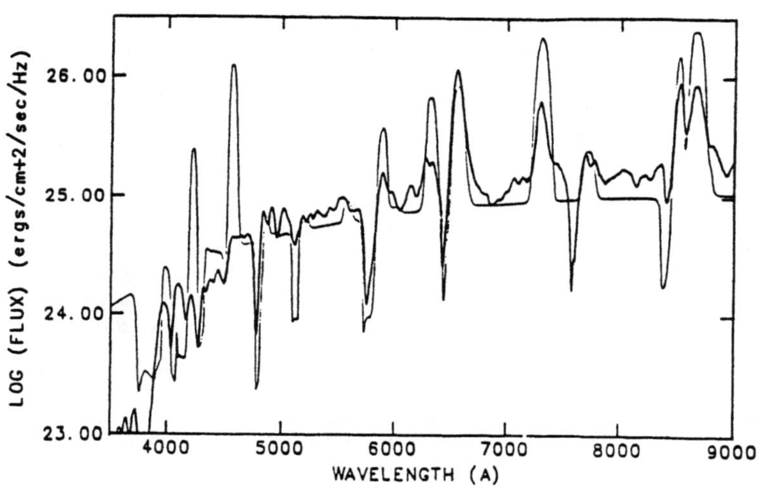

Figure 3: Spectra at 200 days comparing observations (heavy line) to unmixed (top) and mixed composition models.

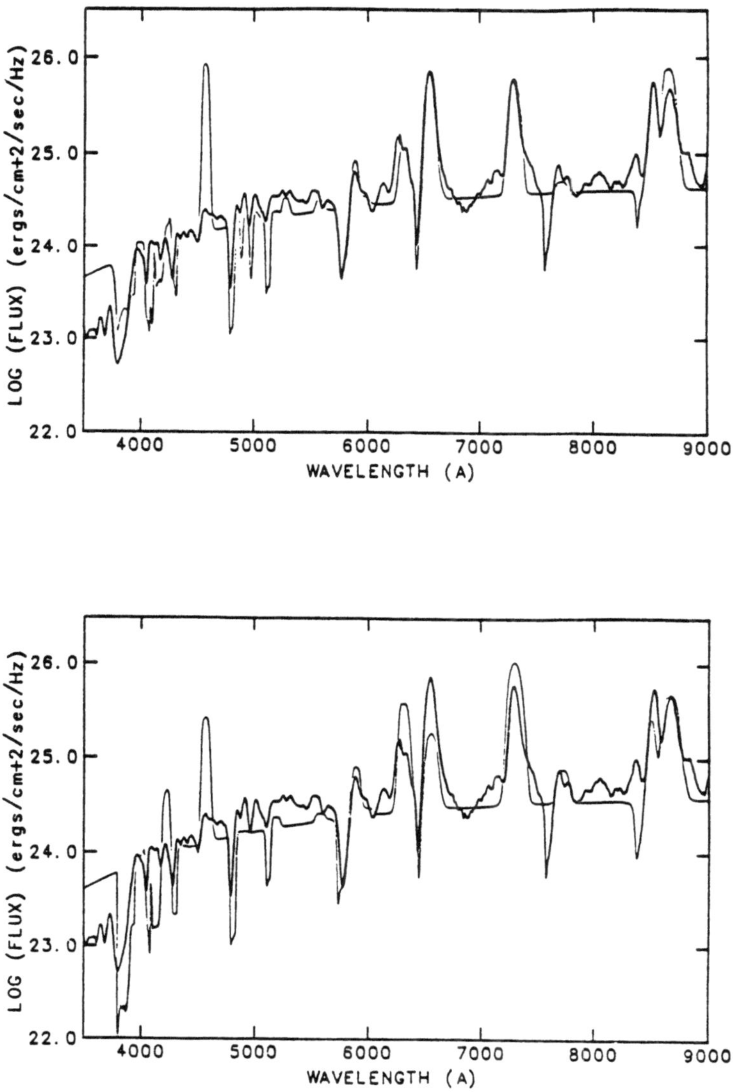

Figure 4: Spectra at 300 days

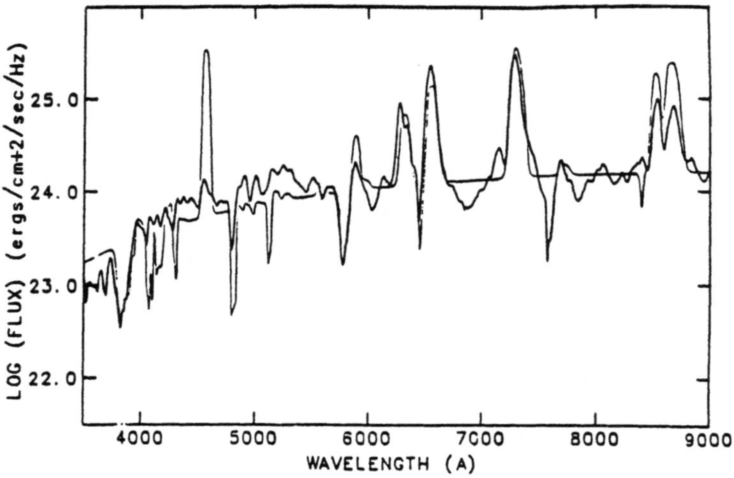

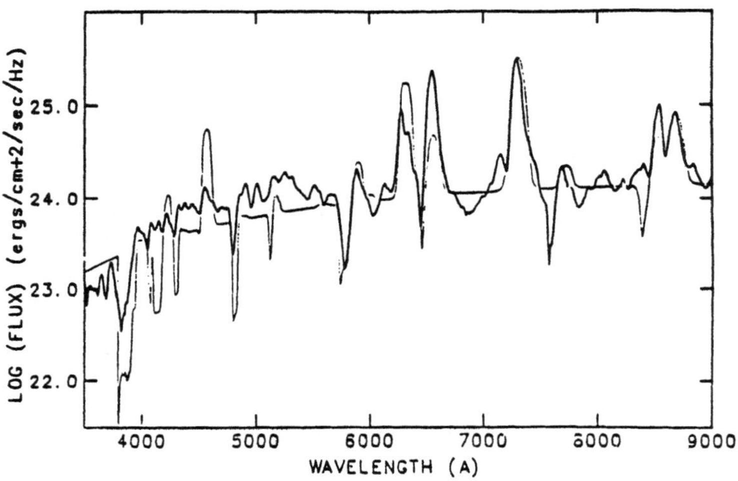

Figure 5: Spectra at 400 days

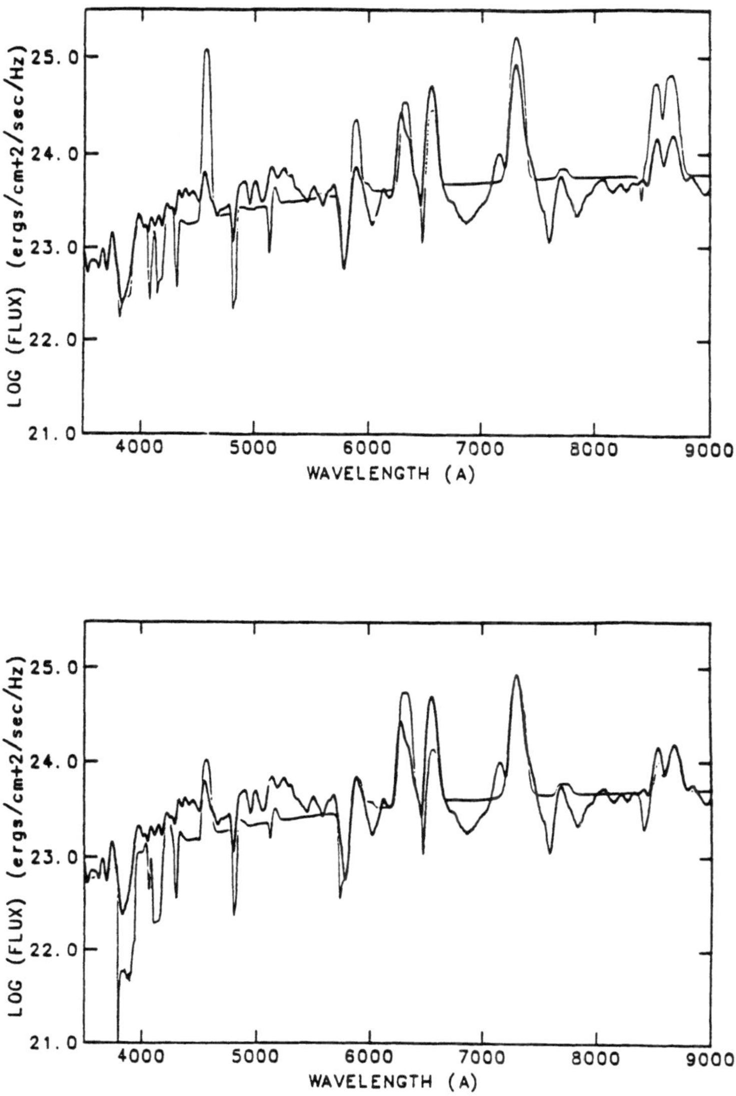

Figure 6: Spectra at 500 days

features near 5000 Å and the spectrum between 6000 and 8000 Å. The CaII infrared feature improves in the mixed model at 400 and 500 days while the stratified model fit to this feature worsens. This is also true of the NaI 5890 Å line. The opposite trend is apparent in the OI 6300 Å CaII H and K, and H_α lines. It is impossible to exclude either of the two composition mixing prescriptions on the basis of these comparisons. The unmixed model, however, provides the best overall fit during the time frame modelled here and is the preferred model in this work.

Supernova SN 1987A is the most thoroughly investigated supernova in history and the conclusions made here are discussed in context with those based on other independent research of this object.

The models assume radioactive decay of ^{56}Co is the sole source of heating in SN 1987A at late times. From observations, and the known distance to the LMC, the total mass of this element (originally ^{56}Ni) is 0.075 $M_\odot$. A sufficient amount of energy is supplied by this material to account for the observed spectrum. A similar conclusion has been reached previously for Type Ia supernovae (AXELROD [1]) and has been inferred from observations of the H_α line profile of SN 1980K (UOMOTO and KIRSHNER [6]). It has long been known that shock heating alone will not maintain the observed optical display into the late-time stage. This work shows, for the first time, that the cobalt decay energy source alone can account for the observed late-time spectrum of a Type II supernovae.

To reproduce the observed optical spectrum of SN 1987A, the cobalt must be distributed throughout a substantial fraction of the atmosphere and cannot reside in the deepest layers where it was created. This result was anticipated from investigations of the γ-ray and X-ray spectral evolution of SN 1987A, but this is the first time mixing of the cobalt has been deduced from models of the optical spectrum. For distant supernovae, the γ-ray and X-ray flux is not detectable (PINTO and WOOSLEY [4]). The optical spectrum, on the other hand, has been observed from many such supernovae. Thus models of distant supernovae using the methods presented here can determine if mixing is occurring in these objects as well or if the phenomenon is unique to events such as SN 1987A.

The novel approach of using a continuum emitting lower boundary in late-time supernova atmosphere models has led to the conclusion that the observed optical line profiles of SN 1987A arise solely from material in the envelope and that the core regions are somehow obscured from direct view (at least until 500 days). Furthermore, the best fit is obtained if the envelope retains its original composition and has not been enhanced, particularly in intermediate mass elements, through a mixing of core and envelope material.

Composition mixing was introduced in the literature in efforts to obtain better fits to the observed light curve. Mixing of ^{56}Co and other iron group elements was specifically invoked to explain the γ-ray and X-ray flux. Mixing of iron group elements into the envelope cannot be discounted in the present analysis due to the oversimplified models of the iron group ions used in this work. Based on the conclusions reached here, it appears that some mechanism exists which has enabled the radioactive material to be mixed into the envelope without disturbing the original stratification of the remaining elements. This conclusion narrows the possibilities of what that mechanism might be. If all the material were mixed, then large-scale Rayleigh-Taylor mixing could be responsible. The fact that at most only a few elements, and these originally located in the deepest layers of the ejecta, have been mixed points toward the formation of energetic Ni-rich clumps which have been able to penetrate to the outer layers.

SN 1987A shows distinctive blue-shifted absorption in the late-time spectra. These features were reproduced by placing a lower boundary in the models at the base of the hydrogen-rich envelope and this implies the existence of broad-band opacity sources which obscure the core region. Other Type II supernovae do not show the blue-shifted absorption but otherwise are very similar to SN 1987A in spectral features and in the time evolution of those features. It has been commonly assumed that the evolution of many of these features, for example the [OI] 6300 Å line, indicates the exposure of deeper layers (namely the oxygen-rich core) as the supernova expands. We have shown that this is not the case for SN 1987A. Instead, this and other features evolve in response to changes in the heating and cooling rates brought about by changes in local conditions in the envelope. That similar evolution of these features occurs in other Type IIs suggests the same mechanism is responsible in those cases as well.

This research is suppored in part by the R. A. Welch Foundation and by NSF Grant 8717166. The computations were done on the Cray X-MP of the University of Texas System Center for High Performance Computing.

References

[1] Axelrod, T.S., (1980a) PhD Thesis Univ. Calif. Santa Cruz

[2] Axelrod, T.S., (1980b) in *Type I Supernovae* ed. J.C. Wheeler (Austin:UT Press), p. 80.

[3] Pinto, P.A., Woosley, S.E., (1988a) Nature **333**,534.

[4] Pinto, P.A., Woosley, S.E., (1988b) Astrophys. J. **329**,820.

[5] Swartz, D.A., Harkness, R.P., Wheeler, J.C., (1989) Nature, **337**,439.

[6] Uomoto, A., Kirshner, R.P., (1986) Astrophys. J. **308**,685.

[7] Witteborn, F.C., Bregman, J.D., Wooden, D.H., Pinto, P.A., Rank, D.M., Woosley, S.E., Cohen, M., (1989) in press.

[8] Woosley, S.E., (1988) Astrophys. J. **330**,218.

Late Time Spectrum of SN 1987A

Yueming Xu & Richard McCray

I. Introduction

At late times ($\gtrsim$ 200 days), when the photosphere of continuum emission has disappeared from the envelope, the radiation of the supernova 1987A is powered by γ-rays resulting from ^{56}Co decays. The spectrum formation at such a late time is a highly non-linear process (XU [1]). The γ-ray photons first lose their energy due to Compton scattering and photoelectric absorption, generating non-thermal energetic electrons. These energetic electrons are then slowed down by Coulomb scattering or ionizing and exciting different ions in the envelope. The ions will recombine and de-excite, radiating away their energy in infrared (IR), optical and primarily ultraviolet (UV) wavebands. Most of the UV photons may be absorbed by photoelectric ionizations or split into photons of longer wavelengths by resonance line scatterings as they propagate through the supernova envelope. These processes continue until most of the internal luminosity is converted into optical or IR photons which can escape without further interactions. The self-consistent modeling of these processes is a coupled problem of the thermal equilibrium, ionization equilibrium and radiative transfer. No satisfactory model is available yet because the necessary atomic data set is not complete and because the envelope dynamics (such as mixing or clumping) is not clear at present.

In this paper we emphasize the importance of charge transfer, line trapping and line splitting in the spectrum formation of late-time supernovae. In §II we show that the observed hydrogen ionization of SN1987A can be explained by the photoionization from the $n = 2$ state, which is over-populated because of the Lyα trapping. Charge transfer is the main process by which hydrogen recombines and heavy elements become singly ionized. In §III we show that the line splitting process is the dominant source of opacity for photons of wavelengths less than 2000Å. Finally, in §IV we discuss the general theory of spectrum formation in late-time supernovae.

II. Envelope Ionization

The infrared continuum ($> 4\ \mu$m) emission of SN1987A in November 1987 requires an emission integral (MOSELEY *et al.* [2]):

$$EM^{+} = \sum_{z} z^{2} \int n_{e} n_{z} dV = 3 \times 10^{64}\ \mathrm{cm}^{-3}, \tag{1}$$

where n_z is the number density of the ions with charge z, and n_e is the electron number density. The infrared hydrogen recombination line emissivities also require a comparable value (XU [1]).

For a homogeneous model with the chemical abundances of the model 10H given by WOOSLEY [3], the emission integral can be written as

$$\int n_e n(H^+)dV = (3.4 \times 10^{68}\mathrm{cm}^{-3})t_2^{-3}\mathrm{M}_{15}^{7/2}\mathrm{E}_{51}^{-3/2}\left(\frac{n(H^+)}{n_H}\right)^2. \qquad (2)$$

where t_2 is the supernova age in units of 100 days, M_{15} is the envelope mass in units of $15\,\mathrm{M}_\odot$, and E_{51} is the total kinetic energy of the supernova in units of 10^{51} ergs. By equating (1) and (2), we estimate the hydrogen ionization fraction at ~300 days:

$$\frac{n(H^+)}{n_H} = 0.05\,\mathrm{M}_{15}^{-7/4}\mathrm{E}_{51}^{3/4}.$$

A temperature of $T \approx 5000$ K would be sufficient to account for this ionization level in LTE, according to the Saha equation; but LTE would require a blackbody radiation field of 5000 K, which is clearly not there.

The rate of impact ionization of hydrogen atoms by thermal electrons at $T =$ 3000K is far too low to explain this ionization level. The ionization cannot be explained by the non-thermal electrons resulting from γ-ray degradation, either. In fact, the total γ-ray energy available is

$$\mathrm{L}_{Co} = (5 \times 10^{53}\ \mathrm{eV\ s}^{-1})\mathrm{M}_{Co}\exp\{-0.88t_2\},$$

where M_{Co} is the initial mass of ^{56}Ni in units of $0.07\,\mathrm{M}_\odot$. A fraction η_i of the energy can be used to ionize hydrogen atoms. $\eta_i \leq 0.3$ for pure hydrogen gas (SHULL and VAN STEENBERG [4]), and it should be smaller for the supernova ejecta which is enriched by heavy elements. However, the hydrogen recombination rate implied by the observed recombination line strength is $R_C = \alpha^{(3)}(T)EM^+ = 4.5 \times 10^{51}T_4^{-0.92}(\ \mathrm{s}^{-1})$ for Case C recombinations (McCRAY [5]; XU [1]). Therefore, the energy available for each hydrogen ionization is at most $\zeta_I = \mathrm{L}_{Co}\eta_i/R_C = 4.2\eta_i(\ \mathrm{eV})$, which is much less than the hydrogen ionization potential 13.6 eV.

The fact that $\zeta_I \leq 4$ eV is a strong hint that the H^+ comes from excited $H^*(n = 2)$ rather than ground state hydrogen atoms. If this is true, the necessary energy per ionization is 3.4 eV, thus the fraction of fast electron energy used for hydrogen ionization must be $\eta_i \gtrsim 0.83$. This cannot happen in a pure hydrogen and helium gas where more than 30% fast electron energy is deposited as heat by Coulomb scattering (SHULL and VAN STEENBERG [4]) and there is no efficient way to turn the heat into hydrogen ionization.

In a metal-rich region, however, the radioactive energy can be used more efficiently to ionize hydrogen atoms. Most of the fast electron energy is deposited first in the excitation and ionization of heavy elements, such as Fe, because these elements have more atomic states and larger cross sections for electron collisions. The metal atoms and ions will give away the absorbed energy by emitting mostly UV photons. These photons cannot escape directly (§III), and will finally be turned into heat or hydrogen ionization. The energy deposited as heat can also excite some UV lines of metals. Thus, it is most likely that H^* is photoionized by near-UV lines ($3.4 < h\nu <$ 5 eV) produced by electron impact excitation of metals such as Mg IIλ2800 (CHUGAI

[6]) and Fe II. This indirect ionization mechanism makes it possible to increase the efficiency η_i of hydrogen ionization; it is more effective than direct electron impact ionization of H^* because the metal ions are more abundant than H^* (by factors $\gtrsim 10^3$) and because the cross-sections for electron impact excitation of metals are greater than the cross-section for electron impact ionization of H^*.

The indirect ionization equilibrium can be approximated with a three-level atomic model, *i.e.*, the ground state, the first excited state (n=2) and the continuum of hydrogen atoms. The ionization equilibrium can be written as

$$\Gamma \frac{1}{\zeta_H} n(H^0) + \xi n(H^*) = \chi_e \alpha(H^+) n(H^+), \tag{3}$$

where $\chi_e = n_e/n_T$ is the fractional number density of electrons, n_T is the total ion number density, $\alpha(H^+)$ cm^3 s^{-1} is the radiative recombination rate coefficient of H^+, ξ is a parameter representing the ratio of the UV radiation field to total density n_T, defined such that $\xi n_T n(H^*)$ is the rate of photoionization of hydrogen from the $n = 2$ state, and Γ is the ratio of energy deposition rate to atomic number density, defined such that the rate of the collisional ionization of fast electrons is $\Gamma n_T n(H^0)/\zeta_H$, ζ_H is the hydrogen ionization potential in units of eV. Assuming that about half of the γ-ray energy is deposited to ionizations, we have the collisional ionization rate coefficient:

$$\Gamma \approx (2.4 \times 10^{-16} \text{ eV cm}^3 \text{ s}^{-1}) t_2^3 \mathrm{M}_{Co} \mathrm{M}_{15}^{-7/2} \mathrm{E}_{51}^{3/2} \exp\{-0.88 t_2\}. \tag{4}$$

The equilibrium equation of hydrogen excitation to the $n = 2$ state can be written as

$$\Gamma \frac{1}{\zeta_H} n(H^0) + \chi_e \alpha(H^+) n(H^+) = (\xi + \xi_c) n(H^*), \tag{5}$$

where the first term is the approximate collisional excitation rate to the $n = 2$ state by fast electrons, the second term is the radiative recombination to the $n = 2$ state, the first term on the right hand side represents the photoionization from the $n = 2$ state, and $\xi_c = \chi_e C_{21} + n_T^{-1}(A_{21}/\tau_{L\alpha} + A_{2\gamma})$ is the rate of collisional and radiative de-excitation from $n = 2$ state, C_{21} is the rate coefficient for collisional de-excitation by thermal electrons, $A_{2\gamma}$ is the two-photon emission rate from the 2s to 1s state of hydrogen atoms (NUSSBAUMER and SCHMÜTZ [7]), and $A_{21}/\tau_{L\alpha}$ is the effective Lyα emission rate corrected for resonant trapping when the Lyα optical depth $\tau_{L\alpha} \gg 1$. By solving (3) and (5), we obtain

$$\frac{n(H^+)}{n(H^0)} = \frac{\Gamma(2\xi + \xi_c)}{\zeta_H \xi_c \chi_e \alpha(H^+)}.$$

As a result, if the radiation field is weak, *i.e.*, $\xi \ll \xi_c$, we have $n(H^+)/n(H^0) \approx 3 \times 10^{-3}$ at 300 days, which is about an order of magnitude smaller than the observations, as we have already seen from the earlier discussion of the energy budget problem. On the other hand, if the UV radiation field is strong ($\xi \gg \xi_c$) and is approximately a diluted black body emission of 5000K, we have $n(H^+)/n(H^0) \approx 0.8W$ at 300 days, where W is the dilution factor. Therefore, in order to explain the hydrogen ionization at 300 days, we need a UV radiation field within the hydrogen-metal mixed region. If this radiation can be fitted by a diluted 5000 K blackbody spectrum, the dilution factor cannot be smaller than $W = 0.06$. We know that Sobolev thick lines at the line center can have the same emissivity as a blackbody ($W = 1$) if they are effectively thick, *i.e.*, if $n_e C_{ul} \tau / A_{ul} > 1$. It is possible that there are enough thick UV lines to build the required ionizing radiation field.

The two-step mechanism of hydrogen ionization described here cannot work without the help of heavy elements. Moreover, the effect of charge transfer of hydrogen with the mixed heavy elements is to exacerbate the already puzzling problem of understanding the hydrogen ionization. Once a hydrogen atom is photoionized, it can be neutralized by charge transfer to heavy elements. If the charge transfer recombination rate is fast enough, the hydrogen ionization will be largely suppressed.

The rates of the charge transfer recombination and ionization of hydrogen can be written $C_M^r n(M) n(H^+)$ and $C_M^i n(M^+) n(H^0)$, respectively. The equation of the ionization equilibrium (3) becomes

$$\Gamma \frac{1}{\zeta_H} n(H^0) + \xi n(H^*) + C_M^i n(M^+) n(H^0) = \chi_e \alpha(H^+) n(H^+) + C_M^r n(M) n(H^+). \quad (6)$$

Thus, if the metal number density $n(M)$ is large, so that $C_M^r n(M) \gg \chi_e \alpha(H^+)$, the hydrogen ionization will be suppressed to a level much lower than would obtain without charge transfer. Even if the ionizing UV radiation field is very strong ($\xi \gg \xi_c$), the hydrogen ionization could saturate to a low value:

$$\frac{n(H^+)}{n(H^0)} = \frac{2\Gamma/\zeta_H + C_M^i n(M^+)}{C_M^r n(M)}.$$

Figure 1 shows some model calculations for the hydrogen ionization. We have assumed that the electron temperature $T = 3000\mathrm{K}$ and $\Gamma = 10^{-15}$ cm^3 s^{-1}. The model without charge transfer shows that the hydrogen ionization increases with the ionizing radiation. The totally mixed homogeneous model 10H, on the other hand, has too much charge transfer, which leads to very low hydrogen ionization, $n(H^+)/n(H^0) \lesssim 2 \times 10^{-4}$. The observed hydrogen ionization can be obtained if hydrogen atoms are mixed with only 10% of the heavy elements in the model 10H. Detailed calculations (XU [1]) show that the mixed heavy elements can be singly ionized for a long time by the charge transfer to hydrogen.

We conclude that a large fraction of the radioactive energy must be absorbed in the hydrogen-rich material, and some heavy elements must be mixed with the hydrogen gas to provide an ionizing UV radiation field, but the mass ratio of heavy elements to hydrogen cannot be larger than 5% in the mixed region. How can this model be possible? According to standard supernova models, most of the γ-ray sources, ^{56}Co nuclei, are within the central region which is enriched in heavy elements. Most of the UV photons ($h\nu > 3.4$ eV) cannot avoid photoabsorption in this metal-rich layer, thus the hydrogen atoms in the outer layers would not be exposed to enough photoionizing flux to maintain the observed hydrogen ionization. The probable solution to this puzzle is that most of the heavy elements (including ^{56}Co) in the supernova ejecta are clumped and the hydrogen is mixed into the interior region; thus γ-rays generated in the ^{56}Co knots can escape directly to the hydrogen-rich gas and deposit energy there by Compton scattering. Actually, the fact that the heavy element knots are observed in some supernova remnants, such as Cas A, support this clumpy model of the supernova ejecta. X-ray observations of this supernova also show properties of a clumpy model (KUMAGAI *et al.* [8]; XU [1]).

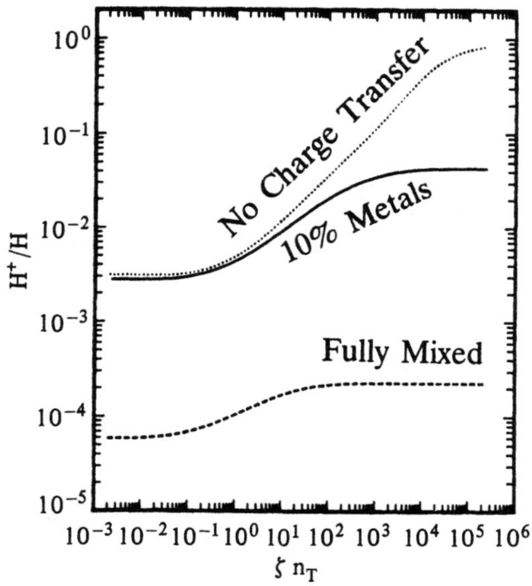

Figure 1: Hydrogen ionization.

III. Emergent Spectrum

In the late-time supernova envelope, strong EUV ($\lambda < 2000$Å) lines, such as He I $\lambda 584$ and H Lyα, are generated by the collisions of fast electrons (XU [1]). These photons cannot escape because the envelope has a large scattering opacity produced by many broad overlapping resonance lines of species such as Fe I and Fe II (McCRAY [5]; XU [1]). In fact, a photon of wavelength λ can be scattered by all lines with wavelength $\lambda' \in [\lambda, \lambda + \lambda_{exp}]$, where $\lambda_{exp} = \lambda v_{exp}/c$ is the redshift at the surface of the expanding envelope. In the Sobolev approximation (MIHALAS [9]), the direct escape probability through the resonant scattering region of line i is $P_i = \exp\{-\tau_i\}$, where τ_i is the Sobolev optical depth of the line i. Thus, the probability for a photon of wavelength λ to escape directly from the envelope can be written as

$$P_0(\lambda) = \exp\left\{ - \sum_{\lambda_i > \lambda}^{\lambda_i < \lambda + \lambda_{exp}} \tau_i \right\}. \tag{7}$$

In general, photons can escape after resonant scatterings if the destruction probability in Sobolev layers is small. Since the effect of the multiple scatterings in a geometrically thin Sobolev layer is the same as that of one scattering at a point in space, we can assume that the effective number of the resonant scatterings by line i is $1 - \exp\{-\tau_i\}$ as far as the spatial diffusion is concerned. Thus, the mean redshift before the next scattering, $\delta\lambda$, of a photon of wavelength λ can be defined such that

$$\sum_{\lambda_i > \lambda}^{\lambda_i < \lambda + \delta\lambda} \left(1 - e^{-\tau_i}\right) = 1. \tag{8}$$

As a result of the diffusion process, the total wavelength redshift $\Delta\lambda$ of an emergent photon can be much larger than λ_{exp}. The redshift $\Delta\lambda$ can be calculated as follows. We use (8) to calculate the mean free path $\delta\lambda_0 = \delta\lambda$ of a photon with initial wavelength $\lambda_0 = \lambda$, and then obtain the mean wavelength after one scattering, $\lambda_1 = \lambda_0 + \delta\lambda_0$. With this procedure, we calculate the wavelength after n scatterings, $\lambda_n = \lambda_{n-1} + \delta\lambda_{n-1}$ for $n = 2, 3$, *etc.* Photons will diffuse out from the envelope when

$$\left(\frac{v_{exp}}{c}\right)^2 = \sum_{i=0}^{n_s}\left(\frac{\delta\lambda_i}{\lambda_i}\right)^2, \tag{9}$$

where n_s is the total number of scatterings. UV radiation becomes transparent only if the total redshift $\Delta\lambda = \sum_{i=0}^{n_s}(\delta\lambda_i) < \lambda$. n_s decreases rapidly with the envelope temperature because the Sobolev depths of the lines corresponding to transitions between excited states decrease exponentially with the temperature. n_s also depends on the initial photon wavelength. The envelope will first become transparent at some wavelength "windows", where scattering lines are neither numerous enough nor thick enough to scatter UV photons.

In Figure 2, we plot the number of resonant scatterings n_s calculated from (9) for temperatures of 3000K and 500K. Here, we considered all the abundant element in the model 10H, but we artificially homogenized the chemical abundances throughout the envelope. We took the ionization fractions from the calculations in the last section except for Fe atoms, which we took the observed value, 50% (OLIVA *et al.* [10]). We used $\sim$ 58,000 lines of the 20 important ions from the line list of ABBOTT and LUCY [11]. But, most of the important lines are from Fe I and Fe II. At 3000K, photons with wavelength less than 5800Å cannot escape without scattering, and thus the fast pulsations of a possible central source (if it exists) would be smeared out. However, the redshift of UV photons is not large enough to shift the photons to the optical waveband. In fact, the scattering number is $n_s \approx 200$, and thus the total redshift is $\Delta\lambda/\lambda \approx \sqrt{n_s} v_{exp}/c \sim 0.1$. At 500K, UV photons can escape directly through the windows at wavelengths $\lambda > 2500$Å.

Photons can be destroyed during the Sobolev scattering. Once an atom is excited by absorbing a UV photon, it may cascade down, splitting into several different photons. We find that the radiative splitting is the dominant destruction process. For simplicity, we assume that scatterings of the surviving photons are coherent in the comoving frame of the expanding envelope. XU [1] proved that the escape probability of a photon from the resonant region of the line i is given by

$$P_i = e^{-(1-\eta_i)s_i}, \tag{10}$$

where s_i is the survival probability of the photon in one scattering. Thus, the total probability for a photon of wavelength λ to escape from an expanding envelope is

$$P_{esc}(\lambda) = \exp\left\{-\sum_{\lambda_i>\lambda}^{\lambda_i<\lambda+\Delta\lambda}(1-\eta_i)\tau_i\right\}. \tag{11}$$

In Figure 3, we plot the total escape probability (11) together with the escape probability (10) of single lines at 3000K. One sees that no photon with wavelengths less than 4500Å can escape directly from the envelope at 3000K. However, the envelope will become more transparent when the temperature decreases. We compared

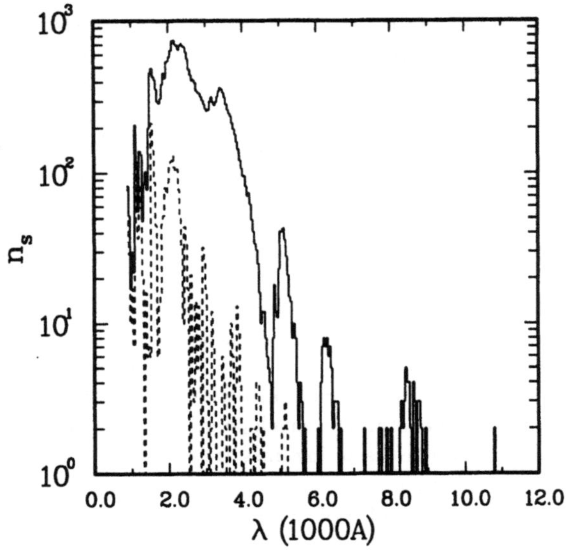

Figure 2: Number of resonant line scatterings in the envelope with expansion velocity $v_{exp}/c = 0.01$. Solid line: T = 3000K, $n_b = 2.4 \times 10^{10} \text{cm}^{-3}$; Dashed line: T = 500K, $n_b = 2.7 \times 10^9 \text{cm}^{-3}$.

P_{esc} with P_0 and found only small differences, which means that photon splitting is a much more common process than coherent line scatterings, *i.e.*, most photons are destroyed by line splitting during resonant scatterings.

We simulated the line scattering and splitting processes in the envelope of the homogenized model 10H by use of the Monte Carlo technique. We assumed that the envelope has a baryon number density of $n_b = 2.4 \times 10^{10} \text{cm}^{-3}$, and the level population of the ions is in LTE at the temperature of 3000K. We believe that these conditions are representative of the inner ejected material of SN1987A at 300 days. In our calculation, the line source function consists of the thermal emission, recombination lines and lines excited by fast electrons. We used the Bethe approximation to estimate the latter (XU [1]). The emergent spctrum (Figure 4) shows no strong source lines at wavelengths less than 2000Å. These lines have split and the energy has been redistributed to the wavelength range $\gtrsim$ 3000Å as a quasi-continuum. The observed spectrum (PHILLIPS [12]) in December 1987 showed a similar quasi-continuum emission in the waveband of 4000Å to 5500Å but was black at the wavelength less than 4000Å. In fact, the photons of wavelength less than 3646Å shown in Figure 4 provide a necessary ionizing UV radiation field for hydrogen ionization as discussed in the last section. In the Monte Carlo calculation, we did not include the continuum absorption due to the photoionization of hydrogen atoms from the $n = 2$ state.

The three predominant lines in the wavelength range 6000Å to 9000Å in Figure 4 correspond to the observed strong lines: [OI]λλ6300 and Hαλ6563, Ca II]λλ7324 and Ca II λλ8662. However, the calculated line ratios do not agree with the observations because our one-zone model is too simple. In fact, since Hα is very thick at 300 days,

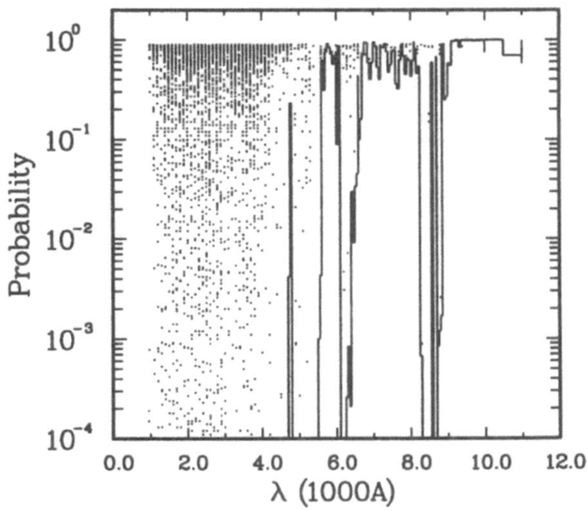

Figure 3: Total escape probability from the envelope with expansion velocity $v_{exp}/c = 0.01$. T = 3000K, $n_b = 2.4 \times 10^{10} \mathrm{cm}^{-3}$. Solid line is the probability from the envelope; dots are for single lines.

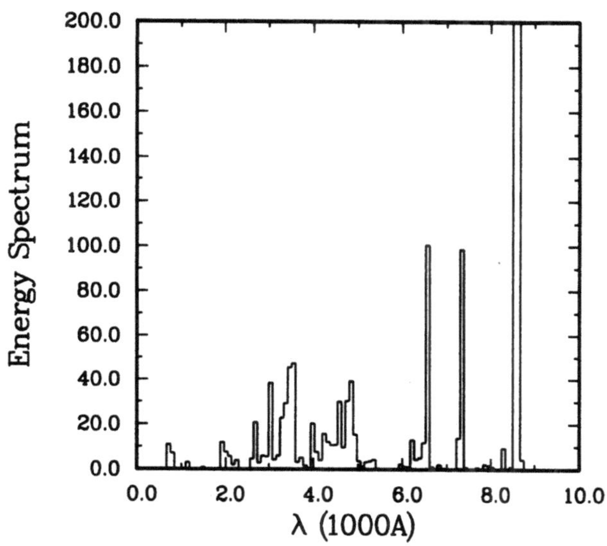

Figure 4: Emergent spectrum calculated for a typical model at 300 days.

it can be emitted from a much larger region than the Ca II lines, and so can be much stronger than in our model calculation. An alternate explanation is that Ca II is not so abundant in the main emission region, so the Ca II lines can be weaker than the model calculation. [OI]$\lambda\lambda$6300 and Ca II]$\lambda\lambda$7324 are forbidden lines connected to ground states. These lines can be thermally excited. Ca II $\lambda\lambda$8662 can also be thermally excited because Ca II H, K lines are trapped, or it can be pumped by the Ca II H, K lines ($\lambda\lambda$3968) or other high energy processes, such as fast electron excitations. If Ca II $\lambda\lambda$8662 is thermally excited, its line ratio to Ca II] $\lambda\lambda$7324 must depend on the electron temperature.

In summary, our simple model calculation has shown the basic features of the observed spectrum. The photon splitting process will destroy all photons of wavelengths less than 2000Å and redistribute the energy to the wavelength range $\gtrsim$ 3000Å. Photons of wavelengths greater than 2000Å cannot be destroyed by splitting because even if a photon splits, it will result in a photon with similar wavelength, and so it must diffuse out if there is no other absorption processes. However, the photoionization of hydrogen atoms from the $n = 2$ state can absorb all photons of wavelengths $\leq$3646Å. The emission of Ca II $\lambda\lambda$8662 is an example of the line splitting.

IV. Discussion

Up to now, the charge transfer and line splitting processes have been neglected in the theory of the spectrum formation in late-time supernovae. We have shown that both of these processes are crucial to the envelope structure and the emergent line spectrum, although our calculations are very preliminary. The current theory of the spectrum formation includes four basic problems: (1) how γ-rays deposit their energy within the supernova envelope; (2) how the subsequent fast electrons deposit their energy among excitations, ionizations and heating; (3) how charge transfer affects the ionization equilibrium; and (4) how the line splitting affects the radiative transfer with non-local coupling.

The former two are basic physics problems, and can be solved separately and accurately. The γ-ray energy deposition can be accurately calculated by the Monte Carlo simulation of the Comptonization (XU *et al.* [13]; SHULL and XU [14]). The key to the accuracy of the γ-ray deposition function is the spatial distribution of ^{56}Co, which can be inferred from the light curves of X-rays and γ-rays. The energy degradation of fast electrons can be calculated from the Spencer-Fano equation (SPENCER and FANO [15]; XU [1]), although the accuracy of the calculation is limited by the lack of knowledge of the cross sections for collisional excitations and ionizations, especially for heavy elements. In future research, more detailed work needs to be done in this field, including the collection of the cross sections.

The thermal and ionization equilibria are closely coupled with the radiative transfer of spectral lines. The charge transfer and line splitting processes add complexity to this already non-linear problem. The charge transfer tends to suppress the hydrogen ionization and to keep the heavy elements singly ionized. It is clear that more accurate atomic data for charge transfer are required to improve the calculation of the ionization equilibrium in supernovae. The incoherent line scattering will destroy all EUV photons ($\lambda <$ 2000Å) and redistribute the energy into longer wavelengths, which makes the radiative transfer harder to calculate accurately. Since there is no continuum radiation field in late-time supernova envelope, the Monte Carlo technique

(ABBOTT and LUCY [11]) is not efficient for solving iteratively the radiative transfer here. RYBICKI and HUMMER [16] developed a Sobolev approach to treat the line transfer with non-local coupling. This approach may be generalized to include the continuous absorption and line splitting processes.

The ultimate goal of the theory of spectrum formation in late-time supernovae is to calculate the emergent spectrum and its evolution from basic principles. However, the purpose of theoretical modeling is not just to fit the observational data, but to understand the physics implied by the observations. The comparison of theories and observations can help us to understand the structure and the chemical abundances of the ejected material. But, given the incompleteness of atomic data and the uncertainties about the interior dynamics of the supernova, we do not think that it would be successful to try a pure theoretical modeling at present. Therefore, in our future research, we should continue to work from observations of line ratios and line profiles to deduce more information about the properties of the supernova interior.

References

1. Y. Xu: *Ph. D. Thesis*, University of Colorado, Boulder (1989).
2. S. H. Moseley, E. Dwek, R. F. Silverberg, W. J. Glaccum, J. R. Graham, R. F. Loewenstein: *Ap. J.*, (1989), in press.
3. S. E. Woosley: *Ap. J.*, **330**, 218 (1988).
4. J. M. Shull, M. E. Van Steenberg: *Ap. J.*, **298**, 268 (1985).
5. R. McCray: In *Molecular Processes in Astrophysics*, ed. by T. H. Hartquist, (Cambridge University Press, 1989), in press.
6. N. N. Chugai: *Soviet Astron. Lett.*, **13(4)**, 282 (1987).
7. H. Nussbaumer, W. Schmütz: *Astron. Astrophys.*, **138**, 495 (1984).
8. S. Kumagai, T. Shigeyama, K. Nomoto, M. Itoh, J. Nishimura, S. Tsuruta: *Ap. J.*, (1989), submitted.
9. D. Mihalas: *Stellar Atmosphere*, (San Franciso: W. H. Freeman and Company, 1978).
10. E. Oliva, A. F. M. Moorwood, I. J. Danziger: In *Proc. 22nd ESLAB Symp. on Infrared Spectroscopy in Astronmy*, ed. by H. Glasse, M. S. Kessler, and R. Gonzales-Riesta, (ESA SP, 1988) p. 270.
11. D. C. Abbott, L. B. Lucy: *Ap. J.*, **288**, 679 (1985).
12. M. M. Phillips: In *Supernova 1987A in the Large Magellanic Cloud*, ed. by M. Kafatos, and A. Michalitsianos, (Cambridge: Univ. Press, 1988), p. 16.
13. Y. Xu, P. G. Sutherland, R. McCray, R. R. Ross: *Ap. J.*, **327**, 197 (1988).
14. J. M. Shull, Y. Xu: In *Supernova 1987A in the Large Magellanic Cloud*, ed. by M. Kafatos, and A. Michalitsianos, (Cambridge: Univ. Press, 1988), p. 371.
15. L. V. Spencer, U. Fano: *Phys. Rev.*, **93**, 1172 (1954).
16. G. H. Rybicki, D. G. Hummer: *Ap. J.*, **219**, 654 (1978).

A Comparison of Carbon Deflagration Models for SN Ia

Robert Harkness

The observational properties of Type Ia supernovae can be qualitatively explained in terms of the explosion of a carbon-oxygen white dwarf. For several years this hypothesis has gained support although there seems to be no clear understanding of how the white dwarf evolves to the appropriate conditions. Furthermore, the details of how degenerate carbon ignition occurs, and how the burning propagates are quite controversial. Recent studies on very fine mass scales tend to show that detonation is inevitable for one-dimensional models. Such a detonation, once formed, incinerates the entire star, leaving no significant quantity of the intermediate mass elements which are required in quantity to reproduce the observed spectra for the first two to three weeks following the explosion.

Carbon deflagration models are attractive because the subsonic burning allows the white dwarf to expand ahead of the burning front so combustion occurs at lower densities. This results in a partially burned layer surrounding the core and can also cause the nuclear burning to cease before reaching the surface of the star.

Four carbon deflagrations are considered here, two models computed by WOOSLEY and WEAVER [1] (models G7 and F7, corresponding to their models 2 and 3, respectively), and two computed by NOMOTO, THIELEMANN and YOKOI [2], (models C6 and W7), incorporating the revised nucleosynthesis calculations of THIELEMANN, NOMOTO and YOKOI [3]. These two sets of models each use a different prescription for the propagation of the deflagration front. In Woosley's models the burning velocity is determined by limiting the rate of increase of the convective luminosity coupling the burning zone and the next unburned zone. This tends to produce a deflagration which propagates at a roughly constant fraction of the sound speed. In the models of Nomoto et al. the flame speed is determined by a time-dependent mixing-length theory, where the mixing length is taken to be a constant fraction of the pressure scale height. This has the consequence that the deflagration wave propagates initially very slowly and accelerates towards the surface of the white dwarf. In both cases the prescription is arbitrary and amounts to no more than a convenient parameterization of the front velocity. Details of the four models are given in the table below.

At first glance it might seem that there is little to choose between these four explosion models, particularly if one examines the abundance distribution as a function of the mass coordinate. If, however, one examines the abundance distribution as a function of velocity (or radius) after the expansion becomes homologous, there are clear distinctions and these are reflected in the emergent spectrum near maximum light (see Figures 1 to 4).

The emergent spectrum is dependent upon not only the abundance distribution, but also on the gamma ray deposition function (which in turn is dependent on the ^{56}Ni distribution, the total mass of ^{56}Ni, the density profile and the expansion velocity). In earlier calculations of the maximum light spectrum of model W7, BRANCH et al. [4] found that their synthetic spectra could be improved if the partially burned matter with expansion velocity greater than about 10,000 km/s was homogenized. Calculations by HARKNESS [5] of model atmospheres based on model W7 reached the same conclusion and similar prescriptions for mixing are considered for each of the models presented here.

Model	**C6**	**W7**	**G7**	**F7**
E_k ($\times 10^{51}$ ergs)	0.91	1.30	1.04	1.73
ρ($\times 10^9$ gm/cm^3)	1.5	2.6	2.1	2.1
$M/M_\odot$	1.366	1.378	1.40	1.40
M_{Ni}	0.48	0.58	0.51	0.89
$M_{C/O}$	0.4	0.08	0.4	0
$M_{Ni/Fe}$	0.66	0.89	0.85	1.25
$V_{partial}$ (km/s)	6500-10000	9000-15000	9500-11000	14300-17000

All of the explosion models provided by their authors were terminated at a few seconds after the explosion. Their subsequent evolution was computed with a Lagrangian hydrodynamics code with flux-limited diffusion, using Rosseland mean opacities calculated from the detailed composition of the model. The gamma ray deposition is calculated from a solution of the radiative transfer equation in co-moving coordinates, assuming the gamma ray opacity is purely absorptive with $\kappa = 0.03$ cm^2/gm.

The spectrum of each model was then computed at 14 days after the explosion using the supernova radiative transfer code described in HARKNESS[5] and WHEELER and HARKNESS [6]. The effects of mixing were simulated by artificially homogenizing the composition above a particular expansion velocity, usually taken to be at the point where the composition changes from the products of nuclear statistical equilibrium (NSE) to the products of partial burning. Approximately 400 of the strongest resonance lines are included in the calculations described here.

Model W7 provides an excellent match to the maximum light spectrum of SN1981B. The combined IUE and McDonald Observatory data (BRANCH et. al [7]) span the wavelength range 1500 - 8500Å and model W7 is in excellent agreement throughout this range (see figure 5). All of the observed spectral features can be accounted for using the original composition profile. When the matter with expansion velocity in excess 11,000 km/s is homogenized the fit in the near ultraviolet is improved, although the optical potion of the spectrum fits slightly less well. This is principally due to the effects of mixing a small amount of iron and cobalt from the NSE region out to high velocity. The observed absorption feature at 3300Å is most easily accounted for by a number of Co II lines. Unfortunately, this feature is located near the ground-based optical cutoff and slightly beyond the long wavelength sensitivity of IUE. Confirmation of this absorption in other SN Ia should be considered a high priority.

Calculations performed after the conference indicate that when sufficient numbers of

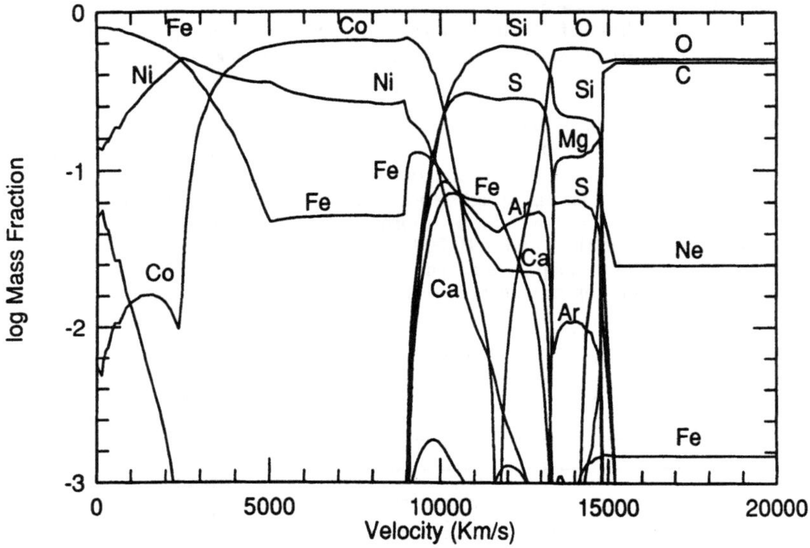

Figure 1: Abundance profile at 14 days for unmixed model W7.

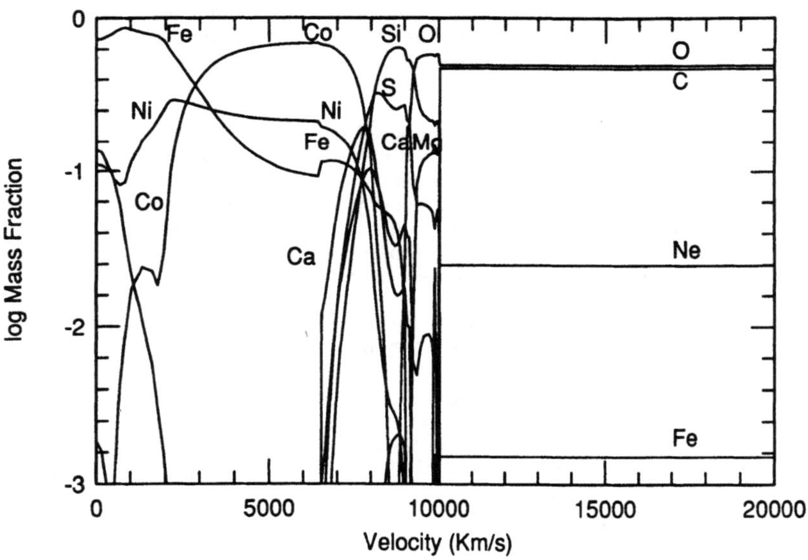

Figure 2: Abundance profile at 14 days for unmixed model C6.

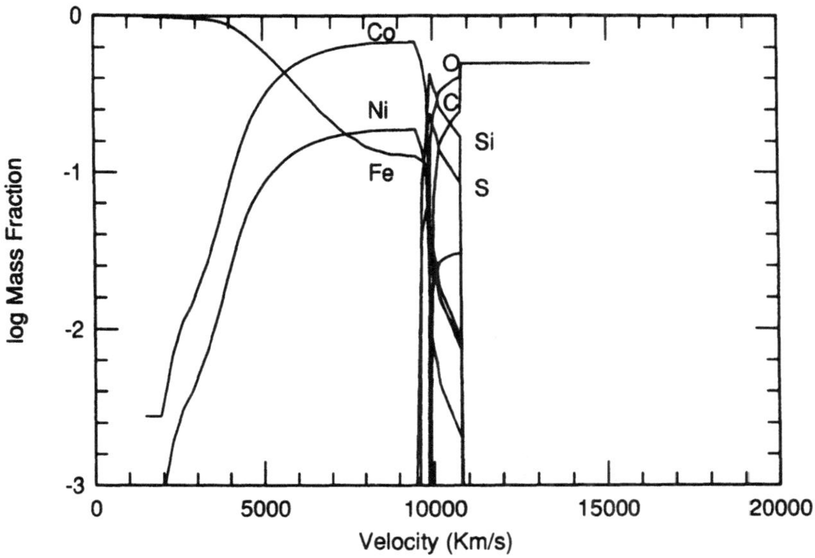

Figure 3: Abundance profile at 14 days for unmixed model G7.

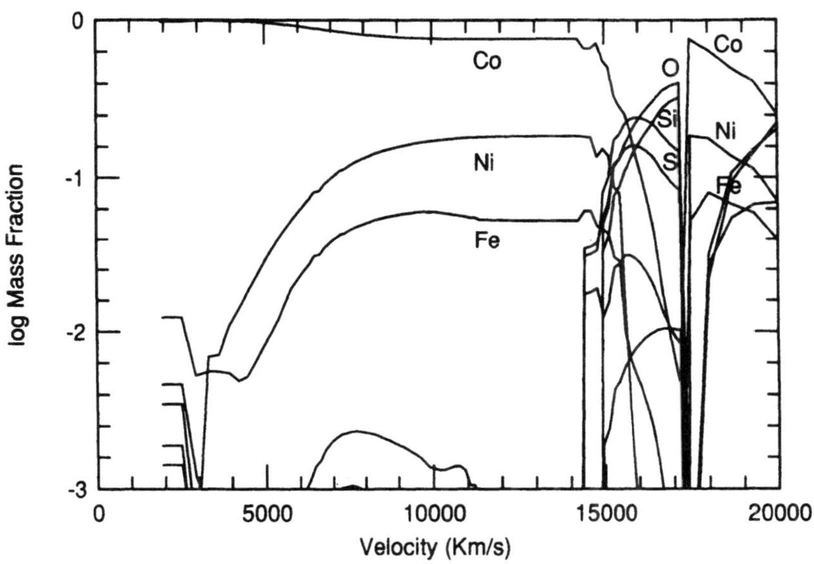

Figure 4: Abundance profile at 14 days for unmixed model F7.

spectral lines are included in the ultraviolet, the emergent spectrum of the *unmixed* W7 model may be superior at all wavelengths to the version which is mixed for velocities greater than 11,000 km/s. This is in contradiction to the previous results of BRANCH et. al [4] and HARKNESS [5].

The spectral evolution of model W7 has been computed in some detail, and will be presented elsewhere. The spectra change slowly from 4 days to around 16 days, when the NSE region is uncovered. The spectrum then changes extremely rapidly to become totally dominated by lines due to Fe II, together with minor features due to Co II and other iron-group elements.

Model C6 gives a good optical spectrum only when the intermediate mass elements are mixed for expansion velocities greater than 6,500 km/s (see figure 6). In the unmixed case the large mass of unburned C/O ($0.4M_{\odot}$) provides a greater number of free electrons per gram and the electron scattering surface is situated in this unburned layer. While the mixed model accounts for the presence of most of the observed spectral features, the lines all occur at longer wavelengths than observed in the case of SN1981B. It is possible that the mixed C6 model could be a decent fit to some of the slow SN Ia.

Comparing figures 2 and 3, one might expect the spectra of models C6 and G7 to be quite similar, but this is not the case. Like C6, model G7 has a large amount of unburned C/O and the unmixed spectrum shows essentially a continuum with a few weak line of oxygen and carbon. The principal differences between C6 and G7 are the density structure and the relative distribution of the Ni^{56}. G7 has a higher density in the unburned layer than C6, and the Ni^{56} is located in a shell immediately below this layer, and the local heating rate is higher. The intermediate mass elements lie in a very narrow shell which is much too restricted in velocity space. Even when the model is mixed from the base of this layer ($V > 9{,}500$ km/s) it provides an unsatisfactory spectrum because the kinetic energy of the outer layers is too low (figure 7).

The final model, F7, provides an interesting constraint. The intermediate mass elements occur at high expansion velocity, which alone might be expected to produce a spectrum incompatible with observations. However, as the shock wave propagated through the outer layers of this model, the deflagration became a detonation, burning the outer layers to NSE at very high expansion velocities ($V > 17{,}000$ km/s). The net result is a high velocity layer of the intermediate mass elements sandwiched between the core NSE region, and a very high velocity surface NSE region. The resulting large mass of Ni^{56} produces the largest gamma ray deposition of the four models. The spectra of the unmixed model and a model mixed from the base of the partially burned region are almost indistinguishable because the spectra are always dominated by iron group elements and Fe II in particular (figure 8). Furthermore, the spectrum evolves far too rapidly to be a viable model for normal SN Ia.

From this comparison of four current deflagration models we can draw the following conclusions:

1. The amount of Ni^{56} produced in the explosion should probably not exceed $\approx 0.8M_{\odot}$, and should not be thoroughly mixed with the partially burned matter. Model W7 also produces an excellent light curve, so around $0.6M_{\odot}$ may be optimum.

2. It seems a major reason for the success of model W7 is that the partial burning extends almost to the surface. As it is, W7 would be even better if the burning DID extend to the surface, producing more magnesium. In this case it appears that the artificial mixing may prove to be unnecessary. Clearly, a large mass of C/O should not remain.

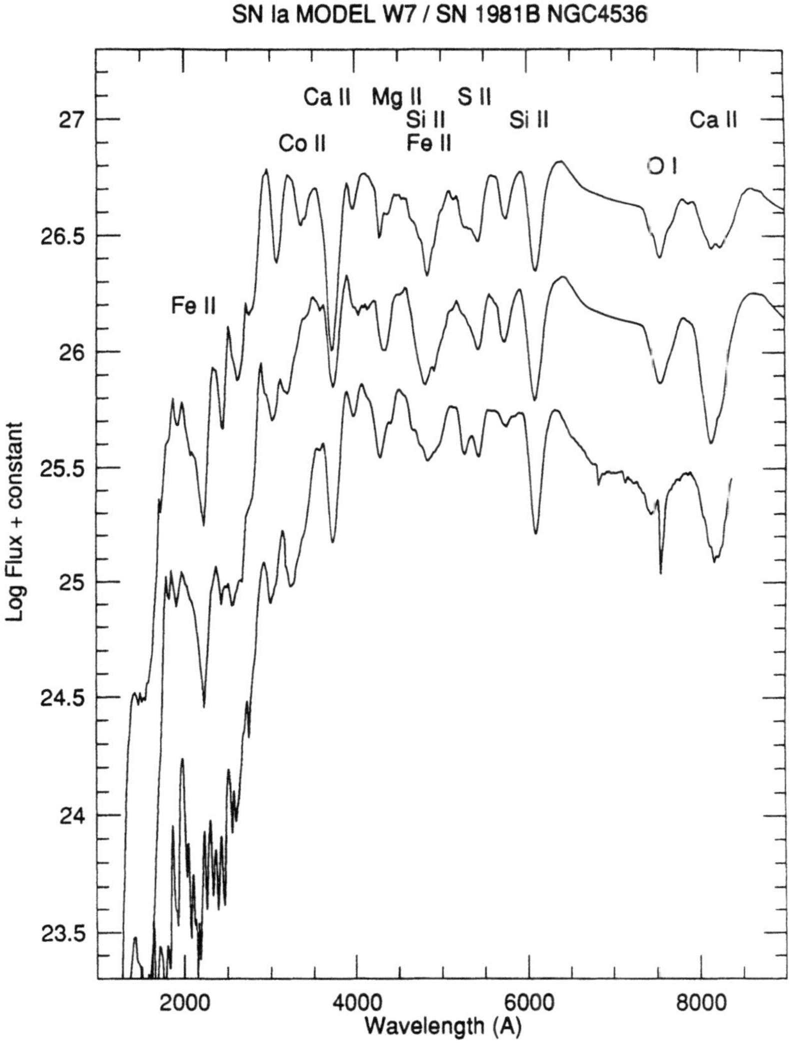

Figure 5: Emergent spectra of model W7 at 14 days, with no mixing (top) and mixing for V > 11,000 km/s (middle) compared with the maximum light spectrum of SN1981B.

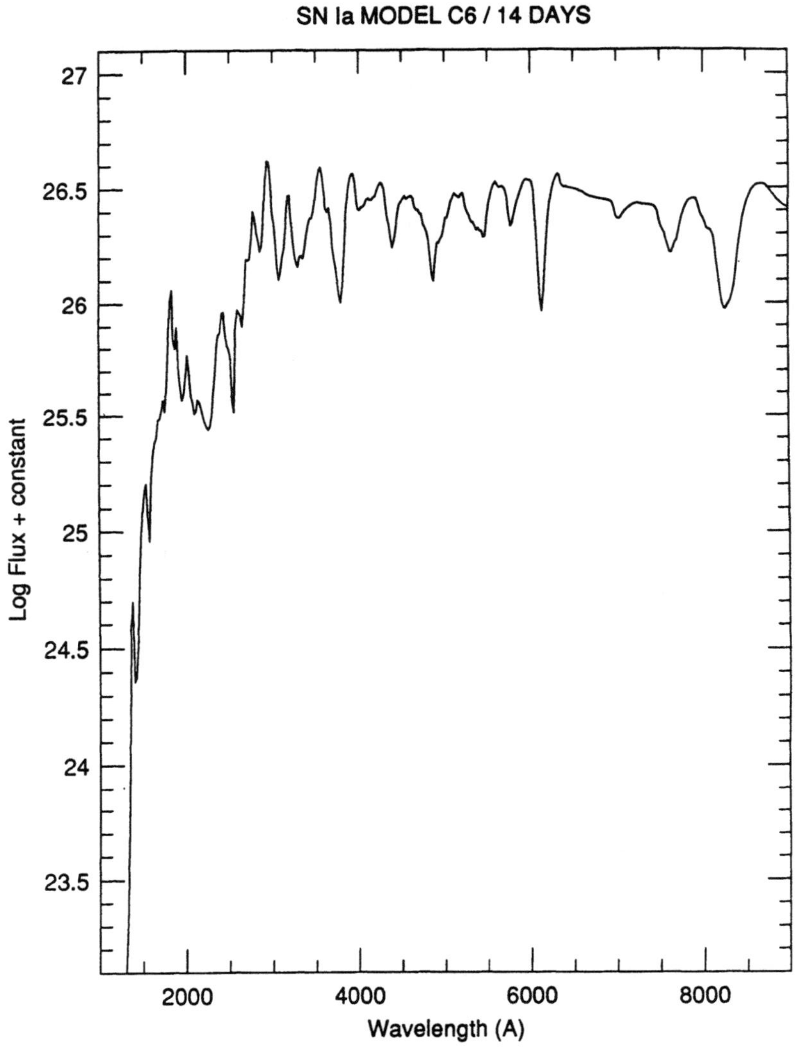

Figure 6: Emergent spectrum of model C6 at 14 days, with mixing for V > 6,500 km/s.

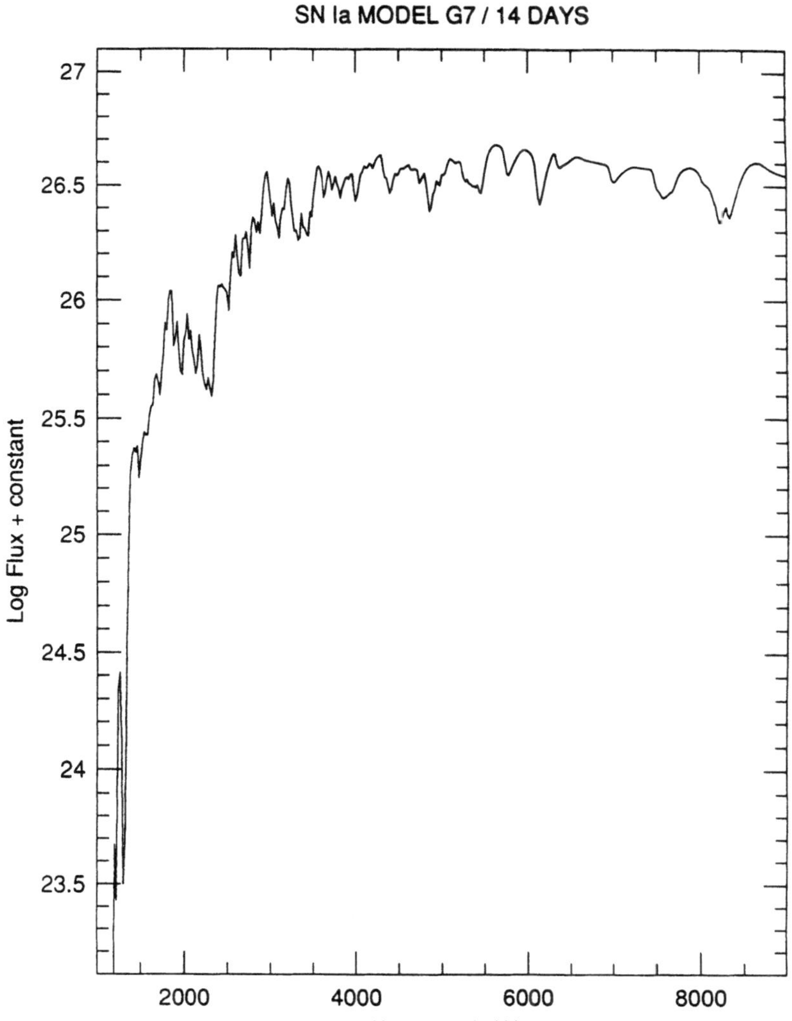

Figure 7: Emergent spectrum of model G7 at 14 days, with mixing for V > 9,500 km/s.

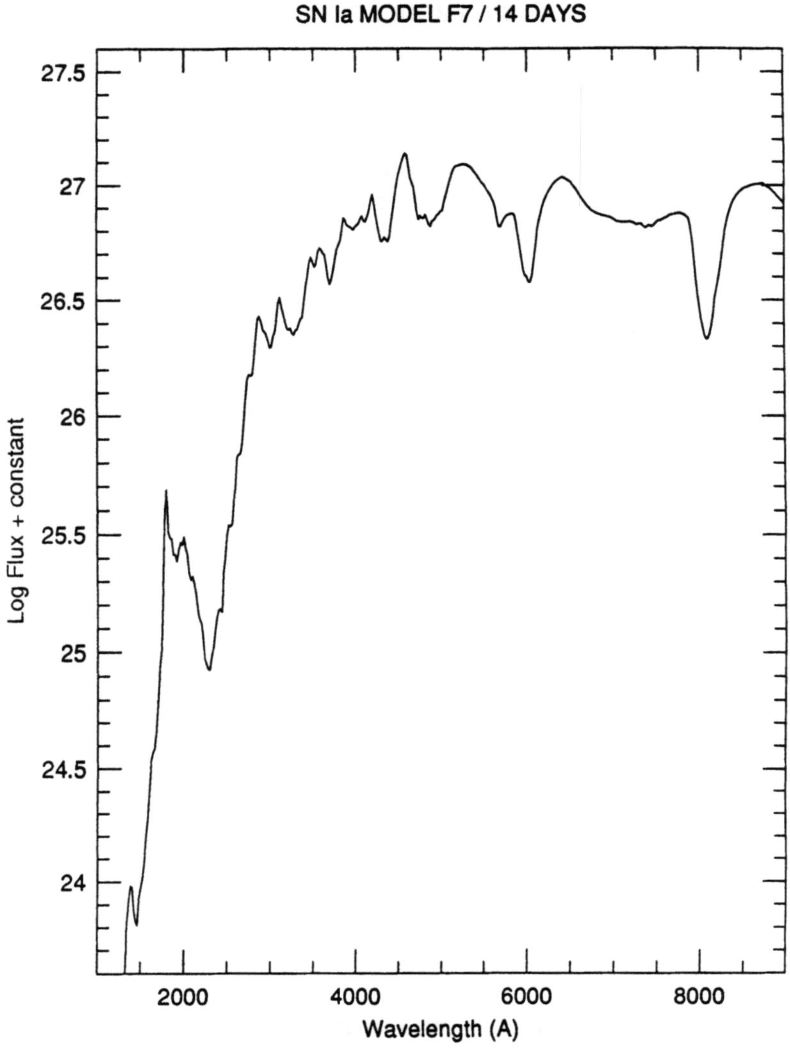

Figure 8: Emergent spectrum of model F7 at 14 days, with mixing for V > 14,000 km/s.

3. The intermediate mass elements resulting from the partial burning must be distributed over a large range of expansion velocity, with minimum velocity V > 8-10,000 km/s. The NSE products should occur at lower velocity. The more gradual transition from NSE seen in models C6 and W7 seems to produce better spectra.

I would like to thank Dr. K. Nomoto and Dr. S. Woosley for providing their explosion models.

All of the computations described here were performed on the Cray X-MP/24 and Cray EA X-MP/14se computers at the University of Texas System Center for High Performance Computing.

References

[1] Woosley, S.E., and Weaver, T.A., in *IAU Colloquium 89, Radiation Hydrodynamics in Stars and Compact Objects*, ed. by D. Mihalas and K.-H.A. Winkler (Berlin:Springer-Verlag),p. 91 (1986).

[2] Nomoto, K., Thielemann, F.-K., and Yokoi, K., Astrophys. J. **286**,644 (1984).

[3] Thielemann, F.-K., Nomoto, K., and Yokoi, K., Astron. Astrophys. **158**,17 (1986).

[4] Branch, D., Doggett, J.B., Nomoto, K., and Thielemann, F.-K., Astrophys. J. **294**,619 (1985).

[5] Harkness, R.P., in *IAU Colloquium 89, Radiation Hydrodynamics in Stars and Compact Objects*, ed. by D. Mihalas and K.-H.A. Winkler (Berlin:Springer-Verlag),p. 166 (1986).

[6] Wheeler, J.C., and Harkness, R.P., submitted to Reports on Progress in Physics (1989).

[7] Branch, D., Lacy, C.H., McCall, M.L., Sutherland, P.G., Uomoto, A., Wheeler, J.C., and Wills, B.J., Astrophys. J. **270**,123 (1983).

SECTION IX

OBSERVATIONS OF RECENT SUPERNOVAE (NOT SN 1987A)

Supernovae: Fabulous Results and Stories

Alexei V. Filippenko

1. Introduction

When I opened the printed program for this workshop and saw that Bob Kirshner, the chairman of the session on "Supernova Classification and Observations of Recent Supernovae (not 87A)," had entitled my talk "Fabulous Results and Stories," I was somewhat concerned that people would think this was the title I personally had chosen. After all, what right do I have to call my own results fabulous? On the brighter side, though, the title (and Bob's introduction — in which he referred to the "tall tales" I would tell) clearly gave me a chance to relate some interesting tidbits that might indeed entertain and amaze you. I love describing supernovae (SNe); they are often so different from one another. I wish Fritz Zwicky were alive today, so that he could see the bizarre spectra our wonderful new detectors make available to us.

2. Type Ib Supernovae: A Distinct New Subclass

It has been clear for many decades [1] that a useful way to classify SNe is to look at their optical spectra, usually near maximum brightness when the spectra are easiest to procure, and check whether there is strong evidence for hydrogen (other than from underlying H II regions) in them. If there's *no* hydrogen the SN is Type I, and if there *is* hydrogen the SN is Type II; that's all there is to it! Sure, one can also look for correlations among SN spectra and light curves, but those are all secondary characteristics. In part, the presence or absence of hydrogen is thought to be of fundamental importance because it naturally leads us to consider two different types of progenitors: white dwarfs (SN I) and supergiants (SN II). The other observed properties of SNe — such as the great homogeneity among the spectra and light curves of most SNe I, and the great heterogeneity among SNe II — seem to corroborate this simple scheme.

Long ago, though, BERTOLA and collaborators [2, 3] had noticed that *some* SNe I (specifically SNe 1962L and 1964L) lack the deep absorption trough generally seen at about 6120 Å within the first month past maximum brightness. For two decades there were few, if any, new examples of such objects, and they were simply labeled as "peculiar SNe I" (SNe Ip). Interest in them was revitalized in the mid-1980s by the studies of several newly-discovered SNe Ip made by UOMOTO and KIRSHNER [4], WHEELER and LEVREAULT [5], SRAMEK, PANAGIA, and WEILER [6], and ELIAS *et al.* [7]. Particularly influential was the thorough, much referenced (but still unpublished) optical and ultraviolet investigation of SN 1983N done by

Nino Panagia and collaborators. As summarized by PORTER and FILIPPENKO [8], SNe Ip seemed to constitute a distinct subclass, characterized by their (*a*) lack of the 6120 Å absorption trough, thought to be blueshifted Si II λλ6347, 6371 in normal SNe I, (*b*) preference for galaxies having Hubble types Sbc or later, (*c*) proximity to H II regions, (*d*) rather low luminosity, typically 1.5 mag fainter than classical SNe I, (*e*) distinct IR light curves having no secondary maximum around 1 month past primary maximum, (*f*) reddish colors, and (*g*) emission of radio radiation within a year past maximum. The subclass was named "Type Ib" [7] to distinguish it from normal SNe Ia. At least one of the earliest studies [5] had concluded that the explosion mechanism might be more closely related to that of SNe II than to SNe Ia, but nobody was really sure because the spectroscopic appearance of SNe Ib near maximum resembled that of somewhat older SNe Ia (~ 1 month past maximum). More clues to the nature of SNe Ib were needed!

On 28 February 1985 UT, during a spectroscopic survey of nearby galaxies with the Hale reflector at Palomar Observatory, Wal Sargent and I "accidentally" discovered SN 1985F near the nucleus of the peculiar SBbc galaxy NGC 4618 [9, 10]. That is, we saw a starlike object close to the end of the central bar of H II regions in NGC 4618, and thought it odd that such a bright nucleus would be present in a late-type galaxy. We guessed that it was either a starburst nucleus or a Galactic foreground star, but its spectrum proved otherwise. In addition to the narrow emission lines produced by the surrounding "fuzz" (the H II regions), we saw very strong, broad emission lines that were later identified as [O I] λλ6300, 6364, Mg I] λ4571, and Na I D. Additional spectra obtained during the next two months revealed emission lines of [Ca II] λλ7291, 7324, the Ca II infrared triplet, O I λ7774, [C I] λλ8727, 9823, 9849, and other low-excitation lines of neutral or singly ionized species (Fig. 1).

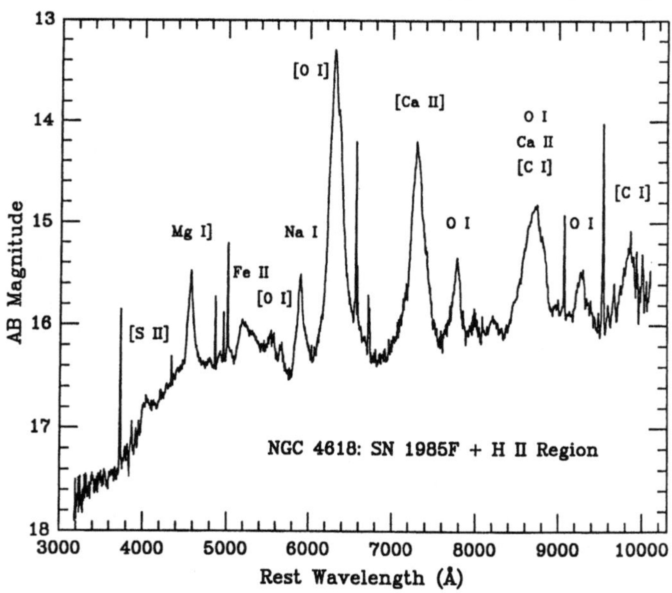

Figure 1: Spectrum of SN 1985F in NGC 4618, obtained with the Palomar Hale (5 m) and Lick Shane (3 m) reflectors in March and April 1985. Only broad (SN) lines are labeled; narrow lines are from the superposed H II regions. AB magnitude $= -2.5 \log f_\nu - 48.6$, where the units of f_ν are erg s^{-1} cm^{-2} Hz^{-1}.

The emission-line spectrum of SN 1985F, dominated by forbidden lines, suggested that we were looking at an old SN, as did the exponential decline of the derived light curve [11]. The complete absence of hydrogen led to a formal classification of SN I, although no known spectra of SNe I at *any* stage of development resembled that of SN 1985F. Given the physical conditions implied by a rough analysis of the emission-line intensity ratios, it was likely that lines of hydrogen would have been visible had hydrogen been present in appreciable quantities [9, 10]. This, along with the dominance of intermediate-mass elements, the crude estimates of at least several solar masses of ejecta, and the apparently close association with an H II region, suggested that we were looking at the explosion of a massive star that had rid itself of hydrogen prior to exploding, somewhat like the progenitor long ago proposed for Cas A by CHEVALIER [12]. The *San Francisco Chronicle* thus reported that we had "peeked at a stripteasing star," although I don't think I actually ever used those terms. Several theorists did more detailed calculations which agreed that the progenitor was probably quite massive [13, 14].

So, here was yet another new puzzle in the overall picture of SNe I, long thought to be nearly homogeneous in their observed properties. On the one hand there were the SNe Ib, which appeared to differ from SNe Ia in several ways, but which had not definitely been linked to massive stars or to a different explosion mechanism. On the other hand there was SN 1985F, spectroscopically distinct from SNe Ia (aside from the absence of hydrogen), and most likely the explosion of a massive star. It seemed that SNe I, far from being all the same, were breaking up into several distinct subclasses.

A crucial "unification" occurred when the Texas group [15] showed that a spectrum of the Type Ib SN 1983N, obtained 8 months past maximum, was very similar to that of SN 1985F at the time of its discovery. Moreover, Bob Kirshner (quoted in CHEVALIER [16]) found that a late-time spectrum of the Type Ib SN 1984L also resembled that of SN 1985F. (His spectrum is now published in [17]). Thus, SN 1985F was probably a SN Ib discovered long after maximum and, conversely, SNe Ib eventually turn into objects whose spectra really are vastly different from those of SNe Ia (Fig. 2). This provided much-needed evidence that SNe Ib constitute a physically separate subclass of SNe I, possibly having a fundamentally different explosion mechanism. It is interesting that CHUGAI [18] had, in fact, already suggested that SN 1985F might be a SN Ib discovered long after maximum.

Based on a comparison between the spectroscopic appearance of SN 1983N and SN 1985F, GASKELL *et al.* [15] deduced that Sargent and I had discovered SN 1985F 8–9 months past maximum. This was later confirmed by TSVETKOV [19], whose inspection of pre-discovery plates taken at the Crimean Station of the Sternberg State Astronomical Institute showed that SN 1985F had reached maximum brightness at $B = 12.1$ mag about 260 days prior to our discovery. In fact, the earliest plate had been obtained when SN 1985F was still on the rise, and it is sad that the object was not recognized at that time. It would have been among the brightest SNe, and certainly the brightest SN Ib, in many years. On the other hand, it is unlikely that I personally would have become so interested in supernovae had Sargent and I not stumbled across SN 1985F. Such is the nature of serendipity in astronomy!

Not having a complete series of spectra of SN 1985F over the course of its development, we must resort to other SNe Ib to see how they evolve. At present, the most complete long-term series of spectra is that of FILIPPENKO, PORTER, and SARGENT [20] for SN 1987M. The data are shown and discussed in the contribution by PORTER [21] in this volume. Of significance is the relatively early emergence of the [O I], [Ca II], and Ca II infrared triplet emission lines; they are already very prominent at $t = 2$ months. It is interesting that at this time, [O I] $\lambda6364$ is approximately as strong as [O I] $\lambda6300$, because both lines are near the intensity of the Planck function at the appropriate temperature.

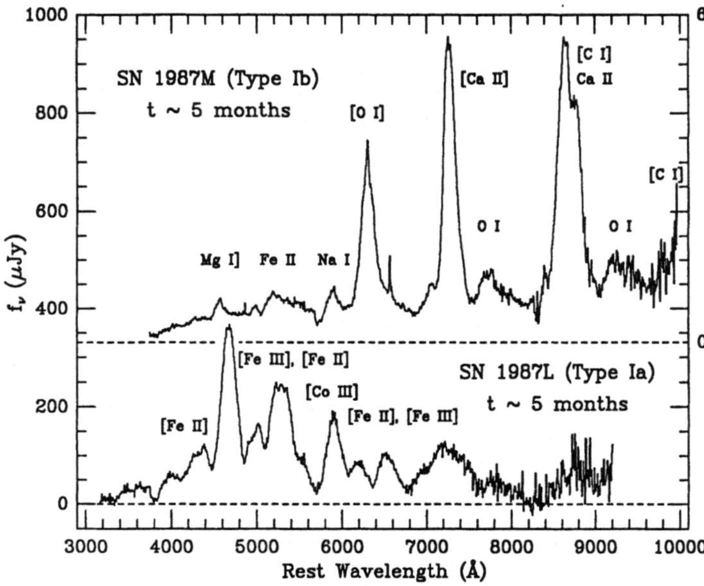

Figure 2: Spectra of SN 1987M (Type Ib) and SN 1987L (Type Ia), each obtained about 5 months past maximum brightness. Although both objects exhibit strong emission lines, the spectra could hardly be more dissimilar.

It is now becoming clear that SNe Ib may constitute a rather heterogeneous subclass, with large variations in the observed strengths of helium absorption lines in spectra obtained around maximum brightness [22, 23]. HARKNESS and WHEELER [24] even suggest that SNe Ib should actually be divided into two separate categories: SNe Ib are those showing strong He lines (e.g., SNe 1983N, 1984L), and SNe Ic are those in which He is weak or absent near maximum (e.g., SN 1987M). Perhaps a larger sample of SNe Ib should be observed before SNe Ic are formally introduced into the nomenclature. It is quite possible, for example, that there exists a continuum of helium relative strengths among SNe Ib, and that He-rich objects are not fundamentally different from He-poor objects — just as B-type main-sequence stars are not fundamentally different from A-type main-sequence stars. If a continuum is indeed found, a satisfactory classification scheme can later be developed.

3. What are the Progenitors of SNe Ib?

During the period 1984–1987 many authors [5, 10, 13, 14, 15, 16, 21, 22, 23, 25] had come to the conclusion that SN 1985F and other SNe Ib were the explosions of massive stars that had peeled off (either by winds or by mass transfer) their outer layers of hydrogen prior to exploding. In this case, the explosion mechanism was generally thought to be core collapse, as in SNe II. One of the most recent, and persuasive, arguments is that of FRANSSON and CHEVALIER [26] (see also [27] and [28]). They show that a 25 $M_\odot$ (zero-age main sequence) star that evolves to a 8 $M_\odot$ helium core develops, 9–10 months after its explosion, an emission-line spectrum remarkably similar to that of SN 1985F. The predicted line intensity ratios are in good agreement with the observed ones, but the calculated line profiles are much too flat-topped, or "boxy." This suggests that mixing of different layers occurs during the explosion. Ample evidence of such mixing has in fact been found in SN 1987A; see the review by ARNETT *et al.* [29].

Despite the evidence for massive SN Ib progenitors, there are problems to be addressed. PANAGIA and LAIDLER [30], for example, claim that SNe Ib generally occur at greater distances from H II regions (as measured by them) than do SNe II (as measured by HUANG [31]). If there really is a significant difference in distance, it implies that the progenitors of SNe Ib are probably less massive, on average, than those of SNe II — they have more time to wander away from their birth places before exploding. This interesting result must be verified with a carefully chosen sample of SNe Ib and SNe II, all measured in a consistent manner. (At least one SN Ib [SN 1985F] directly superposed on a bright H II region seems to have been excluded from the study reported in [30].) Another potential difficulty, discussed by ENSMAN and WOOSLEY [32], is that the light curve produced by iron core collapse in an 8 $M_\odot$ helium core is too broad to be consistent with that of a typical SN Ib. Even 6 $M_\odot$ helium cores yield rather broad light curves [32] (see also [33]); most SNe Ib rise and decline at about the same rate as SNe Ia during the first few months.

The above problem with light curves, however, might not be serious. Recent calculations of evolving stars by LANGER [34] suggest that the final helium core of a very massive star can be substantially smaller than that predicted in earlier studies, if mass loss is properly taken into account. This is at least partially supported by observational studies of Wolf-Rayet stars in binary systems (MOFFAT *et al.* [35]). Moreover, ENSMAN and WOOSLEY [32] point out that clumping of the ejecta can lead to smaller gamma-ray deposition, and hence to a narrower light curve. Evidence for clumping has been found in SN 1987A, of course [29], and more recently in the Type Ib SN 1985F [36]. As shown in Figure 3, both of the [O I] $\lambda\lambda$6300, 6364 lines in SN 1985F showed small-scale structure at the same relative velocities within the line profiles (thereby confirming the reality of the features), and this structure became more pronounced with time.

As emphasized by BRANCH [37], it is nevertheless possible that at least some SNe Ib represent the explosions of white dwarfs [38, 39, 40]. The most widely discussed model, that of BRANCH and NOMOTO [38] (see also [37]), invokes an off-center detonation at the base of the helium layer in a white dwarf that is slowly accreting hydrogen from the wind of a companion star. Naively, I expect that the resulting light curve should decline *faster* than that of SNe Ia, since the detonation only occurs in the outer few tenths of a solar mass; this is supported by the calculations in [41]. In addition, the predicted late-time spectrum [26, 27] does not exhibit sufficiently strong emission lines of intermediate-mass elements, and the relative intensities of certain lines (such as [C I] λ8727 and [O I] $\lambda\lambda$6300, 6364) differ from the observations. More recently, though, WOOSLEY [42] has suggested that a slow-flame carbon deflagration in a white dwarf may explain some SNe Ib; the light curve is fainter and broader, and the late-time spectrum may be consistent with observations. Further study of white dwarf models should prove illuminating.

Of course, the question of low-mass versus high-mass progenitors could be settled if the mass of the SN Ib ejecta were measured accurately. During the late-time "supernebular" phase this might seem to be an easy task with the help of standard nebular diagnostics of temperature and density [13, 17, 25, 26, 43]. In particular, the strength of [O I] λ5577 could be compared with that of [O I] $\lambda\lambda$6300, 6364 to get the electron temperature, and in the high-density limit ($n_e \gtrsim 10^6$ cm^{-3}, almost surely present at $t \lesssim 1$ year) the oxygen mass is easily computed (e.g., [25]). The problem is that the mass depends exponentially on T_e, and T_e is very difficult to measure because the weak [O I] λ5577 line is heavily blended with Fe II and other contaminants; see Figure 1, and the spectra of SN 1987M shown in [20] and [21]. For example, if the distance of SN 1985F is 7.1 Mpc and the measured flux of [O I] λλ6300, 6364 in February 1985 was 2.2×10^{-12} erg s^{-1} cm^{-2} [9], then we find that $M_O = 0.11\ M_\odot$ if $T_e = 10\,000$ K, $M_O = 1.1\ M_\odot$ if $T_e = 5000$ K, and $M_O = 11\ M_\odot$ if $T_e = 3300$ K. Clearly, an accurate measurement of T_e is needed if we are to directly determine the mass of the ejected oxygen!

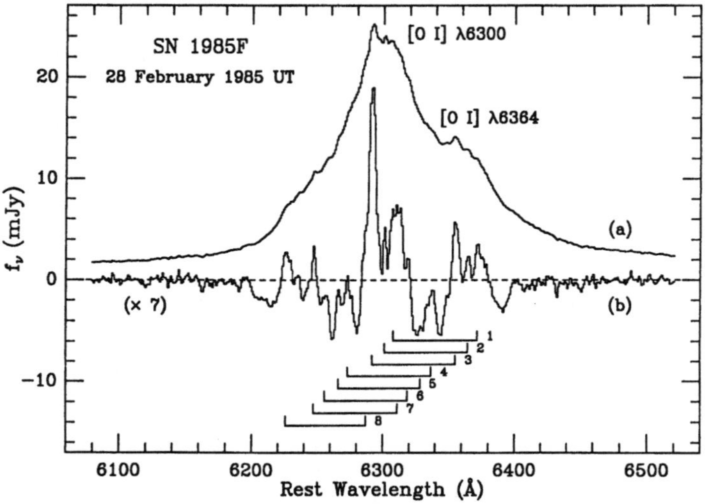

Figure 3: (*a*) [O I] blend in SN 1985F, roughly 9 months past maximum. (*b*) A heavily smoothed version of (*a*) was subtracted from (*a*) to obtain this residual spectrum, scaled by a factor of 7. Pairs of [O I] λλ6300, 6364 "peaks" having the same relative velocity are indicated.

4. The "Missing Link" Between SNe Ib and SNe II?

In late July 1987, a supernova which may play a particularly significant role in our understanding of SNe Ib was discovered [44] by the Berkeley Automated Supernova Search Team [45] in the Virgo spiral galaxy NGC 4651. The first few spectra of SN 1987K showed it to be a SN II, with a P Cygni profile of Hα, and I dutifully reported its spectral type to the *IAU Circulars*. However, Hα did not seem as prominent as in most SNe II having well-developed low-excitation absorption lines elsewhere in the spectrum. Furthermore, the object was atypical in that the relative strength of the Hα emission line did not grow during the next two weeks. Anyone who has tried to observe the Virgo cluster in August knows that it's quite tough to do, as the Sun tends to get in the way; sadly, the object was subsequently lost for a few months.

When I reobserved it in December 1987, I was astonished to find that, unlike previously observed SNe II, it had lost all traces of the broad Hα emission line [46]. Instead, the spectrum exhibited broad [O I] λλ6300, 6364 and [Ca II] λλ7291, 7324 — the defining characteristics of *Type Ib SNe* long after maximum! These results were confirmed by a spectrum obtained in February 1988 (Fig. 4). Such a metamorphosis is unprecedented, but would undoubtedly have pleased Fritz Zwicky.

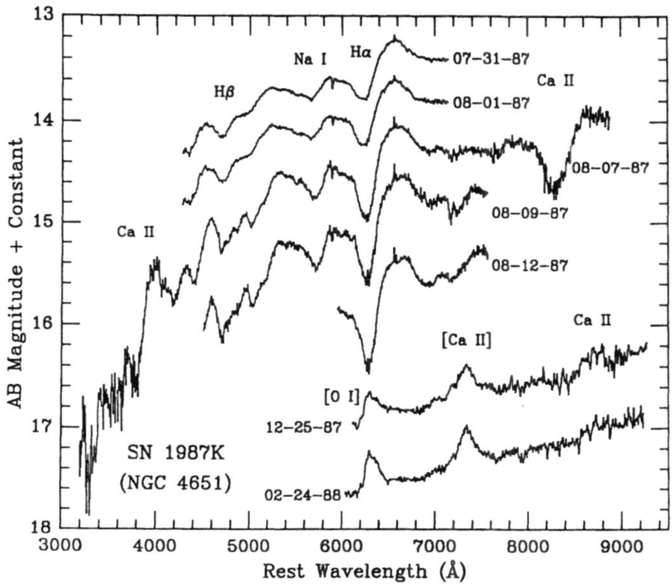

Figure 4: Series of spectra of SN 1987K, showing its transformation from a SN II into a SN Ib. Narrow Hα and other emission lines from the underlying H II region have been excised in the last two spectra, for clarity. Constants added to the spectra before plotting were (top to bottom) −0.85, −0.45, 0.0, 0.15, 0.0, 0.3, and 0.2 mag.

Apparently, Bob Kirshner and Eric Schlegel had not obtained an early spectrum of SN 1987K, but observed it for the first time in April 1988. They, too, found the spectrum to be that of a SN Ib [17]. Not knowing my new results, over which I was still pondering because they were so unexpected, Bob telephoned me and said that I had mistakenly called SN 1987K a SN II, when actually it was a SN Ib. I assured Bob that SN 1987K had undergone the supernova equivalent of a sex-change operation, and that my early classification was indeed correct. Bob, naturally, seemed skeptical, but I think that now, having seen the full set of spectra, he's a believer.

The simplest interpretation of SN 1987K, of course, is that it was a massive star that lost *most*, yet not all, of its outer layer of hydrogen prior to its explosion as a regular core-collapse SN [46, 47]. This could occur if the star were not sufficiently massive (say, 25–35 $M_\odot$?) for its wind to blow away all of the hydrogen and become a helium giant, or if the star were less massive but dumped much of its atmosphere onto a binary companion. If it had a thin hydrogen shell at the time of its explosion, it would masquerade as a SN II for a while, and as the expanding ejecta thinned out the spectrum would become dominated by emission from the deeper and denser layers. A completely different conjecture is that there was plenty of hydrogen in the atmosphere of the star, but at late times it was poorly heated, perhaps because radioactive nuclides were not mixed sufficiently well with the hydrogen.

It is interesting that, in the early models of SN 1987A, WOOSLEY *et al.* [48] suggested that the progenitor might have undergone much mass loss, leaving behind a very thin hydrogen shell. They dubbed such an object a SN IIb — having the spectroscopic properties of SNe II initially, and of SNe Ib later on. In fact, we now know that SN 1987A had a massive hydrogen envelope (5–10 $M_\odot$), but the idea is nevertheless a potentially valid one for other SNe. Notably, VAN DEN BERGH [49] had previously argued that the difference between SNe Ib and SNe II might be only skin deep. Further evidence for this point of view is provided by the detailed analysis of Cas A ejecta done by FESEN, BECKER, and GOODRICH [50].

If the above hypothesis of a thin hydrogen skin is correct, we expect several observable consequences. First, the expansion velocities of the oxygen and calcium ejecta in SN 1987K should be larger than those in typical SNe II many months after core collapse, since SN 1987K didn't have a massive hydrogen envelope tamping the inner layers. This is indeed the case, as shown in Figure 1 of FILIPPENKO [51], where a comparison is made between the late-time line widths in SN 1987K and in the Type II SN 1986I. Second, SN 1987K should have had a *broader* light curve than typical SNe Ib, because of the hydrogen recombination wave traversing the remaining envelope in SN 1987K. The light curve illustrated in [46] seems to confirm this, although the evidence is not conclusive because the data were taken through a red (rather than visual or blue) bandpass. Finally, some SNe Ib should exhibit a *very* weak Hα line near maximum brightness, to complete the mass-loss continuity between SNe Ib and SNe II. In retrospect, we do see local maxima near Hα in the spectra of a few SNe Ib (e.g., SN 1988L [46]), but the mere presence of a local maximum at the right position in a very bumpy spectrum does not constitute proof that hydrogen is present. It would be more convincing if a shallow P Cygni profile were present at the correct position. Such might be the case in SN 1989O [52], as shown in Figure 5. SN 1989O appears to be a SN II, but Hα is very weak, even compared with Hα in SN 1987K. (The Ca II lines in SN 1989O seem even weaker.) Unfortunately, this object was so distant ($z \approx 0.064$) and faint at maximum ($m \approx 18$) that it was not possible to follow for a long time.

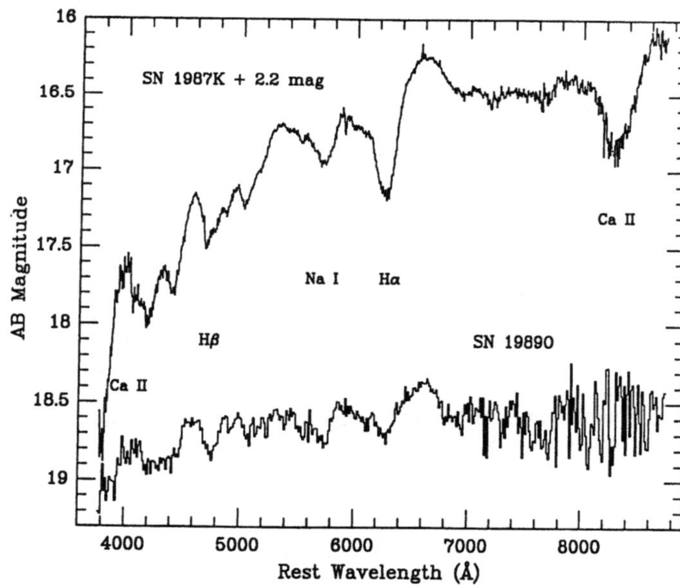

Figure 5: Spectrum of SN 1989O in MCG 6–1–26, obtained at Lick Observatory 5–6 days after discovery, compared with that of SN 1987K. Note the weak Hα and Ca II in SN 1989O, and the overall similarity of the spectra.

Despite the superficial success of my naive interpretation of SN 1987K, what we really need is detailed models which show that the spectrum of a massive star having a thin hydrogen envelope really does resemble that of SN 1987K, and evolves in the observed manner. Robert Harkness and Craig Wheeler are currently attempting to do the requisite calculations, and the preliminary results are not encouraging. The theoretical spectrum seems to show deep features produced by the heavy elements that would naturally be present (with roughly solar abundances) in the hydrogen and helium shells of a massive star, yet these are not observed. I am not yet sure whether this represents a flaw in the calculations, in the specific assumptions, or in the basic hypothesis; further theoretical work is needed to clarify the situation. More observations of SNe like SN 1987K are also crucial, but these objects are apparently quite rare. Aside from the questionable case of SN 1989O, no known SN since SN 1987K has shown similar properties. Given its relatively broad light curve [32], SN 1985F may have actually been a SN II near maximum (like SN 1987K), but we will probably never know. Unlike the case for SN 1987A, the light echoes of SN 1985F are much too faint to observe, let alone get spectra of; in this case we can't "play it again, Sam" (to quote Bob Kirshner).

5. A Distinct Subclass of Type II Supernovae?

I would now like to mention a topic similar to that discussed by Eric Schlegel at this workshop — namely, the gradual emergence of a new, distinct subclass of SNe II. Most SNe II show prominent P Cygni profiles of hydrogen in their spectra, but in some objects the absorption component is either very weak or entirely absent. These SNe exhibit strong Hα emission whose equivalent width grows to astoundingly high values as they age. The Hα line is sometimes superposed on a much broader component of Hα emission. Their continua seem bluer than normal, they stay bright for a very long time, and their late-time spectra ($t \approx 1$ year) indicate extremely high densities in the ejecta. Some of them are also unusually luminous at maximum, compared with other SNe II. Finally, a few show strong evidence of fairly dense circumstellar matter, probably ejected prior to the explosion.

An excellent example of such an object is SN 1987F, which occurred in a bright H II region in a spiral arm of NGC 4615 (FILIPPENKO [53]). When first observed, broad Hα emission was superposed on a luminous ($M_V \approx -19.3$ mag), nearly featureless continuum, but its profile did not have the characteristic P Cygni shape, and its centroid was blueshifted by $\gtrsim 1500$ km s^{-1} with respect to the systemic velocity of the parent galaxy (Fig. 6). The near-maximum spectrum of SN 1988I, also discussed in [53], was similarly dominated by broad hydrogen Balmer lines, but the SN was too faint to monitor for an extended period of time. Many months later, the broad Hα in SN 1987F was more luminous, and had much larger equivalent width; Fe II, Ca II, and O I emission were detected as well. Forbidden lines, normally quite strong at this phase, were very weak. The narrow component of Hα, initially quite luminous, was now much less prominent. At early times it may have been produced by material previously ejected by the progenitor, but this gas was eventually engulfed by the expanding SN ejecta, as in the case of SN 1984E (GASKELL [54]; HENRY and BRANCH [55]).

The derived electron density in the ejected envelope of SN 1987F at late times was $\gtrsim 10^9$ cm^{-3}. This, together with the observed flux of Hα emission, can be used to derive the mass of the emitting hydrogen: $M \lesssim 0.1\ M_\odot$ if $H_0 = 75$ km s^{-1} Mpc^{-1}. (An earlier calculation [53] gave $M \gtrsim 5-30\ M_\odot$, but clumping of the ejecta had unfortunately been neglected. In fact, I find that the filling factor of clumps is $\lesssim 0.01$ if the ejecta are expanding at $v \gtrsim 6000$ km s^{-1}.) Although this does not seem like a large amount of gas, it is likely that the progenitor was massive; the light curve declined very slowly at both early and late times. With a few small but probably significant exceptions, the overall optical spectroscopic properties of SN 1987F closely resemble those of type 1 Seyfert nuclei and QSOs, whose emission-line spectra are thought to be produced by high-density clouds irradiated by a flat ultraviolet continuum.

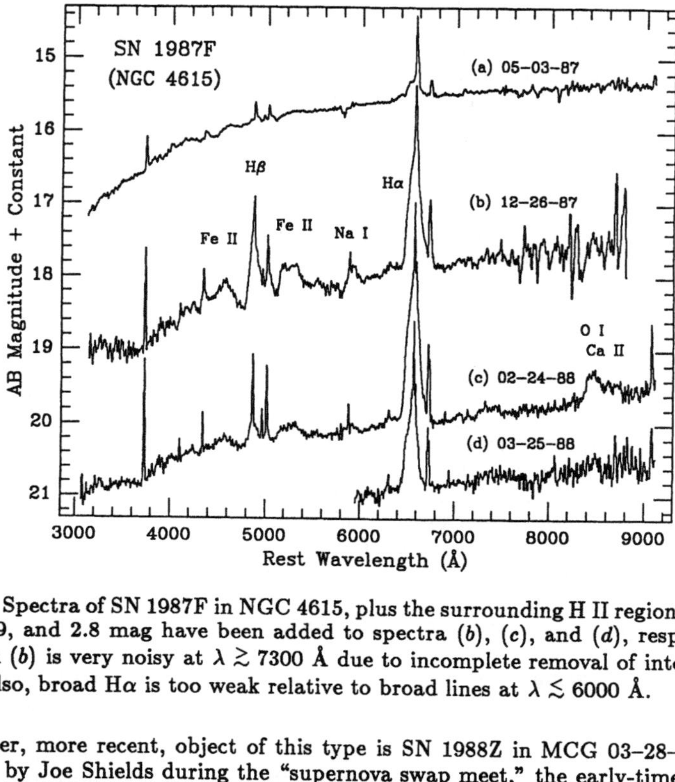

Figure 6: Spectra of SN 1987F in NGC 4615, plus the surrounding H II region. Offsets of 0.2, 1.9, and 2.8 mag have been added to spectra (*b*), (*c*), and (*d*), respectively. Spectrum (*b*) is very noisy at $\lambda \gtrsim 7300$ Å due to incomplete removal of interference fringes; also, broad Hα is too weak relative to broad lines at $\lambda \lesssim 6000$ Å.

Another, more recent, object of this type is SN 1988Z in MCG 03–28–022. As discussed by Joe Shields during the "supernova swap meet," the early-time spectra of SN 1988Z showed very narrow [O III] λ4363 and [O III] $\lambda\lambda$4959, 5007 emission lines whose relative intensities indicate $n_e \gtrsim 10^{6.9}$ cm^{-3}, *regardless* of the electron temperature. The actual electron density may be substantially greater, but in any case it is much higher than that of the normal interstellar medium. Although careful measurements are still in progress, the flux of [O III] λ5007 may have grown with time, at least during the first month. Moreover, the helium abundance of this gas appears to be abnormally high. The obvious conclusion is that the progenitor of SN 1988Z expelled a shell of dense material, and this material became ionized by the flash of ultraviolet radiation emitted by the SN as the shock wave broke through the stellar surface, as in SN 1987A (FRANSSON *et al.* [56]).

Even after several months, SN 1988Z remained remarkably blue. Moreover, as in SN 1987F, the density of the actual SN ejecta was very high: forbidden lines were weak or absent, and very strong lines of Fe II, Ca II, and O I emerged (Fig. 7). The blend of O I λ8446 and the Ca II infrared triplet, in particular, became stronger than the very broad (FWHM $\approx$ 15 000 km s^{-1}) component of Hα, yet little or no [Ca II] $\lambda\lambda$7291, 7324 was present.

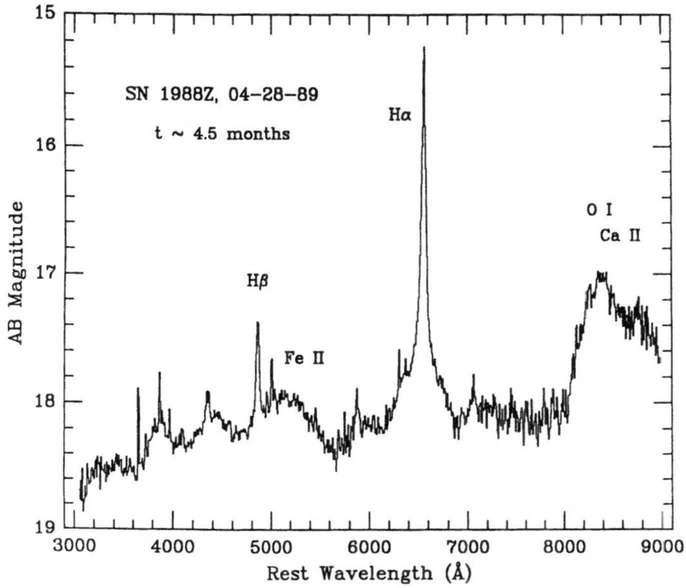

Figure 7: Spectrum of SN 1988Z, obtained with the Shane reflector at Lick Observatory. The narrow [O III] lines reveal the presence of a dense circumstellar shell. Balmer lines of intermediate and great width are also present. Note the broad undulations in the blue-green part of the spectrum, which are probably produced by Fe II.

But SN 1988Z has a lot more in store for us than this! Perusal of the spectra shows that an intermediate-width component of the Balmer lines (FWHM $\approx$ 1600 km s^{-1}) became visible after about 1 month, and its flux steadily grew with time. As shown in Figure 7, the Balmer decrement of this component is very steep, indicating that some sort of collisional excitation might be operating. Is this another shell of matter, ejected at high velocity from the progenitor of SN 1988Z immediately prior to the titanic explosion? Its origin is not yet clear, but analysis of the old spectra — as well as new data — is continuing. SN 1988Z has remained so bright that it is still easily observed, nearly one year after discovery, despite its relatively large distance ($z \approx 0.022$). The new spectra reveal the dominance of the intermediate component; the equivalent width of Hα is enormous. In fact, the optical spectrum resembles that of the radio-loud SN 1986J (RUPEN *et al.* [57]), and an attempt should be made to observe SN 1988Z with the VLA.

6. Future Prospects

We now see that ZWICKY [58] was probably right when he said that there are actually many classes of SNe; with sufficiently good spectra, we are finding all sorts of interesting differences among objects. However, we probably shouldn't further subdivide SNe into new subclasses (other than Ia, Ib, II-L, and II-P) until we have a much more extensive data base. It would be good to choose observable characteristics that are most fundamental from a *physical* point of view. Perhaps, for example, those objects having strong Hα emission, but little or no absorption, always experience pre-supernova winds; on the other hand, SN 1987A went through a prominent phase of gas ejection, yet it developed Balmer lines having normal P Cygni profiles.

To achieve such a data base, it is clear that observers need to start pooling their data — especially their spectra, which do not depend on site-specific bandpasses of various filters. Although obtaining one spectrum of each available SN is still useful for statistical purposes, we can take a major step beyond this by getting relatively full coverage of the spectral evolution of a few well-placed objects. This is difficult for one observer, or even a team of astronomers at one observatory, to achieve because of ever-present time constraints and the rapid fading of most SNe. However, by working together we can obtain high-quality data for many objects other than SN 1987A. Fortunately, recent progress in the use of electronic (computer) mail and standardized software has made our task considerably easier.

I am grateful to the entire local organizing committee of this workshop for making our stay in Santa Cruz so pleasant, and especially for asking Mother Nature to postpone the big earthquake until well after we had departed. Special thanks also go to Stan Woosley for his astonishing patience; my contribution must be the very last one to be submitted. I also thank my collaborators on various projects for allowing me to discuss results prior to formal publication elsewhere. This work is partially funded by a Presidential Young Investigator Award (NSF grant AST–8957063), as well as by the Center for Particle Astrophysics, an NSF Science and Technology Center operated by the University of California at Berkeley under Cooperative Agreement AST–8809616. Support from the California Space Institute (grant CS–41–88) is also appreciated. Lick Observatory receives partial funding from NSF through Core Block grant AST–8614510.

References

1. R. Minkowski: *Pub. A. S. P.*, **53**, 224 (1941).
2. F. Bertola: *Ann. d'Ap.*, **27**, 319 (1964).
3. F. Bertola, A. Mammano, and M. Perinotto: *Contrib. Asiago Obs.*, **174**, 51 (1965).
4. A. Uomoto and R. P. Kirshner: *Astr. Ap.*, **149**, L7 (1985).
5. J. C. Wheeler and R. Levreault: *Ap. J. (Letters)*, **294**, L17 (1985).
6. R. A. Sramek, N. Panagia, and K. W. Weiler: *Ap. J. (Letters)*, **285**, L59 (1984).
7. J. H. Elias, K. Mathews, G. Neugebauer, and S. E. Persson: *Ap. J.*, **296**, 379 (1985).
8. A. C. Porter and A. V. Filippenko: *A. J.*, **93**, 1372 (1987).
9. A. V. Filippenko and W. L. W. Sargent: *Nature*, **316**, 407 (1985).
10. A. V. Filippenko and W. L. W. Sargent: *A. J.*, **91**, 691 (1986).
11. A. V. Filippenko, A. C. Porter, W. L. W. Sargent, and D. P. Schneider: *A. J.*, **92**, 1341 (1986).
12. R. A. Chevalier: *Ap. J.*, **208**, 826 (1976).
13. M. C. Begelman and C. L. Sarazin: *Ap. J. (Letters)*, **302**, L59 (1986).
14. R. Schaeffer, M. Cassé, and S. Cahen: *Ap. J. (Letters)*, **316**, L31 (1987).
15. C. M. Gaskell, *et al.*: *Ap. J. (Letters)*, **306**, L77 (1986).
16. R. A. Chevalier: *Highlights Astron.*, **7**, 599 (1986).
17. E. M. Schlegel and R. P. Kirshner: *A. J.*, **98**, 577 (1989).
18. N. N. Chugai: *Pis'ma Astron. Zh.*, **12**, 461 (*Sov. Astron. Letters* **12**, 192) (1986).
19. D. Yu. Tsvetkov: *Pis'ma Astron. Zh.*, **12**, 784 (*Sov. Astron. Letters* **12**, 328) (1986).
20. A. V. Filippenko, A. C. Porter, and W. L. W. Sargent: submitted (1990).

21. A. C. Porter: These *Proceedings*.
22. J. C. Wheeler, *et al.*: *Ap. J. (Letters)*, **313**, L69 (1987).
23. R. P. Harkness, *et al.*: *Ap. J.*, **317**, 355 (1987).
24. R. P. Harkness and J. C. Wheeler: In *Supernovae*, ed. by A. Petschek (Springer-Verlag, Berlin 1989), in press.
25. A. Uomoto: *Ap. J. (Letters)*, **310**, L35 (1986).
26. C. Fransson and R. A. Chevalier: *Ap. J.*, **343**, 323 (1989).
27. C. Fransson: In *Atmospheric Diagnostics of Stellar Evolution*, IAU Colloquium No. 108, ed. by K. Nomoto (Springer-Verlag, Berlin 1988), p. 385.
28. T. Axelrod: In *Atmospheric Diagnostics of Stellar Evolution*, IAU Colloquium No. 108, ed. by K. Nomoto (Springer-Verlag, Berlin 1988), p. 319.
29. W. D. Arnett, J. N. Bahcall, R. P. Kirshner, and S. E. Woosley: *Ann. Rev. Astr. Ap.*, **27**, 629 (1989).
30. N. Panagia and V. G. Laidler: In *Supernova Shells and their Birth Events*, ed. by W. Kundt (Springer-Verlag, Berlin 1989), in press.
31. Y.-L. Huang: *Pub. A. S. P.*, **99**, 461 (1987).
32. L. M. Ensman and S. E. Woosley: *Ap. J.*, **333**, 754 (1989).
33. K. Nomoto, T. Shigeyama, and M. Hashimoto: In *Atmospheric Diagnostics of Stellar Evolution*, IAU Colloquium No. 108, ed. by K. Nomoto (Springer-Verlag, Berlin 1988), p. 319.
34. N. Langer: *Astr. Ap.*, **220**, 135 (1989).
35. A. F. J. Moffat, *et al.*: *A. J.*, **91**, 1386 (1986).
36. A. V. Filippenko and W. L. W. Sargent: *Ap. J. (Letters)*, **345**, L43 (1989).
37. D. Branch: In *Atmospheric Diagnostics of Stellar Evolution*, IAU Colloquium No. 108, ed. by K. Nomoto (Springer-Verlag, Berlin 1988), p. 281.
38. D. Branch and K. Nomoto: *Astr. Ap.*, **164**, L13 (1986).
39. A. M. Khokhlov and E. V. Ergma: *Pis'ma Astron. Zh.*, **12**, 366 (*Sov. Astron. Letters*, **12**, 152) (1986).
40. I. Iben, K. Nomoto, A. Tornambé, and A. Tutukov: *Ap. J.*, **317**, 717 (1987).
41. S. E. Woosley, R. E. Taam, and T. A. Weaver: *Ap. J.*, **301**, 601 (1986).
42. S. E. Woosley: In *Supernovae*, ed. by A. Petschek (Springer-Verlag, Berlin 1989), in press.
43. N. N. Chugai: *Sov. Astron. Circular* No. 1469 (1986).
44. C. Pennypacker: *IAU Circular* No. 4426 (1987).
45. S. Perlmutter, *et al.*: In *Instrumentation for Ground-Based Optical Astronomy*, ed. by L. B. Robinson (Springer-Verlag, Berlin 1988), p. 67.
46. A. V. Filippenko: *A. J.*, **96**, 1941 (1988).
47. A. V. Filippenko: *Proc. Astr. Soc. Aust.*, **7**, 540 (1988).
48. S. E. Woosley, P. A. Pinto, P. G. Martin, and T. A. Weaver: *Ap. J.*, **318**, 664 (1987).
49. S. van den Bergh: *Ap. J.*, **327**, 156 (1988).
50. R. A. Fesen, R. H. Becker, and R. W. Goodrich: *Ap. J. (Letters)*, **329**, L89 (1988).
51. A. V. Filippenko: In *Particle Astrophysics: Forefront Experimental Issues*, ed. by E. B. Norman (World Scientific, Singapore 1989), p. 177.
52. A. V. Filippenko and J. C. Shields: *IAU Circular* No. 4851.
53. A. V. Filippenko: *A. J.*, **97**, 726 (1989).
54. C. M. Gaskell: *Pub. A. S. P.*, **96**, 789 (1985).
55. R. B. C. Henry and D. Branch: *Pub. A. S. P.*, **99**, 112 (1987).
56. C. Fransson, *et al.*: *Ap. J.*, **336**, 429 (1989).
57. M. P. Rupen, *et al.*: *A. J.*, **94**, 61 (1987).
58. F. Zwicky: In *Stars and Stellar Systems*, Vol. 8, ed. by L. H. Aller and D. B. McLaughlin (Univ. of Chicago Press, Chicago 1965), p. 367.

A Spectroscopic Glance at the CfA Supernovae Atlas

Eric M. Schlegel

Supernovae have traditionally been difficult to study. These objects are transient events, hence unpredictable. The scheduling of telescope time for observations, particularly to cover a sizable portion of the evolution, is impossible. The obvious alternative is to request observations by the available observers, using the available equipment. This approach usually works near maximum, as most spectroscopists are at least somewhat curious about supernova spectra. However, it usually proves difficult to maintain interest, particularly as the supernova fades, which is when the observations become difficult. Consequently, a few, stray spectra of a particular supernova, near maximum, are obtained, and either published or left to rot on a data tape. There are two cures for this problem. First, interested observers can migrate to those institutions which have good facilities, and subsequently make use of them. Second, one can "arrange" to blow up a nearby star, and raise the interest level of many observers. The workshop has demonstrated that SN 1987A produced the second cure. This contribution describes a program which tackles the first.

Data for all supernovae other than SN 1987A are relatively sparse. A program exists at the Center for Astrophysics to observe as many supernovae as possible as often as possible. The observations consist of CCD photometry and spectroscopy, using a variety of telescopes and instruments. The telescopes used include the 0.6m for CCD photometry, the 1.5m Tillinghast reflector, both on Mt. Hopkins, and the Multiple Mirror Telescope. The instruments used include the "Z-machine" (DAVIS and LATHAM [1]) (a Reticon spectrograph built for the CfA redshift survey), occasionally the Faint Object Grism Spectrograph on the MMT, and more recently, the MMT "Red Channel" spectrograph (SCHMIDT *et al.* [2]). The wavelength coverage is generally 4000Å to 7000Å, at about 5-10Å resolution. It should be noted that the Z-machine flux calibration is sometimes poor (too much operator experience with the redshift survey). The desire for good flux calibration indicates a need for good photometry with which to calibrate the spectroscopy.

A supernova data base is slowly being assembled, and will eventually be made available to all. Table 1 lists the objects present to date (this review written in August 1989), with the number of spectra obtained on each object. This type of data base can be as useful as the vast pile of data on SN 1987A, simply because of the wider range of supernova properties sampled. Numerous questions can be addressed with

Table 1
Supernovae Atlas by Type: Spectra Available

SN II		SN Ia		SN Ib		Unknown	
SN 80K	19*	SN 81F	1	SN 83I	1	SN 82D	1
SN 81E	1	SN 81G	3	SN 83N	5	SN 82V	1
SN 82F	3	SN 82C	1	SN 84L	12*	SN 82X	1
SN 84E	6?	SN 85A	9*	SN 85F	4*	SN 82Y	1
SN 85H	2	SN 85B	8*	SN 87K	2	SN 88W	1
SN 85L	1	SN 86A	5	SN 87M	4		
SN 85O	13	SN 86N	5	SN 88L	1		
SN 85R	5	SN 86O	6	SN 89E	1		
SN 86B	1	SN 87D	3				
SN 86E	6	SN 87L	6				
SN 86I	4	SN 87N	5				
SN 86K	7	SN 88B	2				
SN 87B	5	SN 88C	2				
SN 87C	3	SN 88F	2				
SN 87F	2	SN 88V	4x?				
SN 88A	13x	SN 89B	15x				
SN 88H	4	SN 89D	8				
SN 88Q	1	SN 89M	2x				
SN 88S	1						
SN 88Y	1						
SN 88Z	5x						
SN 89A	5						
SN 89C	11x						
SN 89F	2						
SN 89K	1						
SN 89L	3x						

* = data published, or data published by others and donated
x = still following (August 1989)

these data, including the few listed below:

- spectroscopic differences between SN Ia in E, S, and Irr galaxies;
- spectroscopic differences between SN II-L and SN II-P;
- velocity behavior of particular lines, for example, the Si II 6150[Å] absorption

line seen in SN Ia;

– the net H_α flux evolution.

A typical collection of spectra for the SN Ib SN 1984L, at late times, is shown in Figure 1 (next page). Numbers listed in the upper right-hand corner are the age in days past maximum (if known; otherwise, the age in days past discovery), and the vertical offset to separate the data. The offset is the value given by

$$\text{plotted vertical position} = \log(\text{ data value }) + \text{offset}.$$

Note the large [O I] 6300/6363Å emission lines, a characteristic of the late-time spectra of SN Ib. The SN 1984L data indicate the benefit of late-time supernova data: the 385-day spectrum, while noisy (the "technical" term is "herbaceous"), provided the connection between SN 1983N and SN 1985F, thereby establishing the subclass of SN Ib. SN 1983N was relatively well-observed at early times (PANAGIA *et al.* [3]), but was not observed beyond about 100 days. SN 1985F, the Filippenko-Sargent object (FILIPPENKO and SARGENT [4]), was discovered about 9 months past maximum. SN 1984L was well-observed at early times (HARKNESS *et al.* [5]), and was observed at late times. The early-time data on SN 1983N and SN 1984L showed that some supernovae of Type I belonged in a separate subclass, and the late-time data have indicated what types of object may be the progenitors for these supernovae (SCHLEGEL and KIRSHNER [6]). A summary of the late-time data on SN 1984L is included elsewhere in this volume (SCHLEGEL [7]).

Figure 2 (second page after this one) is a spectrum of SN 1988S, discovered by C. Pollas and workers (POLLAS [8]). I include this spectrum, the only one I know of, because it appears to show a bump in the H_α profile at the same location as that seen in SN 1987A, and described there as the "Bochum event" (CRISTIANI *et al.* [9]). I point out this feature in the SN 1988S spectrum for two reasons. First, the Bochum event, described as the photospheric breakout of ^{56}Ni (PHILLIPS and HEATHCOTE [10], LUCY [11], HANUSCHIK *et al.* [12]), could also be studied in distant supernovae, thereby providing a wider range of behavior to aid in understanding what this event is. Second, high-resolution spectroscopy is not a waste of time, if that is what you as an observer can contribute. The only thing which must be considered with high-resolution spectroscopy is what bandpass in which to work. The bandpass will be most useful if it is matched to the supernova type. For a Type II, working near H_α is best.

Figure 3 (third, fourth pages after this one) shows a series of early-time spectra of the recent Type Ia supernova SN 1989B. The SN Ia characteristic Si II 6150Å absorption is present, and can be followed for about the first 25 days. Note that a feature is present in the core of the line, and that it appears to vanish after a few days. If the line persists for a few days in the cores of the Si II lines of all SN Ia, then it could bias the velocity centroid of the line. This *might* lead to the variations of the sort discussed in BRANCH *et al.* [13]. A larger discussion of the Si II line for all SN Ia observations included in the database will be published elsewhere.

Two patterns that I believe are present in the data, and which were discussed at the Workshop as "work in progress", have not been included in these proceedings.

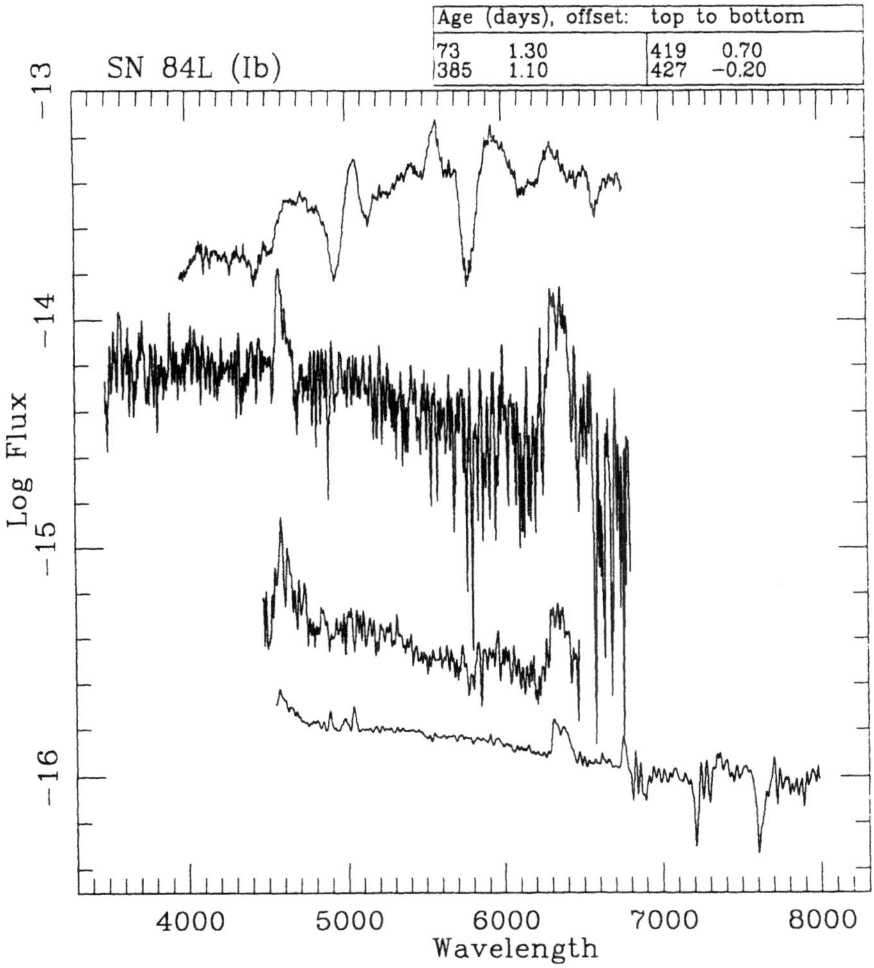

Figure 1: The late-time spectra of the Type Ib SN 1984L. Note the [O I] 6300-6363Å lines, which characterize the late-time emission of this subclass.

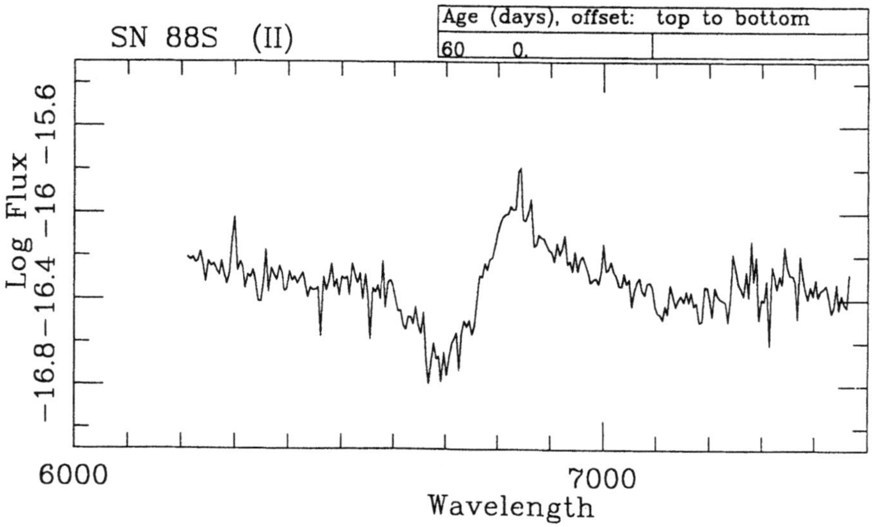

Figure 2: A high-resolution spectrum of the Type II SN 1988S, about 60 days past maximum. Note the feature on the blue edge of the emission component, just redward of 6800Å. A similar feature, called the "Bochum event", was seen in SN 1987A. This raises the possibility that such an event, whatever its cause, can be studied in distant supernovae.

The two patterns, a possible spectroscopic distinction between SN II-L and SN II-P, and a pattern of velocity behavior in the H_α lines, are still being worked on, and will be reported in detail at another time.

Finally, recent data on some Type II supernovae point to the possibility of a new subclass. These supernovae are characterized by having weak H_α profiles (relative to the typical Type II, such as SN 1986I). The profiles are often asymmetric, with the red portion of the profile missing. The broad base generally lies blueward relative to the narrow portion of the line. The narrow lines are often resolved, and appear to be wider than the lines from neighboring H II regions. Table 2 lists the candidate objects, and Figure 4 shows two spectra of SN 1987F. SN 1987F has been described by FILIPPENKO [14], where he labels these objects "Seyfert I supernovae". I prefer the label "SN IIn (n = narrow)", which describes the most apparent feature in the spectrum without invoking a particular model. A somewhat larger discussion of these objects has been submitted for publication (SCHLEGEL [15]). These objects may be related to SN 1984E (e.g., HENRY and BRANCH [16]) and SN 1983K (Niemela *et al.* [17]). Some data on SN 1984E are being resurrected for comparison. The published spectra of SN 1984E appear to be superficially related in that one or two of the spectra look similar to the proposed SN IIn spectra. However, the SN 1984E data show a broader H_β line, while the SN 1983K data show He II 4686Å in emission, and absorption at relatively early times at H_α. Both features are not seen in the proposed SN IIn subclass. Perhaps there is a continuum in some parameter, and SN 1983K and SN 1984E sit a one end, while the proposed SN IIn subclass objects sit at the other end. Further analysis will indicate how likely this idea is.

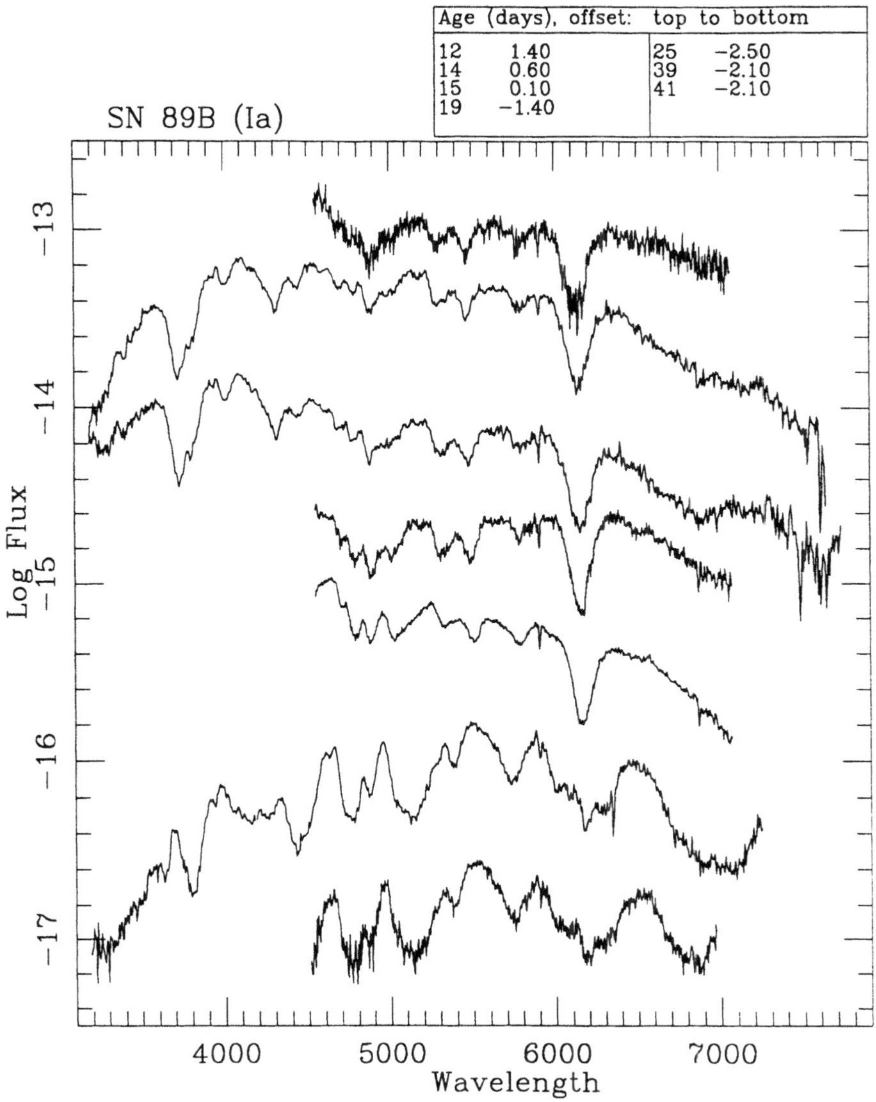

Figure 3. A series of early-time spectra, out to day 100, of the Type Ia SN 1989B. The characteristic 6150Å absorption is present. Note the presence of an emission-like component just redward of the core of the line near day 15. This core emission could distort the velocity behavior attributed to the line. A large database similar to the one being assembled here will investigate such behavior.

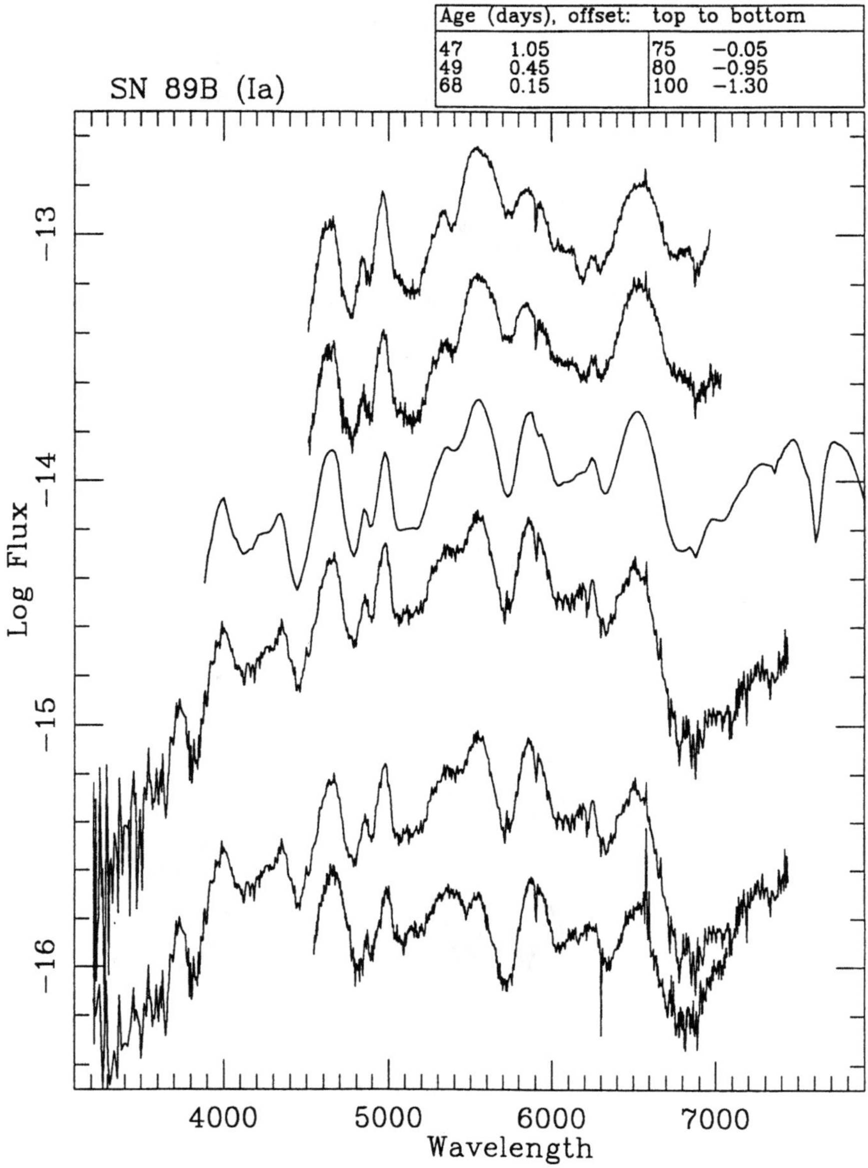

Figure 3. A series of early-time spectra, out to day 100, of the Type Ia SN 1989B. The characteristic 6150Å absorption is present. Note the presence of an emission-like component just redward of the core of the line near day 15. This core emission could distort the velocity behavior attributed to the line. A large database similar to the one being assembled here will investigate such behavior.

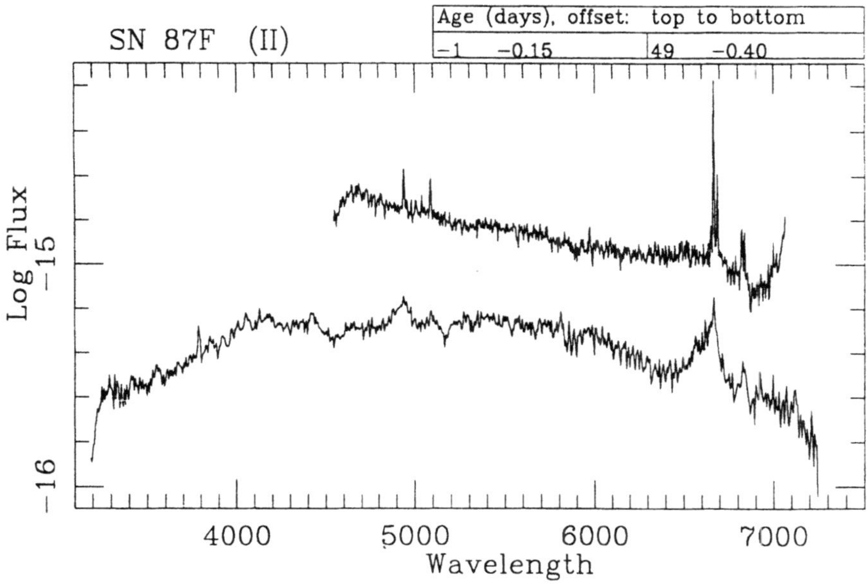

Figure 4: Spectra of SN 1987F, representative of the possible new subclass of Type II supernovae. The narrow lines are prominent in many of the candidate members of this subclass. In addition, the evolution of the spectra is relatively slow.

Table 2
Supernova Members of Proposed Subclass

SN	Galaxy	Type	Offsets (arc secs)
78G	IC 5201	SBcd	W96, N42
87B	NGC 5850	SBb	W75, S145
87C	Mk 90	???	E12.6, S16.8
87F	NGC 4615	Sd	E24, S6
88I	1018+3554	Sc(?)	E6, N1.1
88Z	MCG+3-28-22	Sa(?)	E11, S1
89C	UGC 5249	SBcd	???
89L	NGC 7339	Sbc	E38, N1

Acknowledgments

I thank the following people for contributions to this project: Robert Kirshner, Ed Horine, John Huchra, Ron Marzke, Charles Maxson, April Michel, Doug Mink, Jim Peters, Rudy Schild, Chris Smith, Susan Tokarz, William Wyatt.

References

1. Davis, M. and Latham, D. L. 1979, *SPIE Proceedings*, Tucson.
2. Schmidt, G., Weymann, R., and Foltz, C. 1989, *Preprint.*
3. Panagia, N. *et al.* 1989, *Preprint.*
4. Filippenko, A. V. and Sargent, W. 1985, *Nature*, **316**, 407.
5. Harkness, R., Wheeler, J. C., Margon, B., Downes, R., Kirshner, R., Uomoto, A., Barker, E., Cochran, A., Dinerstein, H., Garnett, D., and Levreault, R., 1987, *Ap. J.*, **317**, 355.
6. Schlegel, E. M. and Kirshner, R. P. 1989, *Astr. J.*, **98**, 577.
7. Schlegel, E. M. 1989, *Proceedings of the Santa Cruz Summer Workshop on Supernovae*, this volume.
8. Pollas, C. 1988, *IAU Circular*, 4651.
9. Cristiani, S., Gouiffes, C., Hanuschik, R., and Magain, P., 1987, *IAU Circular*, 4350.
10. Phillips, M. and Heathcote, S., 1989, *Publ. Astr. Soc. Pac.*, **101**, 137.
11. Lucy, L., 1988, in *Supernova 1987A in the Large Magellanic Cloud*, ed. M. Kafatos and A. Michalitsianos, (Cambridge: Cambridge University Press). p. 74.
12. Hanuschik, R., Thimm, G., and Dachs, J. 1988, *M. N. R. A. S.*, **234**, 41P.
13. Branch, D., Drucker, W., and Jeffrey, D. 1988, *Ap. J. (Letters)*, **330**, L117.
14. Filippenko, A. V. 1989, *Astr. J.*, **97**, 726.
15. Schlegel, E. M. 1989, *Preprint.*
16. Henry, R. B. C. and Branch, D. 1987, *Publ. Astr. Soc. Pac.*, **99**, 112.
17. Niemela, V., Ruiz, M. T., and Phillips, M. 1985, *Ap. J.*, **289**, 52.

Late Time Photometric Behavior of Supernova

Enrico Cappellano, Roberto Barbon, Massimo Della Valle, Sergio Ortolani, Leonida Rosino, & Massimo Turatto

In a previous paper BARBON et al. [1] have shown, from the analysis of the light curves of few well studied supernovae (SNe), that the blue luminosities of SNe in the range 200–400 days after maximum light fade linearly with a decline rate which seems a characteristic for each SN type. The average decline rates in the blue band were found $\gamma_{SNI} = 1.52$, and $\gamma_{SNII} = 0.81$ $[mag/100^d]$. Due to the scanty material available no distinction was made for the different subtypes of SNe, neither was possible to analyze the behaviour in different photometric bands . To improve that work, we started a program of photometric observations of SNe at late stages, using the ESO telescopes at La Silla (DELLA VALLE et al. [2]). In the first run, we observed a sample of ten SNe, i.e. 1985P, 1986G, 1986E, 1986I, 1986L, 1986N, 1986O, 1987D, 1987F, 1987K, that, at the time of observations, were at phase ranging from six months to two years. A full account of the observations and of the reduction technique will be presented elsewhere (TURATTO et al. [3]), while the detailed photometric and spectroscopic study of three SNe of the sample (SN 1986E, SN1987D and SN 1987F) has been given in a previous paper (CAPPELLARO et al. [4]).

In order to study the photometric behaviour of as many SNe as possible, all SN photometry obtained in B and V bands later that 100 days from maximum light was collected from the literature . The complete description of the late time light curves will be given elsewhere ([3]). In the following, we will focus on two general aspects: the late time absolute luminosities and the late time decline rates.

I. Late Time Absolute Luminosity

In Fig. 1 we report the absolute V magnitudes, in the phase range $100 \div 400$ days, of 5 SNe which are representative of different subclasses of SNe.

These objects were chosen because, they have a good coverage and a photometric behaviour typical of their subclass. Absolute magnitudes were calculated using the parent galaxy distance moduli from TULLY [5] and no correction for extinction was applied, since, for all five SNe, it has been estimated to be very low.

Although this should be checked with a more complete sample, Figure 1 shows that, in the considered phase range, the absolute V luminosities of all types of SN have a very little scatter. In particular, we stress the very close match of the light curves of all subclasses of SN II, despite of the very large differences exhibited at earlier

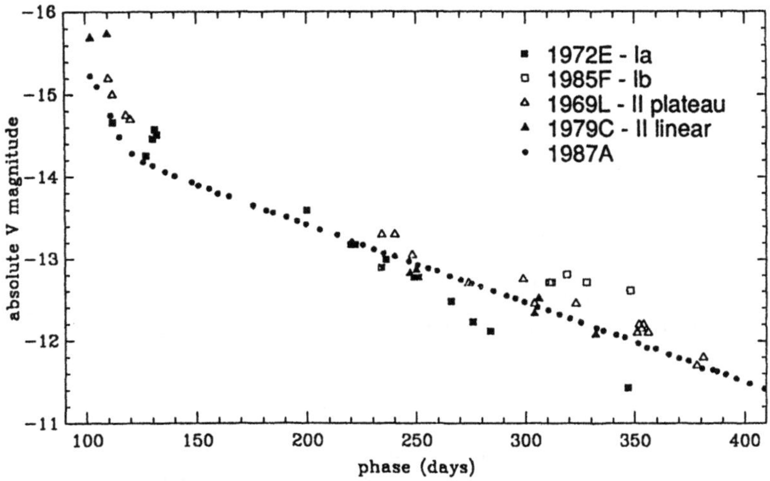

Figure 1: Absolute V light curves of 5 SNe representative of different subclasses in the phase range 100 ÷ 400 days after maximum.

phases, which, in the framework of the radiactive decay input energy model, could argue in favour of the production of a similar amount of ^{56}Ni in the different subclasses of SNII. It appears also that due to the different decline rates, the V luminosity of SN Ia is equal to that of SN II around phase 200. In the radioactive decay context, this is explained assuming that SNI produce a larger amount of ^{56}Ni but, because of a less massive envelope compared to SNII, an increasing fraction of the produced γ-ray escapes from the envelope without being thermalized and giving no contribution to the optical luminosity. Although only few data are available for SNIb, it seems that, while at maximum SNIb are fainter than SNIa due to the slower light curve SNIb are brighter than SNIa after about 300 days.

II. Late Time Decline Rate

Using data collected from the literature together with our observations, we derived the decline rates γ_B, γ_V for all SNe of our sample. The decline rates have been computed with regression lines through all points with phase 150 ÷ 400 days. The lower phase limit was chosen by taking into account two conflicting requirements: to include the largest number of SNe in order to have a fair sample of objects, and, in the meantime, to avoid the early phases in which the light curves show different behaviours peculiar to the different SN types. The upper limit, instead,, was chosen to restrict our analysis to the constant decline regions as shown by SN 1987A (CATCHPOLE et al. [6]).

In Fig. 2 we show the histograms of the observed decline rate in B and V bands, in unit of magnitude per 100 days. In the upper panel also three well studied SNI, for which only γ_{pg} is available, are included [1]. Two more SNII, for which the decline rates have been calculated using photometry at phases later than 400 days are also included. These objects are marked with the symbol * and they have not been used in the computation of the average decline rates reported in Table 1. The Table and Figure confirm that the rate of decline of SNI is in the average higher than for SNII.

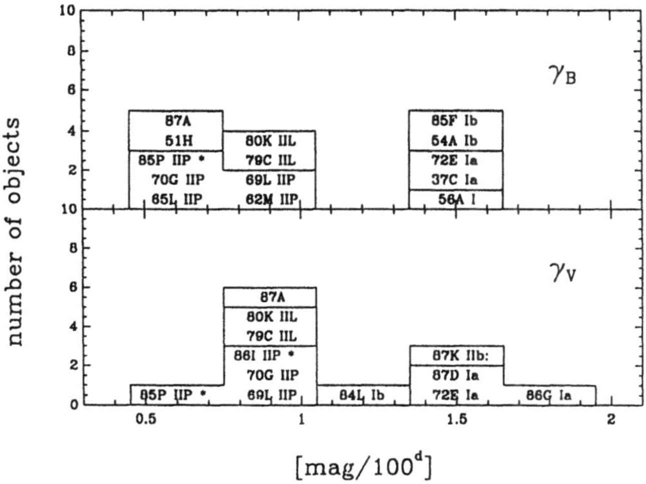

Figure 2: Histograms of the observed decline rates ($[mag/100^d]$) for a sample of SNe of different subclasses, in B (upper panel) and V (lower panel) bands.

A closer inspection, taking into account the different subclasses of SNe, shows that SNIa and SNII *Linear* exhibit the same decline rate in V and B bands, whereas SNII *Plateau*, SN 1987A and SN 1951H (whose late light curve is very similar to that of SN 1987A [3]) appear, in the average, significantly slower in B compared to the V band. For the SNIb it seems that $\gamma_V < \gamma_B$, but we stress again that photometric data on SNIb are scanty. In the lower panel of Fig. 2 the peculiar SN 1987K is classified IIb as suggested by Filippenko [8]. Its late time photometric behaviour appears close to that of SNIa. Finally, it is worth noting that for SN 1987A the V and the bolometric late decline are very close (WHITELOCK et al. [7]), and that the decline rate in the V band for all SNII is very close to that expected from the radioactive decay of ^{56}Co, i.e. $\gamma = 0.976$ $[mag/100^d]$.

Table 1: Average decline rates, $\gamma[mag/100^d]$, in B and V bands for different subtypes of SNe. The number of SNe in each subclass is indicated by n..

	γ_B	n.	γ_V	n.
SNIa	1.52	2	1.59	3
SNIb	1.52	2	1.06	1
SNI total	1.51	5	1.46	4
SNII *Plateau*	0.67	4	0.92	2
SNII *Linear*	0.84	2	0.87	2
1987A-like	0.72	2	0.97	1
SNII total	0.73	8	0.91	5

References

1. Barbon, R., Cappellaro, E. and Turatto, M. 1984, *Astr. Ap.* **135**, 27
2. Della Valle, M., Cappellaro, E., Ortolani, S., Turatto, M. 1988, *ESO Messenger* **52**, 16
3. Turatto, M., Cappellaro, E., Barbon, R., Della Valle, M., Ortolani, S., Rosino, L. in preparation
4. Cappellaro, E., Della Valle, M., Iijima, T., Turatto, M. 1989, *Astr. Ap.* in press
5. Tully, B. R. 1988 *Nearby Galaxies Catalog* Cambridge University Press
6. Catchpole, R. M. *et al.* 1988, *Mon. Not. R. astr. Soc.* **237**, 55P
7. Whitelock, R. M. *et al.* 1988, *Mon. Not. R. astr. Soc.* **234**, 5P
8. Filippenko, A. V. 1988, *Astron. J.* **96**, 194

Mean Evolution and Characteristics of Type I Supernova Spectra

Stefano Benetti & Roberto Barbon

The growing interest on the issue of type I supernova homogeneity (see e.g. BRANCH et al. [1] and references therein) has suggested us to use all the spectroscopic material obtained at Asiago Observatory in the past years, through a fairly homogenous set of instrumentation, to give a contribution to the understanding of this important problem. Our first goal was to produce a temporal sequence of spectra defining a standard evolution for SNeI to be used as reference in search for peculiarities and, afterwards, to correlate these ones with other SN features such as the photometric parameters.

I. Type Ia supernovae

We have examined 140 spectra (40% of which collected at Asiago) for a total of 15 supernovae. The Asiago observations have been made either with the 122 cm reflector (prism spectrograph and RCA image tube) or with the 182 cm telescope (grating spectrograph and ITT/VARO image tube). The plates have been digitized with a PDS microphotometer and intensity calibrated. Wavelength calibration has been made through IHAP commands at the HP working station of Asiago Observatory. Although the intrinsic error of the measurements, as derived by wavelength fits to night sky lines, is of the order of 2-3 Å, the central wavelength of the broad SN features has been given with an error of the order of 5-10 Å. No correction for spectral response of the detector has been applied since it is not essential at this stage of the work.

Using the best Asiago data, we have assembled a spectroscopic Atlas showing the mean spectral evolution of SNeIa from 5 days before maximum to 53 days past maximum. Characteristics phase intervals have been found and the common spectral behaviour has been described, deriving also some useful methods to date the spectra independently of the photometry.

We have also found that some features, as the Fe II bands at 4800-4900 Å, do not fit to the normal scheme since in some supernovae (e.g 1986G) these bands appear quite early, whereas in other objects (e.g. 1968E) they develop some weeks after maximum. A full account on the spectral evolution of SNeIa will be given in a forthcoming paper (BENETTI and BARBON [2]); here we anticipate some of the results.

Following [1] we show in Fig.1, for various SNe, the evolution of the photospheric velocity, corrected for the parent galaxy recession, as derived from the SiII λ6355 absorption. The data plotted in the Figure should be accurate and fairly homoge-

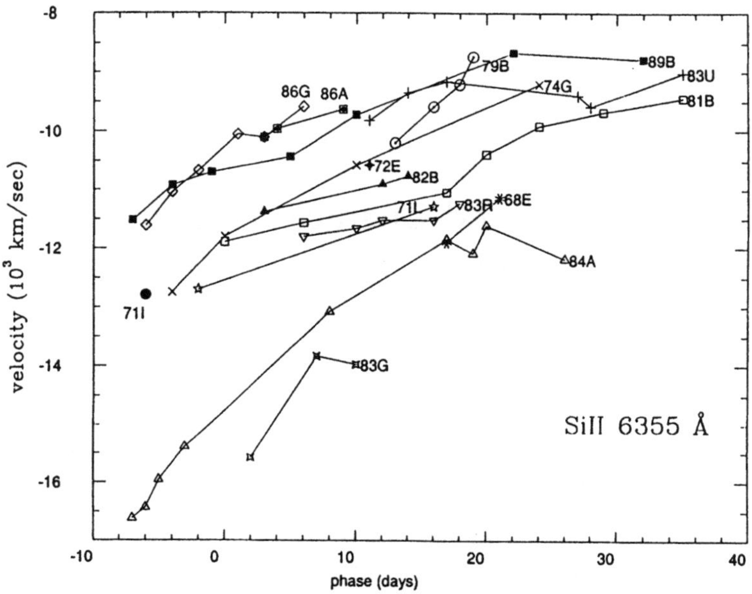

Figure 1: Kinematical evolution, in the frame of the parent galaxy, of the SiII λ6355 absorption feature for 15 supernovae.

neous since they refer to well studied supernovae and half of them come from Asiago spectra. We confirm the velocity evolution found in [1] and moreover, we argue that the inhomogeneity shown by the different objects may decrease with increasing phase. The scatter in Fig.1, from maximum to phase +20 ÷ +25 days, decreases of about 2000 Km/s.

In Fig.2, the same kind of data are given as derived from the MgII λ4481 absorption band. The velocity correlates linearly with phase and, most important, although the error bar is larger than in Fig.1 (670 Km/s against 470 Km/s) the scatter appears smaller. In Fig.2 the kinematically fast SNe 1984A and 1983G behave normally, whereas the data for SN 1989B, which define the low velocity end of our sample, suffer for uncertainty on the date of maximum which has been preliminary set on February 6.

Both Figures 1 and 2 confirm the existence of kinematical differences among SNeIa as recently found, but they also seem to suggest that the amount of this discrepancy depend on the particular spectral feature used and, moreover, that it decreases as the SNe get older.

We have also investigated whether the properties of some spectral features (e.g. phase of appearance, wavelength, etc.) correlate with the photometric parameter β which gives, in mag/100^d, the rate of the initial decline in the B light curve.

Firstly, we checked the correlation between β and the wavelength of the SiII λ6355 absorption at phase $+12^d$, already found by PSKOVSKII [3] and BRANCH [4], using a larger sample of objects and with better values of the parameter β (PSKOVSKII [5]). The data are shown in Fig.3, and no correlation is apparent.

Secondly, we studied the behaviour of the FeII emission bands at 4800 Å and 4900 Å which, as stated above, do not fit to the standard spectral evolution. Relating the phase of appearance of such bands with the same photometric parameter β, we found

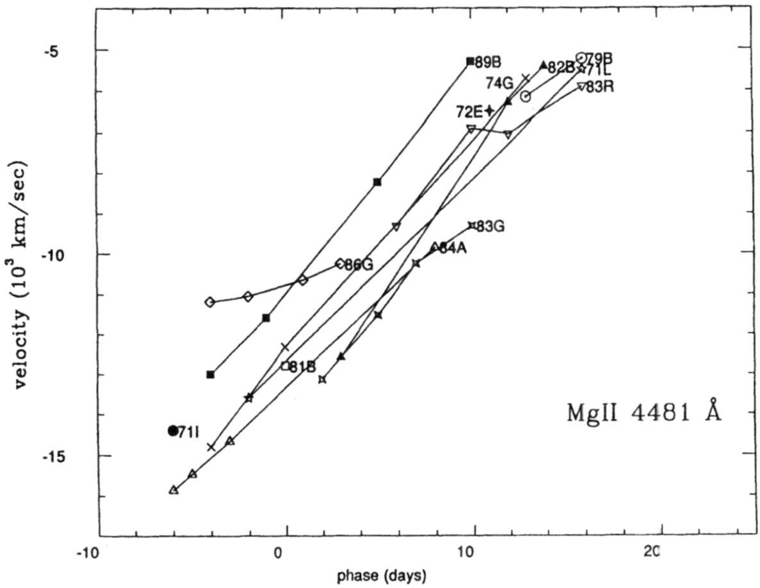

Figure 2: Kinematical evolution, in the frame of the parent galaxy, of the MgII λ4481 absorption feature for 12 supernovae.

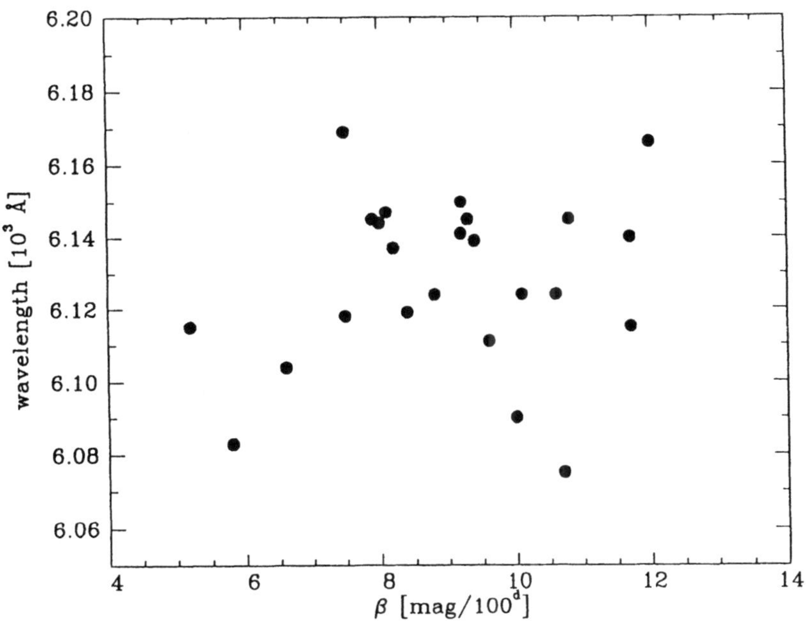

Figure 3: Wavelength of the SiII λ6355 absorption, measured at phase $+12^d$, is plotted against the rate of the luminosity decline β.

that, in the average, SNeIa showing large β values (~ 10) develop such bands at early phases, whereas objects with smaller β (~ 5) show these features appreciably later on. Since the rate of luminosity decline depends on the transparency of the ejected supernova envelope, these findings may suggest that such bands are signatures of (processed?) material in the internal layers of the supernova.

II. Type Ib supernovae

A similar study for SNeIb is not yet possible since few such objects have been thoroughly studied: 1962L, 1964L (observed at Asiago), 1983N and 1984L.

However, one result may be anticipated. Taking the behaviour of SN 1984L and 1983N as representative of the subclass, we noticed, respect to SNeIa, a higher degree of peculiarity, such as the lack of the redshift of the spectral features in 1962L and the anomalous behaviour of the absorption doublet at 6300-6500 Å in the same object as well as in SN 1964L. Due to the small sample, more observation of SNeIb are necessary to confirm the large spectral inhomogeneity of this subclass.

Finally, concerning the discrimination between type Ia at late phases and Ib at early epochs, we found that the latter objects exhibit a much faster shift to longer wavelenghts.

References

1. Branch, D., Drucker, W., and Jeffery, D. J., 1988, Ap. J.(Letters), 330, L117.
2. Benetti, S. and Barbon, R. 1989, in preparation.
3. Pskovskii, Y. P. 1977, Soviet Astr., 21, 675.
4. Branch, D. 1981, Ap. J., 248, 1076.
5. Pskovskii, Y. P. 1984, Soviet Astr., 28, 658.

IUE Observations of Supernovae

Nino Panagia & Roberto Gilmozzi

1. Introduction

The launch of the International Ultraviolet Explorer (IUE) satellite in early 1978 marked the beginning of a new era for SN studies because of its capability of measuring the ultraviolet emission of objects as faint as $m_B = 15$. Moreover, just around that time other powerful astronomical instruments have become available, such as the Einstein Observatory for X-ray measurements, the VLA for observations at radio wavelengths and a number of new telescopes either dedicated to infrared observations (*e.g.* UKIRT and IRTF at Mauna Kea) or equipped with new and highly efficient IR instrumentation (*e.g.* AAT and ESO observatories). As a result, a wealth of new information has become available which, thanks to the coordinated effort of astronomers operating at widely different wavelengths, has provided us with fresh insights as for the properties and the nature of supernovae of both types.

The first supernova observed with IUE was SN 1978G in IC 5021, just toward the end of the first year of IUE operation. It was a type II SN discovered somewhat after maximum so that the observations could not be carried out for any long time.

In 1979 a joint ESA-SERC target-of-opportunity program was started for observing bright supernovae, defined such as *i.e.* $B_{max} < 12$. And sure enough, in April 1979 a bright supernova, SN 1979C in NGC 4321, was discovered, promptly observed and followed with IUE in a collaboration across the Atlantic, for more than three months (PANAGIA *et al.* [1]). Since then, IUE has observed all bright supernovae plus a number of fainter ones: a total of 15, out of which 6 are of Type II, 7 Type Ia and 2 Type Ib. However, only five SNe, namely 1979C, 1980K, 1981B, 1983N and, obviously, 1987A, were bright enough to obtain high quality ultraviolet spectra and/or to follow their time evolution in detail.

A summary of the IUE observations is presented in Table 1. The data of the first 6 SNe have been collected in an atlas compiled by BENVENUTI *et al.* [2]. A paper containing IUE, optical and IR observations of the SNIb prototype 1983N will eventually be completed this year (PANAGIA *et al.*, [3]), soon followed by an analogous paper on SN 1980K. A preliminary discussion of both SNe can be found in PANAGIA [4]. The observations of SN 1987A have already been discussed in part in a number of papers which I am not going to refer to here. The relevant references can be found

Table I

Summary of *IUE* Observations of Supernovae

SN	Type	B_{max}	Galaxy	Observing Period	# SW	# LW	Station
1978G	II	<12.9 V	IC 5201	78/11/30–78/12/11	–	2	G
1979C	II	11.6	NGC 4321	79/04/21–79/08/04	12	19	V,G
1980K	II	11.6	NGC 6946	80/10/30–80/12/09	12	24[a]	V,G
1980N	Ia	~12.5	NGC 1316	80/12/11–81/01/16	1	8	V,G
1981B	Ia	12.0	NGC 4536	81/03/09–81/04/05	2	5	V
1982B	Ia	13.7	NGC 2268	82/02/18	1	2	V
1983G	Ia	12.9	NGC 4753	83/04/08–83/04/25	1	7	V,G
1983N	Ib	11.6	NGC 5236	83/07/04–84/07/31	12	16	V,G
1984J	II	13.2 V	NGC 1559	84/08/13	1	–	V
1985F	Ib	12.1	NGC 4618	85/05/18	–	1	V
1985L	II	<12.5	NGC 5033	85/06/28–85/07/17	1	2	V
1986G	Ia	12.5	NGC 5128	86/05/06–86/05/29	–	6	V,G
1987A	II	4.5	LMC	87/02/24–present[b]	207[a]	509[a]	V,G
1989B	Ia	11.8 V	NGC 3627	89/01/13–89/02/01	–	4	G
1989M	Ia	12.0 V	NGC 4579	89/07/06–89/07/20	–	7	V,G

[a] High resolution observations obtained as well.
[b] As of April '89 for GSFC, July '89 for Vilspa.

in the recent review by KIRSHNER and GILMOZZI [5]. Let me just mention here that the data collected in the first two years will be presented in an extensive atlas in preparation (KIRSHNER *et al.* [6]).

In addition to the obvious interest of observing SNe *per se*, *i.e.* to understand their nature and the mechanisms inducing their explosion, bright supernovae have been used as background sources to study the intervening interstellar medium, most remarkably the galactic haloes, with an accuracy impossible otherwise. Two such supernovae have been observed in the UV at high dispersion, namely SN 1980K in NGC 6946 (PETTINI *et al.* [7]) and, needless to say, SN 1987A in the LMC (DE BOER *et al.* [8], DUPREE *et al.* [9], BLADES *et al.* [10], [11]).

2. Type Ia Supernovae

Although 7 Type Ia SNe have been observed with IUE none of them has been followed for a long time either because of their intrinsic UV faintness or because of satellite pointing constraints. Therefore, for all of them we have observations concentrated around the epoch of their maximum light but we know little about their time evolution.

The UV spectrum is declining quickly with frequency, making it hard to detect any signal at short wavelengths. This aspect is illustrated in Figure 1, which displays the LW spectra of the first three Type Ia SNe observed with IUE, plus SN 1986G. In all cases the observing epoch is within three days from the optical maximum.It also appears that the spectrum is not a smooth continuum but rather consists of a number of "bands" which are observed, with somewhat different strength. The most prominent feature is the emission that peaks at $\sim$ 2950 Å with a half-power width of $\sim$ 100 Å, *i.e.* $\Delta v \simeq 10^4$ $km\ s^{-1}$. Alternatively, this band may be the result of strong absorptions occuring on both sides of the apparent emission, *i.e.* centered at $\sim$ 2840 Å and $\sim$ 3060 Å and having half-power widths of the order of 100 Å. A similarly prominent emission band is seen at $\lambda \sim$ 1890 Å in the only one spectrum obtained at short wavelengths for SN 1981b. Several other absorptions features can be recognized, which are present at all epochs of observation. Although some of them might be identified with multiplets of Fe I, Fe II and Mg II, no detailed study has been made yet for the majority of the absorptions. Nevertheless, the very fact that the spectrum is so similar for the first three SNe and at all epochs when observations were made is already an important result. This support the idea of an overall homogeneity of properties of Type I SNe.

Confirmation of this result has come from observation of all of the other Type Ia SNe with the only clear exception of SN 1986G. This latter deviates from the main path in that the various UV features, such as the one around 2950 Å, are considerably less prominent relative to the continuum as compared with canonical Type Ia spectra (see Fig. 1). It is of interest to note that the rate of decline of the B light curve of SN 1986G was remarkably fast, corresponding to the infrequent Pskovskii class $\beta = 12$ (PHILLIPS *et al* [12]).

3. Type Ib Supernovae

SN 1983N in NGC 5236 (= M 83) is one of the best studied SNe with IUE. Preliminary results can be found in PANAGIA [4], while a complete account of the UV, optical and

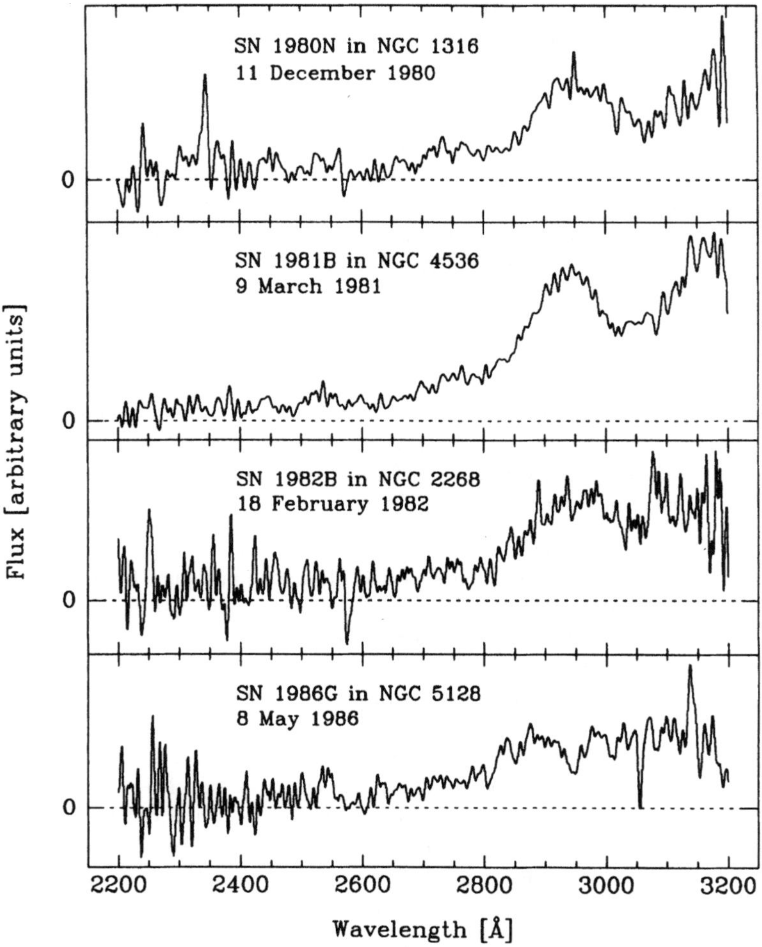

Figure 1 - The LW spectra of SN 1980N, 1981B, 1982B and 1986G at epochs near the optical maximum.

IR observations obtained in the first two months after discovery will be published soon (PANAGIA *et al.* [3]).

Its UV spectrum closely resembles that of Type Ia SNe at comparable epoch and, as such, only a minor fraction of the SN energy is radiated in the UV. In particular, only 13% of the total luminosity was emitted shortward of 3400 Å at the time of the UV maximum. Moreover, there is no indication of any stronger emission in the UV at very early epochs: this implies that the initial radius of the SN, *i.e.* the radius the stellar progenitor had when the shock front reached the photosphere, was definitely much less than 10^{13} *cm* and probably lower than 10^{12} *cm*, and rules out a red supergiant as a possible progenitor of this supernova. A WR star can equally well be excluded according to the stringent arguments summarized by PANAGIA and LAIDLER [13] The most likely scenario is that of a binary system in which both stars are relative massive ($> 5M_\odot$) in which the primary explodes after becoming a massive white dwarf ($M \sim 1.35\ M_\odot$) and accreting matter from the companion while it is a red giant (*cf.* SRAMEK, PANAGIA and WEILER [14]).

SN 1985F was discovered by FILIPPENKO and SARGENT [15] near the nucleus of NGC 4618. It was observed once with IUE, on 1985 May 18, and only with the long wavelength camera. At that time the SN was approximately a year old. The observed spectrum is essentially featureless and flat, with an average flux of 6×10^{14} $erg\ cm^{-2}\ \AA^{-1}\ s^{-1}$. This is approximately one half of the optical continuum level and may entirely be due to the emission of the underlying HII region.

4. Type II Supernovae

Among the five Type II SNe which have been observed with IUE, only two, namely SN 1979C and SN 1980K, were bright enough to allow a detailed study of their properties in the UV.

The main characteristic is that the UV emission is rather strong for quite some time (PANAGIA *et al.* [1]). In particular, for both 1979C and 1980K the UV flux is higher than the extrapolation of the optical spectrum with a black body curve, with a clear excess shortward of $\lambda = 2000$ Å. Such an excess is found at all epochs although it is less pronounced at early times. FRANSSON [16] has shown that the UV excess may be photospheric radiation which has been Compton-scattered by energetic, thermal electrons ($T \sim 10^9\ K$) at the shock front where the ejecta interact with pre-existing circumstellar material. This model can explain both the extra radiation observed at short wavelengths and the high ionization implied by some emission lines (*e.g.* NIV] 1486 Å, CIV 1550 Å, etc.) observed in the spectrum of SN 1979C (PANAGIA *et al.* [1]).

Emission lines in the UV have been detected only for SN 1979C. From a comparison of the line profiles observed in the UV and in the visual with theoretical calculations (FRANSSON *et al.* [17]) concluded that the UV emission lines of highly ionized species are produced in the upper atmosphere, as well as Hα or Mg II λ 2800 Å, although just in the outermost layers where density is lower and the ionizing radiation flux is higher. The line profiles imply an expansion velocity of 8400 $km\ s^{-1}$ (FRANSSON *et al.* [17]), which is only marginally lower than that measured in the optical (*i.e.* 9200 $km\ s^{-1}$; PANAGIA *et al.* [1]). From an analysis of the NV 1240 Å, NIV] 1486 Å, CIV 1550

Å, NIII] 1750 Å and C III] 1909 Å line intensities the abundance ratio of nitrogen to carbon has been estimated to be N/C ~ 8 (FRANSSON *et al.* [17]) *i.e.* ~ 30 times higher than the cosmic value. This strong enhancement of nitrogen relative to carbon suggests that the pre-supernova star was a massive supergiant which had undergone a long period of mass loss, thereby exposing CNO processed material.

A similar overabundance of N has been found in the wind of SN 1987A progenitor (FRANSSON *et al.* [18]). This result suggests that a considerable N overabundance may be a signature of Type II supernovae.

References

1. Panagia, N., *et al.*: M.N.R.A.S., 192, 861, (1980).
2. Benvenuti, P., Sanz Fernandez de Cordoba, L., Wamsteker, W., Macchetto, F., Palumbo, G. C., Panagia, N.: ESA SP-1046 (1982).
3. Panagia, N. *et al.*: in preparation (1989).
4. Panagia, N.: in "Supernovae as Distance Indicators", ed. N. Bartel (Berlin: Springer), p. 14 (1985).
5. Kirshner, R. P., Gilmozzi, R.: in "Exploring the Universe with the IUE Satellite" (2nd edition), ed. in chief Y.Kondo, (Dordrecht: Kluwer), in press (1989).
6. Kirshner, R. P., Panagia, *et al.*: in preparation (1990).
7. Pettini, M. *et al.*: M.N.R.A.S., 199, 409 (1982).
8. de Boer, K., Grewing, M., Richtler, T., Wamsteker, W., Gry, C., Panagia, N.: Astron. Ap., 177, L37 (1987).
9. Dupree, A.K., Kirshner, R.P., Nassiopoulos, G.E., Raymond, J.C., Sonneborn, G.: Ap. J., 320, 597 (1987).
10. Blades, J.C., Wheatley, J.M., Panagia, N., Grewing, M., Pettini, M., Wamsteker, W.: Ap. J. (Letters), 332, L75 (1988).
11. Blades, J.C., Wheatley, J.M., Panagia, N., Grewing, M., Pettini, M.,Wamsteker, W.: Ap. J., 334, 308 (1988).
12. Phillips, M.M., *et al.*: P.A.S.P., 99, 592 (1989).
13. Panagia, N., Laidler, V.C.: this Conference.
14. Sramek, R. A., Panagia, N., Weiler, K. W.: Ap.J. (Letters), 285, L59 (1984).
15. Filippenko, A. V., Sargent, W. L. W.: Astr. J., 91, 691 (1986).
16. Fransson, C.: Physica Scripta, T7, 50 (1984).
17. Fransson, C., Benvenuti, P., Gordon, C., Hempe, K., Palumbo, G. G. C., Panagia, N., Reimers, D., Wamsteker, W.: Astr. Ap., 132, 1 (1984).
18. Fransson, C., Cassatella, A., Gilmozzi, R., Kirshner, R.P., Panagia, N., Sonneborn, G., Wamsteker, W.: Ap. J., 336, 429 (1989).

VLBI Observations of Supernovae and Their Remnants

Norbert Bartel

1. INTRODUCTION

The technique of Very-Long-Baseline Interferometry (VLBI) allows us to contribute uniquely to research of supernovae (SNe) and supernova remnants (SNRs). To illustrate the potential of SN VLBI, let us take as an example the VLA image of the SNR Cas A (BRAUN, GULL, and PERLEY [1]) and consider the wealth of information revealed in the details of the brightness distribution in each segment of the shell. Let us then take as an example any recent SN and imagine that we make a set of detailed maps of the radio brightness distribution. Not only would we have imaged the supernova shortly after the explosion, but we would also have monitored the two-dimensional expansion of the debris over a large fraction of the SN's lifetime and, perhaps, eventually revealed a pulsar nebula in the center of the expanding cloud. However, until recently, the reality of VLBI of SNe and SNRs only trailed our imagination and caused us to be more modest, especially in those cases where we not only failed to detect a SN but also failed to obtain any useful information on that SN. I do not really want to elaborate on these cases.

More rewarding were VLBI observations of SN1987A made only 5.2 d after the neutrino burst. We did not detect the SN, but we obtained important lower bounds on the angular radius and mean angular expansion velocity of the SN's radiosphere.

SN1980K in NGC6946 and an SNR in NGC4449 were detected in VLBI observations and allowed bounds on their angular radii to be determined. In the latter case, such a bound, when combined with a value for the distance of the galaxy and with information from optical spectroscopic data, led to an estimate of the age of the SNR.

SN1979C in M100 in the Virgo cluster of galaxies has been detected in VLBI observations at several epochs. For each epoch, an angular radius was determined. By virtue of the SN's having been monitored for a large fraction of its lifetime, and its relatively large distance from us, SN1979C has been, scientifically, one of the most rewarding supernovae of the ones observed with VLBI. The observations allowed estimates of the supernova's angular expansion velocity and of bounds on any decleration or acceleration of it. Combined with a determination of the radial expansion velocity of the line-emitting and -absorbing gas, the distance to the Virgo cluster and Hubble's constant could be estimated (for the two latter estimates, see BARTEL, this volume).

Of the 23 hot spots in the starburst galaxy M82, six were detected in VLBI observations. For one of these sources, an image could be obtained. The image, combined with an estimate of the expansion velocity of the hot spot and earlier measurements of their luminosities and radio light curves, led to the conclusion that the hot spots are indeed SNe and SNRs. This result led to further conclusions about the SNe/SNRs' ages and about the SN rate in the inner 600 pc of M82.

SN1986J in NGC891 is the first SN imaged with VLBI. By virtue of the SN's relatively large flux density and relatively compact structure and of the VLBI array's recently increased sensitivity, we can now indeed hope to obtain a set of images of the expanding gas and make a movie of an exploding star.

In the remainder, I will review the observations and their results for the six SNe and SNRs investigated with VLBI, in the order of increasing information obtained and obtainable as outlined in this section. An earlier review is given in BARTEL [2].

2. SNe AND SNRs OBSERVED WITH VLBI

2.1 SN1987A: no detection – bound on θ

With an unprecedented promptness, the radio emission of SN1987A reached its maximum of ~ 140 mJy at 1.4 GHz (TURTLE *et al.* [3]) only three days after the neutrino burst (AGLIETTA *et al.* [4]; BIONTA *et al.* [5]; HIRATA *et al.* [6]). The only antennas in the southern hemisphere that were conceivably capable of resolving the expanding shell of SN1987A were one of the antennas in Tidbinbilla, Australia, and the 26-m diameter antenna in Hartebeesthoek, South Africa. After frantic organization, Mark III VLBI observations were made at 2.3 GHz at $t = 5.2$, 6.2, and 7.2 d with the latter antenna and NASA's 34-m diameter DSS42 antenna in Tidbinbilla (SHAPIRO *et al.* [7]). Unfortunately, at the time of the observations, the SN was already too weak and too extended to be detected with the above interferometer. No fringes were found from the supernova's radio emission on any of those three observing days, although fringes were obtained for the calibrator sources on all three days with amplitudes agreeing (to within the corresponding combined standard errors) with those obtained from VLBI measurements made at the same resolution, but five years earlier (PRESTON *et al.* [8]; G. NICOLSON 1987, priv. communication).

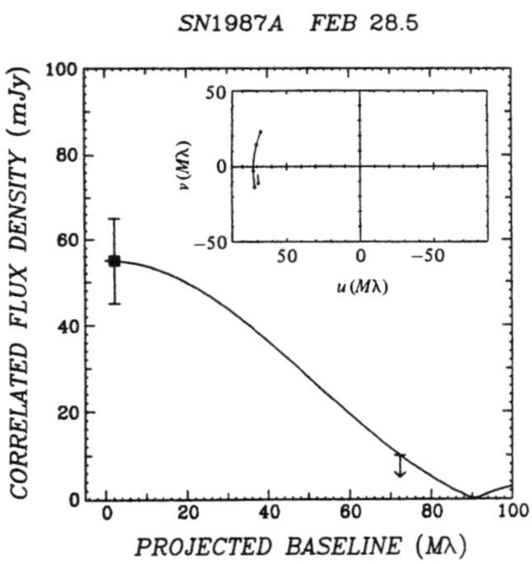

Fig. 1. The total flux density and an upper bound on the correlated flux density of SN1987A on ~ 28.5 Feb. 1987, together with the prediction of a model of an optically thin uniform sphere with a radius of 1.6 mas. The prediction is similar to that from a model of an optically thin shell with an outer angular radius of 1.25 mas. The u-v track for the VLBI observations is plotted in the inset.

The upper bound on the correlated flux density of SN1987A obtained on the first day of the VLBI observations was compared with the corresponding total flux density (Fig. 1) to derive a lower bound on the angular radius of the radiosphere (BARTEL *et al.* [9]; JAUNCEY *et al.* [10]; SHAPIRO *et al.* [7]). For an optically thin shell model with a shell thickness of ~ 15% of the shell's outer radius, the lower bound on the radius is $\theta_{\rm radio} > 1.25 \pm 0.07$ mas. Here and hereafter, the quoted errors are meant to be standard errors (σ), with statistical and systematic contributions combined, unless otherwise stated. The use of any other physically plausible model would result in an up to $\sim 30\%$ larger lower bound (see, e.g., MARSCHER [11]). Given a distance to SN1987A of 50 ± 5 kpc (FEAST and WALKER [12]), the lower bound on the angular radius corresponds to a lower bound (with the combined 1σ error subtracted) on the linear radius, $R_{\rm radio}$, of $R_{\rm radio} > 8.3 \times 10^{14}$ cm $= 12 \times 10^3$ $R_\odot = 55$ AU, at $t = 5.2$ d. If one assumes that the radiosphere expanded linearly from zero size at $t = 0$, the lower bound on the radius corresponds to a lower bound on the expansion velocity, $v_{\rm radio}$, of the radiosphere: $v_{\rm radio} > 19 \times 10^3$ km s^{-1}.

Since the distance to the Large Magellanic Cloud is known to within about 10%, SN1987A allowed for the first time a comparison of the linear radii and the expansion velocities of a supernova's photosphere and radiosphere with those of the supernova's line-emitting and -absorbing regions. In Fig. 2, the lower bounds on the radius and expansion velocity obtained from the VLBI radio data are compared with the corresponding radii and velocities obtained from optical photometric (MENZIES *et al.* [13]) and spectroscopic (HANUSCHIK and DACHS [14]; BLANCO *et al.* [15]) data.

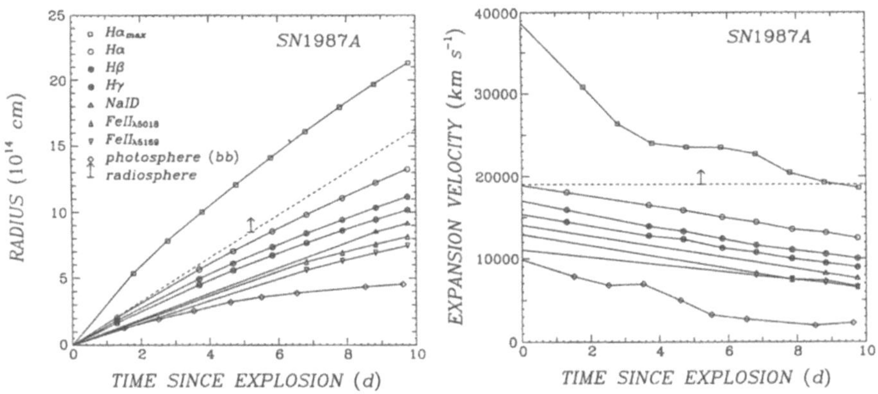

Fig. 2. Lower bounds on the radius and the corresponding (assumed uniform) expansion velocity of the radiosphere of SN1987A (dashed lines), compared with radii and expansion velocities of the blackbody photosphere and the line-forming regions. The radii of the line-forming regions were obtained from an integration of the velocities corresponding to the blueshifts of the absorption minima of the indicated lines and the largest observed blueshift, $H\alpha_{\rm max}$, the latter from the blue edge of the $H\alpha$ absorption trough. The expansion velocities of the photosphere were obtained from a differentiation of the photosphere's radii. The uncertainties of the $H\alpha_{\rm max}$ velocities are ~ 1000 km s^{-1} and those of the $H\alpha_{\rm max}$ radii are smaller than the symbols. Other uncertainties were not given in the original papers but are believed to be not larger than those of the $H\alpha_{\rm max}$ velocities and radii, respectively.

At the time of the VLBI observations, the radius and the expansion velocity of the radiosphere were both at least a factor 2.5 larger than those of the blackbody photosphere and, respectively, at least 10% and 25% larger than the radius and velocity inferred from the $H\alpha$-line absorption minimum. The results not only add to our knowledge of supernovae but are also important in limiting the uncertainties accompanying the use of supernovae as distance indicators (see, e.g., BARTEL [16], [17]; BARTEL *et al.* [18]), if the physical processes responsible for the radio emission from SN1987A are typical for SNe in general (see BARTEL, this volume).

2.2 SN1980K and SNR in NGC4449: detection – bound on Θ

The supernova SN1980K in the galaxy NGC6946 reached its maximum flux density of ~ 2.5 mJy at 1.5 GHz ~ 0.4 yr after the explosion that occurred on, or near, 1980 Oct. 17 (WEILER *et al.* [19]). VLBI observations were made at 2.3 GHz on 1983 May 7 with the sensitive NASA 64-m antennas at Goldstone, CA and Madrid, Spain. The supernova was detected in two out of three adjacent 13-min scans. The visibility amplitudes determined in the two scans, and an upper bound determined in the third scan, are shown in Fig. 3. Since the visibility amplitudes of the three segments are expected to be approximately Gaussianly distributed, we can compute their mean value and standard error and infer that SN1980K was unresolved at the epoch of our VLBI observations, 2.54 yr after the assumed date of 1980 Oct. 17 for the explosion. The value for the angular radius of a shell model is: $\Theta \lesssim 0.5 \pm 0.5$ mas, equivalent to $\Theta < 1$ mas (BARTEL [20]). The prediction from this model is also shown in Fig. 3.

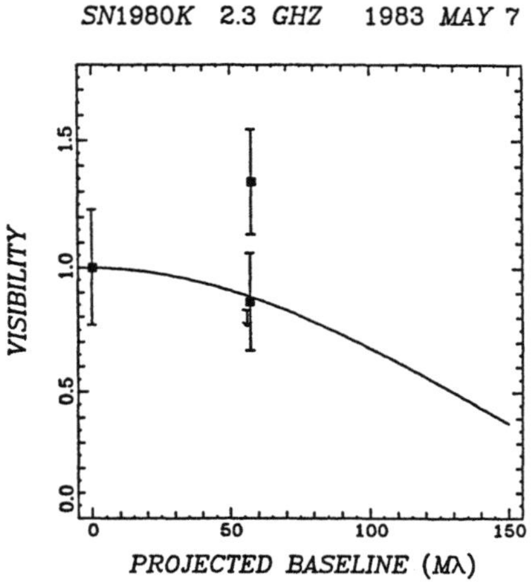

Fig. 3. Visibility amplitudes and the prediction from a shell model. The visibility amplitude at zero spacing was obtained from interpolating between several measurements made with the VLA by WEILER *et al.* [19] at 1.5 and 5 GHz only two days to seven weeks prior to our VLBI observations.

The SNR in the galaxy NGC4449 (SEAQUIST and BIGNELL [21]; BALICK and HECKMAN [22]) at a distance from Earth of a few Mpc shows a declining flux density that was ~ 15 mJy at 1.5 GHz in 1981 (DE BRUYN [23]). So far, only an upper bound on the SNR's angular radius of 38 mas (Cas A morphology assumed) has been determined (DE BRUYN [23]) with VLBI, but new data may some time allow an image of this source to be made (DE BRUYN 1988, priv. communication).

When the upper bound is combined with the galaxy's distance and the SNR's linear expansion velocity (KIRSHNER and BLAIR [24]), an upper bound on the SNR's age of ~ 200 yr can be estimated. There is some optical evidence for a lower bound on the age of ~ 60 yr. If this lower bound could be confirmed, the SNR in NGC4449 would be, together with the SNRs in M82, an important element for supernova VLBI research in bridging the gap of ages between extragalactic SNe with ages of up to ~ 10 yr and galactic SNRs with ages $\gtrsim$ 300 yr.

2.3 SN1979C: detection – $\theta(t)$

The supernova SN1979C in the galaxy M100 in the Virgo cluster reached its maximum flux density at 1.4 GHz of ~ 10 mJy 3.7 yr after the explosion on, or around, 1979 Apr. 1 (WEILER *et al.* [19]). Figure 4 displays a radio map (BARTEL *et al.* [18]) of the galaxy with the supernova located in one of the galaxy's spiral arms.

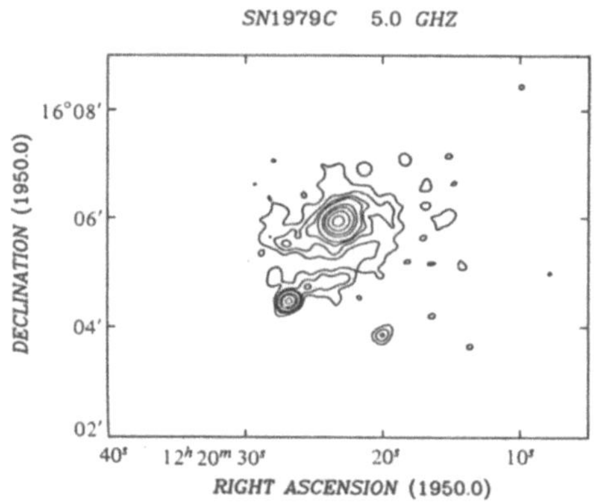

Fig. 4. Radio map of SN1979C, located at the southern edge of a spiral arm of the galaxy M100. The map was made with the VLA on 1982 Dec. 8.

VLBI observations commenced on 1982 Dec. 8 and have resulted in angular radius determinations at 5 GHz at four consecutive epochs. The measured visibility amplitudes and the predictions from a model for the supernova's brightness distribution are shown in Fig. 5.

The angular radius determinations, Θ, are plotted, as a function of the time since explosion, t, in Fig. 6. A weighted least-squares fit of the form $\Theta \propto t^m$ gives $\Theta = 0.42 \pm 0.03$ mas for the time of our first VLBI observations at $t = 3.69$ yr, and $m = 1.03 \pm 0.15$, consistent with uniform expansion.

These data provide the first direct measurement of the expansion of a supernova. The expansion is consistent with being uniform, as predicted on the basis of a fit of the circumstellar interaction model (CHEVALIER [25]) to the radio light curves at two frequencies (CHEVALIER [26]; WEILER *et al.* [19]).

VLBI observations from 1988 are still being analyzed. The results from these observations together with results from anticipated observations in the spring of 1990 could help to reduce the uncertainty in m considerably. A small uncertainty of m could lead to a model-independent distinction between a uniform or decelerated expansion ($m \leq 1$), as expected in the circumstellar interaction model, or an accelerated expansion, as could be expected if a central pulsar powers the nebula.

2.4 SNR41.9+58 and other SNRs in M82: image – $\dot{\Theta}$

The starburst galaxy M82 (Fig. 7) has been known for some time to contain more than ~ 20 radio hot spots in its central region of ~ 600 pc extent (KRONGERG, BIERMANN, and SCHWAB [27], [29]; UNGER *et al.* [30]). We assume the galaxy's distance from Earth to be 3.3 Mpc (TAMMANN and SANDAGE [31]; but see also SANDAGE [32] for a $\sim 50\%$ larger estimate). Since the hot spots' time variability (see, e.g., KRONBERG and SRAMEK [33]) and spectra (e.g., KRONBERG, BIERMANN, and SCHWAB [27]) also resemble those of SNe and since a high star-formation rate with a correspondingly high SN rate of ~ 0.3 yr^{-1} is expected for M82 on the basis of infrared observations (RIEKE *et al.* [34]), the hot spots in M82 were believed to be SNe or young SNRs (e.g., KRONBERG, BIERMANN, and SCHWAB [27], [29]; UNGER *et al.* [30]).

VLBI determinations of the sizes, morphologies, and expansion velocities of the hot spots further helped to clarify their nature and qualified earlier VLBI observations made by GELDZAHLER *et al.* [35] that seemed to support different interpretations of the nature of the hot spots.

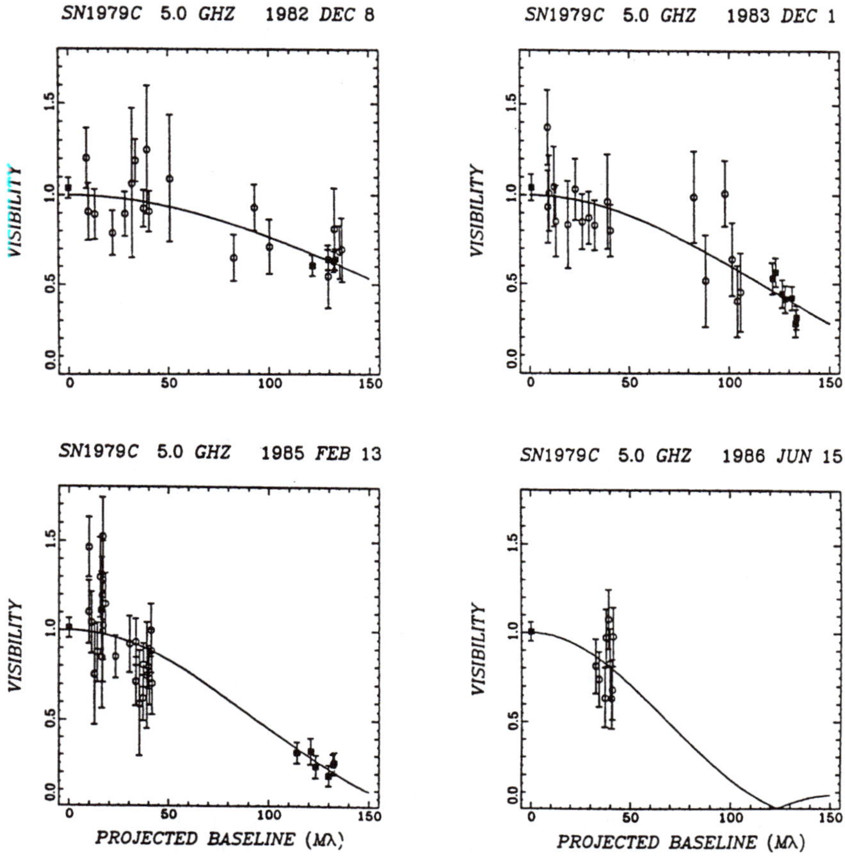

Figure 5. Measured visibility amplitudes at 5.0 GHz and predictions from the fit of a uniform sphere model. The filled squares show the most significant data points, obtained with the VLA alone and with the Bonn-VLA interferometer. At epoch 1986 June 15, the antenna at Bonn was malfunctioning and therefore prevented the recording of data from SN1979C with the transatlantic interferometers.

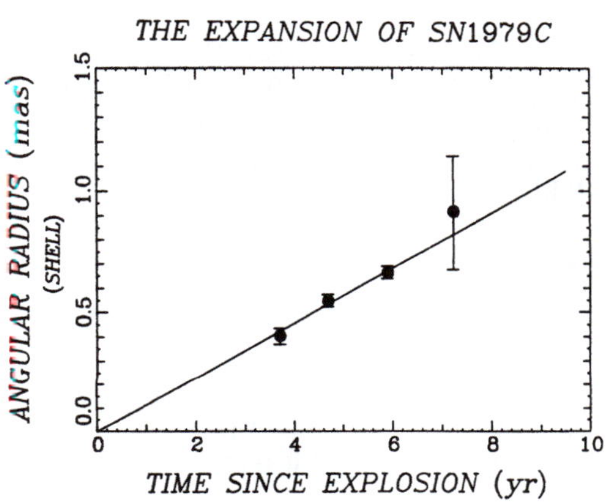

Fig. 6. The angular radius determinations for a shell model. For the extreme models of a ring and a uniform sphere, the ordinate scale has to be multiplied by 0.8 and 1.3, respectively. The solid line represents uniform expansion ($m = 1$), which is consistent with our weighted least-squares solution for m.

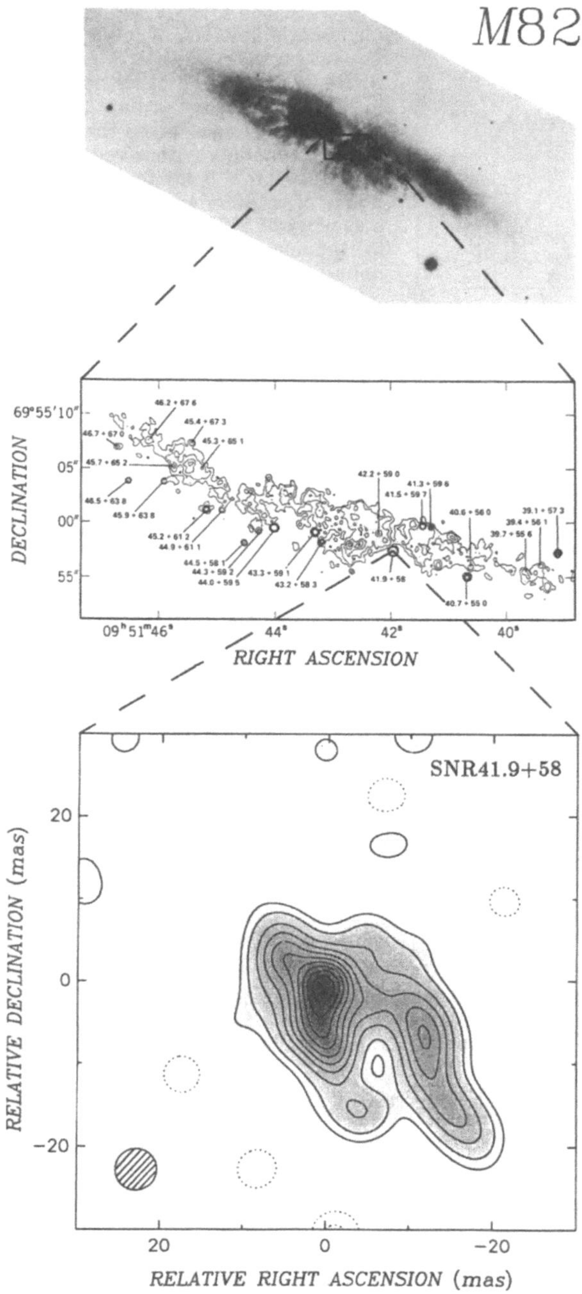

Fig. 7. A hybrid map at 2.3 GHz (lower part) of SNR41.9+58, the strongest compact component in the nuclear region of M82. The contours are at 90, 80, 70, 60, 50, 40, 30, 10, 5, −5% of the peak brightness of 25 mJy per beam area, equivalent to 4.4×10^8 K. The 50% contour of the restoring beam is shown as the striped circle in the lower left corner. The map is shown in relation to a radio map of the inner 600 pc of M82, made by KRONBERG, BIERMANN, and SCHWAB [27] with the VLA at a frequency of 4.9 GHz and with an angular resolution of 0″.34. The contours of the VLA map are shown at 40, 30, 20, 10, and 5% of the peak brightness of 102 mJy per beam area. All the sources we observed with VLBI, and analyzed, are labeled. For results from VLBI observations of the sources other than SNR41.9+58, which are presumably also SNRs, see BARTEL *et al.*[28]. The radio map is juxtaposed to an optical image of M82 taken from Sandage's Hubble atlas of galaxies. For each of the three images, north is up and east to the left.

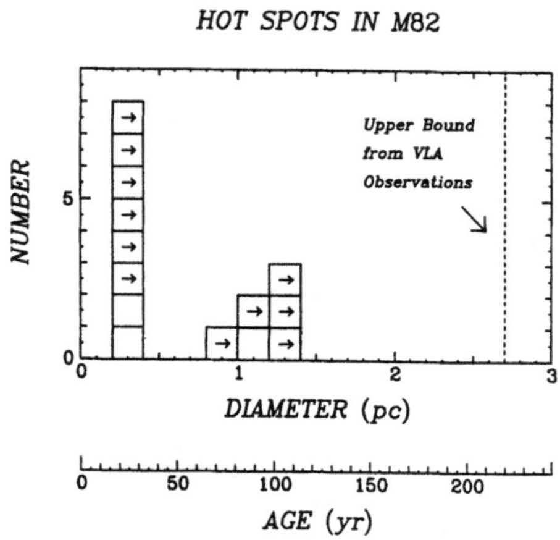

Fig. 8. A histogram of the FWHM-Gaussian diameters, or lower bounds on them (with arrows) of SNR41.9+58 and 13 other hot spots in M82. The second ordinate gives the hot spots' ages, given that they are all expanding with a velocity of 6×10^3 km s^{-1} for the HWHM points of a Gaussian, equivalent to a velocity of 8500 ± 4300 km s^{-1} for the outer parts of a shell. The bound on the diameters from VLA observations (KRONBERG, BIERMANN, and SCHWAB [27]) is between ~ 2.7 and ~ 5.4 pc ($d \equiv 3.3$ pc), depending on the observing frequency and the VLA configuration.

The lower part of Fig. 7 shows a map of the brightest hot spot, 41.9+58 (BARTEL *et al.* [28]; see also WILKINSON and DE BRUYN [36] for an image of 41.9+58 at 5 GHz resembling in several details ours at 2.3 GHz).

The map displays a shell-type morphology. The apparent deviation of the structure from circular symmetry may have been caused by relatively large Rayleigh–Taylor instabilities and by the relatively high pressure of the interstellar medium in the nuclear region of M82, which is estimated to be 30–300 times larger than the equivalent pressure in our Galaxy (LUGTEN *et al.* [37]). Further detailed imaging of 41.9+58 is of particular interest, since it promises to reveal the spectral evolution of each segment of the supernova's shockfront and its expansion into the dense ambient medium.

A value for the expansion velocity was found by fitting Gaussian models to the visibility data at several epochs. The expansion velocity was found by combining estimates of angular radii at epochs 1980 (WILKINSON and DE BRUYN [38]) and 1983 (BARTEL *et al.* [28]): 6000 ± 3000 km s^{-1} along the northeast–southwest axis. This value is equivalent to an expansion velocity of 8500 ± 4300 km s^{-1} for the outer part of a shell-like brightness distribution. Assuming uniform expansion, the date of explosion is 1955^{+10}_{-20}. A backwards extrapolation of the source's light curve to a time around the date of explosion suggests a maximum flux density at 1.5 GHz of the order of 1 Jy. The shell-like morphology of 41.9+58 and its large expansion velocity, combined with the previously found characteristics of 41.9+58 and those of the other hot spots, lead to the conclusion that 41.9+58 and most other hot spots, if not all, in M82 are indeed SNe and young SNRs.

Apart from SNR41.9+58, all other hot spots that were labeled in the VLA image in Fig. 7 were observed with VLBI. For 13 of these sources, sizes, or lower bounds on them, could be determined (Fig. 8). These values combined with upper bounds on their sizes from VLA observations (KRONBERG, BIERMANN, and SCHWAB [27]) indicate a radio SN rate in the inner 600 pc of M82 of 0.1 yr^{-1} (BARTEL *et al.* [28]). The rate for SNe in general could even be larger.

2.5 SN1986J: image – potential for movie

The supernova SN1986J in the galaxy NGC891 (Fig. 9), at 0.56 the distance to the center of the Virgo cluster (AARONSON *et al.* [39]), reached its maximum flux density of ~ 120 mJy at 1.5 GHz in mid 1987 (WEILER, this volume). This epoch occurred about one year after its discovery at the VLA (RUPEN *et al.* [40]) and several years after the (unobserved) explosion. On the basis of the supernova's light curve and a theoretical model for the interaction of a shell of the SN with its circumstellar material ejected in the pre-supernova phase (CHEVALIER [41]; WEILER, this volume), the explosion occurred sometime between 1982 and 1983.

VLBI observations were made at 1.7 GHz on 1986 Sep. 29 (BARTEL, RUPEN, and SHAPIRO [42]), at 10.7 GHz on 1987 Feb. 23, at 5.0 GHz on 1987 May 30 (BARTEL, RUPEN, and SHAPIRO [43]), and at 8.4 GHz on 1988 Sep. 29. Since each of the observations was made at a different frequency, and since only a sparse set of data was obtained for the two earlier observations, our sensitivity for determining any expansion was relatively small. Probably partly as a result, we have not yet detected any significant expansion.

The first three observations did not allow more complicated models than circular and elliptical Gaussians to be fit to the visibility data. For the 5.0 GHz data, we obtained a half-width at half-maximum of the major axis of 0.8 ± 0.1 mas with a position angle of $145° \pm 5°$ and a ratio between the minor and the major axis of 0.62 ± 0.05 (BARTEL, RUPEN, and SHAPIRO [43]).

The last observations so far were the most sensitive and most extensive ones and allowed us to make a VLBI image, the first one ever of a SN. The image of SN1986J is shown in the lower part of Fig. 9 juxtaposed to an image of the host galaxy NGC891.

SN1986J has a complex brightness distribution elongated in the direction indicated by the model in the earlier observations. An elongation could have been caused by a number of processes. First, it is conceivable that the flow pattern of the expanding gas is anisotropic due to rotation of the progenitor star (BODENHEIMER and WOOSLEY [44]) or to an asymmetric explosion (SHKLOVSKII [45]) or both. Second, the flow pattern of the expanding gas could be isotropic, but the combination of the rate of mass loss and wind velocity of the progenitor star could have varied angularly and given rise to an anisotropic density distribution of the circumstellar medium (EMMERING and CHEVALIER [46]). Third, the radio emission might not emanate from the shockfront region but rather from the amorphously shaped relativistic particle plasma emanating from a central pulsar (PACINI and SALVATI [47]). In this case, deviations from spherical symmetry of the brightness distribution of the supernova are a natural consequence. Because of the complexity of the brightness distribution, rotation of the progenitor star cannot be considered as the only cause for the SN's morphology.

Because of the limited angular resolution, the type of the morphology has been difficult to determine. Different mapping techniques give somewhat different results. Although mapping the outer structure with its jetlike extensions appears to be independent of the imaging technique, mapping the inner 2 mas is not quite independent. The clearest indication for a shell or composite structure in the brightness distribution of SN1986J, with a local minimum in its center, comes from the "maximum entropy" map as displayed in Fig. 9. A shell-type morphology would be expected (WEILER and SRAMEK [48]), since the radio spectrum of SN1986J around 8 GHz is steep, having an index $\alpha \sim -0.7$ ($S_\nu \propto \nu^\alpha$) (WEILER, this volume). It remains to be seen whether such a morphology becomes more apparent in SN1986J's further evolution with either mapping technique and how the morphology's complexity is linked to the velocity field of the expanding material. A sequence of detailed VLBI images may answer these and related questions. Such a sequence of images is indeed obtainable for SN1986J.

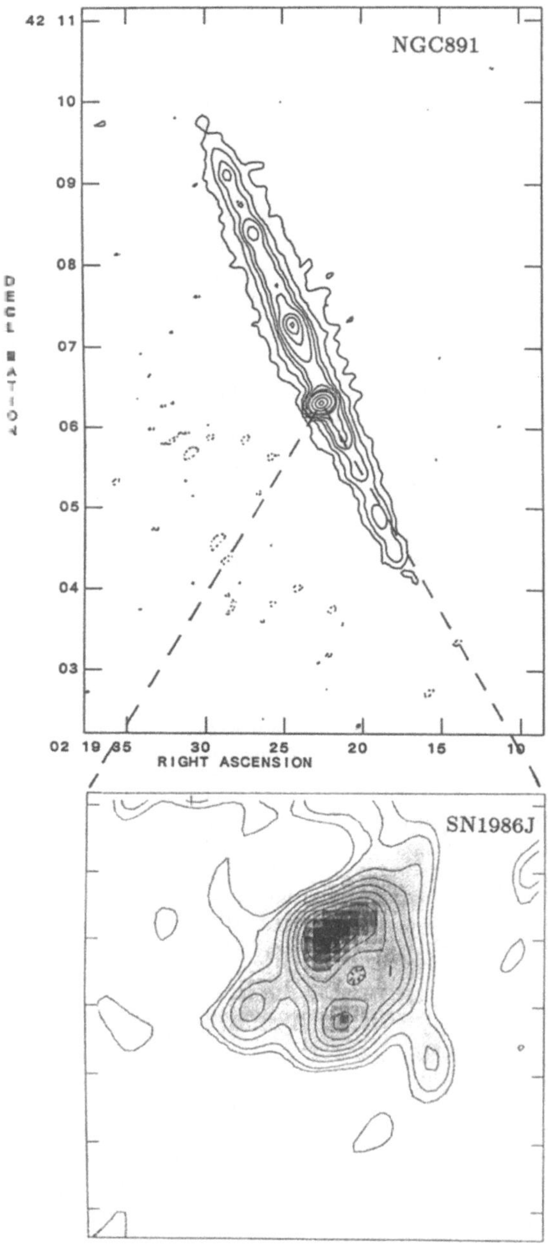

Fig. 9. A "clean" map of the galaxy NGC891 and its pointlike supernova SN1986J made with the VLA at the frequency and epoch of the VLBI observations on 1988 Sep. 29. The contours are −0.2, +0.2, 0.6, 1.1, 1.7, 2.9, 5.7, 8.6, 11.4, 28.6, 57.1, and 85.7% of 53 mJy per beam area. The negative contours are dashed. The coordinates are for epoch B1950.0. The lower part of the figure displays a preliminary VLBI "maximum entropy" image of SN1986J. The contours are 5, 10, 20, 30, 40, 50, 60, 70, 80, 90, and 99% of the peak brightness. Any angle of 1 mas on the sky is indicated by the separation of the tick marks on the left and right side of the figure. North is up and east to the left. There is a 180° ambiguity as to the orientation of the apparently shell-like inner structure of the map. The ambiguity is probably due to the lack of sufficient closure phase information from the intermediate and long baselines. There still remains some uncertainty as to the reality of the shell-like inner structure. For comparison, a "clean" map, made from convolving a set of δ-components, shows similar condensations to those shown here but not smoothly connected to a shell. Only indications of arcs are apparent. The final analysis is pending.

3. CONCLUSIONS

The most important results from VLBI observations of SNe and their remnants have been a) the imaging of a supernova (SN1986J) and a young supernova remnant (SNR41.9+58) and the estimate of the radio SN rate in the central region of the SNR's host galaxy (M82), b) the determinations of the angular expansion rate, and bounds on any acceleration or deceleration of it, of a supernova (SN1979C), c) the estimate of a useful lower bound on the size of the radio shell of a supernova (SN1987A) relative to the sizes of the line-forming regions, and d) estimates of the distance to the Virgo cluster of galaxies and H_0 (see BARTEL, this volume).

More observations of the supernovae discussed here and of supernovae yet to be discovered should allow, at least in some cases, investigations of:

— the detailed dynamic and spectral evolutions, over a large fraction of the SNe's lifetime, of different segments of the shockfront as it expands into the circumstellar medium;

— the relation between the dynamics of the shockfront and those of a) the outer jetlike features and b) the inner pulsar nebula, if the latter exists and can be detected;

— the correlation between radio/optical light curves and the properties derived from the VLBI images; and

— the method of determining distances to galaxies as far away as $\sim$ 40 Mpc.

The VLBA, the new Green Bank 100-m class telescope, and other sensitive antennas together with space-based antennas will aid considerably in the realization of the projects discussed and afford us the intriguing opportunity to make movies, with unprecedented details, of exploding stars.

4. ACKNOWLEDGMENT

This research was supported in part by the NSF under grant No. AST-8902087.

5. REFERENCES

1. Braun, R., Gull, S. F., and Perley, R. A. 1987, *Nature*, **327**, 395.
2. Bartel, N. 1988, in *Supernova Shells and Their Birth Events*, Lecture Notes in Physics, ed. W. Kundt (Springer–Verlag, Berlin), **316**, 206.
3. Turtle, A. J. *et al.* 1987, *Nature*, **327**, 38.
4. Aglietta, M. *et al.* 1987, *Europhys. Lett.*, **3**, 1315.
5. Bionta, R. M. *et al.* 1987, *Phys. Rev. Lett.*, **58**, 1494.
6. Hirata, K. *et al.* 1987, *Phys. Rev. Lett.*, **58**, 1490.
7. Shapiro, I. I. *et al.* 1988, in *IAU Symposium 129, The Impact of VLBI on Astrophysics and Geophysics*, eds. M. J. Reid and J. M. Moran (Reidel, Dordrecht), p. 185.
8. Preston, R. A., Morabito, D. D., Williams, J. G., Faulkner, J., Jauncey, D. L., and Nicolson, G. D. 1985, *A. J.*, **90**, 1599.
9. Bartel, N. *et al.* 1988, in *Supernova 1987A in the Large Magellanic Cloud*, eds. M. Kafatos and A. Michalitsianos (Cambridge Univ. Press, Cambridge), p. 81.
10. Jauncey, D.L. *et al.* 1988, *Nature*, **334**, 412.
11. Marscher, A. P. 1985, in *Supernova as Distance Indicators*, Lecture Notes in Physics, ed. N. Bartel (Springer–Verlag, Berlin), **224**, 130.
12. Feast, M. W., and Walker, A. R. 1987, *Ann. Rev. Astr. Ap.*, **25**, 345.
13. Menzies, J. W. *et al.* 1987, *M. N. R. A. S.*, **227**, 39p.

14. Hanuschik, R. W., and Dachs, J. 1987, *Astr. Ap. Lett.*, **182**, L29.
15. Blanco, V. M. *et al.* 1987, *Ap. J.*, **320**, 589.
16. Bartel, N. 1985, in *Supernovae as Distance Indicators*, Lecture Notes in Physics, ed. N. Bartel (Springer–Verlag, Berlin), **224**, 107.
17. Bartel, N. 1986, in *Highlights of Astronomy*, ed. J. P. Swings (Reidel, Dordrecht), **7**, 655.
18. Bartel, N., Rogers, A. E. E., Shapiro, I. I., Gorenstein, M. V., Gwinn, C. R., Marcaide, J. M., and Weiler, K. W. 1985, *Nature*, **318**, 25.
19. Weiler, K. W., Sramek, R. A., Panagia, N., van der Hulst, J. M., and Salvati, M. 1986, *Ap. J.*, **301**, 790.
20. Bartel, N. 1988, in *IAU Symposium 129, The Impact of VLBI on Astrophysics and Geophysics*, eds. M. J. Reid and J. M. Moran (Reidel, Dordrecht), p. 175.
21. Seaquist, E. R. and Bignell, R. C. 1978, *Ap. J. (Letters)*, **226**, L5.
22. Balick, B. and Heckmann, T. 1978, *Ap. J. (Letters)*, **226**, L7.
23. de Bruyn, A. G. 1983, *Astr. Ap.*, **119**, 301.
24. Kirshner, R. P. and Blair, W. P. 1980, *Ap. J.*, **236**, 135.
25. Chevalier, R. A. 1982, *Ap. J.*, **259**, 302.
26. Chevalier, R. A. 1984, *Ann. NY Acad. Sci.*, **422**, 215.
27. Kronberg, P. P., Biermann, P., and Schwab, F. R. 1985, *Ap. J.*, **291**, 693.
28. Bartel, N. *et al.* 1987, *Ap. J.*, **323**, 505.
29. Kronberg, P. P., Biermann, P., and Schwab, F. R. 1981, *Ap. J.*, **246**, 28.
30. Unger, S. W., Pedlar, A., Axon, D. J., Wilkinson, P. N., and Appleton, P. N. 1984, *M. N. R. A. S.*, **211**, 783.
31. Tammann, A. and Sandage, A. R. 1968, *Ap. J.*, **151**, 825.
32. Sandage, A. 1984, *A. J.*, **89**, 621.
33. Kronberg, P. P., and Sramek, R. A. 1985, *Science*, **277**, 28.
34. Rieke, G. H., Lebofsky, M. J., Thompson, R. I., Low, F. J., and Tokunaga, A. T. 1980, *Ap. J.*, **238**, 24.
35. Geldzahler, B. J., Kellermann, K. I., Shaffer, D. B., and Clark, B. G. 1977, *Ap. J. (Letters)*, **215**, L5.
36. Wilkinson, P. N. and de Bruyn, A. G. 1988, in *IAU Symposium 129, The Impact of VLBI on Astrophysics and Geophysics*, eds. M. J. Reid and J. M. Moran (Reidel, Dordrecht), p. 187.
37. Lugten, J. B., Watson, D. M., Crawford, M. K., and Genzel, R. 1986, *Ap. J. (Letters)*, **311**, L51.
38. Wilkinson, P. N. and de Bruyn, A. G. 1984, *M. N. R. A. S.*, **211**, 593.
39. Aaronson, N. *et al.* 1982, *Ap. J. Suppl.*, **50**, 241.
40. Rupen, M. P., van Gorkom, J. H., Knapp, G. R., Gunn, J. E., and Schneider, D. P. 1987, *A. J.*, **94**, 61.
41. Chevalier, R. A. 1987, *Nature*, **329**, 611.
42. Bartel, N., Rupen, M., and Shapiro, I. 1987, IAU Circ. No. **4292**.
43. Bartel, N., Rupen, M. R., and Shapiro, I. I. 1988, *Ap. J. (Letters)*, **337**, L85.
44. Bodenheimer, P., and Woosley, S. E. 1983, *Ap. J.*, **269**, 281.
45. Shklovskii, I. S. 1970, *Soviet Astr.*, **13**, 562, transl. from *Astr. Zh.*, **46**, 715.
46. Emmering, R. T., and Chevalier, R. A. 1988, *A. J.*, **95**, 152.
47. Pacini, F. and Salvati, M. 1981, *Ap. J. (Letters)*, **245**, L107.
48. Weiler, K. W., and Sramek, R. A. 1988, *Ann. Rev. Astr. Ap.*, **26**, 295.

A Supernova with a Difference, SN 1986J

Kurt W. Weiler, Nino Panagia, & Richard Sramek

I. Introduction

A new radio source was discovered on 21 August 1986 by VAN GORKOM et al. [1] during their study of the nearby, edge-on, spiral galaxy NGC891. Because the source lay in the disk and had HI absorption features characteristic of the galaxy, they concluded that it was a supernova in NGC891 and their report assigned the designation SN1986J. A summary of the known properties of SN1986J is given in Table 1.

Table 1. Properties and references for SN1986J

Property	Value	Reference
Position		
Right Ascension	$02^h19^m22^s.60 \pm 0^s.02$	
Declination	$+42°06'18''.9 \pm 0''.2$	
Distance (NGC891)	12 Mpc	2,3
Explosion Date	13 Sept. 1982 ± ~250 days	
Discovery Date	21 Aug 1986	1
Optical/IR photometry		1,4,5,6
Optical/IR spectroscopy		4,7
Radio flux density meas.		1,4,7,8,9,10,11 12,13,14,15
Radio diameter measurements (Gaussian model, FWHM)	1.6 x 1.0 mas in 5/87	16
Physical diameter	$\geq 2.9 \times 1.8 \times 10^{17}$ cm	
Average expansion velocity	$\geq 9.6 \times 6.1 \times 10^3$ km s^{-1}	
Maximum line width	$\sim 10^3$ km s^{-1}	4
Apparent B magnitude at max.	$> 18^m$?	5
Absolute B magnitude at max.	$>-14^m.9$? for $A_B = 2^m.5$	
Peak flux density (λ6 cm)	~130 mJy	
Average surface brightness at peak flux (λ6 cm)	$\sim 3.5 \times 10^{-11}$ Wm^{-2}Hz^{-1}ster^{-1}	
Peak brightness temp. (λ6 cm)	$\sim 4.5 \times 10^9$ K	
Peak spectral lumin. (λ6 cm)	$\sim 2.1 \times 10^{28}$ erg s^{-1}Hz^{-1}	
Ratio to Cas A	~3,000	3
Ratio to SN1979C	~3	3

Optical spectral observations by RUPEN et al. [4] in 1984 and 1986 of an object at the position of SN1986J reveal the presence of hydrogen lines indicating classification as a Type II, but the lines are rather narrow, being only ~ 1000 km s^{-1} wide. A more detailed discussion of the properties of SN1986J in all wavelength ranges is given in WEILER, PANAGIA, and SRAMEK [17].

Since we are carrying out a long standing program of searching for and monitoring the radio emission from RSNe, we were asked by the collaborators on RUPEN et al. [4] to include SN1986J in our program. We report here on the results of this continuing series of observations and their meaning in terms of the type, nature, and physical properties of SN1986J.

II. Observations

Even though it, unfortunately, was not found optically so that its exact age is not well known, SN1986J is the most powerful RSN ever detected and there are extensive radio observations available. We have added 63 new measurements at the 5 different VLA wavelengths of λλ90, 20, 6, 2, and 1.3 cm over a period of more than two years. Including the measurements available from other observers and from the literature, there are now

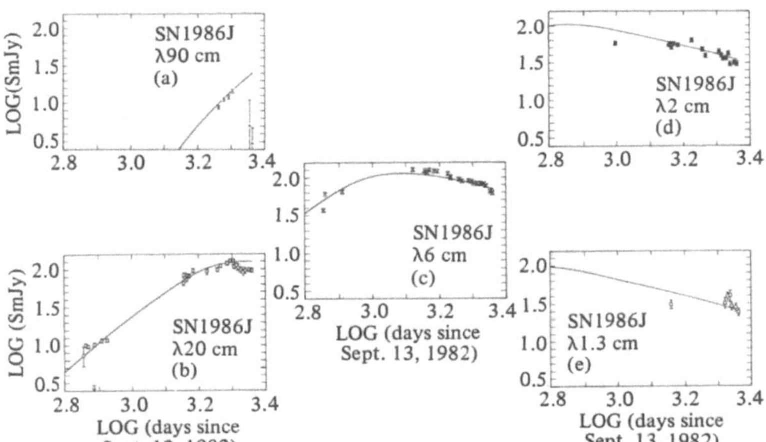

Figure 1. Radio "light curves" for SN1986J in NGC891 for the five wavelengths λλ90, 20, 6, 2, and 1.3 cm. The age of the supernova is measured in days from the best fit date of explosion on 13 September 1982. The solid lines represent the best fit light curves of the form $S(mJy) = K_1[\nu/(5\ GHz)]^{\alpha}[(t-t_0)/(1\ day)]^{\beta}e^{-\tau}(1-e^{-\tau'})\tau'^{-1}$, where $\tau = K_2[\nu/(5\ GHz)]^{-2.1}[(t-t_0)/(1\ day)]^{\delta}$ and $\tau' = K_3[\nu/(5\ GHz)]^{-2.1}[(t-t_0)/(1\ day)]^{\delta'}$ with $\delta = \alpha-\beta-3$ and $\delta' = 5\delta/3$ from the CHEVALIER [18,19] model.

almost 100 flux density measurements available at these 5 wavelengths and these are plotted in Fig. 1.

Examination of Fig. 1 shows some unusual features: a. the emission turns on relatively slowly with time compared with SN1979C and SN1980K (see WEILER et al. [20]) and b. the shortest wavelength for which there is a reasonable data set (λ1.3 cm) shows roughly constant flux density over the measurement interval.

One aspect which is perhaps not unusual, although it might appear so at first sight, is the apparent disappearance of SN1986J at λ90 cm in November and December 1988 after detection by SUKUMAR and ALLEN [11] of strong and rising flux density in the period from September 1987 to March 1988. Given the high optical depth at λ90 cm still present in 1988, (see models below) any fluctuation in this opacity would quickly blot out the λ90 cm radiation since we are seeing such a small fraction of the total emission. Such a flux density fluctuation in the same time interval is also evident in the λ20 cm data. This opacity change can, in fact, be quantified as has been done by WEILER, PANAGIA, and SRAMEK [17] who show that an external absorption variation can explain the observed decrement. Similar short term changes in the opacity of the external absorbing medium were also observed in SN1979C (WEILER et al. [20]) and have been interpreted by LUNDQVIST and FRANSSON [21] as due to variation in the temperature and ionization of the presupernova stellar wind close to the supernova blast wave.

Our extensive data set yields many determinations of the spectral index α ($S \propto \nu^{+\alpha}$) both in time and between different pairs of frequencies. These are plotted in Fig. 2. Examination of Fig. 2 shows that:

a. The spectrum from λ90 cm to λ20 cm through the end of 1988 is "inverted" with a large positive index ($\alpha \sim +1.5$) indicating that SN1986J is still very optically thick to its λ90 cm emission.
b. The spectral index from λ20 cm to λ6 cm changes from a large positive index ($\alpha \sim +1.5$) initially, when the RSN was still optically thick to its λ20 cm emission, to a more "normal" negative index of $\alpha \sim -0.3$ for the most recent data as SN1986J has become almost optically thin at λ20 cm. The form of the change, however, is quite unlike the sharp decline of the spectral index followed by an asymptotic approach to the optically thin value as was seen for

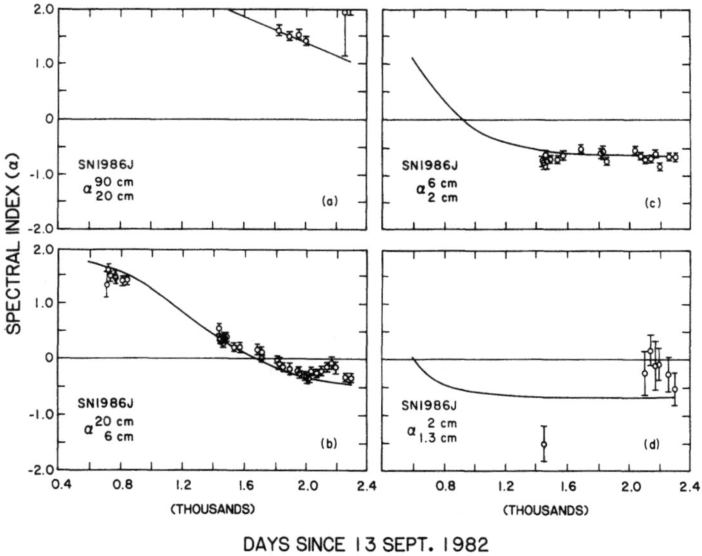

Figure 2. Spectral index α, ($S \propto \nu^{+\alpha}$) evolution for SN1986J between (a) λ90 cm -- λ20 cm, (b) λ20 cm -- λ6 cm, (c) λ6 cm -- λ2 cm and (d) λ2 cm -- λ1.3 cm plotted as a function of time in days since the explosion date on ~13 September 1982. The solid lines are calculated from the best fit theoretical "light curves" shown in Fig. 1 and described below.

SN1979C and SN1980K (WEILER et al. [20]). It is more of plateau at $\alpha \sim +1.5$ followed by a slow decline to an optically thin value which may eventually reach $\alpha \sim -0.7$.

c. At λ6 cm to λ2 cm, for the period where there is data, the RSN is optically thin to its emission at both wavelengths with an average index of $\alpha = -0.7 \pm 0.1$.

d. The spectrum between λ2 cm and λ1.3 cm, even though sparse and hampered by larger errors is, quite surprisingly, flat with an average index of only $\alpha = -0.2 \pm 0.2$.

III. Discussion

1. Light Curves

In their study of a number of RSNe WEILER et al. [20] established several common properties:

a. non-thermal emission with high brightness temperature;

b. turn-on first at shorter wavelengths and later at longer wavelengths;

c. initial rapid increase of flux density with time at each wavelength;
d. power law decline in flux density at each wavelength after maximum is reached;
e. sharp initial decrease in spectral index between any two wavelengths as the longer wavelength goes from optically thick to optically thin, and
f. a final asymptotic approach of the spectral index α to an optically thin, non-thermal, constant negative value ($S \propto \nu^{+\alpha}$).

They showed that such behavior is best described by the change in optical depth of a external, thermal, absorbing screen and ruled out such absorbing processes as synchrotron self-absorption, Razin-Tsytovich effect, and mixed, internal, non-thermal emitting, thermal absorbing gas. They then modelled the radio "light curves" of the RSNe with the simple mathematical form:

$$S(\mathrm{mJy}) = K_1 \left(\frac{\nu}{5\ \mathrm{GHz}}\right)^{\alpha} \left(\frac{t - t_0}{1\ \mathrm{day}}\right)^{\beta} e^{-\tau} \tag{1}$$

with

$$\tau = K_2 \left(\frac{\nu}{5\ \mathrm{GHz}}\right)^{-2.1} \left(\frac{t - t_0}{1\ \mathrm{day}}\right)^{\delta} . \tag{2}$$

This formulation assumes that both the flux density S and the optical depth τ are well described by power-law functions of the supernova age ($t - t_0$), with powers β and δ, respectively; that the absorption is purely thermal, free-free absorption in an ionized medium (frequency dependence $\nu^{-2.1}$) external to the emitting region with a radial dependence of r^{-2} from a constant speed, red supergiant wind; and that the intrinsic emission is due to the nonthermal synchrotron process with an optically thin spectral index α. The quantities K_1 and K_2 are two scaling factors for the units of choice of mJy, GHz, and days, and correspond formally to the flux density (K_1) and optical depth (K_2) at 5 GHz 1 day after the SN explosion.

While SN1986J certainly shares a number of common properties with other RSNe such as Points a, b, d, and f listed above, it is clearly different in its slow turn on at each frequency, early detection at low frequencies, and spectral index evolution. The purely external, thermal absorbing gas described by (1) and (2) provides a poor fit to our data set. Quantitatively, the best fit which could be obtained with the model of (1) and (2) still had a reduced $\chi_{red}^2 > 10$ even though a minimum measurement error of 14% was introduced into the calculations.

Since synchrotron self-absorption and the Razin-Tsytovich effect are still unlikely to be significant factors in a relatively extended emitter such as SN1986J, the most attractive alternative mechanism to describe its radio light curves is to add an additional component consisting of an expanding, mixed, thermal absorbing/non-thermal emitting gas undergoing a transition from optically thick to thin. This can be mathematically expressed by including an additional absorption term in (1) to give

$$S(\mathrm{mJy}) = K_1 \left(\frac{\nu}{5\ \mathrm{GHz}}\right)^{\alpha} \left(\frac{t - t_0}{1\ \mathrm{day}}\right)^{\beta} e^{-\tau} \left(1 - e^{-\tau'}\right) \tau'^{-1} \tag{3}$$

where τ is still the external absorption described by (2) and

$$\tau' = K_3 \left(\frac{\nu}{5\ \mathrm{GHz}}\right)^{-2.1} \left(\frac{t - t_0}{1\ \mathrm{day}}\right)^{\delta'} \tag{4}$$

with τ', K_3 and δ' describing the properties of a thermal absorbing medium which is <u>mixed</u> with the nonthermal emitting gas. To limit the number of free parameters, we again adopt the CHEVALIER [18,19] model which determines $\delta = \alpha - \beta - 3$ and $\delta' = 5\delta/3$.

Searching parameter space for a minimum χ^2 with (2), (3), and (4) yields the values listed in Table 2. These values are then used to calculate the model curves which are shown as the solid lines in Figs. 1 and 2. As can be seen in the Figures, the fit to the data is quite satisfactory, especially considering the simplicity of the model, and describes quite well the changes in the flux density of SN1986J both with time and frequency.

As might be expected, the degree to which the several parameters are determined by the fitting process varies greatly. The spectral index (α) and rate of decline of the emitting process (β) are tightly constrained with the intrinsic intensity of the non-thermal emission (K_1) and the amount of internal absorption (K_3) somewhat less so. The solutions are not very sensitive to the explosion date (t_0), with indeterminacy of ~8 months earlier or later than the best fit date of 13 September 1982, or to the amount of external absorption (K_2). The external absorption

Table 2. Fitting parameters for SN1986J[1]

Parameter	Value	Deviation Range[a]
K_1	6.7×10^5	$(3.8 -- 9.2) \times 10^5$
α	-0.67	-(0.59 -- 0.71)
β	-1.18	-(1.16 -- 1.22)
K_2	3×10^5	$(0 -- 63) \times 10^5$
δ $(=\alpha-\beta-3)$	-2.49	-(2.19 -- 2.69)
K_3	4×10^{12}	$(2 -- 12) \times 10^{12}$
δ' $(=5\delta/3)$	-4.15	-(3.65 -- 4.45)
t_0	13 Sept. 1982	(25 Feb. '82 -- 10 Jul. '83)

Derived Quantities

M_{dot} ~ $1.8 \times 10^{-4}(v/10 \text{ km s}^{-1})$ $M_\odot$ yr^{-1}

$M_{eject} \geq 5.5$ $M_\odot$

[1]With an additional error of 14% of the measured flux density to account for the fact that no simple model can completely describe such a complicated phenomenon, these parameters yield a $\chi^2_{red} = 0.96$.

[a]The Deviation Range is the range in which there is a ~67% probability that the true value lies. This is equivalent to a 1σ range for a one parameter solution.

could, in fact, be omitted entirely ($K_2 = 0$) and still obtain a reasonable description of the available data. However, $K_2 = 3 \times 10^5$ provides the best fit and there are other reasons for expecting an external absorbing medium to be present.

2. Models

As discussed above, SN1986J is more complex than a simple mini-shell model can describe. In addition to the presumed presence of an external absorbing medium surrounding the emission region there is an additional absorbing medium which is mixed with the emission regions [the additional terms in (3) and (4)]. We shall discuss these separately.

a. The external absorber

Because of the dominance of the mixed emitting/absorbing component, the properties of the purely external absorber are relatively poorly determined for SN1986J. Examination of the uncertainty for parameter K_2 in Table 2 (which specifies the initial amount of external absorption) shows that the external absorption could range from $0 < K_2 < 6.3 \times 10^6$ to within 1σ. In other words, the amount of external absorption could range from none to almost as much as was estimated for the next brightest RSN SN1979C ($K_2 = 5.1 \times 10^7$; WEILER et al. [20]).

However, as discussed above, shorter term variations in the optical depth at λλ90 and 20 cm imply the existence of an external absorbing medium of roughly this magnitude. Attributing this external absorber to the presence of a surrounding cocoon of matter established by mass loss from the SN progenitor in the last, presupernova stages of stellar evolution, one can estimate a mass loss rate (see WEILER et al. [20] for a discussion of the assumptions involved).

Following LUNDQVIST and FRANSSON [21], we adopt a temperature of 3×10^4 K for the circumstellar gas prior to the strong cooling episode. We also adopt an average expansion velocity of 7.8×10^3 km s^{-1} derived from VLBI observations in May 1987 (t ~ 1720 days; BARTEL, RUPEN and SHAPIRO [16]). Then, using Equation 16 of WEILER et al. [20], we obtain a mass loss rate of M_{dot}(absorption) = 1.8×10^{-4} (v/10 km s^{-1}) $M_\odot$ yr^{-1}. This value agrees quite well with the estimate one can make from the level of intrinsic emission as determined by the interaction of the SN ejecta with the circumstellar material. Using the simplified formulation of WEILER et al. [3] valid for Type II SNe, we obtain M_{dot}(emission) = 1.3×10^{-4} $M_\odot$ yr^{-1}. Although the excellent agreement of the two estimates may be coincidental, it is clear that the mass loss rate of the SN1986J progenitor was quite high. This implies that it was a red supergiant as massive as possible to lose mass at such a high rate, i.e., M(ZAMS) >> 8 $M_\odot$ which is the limit above which stars are supposed to be able to explode by core collapse. However, the original mass should not be too high, say ≤30 $M_\odot$, because otherwise the progenitor would have first become a Wolf-Rayet star before exploding and would have created a much less dense circumstellar envelope.

b. The mixed absorber

Even though the presence of an external, wind created absorbing medium of density comparable to that observed for SN1979C is indicated both by our parameter fits and by the observed short term light curve variations, the radio properties of SN1986J are dominated by an absorbing medium which is mixed with the non-thermal emitters. Equations (3) and (4) implicitly assume an expanding, filled, volume with uniformly mixed emitters and absorbers.

Adopting for the absorbing material mixed with the non-thermal gas a temperature of 3×10^4 K, the value of $K_3 = 4 \times 10^{12}$ implies an emission measure of $n_e^2 \Delta R = 1.9 \times 10^{26}$ cm^{-5}. In the case of a spherical structure with constant density, this indicates a total absorbing mass of M(internal abs.) ~ 3 $M_\odot$. The mass would be lower if the filling factor is significantly lower than unity but higher if the gas ionization is not complete.

Similarly, we can estimate the properties of the thermal gas which is emitting Hα radiation. From RUPEN et al. [4] the intensity of Hα in September 1986, corrected for an extinction corresponding to $A_V = 2^m$, is $I_{corr}(H\alpha) = 1.9 \times 10^{-13}$ erg cm^{-2} s^{-1} which, for a distance of 12 Mpc, becomes $L(H\alpha) = 3.2 \times 10^{39}$ erg s^{-1}. Since the expansion velocity of the Hα emitting region is about 10^3 km s^{-1}, its radius is about 1/7.8 of the radius of the radio emitting region. Therefore, extrapolating to May 1987, the average density turns out to be $n_{ave}(H\alpha) = 7.3 \times 10^6$ and the emitting mass to be ~2.5 $M_\odot$ (again assuming $T = 3 \times 10^4$ K). Since the density is so much higher than that of the radio emitting region and the radius so much smaller, we have direct evidence for a strong density gradient. Furthermore, it is likely that the two regions are spatially distinct, so that the total mass of the ejecta is essentially the sum of the two masses, i.e., ~5.5 $M_\odot$. In fact, this may even be just a lower limit to the true mass because it is quite possible that, similar to what is found in SN1987A, some dust may have condensed within the emitting gas and be absorbing a sizeable fraction of the Hα radiation.

3. What is SN1986J?

It is clear that SN1986J is different from all previously studied RSNe and may be the defining member of a new subclass of Type II SNe. For illustrative purposes, we sketch an estimate of its basic structure in Fig. 3. Since the origin and role of its apparently elongated radio structure (BARTEL, RUPEN, and SHAPIRO [16]) is not known at the present time, we represent it as spherically symmetric.

Although very speculative at this stage, we suggest that we may be witnessing the birth of a plerion -- a Crab Nebula. The possibly high mass progenitor star, the high abundance of He, the mixed thermal and non-thermal matter, the extensive filamentation, and the elliptical morphology all bear some similarities to the Crab Nebula. HOWEVER, it should be emphasized that emission from a mini-plerion is apparently not presently being observed at radio wavelengths although at optical wavelengths it may be (Chevalier [22]). A defining characteristic of a plerion is its flat spectrum ($\alpha > -0.3$) radio emission while SN1986J has a rather steep (α = -0.67) spectrum. Although there is slight evidence in our data for flattening of the radio spectrum at the short wavelengths (λ1.3 cm) that must still be confirmed. If a mini-plerion is indeed active and just becoming visible at the present time, one expects the radio light curves to begin to deviate from the description of (2), (3), (4) and Table 2 as the "new" emission becomes visible. Since quarterly monitoring of SN1986J is continuing, its full nature and the possible presence of additional emitting components should become clearer in time.

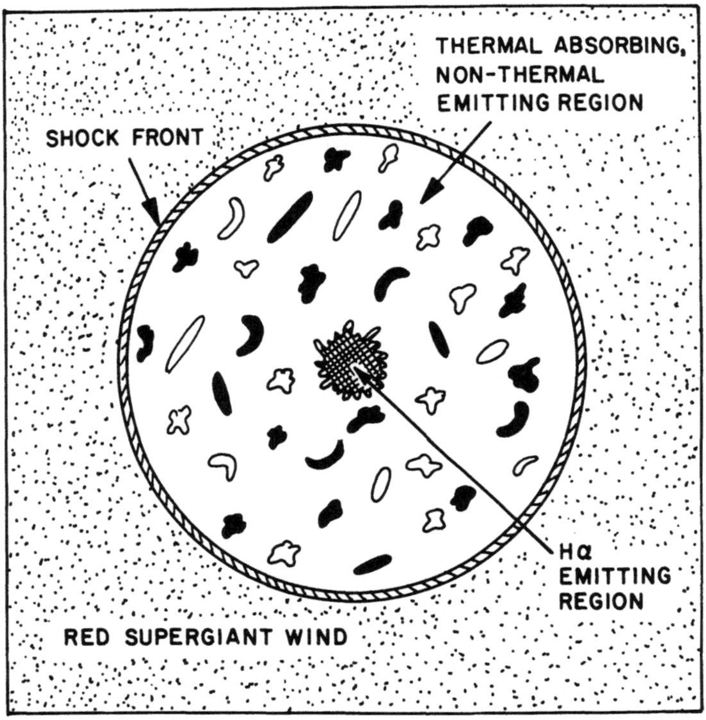

Figure 3. A spherically symmetric cartoon of the possible structure of SN1986J at the present epoch. The circumstellar cocoon from the red supergiant (RSG) wind phase of the presupernova star extends to an unknown distance. The supernova shock front presently forms the boundary between the RSG wind and the thermal absorbing/nonthermal emitting regions. The diameter of this shock front is equal to the VLBI size of the radio emission from SN1986J or $\sim 2.4\times10^{17}$ cm and it is expanding at $\sim$7,800 km s^{-1}. The thermal absorbing/nonthermal emitting region appears to be heavily fragmented into thermal absorbing blobs or filaments co-mixed with numerous nonthermal emitting regions. The high density Hα emitting region is central, is expanding much more slowly ($\sim$1000 km s^{-1}), and is much smaller being $\sim$1/7.8 of the VLBI size or $\sim 3\times10^{16}$ cm.

IV. Conclusions

Analysis of all of the available data for SN1986J leads us to conclude:

1. It is clearly different from previously known Type II RSNe and may represent a new class.
2. Its radio characteristics are dominated by a mixed thermal absorbing/nonthermal emitting medium which is evolving in a regular way with time.
3. An external absorbing medium, although contributing relatively less to the observed properties of the radio emission, is present with a magnitude similar to that seen in previous RSNe such as SN1979C and SN1980K. It is also probably responsible for the short term fluctuations seen in the radio light curves.
4. A presupernova stellar mass loss rate of $M_{dot} \sim 1.8 \times 10^{-4}\ M_{\odot}\ yr^{-1}$ is needed to account for the most likely external absorbing medium.
5. A minimum ejected mass of $M \sim 5.5\ M_{\odot}$ is needed to account for the observed absorption in the mixed emitting/absorbing region and the observed Hα emission.
6. The density of the presupernova stellar wind and the ejected mass which can be accounted for in the mixed thermal absorbing/non-thermal emitting region and the Hα emitting region imply a zero age main sequence stellar mass M(ZAMS) ~ 20 -- 30 $M_{\odot}$.
7. We may be witnessing the birth of a plerion like the Crab Nebula and it has been suggested by Chevalier [22] that pulsar driven optical emission has already been observed. However, the presently detectable radio emission is steep spectrum ($\alpha \sim -0.67$) and is not likely to be plerionic. A new, flat spectrum radio component <u>may</u> be emerging at the highest monitoring frequencies (λλ2 -- 1.3 cm), but must still be confirmed.

In summary, SN1986J is a very interesting and unusual object which has already revealed enough of its nature to allow us to model its physical structure quite well. As it evolves we are continuing to monitor it at regular intervals to confirm and refine those properties already established and to look for the possible development of new characteristics.

References

1. van Gorkom, J., Rupen, M., Knapp, G., and Gunn, J.: IAU Circ., No. 4248 (1986).
2. Aaronson, M., et al.: Ap. J. Suppl. 50, 241 (1982).

3. Weiler, K.W., Panagia, N., Sramek, R.A., van der Hulst, J.M., Roberts, M.S., and Nguyen, L.: Ap. J., 336, 421 (1989).
4. Rupen, M.P., van Gorkom, J.H., Knapp, G.R., Gunn, J.E., and Schneider, D.P.: Astron. J. 94, 61 (1987).
5. Cappellaro E. and Turatto, M.: IAU Circ., No. 4262 (1986).
6. Kent, S. and Schild, R.: IAU Circ., No. 4423 (1987).
7. Gunn, J.: IAU Circ., No. 4258 (1986).
8. Wehrle, A.: IAU Circ., No. 4260 (1986).
9. Sukumar, S.: IAU Circ., No. 4287 (1986).
10. Allen, R.J., Sukumar, S., and Beck, R.: IAU Circ., No. 4595 (1988).
11. Sukumar, S. and Allen, R.J.: Ap. J., 341, 883 (1989).
12. Gioia, I.M. and Fabbiano, G.: Ap. J. Suppl. 63, 771 (1987).
13. Condon, J.J.: IAU Circ., No. 4258 (1986).
14. Fabbiano, G. and Gioia, I.M.: IAU Circ., No. 4258 (1986).
15. van der Hulst, J.M., de Bruyn, A.G., and Allen, R.J.: IAU Circ., No. 4258 (1986).
16. Bartel, N., Rupen, M.R., and Shapiro, I.: Ap. J. Lett., 337, L85 (1989).
17. Weiler, K.W., Panagia, N., and Sramek, R.A.: Ap.J. (1990), in press.
18. Chevalier, R.: Ap. J. 251, 259 (1981).
19. Chevalier, R.A.: Ap. J. 259, 302 (1982).
20. Weiler, K.W., Sramek, R.A., Panagia, N., van der Hulst, J.M., and Salvati, M.: Ap. J. 301, 790 (1986).
21. Lundqvist, P. and Fransson, C.: Astr. Ap. 192, 221 (1987).
22. Chevalier, R. A.: Nature 329, 611 (1987).

Spectrophotometry and Photometry of SN 1987M

Alain Porter

Supernova 1987M was discovered in NGC 2715 by M. LOVAS [1] of Konkoly Observatory on September 21, 1987. Spectroscopy at Lick and Palomar Observatories in the following months showed that it was a Type Ib supernova, and provided the best coverage to date of the start of the supernebular phase in this class.

1. Spectrophotometry

The montage in Fig. 1 shows the spectrum between September 28, 1987, and February 25, 1988. The early spectrum is established as Type I by the absence of hydrogen lines, and as Type Ib by the absence of Si II λ 6355 absorption near 6150 Å. The classification is confirmed by the emergence of strong, broad emission lines of intermediate mass elements (O, Ca, Mg, Na) several months after maximum. The early and rapid onset of this supernebular phase was a bit of a surprise. No [O I] λ6300 emission is visible in the October 20 spectrum, but the calcium emission is already very strong only one month later. However, this behavior was predicted by CHUGAI [2].

Lacking He I λ5876 and λ6678 absorption lines, SN 1987M more closely resembles SNe 1983I and 1983V than the "prototypical" SNe Ib 1983N and 1984L. Wheeler *et al.* [3] give SN 1983I and 1983V a separate classification, "Type Ic." These objects certainly seem to have different abundances from other SNe Ib, and a comparison of published spectra suggests that they have continua which are redder (cooler) in the optical and near infrared region. This may significantly constrain the mechanism and ^{56}Ni mass of the explosion, and photometric and spectrophotometric tests of this possibility should be a top observing priority.

The rapid decrease in density of the remnant with time is traced by the increasing strength of forbidden relative to permitted calcium emission. It also appears that the Ca II H and K (permitted) absorption trough may be filled in by emission in the last spectrum, though by this time the star was becoming difficult to observe. In work in progress (FILIPPENKO *et al.* [4]), we hope to be able to estimate densities in the emitting regions using the calcium emission lines.

2. Photometry

Photometry of the supernova in the Thuan-Gunn g and r passbands was collected by several direct CCD observers using the 1.5 meter Oscar Mayer telescope at Palomar Observatory. Although the data are sparse, the light curve is consistent with a maximum near the discovery date and an inflection point about 30 days later. Except for a zero point shift, the g light curve agrees to within better than 0.3 magnitudes with the average Type Ia V light curve compiled by DOGGETT and BRANCH [5]. However, SN 1987M appears to have been subluminous. Assuming the Revised Shapley-Ames distance modulus of 32.4 magnitudes to NGC 2715, the maximum absolute magnitude in Gunn g (4950Å) was -16.8. Typical SNe I have maximum V

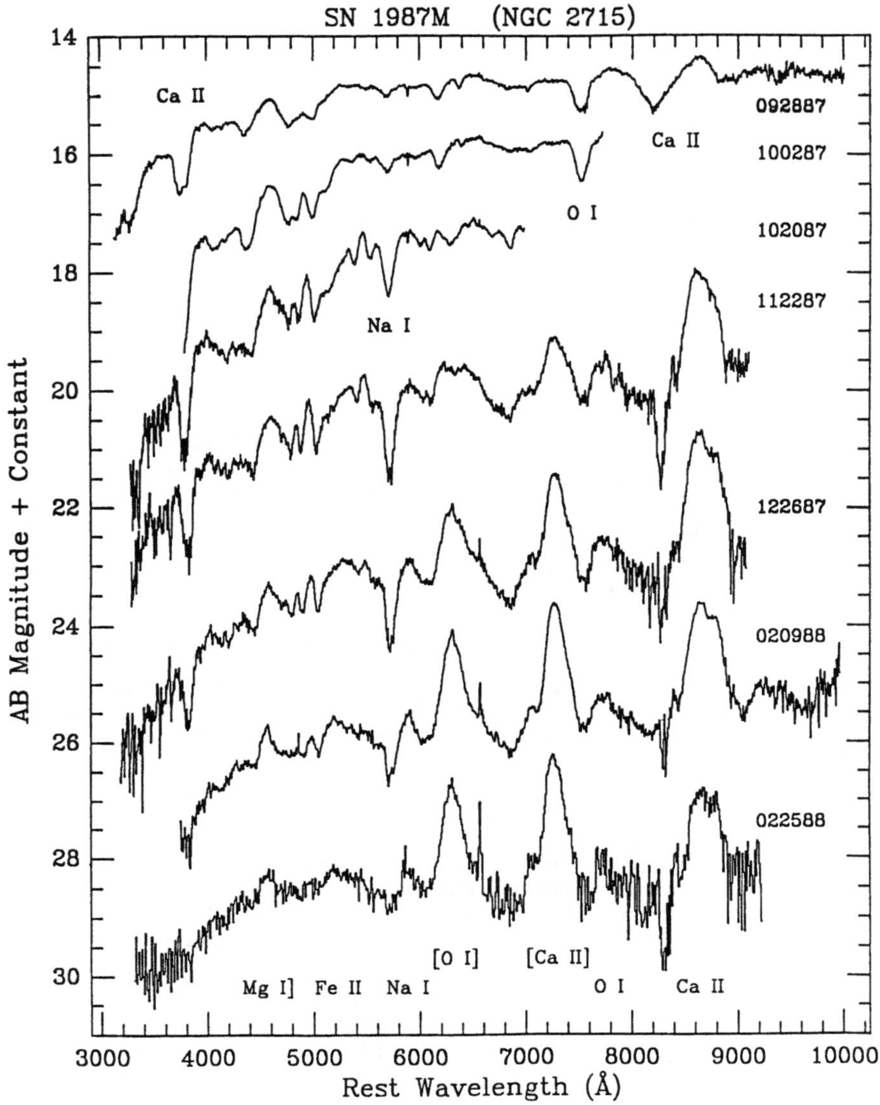

Figure 1. Spectra of SN 1987M.

(5500Å) magnitudes of -19.5 ($H_0 = 50$). It is unlikely that the $g - V$ color of the supernova and error in the distance modulus can account for this difference.

Could the supernova have been heavily obscured by dust? Na D absorption from NGC 2715 is visible in the early spectra, but its equivalent width is only 1.5 Å. PHILLIPS *et al.* [6] and RICH [7] found 3.6 Å of Na D absorption in the line of sight to SN 1986G in Cen A. They agreed closely on a reddening of $E_{B-V} = 0.9$ and a blue extinction of 3.6 magnitudes. ALLEN [8] also quotes a relation $A_V = 3.3E_{B-V}$ for visual extinction. On this basis, the visual extinction to SN 1987M is likely to have been 1.2-1.4 magnitudes. This alone certainly does not account for the star's faintness. Extinction, distance error, and color together *might* explain it, but only with difficulty.

3. Of Progenitors

The similarity of the *shape* of the light curve of SN 1987M to the Ia template poses a problem for a favored model of Type Ib explosions. ENSMAN and WOOSLEY [9] have pointed out that the light curves of exploding Wolf-Rayet stars should have broader, shallower peaks than most SNe I show. On the other hand, white dwarf detonation models are either too luminous to be SNe Ib, or fade away too rapidly (BRANCH [10]). Light curves of large samples of supernovae on *uniform* photometric systems are as necessary as spectroscopy to choose between the models and measure the explosion energies.

SN 1987M differed from most earlier SNe Ib in that it was *not* superposed on a luminous H II/OB complex. This is evident, not only from the template image of NGC 2715, but from the weakness of the narrow Hα emission in the spectra, which is not visible until the end of 1987. PANAGIA and LAIDLER [11] have argued that SNe Ib are less closely associated with H II/OB complexes than are SNe II, and that they therefore have less massive progenitors–as also suggested by their light curves. Careful astrometry, photometry, and spectroscopy of H II regions in host galaxies is needed to solve this problem.

References

1. M. Lovas: I.A.U.Circular # 4451 (1987).
2. N. N. Chugai: Pis'ma Astron. Zh. 12 461 (1986).
3. J. C. Wheeler *et al.:* Astrophys. J. 313 L69 (1987).
4. A. V. Filippenko *et al.:* (1990), in preparation.
5. L. Doggett, D. Branch: Astron. J. 90 2303 (1985).
6. M. M. Phillips *et al.:* Pub. Astron. Soc. Pacific 99 592 (1987).
7. R. M. Rich: Astron. J. 94 651 (1987).
8. C. W. Allen: Astrophysical Quantities (London, Athlone Press 1973).
9. L. Ensman, S. Woosley: Astrophys. J. 333 754 (1988).
10. D. Branch: In Atmospheric Diagnostics of Stellar Evolution, IAU Colloquium # 108, ed. by K. Nomoto (New York, Springer-Verlag 1987).
11. N. Panagia, V. Laidler: preprint (1989).

The Type Ib Supernova SN 1984L in NGC 991

Eric M. Schlegel

SN 1984L in NGC 991 is a temporarily unique object for the Type Ib subclass of supernovae. It remains the only "fully" observed SN Ib to date, where "fully" means photometry *and* spectroscopy at early *and* late times.

The Type Ib subclass was only recognized about 1985, with the discoveries of SN 1983N and SN 1984L (e.g., UOMOTO and KIRSHNER [1], WHEELER and LEVREAULT [2]). The early-time observations of these supernovae showed an overall similarity with the typical SN Ia supernovae, of which the recent prime examples are SN 1981B (BRANCH *et al.* [3]) and SN 1989B (PHILLIPS *et al.* [4]). The obvious difference was the lack of the 6150Å absorption so characteristic of the first 20-30 days of the evolution of SN Ia. The discovery of SN 1985F, with its obviously strange spectrum dominated by [O I] 6300Å (FILIPPENKO and SARGENT [5]), and the late-time spectra of SN 1984L and SN 1983N (GASKELL *et al.* [6]), gave an indication of exactly what the late-time evolution was. Unfortunately, SN 1985F was not observed spectroscopically before about 250 days past maximum. SN 1984L, another Reverend Evans discovery (EVANS [7]), was observed near maximum, and again about 10 months later. It matched the appearance of SN 1983N at early times, and showed the [O I] lines now known to be characteristic of the late evolution of SN Ib. Subsequent Type Ib supernovae have generally been quite faint relative to SN 1983N and SN 1984L. The general properties have been summarized in SCHLEGEL and KIRSHNER [8].

The late-time light curve of SN 1984L was obtained from CCD images of the supernova taken about 300 to 500 days past maximum. An image of NGC 991 obtained at approximately 1250 days past maximum was used to remove the galaxy background. The resulting light curve is shown in Figure 1 (next page). The early-time points are from the literature (see SCHLEGEL and KIRSHNER [8] for the details). The overall light curve behavior does not match any of the standard light curves from DOGGETT and BRANCH [9]. The measured late-time slope for the V light curve falls almost embarrassingly on top of the e-folding time for ^{56}Co decay. Given that this is not a bolometric light curve, the agreement is perhaps surprising. No tricks or antics were used to get the light curve: a straight-forward, cross correlation of the "galaxy plus supernova" CCD frame with the galaxy frame was used to obtain the correctly subtracted image. The remaining pixels at the supernova's location were summed through a 7-arcsec aperture. Perhaps the visible light at 500 days past max-

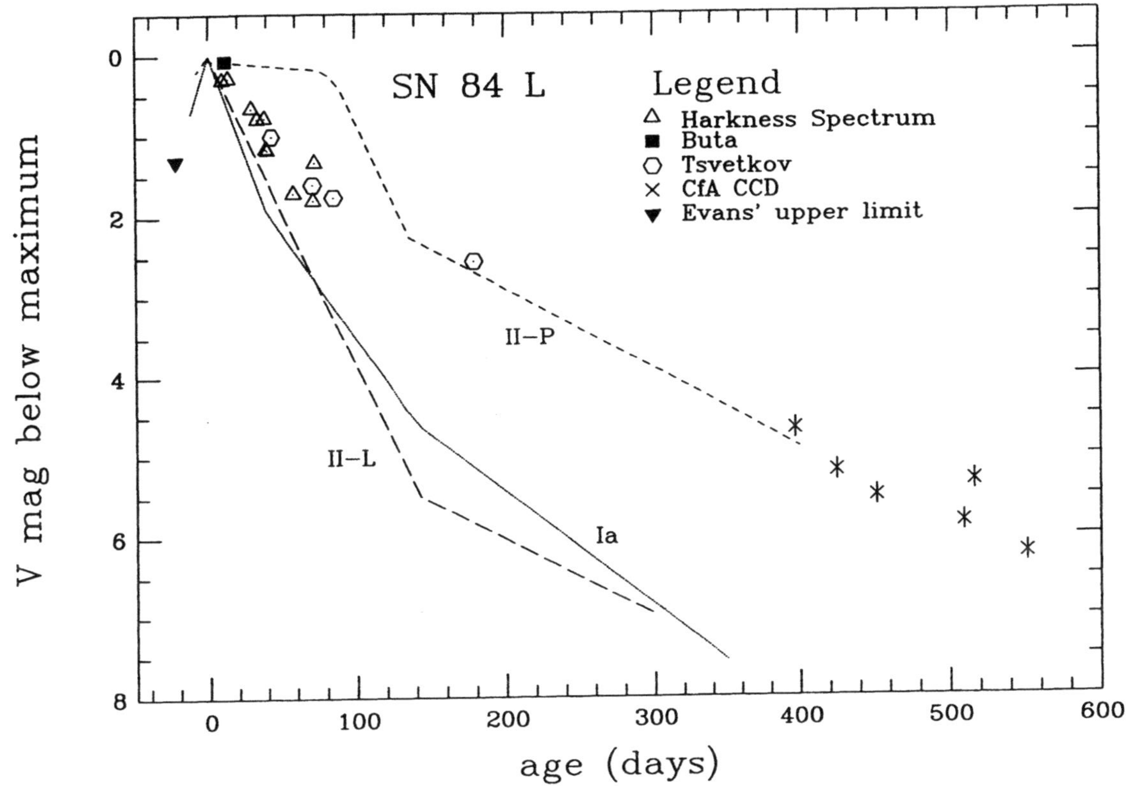

Figure 1. The V light curve for SN 1984L. The early-time points are from the literature. See [8] for citations and details.

imum is a crude approximation to the bolometric flux. Only a future, well-observed SN Ib will tell us.

The late-time spectra of SN 1984L are present in Figure 1 of SCHLEGEL (this volume) [10]. The characteristic [O I] 6300/6363Å emission is present, as well as Mg I] 4571Å emission. The presence of the oxygen emission in the debris implies a massive star has exploded. A spectrum, with poor S/N (presented in [10]), which extends to about 1μ, shows what may be the [C I] 9830Å complex. The ratio of the fluxes of the [C I] 9830Å to [C I] 8727Å provides a limit of about 3000[K] on the temperature of the debris. The number density is greater than about 10^6 cm^{-3}. These numbers lead to an estimate of the ejected oxygen mass of about $1M_{\odot}$. This number is about a factor of 3-5 greater than in the Type II SN 1980K (UOMOTO and KIRSHNER [11]).

Further details may be found in SCHLEGEL and KIRSHNER [5].

References

1. Uomoto, A. and Kirshner, 1985, *Astr. Ap.*, **149**, L7.
2. Wheeler, J. C. and Levreault, R. 1985, *Ap. J. (Letters)*, **294**, l17.
3. Branch, D., Lacy, C. H., McCall, M. L., Sutherland, P. G., Uomoto, A., Wheeler, J. C., and Wills, B. J. 1983, *Ap. J.*, **270**, 123.
4. Phillips, M. *et al.* 1989, *In preparation.*
5. Filippenko, A. V. and Sargent, W. 1985, *Nature*, **316**, 407.
6. Gaskell, C. M., Capellaro, E., Dinerstein, H., Garnett, D., Harkness, R., and Wheeler, J. C. 1986, *Ap. J. (Letters)*, **306**, L77.
7. Evans, R. 1984, *IAU Circular*, 3979.
8. Schlegel, E. M. and Kirshner, R. P. 1989, *Astr. J.*, **98**, 577.
9. Doggett, J. and Branch, D. 1985, *Astr. J.*, **90**, 2303.
10. Schlegel, E. M. 1989, in *Santa Cruz Summer Workshop on Supernovae*, ed. S. Woosley, (Springer-Verlag). in press.
11. Uomoto, A. and Kirshner, R. P. 1986, *Ap. J.*, **308**, 685.

SECTION X

TYPE Ia AND Ib SUPERNOVAE — PROGENITORS AND MECHANISMS

Evolution of Type Ia Progenitors

Ramon Canal, Jordi Isern, Javier Labay, & Rosario Lopez

Type Ia supernova (SNIa) progenitors are widely believed to be mass-accreting white dwarfs, growing toward Chandrasekhar's limit. Explosive ignition of the electron-degenerate material induces mass ejection and creates the radioactive nuclei whose decays power the light curve (COLGATE, PETSCHECK, and KRIESE [1]; ARNETT [2]; WOOSLEY and WEAVER [3]; see also WOOSLEY, this volume). Progenitor requirements can be summarized as follows:

a) According to the spectra, the object has to be devoid of hydrogen at the time of explosion. There is some debate as to the maximum amount of H that would still be compatible with the absence of Balmer lines. A current estimate is $M_H \leq 0.1 M_\odot$ (see WHEELER, this volume).

b) It has to be long-lived (ages up to several Gyr) to account for the occurrence of SNIa in elliptical galaxies, where major activity of star formation stopped long ago, and also in view of the lack of association with HII regions in spiral galaxies (MAZA and VAN DEN BERGH [4]).

c) The explosion has to produce a nickel mass $M_{Ni} \geq 0.5 M_\odot$ in order to explain the light curves (ARNETT [2]; ARNETT, BRANCH, and WHEELER [5]) and late-time spectra (WOOSLEY, AXELROD, and WEAVER [6]).

d) Intermediate-mass elements (O, Mg, Si, S, Ca) must be present in the outer layers, as deduced from the spectra at maximum light (BRANCH et al. [7]; WHEELER, HARKNESS, and CAPPELLARO [8]; see also HARKNESS, this volume).

e) The light curves of the explosions must be fairly homogeneous (CADONAU, SANDAGE, and TAMMANN [9]; see also both BRANCH and LEIBUNDGUT, this volume). Nonetheless, observed inhomogeneities (BRANCH [10]; PHILLIPS et al. [11]) may provide important diagnostics of progenitor characteristics and mechanism of explosion (CANAL, ISERN, and LOPEZ [12]).

f) In addition, of course, the progenitors must die at the observed rates of SNIa (see VAN DEN BERGH, this volume) and produce acceptable nucleosynthesis.

All of the preceding (but maybe the last point) lends strong support to the white dwarf hypothesis, and model explosions based on it do account fairly well for the general characteristics of SNIa outbursts even if important problems remain unsolved (see the above references). Finding a consistent scenario for evolution of the white dwarfs up to explosion is, however, a pending task.

Unless it were possible to revive some long-term mechanism for igniting single white dwarfs (such as the slow electron capture once proposed by FINZI and WOLF [13]), all evidence points towards white dwarfs in binary systems. There the clock is set by the evolution of the system up to the point when mass accretion by the white dwarf starts. All chemical compositions of the white dwarf are possible in principle: helium, carbon-oxygen, or oxygen-neon-magnesium. Most studies, however, have concentrated on C+O white dwarfs. He white dwarfs would almost certainly detonate (burn supersonically) after igniting. Their material would be completely incinerated (no intermediate-mass elements, then), and expansion velocities would be too large. At the other end, O+Ne+Mg white dwarfs have rather been thought to collapse after igniting O at high densities (MIYAJI et al. [14]; MIYAJI and NOMOTO [15], but see below). This leaves C+O white dwarfs, that can ignite carbon at lower densities than O+Ne+Mg ones and may, nonetheless, be able to sustain subsonic burning fronts through either the whole or a fraction of their mass (WOOSLEY [16]).

I. C + O White Dwarfs in Close Binary Systems

C+O white dwarfs can form in binary systems either by Roche-lobe overflow just before or just after ignition of He in *initially close binaries* or by Roche-lobe overflow during early asymptotic giant branch (AGB) phase or thermally pulsing AGB phase in *initially wide binaries*. In the last case, common envelope evolution should allow losing enough orbital angular momentum to change the wide binary into a close one. A problem arises as to which are the typical mass range and the upper mass limit for C+O white dwarfs formed this way. Observation of cataclysmic variables and of classical novae give average masses of $0.75M_\odot$ and of $1.23M_\odot$, respectively, and a recent model for the recurrent nova U Sco gives a mass of $\simeq 1.38M_\odot$ (STARRFIELD, SPARKS, and SHAVIV [17]). But, as we will see, nor novae nor cataclysmic variables in general seem to be the progenitors of SNIa. Besides, the most massive white dwarfs found in those systems may well not be C+O white dwarfs but O+Ne+Mg ones.

Depending on the initial parameters of the system, the companion of the C+O white dwarf may either be:

1) A main-sequence star

2) A subgiant or a red-giant star

3) A He white dwarf (or a He nondegenerate star)

4) Another C+O white dwarf

Scenarios leading to the formation of those systems have been thoroughly studied by IBEN and TUTUKOV [18]. Depending on the type of companion, widely different time intervals may separate the onset of mass transfer from the epoch of white dwarf formation.

II. The Mass Growth Process

IIa. The outer layers

Mass accretion affects both the outer layers and the core of the white dwarf. Concerning the first, for the white dwarf to grow by a few tenths of a solar mass the material has not only to fall onto its surface but also to be incorporated into the degenerate core. This poses many problems and a fully satisfactory scenario has not yet been devised.

Accretion of *H-rich material* (from the two first types of companion listed above) should avoid its *explosive ignition* (as in novae) and also formation of a *red-giant envelope* (eventually engulfing the white dwarf and its companion together) on top of a H-burning shell or direct formation of a *common envelope.* Usually quoted values are:

$\dot{M}_H \leq 10^{-9} M_{\odot} yr^{-1}$ as leading to nova outburst

$\dot{M}_H \geq 10^{-6} M_{\odot} yr^{-1}$ formation of a red-giant envelope

$\dot{M}_H \geq \dot{M}_{Edd} \sim 10^{-5} M_{\odot} yr^{-1}$ direct formation of a common envelope

Especially concerning the first limit, one must note that it is based on calculations assuming spherically symmetric and "soft" accretion. Most likely, material will form a disk around the compact star and the accretion process will thus include angular momentum and kinetic energy dissipation. When those effects are included, the actual range of $\dot{M}_H$ producing nova explosions is ill-defined (SHAVIV and STARRFIELD [19]; SPARKS and KUTTER [20]). What seems clear, however, is that nova outbursts will reduce actual mass growth to, at most, 10% of $\dot{M}_H$ (TRURAN [21]).

Common envelope (either due to accretion above Eddington's limit or to formation of a red-giant envelope) should induce mass loss by the system as a whole and thus inhibit further growth of the white dwarf. Nonetheless, HACHISU, KATO, and SAIO [22] propose a model where stable mass transfer in common envelope is possible (through steady hydrogen burning) for a range of parameters of the binary system. Predicted mass-accretion rates are $\dot{M}_H \geq 10^{-7} M_{\odot}$.

There is good evidence that SNIa are not the descendants of novae nor of cataclysmic variables. One argument comes from the observation that in M31 the galactic bulge is ~ 20 times more efficient than the disk in producing novae whereas, on the contrary, galactic disks in general are more efficient than bulges in producing SNIa (RENZINI [23]). Also, the birthrate of cataclysmic variables (in the solar neighbourhood at least) is one order of magnitude lower than the SNIa rate (RITTER [24]).

Hypothetical close binaries where H were accreted at $10^{-9} M_{\odot} yr^{-1} \leq \dot{M}_H \leq 10^{-6} M_{\odot} yr^{-1}$ might burn it into He steadily or in weak flashes (and would thus not be novae nor cataclysmic variables). Anyway, those systems would be very luminous, and assuming that they had to accrete a few tenths of a solar mass there should now be $\sim 10^4 - 10^5$ of them in the Galaxy to account for one SNIa event every ~ 100 years. No known object is a likely candidate. When igniting H they would appear as bright EUV sources. Between burning episodes their luminosities should equally be high, due to the energy released by accretion. Would they appear as symbiotic stars, or as emission-line variables? If they were long-period variables, observational selection effects would however be against their detection.

A further limitation is set by the fact that the He layer resulting from burning the accreted H will explosively ignite if it is accumulated at a rate $10^{-9} M_{\odot} yr^{-1} \leq \dot{M}_{He} \leq 5 \times 10^{-8} M_{\odot} yr^{-1}$. The lower limit, however, is only valid for initial masses of the C+O core: $M_{core} \leq 1.13 M_{\odot}$ (NOMOTO [25]). It is unclear to which extent detonation of the He layer would propagate into the C+O core, but no such outburst could be a valid model for SNIa: iron-peak instead of intermediate-mass elements would appear in the spectra at maximum light.

The same restriction applies to direct accretion of He from either a nondegenerate or a degenerate companion. Concerning the former, only 3-4 systems are known at present and in all of them the mass of the He star is exceedingly low. C+O plus He white dwarf binaries, on the other hand, would have very short lifetimes, and this is incompatible with SNIa statistics.

In recent years, the idea (Webbink [26]; IBEN and TUTUKOV [18]) that SNIa originate from double C+O white dwarf systems has become popular. Accretion of C+O material onto a white dwarf of the same composition does not pose the problems encountered in the accretion of H or He. Once the two C+O white dwarfs have formed, orbital angular momentum will be lost by emission of gravitational radiation, and the stars will get closer on the corresponding time scale. Roche-lobe overflow by the lower-mass component will eventually occur and have a runaway character: the secondary will be disrupted and its material will likely form a massive disk around the primary. It is not clear what should happen next. Very fast accretion ($\dot{M}_{C+O} \geq 5 \times 10^{-6} M_{\odot} yr^{-1}$) would induce carbon ignition close to the surface and the star would not explode as a SNIa but quasi-hydrostatically burn into a O+Ne+Mg white dwarf (NOMOTO and IBEN [27]), that might then collapse (or explode: see below). Lower accretion rates would allow increase in central density and entropy up to the point of explosive carbon ignition and a SNIa event might result. The actual rate depends on viscous dissipation in the disk (MOCHKOVITCH and LIVIO [28]), and the issue remains undecided (see MOCHKOVITCH, this volume).

An additional problem is the actual frequency of such white dwarf binaries in the Galaxy. Two independent searches (ROBINSON and SHAFTER [29]; BRAGAGLIA et al. [30]; see also BRAGAGLIA, this volume) have yielded none and only one double degenerates, respectively. The one found would need more than a Hubble time to merge and is probably a double He white dwarf system. Taking the two surveys together, the frequency of double degenerates in the white dwarf population would be $\nu_{DD} < 0.04$ (with 90% probability) or $\nu_{DD} < 0.02$ (with 70% probability): far too low to account for SNIa (the required frequency would be $\nu_{DD} \sim 0.1$). Observational bias has favored DA white dwarfs (84 out of 89), and one can argue that double C+O degenerates should rather be non-DA (RENZINI [23]). However, we know that (in single stars at least) non-DA may turn into DA (H finally floating at the surface), and thus the validity of this explanation is questionable.

Indirect evidence favoring the double white dwarf scenario might come from recent measurements of the [O/Fe] ratio in main-sequence dwarfs belonging to the old disk and halo populations of our Galaxy (ABIA and REBOLO [31]): to account for the variation of the ratio with Fe abundance, the start of the main production of this last element (by SNIa) should be delayed by an amount that agrees with the predictions from merging of two white dwarfs by emission of gravitational radiation (ABIA et al. [32]).

IIb. The core

We turn now to core evolution. It begins with the formation of the C+O white dwarf. Prior to mass transfer, it consists only of cooling. If the time interval between formation and the onset of mass transfer from the secondary is long enough (that depends also on the mass of the white dwarf), a first-order phase transition will take place: the star's core will *crystallize*, the ion sites forming a body-centered cubic lattice (bcc). This happens for a critical value of the plasma-coupling constant $\Gamma = Z^2e^2/r_skT$ (where Z is atomic number, e the the electron charge, k Boltzmann's constant, T temperature, and r_s the *ion-sphere radius*: the radius of the sphere containing on average one ion). Monte Carlo calculations for the one-component plasma (OCP) give $\Gamma_{crit} = 155 \pm 5$ (SLATTERY, DOOLEN, and DE WITT [33]). The case of a C+O plasma or of a binary ion mixture (BIM) in general is more complex. We see, from the expression of Γ, that in a core cooling at constant density (constant r_s) Γ_{crit} will be reached first for oxygen. The question arises of whether carbon and oxygen are miscible in the solid phase. Early Monte Carlo simulations by LOUMOS and HUBBARD [34] seemed to support the idea that they actually were and that the crystallization temperature of the mixture (at a fixed density) should be the weighted (by mass fraction) mean of those of pure carbon and pure oxygen. Later, STEVENSON [35] proposed an *eutectic* phase diagram by adopting the *random alloy mixing* (RAM) model for internal energies in the solid phase: carbon and oxygen would separate when crystallizing, solid oxygen accumulating at the center and a pure oxygen core thus progressively growing (MOCHKOVITCH [36]). More recently, however, there have been new calculations of the phase diagram for BIM. On one hand, BARRAT, HANSEN, and MOCHKOVITCH [37], using a density-functional approach for calculating the energies, deduce a diagram of the *spindle* form: a *random* C+O alloy, only slightly more oxygen-rich than the fluid phase would result from crystallization. Transition to an *ordered* alloy of the ClCs type (bcc lattice with carbon and oxygen ion sites forming simple cubic sublattices) is predicted for values of Γ larger than Γ_{crit}. On the other hand, ICHIMARU, IYETOMI, and OGATA [38] find, by Monte Carlo simulations, that the *linear mixing* formula is more accurate than the *random-mixing* one for calculating the energies of both fluid and solid phases. They obtain an *azeotropic* phase diagram. Again, as in BARRAT et al. [37], only a moderate oxygen enrichment of the solid alloy is predicted. In the Monte Carlo simulations of the bcc lattice, ICHIMARU et al. [38] find the equilibrium final states to take *random* bcc-solid configurations. This does not necessarily exclude a transition to an ordered configuration at a value of Γ larger that those considered in these last simulations. Finally, GODON et al. [39] have also recently approached the problem by solid state physics methods (linear Muffin-Tin orbitals, LMTO). They conclude that, within 1% accuracy, microscopic separation requires no energy and that the issue cannot be decided. In the remainder of this discussion we will, however, mainly adopt the view that carbon and oxygen *do not* appreciably separate when freezing, and that they form a *random alloy*. Implications of the other two alternatives (phase separation or ordered crystal) will be pointed out in due course.

When mass accretion starts, whatever of the mass-transfer scenarios that we have previously considered works, the degenerate core will grow in mass and it will be heated by compression and (in the cases of either H or He accretion) also by the inward flow from the He-burning shell (but for the cases $\dot{M}_H \leq 10^{-9}M_\odot yr^{-1}$ or $\dot{M}_{He} \leq 10^{-9}M_\odot yr^{-1}$ and $M_{core} \geq 1.13M_\odot$, where there would only be a H-burning shell in the first case and no shell at all in the second). Compressional heating will also affect the outer layers first and diffuse inward (NOMOTO [40]). The star's center will thus increase both in density and in entropy (the last by heat inflow from the outer core) up to the point of carbon ignition. An exception to this is when either the accretion rate is low ($\dot{M} \leq 10^{-10}M_\odot yr^{-1}$) or both M_{core} and $\dot{M}$ are large (HERNANZ et al. [41]). In the first case, heat inflow is quenched by thermal neutrino losses. In the second one, carbon ignition precedes the arrival to the center of the heat flow. In both cases the central layers of the white dwarf are compressed *adiabatically* (but for thermal neutrino emission). This is an important point, since in these cases a variable fraction of any solid core that forms during the cooling phase survives until carbon ignition.

III. Explosive Carbon Ignition

Most currently available calculations of C+O white dwarf evolution upon mass accretion (NOMOTO, THIELEMANN, and YOKOI [42]; SUTHERLAND and WHEELER [43]; WOOSLEY, AXELROD, and WEAVER [6]) have assumed combinations of initial mass and internal temperature plus accretion rate that lead to central carbon ignition at $\rho_c \simeq 2 - 3 \times 10^9 g\ cm^{-3}$ in an entirely fluid core ($T \simeq 2 \times 10^8 K$ and $\Gamma \ll \Gamma_{crit}$). Prior to runaway, the central layers become convectively unstable. Urca nuclei such as ^{23}Na are produced during carbon burning and electron Fermi energies are high enough at those densities for convectively driven Urca processes (PACZYNSKI [44]) to operate. IBEN [45] found that Urca neutrino losses would then delay thermonuclear runaway to higher densities. More recently, BARKAT and WHEELER [46] (see also BARKAT, this volume) have found that a steady state can be reached where convective Urca losses exactly balance the energy released by carbon burning. Subsequent evolution is unclear: it might lead to collapse or maybe to off-center ignition of carbon.

It must be noted, however, that if one considers the full spectrum of initial white dwarf masses, internal temperatures, and accretion rates, central carbon ignition densities (even without convective Urca losses) can span a range $2-3 \times 10^9 g\ cm^{-3} \leq \rho_c \leq 1.5 \times 10^{10} g\ cm^{-3}$, the highest values corresponding to cases where ignition (in the *pycnonuclear* regime) takes place inside a residual solid core (HERNANZ et al. [41]).

Assuming that a thermonuclear runaway still happens, it can propagate either supersonically, by the shock wave it initiates (*detonation*) or subsonically, by conduction and turbulent mixing (*deflagration*). Deflagration has been favored in recent years (see the above references and also WOOSLEY, this volume), since it simultaneously provides incineration of a fraction of the core (needed to produce the light curves and late-time spectra) and partial burning of another (outer) fraction (needed to produce the spectra at maximum). Its physical soundness has recently been questioned (BLINNIKOV and KHOKHLOV [47]; WHEELER et al. [48]). If an adiabatic (or even flatter) temperature gradient prevails close to the center, spontaneous ignition may simultaneously affect a region large enough to initiate detonation. WOOSLEY [16] disputes this on the basis that once runaway starts no large enough regions can remain isothermal to the required degree. A major objection to the detonation model has been that the star should be completely incinerated and no intermediate-mass elements would be produced. This might not be true if the detonation breaks down for densities $\rho \leq 10^7 g\ cm^{-3}$ (IMSHENNIK and KHOKHLOV [49]; KHOKHLOV [50]).

An unsolved difficulty in eith r case (detonation or deflagration) is that the calculated isotopic abundances for the iron group elements (of which SNIa are thought to be the main producers) do not agree with those measured in the solar system (THIELEMANN, NOMOTO, and YOKOI [51]): ^{58}Ni and ^{54}Fe are overproduced by factors $\simeq 5$ and $\simeq 2$, respectively. This results from incinerating the material at high densities, where a large amount of electron capture happens. In the framework of the deflagration model, conveniently varying the flame propagation speed (having it starting very slowly and later accelerating) might provide an initial expansion of the star that would then allow most incineration to happen at lower densities (WOOSLEY [16], and this volume). All the preceding is based on one-dimensional calculations, that unavoidably resort to parametrization. Two- or three-dimensional hydrodynamic simulations can not, however, be expected to provide reliable answers in the near future (see FRYXELL, MÜLLER, and ARNETT [52]).

Another possible way out of the excess neutronization would be to centrally ignite carbon at lower densities: $\rho_c \leq 1.0 - 1.5\ g\ cm^{-3}$ (NOMOTO [53]), or to leave gravitationally bound, as a result of central ignition in a partially solid core, the material that is incinerated at high densities (ISERN, LABAY, and CANAL [54]).

IV. The Branch – Pskovskii Correlation (?) and Off – Center Carbon Ignitions

In spite of their overall homogeneity, SNIa show some range of variation in their spectroscopic and photometric properties. Clear differences in photospheric velocities at maximum light have recently been confirmed (BRANCH [10]). The same is true of differences in the rate of postmaximum decline of the light curves (PHILLIPS et al. [11]). A correlation had earlier been proposed among maximum luminosity, photospheric velocity, and decline rate (PSKOVSKII [55]; BRANCH [56]): brighter SNIa would expand faster and decline more slowly. Varying only ^{56}Ni mass from explosion to explosion would give just the opposite correlation (NOMOTO, THIELEMANN, and YOKOI [42]). It would however find a natural explanation in models based on phase separation of carbon and oxygen at crystallization (LOPEZ et al. [57]): mass accretion would induce off-center carbon ignition at the bottom of the carbon-rich layers surrounding the central oxygen core and so the total mass involved in the explosion would no longer be fixed. As we have discussed above, recent calculations of the phase diagram for C+O mixtures do not confirm this picture. Off-center ignitions would still be possible, nonetheless, for an *ordered* C+O alloy (where $^{12}C+^{12}C$ reactions are inhibited: see below): they would happen when melting of the initial solid core by heat inflow reached deep enough layers (with densities $\simeq 2-3 \times 10^9 g\ cm^{-3}$). A similar effect would be obtained for central carbon ignition in a (random) solid core at comparatively low densities (the last to avoid core collapse: see below), when the burning front emerged from the solid into the fluid layers (ISERN, LABAY, and CANAL [54]). Another equally speculative possibility is that off-center ignitions result from growth of a central convective core with burning stabilized by Urca losses (BARKAT and WHEELER [46]).

GRAHAM [58] recently proposed varying central ignition density in the "standard" deflagration model: for increasing densities a correspondingly larger "hole" of neutronized, nonradioactive material would form. ^{56}Ni (whose decay accounts for the luminosity peak down to the beginning of the exponential "tail" of the light curve) would thus be closer to the surface and the light curves would decay faster. CANAL, ISERN, and LOPEZ [12] have shown that, if this were the case, both higher luminosities at maximum and larger expansion velocities would correspond to faster decline of the light curve, contrary to the suggested correlation. BRANCH (this volume) now thinks that most likely maximum luminosity variations can be accounted for by differences in the amount of extinction in the parent galaxy. If only variation in photospheric velocities and rate of peak luminosity decline are intrinsic (but see LEIBUNDGUT, this volume), and correlated in the aformentioned way, the last conclusions would nonetheless be valid and some kind of off-center ignition should be involved in SNIa outbursts.

V. Collapse of C + O White Dwarfs

Accretion-induced collapse (AIC) of C+O white dwarfs does most likely not produce SNIa outbursts but its physics is closely related to that of the progenitors of such events. As we already pointed out, a central solid core can subsist up to carbon ignition for a range of initial masses, temperatures, and accretion rates. Carbon ignition then happens in the *pycnonuclear* regime (SALPETER and VAN HORN [59]), at central densities $9.5 \times 10^9 g\ cm^{-3} \leq \rho_c \leq 1.5 \times 10^{10} g\ cm^{-3}$ (HERNANZ et al. [41]). Besides, burning propagates conductively (at velocities of the order of 0.01-0.001 times the local sound speed) until it reaches the boundary of the solid core. This combination (high density plus relatively low propagation of burning) allows electron captures to remove energy and to decrease Chandrasekhar's mass faster than the energy release by carbon burning promotes expansion, and collapse ensues (CANAL and ISERN [60]; ISERN et al. [61]; CANAL, ISERN, and LABAY [62]). A complicating factor, however, are the neutrinos emitted by electron capture in the central, growing incinerated region: they heat up the solid layers ahead of the burning front and can melt them before the start of dynamical contraction, when Chandrasekhar's mass is still lower than the star's mass. The burning front should accelerate in reaching fluid layers, where hydrodynamic instabilities can grow. How severe this limitation on both ρ_c at ignition and on conductive velocities (within their range of uncertainty) is, for producing collapse, depends on the poorly known growth rate of hydrodynamic instability (the initial stages of formation of a turbulent flame front: see WOOSLEY [16]).

Once collapse starts, neutrino heating also induces carbon ignition further out than sole compression would otherwise produce, and it has therefore to be included in collapse models (see ISERN, his volume). The high entropy of these collapsing cores (as compared to the Fe-Ni cores of massive stars) should result in little mass ejection and produce only a "dim" event.

The (by now marginal) possibility of formation of an *ordered* C+O alloy when the fluid C+O mixture crystallizes would delay carbon ignition to still higher densities, since only $^{12}C +^{16} O$ (and $^{16}O +^{16} O$) reactions could take place in the solid (for $x_C \leq x_O$, x_i being fraction by number). It would happen at $\rho_c \simeq 2 \times 10^{10} g\ cm^{-3}$ and be triggered by electron captures on oxygen, provided that the core has not been melted before by heat inflow or off-center carbon ignition has not been induced by this last process (see the preceding paragraph).

Concerning carbon ignition in the random alloy case, the limited oxygen enrichment of the solid phase predicted by both BARRAT et al. [37] and ICHIMARU et al. [38] should be included together with the chemical stratification predicted in models for the evolution leading to formation of the white dwarf (MAZZITELLI and D'ANTONA [63]) (it is, however, dependent on the value of the $^{12}C(\alpha,\gamma)^{16}O$ cross-section). One should equally bear in mind the uncertainties in the *pycnonuclear* reaction rates.

VI. The Case of O + Ne + Mg White Dwarfs

Mass-accreting O+Ne+Mg white dwarfs should, on average, be more massive than C+O ones and thus would have to accrete less material before (in this case) Ne-O ignition at the center. Prediction that such ignition would happen at $\rho_c \simeq 2\times 10^{10} g\ cm^{-3}$ (MIYAJI et al. [14]) was based on the assumption that Schwarzschild's criterion is adequate in determining the onset of convection in those cores. A convective core would already develop due to heating by electron captures on ^{24}Mg and ^{24}Na (at $\rho \simeq 4\times 10^{9} g\ cm^{-3}$). This ignored that electron captures not only heat up the core but also produce a negative Y_e-gradient that has a stabilizing effect. Applying Ledoux's criterion, MIYAJI and NOMOTO [15] find that convection is inhibited up to Ne-O ignition at $\rho_c \simeq 9.5\times 10^{9} g\ cm^{-3}$. Contention that semiconvective instability might develop and produce mixing (as in the "salt finger" instability in the oceans) disregards the fact that extra energy is needed to bring electrons from regions of lower Fermi level into regions of higher Fermi level, and also that the time scale of oscillations is here much shorter than the time scale of heat diffusion. The last ignition density should thus be a good estimate. MIYAJI and NOMOTO [15] predict collapse, but in their calculation they suppress any form of burning propagation. Conductive burning propagation can however not be averted, and both convective and Rayleigh-Taylor instabilities can develop from the beginning since the layers are always fluid (due, in last resort, to previous heating by electron captures on ^{24}Mg and ^{24}Na). When current estimates of hydrodynamic burning propagation velocities (WOOSLEY and WEAVER [3]) are adopted, the outcome is explosion, not collapse. The problem is again how fast hydrodynamic burning propagates in its initial stages. Explosion of O+Ne+Mg white dwarfs is thus an open possibility. Their nucleosynthesis would nonetheless yield excess neutron-rich isotopes. The corresponding outbursts have still to be modelled to see if they would produce viable SNIa (or maybe SNIb).

VII. Conclusions

While mass-accreting C+O white dwarfs are the most likely progenitors of SNIa, no consistent and observationally confirmed evolutionary path leading to the outburst is known. Possible progenitor systems include white dwarf / H-rich star binaries where slowly accreted H would somehow manage to burn into He and remain there, or a similar (well-hidden) pair where H would be accreted fast, or still equally elusive double C+O white dwarf binaries (either non-DA or rapidly merging DA, to be compatible with the statistics). In the last case, accretion from the massive disk resulting from disruption of the secondary should be finely tuned.

Confirmed inhomogeneity in the spectroscopic and photometric characteristics of SNIa would indicate some type of off-center ignition. This might be either due to growth of an Urca-stabilized convective core or to the presence of a central solid core.

The problem of a too neutron-rich nucleosynthesis stands. If it could not be solved by the dynamics of the burning front, this would mean that either white dwarfs manage to ignite at lower densities than it is currently thought or, on the contrary, that part of the ashes remains gravitationally bound. Collapsing models, however, do not seem able to produce viable outbursts.

References

1. Colgate, S.A., Petschek, A.G., and Kriese, J.T. 1980, *Ap. J. (Letters)*, **237**, L81
2. Arnett, W.D. 1982, *Ap. J.*, **253**, 785
3. Woosley, S.E., and Weaver, T.A. 1986, *Ann. Rev. Astron. Ap.*, **24**, 205
4. Maza, J., and van den Bergh, S. 1976, *Ap. J.*, **204**, 519
5. Arnett, W.D., Branch, D., and Wheeler, J.C. 1985, *Nature*, **314**, 337
6. Woosley, S.E., Axelrod, T.S., and Weaver, T.A. 1984, in *Stellar Nucleosynthesis*,

ed. C. Chiosi and A. Renzini (Dordrecht: Reidel), p.263

7. Branch, D., et al. 1982, *Ap. J. (Letters)*, **252**, L61

8. Wheeler, J.C., Harkness, R.P., and Cappellaro, E. 1987, in *Proc. 13th Texas Symposium on Relativistic Astrophysics*, ed. M.P. Ulmer (Singapore: World Scientific), p. 402

9. Cadonau, R., Sandage, A., and Tammann, G.A. 1985, in *Supernovae as Distance Indicators*, ed. N. Bartel (Berlin: Springer Verlag), p. 151

10. Branch, D. 1987, *Ap. J. (Letters)*, **316**, L81

11. Phillips, M.M., et al. 1987, *Pub. Astron. Soc. Pacific*, **99**, 592

12. Canal, R., Isern, J., and Lopez, R. 1988, *Ap. J. (Letters)*, **330**, L113

13. Finzi, A., and Wolf, R.A. 1967, *Ap. J.*, **150**, 115

14. Miyaji, S., Nomoto, K., Yokoi, K., and Sugimoto, D. 1980, *Publ. Astron. Soc. Japan*, **32**, 303

15. Miyaji, S., and Nomoto, K. 1987, *Ap. J.*, **318**, 307

16. Woosley, S.E. 1989, in *Supernovae*, ed. A.G. Petschek (Berlin: Springer Verlag), preprint

17. Starrfield, S., Sparks, W.M., and Shaviv, G. 1989, preprint

18. Iben, I., and Tutukov, A. 1984, *Ap. J. Suppl.*, **54**, 335

19. Shaviv, G., and Starrfield, S. 1987, *Ap. J. (Letters)*, **321**, L51

20. Sparks, W.M., and Kutter, G.S. 1987, *Ap. J.*, **321**, 394

21. Truran, J.W. 1989, private communication

22. Hachisu, I., Kato, M., and Saio, H. 1989, *Ap. J.*, in press

23. Renzini, A. 1989, private communication

24. Ritter, H. 1989, private communication

25. Nomoto, K. 1982, *Ap. J.*, **253**, 798

26. Webbink, R.F. 1979, in *White Dwarfs and Variable Degenerate Stars*, ed. H.M. Van Horn and V. Weidemann (Rochester: Univ. Rochester) p. 426

27. Nomoto, K., and Iben, I. 1985, *Ap. J.*, **297**, 531

28. Mochkovitch, R., and Livio, M. 1989, *Astron. Ap.*, **209**, 111

29. Robinson, E.L., and Shafter, A.W. 1989, in *White Dwarfs*, ed G. Wegner (Berlin: Springer Verlag), p. 492

30. Bragaglia, A., Greggio, L., Renzini, A., and D'Odorico, S. 1989, in *White Dwarfs*,

ed. G. Wegner (Berlin: Springer Verlag), p. 138

31. Abia, C., and Rebolo, R. 1989, *Ap. J.*, in press

32. Abia, C., et al. 1989, in preparation

33. Slattery, W.L., Doolen, G.D., and DeWitt, H.E. 1982, *Phys. Rev. A*, **26**, 2255

34. Loumos, G.L., and Hubbard, W.B. 1973, *Ap. J.*, **180**, 199

35. Stevenson, D.J. 1980, *Jour. Phys. Suppl.*, No 3, **41**, C2-53

36. Mochkovitch, R. 1983, *Astron. Ap.*, **122**, 212

37. Barrat, J.L., Hansen, J.P., and Mochkovitch, R. 1988, *Astron. Ap.*, **199**, L15

38. Ichimaru, S., Iyetomi, H., and Ogata, S. 1988, *Ap. J. (Letters)*, **334**, L17

39. Godon, P., Shaviv, G., Ashkenazi, J., and Kovetz, A. 1989, in *White Dwarfs*, ed. G. Wegner (Berlin: Springer Verlag), p. 85

40. Nomoto, K. 1982, *Ap. J.*, **257**, 780

41. Hernanz, M., Isern, J., Canal, R., Labay, J., and Mochkovitch, R. 1988, *Ap. J.*, **324**, 331

42. Nomoto, K., Thielemann, F.-K., and Yokoi, K. 1984, *Ap. J.*, **286**, 644

43. Sutherland, P., and Wheeler J. C. 1984, *Ap. J.*, **280**, 282

44. Paczynski, B. 1972, *Astrophys. Lett.*, **11**, 53

45. Iben, I. 1982, *Ap. J.*, **253**, 248

46. Barkat, Z., and Wheeler, J. C. 1989, in preparation

47. Blinnikov, S.I., and Khokhlov, A.M. 1986, *Soviet Astron. Lett.*, **12**, 131; 1987, *ibid.*, **13**, 364

48. Wheeler, J.C., Harkness, R.P., Barkat, Z., and Swartz, D. 1986, *Publ. Astron. Soc. Pacific*, **98**, 1018

49. Imshennik, V.S., and Khokhlov, A.M. 1984, *Soviet Astron. Lett.*, **10**, 262

50. Khokhlov, A.M. 1989, *M.N.R.A.S.*, **239**, 785

51. Thielemann, F.-K., Nomoto, K., and Yokoi, K. 1985, *Astron. Ap.*, **158**, 17

52. Fryxell, B.A., Müller, E., and Arnett, W.D. 1989, in *Numerical Methods in Astrophysics*, ed. P.R. Woodward (New York: Academic Press), in press

53. Nomoto, K. 1989, private communication

54. Isern, J., Labay, J., and Canal, R. 1984, *Nature*, **309**, 431

55. Pskovskii, Y.P. 1977, *Soviet Astron.*, **21**, 675; *ibid.*, **28**, 658

56. Branch, D. 1982, *Ap. J.*, **285**, 35

57. Lopez, R., Isern, J., Canal, R., and Labay, J. 1986, *Astron. Ap.*, **155**, 1

58. Graham, J.R. 1987, *Ap. J. (Letters)*, **318**, L47

59. Salpeter, E.E., and Van Horn, H.M. 1969, *Ap. J.*, **155**, 183

60. Canal, R., and Isern, J. 1979, in *White Dwarfs and Variable Degenerate Stars*, ed. H.M. Van Horn and V. Weidemann (Rochester: Univ. Rochester), p. 52

61. Isern, J., Labay, J., Hernanz, M., and Canal, R. 1983, *Ap. J.*, **273**, 320

62. Canal, R., Isern, J., and Labay, J. 1989, in *Timing Neutron Stars*, ed. H. Ögelman and E.P.J. van den Heuvel (Dordrecht: Reidel), p. 631

63. Mazzitelli, I., and D'Antona, F. 1987, in *The Second Conference on Faint Blue Stars*, ed. A.G. Davis, D.S. Hayes, and J.W. Liebert (Schenectady: Davis Press), p. 351

Wolf-Rayet Stars as Supernova Precursors

Norbert Langer

Wolf-Rayet (WR) stars are thought to be massive stars which are burning helium in their centers and have lost all or most of their hydrogen rich envelope, consequently showing products of hydrogen burning (WN stars) or helium burning (WC stars) at their surface. The very intense stellar wind of those objects ($\dot{M}_{WR} > 10^{-5}\,M_{\odot}\,yr^{-1}$) is responsible for the dominance of emission lines in their spectra, which is their most prominent characteristic. All massive single stars with initial masses above a certain limit M_{WR} are supposed to evolve into WR stars, which represents their final hydrostatic evolutionary stage (cf. Chiosi and Maeder, 1986). Note that M_{WR}, which is of the order of $25 - 50\,M_{\odot}$ in our Galaxy (van der Hucht et al., 1988; Humphreys et al., 1985), may depend on the stellar metallicity (cf. Dopita, this volume).

As with massive stars, most WR stars, after core helium exhaustion, undergo central carbon-, neon-, oxygen-, and silicon burning, leading to the formation of an iron core, which eventually collapses. Whether this collapse can be reversed into a supernova (SN) explosion depends upon the maximum iron core mass $M_{\dagger}$, which can be exploded by the core collapse mechanism (cf. Baron and Cooperstein, this volume; Mayle and Wilson, this volume). The iron core mass developed in a given WR star depends on its actual mass at the end of its evolution, which presumably does not change after core helium exhaustion because of the short amount of time left from this point on to core collapse. Therefore it is important that, because of the high mass loss rate of WR stars and of certain WR progenitors (esp. the Luminous Blue Variables, LBVs), the final WR mass is much smaller than the corresponding zero age main sequence (ZAMS) mass, and in most cases even much smaller than the helium core mass at hydrogen exhaustion. Both, $M_{\dagger}$ and the initial-final mass relation for massive stars are rather uncertain at present (see below for latter), and therefore no general conclusion about the final fate of WR stars can be drawn. However, some WR stars (e.g. HD 152270: $5 \pm 2\,M_{\odot}$; St. Louis et al., 1987) are known to have a smaller total mass than the presumed helium core mass of the SN 1987A progenitor ($\sim 6\,M_{\odot}$), which implies that at least those are very likely to explode.

May each WR subclass contain pre-SN stars?

As mentioned above, there are two spectroscopic WR subclasses, the WN and the WC stars. Observationally determined surface abundances agree well with abundances resulting from partial or complete hydrogen burning for the WNs and incomplete helium burning for the WC-stars (cf. Maeder, 1983), which allows one to directly compare stellar models with appropriate surface abundances to stars of the respective WR subclass. Both WR subclasses are further divided into subtypes according to spectroscopic criteria (see van der Hucht et al., 1981). There is some evidence that the so called late WN stars (WNLs) still contain a considerable amount of hydrogen in there atmospheres, in contrast to the early WN stars (WNEs), where hydrogen seems to be very rare or absent (Willis, 1982). The correspondence of WN subtypes to the surface hydrogen abundance seems not to be strict (cf. Hamann, 1989); however, we shall designate hydrogenless WN stellar models as WNE stars and hydrogen containing ones as WNL stars in this work. Also the WC subclass is divided into subtypes (WCE, WCL, and WO), but a corresponding distinction of surface abundances could not be established observationally (Torres, 1988; Smith and Hummer, 1988). Therefore, this division can not be transposed to stellar models.

In any evolutionary scenario for WR stars, a WNL phase must precede the WNE phase. This is so because the hydrogen burning convective core is shrinking with time, i.e. in order to expose ashes of complete H-burning at the surface (= WNE-phase), ashes of incomplete H-burning have to be exposed earlier (= WNL-phase). Similarly, because the He-burning convective core is smaller than the H-burning convective core, a WC phase must be preceded by a WN phase. However, this does not necessarily imply, that the time sequence WNL $\rightarrow$ WNE $\rightarrow$ WC is followed up to the end by all massive stars which enter the WR stage (which would mean that only WC stars could be SN progenitors). Actually, the fact that the WR phase of any massive star has to start with a WN stage indicates, that at least stars with a ZAMS mass closely above the lower ZAMS mass limit for WR formation (M_{WR}) should terminate their evolution as WN stars: their ZAMS masses being closely larger than M_{WR} means that they reach the WR phase just at the end of their evolution, implying that no more time is left to further remove sufficient mass in order to reach the WC stage.

Stellar evolution calculations for the pre-WR phases of massive stars still contain far too large uncertainties (concerning $\dot{M}$ in the red supergiant or LBV phase, convection, etc.; cf. Langer and El Eid, 1986) in order to allow reliable estimates for the relative number of SNe to be expected from the different WR subtypes. However, qualitative conclusions may be drawn, though even most of them have to be regarded as tentative and preliminary.

WNL stars as progenitors of peculiar Type II SNe

The above argument for the existence of stars which terminate their evolution as WN stars seems to imply that stars with M_{ZAMS} closely above M_{WR} might explode as WNL star rather than WNE stars. Though this may be right in principle, the number of stars to which this applies may be small. Since massive stars in the ZAMS mass range considered (i.e. somewhere between 25 and 50 $M_{\odot}$) contain a steep hydrogen gradient above the H-burning shell, the mass range ΔM_r where the H mass fraction changes from 0 up to a value considered too high to apply to a WNL surface (e.g. ~ 0.3; cf. Maeder, 1983) is very small ($< 0.1\,M_{\odot}$). Let $\dot{M}_{WNL}$ be a typical mass loss rate for WNL stars (say $3 \cdot 10^{-5}\,M_{\odot}\,yr^{-1}$), then $\Delta M_r/\dot{M}_{WNL}$ is an upper limit for the duration of the WNL phase τ_{WNL}, which, with above numbers, turns to $\tau_{WNL} < 3000\,yr$. This means that stars with the smallest ZAMS masses considered in this paper may, on becoming WR stars, almost immediately turn into WNE stars.

However, the concept of estimating the duration of the different WR phases on the basis of internal composition profiles and typical mass loss rates may lead to the conclusion, that the most massive stars terminate their lifes as WNL stars (cf. Langer, 1987): the thickness (in mass) of the convective zone above the H-burning shell (i.e. also the thickness of the corresponding hydrogen plateau ΔM_{ICZ}) is an increasing function of the stellar mass. The H mass fraction established in this convective zone, X_{ICZ}, is a decreasing function of the stellar mass. For X_{ICZ} to be sufficiently small in order to correspond to the WNL phase, its duration will be $\tau_{WNL} \simeq \Delta M_{ICZ}/\dot{M}_{WNL}$, which, for sufficiently massive stars, may become larger than the post main sequence lifetime, meaning that the WNL phase is the final evolutionary phase in this case. The lower ZAMS mass limit for stars terminating their evolution as WNL stars according to this scenario is rather uncertain, but $M_{ZAMS} \simeq 100\,M_{\odot}$ may give an order of magnitude (cf. Langer and El Eid, 1986).

$M_{ZAMS} \simeq 100\,M_{\odot}$, on the other hand, is also a limiting mass concerning the final fate of massive stars: while stars of lower initial mass form an iron core, leading eventually to core collapse, higher mass stars perform a collapse due to the formation of $e^{\pm}$-pairs prior to central oxygen ignition, and subsequent explosive oxygen burning is able to disrupt the star, giving rise to the so called pair creation supernovae (PCSN) (cf. Woosley and Weaver, 1986). Towards still higher masses, explosive oxygen burning does not deliberate sufficient energy in order to reverse the collapse into an explosion (at least in nonrotating objects; see Glatzel et al.,1985 ; Stringfellow and Woosley, 1988), and the stars collapse to black holes; however, such massive objects are, if formed at all at the present epoch, extremely rare due to the steep decline of the initial mass function. Note that the fate of stars with masses close to the lower mass limit for PCSNe is still somewhat uncertain, some of them becoming pulsationally pair-unstable (see Woosley, 1986).

In summary, the most massive stars (i.e. $100\,M_{\odot} \lesssim M_{ZAMS} \lesssim 200\,M_{\odot}$) may end their evolution as WNL stars performing a PCSN. Herzig et al. (1990) performed numerical computations for lightcurves of pair-unstable exploding massive WR stars and investigated the dependence of the peak luminosity on the amount of synthesized radioactive nickel. They estimated the fraction of PCSN events among the observed supernovae to be of the order of 1%. This number is small, but taken at face it means, that two or three PCSNe could have already been observed, probably classified as peculiar Type II SNe. Herzig et al. argue, that a PCSN may be distinguished from a usual Type II SN by either very low or very high peak luminosity due to the fact that the amount of nickel synthesized may be somewhere in the range between 0 and several solar masses, depending on the mass of the progenitor star. Furthermore, the high mass loss rate of WR stars should result in a high radio luminosity of PCSNe. Possible candidates for PCSNe are SN 1961v (cf. discussion in Langer, 1987a) and SN 1986j (Rupen et al., 1987), while El Eid and Langer (1986) discussed the possibility of Cas A being a PCSN remnant. All three cases, which undoubtly have massive progenitor stars, have a very high radio luminosity, unusual lightcurves and/or peak luminosities, and show strong indications of considerable hydrogen deficiency and helium and nitrogen overabundance.

Hydrogenless WR stars: SN Ib progenitors?

The bulk of observed WR stars, i.e. most WNE and all WC stars, apparently contains no hydrogen at all. Concerning the stellar structure, it appears to be crucial whether the envelope of a WR star contains some amounts of hydrogen or not: its presence implies a hydrogen burning shell as a second nuclear energy source besides the He-burning stellar center. Consequently, the star has a larger radius and smaller surface temperature compared to hydrogenless stars of similar mass, which is also supported by the high radiative opacity and small mean molecular weight of hydrogen. Furthermore it is known, that a hydrogen containing envelope can prevent the WR star from being unstable due to radial pulsations according to the ϵ-mechanism. Maeder (1985) finds massive WR stars to be stable if their surface hydrogen mass fraction exceeds ~8%.

In this context it is important to note that there is high observational evidence, that the most massive WR stars are WNL stars, while WNE and WC stars are found to have relatively low masses. Niemela (1983) made a statistic of masses of WR stars in binary systems and found a mean mass of $63\,M_\odot$ for WNLs but values of $8\,M_\odot$ and $15\,M_\odot$ for WNE and WC stars, respectively. This is consistent with estimates for the bolometric luminosity of WR stars (cf. Smith and Willis, 1983; Schmutz et al., 1989), indicating high values for WNLs but much lower ones for WNEs and WCs (note that luminosities can readily be transformed into masses for WR stars, since they obey a narrow mass-luminosity relation; see Maeder, 1983; Langer, 1989).

The recent data of Schmutz et al. indicates a mass interval of $17\,M_\odot \gtrsim M_{WNE} \gtrsim 9\,M_\odot$ for WNE stars and of $10\,M_\odot \gtrsim M_{WC} \gtrsim 4.5\,M_\odot$ for WC stars, and they notice the incompatibility of these data with standard stellar evolution calculations like that of Maeder and Meynet (1987). The basic contradiction is that WR stars have to originate from very massive stars, but (except the WNL stars) their masses are very small. This contradiction can be resolved only, when the massive WNE and WC stage is a very short phase during the evolution of a WR star. Langer (1989a), who computed WR evolutionary sequences assuming mass dependent mass loss rates for hydrogenless WR stars (instead of the standard assumption of a constant mass loss rate) found, that this discrepancy disappears for a mass dependence of the mass loss rate to a power of 1 or larger, which is motivated by WR structure calculations (Langer, 1989), and is in agreement with present observations of WR mass loss rates (cf. references in Langer, 1989a; see also Abbott et al., 1986). Moreover, WR evolution computed with mass dependent mass loss rates is found to account for observed WR subtype number ratios, observational indications for average progenitor ZAMS masses for WNE and WC stars, as well as an increased average total WR lifetime, which is required by the high observed number ratio of WR to O-stars in the Galaxy (Conti et al., 1983).

A further result of WR evolution incorporating mass dependent mass loss rates, which is most relevant to the topic of this paper, is **1.** the mean final mass of stars with $M_{WR} \leq M_{ZAMS} \lesssim 100\,M_\odot$ is very small (i.e. below $10\,M_\odot$), and **2.** the scatter in the final masses is very small ($\sim \pm 1\,M_\odot$), implying that stars originating from the above ZAMS mass range end up practically with the same final mass.

The general characteristics of WR evolution with mass dependent mass loss rates outlined above, which have been obtained by performing a large number of simplified evolutionary computations, are confirmed by elaborated calculations using a hydrodynamical stellar evolution code. E.g. a $60\,M_\odot$ population I star — which may be a typical WR progenitor — is found to end its evolution as a $5.8\,M_\odot$ WC star, using a mass loss rate for hydrogenless WR stars of the form $\dot{M}_{WR} \sim M_{WR}^{2.5}$. The final surface abundances, which were found to be similar within most of the considered ZAMS mass range, are 20% helium, and about 40% carbon and 40% oxygen by mass. The thickness of the radiative envelope in final He-burning stages was $\sim 2\,M_\odot$, indicating the amount of mass which contains noticeable amounts of helium.

Note that $5.8\,M_\odot$ is a mass which is certainly able to perform a supernova explosion after core collapse, since it is of the same size as the He-core of SN 1987A. Light curves of exploding low mass WR stars have been calculated by Ensman and

Woosley (1988) and are found to be compatible with observed SN Ib light curves for such low masses, while abundance constraints from SN Ib spectra are matched by the abundances mentioned above.

We conclude that some fraction of Ib supernovae may be related to WR stars. However, many problems remain to be solved. For ZAMS masses close to M_{WR} the envelope composition of the SN progenitor corresponds to WNE stars, i.e. to complete H-burning ashes. The ZAMS mass range, for which evolution ends in the WNE stage is not well known (for $M_{WR} = 30\,M_\odot$ Langer, 1989a, finds a range of $30\,M_\odot \leq M_{ZAMS} \lesssim 35\,M_\odot$). The spectroscopical display of WNE-supernovae is unknown. Furthermore, for M_{ZAMS} close to M_{WR}, the final WR mass is of the order of the He-core mass of a star of mass M_{WR}. For $M_{WR} > 30\,M_\odot$, this may be more than $10\,M_\odot$, which seems to be incompatible with SN Ib light curves (Ensman and Woosley, 1988; see also: this volume). Also, as mentioned above, it is unclear if such massive cores lead to SN explosions at all.

The structure of the presupernova configurations

Observations of WR stars can tell us almost nothing about the internal structure and surface conditions (except the chemical composition) of WR stars at the time of central collapse. The SN 1987a has impressingly demonstrated, that stellar radius and surface temperature, and due to that also the mass loss rate, may change drastically during the last couple of $10^4\,yr$ of the evolution of hydrogenrich supergiants. The Kelvin-Helmholtz timescale for hydrogenless WR stars can be expressed as $\tau_{KH} \simeq 4.8 \cdot 10^4 (M/M_\odot)^{-0.38}$ (Langer, 1989), and is a good approximation for the time interval from central helium exhaustion up to central collapse. This time is too short in order to find more than 4% of the WR stars in post core helium burning phases, but sufficiently long in order to allow a global rearrangement of the stellar structure.

The low mass WC star originating from the $60\,M_\odot$ evolution as described in the previous section, e.g., finishes core helium burning with a luminosity of $\log L/L_\odot = 4.9$, and a surface temperature of $120\,000\,K$. In the following contraction phase, however, the luminosity increases up to $\log L/L_\odot = 5.2$, and the surface temperature increases first up to $165\,000\,K$ and decreases then to $140\,000\,K$. A similar situation is encountered in massive WR models (cf. Langer et al., 1988). Thus, the final values of luminosity, radius, and surface temperature can be predicted by stellar evolution calculations, but they cannot be compared to observations. Moreover, the dependence of the mass loss rate during the final hydrostatic evolutionary phases of WR stars cannot even be estimated theoretically, since there is no theory for WR mass loss.

We have to conclude that a qualitative comparison of wind properties derived from observed radio fluxes of Type Ib or Type II pec. SNe with observed WR wind properties may be meaningless.

Conclusions

The modeling of the evolution of massive stars still involves many large uncertainties (mass loss rates, convection, nuclear reaction rates, ...). Depending on the choice of physics used for stellar evolution calculations, quite different results concerning presupernova configurations may be obtained. The conclusions of this paper are based on stellar models, which were obtained with a physics compatible to that which successfully reproduced the progenitor evolution of SN 1987A (see Langer et al., this volume). A further basic ingredient is the concept of mass dependent mass loss rates for hydrogenless WR stars (Langer, 1989a). Our conclusions may be summarized as follows:

- WR stars of any subtype (WNL, WNE, WC) may be presupernova stars. WR progenitors may therefore give rise to Type II SNe (WNL) as well as to Type I SNe (WNE, WC).
- The most massive stars ($M_{ZAMS} \gtrsim 100\,M_{\odot}$) are supposed to explode during their WNL phase as peculiar Type II SNe due to the PCSN mechanism. About 1% of the observed SNe may be due to these objects.
- The bulk of very massive stars ($35\,M_{\odot} \lesssim M_{ZAMS} \lesssim 100\,M_{\odot}$) is supposed to end their evolution as low mass WC star ($M_{WC} < 10\,M_{\odot}$). Some of them certainly explode due to the core collapse mechanism, possibly being classified as Type Ib SNe.
- The wind properties of WR stars during their final evolutionary phases are unknown, which allows at most a qualitative comparison between observed and predicted radio fluxes of SNe related to those objects.

The above ZAMS mass limits are not very well known and should only be understood as order of magnitude estimates. Even at fixed input physics for stellar evolution calculations they would not be constant but rather a function of the stellar metallicity.

Acknowledgment. I am grateful to S. Woosley for valuable discussions and for his hospitality at Lick Observatory. This work has been supported by the Deutsche Forschungsgemeinschaft (DFG) through grants La 587/1-2 and La 587/2-1, by the Astronomische Gesellschaft through the Ludwig-Biermann award 1989, and by NASA through grant NAGW-1273.

References

Abbott, D.C., Bieging, J.H., Churchwell, E., Torres, A.V.: 1986, Astrophys. J. **303**, 239

Chiosi, C., Maeder, A.: 1986, Ann. Rev. Astron. Astrophys. **24**, 329

Conti, P.S., Garmany, C., de Loore, C., Vanbeveren, D.: 1983, Astrophys. J. **274**, 302

El Eid, M.F., Langer, N.: 1986, Astron. Astrophys. **167**, 274

Ensman, L.M., Woosley, S.E.: 1988, Astrophys. J. **333**, 754

Glatzel, W., El Eid, M.F., Fricke, K.J.: 1985, Astron. Astrophys. **149**, 413

Hamann, W.-R.: 1989, Proc. Hot Star Workshop, Boulder, C. Garmany, ed., in press

Herzig, K., El Eid, M.F., Fricke, K.J., Langer, N.: 1990, Astron. Astrophys., submitted

van der Hucht, K.A., Conti, P.S., Lundström, I., Stenholm, B.: Space Sci. Rev. **28**, 227

van der Hucht, K.A., Hidayat, B., Admiranto, A.G., Supelli, K.R., Doom, C.: 1988, Astron. Astrophys. **199**, 217

Humphreys, R.M., Nichols, M., Massey, P.: 1985, Astron. J. **90**, 101

Langer, N.: 1987, Astron. Astrophys. *Letter* **171**, L1

Langer, N.: 1987a, in: *Nuclear Astrophysics*, Lecture Notes in Physics **287**, W. Hillebrandt et al., eds., Springer, p. 180

Langer, N.: 1989, Astron. Astrophys. **210**, 93

Langer, N.: 1989a, Astron. Astrophys. **220**, 135

Langer, N., El Eid, M.F.: 1986, Astron. Astrophys. **167**, 265

Langer, N., Kiriakidis, M., El Eid, M.F., Fricke, K.J., Weiss, A.: 1988, Astron. Astrophys. **192**, 177

Maeder, A.: 1983, Astron. Astrophys. **120**, 113

Maeder, A.: 1985, Astron. Astrophys. **147**, 300

Maeder, A., Meynet, G.: 1987, Astron. Astrophys. **182**, 243

Niemela, V.S.: 1983, in: Proc. *Workshop on Wolf-Rayet stars*, Paris-Meudon, eds M.C. Lortet, A, Piltaut, p. III.3

Rupen, M.P., van Gorkom, J.H., Knapp, G.R., Gunn, J.E., Schneider, D.P.: 1987, Astron. J. **94**, 61

Schmutz, W., Hamann, W.-R., Wessolowski, K.: 1989, Astron. Astrophys. **210**, 236

Smith, L.F., Hummer, D.G.: 1988, M.N.R.A.S. **230**, 511

Smith, L.J., Willis, A.J.: 1983, Astron. Astrophys. Suppl. **54**, 229

St.-Louis, N., Drissen, L., Moffat, A.F.J., Bastien, P., Tapia, S.: 1987, Astrophys. J. **322**, 870

Stringfellow, G.S., Woosley, S.E.: 1988, preprint, UCRL-98066

Torres, A.V.: 1988, Astrophys. J. **325**, 759

Willis, A.J.: 1982, in: *Wolf-Rayet Stars: Observations, Physics, Evolution*, IAU-Symp. **99**, C. de Loore, A.J. Willis, eds., p. 87

Woosley, S.E.: 1986, in *Nucleosynthesis and Chemical Evolution*, Saas-Fee lecture, B. Hauck, A. Maeder, eds.

Woosley, S.E., Weaver, T.A.: 1986, Ann. Rev. Astron. Astrophys. **24**, 205

Type Ib Supernovae Wolf-Rayet Stars

Lisa Ensman & S. E. Woosley

Since Type Ib supernovae were identified as a class several years ago, the idea that their progenitors are massive stars that have lost their hydrogen envelopes has been quite popular. ENSMAN and WOOSLEY [1] calculated a series of numerical models to test this idea and found that, if SNIb do come from massive stars (as opposed to, say, white dwarfs), those stars must have masses between approximately 4 and 7 $M_{\odot}$ at explosion. For more massive stars, the long diffusion time produces too broad a light curve. Smaller stars, on the other hand, produce negligible ^{56}Ni, making the light curve too dim. It was estimated that *some* SNe Ib, such as SN 1985F, might have progenitors closer to 8 $M_{\odot}$, but the typical SN Ib seems to require a lighter progenitor. Since observed Wolf–Rayet stars have masses much larger than this, the derived mass constraint led us to conclude that the majority of SNe Ib probably do not come from single Wolf–Rayet stars. They might, however, come from hydrogen–stripped stars produced by mass transfer in interacting binaries.

Since [1], some interesting developments in the study of WR stars have led us to be somewhat less pessimistic about single WR stars as SN Ib progenitors. On the observational side, MOFFAT [2, 3] argues, from plots of WR subtype versus mass ratio q ($q = M(\mathrm{WR})/M(\mathrm{O})$, where $M(\mathrm{O})$ is the mass of an OB binary companion), that, as expected on theoretical grounds, WR stars evolve from type WNL to type WNE as they lose mass, then perhaps make a jump over to the WC sequence at some point, finally ending up at type WO. The smallest WR stars (those at the ends of the WN and WC/O sequences) have a mass ratio of about 0.2, implying that they are indeed about 6 to 8 $M_{\odot}$. Note however that this does not mean that all WR stars must evolve through the whole sequence to 6 to 8 $M_{\odot}$. In principle, a supernova explosion could occur at any point, depending on the age and core structure. However, on the theoretical side, Norbert Langer (this volume), has found that with a mass-dependent mass loss rate for the WR stage, one obtains stars which are indeed on the order of 6 to 10 $M_\odot$ at death, independent of the main sequence mass. Furthermore, he finds that the mass loss affects the core evolution such that, even for a very small final mass, some helium may remain on the surface. (The burning volume shrinks leaving partially burned helium behind.) Helium is seen in the spectra of at least some SNe Ib and provides an important constraint on the progenitors. While neither of these works is the last word on the subject, they do both offer hope for the WR–Ib connection. As the WR and Ib observations and theory improve, we will know with more certainty.

As pointed out in [1], there is also the questionable role of clumping and mixing in supernovae of Type Ib. Either would be expected to have a salutory effect on the light curve leading to a narrower peak and greater peak to tail contrast. (The observed drop in magnitude between maximum and the start of the tail could not be matched by the models in [1].) To study the effects of mixing, two 6 $M_\odot$ models have been run – variations on Model 6A of [1]. The first was totally mixed (a homogeneous composition) after shock breakout, while the second was mixed only to 4000 km/s, roughly the base of the helium layer. In addition, these models had only half as much ^{56}Ni as Model 6A (i.e., 0.22 $M_\odot$). The results are shown in Fig. 1.

Although it allowed the rise to maximum to begin immediately, complete mixing of the ejecta had almost no effect on the full width at half maximum. This is probably due to the fact that energy deposition in the outer layers by the decay of ^{56}Ni kept the temperature from falling and hence the electron scattering opacity remained high, offsetting the decrease in diffusion time resulting from mixing the ^{56}Ni closer to the surface. However, the total gamma ray optical depth was diminished, causing the peak to tail contrast to increase, though not enough to match the SN 1983N data, even in this extreme and probably unrealistic case of complete mixing. In the case of partial mixing, the peak in the light curve is sharper and narrower because as soon as the recombination front reached the bottom of the helium layer, radioactive material was exposed, and the radioactive tail, at near 100% deposition, began almost immediately. There was no nickel in the outer layer to keep it ionized so that that *mass did not have any* effect on the width of the peak. But, again, the changes in

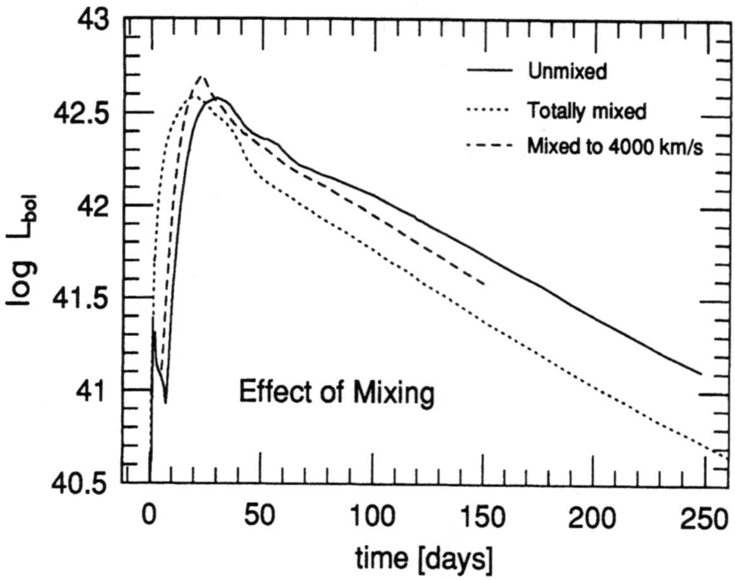

Figure 1. Effect of mixing on the light curve of a 6 $M_\odot$ model.

the light curve were not large enough to improve the fit to the data.

Models which include the effects of clumping have not yet been computed, but "holes" in the ejecta would indeed decrease the diffusion time and allow gamma rays to escape sooner. Whether a large effect on the light curve can be produced with a reasonable amount of clumping remains to be seen. (See [1] though for an extreme case in which the effective gamma ray opacity was decreased by a factor of 10 to roughly simulate the effect of clumping on gamma ray escape.)

References

1. L. Ensman, S. E. Woosley: Ap. J., **333**, 754 (1988).
2. A. Moffat: private communication (1988).
3. A. Moffat: A.J., **91**, 1386 (1986).

On the Nature of Type Ib Supernova Progenitors

Nino Panagia & Victoria G. Laidler

The realization that there is a separate subclass of Type I supernovae (SNe) to be denoted as Type Ib came after the detailed study of the SN 1983N in M83 (PANAGIA *et al.* [1]; see also PANAGIA [2], WHEELER and LEVREAULT [3], UOMOTO and KIRSHNER [4]). It was immediately clear that SN 1983N is distinctly different from the classical variety of Type I SNe (obviously denoted as Type Ia SNe) in a number of important aspects. Since then about a dozen SNe have been classified as Type Ib SNe, some newly discovered and others found just re-examining old spectra or paying due attention to the comments that the observers gave at the time when the original observations were made (*e.g.* BERTOLA [5]). In addition to the necessary condition to be called Type I SNe, *i.e.* the absence of hydrogen lines from the spectrum, the distinctive characteristics of Type Ib SNe can be summarized as follows (cf. PANAGIA *et al.* [6], WEILER and SRAMEK [7]):

- The 6150 Å feature is absent from the spectrum.
- The overall spectral distribution is redder (Δ(B-V)$\sim$ 0.5) and fainter ($\sim$ 1.5 magnitudes) than for Type Ia SNe.
- The optical light curve is essentially "normal", *i.e.* quite similar to that of Type Ia SNe.
- The IR light curve is single-peaked, the maximum occurring a few days after the optical maximum.
- They are strong radio emitters with a steep spectrum and a quick temporal decline.
- They are found only in spiral galaxies.
- They are located in spiral arms.
- They are possibly "associated" with (*i.e.* projected on, or near to) an HII region.

All these properties can be "read" in a rather simple manner and the picture that emerges is that of the explosion of stars which are "compact" (hence ***not*** a red supergiant) and have a small envelope similar to the case of Type Ia SNe (because of the similarity of the light curves), but have a chemical composition different from that of Type Ia progenitors (absence of the 6150 Å band) and a lower amount of ^{56}Ni synthetized in the explosion (redder and fainter emission).

Their radio emission requires the presence of a circumstellar envelope created by mass loss corresponding to $\dot{M}/v_{exp} \sim 3 \times 10^{-7}[M_\odot\ yr^{-1}]/[km\ s^{-1}]$ (WEILER *et al.* [8]). Such a flow would be so opaque as to create a *pseudo-photosphere* in the wind itself. For example, assuming a constant wind velocity, the radius at which the optical depth is of the order of unity would be (PANAGIA and FELLI [9]):

$$R \approx 3 \times 10^{12}\ (\kappa/\sigma_e)\{(\dot{M}/v_{exp})/10^{-7}\}\ [cm]\ >\ 9 \times 10^{12}\ [cm]$$

where κ is the average opacity, σ_e is the electron scattering cross section, $\dot{M}$ is in $M_\odot\ yr^{-1}$ and v_{exp} in $km\ s^{-1}$. It is clear that a radius so large is inconsistent with the hypothesis of a WR stellar wind. On the other hand the value of $\dot{M}/v_{exp} \sim 3 \times 10^{-7}\ [M_\odot\ yr^{-1}]/[km\ s^{-1}]$ is appropriate for a red supergiant: this implies the presence of a relatively massive companion (*i.e.* several solar masses; SRAMEK *et al.* [10]).

The lifetime of their progenitors must be shorter than 3×10^8 years, and, therefore their original mass larger than 5.5 $M_\odot$, in order to satisfy the condition posed by them being located in spiral arms. In fact, *if* Type Ib SNe are intrinsically associated with HII regions, their progenitors should be quite short-lived and, therefore, be much more massive. Since this is the only argument which may favor the idea that Type Ib SNe progenitors be very massive stars, we have considered this point in quite some detail.

First of all we have checked the validity of the "association" of Type Ib SNe with HII regions which is claimed for about 50% of them (WHEELER *et al.* [11]). By overlaying the best positions of Type Ib SNe [among those reported in the BARBON *et al.* Catalog of SNe [12] and those astrometrically determined either in the optical or in the radio] on the galaxy images with the use of the GASP[1] software, we find that 6 ± 1 out of 12 SNe appear to fall within 5" from the image of a knot (*i.e.* presumably an HII region). On the other hand, only in two cases (SN 1981I in NGC 4051 and SN 1985F in NGC 4618) the SN seems to fall on top of an HII region. Therefore, the "association" with an HII region actually means *close proximity*. And since 5" even at a distance as short as 4 Mpc corresponds to 100 pc, such a *proximity* may in fact be just fortuitous. For example, had the LMC been at 4 Mpc instead of 54 kpc, SN 1987A would appear to be *associated* with the 30 Doradus nebula while it is *not*.

A possible way to clarify this issue is to compare these results with a similar statistics made for the case of Type II SNe. Such an analysis is in progress but for the time being let us utilize the results of a similar study done by HUANG [13]. Out of 29 SNe for which there were good position measurements and good galaxy images to make the overlays, 25 objects were found to fall within an average distance of 5" from an HII. This immediately indicates that, independently on whether *any* such association is real in *any* case, the "association" of Type Ib SNe is *much* looser than for Type II

[1]GASP is the Guide Star Astrometric Support Program available at the Space Telescope Science Institute.

SNe. Assuming that this difference is entirely due to "evaporation" of the stars from their birth place, that difference implies lifetimes of the SN Ib progenitors considerably longer (3-10 times) than those of SN II progenitors and, consequently, original masses considerably lower (2-3 times for SN Ib than for SN II). Since Type II SNe are believed to originate from progenitors with masses in the range 8-20 $M_\odot$ (*e.g.* MAEDER [14]; see also VAN DEN BERGH, this Conference), the progenitors of Type Ib must have had masses within the possible range 5-10 $M_\odot$.

Such a range of progenitor's masses can be further narrowed down considering that the frequency of Type Ib explosions in spiral galaxies is about 1/3 that of Type II SNe (BRANCH [15], VAN DEN BERGH, MCCLURE, and EVANS [16]). Therefore, assuming that stars more massive than 8 $M_\odot$ make Type II SNe and adopting an initial mass function proportional to $M^{-2.35}$ the possible mass range for SN Ib progenitors turns out to be about 6.5-8 $M_\odot$. This agrees well with the direct estimate of $M > 6.5$ $M_\odot$ made by SRAMEK *et al.* [10] for SN 1983N on the basis of its radio emission. Also, a relatively "modest" mass for the progenitor can naturally explain why the mass ejected in the explosion (as implied by the "normal" optical light curve) is a few solar masses at most (BRANCH [17]).

We conclude that the only viable scenario to account for Type Ib events is that of a star with original mass around 7 $M_\odot$, which is member of a binary system in which the companion is slightly less massive (say, ~ 5 $M_\odot$). The primary follows its evolution to the end becoming a rather massive degenerate star, which explodes when the secondary has reached the stage of red supergiant. The alternative hypothesis of a very massive progenitor (*i.e.* $M > 20\text{-}30$ $M_\odot$) is ruled out on the basis of the mass loss characteristics required to account for the radio emission, the "light" envelope implied by the behaviour of the optical light curve and the "association" with HII regions which is much looser than for Type II SNe.

References

1. Panagia, N. et al: in preparation (1989).
2. Panagia, N.: in "Supernovae as Distance Indicators", ed. N. Bartel (Berlin: Springer), p. 14 (1985).
3. Wheeler, J.C., Levreault, R.: Ap. J. (Letters), 294, 17 (1985).
4. Uomoto, A., Kirshner, R.P.: Astr. Ap., 149, L7 (1985).
5. Bertola, F.: Ann. Ap., 27, 319 (1964).
6. Panagia, N., Sramek, R.A., Weiler, K.W.: Ap. J. (Letters), 300, L55 (1986).
7. Weiler, K.W., Sramek, R.A.: Ann. Rev. Astr. Ap., 26, 295 (1988).
8. Weiler, K.W., Sramek, R.A., Panagia, N., van der Hulst, J.M., Salvati, M.: Ap. J., 301, 790 (1986).

9. Panagia, N., Felli, M.: in "Wolf-Rayet Stars: Observations, Physics, Evolution", eds. C.W.H. de Loore and A.J. Willis (Dordrecht: Reidel), p. 203 (1982).

10. Sramek, R.A., Panagia, N., Weiler, K.W.: Ap. J. (Letters), 285, L59 (1984).

11. Wheeler, J.C., Harkness, R.P., Cappellaro, E.: Proc. 13th Texas Symposium on Relativistic Astrophysics, ed. M.P. Ulmer, (Singapore: World Scientific), p. 402 (1987).

12. Barbon, R., Cappellaro, E., Ciatti, F., Turatto, M., Kowal, C.T.: Astr. Ap. Suppl., 58, 735 (1984).

13. Huang, Y.-L.: P. A. S. P., 99, 461 (1987).

14. Maeder, A.: ESO Workshop "SN 1987A", ed. I.J. Danziger (Garching: ESO), p. 251 (1987).

15. Branch, D.: Ap. J. (Letters), 300, L51 (1986).

16. van den Bergh, S., McClure, R.D., Evans, R.: Ap. J., 323, 44 (1987).

17. Branch, D.: in Proc. IAU Colloquium No. 108, in press.

The Carbon Explosion Model

Zalman Barkat

The main theoretical characteristics of the carbon explosion model (more often authors use the names carbon deflagration or carbon detonation) are:

a. A carbon/oxygen core grows from below towards the Chandrasekhar mass.

b. In a single star which is able to retain its envelope the growth rate of the core is given by the luminosity-core mass relation and one can show [1][2] that the evolutionary tracks of growing cores converge to a unique track independent of initial mass.

c. In binaries, the growth rate is determined by the companion star and by the geometry. We shall not comment on the reality of scenarios which allow cores to grow. Here we only note that *a priori* there is some freedom for different evolutionary tracks due to different growth rates and the combination of core mass and age (i.e. control temperature). It turns out, however, that for the range of accretion rates $3 \times 10^{-6} \lesssim \dot{M}/M_{\odot} \lesssim 4 \times 10^{-8}$ the possible differences are rather small, especially in regards to the character of the core at carbon ignition, e.g. central density and temperature [3].

d. Note that in the above discussion the initial point from which the mass of the core is growing ("infant mass") is the phase of evolution where the core has already evolved past the point of maximum central temperature and would have been cooling at almost constant density unless forced to grow. One important aspect of this is associated with the fact that if the infant mass is larger than ~ 1.05 $M_{\odot}$, at least some carbon must have already been burnt. Even though the evolutionary tracks of these cores are still almost identical to all the others, the *compositions* of the cores should be different and infant-mass-dependent.

e. Beyond carbon ignition (which is defined as the point where energy generation by nuclear reactions become equal to neutrino energy loss) a convective core forms and grows around the center. Because of the high degeneracy the energy liberated does not affect the pressure so that the usual regulation of burning by expansion does not apply and the temperature increase is accelerated along with the nuclear energy generation rate [4].

f. The only mechanism ever suggested to prevent thermonuclear runaway, i.e. burning on a dynamical timescale, is the convective Urca mechanism, PACZYŃSKI [5]. We shall come back to discuss this issue, but first let us pretend, as has been done in recent years by most authors, that the effect is negligible.

g. One can show that the high sensitivity of the nuclear reactions to the temperature ($\sim T^{20}$) means that the burning proceeds in the form of a narrow front which advances from the center (if this is where it originates) outwards. In general the front can be described by the Hugoniot relations and it can be either a detonation or a deflagration. We shall not discuss here the question whether it is one of these or the other, we only mention that both incinerate matter all the way to nuclear statistical equilibrium as long as the density is higher than $\sim 10^7$ g cm^{-3}. The great advantage of a deflagration is that it is necessarily preceded by a shock which induces matter ahead of the front to expand. This expansion allows the deflagration to "die" at a point beyond which there remain a few tenths of a solar mass whereas in the case of detonation the corresponding amount is negligible. The point is that observations clearly say [6][7] that a significant amount of *partially* burned matter must be present. A detonation could possibly only be allowed if, for some reason, the core is already expanding when the detonation forms. This possibility will be discussed elsewhere.

The attractive thing about the models is the fact that at least some deflagration models [8][9] do produce spectra and light curves which fit observations of Type Ia supernovae amazingly well, and the fact that by their intrinsic nature they must be nearly identical.

The problems associated with the model are:

a. Single stars up to $\sim 7\ M_\odot$ are thought to eject their envelopes in the form of planetary nebulae before their cores grow to the neighborhood of the Chandrasekhar mass [10].

b. If they should appear as Type Ia supernovae they must not possess an excessive hydrogen envelope at explosion.

c. The above two points could possibly be taken care of in a binary system. Again we shall not discuss the reality of such scenarios. We mention, however, that near explosion the luminosity, which is determined by the accretion rate, is, in the range of interest, many thousands of solar luminosities and the supernova progenitor should appear as a blue supergiant. Observationally, it appears that such stars are not abundant enough [11].

d. Single stars having main sequence masses in the range $7 \lesssim M/M_\odot \lesssim 10$ may, depending on some not yet resolved problems in stellar evolution theory, evolve to the point where their cores grow to the Chandrasekhar mass without losing the envelope. If these stars explode they will appear just before the explosion as *red supergiants* with a well defined luminosity ~ 60,000 K (the luminosity-mass relation) and will certainly show hydrogen in the spectrum. Again it is not clear that enough (if any) such stars are observed.

e. From the theoretical point of view, a major problem is associated with the treatment of convective Urca neutrino losses. Current successful models of carbon deflagration [8][9] use as pre-explosion models the structure at carbon ignition, where the central density is 3×10^9 g cm^{-3} and the central temperature ~ 2×10^8 K, on the basis that the runaway is imminent. In 1972 Paczyński has suggested that circulating relevant nuclei, e.g. Na^{23}, by the convective currents past a point in the core (the "Urca shell") where the Fermi energy of electrons matches the threshold energy for capture/emission of electrons produces a cyclic process known as the "Urca process" which constitutes an energy sink, since in both cases a neutrino (or antineutrino) carries away energy. He showed that this sink can become efficient enough to control carbon burning even for nuclei whose abundance is solar, i.e. as low as 10^{-5} (by mass). BRUENN [12], however, drew attention to the fact that in these Urca processes, matter is actually *heating* in spite of the neutrino losses. Over the years since 1972 several authors have discussed this problem which is complicated because of the role which convection is playing. COUCH and ARNETT [13] and IBEN [14][15] have argued that the heating is expended in moving electrons around by the convection, so that the net result of the Urca process is still cooling. Although the detailed treatment is different, both find that carbon burning runaway does not occur at $\rho_c \sim 3 \times 10^9$ g cm^{-3}. As shown in Figure 1, Couch and Arnett find that Urca losses due to the pair Na^{23} – ^{23}Ne delay the runaway until $\rho \sim 6 \times 10^9$ g cm^{-3} while IBEN [16] finds that the temperature first oscillates on a thermal timescale due to alternate cooling and heating phases, and then as the density rises to ~ 4×10^9 g cm^{-3} *drops* sharply due to the appearance of a new Urca shell associated with the pair ^{21}F – ^{21}Ne whose threshold energy is 5.7 Mev which corresponds to $\rho \sim 3.5 \times 10^9$ g cm^{-3}. Oddly enough these results have not been taken into account in the existing carbon explosion models cited above. As can be seen from Iben's work, the problem is indeed quite complicated and the end result unclear.

Recently [17], we have reviewed the problem and believe we can clarify the way it should be treated. We find that the convective core must settle into a quasi-steady state where the Bruenn heating term is canceled by a current of electrons (together with the appropriate nuclei). In this case Urca losses can control the burning as originally suggested. We analyze the stability of the thermal balance and find:

1) The balance is *unstable* towards *cooling*. This is so because the heating depends on central entropy(s) and the cooling on the excursion of the convective zone beyond the Urca shell. These two variables are related in a complicated way which involves the manner in

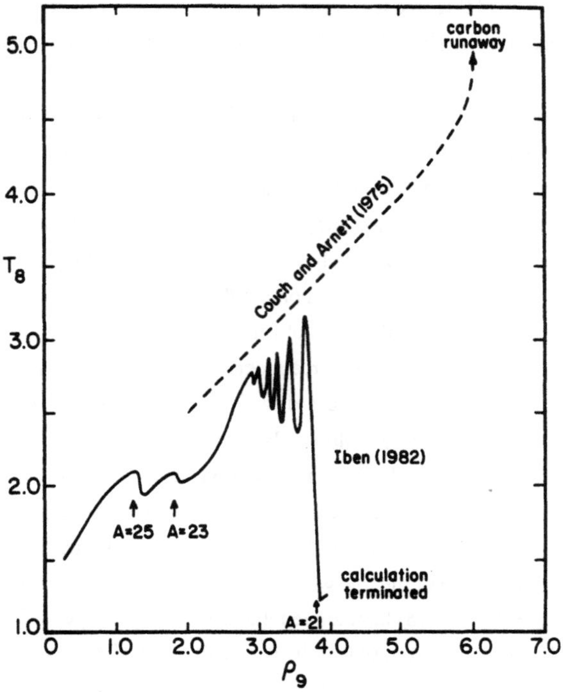

Figure 1. The evolutionary track, on the ρ,T plane, of the central conditions of the core near and following carbon ignition as obtained by IBEN [16] (full line), is compared with the track obtained by COUCH and ARNETT [13] (dashed line).

which the convective boundary retreats upon central entropy decrease on one hand and the outward advance of the Urca shell on the other. The latter is due to core contraction, which is induced by the continuous growth of the core as well as by electron capture on products of carbon burning (e.g. ^{24}Mg).

2) We show that the Urca losses can control burning only up to a limiting value of the nuclear luminosity ~ 10^7 $L_\odot$—the "maximal Urca power." If and when this limit is exceeded, the runaway will be revived.

Our analysis points out what we believe went wrong in Iben's treatment and suggests how to improve on it. At this time we can only conclude that the convective Urca process should not be ignored and speculate on the possible outcome:

1) Runaway at a much higher central density $\gtrsim 6\text{x}10^9$ g cm^{-3}, which may at best lead to an explosion which is quite different from what the existing models give, especially since copious electron capture must occur on nickel-iron group nuclei behind the burning front. In fact, at densities above ~ 10^{10} g cm^{-3} reimplosion can occur [18][3].

2) Runaway off center above a cool inner core. It is not impossible that the result in this case may not be too different from the successful current models.

References

1. B. Paczyński: *Acta Astr.*, **20**, 47 (1970).
2. Z. Barkat: *Ap. J.*, **163**, 433 (1971).
3. K. Nomoto: In *Proceedings of XXIth Rencoutre de Moriond VIth Astrophysics Meeting*, Les Arcs (1986).
4. F. Hoyle and W. A. Fowler: *Ap. J.*, **132**, 565 (1960).
5. B. Paczyński: *Ap. Letters*, **11**, 53 (1972).
6. D. Branch, J. B. Doggett, K. Nomoto, and F.-K. Thielemann: *Ap. J.*, **294**, 619 (1985).
7. R. P. Harkness: In *Radiation Hydrodynamics in Stars and Compact Objects*, ed. D. Michalas and K.-H.A. Winkler (Springer-Verlag, Berlin 1986 [p. 166]).
8. K. Nomoto, F.-K. Thielemann, and K. Yoki: *Ap. J.*, **286**, 644 (1984).
9. S. E. Woosley and T. A. Weaver: In *Radiation Hydrodynamics in Stars and Compact Objects* (Springer-Verlag, Berlin 1986 [p. 91]).
10. Y. Tuchman, N. Sack, and Z. Barkat: *Ap. J. (Letters)*, **225**, L137 (1978).
11. P. G. Sutherland and J. C. Wheeler: *Ap. J.*, **280**, 282 (1984).
12. S. W. Bruenn: *Ap. J. (Letters)*, **183**, L125 (1973b).
13. R. G. Couch and W. D. Arnett: *Ap. J.*, **196**, 791 (1975).
14. I. Iben, Jr.: *Ap. J.*, **219**, 213 (1978a).
15. ________. *Ap. J.*, **226**, 996 (1978b).
16. ________. *Ap. J.*. **253**, 248 (1982).
17. Z. Barkat and J. C. Wheeler: In preparation (1989).
18. S. W. Bruenn: *Ap. J.*, **186**, 1157 (1973a).

Explosion of Massive Wolf-Rayet Stars

Mounib F. El Eid

We have computed the final evolution of two Wolf-Rayet (W-R) stars of 61 $M_\odot$ (model A) and 44.86 $M_\odot$ (model B) through the carbon, neon, and oxygen burning phases using a one dimensional hydrodynamic code, and an extended network of nuclear reactions (El EID and PRANTZOS, [1]) to determine the nuclear energy generation rates, and the nucleosynthesis in such objects. In this contribution, only a brief summary of the main results is given; more details will be published elsewhere.

As described by LANGER and EL EID [2], the core masses above were obtained from evolutionary calculations of a 100 $M_\odot$ pop I star with mass loss and different assumption about convective overshooting (mixing beyond the boundary of the convective core as predicted by the Schwarzschild criterion for convection). According to these computations, model A resembles a W-R star of subtype WC/WO while model B resembles a WN star. Other evolutionary computations (MAEDER and MEYNET [3]) with mass los but a moderate amount of overshooting, show that the core masses above may originate from initial masses of ≈ 115 $M_\odot$ and ≈ 90 $M_\odot$ respectively.

At the end of core helium burning the oxygen mass was 50.67 $M_\odot$ in model A, and 36.36 $M_\odot$ in model B. Before central carbon ignition both models had a relatively short phase (≈ 10^3 years) of gravitational contraction during which the temperatures and densities increased steadily along with the energy losses due to neutrino processes [4]. Due to the small central carbon abundance, and the enhanced neutrino losses, the carbon burning phase proceeded in a radiative core in both models. Towards the end of neon burning, which also occurred in a radiative core, T and ρ have reached values that favour the production of electron-positron pairs ($e^\pm$) by the rediation field. The $e^\pm$ pair creation reduced the entropy of radiation, hence the radiation pressure, and the adiabatic index dropped below the critical value of 4/3. Consequently, both models above got dynamically unstable at the onset of central oxygen burning. A collapse phase followed during which oxygen burnt explosively on a time scale of typically 50 sec (cf. Fig. 1a).

We found that the oxygen burning phase marks the final phases of the present stellar models, but their behaviour during this phase was remarkably different due to their different masses. After oxygen ignition, model A collapsed within 50 sec to a peak central temperature $T_c = 3.73 \cdot 10^9$ K and density $\rho_c = 2.28 \cdot 10^6$ g cm^{-3} and burnt 7.82 $M_\odot$ of oxygen at this time. The nuclear energy release from explosive oxygen burning was sufficient to reverse smoothly (no shock formation) the collapse into explosion, which finally led to total disruption of the star. Since the total amount of burnt oxygen was 8.75 $M_\odot$, the maximum kinetic energy of the explosion attained the value $8.75 \cdot 10^{51}$ erg. The nucleosynthetic yield was dominated by oxygen and its burning products (Si, S, Ar, Ca), and 0.015 $M_\odot$ of ^{56}Ni was synthesized in this model. As we shall see below this amount of nickel has an appreciable effect on the bolometric light curve which may arise from such explosion.

The final evolution of Model B was more complicated, since it encountered the pair instability twice. Fig. 1a shows that the first infall reached a peak central temperature of $3.18 \cdot 10^9$ K and density $1.67 \cdot 10^6$ g cm^{-3} in a time of 50 sec after oxygen ignition. As displayed in Fig. 1b, the inner zones have left the pair instability domain due to the increased density, so that the collapse is halt there. The nuclear energy release from oxygen burning was than sufficient to reverse the collapse, again without shock formation, into explosion. A total amount 1.57 $M_\odot$ of oxygen was burnt, which led to $1.57 \cdot 10^{51}$ erg of explosion kinetic energy. However, during the expansion phase only the outer layers comprising 1.41 $M_\odot$ of unprocessed helium-rich material were ejected, while the total energy of the star remained negative.

After a time of 367 sec (counted from oxygen ignition), the central temperature was reduced to $3.59 \cdot 10^8$ K and the central density to $2.34 \cdot 10^3$ g cm^{-3}. An oscillation phase followed (cf. fig. 1a) with a total duration of about 2500 sec, after which the remaining core became pair unstable for the second time. The ensuing second infall was stronger, and proceeded to $T_c = 4.26 \cdot 10^9$ K and $\rho_c = 4.06 \cdot 10^6$ g cm^{-3} in 30 sec. During this phase, the oxygen mass decreased from 34.70 $M_\odot$ to 24.45 $M_\odot$. This high amount of burnt oxygen led to a stronger explosion which disrupted the star. The final ejected mass of oxygen was 24.01 $M_\odot$ and the maximum kinetic energy turned out to be $8.50 \cdot 10^{51}$ erg. Thus this model has burnt 12.35 $M_\odot$ of oxygen in total.

The composition of the ejecta was dominated by oxygen and its burning products like in model A. However, the ^{56}Ni yield from model B was 0.72 $M_\odot$, remarkably higher than from model A, since model B has achieved higher temperatures and densities during its second collapse phase.

We note that the only other computations which have shown the repeated occurrance of the pair instability were those of WOOSLEY and WEAVER [5] (see also WOOSLEY [6]). They called this process "pulsational pair-instability". According to them, the pair instability was encountered four times during the evolution of a 45 $M_\odot$ W-R star after oxygen ignition. At the end of their calculations, an iron core of 2.2 $M_\odot$ was formed surrounded by 39 $M_\odot$ of outside material. The final fate of such objects is still to be determined [5]. Clearly, this type of evolution is different from that of our model B. The reason is not clear yet. What we know (Woosley, private communication) is, that the initial oxygen of our model B is higher by ≈ 2 $M_\odot$. This may not be the only reason for the discrepancy, though it appears that the final fate of pair unstable stars is very sensitive to the oxygen mass if this mass is close to the limit of 30 $M_\odot$, below which a star may not become pair unstable at all [7]. We leave the clarification of this issue for a forthcoming work.

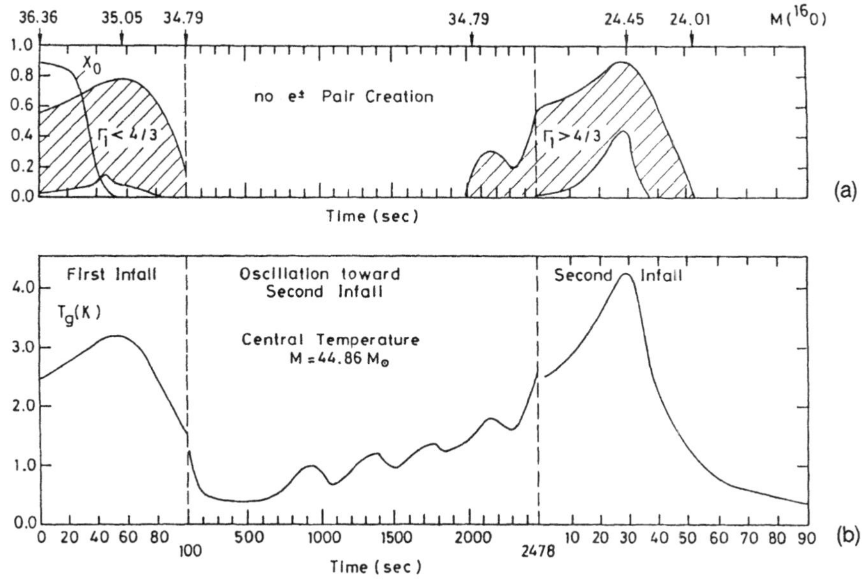

Figure 1. (a) Central temperature as a function of time for the pair unstable Wolf-Rayet star of the indicated mass. Time = 0 is taken at oxygen ignition (note the change of time scale). The Fig. shows that the star encouters the $e^{\pm}$ pair creation instability twice. (b) Development of the pair instability domain with time where the adiabatic index $\Gamma_1 < 4/3$. The variation of the oxygen mass with time is indicated at the top of the Fig. X_O denotes the central abundance (mass fraction) of oxygen.

Finally, we mention that the bolometric light curve, which may be associated with the explosion of the present model A, has recently been calculated at Göttingen (HERZIG [8]; HERZIG et al. [9]). The light curve of model B is currently under study. Only some results will be mentioned here; the details will be given in [9]. In Fig. 2, two light curves of the 61 $M_\odot$ star are shown and compared with the light curve resulting from the explosion of the 8 $M_\odot$ W-R star according to ENSMAN and WOOSLEY [10]. The dotted light curve is calculated with 0.015 $M_\odot$ of ^{56}Ni synthesized in model A as mentioned above, while the dashed curve is obtained when the nickel mass is artificially increased to 0.50 $M_\odot$. The early shape of the light curves is dominated by the energy input from the oxygen recombination, which leads to the first peak, while the late shape is determined by the energy input from the radioactive decay of ^{56}Ni and ^{56}Co. The late influence of the radioactive decay on the light curves is a consequence of the large mass, hence the longer diffusion time scale, especially when nickel is concentrated near the center as in the high mass model. The comparison between the dotted curve and that of the 8 $M_\odot$ star reveals, that the former is less luminous and has a larger width. The larger width is due to the higher core mass, whereas the lower peak luminosity reflects the difference in the nickel mass. The effect of increased nickel mass is demonstrated by the dashed curve in Fig. 2. In this case, the peak luminosities become similar, simply because the nickel masses are now comparable. Thus, the remaining difference between the light curves is due to the different masses. Therefore, it appears possible to distinguish observationally between the two types of explosions on the basis of the bolmetric light curve alone.

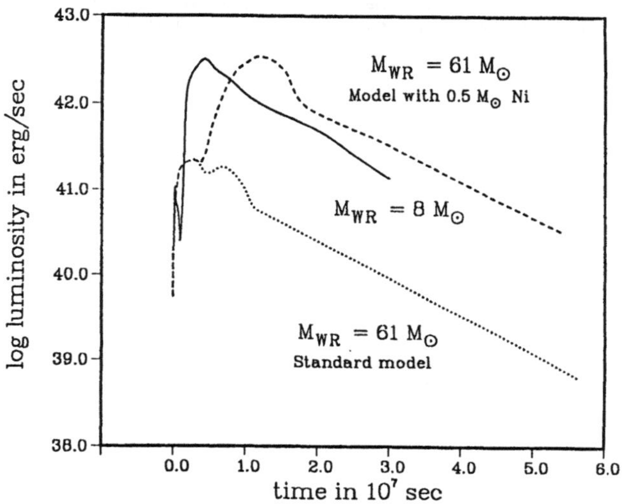

Figure 2. Comparison of the light curves of the exploding high and low mass Wolf-Rayet stars. The effect of the radioactive ^{56}Ni decay on the light curve of the 61 $M_\odot$ star (model A in the text) is illustrated by comparing the dotted and dashed curves. The light curve of the 8 $M_\odot$ is due to ENSMAN and WOOSLEY [10].

References

1. M.F. El Eid, N. Prantzos: in The Origin And Distribution of the Elements, ed. by G.J. Mathews (World Scientific, New Jersey 1988), p. 485
2. N. Langer, M.F. El Eid: Astron. Astrophys. **167**, 265 (1986)
3. A. Maeder, G. Meynet: Astron. Astrophys. **182**, 243 (1987)
4. H. Munakata, Y. Kohyama, N. Itoh: Ap. J. **226**, 197 (1985)
5. S. E. Woosley, T.A. Weaver: in IAU Colloq. No. 89, Radiation Transport and Hydrodynamics, ed. by D. Mihalas and H. Winkler (Reidel, Dordrecht 1986)
6. S. E. Woosley: in Nucleosynthesis and Chemical Evolution, ed. by B. Hack and A. Maeder (Geneva Observatory 1986), p. 1
7. J.R. Bond, W.D. Arnett, B.J. Carr: Ap. J. **280**, 825 (1984)
8. K. Herzig: Diploma Thesis, Univ. of Göttingen, 1988
9. K. Herzig, M.F. El Eid, K.J. Fricke, N. Langer: Theoretical light curves for exploding massive Wolf-Rayet stars, preprint, submitted to Astron. Astrophys., 1989
10. L.M. Ensman, S.E. Woosley: Ap. J. **333**, 754 (1988)

Theoretical Light Curves of Type I Supernovae

K. Nomoto & T. Shigeyama

1. Introduction

Type I supernovae (SN I) are identified from the absence of hydrogen lines in the maximum light spectra. SN I are further subclassified from the helium features in the spectra, namely, SN Ia (no helium), SN Ib (helium-rich), and SN Ic (helium-poor) [34].

The lack of hydrogen lines implies that the progenitors of SN I have lost their hydrogen-rich envelope at the time of explosion. Two cases are possible: (1) tidal mass loss during the evolution of a close binary system, and (2) stellar wind type mass loss. This implies that we have basically two candidates for the progenitor of SN I, i.e., white dwarfs and Wolf-Rayet stars. Because of the complicated mass loss processes, it has not been easy to identify the exact evolutionary origin of SN I. The currently popular models are the carbon deflagration of accreting C+O white dwarfs for SN Ia and the explosion of Wolf-Rayet stars for SN Ib. However, accreting white dwarf models for Ib and especially for Ic may be possible (e.g., helium detonation as speculated in [4]). Here we discuss to what extent these models can account for the observed light curves (§§2–3)

Since the origin of SN Ib, Ic, and variations of Ia have not been quite clear, we reexamine the fate of 8-10 $M_\odot$ stars, O+Ne+Mg white dwarfs, and solid C+O white dwarfs to explore other possible models for SN I (§4).

2. Type Ib and Ic Supernovae

2.1. Exploding Helium Stars

Wolf-Rayet stars form from massive stars by losing their hydrogen-rich envelope and becoming helium stars. Such helium stars eventually undergo iron core collapse like SN II. Because of the lack of hydrogen-rich envelope, the Wolf-Rayet star is too compact to power the light curve by shock heating, thereby requiring the power of radioactive decays to become SN Ib. In fact, the exponential tails of the light curves and the IR emission line of iron at late times [13] show that decays of ^{56}Ni and ^{56}Co also work for SN Ib. The peak luminosity is lower than SN Ia by a factor of $\sim$ 2-4.

We calculated the light curves for the helium star models of masses $M_\alpha = 3.3\ M_\odot$, $4\ M_\odot$ and $6\ M_\odot$ (cooresponding main-sequence masses are $\sim 13M_\odot$, $15M_\odot$, $20M_\odot$, respectively) assuming the production of $0.075\ M_\odot$ ^{56}Ni. the γ-ray opacity of 0.03 cm^2 g^{-1} and the optical opacity are calculated as done for the SN 1987A model [30]. For $M_\alpha = 3.3$, 4, and 6 $M_\odot$, the masses of the ejecta are $M_{\rm ej} = 2.1$, 2.6, and 4.4 $M_\odot$, and the neutron star residue are $M_{\rm ns} = 1.2$, 1.4, and 1.6 $M_\odot$, respectively [5].

The decline rate of the light curve depends largely on how fast γ-rays and X-rays escape from the star. The column depth to the ^{56}Ni-^{56}Co layer at time t is roughly proportional to (mass of the overlying layer)$^2/(Et^2)$. Accordingly the optical light curve declines faster, if the ejecta mass is smaller, ^{56}Ni is mixed closer to the surface, and E is larger.

2.2. Light Curves of Type Ib Supernovae

For SN Ib, two observed light curves are shown in Fig. 1: the slower curve is the visual light curve of SN 1984L [14] and the faster one is the bolometric light curve of 1983N [26]. The latter decline rate is close to that of SN Ia [17].

If all γ-ray energies are deposited in the star, bolometric light curve after the peak declines at the radioactive decay rate as shown by the dotted curve in Fig. 1. With $E = 1 \times 10^{51}$ erg and the original stratified composition structure (i.e., ^{56}Ni being confined in the central region), the tails of the calculated light curves for $M_\alpha = 4$ and $6M_\odot$ are close to the dotted line (see also [29]). This seems to be slower than even the slowest SN Ib, SN 1984L, suggesting a mixing of ^{56}Ni into the outer layers.

The dashed line in Fig. 1 shows a light curve for rather extreme case of mixing where ^{56}Ni is uniformly distributed in the whole ejecta of $M_\alpha = 6\ M_\odot$. It is probably consistent with SN 1984L but its decline is slower than 1983N. The solid line calculated for $M_\alpha = 4\ M_\odot$ ($M_{\rm ej} = 2.6\ M_\odot$) with mixing of ^{56}Ni at $M_r < 2.2\ M_\odot$ is in good agreement with SN 1983N. (If we mix ^{56}Ni closer to the surface, the pre-peak dip in the light curve is too bright to be consistent with 1983N like the solid line in Fig. 2.)

Figure 1 shows that the light curves of SN Ib may be accounted for by the helium star models of $M_\alpha < 6\ M_\odot$ as far as ^{56}Ni is mixed out close to the surface. The variation of the decline rate of the light curve may be ascribed to the variation of the ejecta mass. Mixing of radioactive materials into outer layers might cause a large non-LTE excitation of helium lines [3].

However, the above constraint of $M_\alpha < \sim 6\ M_\odot$ leads to some problems. If the Wolf-Rayet stars are the progenitor of SN Ib, their main-sequence masses should be larger than 20 $M_\odot$ (and $M_\alpha > 6\ M_\odot$) because of the following reason. We know that SN 1987A is the explosion of a 20 $M_\odot$ star ($M_\alpha = 6\ M_\odot$) that produced $\sim 0.07\ M_\odot$ ^{56}Ni. On the other hand, SN Ib should produce 0.15 - 0.3 $M_\odot$ ^{56}Ni, because the peak luminosity of SN Ib is about 1/4 - 1/2 of SN Ia that are powered by the decay of $\sim 0.6\ M_\odot$ ^{56}Ni. Therefore the main-sequence mass of the Wolf-Rayet progenitor of SN Ib should be larger than that of the progenitor of SN 1987A. For smaller mass stars,

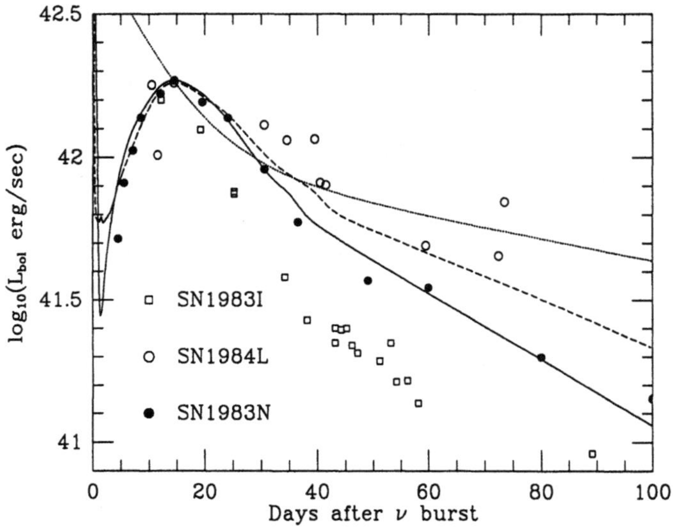

Fig. 1: Observed light curves of SN Ib (bolometric curve of SN 1983N and visual curve of SN 1984L) and SN Ic (visual curve of SN 1983I) are compared with the helium star models of $M_\alpha = 6\ M_\odot$ (with the ejected mass $M_{ej}\ = 4.4\ M_\odot$: dashed curve) and $4\ M_\odot$ ($M_{ej}\ = 2.6\ M_\odot$: solid curve). Both curves assume almost complete mixing of 0.075 $M_\odot$ ^{56}Ni in the ejecta. The dotted curve is the energy generation rate of the ^{56}Ni-^{56}Co decays.

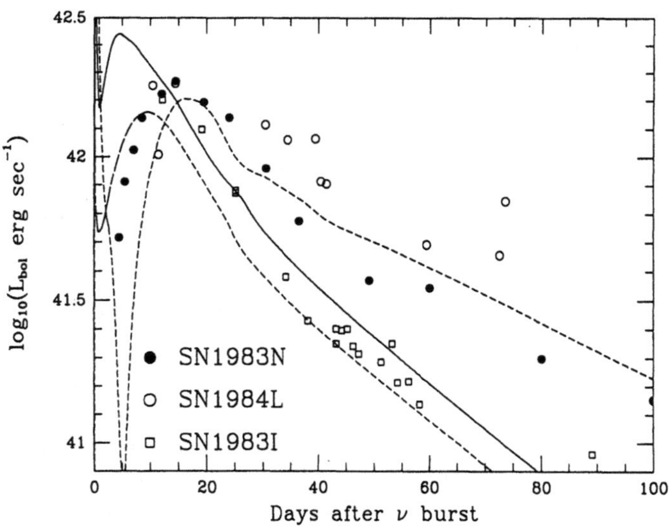

Fig. 2: Same as Fig. 1 but compared with the helium star models of $M_\alpha = 3.3\ M_\odot$ ($M_{ej}\ = 2.1\ M_\odot$) with mixing (solid curve) and $M_\alpha = 4\ M_\odot$ ($M_{ej}\ = 2.2\ M_\odot$) with and without mixing (dashed).

the density of the silicon and oxygen layers is too low to synthesize enough ^{56}Ni. This requirement could meet the constraint from the light curves only for relatively narrow range of stellar mass around main-sequence $\sim$ 20 - 25 $M_\odot$ [33, 9]. If the ejecta are highly clumpy, however, they are more transparent to γ-rays, thus being allowed to be more massive [9].

2.3. Light Curves of Type Ic Supernovae

For SN Ic the fast decline of the luminosity is more difficult to be reconciled with the peak luminosity. In Figures 1 and 2, the visual light curve of SN 1983I is shown since the visual curve is close to the bolometric one. This light curve of SN Ic declines significantly faster than SN Ia and SN Ib. It is clear from Fig. 1 that the two curves for $M_\alpha = 4$ and 6 $M_\odot$ can not be consistent with SN Ic even with extreme mixing. The large difference between the peak luminosity and the tail as well as narrow peak is difficult to reproduce with such a massive star model [9].

To reproduce the SN Ic light curve, the helium star mass should be as small as $M_\alpha = 3.3\ M_\odot$ ($M_{\rm ej} = 2.1\ M_\odot$) as shown by the solid curve in Fig. 2. Almost complete mixing of ^{56}Ni is also necessary for this curve.

The ratio $M_{\rm ej}/M_\alpha$ could be smaller than assumed above, if the helium star loses its helium envelope by mass loss. To explore such a case, we constructed the ejecta model with $M_{\rm ej} = 2.2\ M_\odot$ from the helium star of $M_\alpha = 4\ M_\odot$ by removing the neutron star mass of 1.8 $M_\odot$. The calculated light curves are shown by the two dashed curves in Fig. 2 where the faster and the slower declines correspond to the cases with and without mixing. It is clear that mixing is required to reproduce the light curve of SN 1983I. If $M_{\rm ej}$ is significantly larger than 2.2 $M_\odot$ to be consistent with the ^{56}Ni mass, the effect of clumpiness to make the ejecta transparent to γ-rays should be significant [9].

The evolution of the photospheric velocity $v_{\rm ph}$ observed in SN Ic provides an another important clue to the nature of the progenitor. In SN 1983V, $v_{\rm ph}$ changes from $\sim$ 18000 km s^{-1} [3] to 7500 km s^{-1} [14] in 8 days, i.e., 1200 km s^{-1}/day near maximum light. On the contrary, SN Ib 1983N showed much slower decrease in $v_{\rm ph}$ with 400 km s^{-1}/day [28]. In Fig. 3, the change in $v_{\rm ph}$ is shown for several helium star models as well as the model for SN 1987A. Compared with SN 1987A, the helium stars have smaller envelope mass, thus having the smaller velocity contrast between the envelope and the core. Consequently the velocity gradient in the envelope and the change in $v_{\rm ph}$ are small. This is consistent with SN 1983N but cannot reproduce the observed feature of SN Ic 1983V. The SN Ic feature might better be explained by a model involving accreting white dwarfs (e.g., [4]). Certainly more quantitative studies of theoretical spectra and light curves (including the expansion opacity) are necessary to identify the SN Ib and Ic progenitors.

2.4. Mixing

For massive helium star models of SN Ib and Ic, mixing of ^{56}Ni and formation of clumps are required to explain the observed light curves. For SN 1987A, mixing during explosion

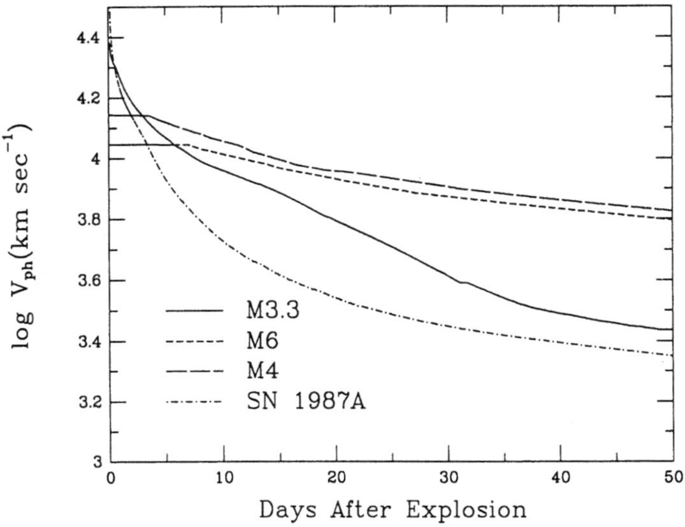

Fig. 3: Evolution of the photospheric velocity for several helium star models and the model for SN 1987A.

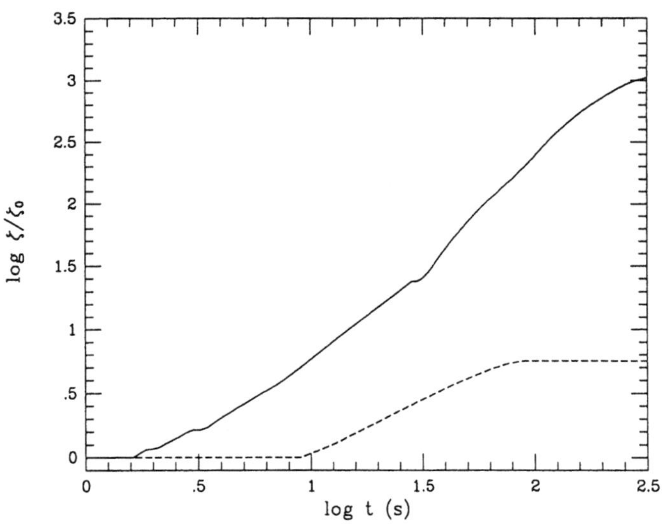

Fig. 4: The growth of the Rayleigh-Taylor instability at the He/C+O interface of helium stars. The amplification factor relative to the initial perturbation is plotted for the explosion of the helium star of $M_\alpha = 6\ M_\odot$ (dashed curve) and $4\ M_\odot$ (solid).

is found to be induced by the Rayleigh-Taylor instability at the composition interface. We have studied whether such a mixing occurs in the exploding helium stars as well by the linear stability analysis [8].

When the shock wave hits the helium envelope, the expansion of the inner core is decelerated. The resulting pressure inversion induces the Rayleigh-Taylor instability at the interface between the core and the helium envelope where the density steeply decreases outward. The growth of the initial perturbations is larger for the smaller mass helium star as shown for $M_\alpha = 3.3\ M_\odot$ (solid) and $6\ M_\odot$ (dashed) in Fig. 4. The reason is as follows:

1) For smaller M_α the mass ratio between the helium envelope and the core (excluding the neutron star mass) is larger (i.e., 2.5, 2.7, 1.0, and 0.45 for M_α = 3.3, 4, 6, and 8 $M_\odot$, respectively) so that the deceleration of the core is larger.
2) Smaller mass stars have steeper density gradient near the composition interface.
3) The stellar radius is larger for smaller M_α so that it takes longer for the shock wave to reach the stellar surface; then the instability grows for a longer time until the rarefaction wave from the surface makes the layer stable.

All these effects result in the shorter timescale of the growth and the larger amplification of the perturbation for the smaller mass helium star (Fig. 4). This implies that a large scale mixing would occur only for stars with masses smaller than a certain limit.

From the ^{56}Ni mass constraint, on the other hand, the mass of the helium star progenitor of SN Ib should be larger than 6 $M_\odot$. Compared with SN 1987A where the Rayleigh-Taylor instability starts to grow at the hydrogen/helium interface, the amplification at the helium/heavy-element interface in the 6 $M_\odot$ helium star is much smaller (dashed curve in Fig. 4). This is because the density gradient is less steep and, moreover, there is not enough time for the instability to grow before the shock wave reaches the surface. On the other hand, some clumpiness in SN Ib is suggested from observations [10]. Further 2D hydrodynamical calculation is necessary to see whether the two constraints of the progenitor mass can be consistent with each other.

3. Type Ia Supernovae

3.1. Carbon Deflagration in Accreting C+O White Dwarfs

The evolution of accreting white dwarfs in close binary systems is determined by the balance between compressional heating and radiative cooling and, thus, strongly depends on the mass accretion rate, $\dot{M}$. For intermediate accretion rates (2.7×10^{-6} $M_\odot$ yr^{-1} > $\dot{M}$ > 4×10^{-8} $M_\odot$ yr^{-1}), the C+O white dwarf can grow up to the Chandrasekhar mass [21, 22]. When the central density reaches $\sim 3 \times 10^9$ g cm^{-3}, explosive carbon burning starts in the white dwarf's center. A carbon burning front then propagates outward on the timescale for convective heat transport, which is called a convective deflagration wave. It takes about 1 sec for the deflagration wave to reach the surface region, during which the white dwarf expands [25].

Behind the deflagration wave, the material undergoes explosive nuclear burning of silicon, oxygen, neon, and carbon. In the inner layer, iron-peak elements, mostly ^{56}Ni, are produced. When the deflagration wave arrives at the outer layers, the density it encounters has already decreased due to the expansion, so that only Ca, Ar, S, and Si are produced. In the intermediate layers, explosive burning of carbon and neon synthesizes S, Si, and Mg. In the outermost layers, the deflagration wave dies and C+O remain unburned. The composition structure after freeze-out is seen in [32].

In the current standard model W7 [25], a carbon deflagration occurs in the white dwarf of mass 1.38 $M_\odot$ and synthesized 0.58 $M_\odot$ ^{56}Ni. The star explodes completely with the kinetic energy of $E = 1.3 \times 10^{51}$ ergs and no neutron star residue. The synthetic spectrum at maximum light is in excellent agreement with the observed optical spectrum of SN 1981B. The agreement implies that both explosion energy and nucleosynthesis in the carbon deflagration model are consistent with the observations near maximum light.

3.2. Optical Light Curve and Distribution of ^{56}Ni

Although most of SN I show similar tails in their light curves, there exists a certain variations in the declining rate as seen in the bolometric light curves of SN 1972E and SN 1981B [11] and the visual light curve of SN 1986G [27] in Fig. 5. SN 1986G shows significantly faster decline than SN 1972E and SN 1981B. We suggest that such a variation would mostly stem from the difference in the distributions of ^{56}Ni.

We calculate the light curves for three different distributions of ^{56}Ni as seen in Fig. 5. Assuming the optical opacity of 0.1 cm^2 g^{-1} and the γ-ray opacity of 0.03 cm^2 g^{-1}.
1) W71 (dashed): ^{56}Ni is confined in the central region of M_r = 0 - 0.58 $M_\odot$.
2) W72 (solid): ^{56}Ni is located at the intermediate region of M_r = 0.28 - 0.86 $M_\odot$ as in the original W7. At $M_r < 0.28$ $M_\odot$, more neutron-rich ^{58}Ni, ^{54}Fe and ^{56}Fe are produced.
3) W73 (dash-dotted): ^{56}Ni is distributed in the outermost layers of M_r = 0.80 - 1.38 $M_\odot$.

The calculated light curves are compared with observations in Fig. 5. The slow light curves of SN 1972E and SN 1981B are in good agreement with W72 (solid, i.e., the original W7), but W71 (dashed) is a little too slow. The fast decline of SN 1986G is well reproduced by W73 (dash- dotted) because γ-rays from the outer layer escape earlier. In order to reproduce SN 1986G with the same ^{56}Ni distribution as in W72, E/M should be 3 times larger than that of W7, which is impossible for carbon deflagration models. In fact, the expansion velocities of SN 1986G is even lower than SN 1981B [5].

This indicates that the light curve shape is more sensitive to the distribution of ^{56}Ni than E/M. This is consistent with the fact that the light curve of SN 1984A is similar to SN 1981B though its velocities are significantly higher [5].

3.3. Variations of Carbon Deflagration Models

Here we discuss the evolutionary origin of the difference in the abundance distribution as well as the effect of the initial metallicity (i.e., population) of the white dwarf. The distribution of ^{56}Ni depends on 1) the size of the neutron-rich core [12] if the carbon deflagration starts from the center and 2) the site of carbon ignition if an off- center deflagration occurs.

The neutronized core could not be very large, because it would lead either to the ejection of too much neutron-rich ^{58}Ni and ^{54}Fe or to the collapse due to rapid electron capture at high central density [22]. Thus as far as the carbon deflagration starts from the center, variation of the light curve would not be so large.

If the off-center carbon deflagration frequently occurs, larger variations are expected. This would take place if a solid oxygen core has formed in the white dwarf [7] or the convective Urca neutrino process efficiently cools down the inner core after carbon ignition [1, 15]. However, the former possibility has been ruled out by the recent findings that chemical separation between carbon and oxygen does not occur [2, 16].

The convective Urca process will not occur if the carbon ignition density, $\rho_{\rm ig}$, is higher than the threshold densities of Urca process, i.e., $\rho_{\rm th} = 1.3 \times 10^9$ g cm^{-3} for ^{25}Na - ^{25}Mg pair and 1.8×10^9 g cm^{-3} for ^{23}Ne - ^{23}Na pair. If the accretion rate is as high as $\dot{M} > \sim 10^{-6}$ $M_\odot$ yr^{-1}, $\rho_{\rm ig}$ would be lower than $\rho_{\rm th}$ for the A = 25 pair. Even for lower $\dot{M}$, the A = 25 Urca shell cooling would not operate if the white dwarf belongs to the old population and has low metallicity. In this case, $\rho_{\rm ig} \sim 1.5 \times 10^9$ g cm^{-3} if $\dot{M}$ corresponds to the steady hydrogen burning as occurs in the AGB star or common envelope ($\sim 8 \times 10^{-7}$ $M_\odot$ yr^{-1}) [19]; this would induce the central deflagration without undergoing convective Urca cooling. For still lower $\dot{M}$, the convective Urca process would operate; however, nova-like explosions might reduce the occurrence frequency of this case.

Our speculation is that the majority of the SN Ia progenitor undergo rapid accretion to avoid nova-like explosions [21]. Then the central deflagration is induced without undergoing convective Urca cooling. Resulting SN Ia form quite a uniform class of light curves. This would be more likely the case for the white dwarfs with low metallicity. The variations of the light curves would result from the off-center carbon deflagration that occurs for slower accretion as well as high metallicity white dwarfs.

4. Possible Alternative Models

As discussed in §§2-3, the origin of SN Ib,c and variations of Ia have not been clarified yet. Thus it is worth exploring possible alternative models, which include O+Ne+Mg white dwarfs, and C+O white dwarfs having high ρ_c. Whether the stars related to these white dwarfs, e.g., 8-10 $M_\odot$ stars and some merging white dwarfs, can produce SN-I like events depends on whether these white dwarfs undergo collapse or explosion.

4.1. 8 - 10 $M_\odot$ Stars and O+Ne+Mg White Dwarfs

The stars in this mass range is distinct from other mass range because they develop a degenerate O+Ne+Mg core. If these stars are in close binary systems, they form O+Ne+Mg white dwarfs [20]. The initial masses of the O+Ne+Mg white dwarfs are as large as 1.2 - 1.37 $M_\odot$ which is favorable feature to produce short period recurrent novae. The abundances of the typical O+Ne+Mg white dwarfs may be consistent with the observed abundance of Neon Novae [35]. Furthermore, such recurrent novae on massive white dwarfs could easily grow the white dwarf mass [31].

When the core mass grows to 1.38 $M_\odot$, the central density reaches 4×10^9 g cm^{-3} and electron captures ^{24}Mg (e$^-$, ν) ^{24}Na (e$^-$, ν) ^{20}Ne and ^{20}Ne (e$^-$, ν) ^{20}F (e$^-$, ν) ^{20}O. The central density at which electron captures ignite a neon/oxygen deflagration could range from 1 - 2.5 $\times 10^{10}$ g cm^{-3} depending on the timescale of material mixing in the electron capture region [18]. If the neon/oxygen deflagration is initiated at $\rho_c \sim 1 \times 10^{10}$ g cm^{-3}, whether it leads to collapse or explosion needs careful study.

4.2. C+O White Dwarfs

The accreting C+O white dwarfs could also either explode or collapse, depending on the conditions of the white dwarfs. Compression of the white dwarf by the accreted matter first heats up a surface layer and, later, heat diffuses inward [19]. If the initial mass of the white dwarf, M_{CO}, is smaller than 1.2 $M_\odot$, the entropy in the center increases substantially due to the heat inflow and carbon ignites at relatively low central density ($\rho_c \sim 3 \times 10^9$ g cm^{-3}). On the other hand, if the white dwarf is more massive than 1.2 $M_\odot$ and cold at the onset of accretion, the central region is compressed only adiabatically and thus is cold when carbon is ignited in the center. In the latter case, the ignition density is as high as 10^{10} g cm^{-3} (e.g., [6]) and the white dwarf may well have a solid core. For such a case, it is important to determine the critical condition for which a carbon deflagration induces collapse rather than explosion.

4.3. Collapse of C+O White Dwarfs Induced by Carbon Deflagration

Nomoto [23, 24] examined the critical condition for which a carbon deflagration initiated at the center of $\rho_c \sim 10^{10}$ g cm^{-3} leads to collapse of the white dwarf. The outcome depends on whether, behind the deflagration wave, nuclear energy release or electron capture is faster. The energy generation rate is determined mainly by the propagation velocity of the deflagration wave, v_{def}, while the electron capture rate depends on the density. If v_{def} is lower than a certain critical speed, v_{crit}, electron capture induces collapse. If v_{def} is sufficiently high, on the contrary, complete disruption results.

First let us start from the *conductive* deflagration in the solid C+O white dwarf. In the old calculation [23, 24], v_{def} obtained from the heat conduction calculation was too low in the very central region because of too coarse mesh points compared with the sphericity. To avoid this problem, we adopt a simpler approach assuming a constant ratio of v_{def}/v_s for conductive deflagration wave, where v_s denotes the local sound velocity.

Figure 6 shows the change in the central density of the C+O white dwarf starting from 9×10^9 g cm^{-3}. Three cases assume $v_{def}/v_s = 0.05$, 0.03, and 0.01, and the latter two slow cases undergo the collapse. This implies that the critical velocity, v_{crit}, that divides collapse and explosion is $v_{crit} \sim 0.03\ v_s$ for $\rho_c \sim 10^{10}$ g cm^{-3}. The realistic value of conductive deflagration speed is $v_{def} \sim 0.01\ v_s$ [36], thereby leading to the collapse of the solid white dwarf.

4.4. Collapse of O+Ne+Mg White Dwarfs Induced by Ne/O Deflagration

Next we discuss the deflagration in O+Ne+Mg white dwarfs. After electron capture on ^{24}Mg starts, resulting entropy production heats up the central region and forms a liquid core even if the white dwarf had initially a solid core [18, 5]. When electron capture on ^{20}Ne ignites a neon flash, therefore, a *convective* neon/oxygen deflagration wave forms. Since the propagation velocity of the deflagration wave, v_{def}, is highly uncertain, we apply a time dependent mixing length prescription with the ratio between the mixing length and the pressure scale height $\alpha = \ell/H_p = 0.7$, 1.4, and 2 [25].

Figure 7 shows the change in ρ_c associated with the propagation of the deflagration wave. The slowest case of $\alpha = 0.7$ goes into collapse, while the case with $\alpha = 1.4$ is marginal. For smaller α, the deflagration speed in the central region is so low that the *conductive* deflagration is faster. Thus the minimum v_{def} is set at $\sim 0.01\ v_s \sim 100$ km s^{-1}. Such a slow propagation is seen from the initial slow increase in M_r at the location of the deflagration front (Fig. 8). Consequently the fate of the convective deflagration wave depends mainly on whether v_{def} exceeds $\sim 0.03\ v_s$ in the central region of $M_r <\sim 0.1$ - $0.2\ M_\odot$.

If $\alpha < 1.4$ is the case, the O+Ne+Mg white dwarf collapses even if $\rho_c \sim 10^{10}$ g cm^{-3}. Since W7 model, which adopts $\alpha = 0.7$ for the carbon deflagration, can nicely account for the observations of SN Ia, $\alpha = 0.7$ may also be preferred for Ne/O deflagration and thus the collapse would be more likely.

If total disruption results from such high central density as 10^{10} g cm^{-3}, such an explosion may not be preferred because too much neutron-rich matter would be ejected. Moreover, the explosion energy is such low as $\sim 10^{50}$ ergs because of large neutrino losses, thereby cannot be consistent with any subclass of SN I.

Though the definite conclusion needs multi-dimensional calculation to determine the propagation speed, the deflagration initiated from $\rho_c \sim 10^{10}$ g cm^{-3} would make a collapse for both O+Ne+Mg and C+O white dwarfs. In other words, such white dwarfs and related stars may not become SN I but become a neutron star. Such an accretion-induced collapse of white dwarf may be necessary to account for the existence of a certain class of binary pulsar and low mass X-ray binaries [24].

K.N. would like to thank Dr. Ramon Canal for useful discussion on the white dwarf collapse.

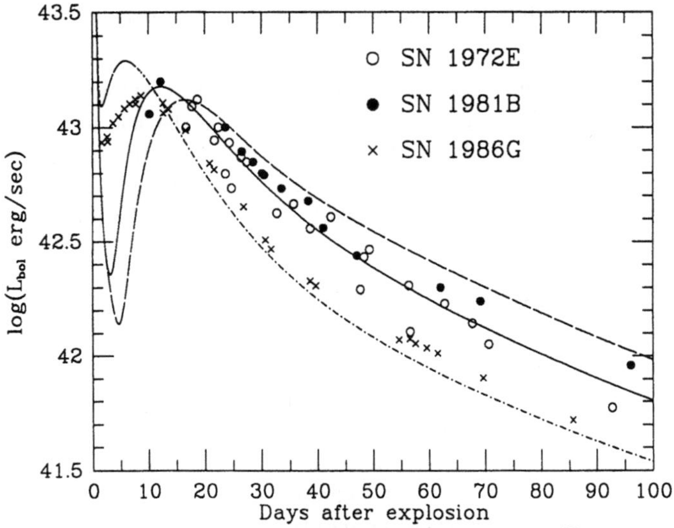

Fig. 5: Bolometric light curves for the different distribution of ^{56}Ni in W7 as compared with the bolometric light curves of SN 1972E and 1981B and the visual light curve of SN 1986G. 1) W71 (dashed): ^{56}Ni is confined in the central region of M_r = 0 - 0.58 $M_\odot$. 2) W72 (solid): ^{56}Ni is located at the intermediate region of M_r = 0.28 - 0.86 $M_\odot$ as in the original W7. 3) W73 (dash-dotted): ^{56}Ni is distributed in the outermost layers of M_r = 0.80 - 1.38 $M_\odot$. The optical opacity of 0.1 cm^2 g^{-1} and the γ-ray opacity of 0.03 cm^2 g^{-1} are assumed.

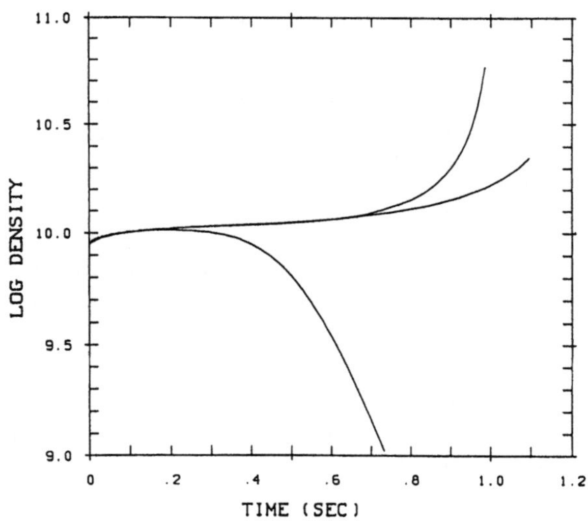

Fig. 6: Change in the central density of the C+O white dwarfs following the propagation of the conductive carbon deflagration wave in the initially solid core. Three cases with $v_{\rm def}/v_s$ = 0.05, 0.03, and 0.01 are shown and the latter two undergo collapse.

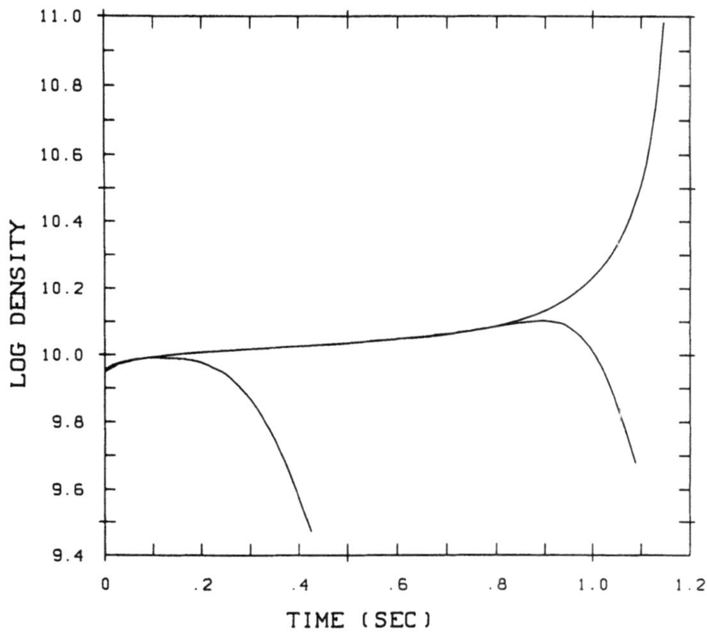

Fig. 7: Same as Fig. 6 but for the convective Ne/O deflagration wave in the O+Ne+Mg white dwarf for three cases with ℓ/H_p = 1.4, 1.0, and 0.7. For the slowest case of ℓ/H_p = 0.7, the white dwarf undergoes collapse.

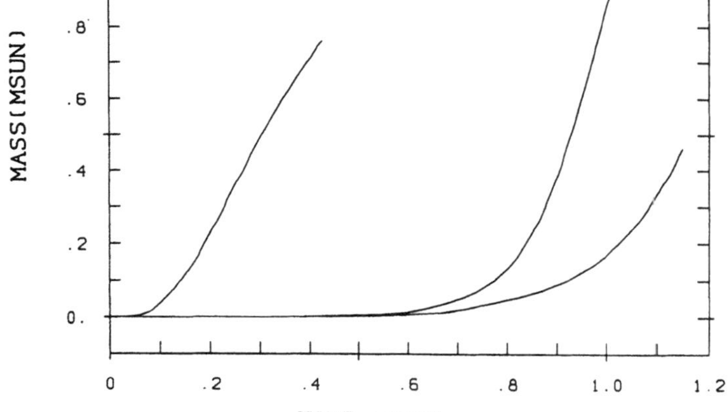

Fig. 8: Propagation of the convective Ne/O deflagration wave in the O+Ne+Mg white dwarf. The location of the deflagration front is plotted as a function of time for three cases with ℓ/H_p = 1.4, 1.0, and 0.7.

References

1. Barkat, Z. 1990, in this volume.
2. Barrat, J.L., Hansen, J.P., and Mochkovitch, R. 1988, *Astr. Ap.*, **199**, L15.
3. Branch, D. 1988, in *IAU Colloquium 108, Atmospheric Diagnostics of Stellar Evolution*, ed. K. Nomoto, *Lecture Notes in Physics*, **305**, 281.
4. Branch, D., and Nomoto, K. 1986, *Astr. Ap.*, **164**, L13.
5. Branch, D., Drucker, W., and Jeffery, D.J. 1988, *Ap. J. (Letters)*, **330**, L117.
6. Canal, R. 1990, in this volume.
7. Canal, R., Isern, J., and Lopez, R. 1988, *Ap. J. (Letters)*, **330**, L113.
8. Ebisuzaki, T., Shigeyama, T., and Nomoto, K. 1989, *Ap. J. (Letters)*, **344**, L65.
9. Ensman, L., and Woosley, S.E. 1988, *Ap. J.*, **333**, 754.
10. Filippenko, A.V., and Sargent, W.L.W. 1989, *Ap. J. (Letters)*, **345**, L43.
11. Graham, J. 1987, *Ap. J.*, **315**, 588.
12. Graham, J. 1987, *Ap. J. (Letters)*, **318**, L47.
13. Graham, J. *et al.* 1986, *M. N. R. A. S.*, **218**, 93.
14. Harkness, R.P. et al.: 1987, *Ap. J.*, **317**, 355.
15. Iben, I.Jr. 1982, *Ap. J.*, **253**, 248.
16. Ichimaru, S., Iyetomi, H., and Ogata, S. 1988, *Ap. J. (Letters)*, **334**, L17.
17. Leibundgut, B. 1988, Ph.D. thesis.
18. Miyaji, S., and Nomoto, K. 1987, *Ap. J.*, **318**, 307.
19. Nomoto, K. 1982, *Ap. J.*, **253**, 798.
20. Nomoto, K. 1984, *Ap. J.*, **277**, 791.
21. Nomoto, K. 1986a, *Ann. NY Acad. Sci.*, **470**, 294.
22. Nomoto, K. 1986b, *Prog. Part. Nucl. Phys.*, **17**, 249.
23. Nomoto, K. 1987, in The Origin and Evolution of Neutron Stars, IAU Symp. 125, ed. D.J. Helfand and J.-H. Huang (D. Reidel), p. 281.
24. Nomoto, K. 1988, in Proc. 13th Texas Symposium on Relativistic Astrophysics, ed. M. Ulmer (World Scientific), p. 519.
25. Nomoto, K., Thielemann, F.-K., and Yokoi, K. 1984, *Ap. J.*, **286**, 644.
26. Panagia, N.: 1987, in *High Energy Phenomena Around Collapsed Stars*, ed. F. Pacini (D. Reidel), p. 33.
27. Phillips, M.M. *et al.* 1987, *P.A.S.P.*, **99**, 592.
28. Richtler, T., and Sadler, E.M. 1983, *Astr. Ap.*, **128**, L3.
29. Schaeffer, R., Casse, M., and Cahen, S. 1987, *Ap. J.*, **316**, L31.
30. Shigeyama, T., and Nomoto, K. 1989, *Ap. J.*, submitted.
31. Starrfield, S. 1990, in this volume.
32. Thielemann, F.-K., Nomoto, K., and Yokoi, K. 1986, *Astr. Ap.*, **158**, 17.
33. Wheeler, J.C., and Levreault, R. 1985, *Ap. J. (Letters)*, **294**, L17.
34. Wheeler, J.C., and Harkness, R. 1990a, in *Supernovae*, ed. A. Petschek (Springer-Verlag), in press.
35. Williams, R.E. *et al.* 1985, *M. N. R. A. S.*, **212**, 753.
36. Woosley, S.E., and Weaver, T.A. 1986, in *IAU Colloquium 89, Radiation Transport and Hydrodynamics*, ed. D. Mihalas and K.H. Winkler, *Lecture Notes in Physics*, **255**, 91.

Expansion Opacity, Type Ia's and the Question of Collapse vs Thermonuclear

Stirling A. Colgate

It is a difficult question whether Type Ia supernova are caused by collapse to a neutron star or a thermonuclear explosion. The density required to ignite carbon burning is only slightly less than that at which collapse occurs due to neutrino emission. In addition there is the question of the URCA shells that may stabilize carbon burning. Finally, there is an argument that a deflagration may turn into a detonation (Woosley 1989). The consequence of a detonation is that the strong shock causes almost the entire star to be synthesized to ^{56}Ni, which consequently overproduces ^{56}Fe and greatly under produces the lower atomic number elements necessary for producing the observed spectrum. On the other than, for the same reason the collapse model must eject the envelope slowly by the hot bubble (Colgate 1989) or overproduce iron relative to lighter elements. Despite this uncertainty, the thermonuclear model has gained credence as the "standard" model (Woosley and Weaver 1986). The principal calculation of the thermonuclear model motivating this perception is that of the light curve. The calculations of the explosion of a $1.4M_\odot$ white dwarf performed using a constant opacity, $\kappa = 0.1cm^2\ g^{-1}$ result in a light curve very close to the observed one. "How could the same kinetic energy in roughly one-third the mass – typical of collapse models ($1M_\odot$ neutron star, $0.4M_\odot$ ejected) result in the same light curve?" The purpose of this note is to show how nearly the same light curve results regardless of the mass or energy. The point is that when the opacity of the expanding nebula is corrected to first order for the expansion opacity (Karp et al. 1977), the optical thickness of the nebula remains nearly constant, contrary to intuition. This means that the simple observation of light maximum must be interpreted as a combination of the exponentially decaying energy source and the extremely complex process of line scattering where a statistically few large gaps between $\sim 10^5$ lines determine the Rosseland mean opacity. Colgate, Petschek and Kriese (1980) and Colgate and Petschek (1980a) have argued that the way to measure the thickness and hence, mass of the SNIa's is by the interpretation of the late light curve, 56-day half life, as due to the escape of the positrons of the ^{56}Co decay. Here "opacity" is due to the less complex and experimentally measured scattering and energy loss of high energy electrons (positrons). In this case the measured expansion velocity from the Doppler shift of the lines and the modification of the 77-day decay of ^{56}Ni to the 56-day observed leads to an interpreted ejected mass of $0.4M_\odot$ rather than the thermonuclear model where $1.41M_{\odot}$ is ejected.

Finally, we point out that this complex behavior of the expansion opacity in the interpretation of the light curves of supernovae is fortunate because recently Kirshner (1989) has published three early light curves of SNIb's that nicely overlap and show a rise time to maximum in 15 days. There is one observation 4^m below (2.6% of) maximum at 14 days before maximum so that there is small error in the estimate of the time (15d) to light maximum. On the other hand, since the "standard model" predicts that the presupernova mass of Type Ib's is 6 to 7 $M_{\odot}$, much larger than Type Ia's, and therefore the time to light maximum at constant opacity should be much larger contrary to observation. The time to light maximum should scale (at constant total energy) as $M^{3/4to1}$. Hence, one would therefore expect the time to light maximum of Type Ia's to be 3 to 5 days or a factor of at least 3 less than many observations. Hence, we need some effect that changes the opacity with mass, i.e. the expansion opacity, to explain Type Ia's versus 1b's regardless of collapse of thermonuclear.

This note will first discuss the effect of the expansion opacity on the light curve. Then the logic of the positron transparency modification of the ^{56}Co (77d) decay in Type Ia's is applied to the corresponding gamma ray transparency and resulting modification of the ^{56}Co (77d) decay in SN1987a.

Expansion Opacity

Karp et al. (1977) showed how the opacity of expanding media was different from static atmospheres due to the monotonic redshift of every photon, sometimes called "tired light." This redshift allows the possibility of a photon to scatter from a sequence of atomic states before Compton scattering from a background of free electrons. Hence, the opacity is enhanced above Compton scattering if the redshift — or expansion — is large enough in the time of one Compton mean free path and the cross section of the states large enough. This requires that the fractional energy gap between states be less than the redshift of the photons in the Compton mean free path time. This fractional redshift is measured as 1/s where s, the expansion parameter is $\kappa_c \rho c t$. Here t is the time since the beginning of the expansion or explosion and κ_c the Compton opacity. The calculations of Karp et al. used a set of 2.6×10^5 lines for temperatures between 6000 K and 30,000 K. The opacity enhancement became significant for $s \sim 10^3$ or fractional redshift between scatterings of $\sim 10^{-3}$. This is a measure of the significant gaps in the Rosseland mean in the 2.6×10^5 lines. When s is small, $100 < s < 500$, and over a limited range of temperature (6000 to 15,000 K) then the expansion opacity scales like $1/s$. This implies that as the density gets lower, the increase in fractional redshift of a photon between Compton scatterings just compensates by increased line scattering the reduction in Compton opacity. A larger line list would broaden in temperature and range of s the conditions at which this approximation is valid. In addition, Karp et al. used a solar mixture depleted in hydrogen, but still with helium 10 times that of the heavier elements. The ejecta of SNIa's is likely to contain less helium. Recognizing these uncertainties and that the $1/s$ approximation applies over a very limited range of parameter space, the scaling of the optical thickness of the supernova nebula can be estimated.

If we express s in terms of the ejected mass of a supernova, M_{eject}, and kinetic energy, E_{51}, as 10^{51} ergs, time t_6 in 10^6 seconds and atomic weight A, per free electron, we obtain

$$s = 2.4 \times 10^3 M_{eject}^{5/2} E_{51}^{-3/2} t_6^{-2} A^{-1}. \tag{1}$$

Hence, for a mean atomic weight of once ionized He of 4, at time of light maximum of 14 days, $M_{eject} = 1.41$, and $E_{51} = 1.6$, characteristic of thermonuclear models, we obtain $s = 420$ just at the boundary where the $1/s$ approximation to the opacity applies. On the other hand for a Type Ib, of $7M_\odot$, 10^{51} ergs and $t = 15$ days, $s = 2.7 \times 10^4$ well within Compton scattering. The collapse model would give $s = 37$, below the region of the $1/s$ approximation. Nevertheless, if we use for the smaller models, $\kappa = \kappa_c(s_o/s)$, then the optical thickness of the expanding nebula, τ_s becomes

$$\tau_s = \kappa\rho R = \frac{\kappa_c \rho R s_o}{\kappa_c \rho c t} \tag{2}$$

or

$$\tau_s = s_o(v/c) = s_o((10/3)E/M)^{\frac{1}{2}}/c \tag{3}$$

where R and v are the outside radius and velocity respectively. Hence, in this rough approximation the optical thickness is independent of time provided $s < s_o$ and becomes even greater with smaller mass! Here $s_o \simeq 700$, that is a value of s where the expansion opacity becomes roughly 2 times Compton opacity. This means that simple intuition is too simple indeed. Hence, the time to light maximum is a convolution of the optical thickness and a decaying exponential heat source. Since $s_o \simeq 700$ and $v/c \simeq 1/25$, $\tau_s is \simeq 28$, the same as calculated for light maximum in the past (Colgate and McKee, 1979; Arnett 1982). However in this simple approximation, the collapse model will have a slightly longer time to light maximum than the thermonuclear. This will be difficult to calculate accurately because of the extreme sensitivity of the expansion opacity to composition, state conditions, and the completeness of the line list. By way of comparison, the optical thickness for Compton opacity alone and $A = 4$ becomes:

$$\tau_c = 26 M_{eject}^2 E_{51}^{-1} t_6^{-2}. \tag{4}$$

This is a factor of 2 smaller than the expansion opacity thickness for the two smaller mass models, but for Type Ib's the Compton opacity will dominate, and then the actual value of A becomes more critical. In general uncertainties in the composition and the line list will make this a continuing difficult problem so that it is difficult to use the light curve as a measure of the ejected mass.

The Gamma Ray Deposition Function Applied to 1987a

Even though SN1987a is a Type II supernova, the method for calculating the late light curve of Ia's can be applied to a modified late light curve of 1987a. Colgate, Petschek, and Kriese (1980a) calculated the gamma ray deposition from the ^{56}Ni decay in an expanding nebula of several models (mixed and unmixed) using an extensive Monte Carlo program. The emergent spectrum showed the early large x-ray flux expected from the gamma ray Compton scattering, which later was observed in 1987a. The numerical results were parameterized in a simple formula that was valid for any spherical expanding absorbing mass and an exponential path length radiation source. This generalization allowed the same formulation of deposition to be applied to positron energy deposition (from ^{56}Co decay) at a later time in the expansion because the peculiar combination of beta ray spectrum, multiple scattering and ionization loss leads (experimentally) to an exponential path length distribution. Hence, by scaling the effective range ($0.1\ g\ cm^2$) to that of the gamma rays ($35.5 g\ cm^{-2}$) the same deposition function could be used for the positrons as used earlier for the gamma rays. This assumed that any magnetic field was combed purely radially and/or that the initial magnetic field was less than 10^4 gauss.

When this deposition function is applied to an exponentially decaying source, a family of curves results of various differing decay rates from that of the original exponential, Fig. 1. The single parameter of each curve is the ratio of the expansion time, t_o, at which the deposition is one half (total thickness, $\tau = 1$) to the half life time t_1 (Colgate, Petschek, and Kriese 1980b). This ratio for our best fit model of SNIa's was 4.9 leading to an exponential decay time of the deposited fraction of the energy of 73% of the unmodified decay time or $0.73 \times 77d = 56.5d$. Transparency time is then $4.9 \times 77d = 377d$ at which $[\rho R]$ of the expanding nebula is $0.1\ g\ cm^{-2}$.

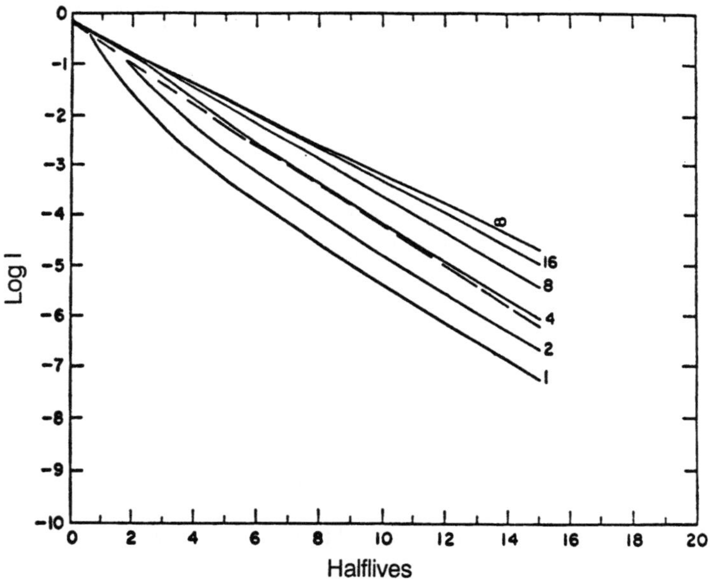

Fig. 1. We plot $D\ e^{-t\ln 2/t_1}$ for various values of t_0/t_1 where t_0 is the time at which deposition is $\frac{1}{2}$ (Colgate, Petschek, and Kriese 1980b)

For an expansion velocity of 1.4×10^9 $cm\ s^{-1}$, this corresponds to a mass of $0.42 M_\odot$. Hence, our claim that Type Ia's are collapse rather than thermonuclear.

The corresponding ratio for the gamma rays of 1987a where one half deposition of the gamma rays occurred at 660 days from Pinto, Woosley and Ensman (1988), is then $t_o/t_1 = 8.57$. From Fig. 1 the modified deposition function has a slope of 82.5% of the 77-day half life or 63.5 days. Figure 2 shows this drawn on the current bolometric light curve (Whitelock 1989). This early agreement is close enough to be a good argument for this interpretation of the 56-day decay observed in SNIa's. Of course later ^{57}Co and a possible pulsar will alter this agreement (Pinto et al. 1988).

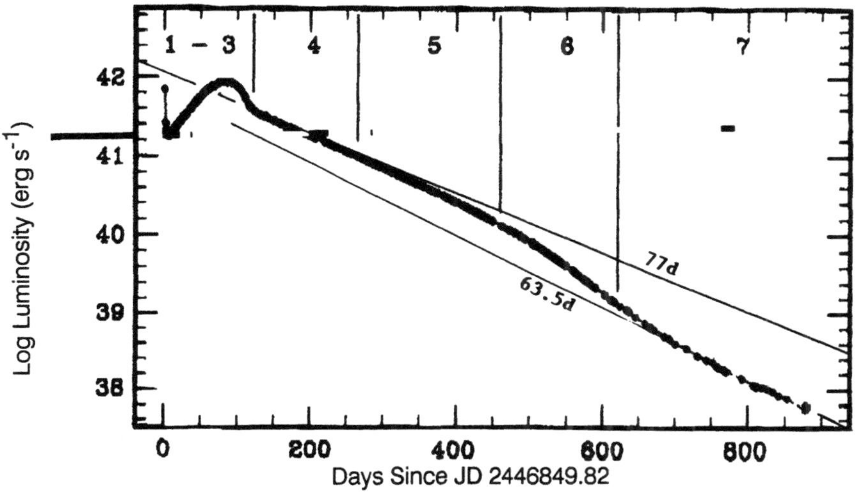

Fig. 2. Bolometric light curve from U to M photometry[6].
Courtesy of P. Whitelock, This Volume.

Since the light curve at this stage of calculation cannot be relied upon for a unique interpretation of the ejected mass because of the expansion opacity (Harkness (1989) expects to do such calculations), meanwhile it seems reasonable to use the positron range as a better measurement. The infrared emission was used by Axelrod (1980) to modify the late SNIa optical decay below the 77-day exponential, but this shows a "guillotine" effect or sharp cut off in time as opposed to the long exponential observed by Kirshner and Oke (1975) of over ten half lives.

Acknowledgements

I am indebted to discussions with Albert Petschek. This work was supported by the DOE.

References

1. S. E. Woosley: In Supernovae, ed. A. G. Petschek (Berlin: Springer-Verlag) (1989).
2. S. A. Colgate: Santa Cruz Workshop on Supernovae, ed. S. E. Woosley (1989).
3. S. A. Colgate, A. G. Petschek and J. T. Kriese: Ap. J. 239, L81 (1980a).
4. S. A. Colgate, A. G. Petschek, and J. R. Kriese: AIP Conf. Proc. 63 *Supernova Spectra*, ed. R. E. Meyerott, LaJolla Inst., LaJolla, CA, p.7 (1980b).
5. P. Whitelock: Santa Cruz Workshop on Supernovae, ed. S. E. Woosley (1989).
6. S. A. Colgate and A. G. Petschek: *Nature* 296, 804 (1982).
7. S. A. Colgate and C. McKee: Ap. J. 157, 623 (1969).
8. A. H. Karp, G. Lasher, K. L. Chan, and E. E. Salpeter: Ap. J. 214, 161 (1977).
9. R. V. Kirshner: In Supernovae, ed. A. G. Petschek (Berlin: Springer-Verlag) (1989).
10. S. E. Woosley and T. A. Weaver: Ann. Rev. Astron. and Ap. 24, 205 (1986).
11. W. D. Arnett: Ap. J. 253, 785 (1982).
12. P. A. Pinto, S. E. Woosley and L. M. Ensman: Ap. J. 331, L101 (1988).
13. T. S. Axelrod: Thesis, Univ. Calif., Berkeley and T. A. Weaver, T. S. Axelrod, and S. E. Woosley: In "Type I Supernova" ed. J. C. Wheeler, (Austin: Univ. of Texas Press) p. 113 (1980).
14. R. V. Kirshner and J. B. Oke: Ap. J. 200, 574 (1975).
15. Harkness, private communication (1989)

SN Ia Models by Coalescence of Two Carbon-Oxygen White Dwarfs

Robert Mochkovitch & Mario Livio

White dwarf coalescence models for type Ia supernovae (SNe Ia) have been recently proposed [1,2] as an alternative to the standard scenario in which a white dwarf accretes hydrogen rich material from a companion star. The need for new models appeared after it was realized that it might be difficult to increase the white dwarf mass up to the Chandrasekhar limit by accretion, at least in a sufficiently large number of systems to account for the SN Ia rate in the Galaxy. Indeed, mass loss during hydrogen and helium flashes and off-center helium detonation considerably restrict the range of values of the accretion rate, white dwarf and companion masses for which central carbon ignition is possible. Another difficulty is the risk of contamination of the spectrum by stripped hydrogen from the companion. Even if a small window of parameters giving a SN Ia explosion may remain (the theoretical modelling of all the processes involved in the presupernova evolution is too uncertain to draw definite conclusions) the search for a new class of progenitors may turn to be more promising.

I. Close Binary White Dwarfs

Very close white dwarf pairs result from the evolution of close binary systems of stars of intermediate mass after one or two episodes of non conservative mass transfer. If the two white dwarfs are formed with a separation of less than about 2 $R_\odot$, they will merge within a Hubble time due to gravitational radiation losses. For two carbon-oxygen (C-O) white dwarfs, the total mass of the system will be generally larger than the Chandrasekhar limit and some kind of violent outcome seems unescapable. Whether it will fit the observable properties of a SN Ia is however still a matter of discussion. Another problem concerns the statistics of the binary white dwarf (BWD) population. WEBBINK [1] and IBEN and TUTUKOV [2] estimated the birthrate of C-O/C-O pairs to be comparable to the SN Ia rate in the Galaxy. This was confirmed in a recent study by TORNAMBE [3] as long as the efficiency of the common envelope phase to reduce the orbital separation is not too large. The long timescale for merging by the emission of gravitational radiation then implies that BWDs should be quite common in the Galaxy. Several groups [4,5,6] have tried to detect them but the two objects which have been found until now, L 870–2 [4] and WD 0957 –666 [5] have periods larger than one day which corresponds to a timescale for merging much larger than 10^{10} years. Moreover, ROBINSON and SHAFTER [6] did not detect any BWD with a period between 30 *s* and 3 *h* in a sample of 44 catalogued white dwarfs. They concluded that the resulting upper limit for the space density of BWDs in the Galaxy is too small to account for the SN Ia rate. The origin of the present disagreement between the theoretical predictions and the observational status is unclear. We shall not discuss this point and rather focus on the evolution which follows the coalescence of the two white dwarfs. This problem will naturally remain of interest even if BWDs are not SN Ia progenitors (if at least one BWD able to merge within a Hubble time is finally discovered!).

II. Post-Coalescence Models

When the less massive white dwarf overfills its Roche lobe coalescence, i.e. dynamical merging occurs if its mass is not too different from that of the primary ($M_2 > 0.5$ $M_\odot$ for $M_1 = 1$ $M_\odot$). In a few orbital periods the secondary is tidally disrupted and forms a thick disk orbiting around the primary, as illustrated by the beautiful 3D hydrodynamical calculations of BENZ et al. [7]. We have used the self-consistent field (SCF) method developed by OSTRIKER and MARK [8] to compute the structure of the resulting configuration. Since we do not follow the evolution during the merging process, some parametrization for the final distribution of angular momentum in the disk must be assumed. We also constraint the pressure to be a function of the density only and use the results of [7] as a guide for the parametrization. The SCF method therefore relies on inputs from the 3D calculations but has the advantage of being more accurate in the outer parts of the disk where the density contrast relative to the center becomes large.

The post coalescence configuration is computed with conservation of the total mass, angular momentum and energy (less than 3% of the total mass is ejected in the case studied by [7]). The distribution of angular momentum in the disk is given by

$$j(m) = j_0 \frac{(m - M_1)}{(M_1 + M_2 + b(m - M_1))} \quad \text{for } m > M_1 \text{ and } j(m) = 0 \text{ for } m < M_1 \,, \qquad (1)$$

where M_1 and M_2 are the original white dwarfs masses ($M_1 > M_2$) and m is the mass within axial cylinders; j_0 is determined by the value of the total angular momentum and b is a parameter. The pressure $P_e(\rho)$ of degenerate electrons is used until a transition density ρ_{tr}, below which the thermal pressure P_{th} becomes important. At $\rho < \rho_{th}$ a polytropic relation $P = K\rho^\gamma$ is adopted and $P_{th} = P - P_e$.

We computed the structure of the configuration resulting from the coalescence process considered in [7]. The central white dwarf of mass 1.2 $M_\odot$ is surrounded by a disk of 0.9 $M_\odot$. The transition density is $\rho_{tr} \approx 10^7$ g.cm^{-3}, $\gamma \approx 1.4$ and $b = 9.5$. The model isodensities are represented in Fig.1. The central density, polar and equatorial radii are $\rho_c = 1.75\ 10^8$ g.cm^{-3}, $R_p = 3500$ km and $R_e = 26500$ km. As already discussed by MOCHKOVITCH and LIVIO [9] for a different model and in agreement with [7] we find that the temperature at the white dwarf–disk boundary is 5 – 10 10^8 K, i.e. possibly larger than the carbon ignition temperature. Off-center carbon burning would strongly affect the subsequent evolution which also critically depends on the viscosity in the disk.

III. Post-Coalescence Evolution

Let us first suppose that carbon was not ignited during merging or that carbon burning was rapidly quenched, for example by material expansion. The disk evolution will then be determined by the balance between heating by viscous dissipation and cooling by neutrino and radiation losses. If molecular viscosity is the only way to carry angular momentum in the disk, heating will be negligible and cooling will occur on a Kelvin-Helmholtz timescale (a few 10^6 years). The structure will become completely degenerate and there will be a slow increase in central density as rotational support is

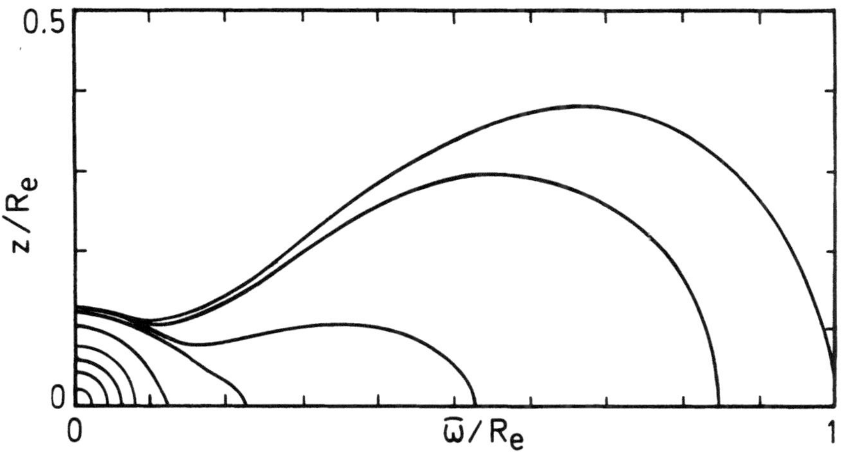

Figure 1. Isodensity contours (ρ/ρ_c = 0.8, 0.5, 0.2, 0.1, 10^{-2}, 3 10^{-3}, 10^{-3}, 10^{-4} and 0) for a model with a central 1.2 $M_\odot$ white dwarf and a 0.9 $M_\odot$ disk.

progressively removed by transport of angular momentum (with a very long timescale $\sim 10^{10}$ years). Finally, central carbon ignition will take place and one may expect that most of the results of the carbon deflagration model [10] will be recovered. However, the high shear in the disk (even if it satisfies the Rayleigh criterion for stability) is likely to induce turbulence as indicated by the value of the Richardson number $Ri < 1/4$ in most of the disk ($Ri > 1/4$ is a sufficient condition for stability; $Ri < 1/4$ is a strong indication for turbulence even if it does not imply it). The α prescription for the viscosity then gives values which can be larger than 10^{12} cm^2.s^{-1}, producing a dissipation rate of 10^{13} erg.g^{-1}.s^{-1} or more in the boundary layer. This certainly leads to carbon ignition, even if it did not occur during the merging process itself.

As a result of carbon burning the object becomes more spherical as its radius increases (the "eccentricity" $e \propto R^{-1/2}$) and eventually adopts a giant-like structure with a carbon burning shell as in the models computed by KAWAI et al. [11]. The luminosity of such models is very close to the Eddington limit, with therefore the possibility of extensive mass loss. Another complication would be the inward propagation of a carbon burning front, incinerating the central white dwarf to O-Ne-Mg composition [12]. One can only speculate on the final outcome of the evolution. Will the central white dwarf finally collapse or explode? Or will mass loss be sufficiently important to reduce the total mass below the Chandrasekhar limit?

In conclusion, white dwarf coalescence still appears far from being a fully conclusive alternative to the standard model for SNe Ia. On the observational side, no BWD able to merge within a Hubble time has been discovered yet. In the theoretical models, it seems difficult to prevent off-center carbon ignition during coalescence or later by viscous heating in the disk. Nuclear burning in semi to non-degenerate material cannot directly produce a SN Ia and the evolution following carbon ignition is extremely uncertain. Clearly, much work will be needed either to confirm or eliminate the coalescence scenario.

References

1. R. F. Webbink: Ap. J. 277, 355 (1984).
2. I. Jr. Iben, A. V. Tutukov: Ap. J. (Suppl.) 54, 335 (1984).
3. A. Tornambe': M.N.R.A.S. 239, 771 (1989).
4. R. A. Saffer, J. Liebert, E. W. Olszewski: Ap. J. 334, 947 (1988).
5. A. Bragaglia, L. Greggio, A. Renzini, S. D'Odorico: Messenger 52, 35 (1988).
6. E. L. Robinson, A. W. Shafter: Ap. J. 322, 296 (1987).
7. W. Benz, R. L. Bowers, A. G. W. Cameron, W. H. Press: preprint (1989).
8. J. P. Ostriker, J. W. K. Mark: Ap. J. 151, 1075 (1968).
9. R. Mochkovitch, M. Livio: Astr. Ap. 202, 211 (1989).
10. K. Nomoto, F. K. Thielemann, K. Yokoi: Ap. J. 286, 644 (1984).
11. Y. Kawai, H. Saio, K. Nomoto: Ap. J. 328, 207 (1988).
12. K. Nomoto, I. Jr. Iben: Ap. J. 297, 531 (1985).

Collapse or Explosion of C+O White Dwarfs

J. Isern, R. Canal, D. Garcia, & J. Labay

The existence and the orbital parameters of low-mass binary x-ray sources suggest that a fraction at least of neutron stars have been produced by the accretion induced collapse (AIC) of a white dwarf (TAAM and VAN DEN HEUVEL [1]). Two types of candidates have been proposed thus far: carbon-oxygen white dwarfs (CANAL and SCHATZMANN [2]; CANAL and ISERN [3]; CANAL, ISERN and LABAY [4]), and oxygen-neon-magnesium white dwarfs (MIYAJI et al [5]; MIYAJI and NOMOTO [6]).

In the firt case, however, there is ample consensus that such stars are the progenitors of Type Ia supernovae. Current models (NOMOTO, THIELEMAN and YOKOI [7]; SUTHERLAND and WHEELER [8]; WOOSLEY and WEAVER [9]) do not predict the formation of any compact remmnant. A possible way out of this dilemma comes from consideration that white dwarfs are, in general, at least partially solid objects. Solidification has two main consequences: 1) Nuclear reaccions start in the pycnonuclear regime, this delaying the onset of thermonuclear runaway up to high densities ($\rho \simeq 10^{10}$ g/cm^3) (HERNANZ *et al* [10]) and 2) The propagation velocity of a burning front inside a solid likely corresponds to conduction only, that implying lower velocities ($v_{BF} \simeq 10^{-2} c_s$; c_s local sound speed) than in a fluid since Rayleigh-Taylor instabilities cannot develop, that allowing the energy loss by electron captures behind the front to win (CANAL & ISERN [3]) over the energy release by propagation of the burnig front.

It has been shown by ISERN et al [11] and by HERNANZ *et al*

[10] that central termonuclear runaway can indeed be delayed up to densities $\rho_c \geq 10^{10}$g cm^{-3} in mass-acreting C+O white dwarfs, provided that they are initially cold and massive and the accretion rates are either $\dot{M} \geq 10^{-7}$ $M_\odot$ yr^{-1} or $\dot{M} \geq 10^{-9}$ $M_\odot$ yr^{-1}. Central densities for which runaway happens inside a central solid core span the range 9.5×10^9g cm^{-3} $\geq \rho_c \geq 1.5 \times 10^{10}$g cm^{-3}.

We have performed a parameter study of the dynamical evolution of a C+O white dwarf for a central ignition starting at $\rho_c = 1.13 \times 10^{10}$g cm^{-3}. Burning front velocities are 0.1, 0.03, 0.01, and 0.005 times c_s. The last velocity roughly corresponds to the estimate of conductive front velocities by WOOSLEY and WEAVER [9]. Electron capture rates ere calculated by means of EPSTEIN and ARNETT's [12] expressions, recalibrated by comparison with the rates of FULLER, FOWLER, and NEWMAN [13] at selected points in (ρ, T, Y_e) space. We see that transition from explosion to collapse takes place between 0.03 and 0.01 times c_s. Calculations were stopped when $\rho \simeq 10^{11.5}$ g cm^{-3} (in the cases of increasing density), and the star was already contracting homologously in a hydrodynamic time scale. Collapse to nuclear densities is thus the only possible issue. In a second series of calculations we have taken into account the change in burning propagation velocity that must follow from the emergence of the burning front outside of the central solid core into fluid layers. In this case the size of this core plays an essential role. We adopted, for the conductive velocities in the solid core, those derived from WOOSLEY and WEAVER's [9] expression. In the fluid layers , we switched to a turbulent front velocity given by SUTHERLAND and WHEELER's [8] prescription for the propagation of Rayleigh-Taylor instabilities. The solid core's size, however, does not stay at its initial value at runaway but changes in the course of burning propagation. A major effect in this is the absorption of neutrinos coming from the electron captures in the central, incinerated regions. The rates of energy deposition in the layers surrounding the central incinerated core are from CHECHETKIN *et al* [14], and IBAÑEZ *et al* [15].

Taking into account the latent heat, this typically gives ≃ 1 s for survival of the solid layers. That, in turn, poses the most severe lower (upper) limits to central ignition density (burning propagation velocity). For a model with ρ_{ign} = 1.13 x 10^{10} g cm^{-3} and initial solid core of $M_{core} \simeq$ 0.6 M_{star}, and the burnig front velocity prescriptions given above, collapse also ensues, but the behaviour is quite diferent: whihout inclusion of the neutrinos, the incinerated mass when $\rho_c = 10^4$ g cm^{-3} would be $M_{inc} \simeq$ 0.07 M_{star} while with neutrinos included it is $M_{inc} \simeq$ 0.86 M_{star} when reaching the same density. The larger mass of the incinerated core in the second case partially arises from switching from conductive to turbulent (hidrodynamic) velocities when reaching the edge of the solid core, but to a much larger extent from ignition of overlying material by neutrino heating when the core has already entered dynamic contraction.

We thus conclude that AIC of C+O white dwarfs is possible but only for a limited range of ignition densities and solid core sizes. Minimum solid core size at ignition should be ≃ 0.04 M_{star} that depending on the still uncertain values of conductive velocities. This, in turn, sets narrow limits to initial mass and temperature (at the star on accretion) and to mass accretion rates.

References

1. R.E.Taam and E.P.J. van den Heuvel: Ap.J. 305, 235 (1986)
2. R.Canal and E.Schatzmann: Astr. Ap.46, 229 (1976)
3. R.Canal and J.Isern: in White Dwarfs and Variable Degenerate Stars, ed. by H.M. van Horn and V.Weidemann (Rochester: Univ. Rochester Press 1979), p.53
4. R.Canal, J.Isern and J.Labay: Ap.J. (Letters) 241, L33 (1980)
5. S.Miyaji, K.Nomoto, K.Yokoi and D.Sugimoto: Publ. Astr. Soc. Japan 32, 303 (1980)
6. S.Miyaji and K.Nomoto: Ap.J. 318, 307 (1987)
7. K.Nomoto, F.Thielemann and K.Yokoi: Ap.J. 286, 644 (1984)
8. P.G.Sutherland and J.C.Wheeler: Ap.J. 280, 282 (1984)
9. S.E.Woosley and T.A.Weaver: Ann. Rev. Astr. Ap. 24, 205 (1986)
10. M.Hernanz, J.Isern, R.Canal, J.Labay and R.Mochkovitch:

Ap.J. 324, 331 (1988)

11. J.Isern, J.Labay, R.Canal and M.Hernanz: Ap.J. 273, 320 (1983)
12. R.Epstein and D.Arnett: Ap.J. 201, 202 (1975)
13. G.M.Fuller, W.F.H.Fowler and M.J.Newman: Ap.J.Suppl. 48, 279 (1982)
14. V.M.Chechetkin, Gershtein S.S., Imshennik V.S., Ivanova L.N., M.Yu. Khlopov: Ap. Space Sci. 67, 61 (1980)
15. J.M.Ibañez, J.A.Miralles, R.Canal, J.Isern and J.Labay: In preparation (1989)

Searching for Double Degenerates: One Out of Fifty

Angela Bragaglia, Laura Greggio, Alvio Renzini, & Sandro D'Odorico

1 – Observations and Results

Motivated by the theoretical suggestion that the precursors of type I supernovae may be binary white dwarfs (WD), also called double degenerates (DD), in 1984 we started a systematic search for binarity among spectroscopically confirmed WDs taken from the existing catalogues. If DDs are Type I SN precursors, we should find some systems which are close enough to merge in less than one Hubble time, i.e. with separations less than 3 $R_\odot$, implying orbital periods less than $\sim 10^h$ and orbital velocities $\gtrsim 300$ kms^{-1} (IBEN & TUTUKOV [1]). This holds, however, only if the bulk of DD are not initially formed much closer than 3 $R_\odot$.

In four observing runs at the ESO 3.6m telescope we have then obtained 155 useful spectra for a total of 49 WDs, all but one of the DA variety. This choice was primarily dictated by the larger number of suitable lines in the spectra of DAs that could be used to detect evidence of binarity by comparing 2÷4 spectra of the same WD with a cross-correlation algorithm, and then looking for a radial velocity variation Δv_r or for a change in line profile, if any. Apart from this, no other selection criterium for the target WDs was imposed (if one excludes apparent brightness, of course). So our sample was not biased towards WDs having *a priori* a larger chance of being double (e.g. exceptionally low gravity).

Details of the observational and reduction techniques are given elsewhere (BRAGAGLIA *et al.* [2]). The 1 σ error of our Δv_r determinations is ~ 20 km/s in the best cases; the whole set of Δv_r's (excluding 3 special cases) is well fitted by a gaussian with $\sigma \simeq 40$ km/s. Therefore, even in the worst cases we should have been largely able to detect a Δv_r as high as the minimum required for a type I SN precursor (a 15 σ effect!).

Till now we have found only one DD system (WD0957-666) and two WD + RD pairs (WD0034-211 and WD0419-487). We have also 2 dubious cases ($\Delta v_r \simeq 2\ \sigma$) that we intend to reobserve.

For the confirmed DD we have till now obtained 17 spectra over a baseline of 15 months. Its radial velocity amplitude is 220 km/s, or $\sim 5\ \sigma$. Of course we are trying to determine the period of this object. The available data indicate $P \simeq 1^d.17$, but alternative periods as long as $1^d.9$ cannot be excluded, as well as the possibility of the unseen companion being a red or brown dwarf. We plan to obtain several new spectra of this object during our next run (January 1990), so as to remove any residual ambiguity. We note that WD0957-666 is the object with the smallest gravity for which KOESTER *et al.* [3] have derived a mass: they found only $0.13 M_\odot$, while GUSEINOV *et al.* [4] obtained 0.16 $M_\odot$. Both groups assumed a mass-gravity relation for completely degenerate configurations. While this may actually result in a sizable underestimate of the mass (cf. Fig. 4 in IBEN & TUTUKOV [5]), still the extreme properties of this object were suggestive of a binary nature that we have now confirmed. Yet, we emphasize that the object was not included on purpose in the sample, and therefore it can be used for statistical considerations. Instead, in the statistics we will not consider WD0153-052 (SAFFER *et al.* [6]) as this latter DD was indeed observed specifically for its *special* properties.

2 – Discussion

Our result can be compared and combined with that of ROBINSON & SHAFTER ([7]; RS). They examined 44 WDs (40 DAs and 4 non-DAs). With their experimental set-up they could detect only short-period binaries (with $30^s < P < 3^h$); they found none. We are sensitive even to much longer periods: from about 20 minutes (twice our typical exposure time) to several days; very short period systems that we could miss must be very rare anyway, because of very rapid gravitational wave radiation shrinking of the orbit.
On the base of their data RS calculated an upper limit for the frequency of DDs among WDs:

$$\nu < 0.05 \qquad \text{(90\% probability)} \tag{1}$$

and the corresponding upper limit of field DDs:

$$N_{\rm DD} < 3.0 \cdot 10^{-5}{\rm pc}^{-3} \qquad \text{(90 \% probability).} \tag{2}$$

We can now use a larger set of WDs by co-adding the two samples: we have a statistics on 89 WDs (4 are in common between the two sets) and the results are:

$$\nu < 0.04 \qquad \text{(90\% probability)} \tag{3}$$

$$N_{\rm DD} < 2.4 \cdot 10^{-5}{\rm pc}^{-3} \qquad \text{(90 \% probability)} \tag{4}$$

If instead we use only our data (49 WDs) we obtain:

$$\nu < 0.07; \qquad N_{\rm DD} < 4.3 \cdot 10^{-5}{\rm pc}^{-3} \qquad \text{(90 \% probability)} \tag{5}$$

The *a posteriori* probability of finding a DD with $P < 3^h$ is less than 4 % and with $P <$ few days is less than 7 % at the 90 % confidence level .

Is this consistent with the hypothesis that DDs are Type I SN progenitors? RS argue that the minimum space density of DD required to produce the observed SNI rate is $2.8 \cdot 10^{-5}{\rm pc}^{-3}$, just about what they find for the *upper limit* to $N_{\rm DD}$, so *all* DDs should produce a SNI. This seems impossible, since most DDs are not enough massive. They conclude that DDs *are not* the dominant progenitors of Type I SNe unless they form with a very short period and quickly merge.
Adding our data we *reinforce* this conclusion. The situation is only marginally more favorable if we choose to consider only our data, on the fact that RS's method is sensible only to very short periods.

3 – DAs or not DAs?

At this point it looks hard to escape the conclusion that type I SN precursors are not DD binary systems contained in catalogues of DA white dwarfs. This negative result has prompted us to give a closer look to the DD scenario, and indeed another possibility yet remains to be observationally tested: in fact this negative result does not exclude the possibility that non-DAs may do it! Only 5 out of 89 observed WDs are non-DAs and therefore very little statistics is available for them at this stage. Are there any plausible reasons why SNI precursors should in case be non-DA's? We now believe that actually there are rather convincing theoretical arguments supporting this notion. Keeping in mind that in case we presently observe the WD produced by the secondary binary component (the primary WD is most likely fainter), we distinguish four main cases concerning the initial mass of the secondary – M_2 – and its evolutionary phase at its final Roche-lobe contact.

Case 1 : $M_2 \lesssim M_{\rm HeF} \simeq 2.2 M_\odot$, contact of type **B**, i.e. during the shell hydrogen-burning phase of the secondary. Objects of this kind have a degenerate helium core at Roche-lobe contact; removing most of the envelope by Roche-lobe overflow will result in a helium WD with $M_{WD} \lesssim 0.5 M_{\odot}$ and still some hydrogen at the surface (IBEN & TUTUKOV [5]), i.e. a DA white dwarf.

Case 2 : $M_2 \lesssim M_{\rm HeF} \simeq 2.2 M_{\odot}$, contact of type **C**, i.e. during the AGB phase of the secondary. The contact can take place either during the early AGB phase (prior to the first thermal pulse) or just during one thermal pulse, when the star significantly expands in response to the pulse. In both cases Roche-lobe overflow leaves a star whose luminosity is powered by helium burning. This should ensure that the residual hydrogen-rich envelope (with an initial mass $\lesssim 10^{-3} M_{\odot}$) can be lost during the bright Post-AGB phase, thereby leaving a bare CO-WD, with a pure helium atmosphere (RENZINI [8]; IBEN [9]), i.e. a non-DA WD with $M_{WD} \gtrsim 0.5 M_{\odot}$.

Case 3 : $M_2 \gtrsim M_{\rm HeF} \simeq 2.2 M_{\odot}$, contact of type **B**. After most envelope is removed by Roche-lobe overflow, the NON-degenerate helium core is still able to ignite helium, thereby producing an almost bare core helium-burning star (envelope mass $\sim 10^{-3} M_{\odot}$, IBEN & TUTUKOV [10]). Even a modest mass loss rate during the helium burning phases is then able of removing the last vestiges of the hydrogen envelope, thus ultimately leaving a CO-WD ($M_{WD} \gtrsim 0.5 M_{\odot}$) without hydrogen layers, i.e. a non-DA dwarf. Finally,

Case 4 : $M_2 \gtrsim M_{\rm HeF} \simeq 2.2 M_{\odot}$, contact of type **C**. The case is practically very similar to Case 2, and therefore a non-DA C-O dwarf is left.

We conclude that in all cases but 1. non-DA WDs are most likely produced. However, in this first case a helium WD is formed, and therefore if SNIs are the product of the merging of two CO-WDs, then these should be searched ONLY among non-DAs. Therefore, at this stage we intend to discontinue our search of DDs among DAs, and will now shift only to non-DAs, in order to have a secure statistics also for this kind of degenerate dwarfs. In closing, we note that the three binary WDs that we have found among DAs (one DD and two WD+RD pairs, cf. [2]), and also the DD observed by SAFFER *et al.* [6], are all probably the result of Case 1-like evolution, as indicated by the small mass of the observed WD ($M_{WD} < 0.5\ M_{\odot}$).

In any event, on the base of the above arguments we predict that the He-WDs formed in interacting binaries should all be of the DA type, while the overwhelming majority of CO-WDs formed in such systems should be of the non-DA type. This is a prediction now subject of observational check.

References

1. I. Iben, A.V. Tutukov: *Ap. J. Suppl.* **54**, 335 (1984).
2. A. Bragaglia, L. Greggio, A. Renzini, S. D'Odorico: *The Messenger* **52**, 35 (1988).
3. D. Koester, H. Schulz, V. Weidemann: *Astron. Astrophys.* **76**, 262 (1979).
4. O.H. Guseinov, H.I. Novruzova, Y.S. Rustamov: *Astrophys. Space Sc.* **96**, 1 (1983).
5. I. Iben, A.V. Tutukov: *Ap. J.* **311**, 742 (1986).
6. R.A. Saffer, J. Liebert, E.W. Olszewski: *Ap. J.* **334**, 947 (1988).
7. E.L. Robinson, A.W. Shafter: *Ap.J.* **322**, 296 (1987).
8. A. Renzini: in Stars and Star Systems, ed. B. Weserlund (Dordrecht, Reidel), p. 155 (1979)
9. I. Iben: *Ap. J.* **277**, 333 (1984).
10. I. Iben, A.V. Tutukov: *Ap. J. Suppl.* **58**, 661 (1985).

Neon Novae, Recurrent Novae, and Type I Supernovae

S. Starrfield, W. M. Sparks, J. W. Truran, & G. Shaviv

Over the past few years, we have been investigating the effects of accretion onto massive white dwarfs and its implications for their growth in mass toward the Chandrasekhar limit, in attempts to identify a possible relationship between SN I and novae. In our studies we have considered accretion at various mass accretion rates onto a variety of different white dwarf masses. We have found that there is a critical white dwarf mass above which a significant fraction of the accreted mass can remain on the white dwarf after the outburst. Below this value of the white dwarf mass, all of the accreted mass, plus core material dredged up into the envelope, is ejected as a result of the explosion. Our latest results include accretion and boundary layer heating produced by the infalling material. From these studies, we have identified some members of the class of recurrent novae, those involving a thermonuclear runaway, as the novae that are occurring on very massive white dwarfs and evolving toward a SN I explosion. One of the outgrowths of our UV studies of novae in outburst has been the identification of a class of novae which eject material that is very rich in the elements from oxygen to aluminum. We have shown that these outbursts occur on ONeMg white dwarfs, which are necessarily very massive white dwarfs.

The assumption commonly made is that a classical nova system is a close binary with one member a white dwarf and the other member a cooler star that fills its Roche lobe. Because it fills its lobe, there is a flow of gas through the inner Lagrangian point into the lobe of the white dwarf. The high angular momentum of the transferred material causes it to spiral into an accretion disk surrounding the white dwarf. Some (unknown) viscous process transfers mass inward and angular momentum outward through the disk, so that a fraction (also unknown) of the material lost by the secondary ultimately ends up on the white dwarf. The accreted layer grows in thickness and is heated by compression until it reaches a temperature that is high enough for thermonuclear burning of hydrogen to begin at the bottom. The simulations show that, if the material is degenerate, a thermonuclear runaway (hereafter: TNR) occurs and the temperature in the accreted envelope can grow to values exceeding 10^8K.

The further evolution of nuclear burning on the white dwarf now depends upon the mass and luminosity of the white dwarf, the rate of mass accretion, and the chemical composition of the reacting layer (Truran 1982; Starrfield 1989; Starrfield, Sparks, and Shaviv 1988; and references therein). Observations of novae ejecta also imply that there is mixing of core material into the accreted layer, so that the chemical composition of the material that is ultimately ejected by the outburst reflects a combination of core plus accreted material (Sparks *et al.* 1988).

It was proposed some years ago that accretion onto white dwarfs in close binary systems could lead to SN I (Truran and Cameron 1971; Whelan and Iben 1973). There have always been difficulties with including novae in this scenario, however, since they have been thought to eject most of their accreted mass. Unfortunately, although there have been other proposals for close binary progenitors of SN I, none of these proposals has yet received observational confirmation. It is the purpose of this paper to show that there is a class of nova-like events in which the white dwarf component is probably increasing in mass and is already close to the Chandrasekhar limit. We also note that the core composition of this class of novae has been identified.

I. RECURRENT NOVAE AND "NEON" NOVAE

Over the past few years, there has been intensive study of a class of novae that is referred to as "recurrent" because they have been *observed* to go through more than one outburst. They experience low amplitude outbursts which repeat on time scales of 20 to 50 years although the shortest known recurrence time is that of U Sco, which underwent outbursts in 1979 and 1987. Recurrent novae eject very small amounts of material, $\leq 10^{-7}$ $M_\odot$, which represents only a *small* fraction of the amount of envelope mass necessary to initiate a TNR on a massive white dwarf. Presumably, the rest of the accreted material remains on the white dwarf.

The short recurrence times for this kind of outburst demands that the mass of the white dwarf exceed $1.3M_\odot$ and the closer to $1.4M_\odot$ the better (Starrfield, Sparks, and Truran 1985; Starrfield, Sparks, and Shaviv 1988; Truran *et al.* 1988). The simulations also show that the accreted envelope is burned to helium very rapidly, ensuring that the entire accreted envelope does not have to be ejected in order for the system to return to quiescence. In addition, abundance analyses of the ejecta of recurrent novae indicate that the heavy element abundances are not enhanced relative to a solar mixture (c.f., Williams *et al.* 1981; Starrfield 1988). We interpret these results as implying that this class of novae is neither mixing core material into the accreted envelope nor ejecting all of the accreted material. Therefore, it seems likely that the mass of the white dwarf, *in recurrent novae systems*, is growing as a result of the nova outburst. We have identified U Sco and V394 CrA as two recurrent novae which contain white dwarfs already close to the Chandrasekhar limit (Starrfield, Sparks, and Shaviv 1988). Finally, because the observed recurrence time scales require high $\dot{M}$ and because most of the material remains on the white dwarf, we predict that their mass must be growing at a very high rate ($\dot{M} > 10^{-7}M_\odot yr^{-1}$).

It is also possible to speculate on the internal composition of the white dwarfs in these systems. The basis of this speculation is the UV studies of novae done with the **IUE** satellite (Starrfield 1988) which show that many novae are ejecting significant amounts of core material into space as a consequence of the outburst. These are the classical novae that show large enhancements of helium and heavier nuclei over a solar mixture (Truran and Livio 1986). In this context, we have been able to identify a sub-class of novae that must be occurring on ONeMg white dwarfs based upon abundance analyses of the ejected material. Currently, we have positive identifications for four such novae (V1370 Aql, V693 CrA, QU Vul, and Nova Vul 1987) and in all cases the intermediate elements are a major fraction of the ejecta.

Observations allow us to estimate what fraction of the ejected material is core material and they show that, in some cases, the fraction is large (Truran and Livio 1986; Sparks *et al.* 1988). In the most extreme case, that of V1370 Aql, the efficiency of this process is very high and it appears that most of the ejecta is core material.

The straightforward interpretation is that the white dwarfs in these nova systems are being eroded in mass as a result of the nova outburst. Therefore, ***novae that show large enrichments of intermediate mass nuclei in their ejecta cannot be the progenitors of SN-like events.*** We note that they must have been formed as high mass white dwarfs in order for the pre-white dwarf to survive carbon burning.

II. CONCLUSIONS

We draw the following conclusions:

(1) Those recurrent novae that are powered by thermonuclear runaways, such as U Sco and V394 CrA, must involve a very massive white dwarf, close to the Chandrasekhar limit. This is demanded by the short recurrence times characteristic of these systems and the fact that only massive degenerate dwarfs can accrete sufficient matter on the observed recurrence time-scales to initiate a runaway.

(2) The white dwarfs in recurrent nova systems appear to be growing in mass since the observed masses of the ejecta are less than the envelope masses necessary to trigger a TNR. This result is also consistent with the findings that the ejecta of recurrent novae show no evidence for high concentrations of heavy nuclei, so that dredge-up and ejection of white dwarf core matter cannot have occurred.

(3) The massive white dwarfs in these systems are most likely to be ONeMg white dwarfs. This prediction is consistent with stellar evolution calculations (Iben, private communication) which suggest that it is difficult to form CO white dwarfs in the mass range $M > 1.3M_\odot$ necessary to understand the recurrent novae and is also consistent with observational evidence for the presence of ONeMg systems among the classical novae.

(4) Typical classical novae are very unlikely to be progenitors of supernovae, since it appears that the compact component in these systems is systematically being reduced in mass as a consequence of successive outbursts.

It thus seems reasonable that the recurrent nova systems, which experience TNR's, may indeed see their white dwarf components evolve to the Chandrasekhar limit, setting the stage for the occurrence of a more violent event. The explosion of an ONeMg white dwarf in a close binary system should be investigated to determine whether the formation of a neutron star in a close binary can be achieved in this manner. Williams(private communication) has suggested a possible relation between recurrent novae and low mass x-ray binaries, since both classes show evidence for helium enrichments in the material being transferred from the secondary. Studies of the long term evolution of recurrent nova systems and hydrodynamic studies of the evolution of the resulting ONeMg white dwarf at the Chandrasekhar limit can hopefully provide a clear statement regarding this matter.

We would like to express our thanks for many useful discussion to Drs. G. S. Kutter, K. Nomoto, S. Shore, E. M. Sion, L. Stryker, G. Sonneborn, R. Wade, R. M. Wagner, and R. E. Williams. S. Starrfield is grateful to Drs. S. Colgate, A. N. Cox, C. F. Keller, M. Henderson, and K. Meyer for the hospitality of the Los Alamos National Laboratory and a generous allotment of computer time. This work was supported in part by NSF Grants AST85- 16173 and AST88-18215 to Arizona State University and AST 86-11500 to the University of Illinois, by the Institute of Geophysics and Planetary Physics at Los Alamos, by NASA grants to Arizona State University and to the University of Colorado, and by the DOE.

REFERENCES

Sparks, W.M., Starrfield, S., Truran, J. W., and Kutter, G. S. 1988, in *Atmospheric Phenomena in Stellar Explosions*, ed. K. Nomoto, (Springer-Verlag: Heidelberg), p. 234.

Starrfield, S. 1988, in *Multiwavelength Astrophysics*, ed. F A. Córdova, (Cambridge: University Press), p. 159.

Starrfield, S. 1989, in *The Classical Nova*, ed. N. Evans, and M. Bode, (New York: Wiley), p. 39.

Starrfield, S., Sparks, W. M., and Shaviv, G. 1988, *Ap. J. Lett.*, **326**, L35.

Starrfield, S., Sparks, W. M., and Truran, J. W. 1985 *Ap. J.*, **291**, 136.

Truran, J. W. 1982, in *Essays in Nuclear Astrophysics*, ed. C.A. Barnes, D.D. Clayton, and D. Schramm (Cambridge: Cambridge U. Press), p. 467.

Truran, J. W., and Cameron, A. G. W 1971, *Astrophys. Space Sci.*, **14**, 179.

Truran, J. W., and Livio, M. 1986, *Ap. J.*, **308**, 721.

Truran, J. W., Livio, M., Hayes, J., Starrfield, S., and Sparks, W. M. 1988, *Ap. J.*, **324**, 345.

Whelan, J. A. J., and Iben, I. 1973, *Ap. J.*, **186**, 1007.

Williams, R.E., Sparks, W.M., Gallagher, J.S., Ney, E.P., Starrfield, S., and Truran, J. W. 1981, *Ap. J.*, **251**, 221.

REFERENCES

SECTION XI

NUCLEOSYNTHESIS IN SUPERNOVAE

Explosive Nucleosynthesis in Type I and Type II Supernovae

Friedrich-Karl Thielemann, Masa-aki Hashimoto, Ken'ichi Nomoto, & Koichi Yokoi

1. Introduction

There exist many original and review articles about the mechanisms of type I and type II supernovae (SNI and SNII, e.g. Nomoto, Thielemann, Yokoi 1984; Nomoto 1986; Bruenn 1989; Cooperstein and Baron 1989; Wilson et al. 1986; Woosley and Weaver 1986) and a number of contributions to this conference (Nomoto, Woosley, Cooperstein and Baron, Mayle and Wilson), so that we do not intend to repeat this discussion here. We rather want to concentrate on the accompanying nucleosynthesis processes. This introduction contains a general presentation of the nucleosynthesis processes, while the application to both types of supernova events is given in sections 2 and 3. Section 4 includes a comparison of both contributions to the enrichment of heavy elements in the interstellar medium and a general conclusion. For a discussion of the nuclear physics input in the present nucleosynthesis calculations see Thielemann (1989) and Thielemann, Hashimoto, Nomoto (1990). One of the major free parameters in stellar evolution is the still uncertain $^{12}C(\alpha,\gamma)^{16}O$ reaction (see Filippone, Humblet, Langanke 1989; Caughlan et al. 1985; Caughlan and Fowler 1988). The present calculations were performed with the rate of Caughlan et al. (1985).

1.1 Explosive Si-Burning

Zones which experience temperatures in excess of 4.0–5.0$\times 10^9$K undergo explosive Si-burning. Temperatures beyond 5×10^9K lead to complete Si-exhaustion and produce only Fe-group nuclei. Explosive Si-burning can be devided into three different regimes: incomplete Si-burning and complete Si-burning with either a normal or alpha-rich freeze-out. Which of the three regimes is encountered depends on the peak temperatures and densities attained during the passage of the shock front (see Fig.20 in Woosley, Arnett, and Clayton 1973, and Fig.5 in Thielemann, Nomoto, Yokoi 1986). The most abundant nucleus in the normal and alpha-rich freeze-out is ^{56}Ni, in case the neutron excess is smaller than 2×10^{-2} or Y_e is larger than 0.49. For the less abundant nuclei the final alpha-capture plays a dominant role transforming nuclei like ^{56}Ni, ^{57}Ni, and ^{58}Ni into ^{60}Zn, ^{61}Zn, and ^{62}Zn in an alpha-rich freeze-out where also trace abundances of ^{40}Ca, ^{44}Ti, ^{48}Cr, and ^{52}Fe are obtained.

Incomplete Si-burning is characterized by peak temperatures of $4-5\times 10^9$K. Temperatures are not high enough for an efficient bridging of the bottle neck above the proton magic number Z=20 by nuclear reactions. Besides the dominant fuel nuclei ^{28}Si and ^{32}S we find the alpha-nuclei ^{36}Ar and ^{40}Ca being most abundant. Partial leakage through the bottle neck above Z=20 produces ^{56}Ni and ^{54}Fe as dominant abundances in the Fe-group. Smaller amounts of ^{52}Fe, ^{58}Ni, ^{55}Co, and ^{57}Ni are encountered.

1.2 Explosive O-burning

Temperatures in excess of roughly 3.3×10^9K lead to a quasi-equilibrium among nuclei in the range $28 < A < 45$ in mass number (Woosley, Arnett, Calyton 1973). These conditions are accomplished in explosive O-burning. The main burning products are ^{28}Si, ^{32}S, ^{36}Ar, ^{40}Ca, ^{38}Ar, and ^{34}S. With mass fractions less than 10^{-2} also ^{33}S, ^{39}K, ^{35}Cl, ^{42}Ca, and ^{37}Ar show up. In zones with temperatures close to 4×10^9K there exists still a contamination by the Fe-group nuclei ^{54}Fe, ^{56}Ni, ^{52}Fe, ^{58}Ni, ^{55}Co, and ^{57}Ni.

1.3 Explosive Ne and C-burning

The main burning products of explosive neon burning are ^{16}O, ^{24}Mg, and ^{28}Si, synthesized via the reaction sequences ^{20}Ne$(\gamma, \alpha)^{16}$O and ^{20}Ne$(\alpha, \gamma)^{24}$Mg$(\alpha, \gamma)^{28}$Si, similar to the hydrostatic case. The mass zones in question have peak temperatures in excess of 2.1×10^9K. They undergo a combined version of explosive neon and carbon burning. Besides the major abundances, mentioned above, explosive neon burning supplies also substantial amounts of ^{27}Al, ^{29}Si, ^{32}S, ^{30}Si, and ^{31}P. Explosive carbon burning contributes in addition the nuclei ^{20}Ne, ^{23}Na, ^{24}Mg, ^{25}Mg, and ^{26}Mg.

1.4 r-Process

The operation of an r-process is characterized by the fact that 10 to 100 neutrons per heavy nucleus have to be available for the onset of substantial neutron capture. Such conditions are only existent after the freeze-out of charged particle reactions, in matter which experienced nuclear statitical equilibrium which was compressed to densities of $10^{11} - 10^{12}$g cm^{-3}, undergoing electron captures until a beta equilibrium is attained (Cameron 1989). A different situation surfaces when the maximum temperatures are below freeze-out conditions for charged particle reactions with Fe-group nuclei. Then reactions among light nuclei which release neutrons, like (α, n) reactions on ^{13}C and ^{22}Ne, can sustain a neutron flux. The constraint of having 10-100 neutrons per heavy nucleus, in order to attain r-process conditions, can then be met by small abundances of Fe-group nuclei. Such conditions were expected when the shock front passes the He-burning shell and enhances the ^{22}Ne(α, n) reaction by orders of magnitude. However, Blake et al. (1981) and Cowan, Cameron and Truran (1983) could show that this neutron source is not strong enough for an r-process in realistic stellar models. Recent research based on additional neutron release via inelastic neutrino scattering (Epstein, Colgate, and Haxton 1988) can also not produce neutron densities which are required for such a process to operate (see also Woosley et al. 1989).

2. Type I(a) Supernovae

In the following we want to discuss in detail the burning conditions as they occur in a SNIa. We take the model W7 by Nomoto, Thielemann, and Yokoi (1984) and Thielemann, Nomoto, and Yokoi (1986) as a typical example for that class of exploding C-O white dwarfs in a binary system. There is still considerable uncertainty in the physics of propagating flame fronts (see e.g. Woosley and Weaver 1986b, Müller and Arnett 1986, Zeldovich et al. 1985, and Nomoto, Woosley, Canal et al. this volume) and open questions remain. The propagation of the burning front after central C-ignition has in published calculations only been treated in a parametrized way, fitting the mixing-length parameter in the time-dependent mixing-length theory of Unno (1967) to the observed supernova energy.

Temperatures in the burning front are increased by about a factor of 10, in comparison to the intitial values, leading to explosive burning of the C-O fuel. Fig.1

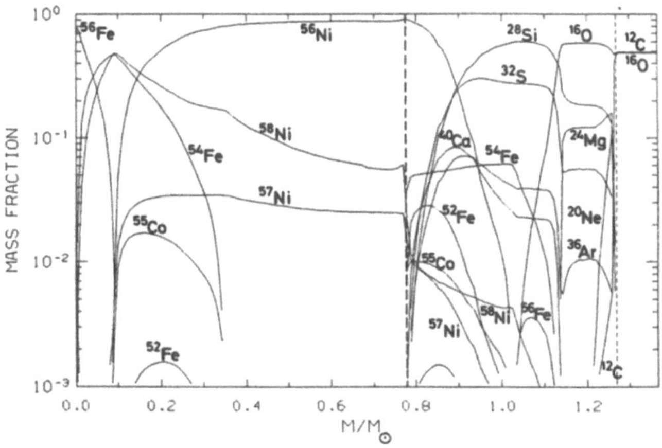

Fig.1: Major abundances after explosive processing. In the inner 0.8 $M_\odot$ only Fe-group nuclei are produced. The inner dashed line shows the transition to incomplete Si-burning, followed further out by products of explosive O, Ne, and C-burning. The outer dashed line marks the quenching of the burning front.

displays the mass fractions of a few major nuclei. We see that the outer zones experience explosive C and Ne-burning, where first the carbon fusion produces ^{20}Ne which photodisintegrates back to ^{16}O. Further towards the center, zones undergo explosive O-burning where also ^{16}O is burned by fusion reactions to ^{28}Si and ^{32}S. Even higher temperatures lead to the burning of Si. In incomplete Si-burning doubly-magic ^{56}Ni is produced together with intermediate nuclei like ^{40}Ca. Inside $0.8 M_\odot$ only "Fe-group" nuclei exist with the dominant abundance of ^{56}Ni, which has the highest binding per nucleon for N=Z nuclei. The situation only changes towards the very center where the densities become high enough to cause Fermi energies of the electron gas in excess of several MeV, which enables appreciable amounts of electron captures on free protons (and to a minor extent on heavy nuclei $\approx$40%). This changes the total proton to neutron ratio and the most abundant nuclei become first ^{54}Fe and ^{58}Ni and finally ^{56}Fe. The region of complete Si-burning is devided into an inner zone of $0.35 M_\odot$ with a normal freeze-out and an outer region with an alpha-rich freeze-out. We see the strong decline of ^{54}Fe and the dominance of ^{58}Ni in the alpha-rich freeze-out.

One of the major aspects of nucleosynthesis calculations is to understand the chemical evolution of galaxies and especially the present abundances in our galaxy, assuming that the solar system abundances give a good representation. Fig.2 shows the ratio of abundances produced in a SNI event to solar abundances. Displayed are abundance ratios after decay of unstable nuclei, normalized to unity for ^{56}Fe. If a SNI event always starts from the same configuration (a white dwarf with $M = M_{Ch}$), the same nucleosynthesis products are expected from each event and the comparison with solar abundances is actually meaningful without averaging over a complete sample. It is obvious from Fig.2 that the production of Fe-group nuclei in comparison to their solar values is a factor of 2 larger than the production of intermediate nuclei from Si to Ca. This shows that SNIs are the dominant production sites of Fe-group nuclei, while SNIIs have to fill in the intermediate nuclei.

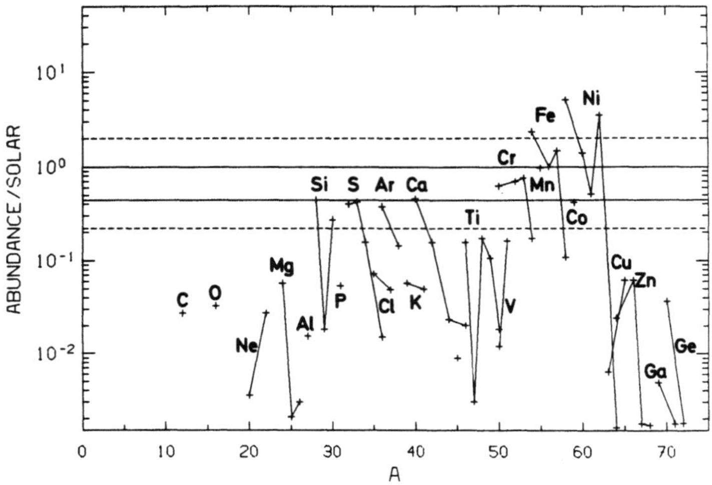

Fig.2: The abundances of stable isotopes formed in a SNI event are shown relative to their solar values. The ratio is normalized to ^{56}Fe. Note the strong overabundances of ^{58,62}Ni and ^{54}Fe.

One undesirable aspect is the large scale of deviations from solar abundances within the Fe-group, when SNIs are the main contributors of these nuclei to the interstellar medium. We notice here especially ^{54}Fe, ^{58}Ni, and ^{62}Ni (originating from ^{62}Zn-decay). All these nuclei come from a chain in the nuclear chart which is displaced by two units to the neutron-rich side from the N=Z chain and measures therefore the neutron excess of the material. Outside of $0.3M_{\odot}$, where electron capture is not effective, the neutron excess is only determined by the ^{22}Ne admixture to ^{12}C and ^{16}O in the original white dwarf, coming from ^{14}N in He-burning, which in turn originated from all CNO-nuclei in H-burning, i.e., the metallicity. Using time-averaged metallicities would reduce the overproduction of these nuclei by 25-40%. Probably more important is that the propagation of the burning front is not fully understood yet. A burning front which starts with a small velocity and then accelerates (Woosley and Weaver 1986b), could reduce the amount of material which is displayed in the mass zones between 0.05 to $0.3M_{\odot}$, where ^{54}Fe, ^{58}Ni, and ^{62}Ni are produced predominantly.

3. Type II Supernovae

While there exists encouraging progress in the understanding of the explosion mechanism of SNIIs (Mayle and Wilson, this volume), we still lack a complete understanding of self-consistant models for collapse and explosion, predicting the mass cut between neutron star and ejecta reliably (Cooperstein and Baron 1989, Myra and Bludman 1989, Bruenn 1989, Mönchmeyer 1989). Despite these open questions it is still possible to model supernova light curves with an artificially induced shock wave of the appropriate energy (see Arnett 1987; Shigeyama et al. 1988; Woosley, Pinto, Ensman 1988 for the case of SN1987A). The same approach has been taken in the past to calculate explosive nucleosynthesis in SNII explosions (Woosley and Weaver 1986a). The only assumption, made implicitely by running the shock wave through the initial model, is that matter which is finally ejected, did not experience significant changes between the onset of the collapse of the inner core and the arrival of the shock wave.

This is well justified for prompt explosions with time scales of 30 ms, but might be questionable for delayed explosions with time scales of seconds.

We want to discuss the general behavior at the example of a $20M_\odot$ star (Nomoto and Hashimoto 1988). The explosion energy used corresponds to a supernova energy of 10^{51}erg. As mentioned before, this treatment cannot predict the position of the mass cut between neutron star and ejecta, but the observation of $0.07 \pm 0.01M_\odot$ of ^{56}Ni in SN1987A (a $20M_\odot$ star) gives an important constraint, because ^{56}Ni is produced in the innermost ejected zones. The explosive nucleosynthesis due to burning in the shock front is shown in Fig.3 for a few major nuclei. Beyond $1.6M_\odot$ all Fe-group nuclei are produced in ***explosive*** Si-burning during the SNII event. At $1.63M_\odot$ Y_e changes from 0.494 to 0.499 and leads to a smaller ^{56}Ni abundance further inside, where more neutron-rich Ni-isotopes share the abundance with ^{56}Ni.

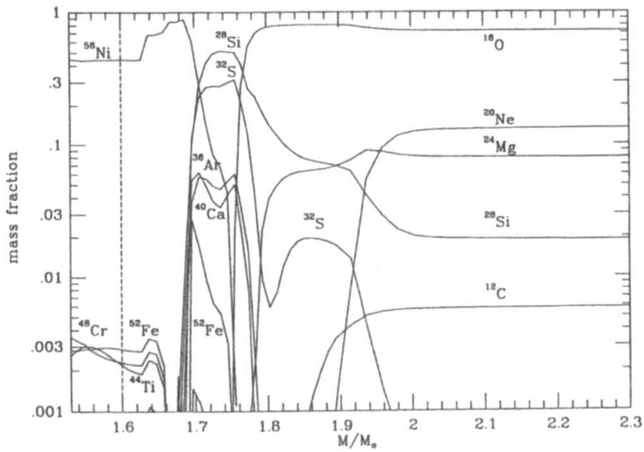

Fig.3: Mass fractions of a few major nuclei after passage of the supernova shockfront. Matter outside $2M_\odot$ is essentially unaltered. Mass zones further in experience explosive Si, O, Ne, and C-burning. In order to eject $0.07M_\odot$ of ^{56}Ni the mass cut between neutron star and ejecta is required to be located at $1.6M_\odot$.

Only alpha-rich freeze-out and incomplete Si-burning are encountered. Contrary to SNIs, densities in excess of 10^8gcm^{-3}, which would result in a normal freeze-out, are not attained in the ejecta of this $20M_\odot$ star. The most abundant nucleus in the normal and alpha-rich freeze-out is ^{56}Ni. For the less abundant nuclei the final alpha-capture plays a dominant role transforming nuclei like ^{56}Ni, ^{57}Ni, and ^{58}Ni into ^{60}Zn, ^{61}Zn, and ^{62}Zn. The region which experiences imcomplete Si-burning starts at $1.69M_\odot$ and extends out to $1.74M_\odot$. In the innermost zones with temperatures close to 4×10^9K there exists still a contamination by the Fe-group nuclei ^{54}Fe, ^{56}Ni, ^{52}Fe, ^{58}Ni, ^{55}Co, and ^{57}Ni. Explosive O-burning occurs in the mass zones up to $1.8M_\odot$. The main burning products are ^{28}Si, ^{32}S, ^{36}Ar, ^{40}Ca, ^{38}Ar, and ^{34}S. With mass fractions less than 10^{-2} also ^{33}S, ^{39}K, ^{35}Cl, ^{42}Ca, and ^{37}Ar are produced. Explosive Ne-burning leads to an ^{16}O-enhancement over its hydrostatic value in the mass zones up to $2M_\odot$.

Traditionally the r-process was assumed to occur close to the mass cut between neutron star and ejecta in matter with very high neutron excess. Using the ^{56}Ni constraint and assuming a spherical explosion led to a mass cut at $1.6M_\odot$. The most neutron-rich zones of the ejecta are located at this inner boundary with a Y_e of 0.494 which corresponds to a neutron excess $\eta = \sum_i (N_i - Z_i)Y_i / \sum_i A_i Y_i = 1 - 2Y_e$ of 1.2×10^{-2}. These zones which experience temperatures in excess of 5×10^9K, produce predominantly nuclei in the mass range 50-60. The quoted value of η therefore indicates that each nucleus has about 0.5 more neutrons than protons. For that mass region this corresponds to a nucleus being still about 1.5 mass units ***more*** proton-rich than the stability line.

No r-process material is ejected from these zones close to the mass cut between neutron star and ejecta. This conclusion relies on the assumption that rotation is not strong enough to violate spherical symmetry, which could cause jet-like ejection (LeBlanc and Wilson 1970; Symbalisty, Schramm, and Wilson 1985). For reasonable ratios of rotational to gravitational energy of 1% before collapse, Mönchmeyer (1989) finds small jet-like circulations at the poles when the shock front is still close to the collapsed core, but obtains an almost spherical symmetry when the shock front reaches the Si-zone. This indicates that a $20M_\odot$ star does not eject r-process nuclei in its final supernova explosion, although SNIIs are strongly expected to be the dominant r-process source. SNIIs with smaller masses, however, could contribute r-process nuclei (Mathews and Cowan 1989).

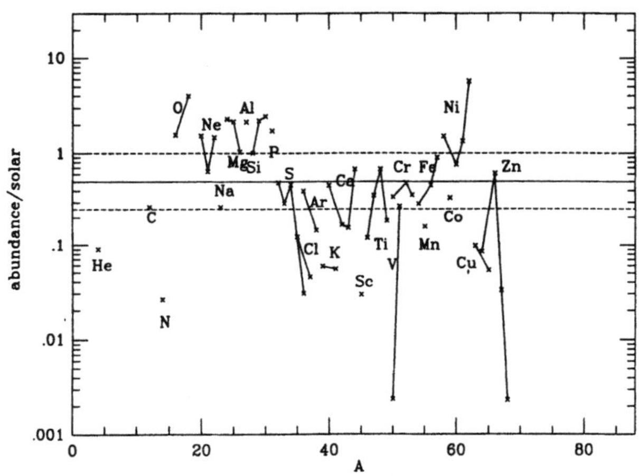

Fig.4: Composition of the supernova ejecta in comparison to solar abundances, normalized to ^{28}Si. Elements lighter than P have strong contributions from hydrostatic C and Ne-burning. Heavier nuclei originate from explosive processing and show similar ratios within a factor of 2 to 3. The ^{58,61,62}Ni abundances are strongly dependent on the mass cut, i.e., the neutron excess of ejected matter

In order to get a general feeling about the nucleosynthesis production in a $20M_\odot$ SNII we display the abundance ratio over solar (normalized to ^{28}Si) in Fig.4. Nuclei heavier than Si and P are on average produced by a factor of 2 to 4 less than ^{28}Si. P, Si, Al, Mg, Na, and Ne, while also produced in explosive burning, have large contributions from the zones of hydrostatic C and Ne-burning, which are unaltered during the explosion (see Fig.3). The reason is the existence of an extended shell of combined C and Ne-burning, ranging from 1.8 to $3.7 M_{\odot}$ in the progenitor star (see. Fig.10 in Nomoto and Hashimoto 1988). Essentially all heavier elements originate from explosive processing. Thus, the ratio between elements heavier than Si and P to lighter elements reflects mainly the size of hydrostatic zones to the explosively processed ones and is a function of stellar mass (and the methods used in stellar evolution calculations). This is evident when comparing to Woosley, Pinto, and Weaver (1988) who use a different treatment of convection (Ledoux vs. Schwarzschild) and have smaller C-burning cores. The amount of ^{16}O is closely linked to the "effective" $^{12}C(\alpha,\gamma)^{16}O$ rate during core He-burning. This effective rate is determined by three factors: 1. the actual nuclear rate, 2. the amount of semiconvection and overshooting, mixing fresh He-fuel into the core at late phases of He-burning, when the temperatures are relatively high and favor alpha-capture on ^{12}C, and 3. the stellar mass which determines the central temperature during He-burning. Our model calculations predict $1.48 M_{\odot}$ of ejected ^{16}O. Hopefully an improved analysis of observations for SN1987A can give better estimates for abundances of the elements discussed above and therefore help to put tighter constraints on the pre-collapse models.

The products of explosive burning from S to Cu originate mainly from mass zones up to $1.8 M_{\odot}$. They fall well along a line of constant overproduction (within reasonable errors). The nuclei ^{58,61,62}Ni, which show large overabundances, are produced in form of the neutron-rich species ^{58}Ni and ^{61,62}Zn. Their production is strongly dependent on Y_e and varies therefore with the position of the mass cut between ejected matter and the remaining neutron star. Especially for the Ni-abundances the position of the mass cut is crucial and one would expect lower mass SNIIs to eject more of these neutron-rich species. This could explain the observed very high Ni abundances in some supernova remnants like the Crab (Henry and Fesen 1988), if they are not explained away by atomic or other effects.

A few nuclei like ^{36}S, ^{37}Cl, ^{40}Ar, ^{58}Fe, and possibly some odd-Z nuclei, which are underabundant in Fig.4, are mainly produced by the weak s-process during core He-burning (Arnett and Thielemann 1985). They were not included in the 30 nuclei network for hydrostatic burning stages and consequently their s-process contribution is neglected. Haxton (1988), Woosley and Haxton (1988), and Woosley et al. (1989) examined the possible effect of inelastic neutrino scattering on explosive nucleosynthesis, an idea which was already introduced earlier by Domogatsky and Nadyozhin (1977). Inelastic neutrino scattering can populate excited states which are unstable against particle emission and produce neighboring nuclei. Outside the neutrino-sphere the scattering events will be rare and therefore this process will be mostly of importance for nuclei with very small abundances, which are not produced otherwise. We did not include this effect in the present calculations.

4. SNI and SNII Contributions to Nucleosynthesis

In order to compare the different contributions of SNIs and SNIIs to the interstellar medium, we display in Table 1 the masses involved in the different types of explosive burning for the standard model of SNI (W7) and a typical (?) representative for SNIIs, the $20 M_{\odot}$ star from section 3. The main difference is seen in the vastly differing amounts involved in explosive Si-burning, 1.03 vs. $0.15 M_{\odot}$. As the major products of Si-burning are Fe-peak elements, it is eminently clear that they are mainly produced by SNIs, as long as the frequencies of both events are comparable. Another difference is given by the fact that only SNIs seem to eject matter which experienced a normal freeze-out in complete Si-burning. This ensures also that the dominant abundances of e.g. ^{54}Fe and ^{55}Mn (from ^{55}Co-decay) come from SNIs, while their counterparts ^{58}Ni and ^{59}Co (from ^{59}Cu-decay) stem from a distorted Fe-peak composition in an alpha-rich freeze-out.

Due to the fact that the innermost matter from SNIs experienced strong electron captures and attained values of Y_e=0.46-0.47, while the $20M_\odot$ star only ejected matter as neutron-rich as $Y_e = 0.498$, the only stable ^{56}Fe – not being a decay product of ^{56}Ni – comes from SNIs. Similarly the contributions to ^{54}Fe and ^{55}Co, both nuclei with $N/Z > 1$, are enhanced. This statement is however very tentative, as we assume that less massive SNIIs can actually eject more neutron-rich material and even r-process nuclei, and the Y_e-values in SNIs depend strongly on the propagation speed of the deflagration front, which governs the amount of electron captures and is still somewhat uncertain.

Zones further out in SNIs and SNIIs have a Y_e determined by the original metalicity in form of ^{22}Ne, which was produced in hydrostatic He-burning from CNO-nuclei. Thus for the same metallicities the burning conditions barely differ but only the masses involved. For explosive O and Ne-C-burning those numbers lie within a factor of 2. While SNIs, consisting initially only of C and O, eject only products of ***explosive*** Ne and C-burning, SNIIs can have extended convective shells of C and Ne-burning which will only be partially processed explosively (see discussion in section 3) and thus a large amount of hydrostatic C and Ne-burning material will also be contributed from SNIIs.

TABLE 1

MASSES IN EXPLOSIVE BURNING

	SNI (TNY 1986)	SNII (THN 1990)
Si-burning	1.03	0.15
complete	0.77	0.10
normal	0.33	
alpha-rich	0.44	0.10
incomplete	0.26	0.05
O-burning	0.11	0.06
Ne+C-burning	0.13	0.20

If we take the very premature approach of assuming that the $20M_\odot$ mass star can represent a typical SNII, one can try to derive a ratio of SNI to SNII events, necessary to produce a solar Si/Fe ratio in total

$$\frac{M(Si)}{M(Fe)}_\odot = \frac{R_I M_I(Si) + R_{II} M_{II}(Si)}{R_I M_I(Fe) + R_{II} M_{II}(Fe)}. \tag{1}$$

Here M denotes the mass involved, R the supernova rate, I and II the types of supernovae. From here we can derive a ratio

$$\frac{R_I}{R_{II}} = -\frac{M_{II}(Si) - M(Si)/M(Fe)_\odot M_{II}(Fe)}{M_I(Si) - M(Si)/M(Fe)_\odot M_I(Fe)} = 0.22 \pm 0.03, \tag{2}$$

if we adopt the appropriate values for M_I and M_{II} from Thielemann, Nomoto and Yokoi (1986) and Thielemann, Hashimoto, Nomoto (1990) [$M_{II}(Fe) = 0.07 \pm 0.01 M_\odot$], as well as $M(Si)/M(Fe)_\odot = 0.559$ (Cameron 1982). Such a ratio is quite uncertain, due to the fact that differences of uncertain numbers are involved. The results for SNIs seem, however, quite reliable since a number of tests comparing theoretical and observational light curves and spectra have been performed. The $20M_\odot$ star is not necessarily typical for SNIIs, but its Si/Fe ratio seems to be (see Table 5 in Thielemann, Hashimoto, Nomoto 1990). R_I/R_{II} reflects the ratio of SNIs to all core collapse events. If SNIbs are interpreted as core collapse events of massive Wolf-Rayet stars (e.g. Wheeler and Levreault 1985, Ensman and Woosley 1988, Nomoto this volume) and one uses the respective observational rates for SNIa, SNIb, and SNII (van den Bergh, this volume), a remarkable agreement is obtained $(R_{Ia}/(R_{Ib} + R_{II}) = 0.214)$.

When we take the value for R_I/R_{II} from Eq.(2), we predict ^{57}Fe/^{56}Fe=0.016-0.023, the errors being due to the uncertainties of the mass cut in SNIIs between neutron star and ejecta. This compares well to a solar value of 0.024 and the ^{57}Fe contribution is dominated by SNIIs. Using the same ratio also results in ^{55}Mn/^{56}Fe=$(8.6 - 9.3) \times 10^{-3}$ which compares well with the solar value of 1.1×10^{-2}. In this case the ^{55}Mn contribution is dominated by SNIs. Similarly we also obtain that the intermediate mass elements from Si to Ca (Si being representative) are produced to roughly 28% in SNIas, while Fe, i.e. ^{56}Fe, is produced to 66% in SNIas. It will be interesting to test similar predictions in the future also for other elements. In order to obtain, however, really meaningful results it will be necessary to have predictions of explosive nucleosynthesis in SNIIs also for masses different from $M = 20M_\odot$ and to perform an integral over the IMF. This is underlined by the deficiency of overabundances in [S/Fe] through [Ca/Fe] for the $20M_\odot$ star in comparison to average values for SNIIs (see Table 5 in Thielemann, Hashimoto, Nomoto 1990).

This research was supported in part by NASA grant NGR 22-007-272, NSF grant AST-8612647 and the National Center for Supercomputer applications at the University of Illinois (AST 890009N).

References

Arnett, W.D. 1987, ***Ap. J.* 319**, 136

Arnett, W.D., Thielemann, F.-K. 1985, ***Ap. J.* 295**, 589

Blake, J.B., Woosley, S.E., Weaver, T.A., Schramm, D.N. 1981, ***Ap. J.* 248**, 315

Bruenn, S.W. 1989, ***Ap. J.* 340**, 955

Cameron, A.G.W. 1982, in ***Essays in Nuclear Astrophysics***, eds. C.A. Barnes, D.D. Clayton, D.N. Schramm, (Cambridge Univ. Press), p. 23

Cameron, A.G.W. 1989, in ***Cosmic Abundances of Matter***, ed. C.J. Waddington, AIP Conf. Proc. 183, p.349

Caughlan, G.R., Fowler, W.A. 1988, ***At. Nucl. Data Tables* 40**, 283

Caughlan, G.R., Fowler, W.A., Harris, M.J, Zimmerman, G.E. 1985, ***At. Nucl. Data Tables*** **32**, 197

Cooperstein, J., Baron, E. 1989, in ***Supernovae***, ed A. Petschek, (Springer-Verlag, New York), in press

Cowan, J.J., Cameron, A.G.W., Truran, J.W. 1983, ***Ap. J.*** **265**, 429

Domogatsky, G.V., Nadyozhin, D.K. 1977, ***M.N.R.A.S.*** **178**, 33p

Ensman, L., Woosley, S.E. 1988, ***Ap. J.*** **333**, 754

Epstein, R.I., Colgate, S.A., Haxton, W.C. 1988, ***Phys. Rev. Lett.*** **61**, 2038

Filippone, B.W., Humblet, J., Langanke, K. 1989, ***Phys. Rev.*** *C*, in press

Haxton, W.C. 1988, ***Phys. Rev. Lett.*** **60**, 1999

Henry, R.B.C., Fesen, R.A. 1988, ***Ap. J.*** **329**, 693

LeBlanc, J.M., Wilson, J.R. 1970, ***Ap. J.*** **161**, 541

Mathews, G.J., Cowan, J.J. 1989, in ***Heavy Ion Physics and Nuclear Astrophysical Problems***, eds. S. Kubono, M. Ishihara, T. Nomura, (World Scientific, Singapore), p.143

Mönchmeyer, R. 1989, Ph. Thesis, TU Munich, unpublished

Müller, E., Arnett, W.D. 1986, ***Ap. J.*** **307**, 619

Myra, E.S., Bludman, S. 1989, ***Ap. J.*** **340**, 384

Nomoto, K. 1986, ***Prog. Part. Nucl. Phys.*** **17**, 249

Nomoto, K., Hashimoto, M. 1988, ***Phys. Rep.*** **163**, 13

Nomoto, K., Thielemann, F.-K., Yokoi, K. 1984, ***Ap. J.***, **286**, 644

Shigeyama, T., Nomoto, K., Hashimoto, M. 1988, ***Astron. Astrophys.*** **196**, 141

Symbalisty, E.M.D., Schramm, D.N., Wilson, J.R. 1985, ***Ap. J. Lett.*** **291**, L11

Thielemann, F.-K. 1989, in ***Nuclear Astrophysics***, ed. M. Lozano, M.I. Gallardo, J.M. Arias (Springer: Berlin), p.106

Thielemann, F.-K., Hashimoto, M., Nomoto, K. 1989, ***Ap. J.*** **348**, in press

Thielemann, F.-K., Nomoto, K., Yokoi, K. 1986, ***Astron. Astrophys.***, **158**, 17

Unno, W. 1967, ***Publ. Astron. Soc. Japan*** **19**, 140

Wheeler, J.C., Levreault, R. 1985, ***Ap. J. Lett.*** **294**, L17

Wilson, J.R., Mayle, R., Woosley, S.E., Weaver, T.A. 1986, in ***Proc. 12th Texas Symp. on Relativistic Astrophysics, Ann. N. Y. Acad. Sci.*** **470**, 267

Woosley, S.E., Arnett, W.D., Clayton, D.D. 1973, ***Ap. J. Suppl.*** **26**, 231

Woosley, S.E., Hartmann, D., Hoffman, R.B., Haxton, W.C. 1989, preprint

Woosley, S.E., Haxton, W.C. 1988, ***Nature*** **334**, 45

Woosley, S.E., Pinto, P.A., Ensman, L. 1988, ***Ap. J.*** **324**, 466

Woosley, S.E., Pinto, P.A., Weaver, T.A. 1988, ***Proc. Astron. Soc. Australia*** **7**, 355

Woosley, S.E., Weaver, T.A. 1986, ***Ann. Rev. Astron. Astrophys.*** **24**, 205

Woosley, S.E., Weaver, T.A. 1986, ***in Radiation Hydrodynamics***, IAU Colloq. No 89, eds., D. Mihalas, K.H. Winkler, (Reidel, Dordrecht), p. 91

Zeldovich, Ya.B., Baerenblatt, G.I., Librovich, V.B., Makhviladze, G.M. 1985, ***The Mathematical Theory of Combustion and Explosions***, (Plenum, New York)

On Supernovae Rates, Oxygen and Iron Abundances

F. X. Timmes

Recently the argument was advanced by Arnett, Schramm and Truran (1989) that there is $\sim 10^9$ $M_\odot$ of oxygen in the $\sim 10^{10}$ year old Galactic disk; that each Type II supernovae ejects ~ 1 $M_\odot$ of oxygen; and hence that Type II supernovae have occurred on the average of once every ten years. By further assuming that all Type II produce 1987A amounts of iron, they conclude that the bulk of iron in the Galactic disk originates from the core collapse of young massive stars. Their conclusions regarding the Type II supernovae rate and Type IIs being the dominant source of iron, even with a gratuitous factor of two, represents an excellent example of extrapolation beyond bound.

Type II supernovae events are the prime nucleosynthetic site for the mid atomic number elements. As the mass of the progenitor star is turned up, calculations by Arnett (1978) and more recently by Woosley (1986) demonstrate that the mass fraction of oxygen produced increases rapidly, soon becoming the dominant element produced. The mass fraction of iron produced by Type II supernovae is relatively small. On the other hand, the carbon deflagration wave front in an accreting carbon + oxygen white dwarf model of Type I supernovae is a copious nucleosynthetic site for the iron peak elements. The computations by Nomoto, Thielemann, and Yokoi (1984) and Woosley (1989) indicate that the mass fraction of iron peak elements produced by this model of Type I supernovae is about 0.7 $M_\odot$ while the mass fraction of oxygen produced is minimal. This suggests that Type I supernovae produce a significant fraction of the iron peak elements in the Galactic disk, which is quite complimentary to the nucleosynthesis of Type II supernovae.

The yield of any element produced by an entire generation of Type I or Type II supernovae is dependent of the birth rate of stars at a previous epoch, the initial mass function, the stellar mass-lifetime relationship, the nucleosyntheses prescriptions and the accretion of primordial material. It is not clear that all Type II events, on average, occur once every ten years and that they all produce 1987A amounts of oxygen and/or iron. To facilitate an understanding of the temporal evolution of the chemical elements and supernovae rates, the coupled, integro-differential system of equations governing the single zone was integrated, assuming the production matrix formalism of Talbot and Arnett (1971, 1973, 1975). The computational algorithms were derived from Press *et al.* (1986). The nucleosynthetic yields of the core collapse of massive stars follows Woosley (1986) while the prescription for intermediate mass stars follows Renzini and Voli (1981). The carbon deflagration model of Type I nucleosynthesis was taken from Nomoto *et al.* The methodology developed by Matteucci and Greggio (1986) for incorporating multiple types of supernovae into the basic equation of chemical evolution was adopted. Further assumptions include a Schmidt (1959) n=2 surface density birth rate, a Miller-Scalo (1974) initial mass function and Chiosi and Matteucci's (1982) functional form for the infall of primordial material.

A typical computation of the Type II and Type I supernovae rates are shown in Figure 1. The models indicate that over a physically reasonable range of model parameters that Type II core collapse events occur at present rate of about two every century and that the present rate of Type I supernovae is about half the rate of core collapse events. These results are in accord with the observational evidence summarized by Van der Bergh (this conference). Note that averaged over the age of the Galaxy, single zone models of chemical evolution yield a core collapse rate of about one every 25 years. As the Galaxy ages, a larger and larger percentage of the iron is attributable to Type I supernovae; the computations suggesting about 65 percent in the present epoch ($\sim$ 12Gy). A similar conclusion was reached by Matteucci and Greggio (1986). It is precisely for this reason, namely, that the bulk of the Galactic oxygen is produced by short lifetime massive stars and that the bulk of Galactic iron is produced by Type I's that produces the [O/Fe] versus [Fe/H] trends observed in F and G main sequence stars by Sneden, Lambert, and Whitaker (1979) and Clegg, Lambert, and Tomkin (1981).

Based on the relatively standard models of chemical evolution adopted here one would conclude that Type II supernovae occur at a present rate of about 2 per century and have occurred at an average rate of about 4 per century over the lifetime of the Galactic disk. Furthermore, Type II supernovae produce the majority of oxygen and other mid atomic weight elements while Type I supernovae produce the bulk of iron peak elements in the Galactic disk.

References

Arnett, W. D. 1978, Ap. J., **219**, 1008.

Arnett, W. D., Schramm, D. N., Truran, J. W. 1989, Ap. J., **339**, L25.

Chiosi, C., Matteucci, F. M., 1982, Astron. Astrophys., **105**, 140.

Clegg, R. E. S., Lambert, D. L., Tomkins, J. 1981, Ap. J., **250**, 262.

Matteucci, F. M., Greggio, L. 1986, Astron. Astrophys., **154**, 279.

Miller, G. E., Scalo, J. M. 1979, Ap. J., **41**, 513.

Nomoto, K., Thielemann, F. K., Yokoi, K. 1984, Ap. J., **286**, 644.

Press, W. H., Flannery, B. P., Teukolsky, S. A., Vetterling, W. T. 1986 Numerical Recipes: *The Art of Scientific Computing*, (Cambridge Univ. Press; Cambridge).

Renzini, A., Voli, M. 1978, Ap. J., **219**, 1008.

Schmidt, M. 1959, Ap. J., **129**, 243.

Sneden, C., Lambert, D. L., Whitaker, R. W. 1979, Ap. J., **234**, 964.

Talbot, R. J., Arnett, W. D. 1971, Ap. J., **170**, 409.

Talbot, R. J., Arnett, W. D. 1973, Ap. J., **186**, 151.

Talbot, R. J., Arnett, W. D. 1975, Ap. J., **197**, 551.

Woosley, S. E. 1986, *Nucleosynthesis and Chemical Evolution*, Hauk, B., Maeder, A., eds. (Geneva Obs.: Geneva), 1-195.

Woosley, S. E. 1989, *Supernovae*, Petschek, A. G., ed. (Springer-Verlag: Berlin), 1-25.

Van den Bergh, S. 1989, This Conference, preprint.

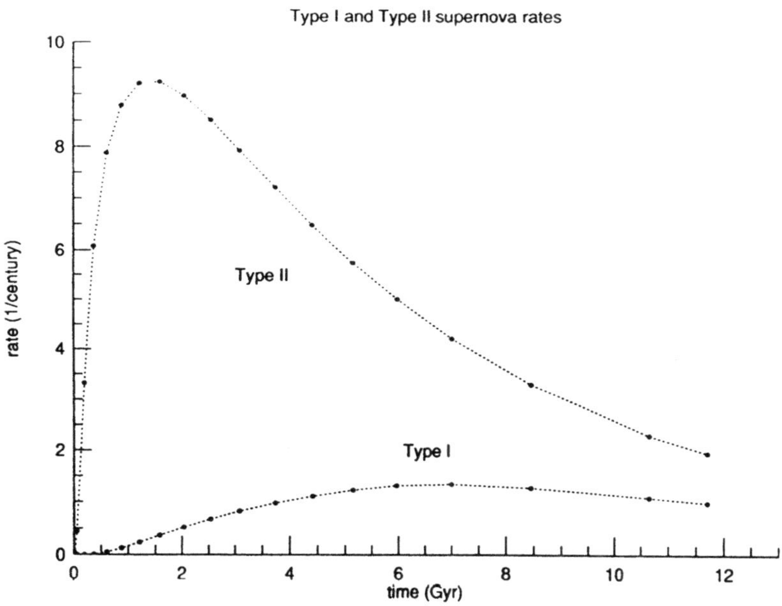

Fig. 1 - The evolution of the number of Type I and Type II supernovae events per century. The average Type II rate is between 3.5 and 5.0 per century, depending on the model parameters, while the average Type I rate is between 2.5 and 1.0 per century.

The p-Process in a Realistic Supernova Model

*N. Prantzos, M. Hashimoto, M. Rayet, &
M. Arnould*

The astrophysically most plausible site for the synthesis of the ($Z \geq 34$) neutron deficient isotopes (the "$p-$nuclei") seems to be the deep interior of highly evolved massive stars. There, (γ, n) photodisintegrations of preexisting more neutron rich ($s-$ and $r-$) species, possibly followed by cascades of (γ, p) and (γ, α) reactions, lead to the production of $p-$nuclei for temperatures ranging roughly from 2 to 3.2×10^9 K. Such a $p-$process is expected to occur in O/Ne layers during a type II supernova explosion (Woosley and Howard, 1978), as well as during a presupernova hydrostatic oxygen burning phase (Arnould, 1976), where radiative proton captures might also contribute to the production of some light $p-$nuclei.

Recently, Rayet, Prantzos and Arnould (1989; henceforth RPA) reinvestigated the $p-$process in the oxygen rich layer of an exploding massive star using a parametrized model where the explosion is simulated by a sudden increase of temperature and density to peak values, followed by an exponential decrease. They used an extended network providing a reliable basis for a self-consistent calculation of a long suite of $12 \leq A \leq 210$ nuclides (see RPA for details) as well as background neutron, proton and helium concentrations. The seed nuclei abundance distribution was obtained by an $s-$process calculation during core helium burning (see e.g. Prantzos et al. 1987). With the simple assumption of six shells of equal mass and constant peak density of 10^6 g cm^{-3}, heated to peak temperatures between 2.2 and 3.2×10^9 K, RPA obtained a $p-$nuclei abundance pattern comparable to the solar system distribution within a factor 3, for 60% of the 35 $p-$isotopes.

In this work, we consider the $p-$process in a realistic type II supernova model recently proposed by Hashimoto et al.(1989) in order to compare their explosive nucleosynthetic yields with spectroscopic observations from SN1987A. The presupernova model, fully described in Nomoto and Hashimoto (1988), consists of an evolved 6 $M_\odot$ helium star, corresponding to the helium core of a 20 $M_\odot$ star, and containing in particular an oxygen/neon rich layer at 1.67 $M_\odot < M_r \leq 3.66\ M_\odot$. An explosion energy of 10^{51} erg is deposited at the inner edge of the ejecta and the changes in temperatures and densities, due to the shock wave propagation and nuclear burning, are calculated hydrodynamically (Shigeyama et al., 1988).

Inspired by previous $p-$process calculations, we select a zone in this layer, where the material is heated during the explosion to a peak temperature T_p ranging from 3.2 to 2.0×10^9 K. Such a zone is located towards the inner edge of the O/Ne layer, at 1.80 $M_\odot \leq M_r \leq 2.05\ M_\odot$, with peak densities ρ_p ranging from 7.2 to 1.9×10^5 g cm^{-3}. A $p-$process calculation was performed in this region with the same input physics as in RPA. For consistency however, seed nuclei abundances were first obtained by an $s-$process calculation during the He burning phase of the 6 $M_\odot$ helium core. Another difference with RPA comes from the use of realistic temperature and density profiles, $T(t)$ and $\rho(t)$. In RPA, T and ρ decayed with a constant timescale given (for ρ) by the hydrodynamic free expansion time $446 \times \rho_p(\text{g cm}^{-3})^{-1/2}$ s. Here

$T(t)$ and $\rho(t)$ decay also in an almost exponential manner, but with timescales which, systematically, are by a factor 2 smaller than the free expansion values obtained with the corresponding values of ρ_p. On the other hand, since ρ_p is as small as $2 \times 10^5\,\mathrm{g\,cm^{-3}}$ in our outermost shells (5 times smaller than in RPA), the decay times in both calculations are in fact comparable in the regions with small values of T_p, while in the inner region ($T_p \geq 2.4$) they become significantly smaller than in RPA.

Figure 1 shows the overproduction factors obtained for a sample of $p-$nuclei from ^{74}Se to ^{190}Pt as a function of T_p. It is seen that each species is produced in a very narrow range of temperature, lower mass nuclei requiring for their synthesis higher values of T_p than the heavier ones. The corresponding mass range involved in the O/Ne layer is also shown on Figure 1. The average overproduction factors $< X > /X_\odot$, where $< X >$ is the mass fraction averaged over the whole 0.25 $M_\odot$ considered layer, are plotted in Figure 2 for the 35 species usually classified as either pure or dominantly $p-$isotopes. This abundance pattern, obtained in realistic astrophysical conditions, essentially confirms the results obtained in the parametrized model of RPA. However, the mean overproduction factor for the 35 $p-$nuclei is only 26, which is small compared to the values 100 – 200 obtained in RPA. This can be explained by the use of the LMC metallicity for calculating the $s-$process seed nuclei as well as by some dilution effect due to the inclusion in this calculation of a relatively thick layer with $2 \leq T_p \leq 2.2\,10^9$ K, where very few $p-$nuclei are produced.

On the other hand, ^{180}Ta is abundantly produced in these layers, which is the first successful attempt to synthesize this especially fragile odd-odd isotope in comparatively large quantities by photonuclear processes (see e.g. Woosley and Howard, 1978). Further investigations are needed to examine the role of the rather large background neutron concentrations obtained with our network in the present model.

Considering the underproduced species of Figure 2, our conclusions are the same as in RPA where their case was discussed in detail. Summing up, ^{113}In, ^{115}Sn, ^{152}Gd and ^{164}Er can in fact be considered as $s-$isotopes, some with slight $p-$process contributions, while other mechanisms are usually invoked to explain the synthesis of ^{138}La. The underproduction of the Mo and Ru $p-$isotopes, the most abundant in the solar system, remains, as expected, unsolved.

The production of ^{146}Sm in the $p-$process has been discussed several times (see e.g. Woosley and Howard, 1978), owing to its relation with the much debated existence of ^{142}Nd/^{144}Nd anomalies in meteorites. The interest in ^{146}Sm has recently been revived by Printzhofer et al. (1989) who have obtained clear evidence of such anomalies in 2 meteorites, from which they conclude to an abundance ratio $P = {}^{146}\mathrm{Sm}/{}^{144}\mathrm{Sm} = 0.015$ at the time of formation of the solar system. Our $p-$process calculation gives a ratio $P = 0.14$, where most of the contribution to ^{146}Sm comes from its short-lived, semi magic ($N = 82$), progenitor ^{146}Gd. Comparison of these two numbers has far reaching implications for the nucleosynthetic history of the galaxy, that will be discussed elsewhere. We only want to note here that, from RPA's calculations, a ratio $P = 0.56$ is obtained, which indicates the sensitivity of P to the detailed astrophysical conditions in which the $p-$process takes place.

We must finally mention that in order to evaluate the enrichment of the supernova ejecta in $p-$nuclei, one must consider the dilution of the latter in the 1.48 $M_\odot$ of oxygen emitted in Hashimoto et al. (1989) model. Since the $p-$nuclei are only produced inside a thin layer of 0.25 $M_\odot$, their mean overabundance will become as small as 0.03 when normalized to ^{16}O. This difficulty, which also arises whith the same acuity for several light elements (see Figure 4 of Hashimoto et al., 1989) might be much alleviated in the explosion of stars with, 1) lower mass (13 to 15 $M_\odot$), which develop relatively smaller oxygen layers (see e.g. Nomoto and Hashimoto, 1988) and, 2) higher metallicity. The $p-$process could occur in such conditions within a larger fraction of the oxygen layer and with a higher efficiency.

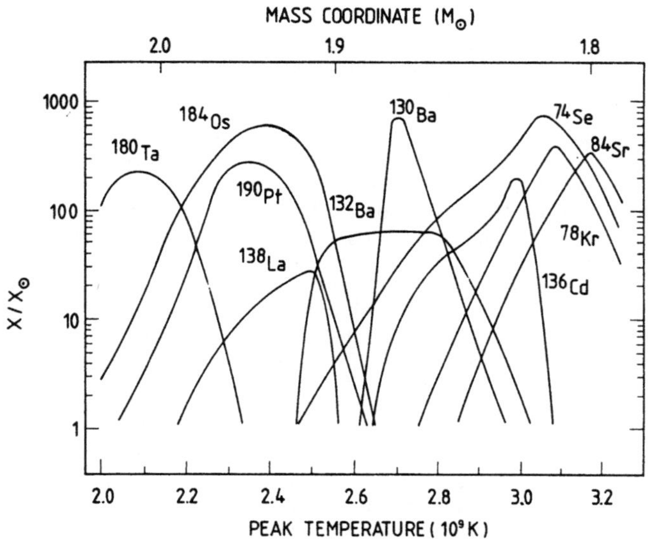

Fig. 1.–Overproduction factors for several p–nuclei as a function of T_p

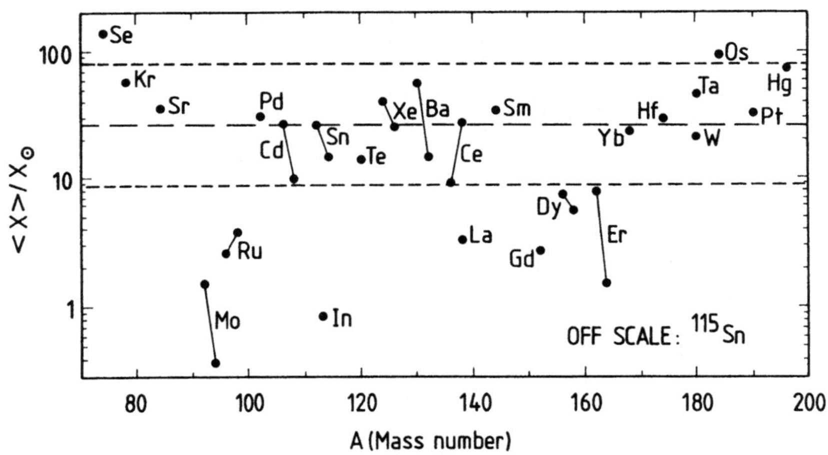

Fig. 2.–The p–nuclei overproduction factors averaged over the whole considered layer. The mean overproduction is shown by the long-dashed line. Solid lines join p–isotopes of the same element

References

Arnould, M. 1976, *Astr. Ap.*, **8**, 436.
Hashimoto, M., Nomoto, K., and Shigeyama, T. 1989, *Astr. Ap.*, **210**, L5.
Nomoto, K., and Hashimoto, M. 1988, *Phys. Rept.*, **163**, 13.
Prantzos, N., Arnould, M., and Arcoragi, J.-P. 1987, *Ap.J.*, **315**, 209.
Prinzhofer, A., Papanastassiou, D.A. and Wasserburg, G.J. 1989, to appear in *Ap.J.(Letters)*
Rayet, M., Prantzos, N., and Arnould, M., 1989, to appear in *Astr. Ap.*
Shigeyama, T., Nomoto, K., and Hashimoto, M., 1988, *Astr. Ap.*, **196**, 141.
Woosley, S.E., and Howard, W.M. 1978, *Ap.J.Suppl.*, **36**, 285.

Neutrino Nucleosynthesis in Massive Stars

Dieter Hartmann

Nucleosynthesis induced by the neutrino burst accompanying core collapse and neutron star formation in massive stars was first studied by DOMOGATSKII ET AL. [1] and WOOSLEY [2]. Since these pioneering studies significant improvements in our understanding of Type II supernovae and weak interactions as well as the direct observation of SN 1987A in ν–light have substantiated the arguments for this ν–process. Recent work by EPSTEIN, COLGATE, and HAXTON [3] and WOOSLEY and HAXTON [4] suggest that a large number of elements could owe their existence in nature to ν–reactions in supernovae. A study of this process including shock wave propagation was carried out by WOOSLEY ET AL. [5;WH^3] for selected zones of a 20 $M_\odot$ star. This paper describes essential features of the ν–process and discusses some of the results obtained by WH^3.

I. Core Collapse Neutrino Burst

The dynamic collapse of massive stars is initiated by photodissociation of iron peak nuclei and electron capture on free protons and heavy nuclei. The electron neutrinos generated by the capture process are trapped inside the collapsing core when $\rho \sim 10^{12}$ g cm^{-3}. In analogy to stellar photospheres one can define an energy–dependent neutrinosphere at which the total 'optical depth" is of order unity. This surface, located at r $\sim$ 50–100 km, radiates $\sim 10^{51}$ ergs during infall ($\sim$ 100 ms). When the core density exceeds nuclear matter density the repulsive nuclear force causes core bounce which drives a strong shock wave through the collapsing envelope [6]. When the shock front reaches the neutrinosphere a burst of $\nu\bar{\nu}$–pairs of all flavors is released. This spike ($\sim$ 4 ms), followed by a cooling tail ($\sim$ 10–20 ms), carries $\sim 3\ 10^{51}$ ergs. However, the total amount of energy lost due to neutrino emission during infall and shock breakout is small compared to the total binding energy of the neutron star that has to be released in the supernova event. Depending on the details of the nuclear matter equation of state the binding energy is approximately $3\ 10^{53}$ ergs (e.g. [7]). The bulk of this energy is emitted during the Kelvin-Helmholtz cooling phase of the proto neutron star lasting several seconds [8]. The energy is equipartitioned between all neutrino flavors. WH^3 assumed the neutrino energy spectrum to be of Fermi-Dirac form with an effective temperature T_ν and zero chemical potential. The mean neutrino energy is then related to the temperature by $\bar{\epsilon}_\nu = 3.15\ T_\nu$. Because the heavy lepton neutrinos (ν_μ, ν_τ) interact with matter only via neutral current interactions, their mean free path is much larger than that of the electron neutrinos. Consequently their neutrinospheres are located further inside the star resulting in a larger neutrino temperature. Expected values are $T(\nu_e) \sim$ 4–5 MeV and $T(\nu_\mu) \sim T(\nu_\tau) \sim 2\ T(\nu_e)$. Recent calculations of neutrino transport in Type II supernovae by JANKA and HILLEBRANDT [9] show that the mean neutrino energy remains approximately constant during the first $\sim$ 10 s following shock breakout, but indicate that the energy distribution is non-thermal. The total energy emitted in all neutrino flavors ($\sim 3\ 10^{53}$

ergs) can be described by an exponentially decaying luminosity with time constant $\tau_\nu \sim 3$ s. At a radius r = 10^9 r_9 cm inside the star the neutrino flux is then given by

$$\Phi_\nu \sim 2.5\, 10^{38}\, r_9^{-2}\, E_{53}\, T_\nu\, \tau_\nu\, exp(-t/\tau_\nu)\; cm^{-2} s^{-1} \,. \tag{1}$$

Consider now ν–interactions in the various shells of heavy elements in the star.

II. Explosive Neutrino Nucleosynthesis

Because of their higher temperatures, μ– and τ–neutrinos can efficiently excite nuclei to particle unbound states by inelastic neutral current scattering: (Z, A) + $\nu \rightarrow$ (Z,A)* + ν'. The excited nuclei then decay by n, p, d, t, α, or multiple particle emission. WH[3] also include charged current cross sections in their study. Nucleons and nuclei generated by neutrino spallation then continue the nucleosynthesis by reacting in the supernova environment. EPSTEIN ET AL. [3] suggest that neutron spallation off ^{4}He in helium-rich zones could significantly contribute to, or perhaps even generate, the elusive r-process. A realistic simulation of the ν–process must include full nuclear network calculations to take reactions due to ν–induced particles into account. Furthermore, a consistent treatment of the ν–irradiation and nuclear flows must also include effects due to the passage of the shock wave. Some stellar zones experience the neutrino flux during the cooling phase following shock ejection while in other zones the ν–induced modifications are already completed when the shock arrives. Explosive nucleosynthesis can be simulated assuming adiabatic expansion on the hydrodynamic time scale $\tau_{dyn} \sim 446/\sqrt{\rho}$ s. In the case of Type II supernovae, WEAVER and WOOSLEY [10] found that, at a time when the shock wave is traversing the mantel of heavy elements, the temperature jumps to its peak value

$$T_p \sim 2.4\ \mp \gamma 10^9\, E_{51}^{1/4}\ r_9^{-3/4}\ K \tag{2}$$

where E_{51} is the total kinetic energy of the supernova in units of 10^{51} ergs. The peak temperature can also be estimated solving the Rankine-Hugoniot equations assuming, as detailed models suggest, that the portion of the supernova behind the shock is nearly isothermal and radiation dominated. Due to the radial expansion (v $\sim$ 5 10^3 km s^{-1}) the neutrino flux during the post-shock phase declines more rapidly than the assumed exponential decay. The neutrino reaction rates after shock arrival are thus given by

$$\lambda_\nu = \sigma(T_\nu)\, \Phi_\nu\, b_c \left(1 + \frac{v_9}{r_9(0)}(t - t_0)\right)^{-2}, \tag{3}$$

where the shock arrival time t_0 can be estimated using the Sedov solution

$$t_0 \sim 0.7\, r_9 \left(\frac{M_{env}}{E_{51}}\right)^{1/2} s, \tag{4}$$

with M_{env} being the Lagrangian mass coordinate at radius r_9 (0) minus the remnant mass (1.4 $M_\odot$). For details of the calculation of the inelastic neutrino scattering across section $\sigma(T_\nu)$ and the branching ratio for emission of particle c we refer the reader to WH[3] and HAXTON [11]. The qualitative features of the nuclear response to neutrinos of energies up to $\sim$ 150 MeV are well understood. The threshold response is determined by allowed Gamow-Teller transitions. The allowed response should be weak for closed-shell nuclei like ^{4}He, ^{16}O, and ^{40}Ca, where such transitions are Pauli blocked, but relatively strong for nuclei located between closed shells. Because of first-forbidden operators the higher energy μ– and τ–neutrinos can strongly excite the supermultiplet of giant resonances. These transitions are direct analogs of the giant E1 resonance observed in nuclear photoabsorption. WH[3] carried out full shell model calculations of neutral and charged current cross sections for a set of 18 "node nuclei" that included most of the abundant network spescies between ^{4}He and ^{80}Zr. Cross sections for non-node nuclei were obtained by interpolation. Theoretical uncertainties in the cross section calculations are estimated to be less than $\sim$ 10%.

III. Results

When one considers the declining number of stars of increasing mass and the fact that more massive stars exhibit greater absolute yields of heavy elements one arrives at a representative Type II supernova with M $\sim$ 25 $M_\odot$. Together with the relevance of SN1987A for determining the ν–properties of Type II supernovae a 10 $M_\odot$ star is motivated for a study of the dominant features of the ν–process. WH^3 adopted a model of WOOSLEY and WEAVER [12] that was evolved without mass loss. Because the stellar model carries only 13 major species a separate calculation using a much larger reaction network has to be carried out "off-line" to determine initial compositions for the various zones under consideration. Population I and II abundance sets were calculated using a scaled solar abundance distribution.

Consider the nucleosynthetic signature of the ν–process when various stellar zones of given initial composition are subjected to the thermodynamic conditions and neutrino flux described in previous sections. Here we emphasize only those elements that are produced by the ν–process in large enough amounts to account for their solar abundance. A large number of species experience significant production enhancements, such as the γ–ray line isotopes ^{26}Al and ^{22}Na, but an absolute calibration of their yield requires more accurate and complete stellar model calculations that include effects of convection in the presupernova star. Typically the ratio of the mass fraction of major elements in the ejecta to their solar mass fraction fall in the range P $\sim$ 5–10 ([10]). Thus for an isotope to be produced by the ν–process in significant amounts one requires

$$\int \frac{X}{X_\odot}(m)dm \sim P\, M_{ej}. \tag{5}$$

For example, an isotope created with a production factor $X/X_\odot$ = 200 in 0.5 $M_\odot$ of the star would be equivalent to a production factor of $\sim$ 5 in the entire star and would therefore be counted as a "success." Successful production of mass 7 and 11 elements results in the arguably most significant change of current nucleosynthesis paradigms. WH^3 find that ^{7}Li and ^{11}B (but not ^{6}Li, ^{9}Be, and ^{10}B) are produced by $^4He(\nu,\nu'n)^3He(\alpha,\gamma)^7Be(\alpha,\gamma)^{11}C$ and $^4He(\nu,\nu'p)^3H(\alpha,\gamma)^7Li(\alpha,\gamma)^{11}B$. The nuclei ^{7}Be and ^{11}C decay to ^{7}Li and ^{11}B, respectively. Mass 7,11 production requires high ^{4}He seed abundances. In the 20 $M_\odot$ model studied by WH^3 this occurs either in the helium shell or in the innermost shells to be ejected from the supernova which expand so rapidly that the freeze-out from NSF is rich in α particles. Lithium has been regarded as either a product of cosmic ray spallation (CRS) or Big Bang nucleosynthesis [13]. Boron is also traditionally a CRS product. However, ^{11}B is not well produced in past studies of CRS which has led to the hypothesis that cosmic rays might have an (unobservable) alternate low-energy component [13]. The results of WH^3 argue against such a contrivance.

Another isotope successfully produced by supernova neutrinos appear to be ^{19}F whose nucleosynthetic origin has traditionally been very uncertain. The results of WH^3 confirm those of WOOSLEY and HAXTON [4] who showed that fluorine is chiefly a product of the ν–process operating in the neon shell. Although fluorine can also be produced in the presupernova star by partial helium burning recent studies of the ν–process confirm the conclusion of WH^3 that ν–production of ^{19}F predominates in massive stars.

Perhaps the most intriguing implication of the ν–process is the possibility that spallation of neutrons off ^{4}He in the helium zone could provide the elusive site of the r-process [3]. WH^3

explored many circumstances in this respect but did not find a successful site in which the solar r-process might be produced. Although some simulations gave r-process like neutron exposures, the neutrons liberated in those cases were not due to ν–spallation but rather reflect the occurrence of explosive helium burning as the shock passes through a mixture of helium, iron, and ^{18}O or ^{22}Ne. However, for very small metallicities (less than 0.001 solar), compact stellar progenitors (helium shell at $r \sim 10^9$ cm), and small neutron capture cross sections on major poisons (such as ^{12}C) a ν–induced r-process contribution could be significant. While these results suggest that a ν–r-process from low mass stars in the very early Galaxy might be observable in extremely metal deficient stars, the origin of the present day r-process must invoke other causes.

References

1. G. V. Domogatskii, R. A. Eramzhyan, and D. K. Nadozhin 1978, *Ap. Space Sci..*, **58,** 273.
2. S. E. Woosley 1977, *Nature,* **269,** 42.
3. R. I. Epstein, S. Colgate, and W. Haxton 1988, *Phys. Rev. Lett.*, **61,** 2038.
4. S. E. Woosley and W. Haxton 1988, *Nature,* **334,** 45.
5. S. E. Woosley, D. Hartmann, R. Hoffman, and W. Haxton 1990, *Ap. J.*, in press.
6. E. Baron, these proceedings.
7. J. Cooperstein 1988, *Phys. Rep.*, **163,** 95.
8. A. Burrows and J. Lattimer 1986, *Ap. J.*, **307,** 178.
9. H.-T. Janka and W. Hillebrandt 1989, *Astr. Ap. Suppl.*, **78,** 375.
10. S. E. Woosley and T. A. Weaver 1986, *Ann. Rev. Astr. Ap.*, **24,** 205.
11. W. C. Haxton 1988, *Phys. Rev. Letters,* **60,** 1999.
12. S. E. Woosley and T. A. Weaver 1988, *Phys. Rep.*, **163,** 79.
13. M. Arnould and M. Forestini 1989, in *Nuclear Astrophysics,* eds. M. Lozano, M. I. Gallardo, and J. M. Arias (Springer Verlag Berlin: Heidelberg), p. 48.

Possible Gamma-Ray Signatures of an r-Process Event

Bradley S. Meyer & W. Michael Howard

The r-(or rapid) process of neutron capture nucleosynthesis is responsible for the formation of roughly half of the abundances of the nuclei with mass greater than $A \approx 80$ and for all of the actinides. It is known that the r-process occurs in a hot ($T > 10^8 K$), high neutron number density ($n_n > 10^{20} cm^{-3}$) environment where neutron captures typically occur more ***rapidly*** than β-decays. The nuclei thus tend to follow a path in the neutron number-proton number plane many neutrons rich of the β-stability line. At closed neutron shells ($N = 50, 82$, and 126), the β-decay rates are especially slow so that the nuclear abundances build up at these points along the r-process path. Once the r-process has stopped, the nuclei decay from the r-process path to the stability line, thereby giving the final abundances of the r-process. The large abundances built up at the closed neutron shells then yield the three well-known abundance peaks in the r-process abundance distribution at $A = 80$, 130, and 195. For a general review of the r-process, see SCHRAMM [1] and MATHEWS and WARD [2].

Although the general features of the r-process are well known, the actual astrophysical environment or event in which the r-process occurs remains a great mystery. Supernova explosions are perhaps the most likely event, but there has as yet been no direct evidence to support this belief. In this paper we suggest several γ-ray lines as possible signatures of a recent r-process. If these lines were observed in the ejecta of SN 1987A, we would have direct evidence for supernovae as r-process events.

The question of γ-rays from cosmic radioactivity is well-studied (see, for example, CLAYTON [3] and WOOSLEY, PINTO, and HARTMANN [4]). Apart from ^{125}Sb [3] and some transbismuth elements (CLAYTON and CRADDOCK [5]), however, r-process nuclei have not been considered as sources of cosmic γ-rays. We consider in this paper γ-rays from r-process nuclei lighter than bismuth. Since the abundance of these nuclei is fairly well-known for the solar system, we can make estimates of the γ-ray flux resulting from their decay. If the r-process producing these nuclei is greatly different from the nucleosynthesis processes producing the solar system r-process nuclei, the flux estimates will be off. The sharpness of the solar system r-process peaks argues against strong variation among r-process events, however, so we expect flux estimates made from the solar system elemental abundance distribution to be reasonably reliable.

The γ-rays that would serve as r-process signatures would result from the β-decay of nuclei from the r-process path to the β-stability line. Near the r-process path, β-decay life-times are on the order of milliseconds. As the decaying nuclei approach the stability line, however, the decay life-times become greater, reaching in some cases tens of years to thousands of years or more. Since decaying parent nuclei in many cases leave daughter nuclei in excited states, the daughter nuclei will decay to their ground states by emitting γ-rays of specific energy, thus producing signatures of a recent r-process. The count rate at time t after the r-process for a γ-ray line of energy E from such a decay is given by

$$N(E) = N_A \lambda_A \Gamma(E) \exp(-\lambda_A t), \tag{1}$$

where N_A is the total number of nuclei of mass A produced in the r-process, λ_A is the β-decay rate of the relevant decaying A-chain nuclide, and $\Gamma(E)$ is the branching ratio of β-decay to a level at excitation energy E in the daughter nucleus followed by γ-deexcitation to the daughter nucleus ground state. Equation (1) assumes that only one nuclide in a given A-chain has a long life-time and has a daughter that emits a γ-ray. If a long-lived nucleus does not emit a γ-ray upon β-decaying but a later, shorter-lived β-decay in that A-chain does emit a γ-ray, the count rate becomes

$$N(E) = N_A \lambda_{A_1} \Gamma_{A_2}(E) \exp(-\lambda_{A_1} t), \tag{2}$$

where λ_{A_1} is the decay rate of the long-lived nuclide A_1 and $\Gamma_{A_2}(E)$ is the branching ratio for β-decay of the short-lived nuclide A_2 to an excitation energy E in its daughter nucleus followed by γ-deexcitation to the daughter nucleus ground state.

We now consider specific candidates for signatures of a recent r-process. We choose decay A-chains that have nuclides with life-times between 200 days and 100 yrs. Gamma-rays from decays with much shorter life-times will probably be obscured by the ejecta of the r-process event. Decays with much longer life-times would produce too low a flux of γ-rays due to the small decay rate. We present the candidate γ-ray decay lines in Table 1. In this table, we identify the relevant decay, the γ-ray energy, the $\tau_{1/2}$ for that line (i.e. $\ln 2/\lambda_A$ in equation (1) or $\ln 2/\lambda_{A_1}$ in equation (2)), and the coefficient B in the following expression for the γ-ray flux at Earth of r-process γ-rays, scaled to the distance of SN 1987A:

$$F(E) = B \exp(-t \;\ln 2/\tau_{1/2}) \left(\frac{50\ kpc}{d}\right)^2 \left(\frac{M_r}{10^{-4} M_\odot}\right) cm^{-2}\ s^{-1}. \tag{3}$$

In equation (3), d is the distance to the r-process event and M_r is the mass of the r-process material produced in the event. In order to compute B, we used life-times and branching ratios given in LEDERER and SHIRLEY [6] and the solar system r-process abundance distribution in HOWARD *et al.* [7]. To get an estimate of M_r, we consider the case that all r-process material was formed in all supernovae. There is approximately $3 \times 10^4\ M_\odot$ of r-process matter in the Galaxy (inferred from the data in HOWARD *et al.* [7] and ANDERS and EBIHARA [8]). Since the Galaxy is roughly 10^{10} years old (for example, see FOWLER and MEISL [9]) and the supernova rate is probably between $0.1 - 0.01\ yr^{-1}$ in our Galaxy (ARNETT, SCHRAMM, and TRURAN [10] and VAN DEN BERGH, MCCLURE, and EVANS [11]), we may estimate that M_r is roughly $10^{-4}\ M_\odot$. We note that we have assumed that all γ-rays produced escape the ejecta of the r-process event. This may not be true at early time but becomes a better and better approximation as the supernova nebula density falls due to expansion. Finally, we also note that we have not included in Table 1 lines from trans-lead α-decays and β-decays since we only have theoretical predictions of the abundances of these trans-lead nuclei in an r-process event. The theoretical predictions of the abundances of these nuclei are typically less than or roughly equal to the abundances of the nuclei in Table 1, so γ-ray fluxes from these trans-lead nuclei should be roughly comparable to or less than the fluxes in Table 1. Interesting cases are the decay A-chains 210, 227, 228, 241, 252, and 257. See also [5].

From Table 1 we see that there are a number of γ-ray lines which may serve as signatures of a recent r-process. We may now consider whether these lines might be observable from SN 1987A with current observational techniques. The Oriented Scintillation Spectrometer Experiment (OSSE) of the Gamma Ray Observatory is expected to have an experimental sensitivity of about $2 \times 10^{-5} cm^{-2}\ s^{-1}$ for 0.1-10 MeV γ-rays in a 10^6 second observing session [12]. The best rate we would expect for the lines in Table 1 in the 0.1-10 MeV energy range is about $10^{-8}\ cm^{-2}\ s^{-1}$. Clearly at present the best one can probably do with GRO is to set a rather large upper limit on the amount of r-process material that could have been produced by SN 1987A, namely, $M_r < 0.2\ M_\odot$. This limit might help confirm or rule out an exotic

Table 1. Possible γ-ray r-process signatures.

Decay	E (keV)	$\tau_{1/2}(yr)$	B
$^{90}Sr \rightarrow ^{90}Y$			
$\rightarrow ^{90}Zr$	1761	28.8	8.9(-10)
$^{106}Ru \rightarrow ^{106}Rh$			
$\rightarrow ^{106}Pd$	622.2	1.0	1.6(-8)
	511.9	1.0	3.3(-8)
$^{125}Sb \rightarrow ^{125}Te$	428.0	2.7	2.0(-8)
	35.5	2.7	5.7(-8)
$^{137}Cs \rightarrow ^{137}Ba$	661.6	30.2	1.5(-9)
$^{144}Ce \rightarrow ^{144}Pr$			
$\rightarrow ^{144}Nd$	133.5	0.8	2.6(-8)
$^{151}Sm \rightarrow ^{151}Eu$	21.5	90	3.1(-12)
$^{155}Eu \rightarrow ^{155}Gd$	105.4	4.9	3.2(-9)
	86.5	4.9	1.7(-9)
$^{171}Tm \rightarrow ^{171}Yb$	66.7	1.9	2.3(-10)
$^{194}Os \rightarrow ^{194}Ir$	43	6.0	1.7(-8)
$^{194}Ir \rightarrow ^{194}Pt$	328.5	6.0	4.9(-9)

r-process associated with SN 1987A, such as the Mystery Spot being a massive chunk ($M \sim 0.1 M_{\odot}$) of the core that was ejected and then underwent some r-processing (PETRICH [13]). Better limits will require improvements in experimental sensitivity and, consequently, increases in detector size. Moreover in order even to see γ-ray lines from r-process nuclei in the ejecta of SN 1987A one would probably require at least a few thousand-fold increase in sensitivity. Such an improvement would be technologically extremely challenging, if not impossible. Nevertheless, let us point out that ^{137}Cs and the $A = 90$ decay chain have half-lifes of about 30 years, so there is time to try.

We thank Charles Dermer, Mark Leising, Loren Petrich, Phil Pinto, and Stan Woosley for useful discussions and comments. This work was performed under the auspices of the U. S. Department of Energy by the Lawrence Livermore National Laboratory under Contract No. W-7405-ENG-48.

References

1. D. N. Schramm: in Essays in Nuclear Astrophysics, ed. by C. N. Barnes, D. D. Clayton, and D. N. Schramm (Cambridge University Press, Cambridge 1983), p.325.
2. G. J. Mathews, R. A. Ward: Rept. Prog. Phys. 48, 1371 (1985).
3. D. D. Clayton: in Essays in Nuclear Astrophysics, ed. by C. N. Barnes, D. D. Clayton, and D. N. Schramm (Cambridge University Press, Cambridge 1983), p. 401.
4. S. E. Woosley, P. A. Pinto, D. Hartmann: Ap. J. (1989), in press.
5. D. D. Clayton, W. Craddock: Ap. J. 142, 189 (1965).
6. Table of the Isotopes, ed. by C. M. Lederer and V. S. Shirley (John Wiley and Sons, New York, 1978).
7. W. M. Howard, G. J. Mathews, R. A. Ward, K. Takahashi: Ap. J. 309, 633 (1986).
8. E. Anders, M. Ebihara: Geochim. Cosmochim. Acta 46, 2363.
9. W. A. Fowler, C. C. Meisl: in Cosmogonical Processes, ed. by W. D. Arnett, C. J. Hansen, J. W. Truran, and S. Tsuruta (VNU Press, Singapore 1986).
10. W. D. Arnett, D. N. Schramm, J. W. Truran: Ap. J. Letters 339, L25 (1989).
11. S. Van den Bergh, R. D. McClure, R. Evans: Ap. J. 323, 44 (1987).
12. Gamma-Ray Observatory Science Plan, March 1985.
13. L. Petrich: private communication.

SECTION XII

SUPERNOVA REMNANTS AND INTERACTION WITH THE ISM

Recent Optical Studies of Supernova Remnants

Robert A. Fesen

Although supernova remnants (SNRs) have been studied the longest time at optical wavelengths, such investigations continue to provide new information and insights into the properties of Galactic and extragalactic SNRs. One reason for this is that optical SNR research can provide data not easily obtainable at other wavelengths. For example, optical studies of SNRs can yield kinematic data (proper motions and radial velocities), gas temperatures and densities, plus elemental abundances relative to hydrogen and helium - information that is at present either difficult or impossible to derive from radio, infrared, or X-ray data. Optical studies also permit the investigation of SNRs over a wide-range of ionization states; from Fe I in S Andromedae (SN 1885; FESEN, HAMILTON, and SAKEN [17]) to [Fe X] and [Fe XIV] in Puppis A, IC 443, and the Cygnus Loop (TESKE and PETRE [37,38]; BROWN, WOODGATE, and PETRE [3]; TESKE and KIRSHNER [36]). Finally, while it is true that only about 25% of the 160 known Galactic SNRs have been detected optically, this list does include the six youngest Galactic remnants known: SN 1006, SN 1054 (the Crab Nebula), SN 1181 (3C58), SN 1572 (Tycho's SN), SN 1604 (Kepler's SN), and Cas A ($\approx$ 1680).

Recent optical research on SNRs has included work on: (1) young Galactic remnants having high-velocity, nonradiative shock emission, (2) young Galactic ejecta-dominated remnants such as the Crab Nebula and Cas A, (3) extragalactic remnant surveys and studies on individual objects, and (4) detection of young remnants associated with historical observed extragalactic supernovae (SNe). Below, I briefly review some of the recent work done in these areas.

I. Young Galactic SNRs with Nonradiative Shock Emission

Optical emission seen in the remnants of SN 1006 and Tycho's SN (SN 1572) consists of thin and very faint filaments which exhibit only hydrogen Balmer-line emission. Such Balmer-dominated shock emission is often referred to as nonradiative shock emission and is interpreted as being the result of a high-velocity shock moving through a low-density medium leading to the production of both broad and narrow emission-line components (CHEVALIER, KIRSHNER, and RAYMOND [4]). A narrow component is produced by the collisional excitation of neutral atoms passing through the collisionless shock front, while a broad component results from the charge exchange with high-velocity protons. The broad component has a width corresponding to the postshock temperature and thus can be directly related to the shock's velocity.

Recently, improved observations of Tycho's SNR as well as the detection of a broad Hα emission component in SN 1006's faint optical filaments have been carried out by KIRSHNER, WINKLER, and CHEVALIER [26]. These new data have significantly improved our knowledge of these remnants' distances and shock velocities. In Tycho's remnant, a bright eastern filament was observed and found to have a broad Hα component with a width of 1800 ± 100 km s^{-1} (FWHM). This, when combined with nonradiative model calculations [4] and filament proper motion studies (HESSER and VAN DEN BERGH [23]), suggests a shock velocity of 1930 – 2670 km s^{-1} and a kinematical distance of 2.0 – 2.8 kpc. SMITH and KIRSHNER [41] have obtained now even better spectroscopic data on Tycho's nonradiative emission filaments which indicate that the broad component and thus the remnant's derived shock velocity may vary among filaments, presumably as a function of filament density. This may mean that an accurate kinematic distance to the remnant requires care in obtaining both proper motion and spectroscopic data on the same filamentary position.

Kirshner, Winkler, and Chevalier were able also to successfully detect the faint broad 2600 km s^{-1} (FWHM) emission component in SN 1006's filaments at a level just below that of the earlier data of LASKER [27]. They derive a shock velocity of 2800 – 3870 km s^{-1} and a distance of 1.5 – 2.1 kpc using the proper motion value of $0.39 \pm 0.06''$ yr^{-1} reported by HESSER and VAN DEN BERGH [23]. A somewhat smaller but more accurate value for SN 1006's proper motion has recently been measured by LONG, BLAIR, and VAN DEN BERGH [30] who find $0.30 \pm 0.04''$ yr^{-1}. This revised value implies a larger SN 1006 distance of 1.7 – 3.1 kpc. A distance of around 1.7 kpc is indicated by an analysis of UV observations of a sdOB star lying behind the remnant (HAMILTON and FESEN [21]).

High-velocity, nonradiative emission has now been detected in Kepler's SNR (SN 1604). This remnant had previously been known to only exhibit bright radiative emission, meaning that it's emission knots show a variety of emission lines and ionization states characteristic of dense, cooling shocks like those commonly found in older remnants. Faint emission along the remnant's northern limb first detected by D'ODORICO et al. [9] has now been shown to be nonradiative emission (FESEN et al. [13]). This emission region exhibits both narrow and broad Hα components with the broad one suggesting a shock velocity in the range 1600 – 2800 km s^{-1}. Besides yielding a direct estimate of Kepler's shock velocity, this result also means that it is possible to directly measure shock velocities in SNRs having largely radiative filaments, and raises questions about remnant classification schemes based upon the presence or absence of nonradiative emission filaments (VAN DEN BERGH [43]).

Even more recent spectroscopic studies of Kepler's optical emission obtained by BLAIR and LONG [2] indicate that not only is the nonradiative emission not confined to just the remnant's northern section but also is emitted by several knots located near the remnant's projected center. Until now, all nonradiative shock emission, whether in young or old remnants like the Cygnus Loop (see HESTER, RAYMOND, and DANIELSON [24] and references therein) has exhibited a thin, faint filamentary morphology located along the tangential edge of a remnant. Blair and Long's new results may imply that the suspected circumstellar medium around Kepler consists of a wide range of densities. The lowest density material produces the northern thin faint nonradiative filaments like present in SN 1006 and Tycho, while slightly higher density clumps produce the observed knotty nonradiative emission, with the densest clumps producing the bright, radiative emission that has been long observed in the remnant. If this is correct, then it lends support to the idea that the medium surrounding Kepler contains circumstellar mass loss material from the SN progenitor (see BANDIERA [1] and references therein).

II. Young Galactic Ejecta Dominated Remnants

The Crab Nebula: As we approach the 1990's, the Crab Nebula remains the subject of much active optical research. Indeed, several recent studies have produced important and somewhat surprising results, running contrary to the widespread feeling that the Crab is a well understood remnant.

DAVIDSON [7] obtained an integrated spectroscopic scan of the whole nebula and found that estimates of the [O III]/continuum flux ratio taken over the last 25 yr are discrepant. He suggested that either the visual continuum of the Crab may be changing rapidly at about 2% yr^{-1} or that the nebula's [O III] flux may have changed. VÉRON-CETTY and WOLTJER [45] have questioned Davidson's results using CCD images taken in 1986 and 1988 which covered the central portion of the nebula. They place an upper limit to a decrease of 0.5% on the continuum's flux and suggest the real decrease is closer to 0.3%. According to Véron-Cetty and Woltjer, this last value would indicate the pulsar resupplies about half of the energy lost in the nebula's expansion, $\approx 10^{38}$ erg s^{-1}. However, DAVIDSON [8] maintains that, since Woltjer and Véron-Cetty's results were based upon imaging of only the central portion of the nebula, the question regarding possible rapid changes in the remnant's optical brightness remains open.

In an extensive new optical study of the remnant, MACALPINE et al. [31] present spectra and images of the Crab Nebula from which they make important new claims about the remnant. They interpret N-S oriented, long-slit spectra as suggesting bipolar 'bubbles' in the Crab's filamentary structure both above and below a band of helium-rich filaments which run across the center of the remnant. They find that many of the brighter, central filaments are composed almost purely of He, as much as 95% by mass. In addition, CCD images taken using an interference filter centered on the [Ni II] 7378 Å emission line show that the remnant's Ni II emission is not distributed like that of other lines but instead is stronger primarily in the northern filaments, coincident with the low He filaments. Strong [Ni II] line strengths can be interpreted as indicating an Ni abundance as much as 55 times over solar, although the true nature of the Crab's unusually strong [Ni II] is uncertain (HENRY and FESEN [22]). Furthermore, they propose that the high Ni and low He found in the northern filaments indicate an interaction of nickel-rich ejecta with an ambient interstellar cloud in this region. MacAlpine et al. conclude that the mass of the filaments + pulsar is closer to 8 - 9 $M_{\odot}$ rather than the 2 - 3 $M_{\odot}$ of previous estimates, thereby suggesting the precursor's intial mass was around 20 – 30 $M_{\odot}$. If true, this would have important implications regarding the mass range of SN II progenitors since such massive stars have been implicated as possible progenitors of the hydrogen-deficient SN Ib (VAN DEN BERGH [43]; WEILER and SRAMEK [46]).

Recent optical data is beginning to resolve some of the general properties of the Crab's peculiar northern 'jet'. WOLTJER and VÉRON-CETTY [47] have reported the detection of faint optical continuum while new kinematic studies of the Crab Nebula's jet by MARCELIN et al. [32] indicate it formed around the same time as the rest of the remnant. Marcelin et al. obtained long-slit spectra which show the jet to be expanding at a velocity of 260 km s^{-1} with random motions of 60 km s^{-1}. This, when combined with improved proper studies of the jet made by STAKER and FESEN [42] which show the jet's average proper motion is $0.27 \pm 0.04''$ yr^{-1}, suggest the jet formed coeval with the rest of the nebula and both have a homologous-like expansion. Marcelin et al. support the idea that the jet formed out of an instability in the Crab's filamentary shell rather than the plasma beam models of SHULL et al. [40] or the 'shadowed flow' model of MORRISON and ROBERTS [35].

Finally, although the presence of dust in the Crab had been suggested earlier from infrared observations (GLACCUM et al. [18]; MARSDEN et al. [33]; MEZGER et al. [34]), optical data now clearly supports the idea of dust inside the denser filaments. In 1987, WOLTJER and VÉRON-CETTY [47] reported evidence of optical absorption in one bright [O III] filament. More recently, FESEN and BLAIR [16] have found that dusty filaments are present throughout the remnant and coincide not with [O III] bright filaments but ones particularly strong in low-ionization emission lines, e.g. [S II], [O I], and [C I]. Dust condensation in the Crab probably occurred early in the evolution of the remnant when the filaments were denser and had cooled to below 1000 K. The presence of dust in the Crab indicates that it can form and survive in a relatively hostel environment around an active pulsar, and may help to explain the recent discovery of molecular hydrogen in some of the Crab filaments (GRAHAM, WRIGHT, and LONGMORE [20]).

Cassiopeia A: New deep optical CCD imaging and spectra of Cas A suggests that the Cas A progenitor star may have been a Wolf-Rayet star up to the moment it exploded. FESEN, BECKER, and BLAIR [14] and FESEN, BECKER, and GOODRICH [15] have discovered more than a dozen faint emission-line knots that lie outside of the remnant's main radio and X-ray emission shells. The new ejecta knots show strong [N II] emission lines, weak or no Hα emission, and possess ejecta in velocities between 7000 and 8600 km s^{-1}. These knots are called 'fast-moving flocculi' (FMFs) and indicate that the Cas A supernova had a maximum expansion velocity of at least 8500 km s^{-1}, a thin nitrogen-rich but hydrogen-deficient envelope, an explosion date around AD 1680, and a likely WN Wolf-Rayet star progenitor. The supernova explosion apparently left no visible stellar remnant or surviving binary companion behind. Deep images by VAN DEN BERGH and PRITCHET [44] taken under excellent seeing conditions have set better magnitudes limits on the presence of a stellar remnant of 23.5 in I and 24.8 in R corresponding to $M_I \gtrsim +8.9$ and $M_R \gtrsim +9.3$. The region around Cas A's center of expansion appears void of any detectable stars.

Recent follow-up images and spectroscopic data (FESEN and BECKER [11]) continue to support the notion of a thin hydrogen and nitrogen-rich layer on the progenitor at time of SN outburst. They report the discovery of an ejecta knot lying near the base of the remnant's NE jet which possesses both the hydrogen and nitrogen emission seen in the remnant's QSFs and FMFs but also shows strong [O I], [O III], and [S II] emissions like those seen in the remnant's oxygen and sulfur-rich FMKs (see Fig. 1). In addition, new images of Cas A (FESEN [10])

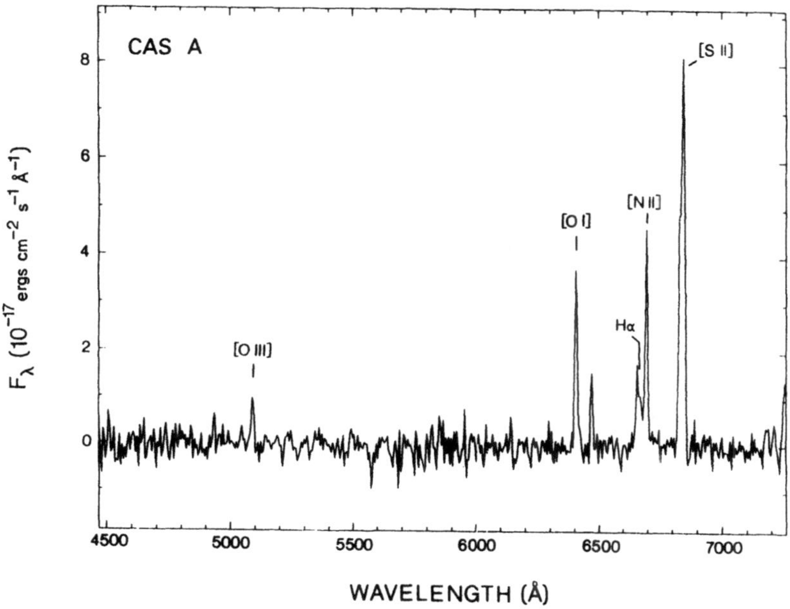

Figure 1: Optical spectrum of a 'mixed-ejecta knot' in Cas A. Knot exhibits the Hα and [N II] line emission like that of the remnant's QSFs and FMFs while also showing strong lines of [O I] 6300,6364 [O III] 4959,5007 and [S II] 6717,6731 like that seen in the FMKs.

show the remnant's NE jet of ejecta stretches out to at least a radial distance of 280″ from the center of expansion. If the explosion date is taken to be 1680 AD, then this angular separation implies an transverse velocity of nearly 13 000 km s^{-1}, assuming no deceleration. This, in turn, means that the jet, which consists largely of S and O-rich knots, was somehow ejected at velocities significantly greater than even those experienced by the progenitor's photosphere. Deep images indeed give one the impression of the jet as an eruption from the main optical shell. The origin of the jet is unknown but what is clear is that it represents an asymmetrical ejection of underlying material up through the progenitor's original surface layers. Finally, one notes that Cas A's jet, like that of the Crab Nebula, appears to lie very close to the plane of the sky thereby permitting us an excellent viewing angle.

III. Extragalactic Remnant Surveys

Although more difficult to study than Galactic SNRs due to increased distances, extragalactic remnants offer distinct advantages in the optical study of SNRs. Due to less extreme extinction variations particularly in face-on systems, it is easier to study an entire galaxy's SNR population. Moreover, while galactic remnants often have highly uncertain distances making meaningful comparisons of their optical morphologies difficult, remnants in a given galaxy are all located at approximately the same distance. Several studies of the LMC, SMC, and local group galaxies have identified now almost 100 optically visible SNRs. These have been detected largely based upon a comparison of [S II] and Hα interference images on which SNR shock emission produces larger [S II]/Hα ratios than for photoionized H II regions. Advances in the optical study of extragalactic SNRs promise to have important implications on general remnant studies.

CHU and KENNICULT [5] obtained long-slit Echelle spectra on nine LMC and SMC remnants from which they derive crude integrated velocity profiles and then use them to distinguish foreground nebula contamination from SNR spectra. They also outline a method to distinguish between SNRs and stellar wind-blown bubbles using kinematic information alone. KIRSHNER et al. [25], using spectra of the oxygen-rich LMC remnant 0540-69.3, find emission-line widths of 2735 km s^{-1} (FWZI) suggesting an expansion age of just 760 ± 50 yr. This would make it perhaps the second youngest SNR in the LMC behind SN 1987A. While this remnant is similar to Cas A in that is shows strong lines of O and S, 0540-69.3 exhibits emission of Fe, Ni, and probably Hα as well. They propose that 0540-69.3 is the young remnant formed by the core collapse of a massive star akin to SN 1987A.

One of the best galaxies to search for SNRs is M33, and LONG et al. [28] have recently completed an atlas of optically identified remnants in this galaxy. Long et al. mainly utilized the optical selection criterion of large [S II]/Hα ratios to identify remnant candidates. Their survey, which consisted of both images and spectra covering the galaxy's inner 15 arcmin region, found a total of 50 SNRs. This suggests that M33's total remnant count could be around 80. The cumulative number vs. diameter relation found for M33's SNRs appears to obey a $N(\leq D) \propto D^{2.1}$ function, substantially steeper than that indicated in previous M33 studies but suggesting an expansion law reasonably close to that of a Sedov expansion. Spectroscopic follow-up on these objects will yield electron densities and abundances which can be used with diameters to further study evolutionary effects and galactic abundance gradients.

IV. The SN–SNR Connection

Until very recently, no supernova had been observed past about 700 days after optical maximum. This led to severe difficulties in relating remnants to their parent SNe. Simply put, we observe extragalactic SNe with ages of 0 – 2 yr yet study young Galactic SNRs with ages of 300 - 2000 yr. Today this situation has changed: there are now 6 SN/SNRs known. These are SN1987A in the LMC, SN1986J in NGC 891, SN1980K in NGC 6946, SN1961V in NGC 1058, SN1957D in M83, and SN1885 in M31. The opening of the observational connection between SN and their remnants promises to be one of the most useful and active fields of SN/SNRs research.

SN 1980K: Recent optical CCD images of the SN site in NGC 6946 by FESEN and BECKER [12] revealed faint Hα emission nearly 8 years after initial outburst. This emission has remained constant to ± 0.2 mag between 1987.5 and 1988.6. A low dispersion 1988 spectrum shows both narrow and broad Hα line emission, broad [O I], [O III], [Fe II], and [Ca II] and/or [O II] emission, and a faint, underlying blue continuum at ≈ 24 mag (see Fig. 2). All broad emission lines appear to have

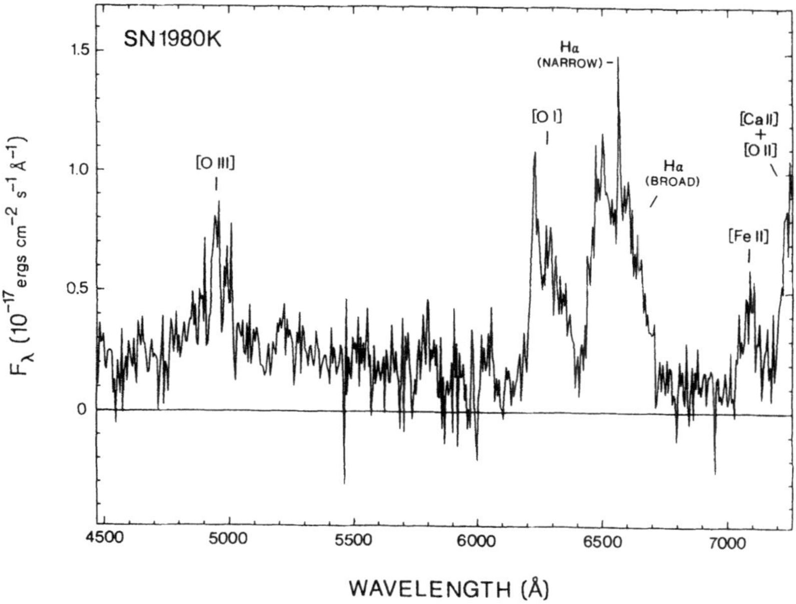

Figure 2: August 1988 spectrum of SN 1980K (Day 2844); from FESEN and BECKER [12].

mean blueshifted line centers possibly indicating the formation of dust within SN 1980K's ejecta. Expansion velocities suggest two distinct emission regions; i.e., a hydrogen-rich zone ($V_{exp} \approx 6900$ km s^{-1}) plus an oxygen- and iron-rich zone ($V_{exp} \approx 5300 - 5500$ km s^{-1}). The narrow Hα emission likely originates in the SN's circumstellar material with the faint blue continuum arising from the pre-SN's parent OB association. Detection of optical emission from the narrow Hα emission likely originates in the SN's circumstellar material with the faint blue continuum arising from the pre-SN's parent OB association. The answer to the key question of the origin of SN 1980K's late-time energy source may be the interaction of the SN's ejecta and shock wave with this circumstellar matter.

SN 1961V: Exhibiting perhaps one of the most unusual SN light curves, SN 1961V in NGC 1058 was classified as the prototype for Type V SNe by Zwicky. It showed narrow H emission lines ($V_{exp} = 2000$ km s^{-1}), and is the only SN other than SN 1987A to have been observed prior to outburst – in this case as an 18th mag star at least back to 1937. Spectra of the site reveals broad Hα emission (FWHM = 2100 km s^{-1}) with a luminosity of 2×10^{36} erg s^{-1} (GOODRICH et al. [19]). Goodrich et al. argue that SN 1961V was not a true SN event (i.e. the explosion of a massive star ending its life) but instead was an eta Car like outburst of a very massive star with the progenitor surviving having faded optically due to the formation of a dusty 1 - 10 $M_\odot$ circumstellar shell. They argue that the progenitor must be one of the most massive and luminous stars known with $M_{ZAMS} \gtrsim 240\ M_\odot$ and believe it is currently of type Of/WN.

SN 1957D: LONG, BLAIR, and KRZEMINSKI [29], using deep interference filter images of the inner portion of M83, have detected the optical remnant of SN at the exact position where COWAN and BRANCH [6] had detected nonthermal radio emission. Spectra show broad [O III] emission (FWHM $\approx$ 2500 km s^{-1}) suggesting the formation of an oxygen-rich, Cas A-like remnant. Improved follow-up spectra have been obtained by TURRATO and DANZIGER [39] which reveal [O I] 6300 and possibly Hα and [S II] emission lines in addition to the [O III] line emission. Unfortunately, no spectra near outburst were taken, leaving its SN classification type unknown.

SN 1885: The first ever observed extragalactic SN, commonly known as S And, occurred in the nuclear bulge of M31, just 16″ from the nucleus. Because this SN showed no bright hydrogen lines near outburst, the object has been classified as a SN I pec. If really a Type I and thus the probable explosion of a wd, then its remnant could be Fe-rich. Also, if S And occurred on the near side of M31's bulge, then its remnant might produce an observable patch of obscuration silhouetted against M31's bright nuclear bulge. FESEN, HAMILTON, and SAKEN [17] have successfully detected the remnant as an unresolved spot of obscuration located within 1″ of S And's reported position by using a interference-filter centered on the Fe I resonance line at 3860 Å. This detection suggests SN 1885's remnant is iron-rich, relatively cool, expanding at $\approx$ 5000 km s^{-1} and is located on near side of the galaxy's bulge.

V. Summary

Future optical SNR research seems promising. In the study of Galactic remnants, further work on young objects such as Puppis A, 3C58, and G292.0+1.8 may provide valuable clues as to the nature of their progenitors and thus indicate the SN type involved. A more complete study of the properties of all young Galactic SNRs will help better define their differences and thus lead to an improved understanding on the variety of SN types.

The field of extragalactic SNR research has taken on a new importance in light of the recent detections of remnants associated with historic SNe thereby finally allowing us to directly connect SN events with SNR types. Although preliminary work indicates that such detections may be rare, the presence of optical emission from the quite normal Type II SN 1980K nearly 8 years after outburst indicates the formation of a young, bright supernova remnant. This raises hope that similarly detectable levels of optical emission may exist for at least some other young extragalactic SNe particularly those with detectable nonthermal radio emission.

REFERENCES

1. R. Bandiera: Ap. J. 319, 885 (1987).
2. W. P. Blair and K. S. Long: in preparation (1989).
3. L. W. Brown, B. E. Woodgate, and R. Petre: Ap. J. 334, 852 (1988).
4. R. A. Chevalier, R. P. Kirshner, and J. C. Raymond: Ap. J. 235, 186 (1980).
5. Y.-H. Chu and R. C. Kennicutt: A. J. 95, 1111 (1988).

6. J. J. Cowan and D. Branch: Ap. J. 293, 400 (1985).
7. K. Davidson: A. J. 94, 964 (1987).
8. K. Davidson: Sky and Telescope 78, 341 (1989).
9. S. D'Odorico, R. Bandiera, J. Danziger, and P. Focardi: A. J. 91, 1382 (1986).
10. R. A. Fesen: in preparation (1989).
11. R. A. Fesen and R. H. Becker: in preparation (1989).
12. R. A. Fesen and R. H. Becker: Ap. J. in press (1989).
13. R. A. Fesen, R. H. Becker, W. P. Blair, and K. S. Long: Ap. J. (Letters), 338, L13 (1989).
14. R. A. Fesen, R. H. Becker, W. P. Blair: Ap. J. 313, 378 (1987).
15. R. A. Fesen, R. H. Becker, R. W. Goodrich: Ap. J. (Letters) 329, L89 (1988).
16. R. A. Fesen and W. P. Blair: Ap. J. submitted (1989).
17. R. A. Fesen, A. J. S. Hamilton, and J. M. Saken: Ap. J. (Letters) 341, L55 (1989).
18. W. Glaccum, D. A. Harper, R. F. Loewenstein, R. Pernic, and F. L. Low: Bull. A.A.S. 14, 612 (1982).
19. R. W. Goodrich, G. S. Stringfellow, G. D. Penrod, and A. V. Filippenko: Ap. J., in press.
20. J. R. Graham, G. S. Wright, and A. J. Longmore: preprint (1989).
21. A. J. S. Hamilton and R. A. Fesen: Ap. J. 327, 178 (1988).
22. R. B. C. Henry and R. A. Fesen: Ap. J. 329 693 (1988).
23. J. E. Hesser and S. van den Bergh: Ap. J. 251, 549 (1981).
24. J. J. Hester, J. C. Raymond, and G. E. Danielson: Ap. J. (Letters) 303, L17 (1986).
25. R. P. Kirshner, J. A. Morse, P. F. Winkler, and W. P. Blair: Ap. J. 342, 260 (1989).
26. R. P. Kirshner, P. F. Winkler, and R. A. Chevalier: Ap. J. (Letters) 315, L135 (1987).
27. B. M. Lasker: Ap. J. 244, 517 (1981).
28. K. S. Long, W. P. Blair, R. P. Kirshner, and P. F. Winkler: Ap. J. in press.
29. K. S. Long, W. P. Blair, and W. Krzeminski: Ap. J. (Letters) 340, L25 (1989).
30. K. S. Long, W. P. Blair, and S. van den Bergh: Ap. J. 333, 749 (1988).
31. G. M. MacAlpine, S. S. McGaugh, J. M. Mazzarella, and A. Uomoto: Ap. J. 342, 364 (1989).
32. M. Marcelin, M.-P. Véron-Cetty, L. Woltjer, J. Boulesteix, S. D'Odorico, and E. Lecoarer: preprint (1989).
33. P. L. Marsden, F. C. Gillett, R. E. Jennings, J. P. Emerson, T. De Jong, and F. M. Olnon: Ap. J. (Letters) 278, L29 (1984).
34. P. G. Mezger, R. J. Tuffs, R. Chini, E. Kreysa, and H.-P. Gemund: Astr. Ap. 167, 145 (1986).
35. P. Morrison and D. Roberts: Nature 313, 661 (1985).
36. R. G. Teske and R. P. Kirshner: Ap. J. 292, 22 (1985).
37. R. G. Teske and R. Petre: Ap. J. 314, 673 (1987).
38. R. G. Teske and R. Petre: Ap. J. 318, 370 (1987).
39. M. Turrato and I. J. Danziger: ESO Messeger, June 1989.
40. P. Shull, U. Carsenty, M. Sarcander, and T. Neckel: Ap. J. 285, L75 (1984).

41. R. Smith and R. P. Kirshner: in preparation (1989).
42. B. Staker and R. A. Fesen: in preparation.
43. S. van den Bergh: Ap. J. 327, 156 (1988).
44. S. van den Bergh and C. J. Pritchet: Ap. J. 307, 723 (1986).
45. M.-P. Véron-Cetty and L. Woltjer: Astr. Ap. 201, L27 (1988).
46. K. W. Weiler and R. A. Sramek: Ann. Rev. Astr. Ap. 26, 295 (1988).
47. L. Woltjer and M.-P. Véron-Cetty: Astr. Ap. 172, L7 (1987).

Supernova Remnants and Candidates in M33

R. Chris Smith, Robert P. Kirshner, P. Frank Winkler, Knox S. Long, & William P. Blair

I. Introduction

Supernova remnants (SNRs) provide fundamental information about stellar evolution and the properties of the interstellar medium. The galactic sample of optical SNRs is large, but severely limited by obscuration, and distances and sizes for these remnants remain uncertain. Extragalactic surveys provide samples for which sizes are more easily determined and for which obscuration is less vexing. Over 70 SNRs have previously been identified in local group galaxies; the largest sample consisting of the 25 remnants in the LMC (MATHEWSON *et al.* [1]). Here we describe the results from the first stage of our survey of M33. These results are based on the atlas of SNRs and candidates by LONG *et al.* [2].

Optical SNR surveys employ the large [S II] to Hα ratio found in typical SNRs compared to that in H II regions. Both models and observations show that SNRs usually have [S II]/Hα ratios greater than 0.4 (RAYMOND [3], FESEN, BLAIR, and KIRSHNER [4]), while the ratio in H II regions is usually $\sim$ 0.1. M33 is ideal for such a survey because it is nearby (720 kpc, DE VAUCOULEURS [5]) and relatively face on (i=57°, CONSIDERE and ATHANASSOULA [6]). Previous optical surveys (see D'ODORICO, DOPITA, and BENVENUTI [7] and references therein) have identified 13 confirmed SNRs and several candidates from interference filter surveys using photographic plates. Less than eight of these SNRs have been detected at radio and X-ray wavelengths (GOSS and VIALLEFOND [8], TRINCHIERI, FABBIANO, and PERES [9], and references therein). Surveys in these bands have not yielded significant numbers of new SNRs in M33, although promising work is underway at the VLA (DURIC [10]), and should be feasible with ROSAT. We have therefore begun an optical survey of M33 using CCDs, taking advantage of their linearity and high quantum efficiency to search for fainter, more diffuse SNRs.

II. Observations and Remnant Identification

We have surveyed the inner 15 square arcmin of M33 with narrow-band interference filter CCD images using the 4m Mayall telescope at Kitt Peak National Observatory with the TI-2 CCD at prime focus. A grid of 18 overlapping 3.5 arcmin fields was used to map the nucleus and inner spiral arms of M33. For each field, images were taken in the light of Hα, [S II] $\lambda\lambda$6717, 6731, [O III] λ5007, and a continuum band at 6100 Å. The Hα, [S II], and continuum images distinguish ISM dominated

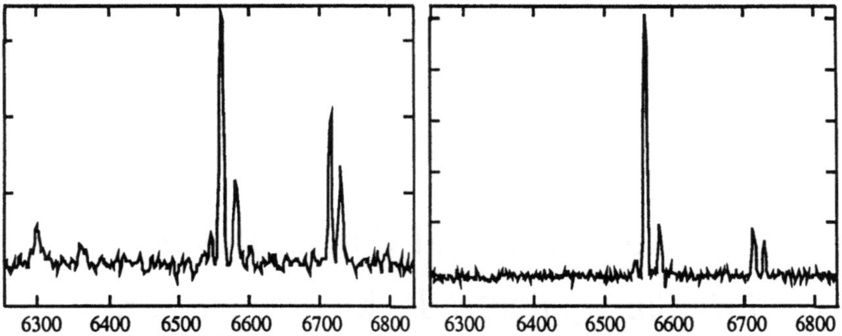

Figure 1: Spectra of a newly identified SNR (left) and a H II region (right) in M33. Note the high [S II]/Hα ratio in the SNR relative to the H II region, as well as the [O I] 6300,6363 emission in the SNR.

SNRs from H II regions and stars, and the [O III] images are used to identify young ejecta-dominated SNRs (see LONG, BLAIR, and KRZEMINSKI [11]).

The SNR candidates were identified in two steps. Objects were initially selected by blinking the $H\alpha$, [S II], and continuum images to identify nebulae which appeared to have high [S II]/$H\alpha$ ratios. These candidates were then verified by measuring background subtracted fluxes from smoothed $H\alpha$ and [S II] images to derive a quantitative [S II]/$H\alpha$ ratio. Of more than 50 candidates identified by blinking, 42 were found to have [S II] to $H\alpha$ ratios greater than 0.4, the dividing line between H II regions and SNRs. This sample includes every one of the 10 spectroscopically confirmed SNRs in our survey area, as well as 2 previously identified candidates which have not been confirmed spectroscopically. An atlas of the 42 objects, with positions, descriptions, and images, can be found in LONG *et al.* [2].

The next stage of our survey is spectroscopic examination of the candidates. We use the Multiple Mirror Telescope with a long slit CCD spectrograph to obtain ~ 5 Å resolution spectra of all of the survey objects. Preliminary reductions show that over 3/4 of the candidates (all those for which we have spectra) exhibit [S II] to $H\alpha$ ratios $\gtrsim 0.4$, and most show [O I] $\lambda\lambda 6300, 6363$ emission, which is often present in SNRs but is weak in photoionized regions (see Fig. 1).

III. Results

In this survey, we have detected 30 NEW candidates in the central region of M33, more than doubling the number of candidates from previous surveys of this region. This confirms the advantages of CCD surveys over photographic techniques, and opens the possibility of pushing optical surveys for SNRs beyond the local group. Figure 2 shows the surface brightness of the 42 objects in our survey, plotted as a function of estimated diameter. Filled points are previously identified SNRs and candidates, and the open points are the newly identified objects. The new candidates are larger and have lower surface brightnesses than previous objects. The spectra show that most, if not all, of these candidates are SNRs.

With this sample, we can begin to study the evolution of SNRs. The distribution of remnants with diameter provides information on the expansion law for remnants.

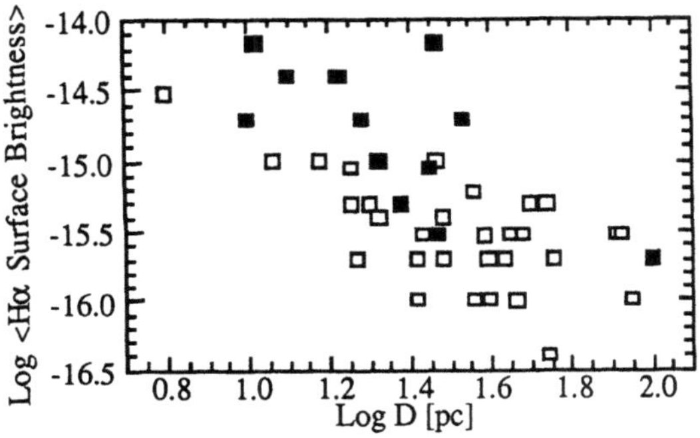

Figure 2: Surface brightness vs. diameter for our sample of SNRs and candidates. Filled squares are previously identified SNRs and candidates; open squares are new candidates. In general, we have identified larger and lower surface brightness objects than previous surveys.

The SEDOV [12] solution predicts $N(<D) \propto D^{2.5}$, in sharp contrast to the results of previous surveys of the LMC (MATHEWSON *et al.* [1]) and M33 (BLAIR and KIRSHNER [13]), which revealed exponents near 1 for this relation, consistent with free expansion of the remnants to their present size. The cumulative number vs. diameter relation for the objects in our survey is plotted in Fig. 3, which shows that the slope in the log-log relation is roughly 2 for diameters smaller than 30 pc in our sample. This slope is in reasonable agreement with the Sedov solution, although a more detailed analysis is required (GREEN [14], BERKHUIJSEN [15])

If the expansion law for remnants is known, we can estimate the supernova rate in M33. We assume that there are as many remnants outside our survey area as there are inside it. With 24 (candidate) remnants less than 30 pc in diameter in our survey, this would give 48 in the whole galaxy. Using $E \sim 10^{51}$ ergs, $n \sim 1\ cm^{-3}$, and Sedov expansion, we obtain a rate of 1 SN every 300 years. This corresponds to 0.9 supernovae per 100 years per $10^{10} L_{B\odot}$, which agrees with the most recent estimates of the total SN rate for late type spirals by EVANS, VAN DEN BERGH, and MCCLURE [16] of $1.9h^2$, for H_0 of $\gtrsim 65$ km s^{-1} Mpc^{-1}.

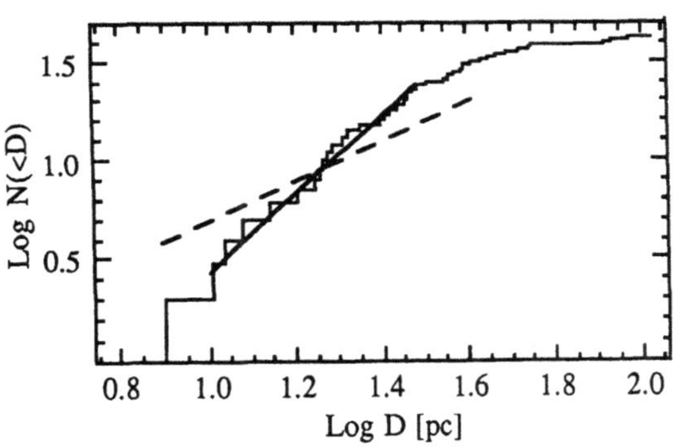

Figure 3: Cumulative number vs. diameter relation for our new sample of candidates and SNRs. The solid line, with a slope 2, is the best fit to the data for objects less than 30 pc in diameter. The dashed line has a slope of 1, the value previous samples gave for this relation.

These results are preliminary, as we are analyzing the images and spectra to quantify the completeness of the sample and perhaps discover a few candidates that were missed by the blink comparison method. The spectra will be used estimate abundances and to derive electron densities, which can be combined with the observed diameters to determine the thermal energy of each remnant (BLAIR, KIRSHNER, and CHEVALIER [17]). In addition, we plan to extend our survey to the whole of M33 with the new large format CCDs that are becoming available. Such a survey will provide a better estimate of the SN rate and facilitate the study of the relation of SNRs (and SNe) with spiral arms and giant star forming regions.

We would like to thank the staffs at Kitt Peak National Observatory and the Multiple Mirror Telescope Observatory for the considerable assistance in the execution of this project, and Tammy Smecker for assistance in the reduction of the image data. RCS and RPK acknowledge support from NSF grant AST 85-16537. PFW's research is supported by NSF grant AST 85-20557, and KSL and WPB acknowledge the support of the Johns Hopkins University Center for Astrophysical Sciences.

References

1. Mathewson, D.S., Ford, V.L., Dopita, M.A., Tuohy, I.R., Long, K.S., and Helfand, D.J. 1983, *Ap. J. Suppl.*, **51**, 345.
2. Long, K.S., Blair, W.P., Kirshner, R.K., and Winkler, P.F. 1990, *Ap. J. Suppl.*, in press for 1 Jan 1990.
3. Raymond, J.C. 1979, *Ap. J. Suppl.*, **39**, 1.
4. Fesen, R.A., Blair, W.P., and Kirshner, R.P. 1985, *Ap. J.*, **292**, 29.
5. de Vaucouleurs, G. 1978, *Ap. J.*, **223**, 730.
6. Considere, S., and Athanassoula, E. 1988, *Astr. Ap. Suppl.*, **76**, 365.
7. D'Odorico, S., Dopita, M.A., and Benvenuti, P. 1980, *Astr. Ap. Suppl.*, **40**, 67.
8. Goss, W. M., and Viallefond, F. 1985, *J. Astrophys. Astr.*, **6**, 145.
9. Trinchieri, G., Fabbiano, G., and Peres, G. 1988, *Ap. J.*, **325**, 531.
10. Duric, N. 1988, in *IAU Coll. 101: Supernova Remnants and the Interstellar Medium,* ed. by R.S. Roger and T.L. Landecker (Cambridge: Cambridge University Press), p. 289.
11. Long, K.S., Blair, W.P., and Krzeminski, W. 1989, *Ap. J. (Letters)*, **340**, L25.
12. Sedov, L.I. 1959, *Similarity and Dimensional Methods in Mechanics*, (New York: Academic Press).
13. Blair, W.P., and Kirshner, R.P. 1985, *Ap. J.*, **289**, 582.
14. Green, D.A. 1984, *M. N. R. A. S.*, **209**, 449.
15. Berkhuijsen, E.M. 1987, *Astr. Ap.*, **181**, 398.
16. Evans, R., van den Bergh, S., and McClure, R.D. 1989, *Ap. J.*, in press.
17. Blair, W.P., Kirshner, R.P., and Chevalier, R.A. 1981, *Ap. J.*, **247**, 879.

The Optical Structure of Cassiopeia A

Jeri E. Reed, A. C. Fabian, & P. F. Winkler

At a distance of 3 kpc and age of about 330 years, Cas A is one of the premier laboratories of the kinematic and chemical structure of supernova remnants. The material in its fast-moving ejecta is unusually rich in oxygen and the silicon-group elements and contains little or no hydrogen (PEIMBERT and VAN DEN BERGH [1], CHEVALIER and KIRSHNER [2]). Although photographic studies of Cas A have been continuing for over 30 years, in this work we present the first complete spectroscopic map of its optical structure.

The optical emission from Cas A consists of a faint, incomplete shell and a jet-like feature of unclear origin. Two major types of optical knots are seen: 1) slow-moving, 'quasi-stationary flocculi' (qsfs) containing nitrogen and hydrogen, which is thought to be circumstellar material lost prior to the explosion; and 2) the very fast-moving ejecta (4000-8000 km/s) containing heavier elements. Asymmetry in Cas A is implied by the existence of its jet, by asymmetric X-ray Doppler shifts (MARKERT *et al.* [3]), and by the lack of correlation between velocity and composition in optical data. Studies of a few bright knots indicated that some heavy material derived from the inner atmosphere is moving faster than the lighter oxygen-rich knots from the outer layers of the star (KAMPER and VAN DEN BERGH [4], CHEVALIER and KIRSHNER [5]), implying that radial mixing has occurred among the layers of the star. This inversion might be principally due to instabilities in the explosion itself, or could reflect interaction with the surrounding material.

In an effort to build a comprehensive map of the kinetic and chemical structure of Cas A, we have obtained optical observations of the face of the nebula and the inner parts of its jet using a CCD spectrograph on the 2.5m Isaac Newton Telescope. The frames were separated by 4 arcsec and covered the wavelength range from 6250 to 7600 Å. The slit width was 2.5″, the resolution along the slit was 0.64 arcsec/pixel and the spectral resolution was 100 km/s/pixel, sufficient to resolve the lines of [SII] and [OII]. The CCD frames were reduced using the FIGARO package, and the spectra corrected for a reddening of $A_V = 4.3$ magnitudes (SEARLE [6]). The complete data set contains about 25,000 spectra. We developed an automated method which allowed us to extract velocities and line strengths for each of the components present in each spectrum of the data set. In this first pass through the data we selected for analysis only lines of 4σ significance or larger. The flux in each line was estimated as the FWHM of the cross correlation peak times the peak line flux. In this way we have found the radial velocity, velocity width and position of every optical knot, and the fluxes of the most important lines needed to determine abundances. We have used this data to create the most complete map of Cas A's optical structure, and the first to show its overall three-dimensional chemical and kinematic structure.

In Fig. 1a we illustrate the correlation of velocity of the fast-moving knots with radial distance from the centre of expansion. It can be seen that the signature of a uniformly expanding shell is present with an expansion velocity of $\sim$ 5000 km/s. Closer inspection of Fig. 1a shows a double arc structure which is also suggested by the radio and X-ray maps of Cas A. The arcs lie along a shell of radius 1.7 arcmin with a velocity centre red shifted by about 900 km/s. The quasi-stationary systems do not appear to be distributed in a shell (Fig. 1b). An expansion velocity of about 500 km/s leads to a 3-D map having major features which cover roughly the same extent as those in maps of the fast knots. A 'cloud' of faster qsfs seems to surround these features.

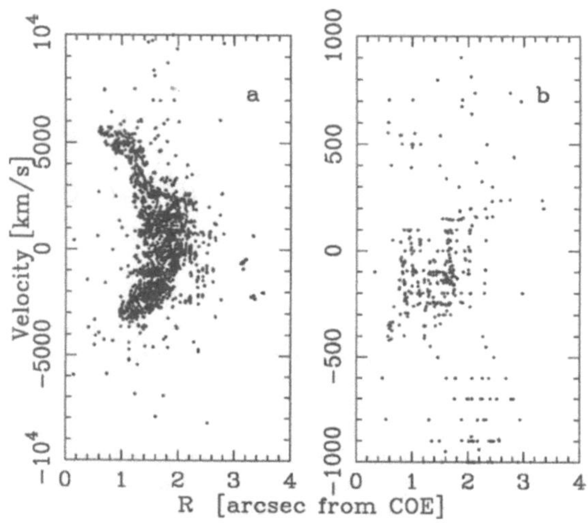

Figure 1. Radial velocity vs. distance from COE for *a)* fast-moving knots; and *b)* quasi-stationary flocculi.

Figure 2 shows the positions of the fast-moving optical knots, where the velocity has been scaled to be 5000 km/s at the radius of the X-ray ring. At least two emission lines had to be detected at 4σ or better for the emission feature to be included in these plots. For comparison, Fig. 2a is a CCD image of Cas A taken through an [OII] filter. Figure 2b is our map of the optical emission in the plane of the sky. The cross-hairs and dotted circle show the centre of expansion and the position of the main ring of X-ray emission at a radius of 1.7 arcmin. The linear appearance of the emission features simply reflects the slit positions on the sky. Figure 2c is the projected view from 'above' the remnant, and Fig. 2d is the projected view from the 'side'. In Fig. 2c and 2d, no correction has been made for the mean velocity of Cas A. Blue shifted velocities have negative values and the dotted line represents the projected X-ray ring as above. The plots do not distinguish between knots of different composition, but the spectra contain lines of O, S, A, and some Ca which we have tabulated for each knot. Lines of Fe and Ni have not been found at the 4σ level, but may be found when we combine the spectra of brighter knots.

The three-dimensional structure is quite asymmetric and seems to consist of a complex of interacting rings. The filaments near the low density region to the northeast, which include the prominent Baade and Minkowski's Filament 1 , appear to be one-dimensional in the plane of the sky. When seen in Figs. 2c and 2d, however, they resolve into a fan of emission from which the jet itself emerges. Our data extends only to the nearest portion of the jet, but this also seems to be spreading in radial velocity with distance from the nebula. The chemical abundance distribution is confused, showing that mixing of knots did occur, but some patterns can be seen. For example, there is a large, blue shifted concentration of oxygen-rich material just east of the centre at a declination between 1 and 2 arcmin above the COE. The large ring seen in Fig. 2c seems to show a gradient from sulfur-rich materials inside towards oxygen-rich at the outer edge.

We are currently searching the data for evidence of very fast moving knots containing nitrogen and hydrogen, which have previously been reported in the outlying portions of the remnant and its jet, but not within the main shell (FESEN *et al.* [7]). (The existence of such knots might imply that material from the outer portions of the star was carried along with the exploding ejecta.) We are also examining the density structure of the nebula via the [SII] lines. Our comprehensive maps of the optical emission and density changes as a function of position and velocity promise to clarify the origin of the asymmetry in Cas A.

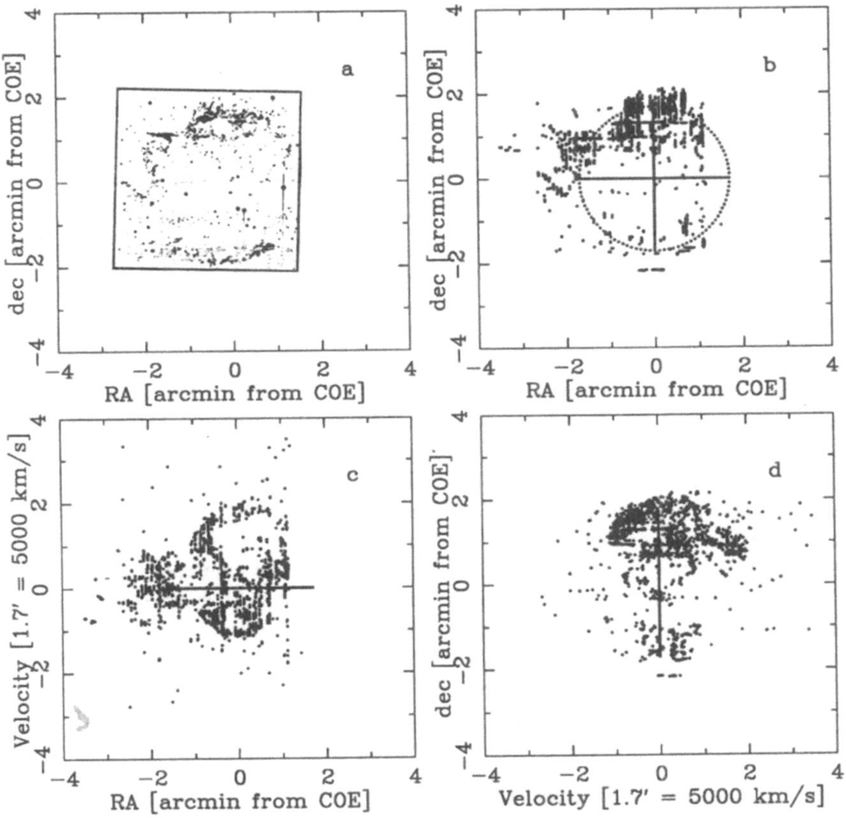

Figure 2. The positions of the fast-moving optical knots. Figure 2*a* is a comparison CCD image of Cas A in the light of [OII] 7325 Å, obtained at the prime focus of the KPNO 4m. The filter bandwidth is 200 Å, sufficient to include the entire velocity range of the Cas A knots. North is up, east to the left. Figure 2*b* represents the positions of the optical knots in the plane of the sky. The cross-hairs indicate the centre of expansion. In Figures 2*c* and 2*d* we have used the appropriately scaled velocity of the knots as a third dimension. Figure 2*c* shows the view from 'above' and Figure 2*d* the view from the 'side' of Cas A in the resulting space. The dotted circle or line represents the position of the X-ray shell at 1.7′.

References

1. M. Piembert and S. van den Bergh: *Astrophys. J.*, **259**, 198 (1971).
2. R.A. Chevalier and R.P. Kirshner: *Astrophys. J.*, **219**, 931 (1978).
3. T.H. Markert, C.R. Canizares, G.W. Clark, P.F. Winkler: *Astrophys. J.*, **268**, 134 (1983).
4. K. Kamper and S. van den Bergh: *Astrophys. J. Suppl.*, **32**, 351 (1976).
5. R.A. Chevalier and R.P. Kirshner: *Astrophys. J.*, **233**, 154 (1979).
6. L. Searle: *Astrophys. J.*, **168**, 41 (1971).
7. R.A. Fesen, R.H. Baker, W.P. Blair: *Astrophys. J.*, **313**, 378 (1987).

Spectrophotometry of Cas A: Implications for Nucleosynthesis in Massive Stars

P. Frank Winkler, Peter F. Roberts, & Robert P. Kirshner

Cas A is the prototype of supernova remnants (SNRs) in which fast-moving, chemically peculiar debris is found. The fast-moving knots have velocities 4000–8000 km/s and are composed entirely of products from advanced stages of nucleosynthesis, with essentially no hydrogen (PEIMBERT and VAN DEN BERGH [1], CHEVALIER and KIRSHNER [2]). Based on this evidence, the fast knots are generally believed to represent uncontaminated ejecta from the core of a massive star. Recent studies by FESEN *et al.* [3] find that the fastest knots in Cas A are very nitrogen-rich, which suggests the more specific identification of the progenitor as a WN star, and that the super-fast nitrogen knots are material from its surface. Along with the handful of remnants with similiar but less extreme properties, Cas A provides one of the few arenas where theoretical models for the evolution and explosion of massive stars can confront observational evidence VAN DEN BERGH [4]. Cas A has become all the more interesting in the wake of SN1987A, for the two objects appear to have many similarities.

We have used the Cryogenic Camera with the KPNO 4-m telescope to obtain long-slit spectra at several positions in Cas A. Three different spectrograph set-ups were used; we have combined the data to give composite spectra for some 20 knots covering the range 4500-10500 Å with approximately 15 Å resolution. Figure 1 shows one long-slit spectrum, taken with the slit oriented nearly E-W across the northern portion of the remnant shell. Velocity variations of several thousand km/s are apparent, even in material which is not resolved as separate knots in direct images. Equally dramatic (though less obvious in this display in which the strong lines are saturated in order to show the faint lines) are large variations in the relative line strengths from one knot to another. Lines from several elements are identified here for the first time in Cas A: [C I] 8727, 9824, and 9850, [Cl II] 8579, [Ni II] 7378, and numerous faint [Fe II] lines.

Figure 2 shows extracted 1-dimensional spectra for two knots of widely varying composition. Lines of oxygen in three different ionization states dominate the spectrum of Fig. 2*a*, and the products of oxygen burning: S, Ar, and Ca manifest themselves only in minute traces, if at all. Furthermore, permitted lines of O I, 7774 and 8446, are seen in addition to the usual forbidden lines. The presence of these lines in the ratio expected for O^+ recombination further confirms an extreme oxygen abundance (WINKLER and KIRSHNER [5]). Oxygen lines are also apparent in the spectrum of Fig. 2*b*, but here the S and Ar lines are far more prominent. The contrasting pair of spectra typifies the spectroscopic range for

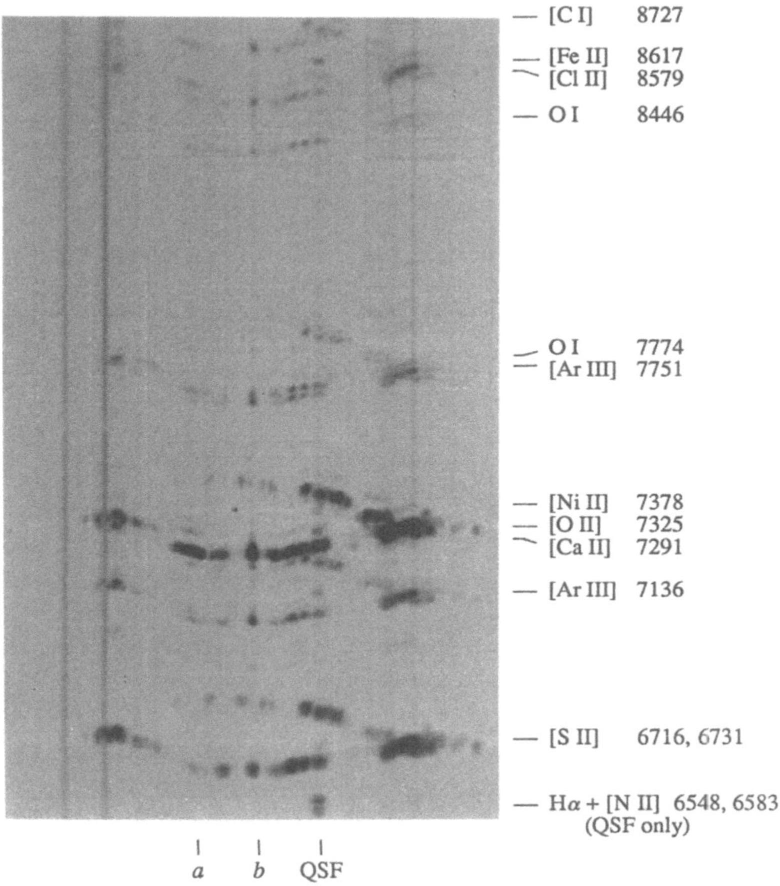

Figure 1. The near-infrared spectrum of Cas A, taken with a long slit crossing the northern portion of the remnant. Knots at the approaching and receding edges of the expanding shell differ in velocity by more than 8000 km/s. Line identifications indicate zero-velocity emission, which is approximately the case near the ends of the slit. Strong lines appear saturated in this display, in order to emphasize the fainter lines. Knots *a* and *b* refer to those shown in Fig. 2. Note the [Fe II] and [Ni II] lines, which are best seen in knot *a*. One quasi-stationary flocculus (QSF) lies on the slit as indicated.

the fast-moving knots in Cas A, from nearly pure oxygen to a mixture in which oxygen-burning products are most evident.

Does the spectroscopic contrast indicate the chemical make-up of these knots, or might unusual physical conditions masquerade as chemical differences? To avoid ionization effects we look at lines from different species that should coexist in the same physical volume, *i.e.*, ions with similar ionization potentials. We have compared the sulfur:oxygen line ratio in different knots for neutral and singly ionized species (Fig. 3*a*). While each ratio varies by almost 2 orders of magnitude among the 20 knots, there is an excellent correlation

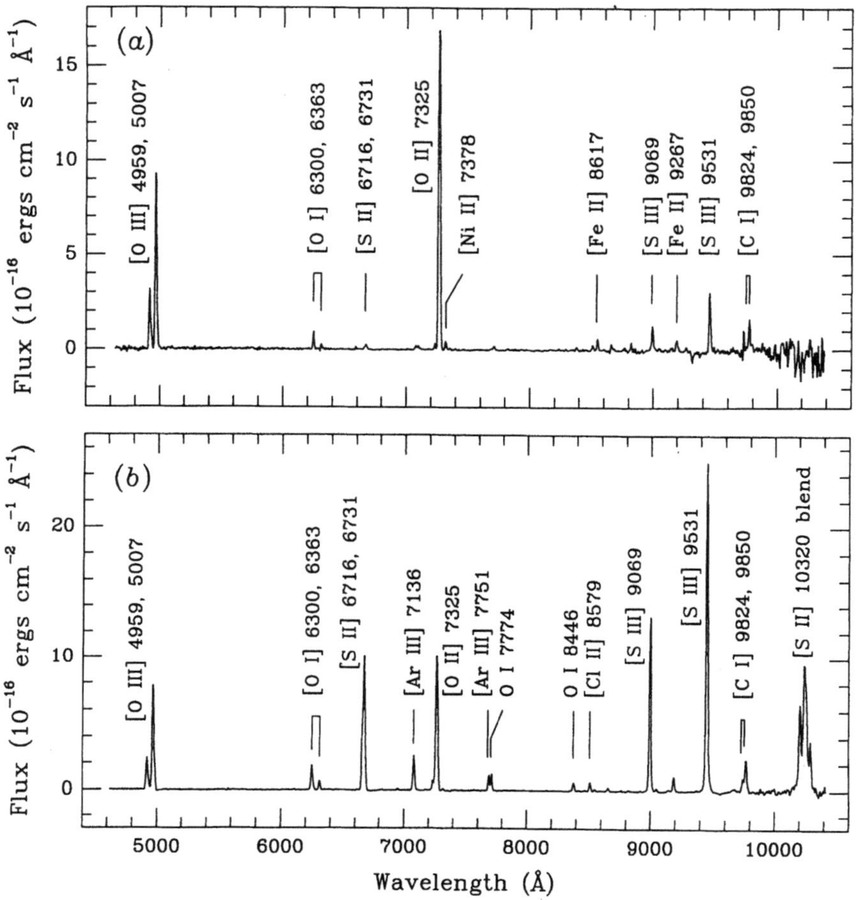

Figure 2. Optical-IR spectra of two fast knots in Cas A, illustrating differences in chemical composition: (*a*) A knot dominated by oxygen lines; note, however, the presence of iron and nickel. (*b*) A knot composed primarily of oxygen and oxygen-burning products. Note the absence of hydrogen emission in both knots.

between the two. A comparison of [S III]/[O II] vs [S II]/[O I] shows a similar correlation. Surely we are seeing true composition variations by large factors. By contrast, we have plotted the [Ar III]/[S III] line ratio vs [S II]/[O II] in Fig. 3*b*. The Ar/S ratio is roughly constant (to within a factor of 3), indicating that Ar and S are formed together in appproximately constant ratio during oxygen burning, while the S/O ratio variations by a factor of 100 represent different mixtures of oxygen and oxygen-burning products – a conclusion tentatively arrived at by CHEVALIER and KIRSHNER [6] based on only a few knots. The measured line strengths indicate that the relative *ionic* abundances of Ar^{++}/S^{++} are in the ratio ~1/8, a value which likely reflects the true *elemental* abundances.

The iron and nickel, though faint, are nevertheless noteworthy. These are most prominent in the "oxygen knot" of Fig. 2*a* (see also Fig. 1), and have the same velocity as the oxygen lines. The same lines are present in many other knots, but with strengths that are

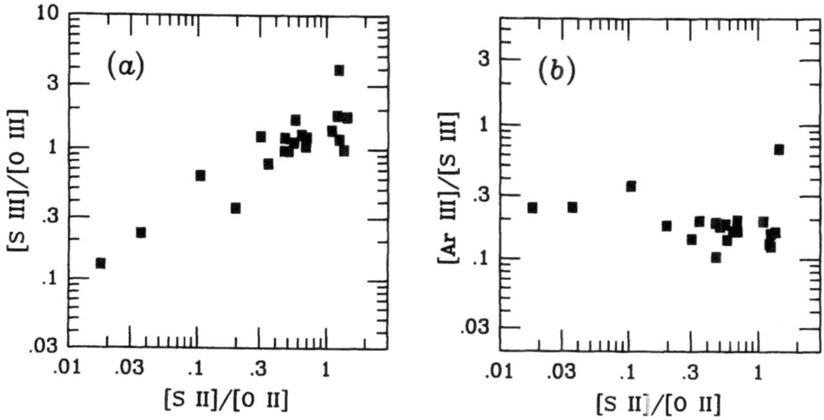

Figure 3. Correlations between flux ratios for pairs of lines as measured in 20 knots of Cas A. (*a*) [S III] 9069 Å : [O III] 5007 Å (vertical) vs [S II] 6716, 6731 Å : [O II] 7325 Å (horizontal). (*b*) [Ar III] 7136 Å : [S III] 9069 Å (vertical) vs [S II] 6716, 6731 Å : [O II] 7325 Å (horizontal).

much weaker relative to oxygen or sulfur. Thus iron-peak elements appear in varied concentrations both in material that is primarily oxygen and that composed predominantly of oxygen-burning products. These observations suggest that some of the iron-peak inner core of the progenitor was mixed by the supernova with overlying material, at least as far out as the oxygen zone. Similar upward mixing of an iron-peak core in SN1987A is widely invoked to explain the early escape of X- and γ-ray emission, as has been discussed by several participants in this workshop. The oxygen-rich SNR 0540-69.3 in the LMC also has iron and nickel at high velocities similar to those of oxygen, suggesting that here too the iron-peak elements are mixed with oxygen in the debris (KIRSHNER *et al.* [7]). Perhaps such mixing is a characteristic feature of massive supernovae.

We are grateful to the organizers of the UCSC workshop on Supernovae for a most interesting meeting, and to participants too numerous to mention individually for stimulating discussions. PFW and PFR acknowledge the support of NSF grant AST 85-20557, NASA grant NAG8-735 and the Middlebury College Faculty Research Fund. RPK acknowledges support from NSF grant AST 85-16537.

References

1. M. Peimbert, S. van den Bergh: *Ap. J.* **167**, 223 (1971).
2. R.A. Chevalier, R.P. Kirshner: *Ap. J.* **219**, 931 (1978).
3. R.A. Fesen, R.H. Becker, R.W. Goodrich: *Ap. J. (Letters)* 329, L89 (1988); also talk by Fesen at this meeting.
4. S. van den Bergh: *Ap. J.* **327**, 156 (1988).
5. P.F. Winkler, R.P. Kirshner: *Ap. J.* **299**, 981.
6. R.A. Chevalier, R.P. Kirshner: *Ap. J.* **233**, 154 (1979).
7. R.P. Kirshner, J.A. Morse, P.F. Winkler: *Ap. J.* **342**, 260 (1989).

Photoionization Models of Iron-Rich Ejecta in SN 1885

A. J. S. Hamilton & R. A. Fesen

1. Introduction

The recent discovery by Fesen *et al.* [1] of the remnant of S Andromedae, the historic supernova of 1885 observed a mere 16″ from the nuclear center of M31, brings to two the number of supernovae whose ejecta have been observed successfully in absorption: SN 1006 (Type Ia; Wu *et al.* [2]; Fesen *et al.* [3]; Hamilton and Fesen [4]) and now SN 1885 (Type I peculiar; de Vaucouleurs and Corwin [5]). In the discovery image, SN 1885 appears as an unresolved spot of Fe I absorption silhouetted against the starry blaze of M31's bulge. From the observed absorption contrast, Fesen *et al.* [1] estimate that SN 1885 has an Fe I diameter of 0.3″, corresponding to an expansion velocity of $\pm 5{,}000\,\mathrm{km\,s^{-1}}$.

Absorption line observations of supernova ejecta can be tremendously powerful, since observed absorption line profiles yield directly the density and composition as a function of radius in the freely expanding ejecta. However, absorption observations are obviously limited to those young remnants which happen to have a suitable background source behind them. This is why most remnants have not been observed in absorption: a suitable background source just does not exist, or at least has not yet been noticed. *HST* may change this situation. In the case of SN 1006, the background object is a single subdwarf OB star (Schweizer and Middleditch [6]). For its part, SN 1885 occults some 10^5 bulge stars in M31.

SN 1885 promises some advantage over SN 1006 in regard to absorption observations. In SN 1006, a good fraction of the ejecta has been shocked, and of the unshocked ejecta only about $0.015\,M_\odot$ remains as observable Fe II, the rest of the unshocked ejecta having been photoionized mainly to Fe III to Fe V mainly by UV and x-ray emission from the shocked ejecta (Hamilton and Fesen [4]). The resonance lines of Fe III and higher ions unfortunately all lie in the far UV inaccessible to current telescopes. By contrast, as discussed below, SN 1885 has probably collided with little ambient gas, and we expect that most of the ions in SN 1885 are still neutral and singly ionized. Thus almost all the ejecta in SN 1885 should still be observable in absorption. Such is the advantage of youth.

2. The Medium Around SN 1885

The absence of detectable emission from the site of SN 1885 in the radio (Dickel and D'Odorico [7]), in the optical (Ciardullo *et al.* [8]; Jacoby *et al.* [9]), or in x-rays (Van Speybroeck *et al.* [10]) implies that SN 1885 has not yet swept up a large mass of ambient gas. The tightest constraint comes from VLA observations (Dickel and D'Odorico [7]) which set an upper limit on the radio luminosity of SN 1885 at less than half the luminosity of Tycho's SNR. If the fractional energy density in relativistic particles and magnetic fields is the same in SN 1885 as in Tycho, then the ambient density must be less than $\sim 2\,\mathrm{cm}^{-3}$ and the mass of shocked gas must be less than $\sim 0.2\,M_\odot$.

A low density of gas surrounding SN 1885 is consistent with what is known about conditions in the bulge of M31. The optical bulge of M31 is dominated by starlight from an old but metal-rich population (Walterbos and Kennicutt [11]; Roger *et al.* [12]; Kent [13]; Faber and French [14]) with a dynamically inferred mass of $\sim 3\times 10^9\,M_\odot$ within the central 5′ (= 1 kpc) diameter (McElroy [15]). There are no OB stars (Welch [16]; Wirth *et al.* [17]), no Wolf-Rayets (Moffat and Shara [18], [19]), nor any other sign of recent star formation. Twenty-one centimeter (Bajaja and Shane [20]) and IRAS observations (Soifer *et al.* [21]) indicate only $\sim$1-2$\times 10^5\,M_\odot$ of neutral hydrogen gas within the central 5′ diameter. This gas is presumably associated with the dust extinction lanes observed optically (McElroy [15]; Ciardullo *et al.* [8]; Nieto *et al.* [22]), but no dust lane is evident at the position of SN 1885 (Fesen *et al.* [1]). X-ray observations (Fabbiano *et al.* [23]; Van Speybroeck *et al.* [10]) set upper limits of $\lesssim 0.02\,\mathrm{cm}^{-3}$ and $\lesssim 10^4\,M_\odot$ to the density and mass of diffuse hot gas at temperatures $\gtrsim 10^6$ K in the central 5′ diameter. Diffuse nonthermal radio emission is nevertheless observed (Walterbos and Gräve [24]; Hjellming and Smarr [25]) with a minimum energy density $\gtrsim 2\,\mathrm{eV\,cm}^{-3}$ in relativistic particles and magnetic fields, interestingly close to the maximum energy density of $\lesssim 2\,\mathrm{eV\,cm}^{-3}$ of hot x-ray emitting gas. These numbers pose the novel possibility that the ambient pressure around SN 1885 may be primarily relativistic, in cosmic rays and magnetic fields.

3. Ambient UV Starlight

Fe I has an ionization potential of 7.87 eV, which is below the Lyman limit, so is relatively easily photoionized by ambient UV starlight. Fe I's photoionization cross-section (Lombardi *et al.* [26]; Hansen *et al.* [27]) averages about $3\times 10^{-18}\,\mathrm{cm}^2$ between threshold (1575 Å) and Lyman α (1216 Å), with a broad peak from many resonances between 1400 Å and 1250 Å.

IUE observations (Johnson [28]; Welch [16]) give a direct handle on the level of ambient UV light, although the actual UV flux experienced by SN 1885 depends on its geometry relative to the nucleus of M31. Between Fe I threshold (1575 Å) and Lyman α (1216 Å) the dereddened flux observed with *IUE* is more or less flat at about $2\times 10^{-14}\,\mathrm{erg\,cm^{-2}\,s^{-1}\,\AA^{-1}}$ through the 178 arcsec^2 race-track-shaped aperture of *IUE*, centered on the nucleus. The flux is about half this at the position of SN 1885. The spectrum and spatial distribution of the observed UV light suggests that it probably arises from blue horizontal branch stars in the bulge of M31 (Welch [16]). If SN 1885 is half way through the bulge, that is, at a position closest to the nucleus, then it is experiencing a photoionizing flux of ~ 5×10^8 phot cm^{-2} s^{-1} between 1575 Å and 1216 Å, summed over all directions. The photoionizing flux is smaller if SN 1885 is further from the nucleus.

Combining the UV flux with the photoionization cross-section of Fe I yields an ionization time of ~20 years, or more if SN 1885 is further than the minimum distance from the nucleus, for Fe I exposed to light from the bulge of M31. Optically thick inner layers of ejecta are however shielded, and remain neutral.

4. Photoionization Models of SN 1885

The estimates of the previous two sections suggest that the main source of ionization of ejecta in SN 1885 is photoionization by ambient UV starlight. Currently, inner layers of

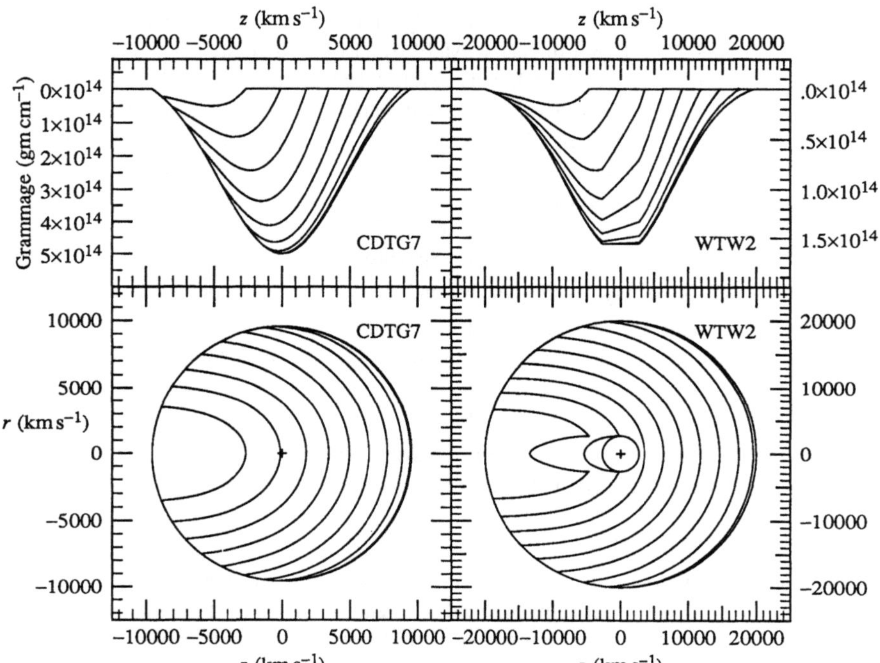

Figure 1. Lower panels show positions of Fe I to Fe II photoionization front in SN 1885, driven by one-sided UV stellar illumination coming from the right. The computations are based on the carbon-deflagrated white dwarf model CDTG7 of Woosley [29], [30], and the off-center helium detonation model 2 of Woosley, Taam and Weaver [31]. The photoionization front is plotted at equal intervals of time, and moves from right to left. Currently the front is about half way through the remnant. Upper panels show the resulting Fe I absorption line profiles expected in spatially unresolved observations, from the point of view of an astronomer observing from the left. As the photoionization front moves through the remnant, the Fe I absorption profile weakens and becomes more blueshifted. The center of WTW2 is composed of unburnt CO.

iron-rich ejecta should be mainly Fe I, but outer layers should have been photoionized to Fe II. One of the intriguing aspects of this photoionization is that SN 1885 is experiencing a one-sided tan — it is being photoionized mainly on the side nearest the bulge (see Fig. 1). If SN 1885 occurred on the near side of the bulge of M31, as the relatively high absorption contrast suggests (Fesen *et al.* [1]), then this one-sided tan should show up as a blue-shift of Fe I lines (Fig. 1), and a corresponding red-shift of Fe II lines. The observation of such a shift would in fact give an indication of SN 1885's position along the line of sight within the bulge of M31.

We have run some idealized models which illustrate the non-spherical ionization structure of SN 1885 which develops as a result of the one-sided photoionization. As examples, we took the carbon-deflagrated white dwarf model CDTG7 of Woosley [29], [30], and, since SN 1885 was after all a peculiar Type I, an off-center helium detonation model, model 2, hereafter WTW2, of Woosley, Taam and Weaver [31]. To simplify the calculations, we assumed (a) that SN 1885 is illuminated by a single UV point source at infinity, and (b) that this causes a sharp Fe I to Fe II photoionization front to propagate through the remnant.

Figure 1 shows the position of the Fe I to Fe II photoionization front at equally spaced intervals of time in models CDTG7 and WTW2 of SN 1885. Currently the photoionization front should be about half way through SN 1885's ejecta. The precise position of the front depends on how close SN 1885 is to M31's nucleus: the front moves faster if SN 1885 is closer to the nucleus, where the UV flux is fiercer.

Figure 1 also shows the resulting Fe I absorption line profiles expected for spatially unresolved (i.e. ground-based) observations of SN 1885. The profiles are shown from the point of view of an observer located the opposite side of the remnant from the photoionizing UV source, which would be a reasonable approximation if SN 1885 is in fact well to the near side of M31's bulge. The depth of the line profile is determined by the "grammage" of intervening Fe I, which is the Fe I density integrated over the transverse directions in the remnant. The off-center helium-detonation model WTW2 differs from the carbon-deflagration model CDTG7 in that the detonation model (a) shows higher expansion velocities, (b) has a lower overall density at any given time, and (c) has Fe in outer layers of ejecta, and unburnt CO on the inside, whereas it's vice versa in the deflagration model.

5. Conclusions

The main message of this paper is that we will learn a lot about the Type I peculiar supernova of 1885 from absorption line observations. Specifically, from observed absorption line profiles we will learn the density and composition of the ejecta as a function of radius. Unlike SN 1006, the only other supernova whose ejecta have so far been observed in absorption, most of the ejecta in SN 1885 are probably still freely expanding and in a low ionization state, mostly neutral and singly ionized. Currently, UV starlight from M31's bulge is driving a lop-sided Fe I to Fe II photoionization front, which has reached about half way through SN 1885's ejecta.

SN 1885's small size, about 0.5″ in diameter, means that observations with *HST* will be essential to realize the full potential of absorption line observations of SN 1885.

References

1. R. A. Fesen, A. J. S. Hamilton, J. M. Saken: *Ap. J. (Letters)*, **341**, L55 (1989).
2. C.-C. Wu, M. Leventhal, C. L. Sarazin, T. R. Gull: *Ap. J. (Letters)*, **269**, L5 (1983).
3. R. A. Fesen, C.-C. Wu, M. Leventhal, A. J. S. Hamilton: *Ap. J.*, **327**, 164 (1988).
4. A. J. S. Hamilton, R. A. Fesen: *Ap. J.*, **327**, 178 (1988).
5. de G. Vaucouleurs, H. G. Corwin, Jr.: *Ap. J.*, **295**, 287 (1985).
6. F. Schweizer, J. Middleditch: *Ap. J.*, **241**, 1039 (1980).
7. J. R. Dickel, S. D'Odorico: *M. N. R. A. S.*, **206**, 351 (1984).
8. R. Ciardullo, V. C. Rubin, G. H. Jacoby, H. C. Ford, W. K. Ford, Jr.: *Astr. J.*, **95**, 438 (1988).
9. G. H. Jacoby, H. Ford, R. Ciardullo: *Ap. J.*, **290**, 136 (1985).
10. Van L. Speybroeck, A. Epstein, W. Forman, R. Giacconi, C. Jones, W. Liller, L. Smarr: *Ap. J. (Letters)*, **234**, L45 (1979).
11. R. A. M. Walterbos, R. C. Kennicutt, Jr.: *Astr. Ap. Suppl.*, **69**, 311 (1987).
12. C. M. Roger, J. P. Phillips, C. S. Magro: *Astr. Ap.*, **161**, 237 (1986).
13. S. M. Kent: *Ap. J.*, **266**, 562 (1983).
14. S. M. Faber, H. B. French: *Ap. J.*, **235**, 405 (1980).
15. D. B. McElroy: *Ap. J.*, **270**, 485 (1983).
16. G. A. Welch: *Ap. J.*, **259**, 77 (1982).
17. A. Wirth, L. L. Smarr, T. L. Bruno: *Ap. J.*, **290**, 140 (1985).
18. A. F. J. Moffat, M. M. Shara: *Ap. J.*, **273**, 544 (1983).
19. A. F. J. Moffat, M. M. Shara: *Ap. J.*, **320**, 266 (1987).
20. E. Bajaja, W. W. Shane: *Astr. Ap. Suppl.*, **49**, 745 (1982).
21. B. T. Soifer, W. L. Rice, J. R. Mould, F. C. Gillett, M. Rowan-Robinson, H. J. Habing: *Ap. J.*, **304**, 651 (1986).
22. J.-L. Nieto, F. D. Macchetto, M. A. C. Perryman, S. di Serego Alighieri, G. Lelièvre: *Astr. Ap.*, **165**, 189 (1986).
23. G. Fabbiano, G. Trinchieri, L. S. Van Speybroeck: *Ap. J.*, **316**, 127 (1987).
24. R. A. M. Walterbos, R. Gräve: *Astr. Ap.*, **150**, L1 (1985).
25. R. M. Hjellming, L. L. Smarr: *Ap. J. (Letters)*, **257**, L13 (1982).
26. G. G. Lombardi, P. L. Smith, W. H. Parkinson: *Phys. Rev. A*, **18**, 2131 (1978).
27. J. E. Hansen, B. Ziegenbein, R. Lincke, H. P. Kelly: *J. Phys. B.*, **10**, 37 (1977).
28. H. M. Johnson: *Ap. J. (Letters)*, **230**, L137 (1979).
29. S. E. Woosley: private communication (1987).
30. S. E. Woosley, T. A. Weaver: In *Proc. IAU Colloquium 89, Radiation Hydrodynamics in Stars and Compact Objects* ed. D. Mihalas, K.-H. A. Winkler (Springer-Verlag, Berlin 1987), p. 91.
31. S. E. Woosley, R. E. Taam, T. A. Weaver: *Ap. J.*, **301**, 601 (1986).

X-Ray Spectroscopy of Young Supernova Remnants: Mixing in the Ejecta of Type I and II Supernovae

John P. Hughes

I. Introduction

Supernova remnants (SNRs) are the product of stellar explosions. From them we can learn much about the late stages of stellar evolution, mass-loss processes in stars, and explosive nucleosynthesis. The events involve the ejection of several solar masses of material at velocities reaching 10^4 km s^{-1} into the circumstellar and interstellar medium. It is possible to observe radio, optical, and X-ray emission from SNRs with ages ranging from hundreds to tens of thousands of years old. Remnants represent a significant input of energy and matter into the galaxy as a whole and thus are important contributors to the overall dynamical and chemical evolution of the galaxy.

The X-ray spectra of SNRs are quite rich: they show strong emission lines from many elements (among which are oxygen, neon, silicon, sulfur, argon, calcium, and iron) with abundances that are often decidedly enhanced relative to the abundances in the sun or the interstellar medium (see Figs. 1, 2, and 4). The continuum emission in some cases is best described by a multi-temperature distribution. Furthermore the rapid evolutionary timescales for these objects influences the observed spectra in subtle ways, through nonequilibrium effects on the ionization states of the various elemental constituents and on the energy transfer between electrons and ions. These effects make SNR X-ray spectra difficult to interpret.

Many previous studies of SNR X-ray spectra have used detailed models for the underlying dynamical evolution to compare to the observational data (ITOH [1,2,3]; GRONENSCHILD and MEWE [4]; SHULL [5]; NUGENT *et al.* [6]; HUGHES and HELFAND [7]; HAMILTON, SARAZIN, and SZYMKOWIAK [8,9]). In general these models have been quite complicated with many parameters and assumptions; unfortunately this has tended to obscure some of the underlying important astrophysics. The models have been computationally expensive and as a result, full exploration of the multidimensional parameter space available has not been attempted. Furthermore no group has been able (or willing) to investigate more than one or two objects.

The limitations in previous analyses discussed above have motivated this investigation of SNR X-ray spectra. The intention was to carry out a consistent spectral analysis of data from several galactic SNRs using simple parametric models for the line and continuum emission. It was planned to utilize data from several satellites in order to have the broadest possible energy band available. The ultimate goal of the project is to obtain, for a number of SNRs, reliable values for the elemental abundances and information on the ionization and thermal state of the plasma. As we show below, some results have already been obtained for three young SNRs: Cassiopeia A, Tycho's SNR, and the remnant of SN1006.

The conclusions of this study are that the ejecta of the Type I remnants (Tycho and SN1006) have retained much of the stratification in their elemental composition to the present day. However the Type II remnant (Cassiopeia A) is much more completely mixed. These results are based on the measurements of line centroids and comparison to a simple nonequilibrium ionization model, which I describe below. I discuss these results in light of recent work on mixing in the ejecta of both Type I and II supernovae (SNe).

II. Nonequilibrium Ionization Model

The observable effects of nonequilibrium ionization (NEI) are most prominent in the comparison between line emission and continuum emission in the X-ray spectra of SNRs. The continuum arises principally from bremsstrahlung emission of electrons on highly ionized atomic species. It depends strongly on the electron temperature in the plasma and very weakly on the atomic species and the ionization state, through the Gaunt factor. In contrast, the line emission depends strongly on the latter; an increase or decrease in elemental abundance or ionization state will cause a proportional change in line intensity. Additionally, the observed line energy for a given transition depends on the ionization state. For example, the energy of the Kα line of iron, in which the principal quantum number changes from n = 2 to n = 1, varies from 6.96 keV for Fe XXVI, to approximately 6.66 keV for Fe XXV, to as low as 6.4 keV for ionization states below about Fe XIX, a consequence of the "screening" of the nuclear charge by innershell electrons. Clearly the line energy of a given electronic transition is a sensitive indicator of the ionization state of that atomic species. However, the ionization state also depends on the electron temperature, as well as whether the plasma is in equilibrium and not. Indeed the line energies of a lower temperature plasma in equilibrium may mimic the true line energies in a higher temperature plasma which is in the ionizing phase. However, given a broad energy band and detectors with sufficient energy resolution to isolate the line emission, it is possible to determine the electron temperature from the continuum emission.

The complications due to NEI effects make it clear that the ionization calculation must be included properly within the context of the evolution of supernova remnants in general. The model to be used in this study assumes a single component plasma that has been recently heated to a temperature of several keV by the passage of a strong shock. The time since the passage of the shock (actually the density times this time, called the ionization timescale) and the temperature of the plasma (assumed constant) become the two relevant parameters. I refer to this as the single temperature – single timescale model. Although in an actual remnant there is a range of temperatures and ionization timescales, this model should be an ideal first approximation to the average state of the plasma (HAMILTON and SARAZIN [10]).

I have coupled my matrix-based solution for the NEI fractions (see [7]) to the RAYMOND and SMITH [11] optically-thin plasma emission code. Additional emission lines appropriate to nonequilibrium conditions were included from MEWE and GRONENSCHILD [12]. For comparison to the fitted Kα line energies, I took the weighted average of all $n = 2$ to $n = 1$ transitions to determine a mean line energy for the NEI model corresponding to a given ionization timescale and electron temperature. This was done separately for each of the astrophysically abundant elements silicon, sulfur, argon, calcium, and iron.

III. Data Analysis

The data used in this project came from the solid state spectrometer (SSS) on the ***Einstein Observatory*** and the gas scintillation proportional counters (GSPCs) on the ***Tenma*** satellite. The SSS data covered the energy range from approximately 0.5 keV to 4.0 keV and the ***Tenma*** data covered the range from 1.5 keV to beyond 10 keV. The combined analysis of these datasets is a powerful technique for the study of line and continuum emission processes in SNRs. As a consequence of the relatively high resolving power of the ***Tenma*** GSPCs (a factor of two better than previous instruments in this energy range), the iron Kα line (~6.5 keV) can be studied to higher precision than ever before. In addition the line and continuum emission can be more easily separated to allow for a good determination of the electron temperature. However the ***Tenma*** GSPCs were not well suited to studying the silicon and sulfur Kα line region (~2 keV) or searching for low temperature (<1 keV) emission components. The ***Einstein*** SSS obtained some of the highest resolution data on SNRs to date, with energy resolution better than about 200 eV. The best data was obtained from the silicon and sulfur Kα line regions, while beyond about 4 keV the ***Einstein Observatory*** X-ray telescope cut-off and below about 1 keV there was the well-known problem of icing on the surface of the SSS. It was difficult to determine continuum temperatures above about 2–3 keV from the SSS data alone.

Thus these two types of data complement each other and joint fits allow for study of the silicon, sulfur, argon, calcium, and iron Kα lines. As discussed below, the average line energies are a sensitive probe of the ionization state of the plasma. Comparison among the different elemental species allows for investigation into the relative ionization histories. Joint fits also allow for a determination of the continuum temperature and searches for multiple components, such as the blast wave in the interstellar medium or the reverse shock in the metal rich stellar ejecta.

In this project I have followed closely the analysis of TSUNEMI *et al.* [13]. The approach is as model independent as possible with the use of parameterized forms for the continuum and line emission. I fitted the continuum emission with an exponential (including Gaunt factor) of two parameters: the normalization and kT. Emission lines were modeled as gaussian profiles with the parameters normalization, central energy, and energy width. The width was held constant at a value $\sigma = 0.050$ keV. Lines were included at energies appropriate to the Kα emission complexes of the elements Ne, Mg, Si, S, Ar, Ca, and Fe; Kβ lines were also included for Si and S; and, when apparently necessary, lines were included in the 1 keV region of the spectra to represent Fe L-shell emission. Absorption due to intervening material in the galaxy along the line of sight was also included. Due to the problem of ice absorption on the SSS, the column density was allowed to vary separately for the SSS and *Tenma* datasets. A relative normalization between the two datasets was also included, in part to account for differences in the fields of view. The field of view of *Tenma* was several degrees, while that of the SSS was only $3'$ in radius, smaller than the sizes of the three SNRs I studied. I assume that the SSS spectra are representative of the entire remnant. Multiple SSS observations of Tycho and SN1006 were done and since in each case the spectra were similar, I have averaged them.

Before carrying out the joint analysis, an initial program of fits to the individual datasets was done. Differences in the silicon and sulfur line energies between the SSS and ***Tenma*** were found for Tycho and Cassiopiea A. This is attributed to an error in the ***Tenma*** energy scale, which was nominally accurate to 30 eV. Assuming that the SSS data was more precise, I shifted the energy scale for the Tycho *Tenma* data up by 15 eV and shifted the Cassiopeia A data down by 70 eV. The shift for Cassiopeia A was uncomfortably large, but probably arose from confusion between one of the calibration lines at 8.04 keV and the source emission at this energy. The result of these shifts is to change the fitted emission line energies from the *Tenma* data, of which the most significant is the iron Kα line. Where appropriate, I discuss how this shift in energy scale influences the conclusions.

My conclusions differ from TSUNEMI ***et al.*** for reasons which arise purely from the inclusion of the SSS data and the shift of the ***Tenma*** energy scale as mentioned above. I was able to reproduce the fitted values of TSUNEMI ***et al.*** for line centroids and continuum temperatures using my analysis system and the ***Tenma*** data alone. The differences in conclusions can be attributed mainly to the improved errors on the silicon and sulfur line centroids, the recognition of multiple temperature components, and the different derived continuum temperatures.

IV. Results for Individual Remnants

SN1006

The remnant of the supernova (SN) explosion in 1006 shows a shell-like morphology in the X-rays with a radius of about $15'$ (PYE ***et al.*** [14]). For a time the X-ray spectrum was thought to be nonthermal, due to a lack of observed emission lines and a single power-law spectral form from 0.5 keV to beyond 10 keV (BECKER ***et al.*** [15]), in seeming contradiction with the imaging data. However an analysis of *Tenma*

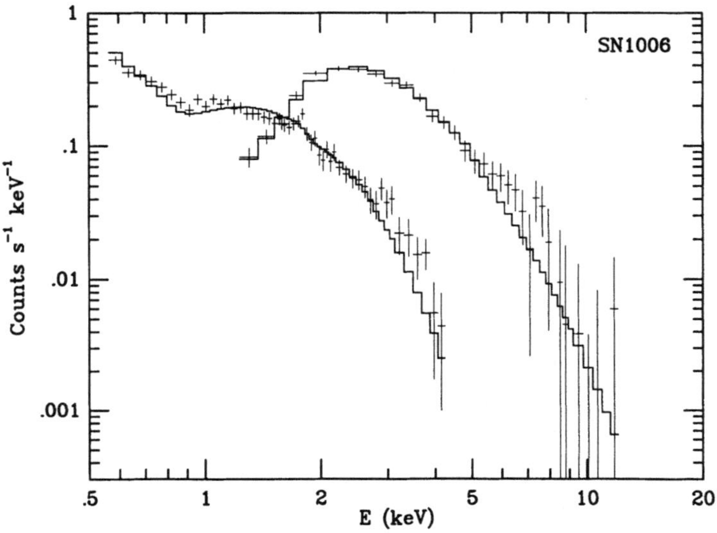

Figure 1. ***Einstein*** **SSS and** ***Tenma*** **X-ray spectra for SN1006 and best-fit single temperature continuum model.**

data (KOYAMA *et al.* [16]) showed the X-ray background emission in the vicinity of SN1006 to be dominated by a 7.5 keV thermal spectrum from the Lupus Loop (an older SNR). Since the 2-10 keV emission from SN1006 is rather weak, proper subtraction of this background component is essential. As I show below this ***Tenma*** spectrum, when analyzed with the SSS data, is more consistent with a single temperature thermal model than the earlier nonthermal, power-law form. In addition, VARTANIAN, LUM, and KU [17] discovered an oxygen emission line from SN1006, adding additional weight to a thermal interpretation of its X-ray emission.

The SSS and ***Tenma*** data are shown in Figure 1, with the best-fit model superposed. This is a pure thermal continuum model with $kT = 1.8$ keV and hydrogen column density of 4×10^{20} atoms cm^2. The χ^2 was 130 for 84 degrees of freedom. Introduction of a gaussian line at an energy of 1.86 keV improved the fit slightly, reducing χ^2 to 121. The best fitting power-law yielded a higher χ^2 value: 151 and in this case the column density was 24×10^{20} atoms cm^2. The 21 cm H I emission survey of HEILES and CLEARY [18] gives a column density of $6.5 \pm 0.6 \times 10^{20}$ in the direction of SN1006, a value which is more consistent with the thermal spectrum than the power-law one.

In the standard model of explosive nucleosynthesis, an iron-rich core is surrounded by layers of progressively lighter elements in an "onion-skin" structure. As the SNR evolves and the ejecta decelerates, a so-called reverse shock begins to propagate into the ejecta, heating the material to X-ray emitting temperatures. The shock travels backwards through the ejecta, progressing into material with increasing atomic number, until the iron core is reached. Much of this spatially stratified structure has been preserved to the present in SN1006. It is known that there are significant amounts of unshocked iron in the interior of the SNR from the presence of broad Fe II absorption lines seen in absorption against the remnant (WU *et al.*

[19]). Yet the only evidence for X-ray line emission is an oxygen line and some hint of a weak silicon line, implying that the reverse shock in this SNR has only recently begun to propagate into the silicon-rich layer of the ejecta.

Tycho

Tycho's SNR is the product of the supernova in 1572, which is widely believed to have been a Type I explosion. The X-ray image of the remnant shows a beautiful shell of emission about 8′ in diameter (SEWARD *et al.* [20]). Three emission components were identified in the *Einstein* high resolution image: an outer shell of swept-up interstellar medium, a diffuse component of ejecta, and clumpy ejecta. The emission measures of these three components were determined to be in the ratios 1:0.25:0.61 and the temperatures were estimated to be 7 keV for the first two and 2 keV for the clumpy ejecta.

The X-ray spectra from the *Einstein* SSS and *Tenma* are shown in Fig. 2. The best fit model shown in the figure includes a number of lines as well as two thermal continuum components. The principle continuum component has a temperature of 1.9 keV; the second component, at a temperature of about 6 keV, has an emission measure nearly 20 times lower than the first. The statistical significance of the second component was not high, since the reduction in χ^2 was modest ($\Delta\chi^2 \sim 25$) when it was included. Taken together these results imply that the SNR is nearly isothermal over the energy band from 0.5 keV to 10 keV. Although the values of fitted temperatures are in agreement with the imaging analysis, SEWARD *et al.* require that the higher temperature component have about twice the emission measure of the cooler component. It would seem that the spectral analysis does not support this picture. I suggest that the emission measure of the hotter component, i.e., the blast wave and the diffuse ejecta, was overestimated in the imaging analysis. For example, the so-called diffuse ejecta component could just as easily be emission clumped at a spatial scale below the resolution of the X-ray image. Further study of the X-ray images from *Einstein* and new observations with ROSAT will help address this point.

With the electron temperature fixed at a value of 1.9 keV, I used the NEI model to determine the expected average Kα line energies for each elemental species as a function of ionization timescale. These are drawn as the smooth curves in Fig. 3; the boxes superposed on each curve are from the fits to the data and are at the 90% confidence level. The errors on the energy centroids for silicon and sulfur are of course quite small due to the excellent spectral resolution of the SSS data. The results on iron are also good because the line is relatively strong and isolated. However the argon and calcium lines are weaker and in the case of the SSS data, fall near the high energy cut-off of the *Einstein* X-ray telescope. As a consequence, the errors on the fitted line centroids are rather large.

It is clear that the ionization timescale derived from the iron line centroid is less than the timescale derived from the other elemental species. In fact the difference between the iron ionization timescale and that of silicon is significant at greater than 99% confidence. Note that this difference is not the result of the energy scale shift of the *Tenma* data. Without the energy shift the iron line centroid would be 0.015 keV ***smaller***, which would ***increase*** the discrepancy. I believe that the difference is a result of the stratification of the ejecta into zones of different elemental composition. As was the case for SN1006, Tycho's SNR has also retained its stratified ejecta, while the main difference is that the reverse shock in Tycho's SNR has begun to enter the iron-rich core.

Cassiopeia A

With an age of approximately 300 years, Cassiopeia A is the youngest known supernova remnant in the Galaxy. The X-ray image (MURRAY *et al.* [21]) shows emission from an inclined ring geometry, which can be identified with the stellar

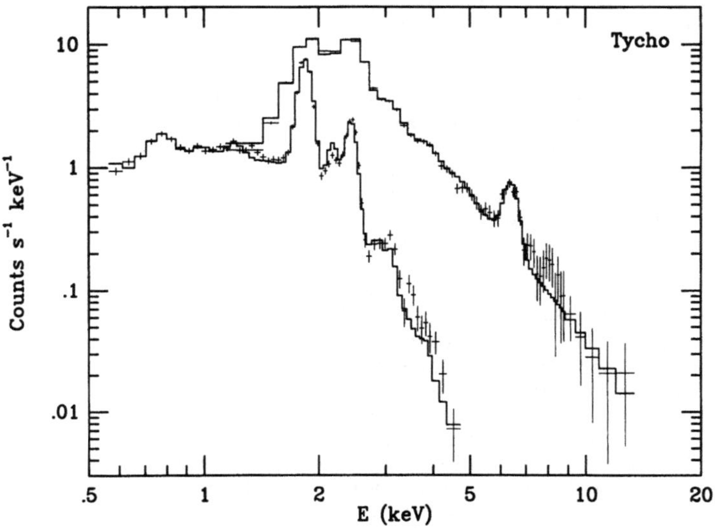

Figure 2. *Einstein* SSS and *Tenma* X-ray spectral data for Tycho's supernova remnant. Prominent lines at 1.9, 2.5 and 6.4 keV are emission from silicon, sulfur, and iron, respectively.

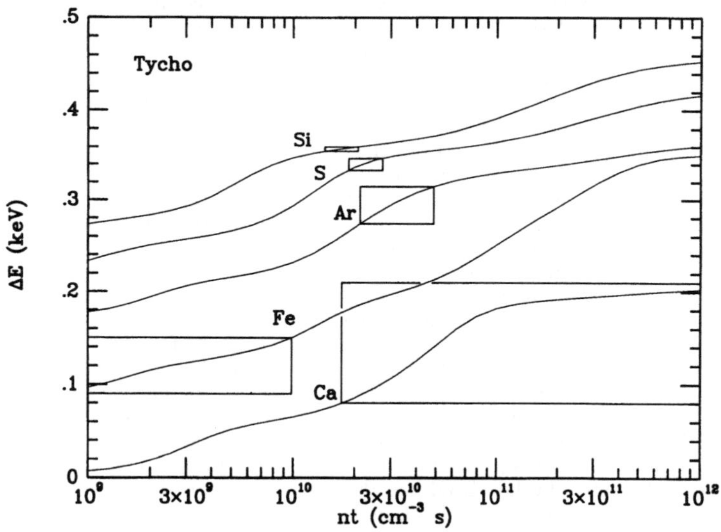

Figure 3. The variation of mean line energy for the Kα lines of silicon, sulfur, argon, calcium, and iron as a function of ionization timescale for Tycho's SNR. The electron temperature was fixed at a value of 1.9 keV. The boxes show the allowed timescales (at the 90% confidence level) derived from the fitted line energies. The reference energies corresponding to the zero of ΔE are: 1.5 keV (Si), 2.1 keV (S), 2.8 keV (Ar), 3.7 keV (Ca), and 6.3 keV (Fe).

ejecta, and an emission region surrounding that, arising from the expansion of the blast wave into the ambient circumstellar medium (CSM). The remnant is believed to be the explosion of a massive star.

The X-ray spectra from the ***Einstein*** SSS and ***Tenma*** are shown in Fig. 4 along with the best fit model which includes a number of lines as well as two thermal continuum components. The presence of two continuum emission components is highly significant. First, the χ^2 decreased by a factor of 3 when two components were included. Second, separate fits to the SSS and ***Tenma*** data using single temperature continua yielded very different temperatures: 1.4 keV (SSS) and 3.8 keV (***Tenma***). Finally, in these separate fits the fluxes for the silicon and sulfur $K\beta$ lines derived from the ***Tenma*** data were too large in comparison to the SSS data (where these lines are resolved). This "extra" flux from the $K\beta$ lines mimics a low temperature thermal component because of the poor spectral resolution of the ***Tenma*** detectors in the low energy range.

The derived continuum temperatures are $kT = 1.2$ keV and $kT = 4.7$ keV and there is no evidence for a significantly hotter (> 10 keV) emission component. The low temperature component has an emission measure which is 7.7 times larger than the high temperature one. I associate the low temperature component with the reverse-shocked ejecta and the high temperature component with the blast wave in the CSM. The ejecta is surely more dense than the CSM, and so the requirement for pressure-equilibrium through the contact interface implies that the ejecta emit at a lower temperature relative to the blast wave.

A deprojection of the high resolution X-ray image (FABIAN ***et al.*** [22]) showed that the emission measure of the ejecta region was about 6.2 times that of the outer plateau region, which is certainly in adequate agreement with the results quoted here from the spectral analysis. Although these authors do find that the plateau region is hotter than the ejecta region, there is no obvious bimodality in their temperature distribution, but rather an almost uniform distribution of emission measures from about 0.5 keV to 4.0 kev, with an additional few temperature values up to about 8 keV. Their analysis neglected nonequilibrium ionization effects and clumping of the ejecta, which should be significant. Nevertheless in future analyses I intend to investigate whether a continuous distribution of thermal continuum models can fit the X-ray spectra as well as the bimodal distribution presented here.

Comparison to the NEI model for this SNR is shown in Fig. 5. I assume that the line emission is produced almost entirely in the low temperature, ejecta component so I fixed the electron temperature at the derived value of $kT = 1.2$ keV. In contrast to Tycho's SNR (Fig. 3), the line centroids for the different elemental species are in good agreement with a single value of the ionization timescale $nt \sim 10^{11}$ cm-1 s. This is about a factor of 5 higher than the silicon and sulfur ionization timescales for Tycho, and thus consistent with the higher densities expected for the massive star progenitor of Cassiopeia A. (Note that if I use the higher temperature value of $kT = 4.7$ keV in the NEI model, the numerical value of the ionization timescale decreases, but the level of agreement among the different elemental constituents remains about the same.) If the shift of the ***Tenma*** energy scale were removed, then the iron line centroid would move to higher values and the ionization timescale would increase by almost an order of magnitude for the full 70 eV shift. Although I consider this unlikely, further analysis of the ***Tenma*** energy scale calibration is essential to eliminate this source of uncertainty.

What can one say about the structure of the supernova ejecta from the fact that a single ionization timescale seems to fit the data for Cassiopeia A? Perhaps it implies that the ejecta are stratified but that the reverse shock has passed through most of it or at least has propagated deeply into the iron-rich core. There are two objections to this picture: (1) it isn't clear that the ionization timescales will be the same since the layers of differing elemental composition would have been shocked at

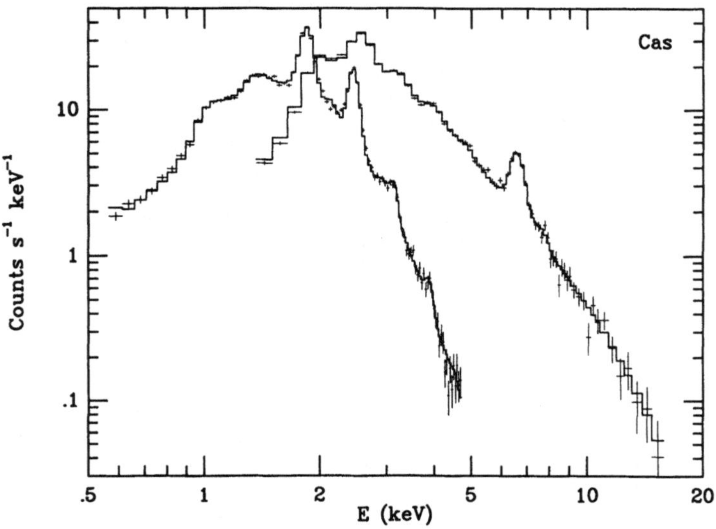

Figure 4. *Einstein* SSS and *Tenma* X-ray spectral data for Cassiopeia A. Prominent lines at 1.9, 2.5 and 6.4 keV are emission from silicon, sulfur, and iron, respectively.

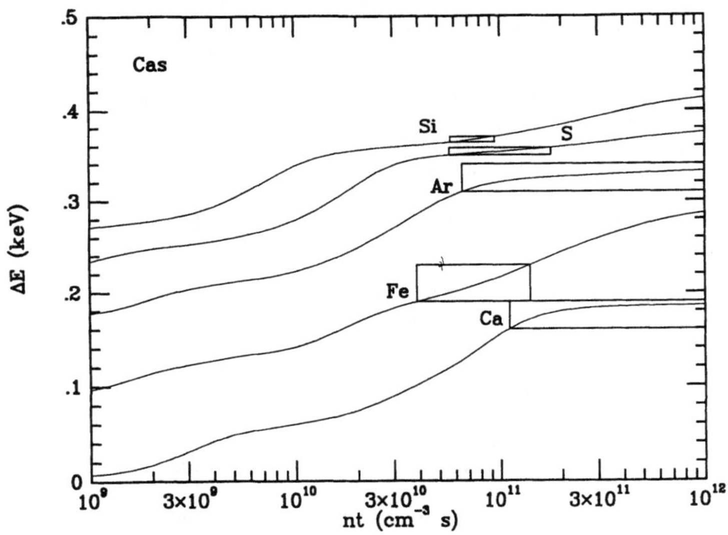

Figure 5. The variation of mean line energy for the Kα lines of silicon, sulfur, argon, calcium, and iron as a function of ionization timescale for Cassiopeia A. The electron temperature was fixed at a value of 1.2 keV. The boxes show the allowed timescales (at the 90% confidence level) derived from the fitted line energies. The reference energies corresponding to the zero of ΔE are: 1.5 keV (Si), 2.1 keV (S), 2.8 keV (Ar), 3.7 keV (Ca), and 6.3 keV (Fe).

different times and densities, and (2) the general appearance of the remnant is more consistent with incomplete thermalization of the ejecta, *i.e.*, more consistent with the remnant being in the free expansion phase than the Sedov phase. Consequently I prefer an alternative explanation wherein the iron-rich core was originally mixed at least as far as the silicon-group layer and the reverse shock is now propagating through a region of nearly homogeneous composition.

V. Conclusions

The two remnants of Type I supernovae which I have discussed in this paper, SN1006 and Tycho's SNR, show strong evidence for spatial stratification of their ejecta. In both cases an iron-rich core is separate from the silicon-group layer, and in the case of SN1006 another distinct region of oxygen or oxygen-group elements may be present. On the other hand, Cassiopeia A, the product of a Type II explosion, seems to have a spatially homogeneous composition.

Our best evidence for mixing in the ejecta of supernovae comes from studies of SN1987A in the Large Magellanic Cloud. These lines of evidence include the early appearance of soft X-rays and γ-rays and fits to the bolometric light curve (see numerous references in this volume). SN1987A was a Type II SN explosion. Evidence for mixing in other SNe is less well established. FILIPPENKO and SARGENT [23] recently observed evidence for inhomogeneities in SN1985F, a Type Ib. The presence of inhomogeneities may indicate the formation of Rayleigh-Taylor instabilities in the ejecta, which would imply significant mixing. However other explanations for clumping are also possible. BRANCH *et al.* [24] require mixing in the outer half of the ejecta of a Type I SN (SN1981B) to explain the early-time optical spectrum. In this case mixing was required to produce the observed velocity-broadening of several Si, S, Ca, and Co lines and to match their blueshifts and absorption depths. However, only about 16% (by mass) of the total iron-peak elements in the ejecta were mixed to higher velocities; the bulk of the iron-rich core was left undisturbed.

Recently two groups (ITOH, MASAI, and NOMOTO [25]; BRINKMANN *et al.* [26]) have compared compositionally-stratified models for a Type I explosion to the X-ray spectra of Tycho's SNR. Both groups started from the carbon deflagration model W7 of NOMOTO, THIELEMANN, and YOKOI [27] and evolved the hydrodynamic and ionization state of the remnant to the present age of Tycho. They found that significant amounts of iron must be mixed out from the interior in order to fit the equivalent width of the iron Kα line. However, neither group used the energy centroids of the lines to establish the relative ionization state between the silicon- and iron-group zones, as I have done here. (This was largely because they did not have access to the high spectral resolution ***Einstein*** SSS data.) Numerous effects can influence the equivalent width of the lines: clumping, thermal conduction, additional heating of electrons by collective plasma interactions, and not least of all, details of the initial conditions, in this case the W7 model. Further research along some of these lines may be necessary to reproduce accurately the broadband X-ray spectrum of young supernova remnants.

This research on the X-ray spectra of the remnants of historical SNe suggests that substantial amounts of mixing cannot have occurred in the ejecta of Type I SNe, while mixing is essential for Type II SNe. This conclusion is consistent with our present understanding of SN1987A as well as other supernovae. Further investigation along the following lines of research are indicated: (1) additional searches for evidence of mixing and/or inhomogeneities in the optical spectra of both Type I and II SNe, and (2) the extension of this analysis to the X-ray spectra of other historical remnants, such as Kepler's SNR, using existing X-ray data. Finally, a significant improvement in our ability to probe questions of the stratification of SN ejecta will come with the planned X-ray astronomy missions of the 90's.

VI. Acknowledgments

I would like to thank K. Koyama and H. Tsumeni for supplying the *Tenma* data and A. Szymkowiak for the SSS data. Special thanks are due to D. Yoon and S. Nolen who carried out much of the data analysis. This research was supported by NASA SADAP grant NAG8-670 and Smithsonian Institution funds.

VII. References

1. H. Itoh, ***Pub. Astr. Soc. Japan***, **29**, 813 (1977).
2. H. Itoh, ***Pub. Astr. Soc. Japan***, **30**, 489 (1978).
3. H. Itoh, ***Pub. Astr. Soc. Japan***, **31**, 541 (1979).
4. E. H. B. M. Gronenschild and R. Mewe, ***Astr. Ap. Suppl.***, **48**, 305 (1982).
5. J. M. Shull, ***Ap. J.***, **262**, 308 (1982).
6. J. J. Nugent, S. H. Pravdo, G. P. Garmire, R. H. Becker, I. R. Tuohy, and P. F. Winkler, ***Ap. J.***, **284**, 612 (1984).
7. J. P. Hughes and D. J. Helfand, ***Ap. J.***, **291**, 544 (1985).
8. A. J. S. Hamilton, C. L. Sarazin, and A. E. Szymkowiak, ***Ap. J.***, **300**, 698 (1986).
9. A. J. S. Hamilton, C. L. Sarazin, and A. E. Szymkowiak, ***Ap. J.***, **300**, 713 (1986).
10. A. J. S. Hamilton and C. L. Sarazin, ***Ap. J.***, **284**, 601 (1984).
11. J. C. Raymond and B. W. Smith, ***Ap. J. Suppl.***, **35**, 419, (1977).
12. R. Mewe and E. H. B. M. Gronenschild, ***Astr. Ap. Suppl.***, **45**, 11 (1981).
13. H. Tsunemi, K. Yamashita, K. Masai, S. Hayakawa, and K. Koyama, ***Ap. J.***, **306**, 248 (1986).
14. J. P. Pye, K. A. Pounds, D. P. Rolf, F. D. Seward, A. Smith, and R. Willingale, ***M.N.R.A.S.***, **194**, 569, (1981).
15. R. H. Becker, A. E. Szymkowiak, E. A. Boldt, S. S. Holt, and P. J. Serlemitsos, ***Ap. J. (Letters)***, **240**, L33, (1980).
16. K. Koyama, H. Tsunemi, R. Becker, and J. P. Hughes, ***Pub. Astr. Soc. Japan***, **39**, 437, (1987).
17. M. H. Vartanian, K. S. K. Lum, and W. H.-M. Ku, ***Ap. J. (Letters)***, **288**, L5, (1985).
18. C. Heiles and M. N. Cleary, ***Australian J. Phys., Ap. Suppl.***, **47**, 1, (1979).
19. C. C. Wu, M. Leventhal, C. L. Sarazin, and T. R. Gull, ***Ap. J. (Letters)***, **269**, L5, (1983).
20. F. Seward, P. Gorenstein, and W. Tucker, ***Ap. J.***, **266**, 287, (1983).
21. S. S. Murray, G. Fabbiano, A. C. Fabian, A. Epstein, and R. Giacconi, ***Ap. J. (Letters)***, **234**, L69, (1979).
22. A. C. Fabian, R. Willingale, J. P. Pye, S. S. Murray, and G. Fabbiano, ***M.N.R.A.S.***, **193**, 175, (1980).
23. A. V. Filippenko and W. L. W. Sargent, ***Ap. J. (Letters)***, **345**, L43, (1989).
24. D. Branch, J. B. Doggett, K. Nomoto, and F.-K. Thielemann, ***Ap. J.***, **294**, 619, (1985).
25. H. Itoh, K. Masai, and K. Nomoto, ***Ap. J.***, **334**, 279, (1988).
26. W. Brinkmann, H. H. Fink, A. Smith, and F. Haberl, preprint, (1989).
27. K. Nomoto, F.-K. Thielemann, and K. Yokoi, ***Ap. J.***, **286**, 644, (1984).

Infrared Knots Around SS433 – Results of Jets

Zhenru Wang, Yang Chen, Richard McCray, & Qinyue Qu

1. INTRODUCTION

SS433 is a famous exotic object in the center of radio source W50. It has been widely observed at various wavelengths and has attracted more and more attention since 1978. The most unusual feature of SS433 is two relativistic jets ejected at a velocity of 0.26c in opposite directions (MARGON [1]). This feature can explain the simultaneous presence of red and blue shifts of strong optical emission lines varying periodically as well as the two bright X-ray lobes stretched in the direction of the jets (SEWARD et al. [2]). In 1987, six infrared knots around SS433 were discovered from IRAS data by BAND [3]. They have similar infrared spectra except for the fifth one. Band suggested that the fifth one is a different kind of object [3], and we agree. But for the other five knots, we suggest a different model. We don't think it is accidental that five knots are all located in the direction of the jets and are within, or adhere to, the radio shell of W50. It is very probable that they result from the interaction between the jets and the interstellar medium. We assume that the knots consist mainly of dust grains and gas. Both are continuously heated by collisions of particles in the jets. Their spectra are explained by the thermal radiation of dust grains and the free-free emission of partly ionized hydrogen gases.

2. THE FITS OF SPECTRA

Suppose the thermal emission of dust is the main component of the steep spectra (from 100 to 60 μ). The dust radiation drops so quickly with frequency that the radiation between 25 and 12 μ can be considered as free-free emission only. The results of fitted parameters, electron density n_e (or n_p), and electronic temperature T_e for the five knots are listed in Table 1. The observed flux between 60 and 100 μ should contain two kinds of radiation. Subtracting the free-free emission contribution from the ob-

Table 1. Physical parameters of the knots from the spectra fits.

knot	1	2	3	4	6
$n_e [10^2\ cm^{-3}]$	3.0	1.5	1.7	1.9	0.55
$T_e [10^3\ K]$	3.3	1.1	0.67	2.1	0.81
$n_d\ [10^{-10}\ cm^{-3}]$	24	1.2	0.62	1.5	0.91
$T_d\ \{K]$	23	25	32	27	22

served fluxes, the fluxes of dust radiation at 100 and 60 μ are easily obtained from the observed data. The flux of dust grains is:

$$F_d(\nu) = 4\pi R^3\ j_d(\nu)/3d^2 = 4\pi R^3\ n_d\ \sigma_d\ Q_a\ B_\nu(T_d)/3d^2 \tag{1}$$

where d is the distance of SS433, 5 kpc, R is the radius of the knots [3], $j_d(\nu)$ is the emissivity of the dust grains (SPITZER [4]), n_d is the number density of dust grains, σ_d is the geometrical cross section of a single grain, $\sigma_d = \pi a^2$, a is the grain radius. T_d is the temperature of dust grains and Q_a is the efficiency factor. Taking 2a = 0.3 μ and the index of refraction m = 1.3 - 0.02i (ALLEN [5]), then $Q_a = 1.45 \times 10^{-16}\ \nu$. Using (1) to fit the steep spectra, the values of n_d and T_d in the knots are obtained and also listed in Table 1. Figure 1 shows the infrared spectrum of knot 1 and its spectrum fit. Other knots are similar.

3. THE CONSISTENCY OF PHYSICAL PARAMETERS

We will now discuss whether the physical parameters of the knots are reasonable.

The hydrogen atoms in the knots are ionized by the bombardment of relativistic particles in the jets. The equation of ionization-recombination equilibrium is:

$$n_H(n_{je}+n_{jp})\ v_j\ \sigma_{1c} = n_e n_p \alpha(2) \tag{2}$$

where n_H is the number density of neutral hydrogen atoms of the knots, n_{je} and n_{jp} are the number densities of relativistic electrons and protons in the jets respectively, $n_{je} = n_{jp} = n_j$, $\alpha(2)$ is the recombination coefficient to all states except the ground state which is a function of T_e [4]. Let $v_j = 0.26c$, and σ_{1c} is the cross section for collision ionization of hydrogen atoms from ground state (LOTZ [6]). From (2) we can easily get the values of $n_H n_j$ for the knots as ~1 cm^{-6}.

Let us now discuss the cooling and heating of dust. The cooling rate of the dust is

$$\Lambda_d = 4\pi \int_0^\infty n_d\sigma_d Q_a B_\nu(T_d)d\nu = 2.64\times10^{-9}\ T_d^5\ n_d\sigma_d\ [\text{ergs cm}^{-3}\ \text{s}^{-1}]. \tag{3}$$

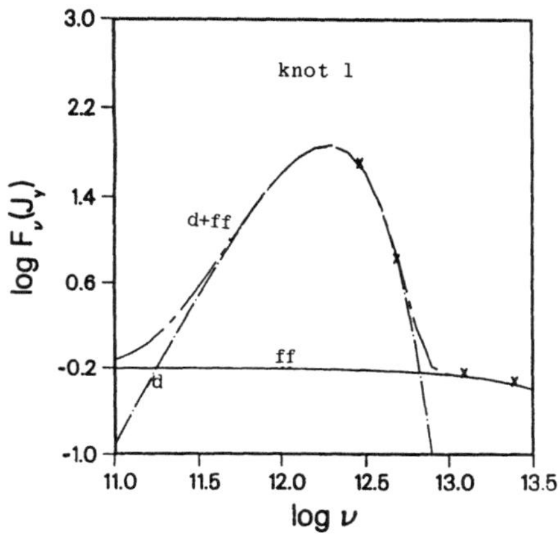

Fig. 1 The infrared spectrum of knot 1. Curve d is the thermal emission of the dust. Curve ff is the free-free emission of hydrogen. The observed data are shown by crosses.

The heating rate of the dust (DWEK [7]) by the jets' electrons is:

$$\Gamma_d^j = n_d \sigma_d n_j v_j [E_e\ \zeta(E_e) + E_p\ \zeta(E_p)] \tag{4}$$

where E_e and E_p are the energy of impinging electrons and protons respectively, $\zeta(E)$ is the fraction of E that is deposited in the dust grain [8]. In our case, $\zeta(E_e) = 0.16$ and $E_p\ \zeta(E_p) \approx E_e\ \zeta(E_e)$. Then

$$\Gamma_d^j = 70\ n_d \sigma_d n_j\ [\text{ergs cm}^{-3}\ \text{s}^{-1}] \quad . \tag{5}$$

The heating rate of the dust by thermal electrons can be obtained by

$$\Gamma_d^T = n_d \sigma_d n_e \int_0^\infty v f(v) \frac{1}{2} m_e v^2 dv = n_d \sigma_d n_e 2(\pi m_e)^{-1/2} (2kT_e)^{3/2} \tag{6}$$

where f(v) is the Maxwell velocity distribution function. From the equation of energy equilibrium of the dust grains $\Lambda_d = \Gamma_d^j + \Gamma_d^T$, we derive values of n_j at the positions of knots ~10^{20} cm from SS433 that range from 10^{-4} to 1.3×10^{-3} cm^{-3}. Compared with previous results (WATSON [8]), these results are consistent with $n_j \propto 1/r^2$. Further, the neutral hydrogen atom number density n_H, the ionization degree χ, and the gas-to-dust ratio $(n_p+n_H)/n_d$ are derived (see [10]).

As to the energy equilibrium of the gas there are two cooling mechanisms. One is thermal bremsstrahlung, the other is due to collision excitation of radiative transitions of trace elements by thermal electrons and atoms. They are represented by $\Lambda_{ff}(T_E)$ and Λ_{cc} (DALGARNO and MCCRAY [9]) respectively

$$\Lambda_{ff}(T_E) = 1.426 \times 10^{-27}\ n_e n_p T_E^{1/2}\ \langle g_{ff} \rangle\ \ [\text{ergs cm}^{-3}\ \text{s}^{-1}] \tag{7}$$

$$\Lambda_{cc} \simeq 4 \times 10^{-24}\, n_e n_H \; [\text{ergs cm}^{-3}\ \text{s}^{-1}] \quad . \tag{8}$$

The heating role is played by the net ionized electrons, and then

$$\Gamma(T_E) = 2 n_H n_J v_J \sigma_{1c} \bar{E}_2 - n_e n_p \alpha(2) \frac{3}{2} kT_E \tag{9}$$

where $\bar{E}_2$ is the mean kinetic energy of electrons released from the collision ionization. The average energy loss of the relativistic particles during a collisional ionization is $E_\ell = \bar{E}_2 + 13.6$ eV. Using the energy equilibrium equation $\Lambda_{ff}(T_E) + \Lambda_{cc}(T_E) = \Gamma(T_E)$ and (2) for the ionization equilibrium together with $T_E \simeq T_e$, we find that the value of E_ℓ is physically reasonable (see [10]).

4. CONCLUSION

Based on the infrared data of knots and the analysis here, it is probable that the five knots are physically related to SS433. The knots are mainly composed of dust and partly ionized hydrogen gas. Their infrared spectra are well fitted with the thermal emission of dust and the free-free emission. The knots are in a state of ionization-recombination equilibrium and energy equilibrium. The physical parameters derived from these physical processes are reasonable and consistent. In another paper (WANG et al. [10]), we further consider the stochastic heating of dust grains and a distribution of grain sizes. We find no important changes in the results, because for the interaction between the jets of SS433 and the interstellar medium, dust grains of larger size are more important and they are nearly in a state of equilibrium. The upper limit on surface brightness of H_α for the knots expected from our model is consistent with Becker's observation ([3],[10]).

This work is supported by the National Science Foundation in the U.S. and the National Natural Science Foundation of China.

References

1. B. Margon: Ann. Rev. Astr. Ap. 22, 507 (1984).
2. F. Seward, J. Grindlay, E. Seaquist, W. Gilmore: Nature 287, 806 (1980).
3. D. L. Band: Publ.A.S.P. 99, 1269 (1987).
4. L. Spitzer: In Physical Processes in the Interstellar Medium (Wiley, 1978).
5. C. W. Allen: Astrophysical Quantities (William Clowes, 1973).
6. W. Lotz: Ap. J. Suppl. 14, 207 (1967).
7. E. Dwek: Ap. J. 322, 812 (1987).
8. M. Watson, R. Willingale, J. Grindlay, F. Seward: Ap. J. 273, 688 (1983).
9. A. Dalgarno, R. McCray: Ann. Rev. Astr. Ap. 10, 375 (1972).
10. Z. R. Wang, Y. Chen, R. McCray, Q. Y. Qu: Astr. Ap. submitted.

The Radio Spectra of Supernova Remnants

John Dickel

1. General Characteristics

The radio emission from supernova remnants (SNRs) is synchrotron radiation by relativistic electrons in magnetic fields. One electron will produce radiation in a series of harmonics of its cyclotron frequency around a magnetic-field line but these harmonics will smear into a continuum with the frequency of peak emission dependent upon the energy of the electron. The radio frequencies of interest are generally well above the peak frequency. A collection of electrons with a power law distribution in energy, $N(E) = CE^{-\gamma}dE$ where $N(E)$ is the number of electrons with energy E, will then produce continuum radiation with a power-law spectrum of the form $S_f = \text{const}\ f^{\alpha}$, where S_f is the observed flux density (in Janskys where 1 Jy = 10^{-26} w m^{-2} Hz^{-1}) at frequency f and the power, α, is called the spectral index where $\alpha = -(\gamma-1)/2$. Typical values of α range between -0.8 and -0.3 for shell SNRs. This means that they are brighter at lower frequencies.

The detailed formula for the intensity of the synchrotron emission is

$$I_f = \frac{A\ e^3}{mc^2}\left(\frac{3e}{4\pi m^3 c^5}\right)^{(\gamma-1)/2} B^{(\gamma+1)/2}\ L\ C\ f^{-(\gamma-1)/2} \qquad (1)$$

where e is the charge of the electron, m its mass, c the speed of light, B the magnetic field strength, f the frequency, γ the energy index, L the depth of the radiating region along the line of sight, C the constant in the energy distribution given above is related to the density, and A is a complex function with a numerical value near 0.1 containing Γ functions of γ (GINZBURG and SYROVATSKI [1]). Evaluation of the spectrum will require knowledge of the magnetic fields and the relativistic electron distributions. The field can be diluted by expansion of the remnant but also compressed by shocks and thermal instabilities or amplified by turbulence. Electrons can suffer losses by radiation and work under adiabatic expansion but also gain energy by Fermi acceleration – first order by crossing shocks and second order in turbulent eddies created by expansion into an irregular medium with Rayleigh-Taylor and Kelvin-Helmhotz instabilities (DICKEL *et al* [2]). Crab-type or filled remnants appear to have continuous injection of particles from a neutron star and values of α near 0.0 (WEILER [3]).

2. Time Changes

How do the radio spectra of shell-type SNR behave under the various competing mechanisms? Observational data indicate that older remnants appear to have flatter spectra than young ones. Figure 1 might suggest that the change in slope is most rapid at young ages but the sample is too small to specify a trend. In addition, Cas A, the youngest known SNR in the Milky Way, is flattening with time (BAARS *et al* [9]); the annual decrease, d, in flux density of Cas A as a function of frequency:

$$d(f)\ [\text{in}\ \%/\text{year}] = 0.97(\pm 0.04) - 0.30(\pm 0.04)\ \log f\ [\text{in GHz}]. \quad (2)$$

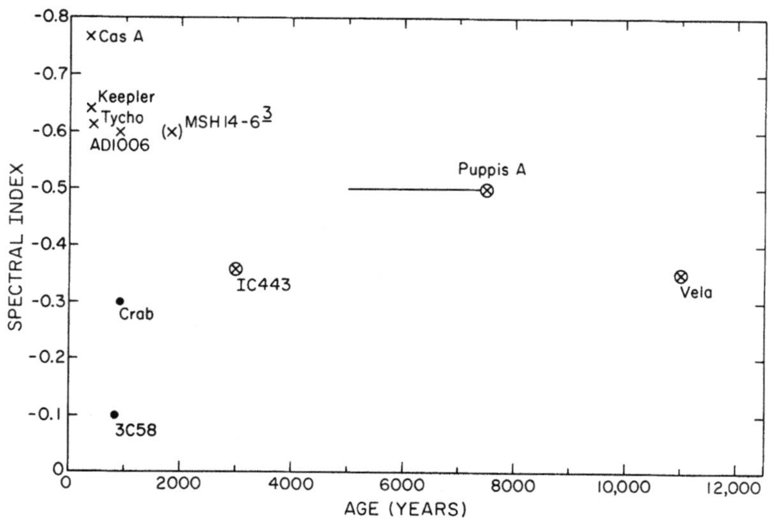

Fig. 1. Radio spectral indices of supernova remnants versus age. Spectral indices are taken from GREEN'S [4] list and references for the ages are: historical remnants (CLARK and STEPHENSON [5]; Puppis A (ARENDT [6]); IC 443 (PETRE *et al* [7]); and Vela (MANCHESTER and TAYLOR [8]).

It is difficult to include more galactic objects in the sample because of inhomogeneities in the database. We don't know how to obtain ages for most remnants and determination of consistent spectral indices is very difficult because of varying beam sizes, sidelobe responses, background removal, etc. at the different frequencies. To obtain a reasonably homogeneous sample we have used a survey of 19 SNRs in the Large Magellanic Cloud (MILNE *et al* [10]). All of these remnants were identified in the radio at 5 GHz and the data used for the spectral index evaluations extend from 0.4 to 14.7 GHz. Although the list does not contain all the SNRs known in the LMC, it should be an unbiased sample of objects. Because we do not know the ages of these objects, we have adopted the 1 GHz surface brightness, Σ, as an indicator of relative evolutionary age. The surface brightness should decrease with time because there is more acceleration at early epochs when the velocities and shocks are faster and, by later times, adiabatic expansion will cause a decrease in density and particle energy. The results, shown in Fig. 2, indicate a rather weak correlation with a value

$$\alpha = -2.25 - 0.09 \log \Sigma \quad (3)$$

and a correlation coefficient of 0.62. The average index of the 19 remnants is -0.44. Thus the spectra appear to be steeper at greater surface brightness which implies they are flattening with age. This is contrary to what one would expect for energy losses alone as the most

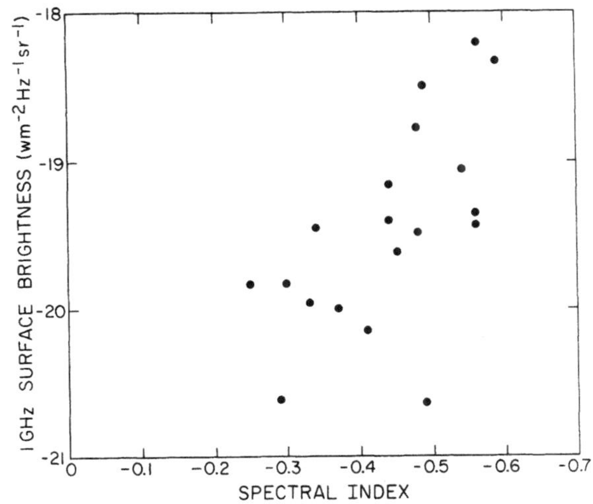

Fig. 2. Radio spectral index versus 1 GHz surface brightness of SNRs in the LMC. Brighter (younger) remnants have more negative (steeper) spectral indices so remnants should evolve toward the lower left.

energetic particles should lose energy fastest. However, COWSIK and SARKAR ([11]) have shown that continued stochastic acceleration of relativistic particles in turbulent regions will tend to decrease the number of particles near the peak in the energy distribution with some going to higher energies and some to lower energies. This will flatten the overall energy distribution and thus the spectrum of the observed radiation with time. The fact that this process continues at least well into the adolescent phase of SNR evolution after the blast has swept up many times the ejected mass, shows that acceleration processes are still important at these later times.

3. Spectral Variations Across a Remnant?

Because particle acceleration depends on turbulent instabilities and shock strength, changes in spectral index might be expected across a remnant as varying conditions are encountered. This does not appear to be the case, however, as there are no conclusive data on changes in spectral index across any shell remnant. Some early indications of such variations have been shown to be erroneous after careful removal of the confusing galactic background, and correction for differing beam resolution near the edges of some remnants, e.g. for IC 443 (ERICKSON and MAHONEY [12]) and Vela (MILNE and MANCHESTER [13]). It thus appears that the spectrum does not depend upon the magnitude of the particle acceleration and field amplification but only on the length of time the stochastic processes have been occurring.

Remnants which do show spectral variations are the composite objects consisting of a filled or Crab-type remnant with a flat spectrum and sharp boundary surrounded by a shell type one with a steeper spectrum, e.g. MSH 15-56 (MILNE *et al* [14]). Also, ERICKSON and PERLEY ([15]) reported a short lived "flare" in the brightness of Cas A at 38 MHz. They could not resolve the feature but its rapid decay required it to be small and thus a small-scale change in spectrum.

4. Changes in Spectral Index With Frequency?

Synchrotron radiation should have a constant spectral index unless the electrons themselves have a changing distribution with energy. Synchrotron losses will steepen a spectrum as they affect the highest energy electrons most significantly but the lifetime for such losses is much greater than 10^5 years, the age of the very oldest observed remnants. Spectra of the old remnants HB9 and the Cygnus Loop, however, suggest kinks near 1GHz with a steeper slope at high frequencies (DE NOYER [16]). These results need confirmation but, if true, might represent a balance between the reduced acceleration processes and compression of previously present ambient particles which are no longer being accelerated. The general galactic background has a change in its spectrum near 200 MHz (BRIDLE [17]) which could be shifted upward in frequency to near 1GHz by the compression. Finally, a number of remnants near the galactic plane show a decrease in brightness at frequencies $\lesssim$ 100 MHz. This can be attributed to free-free absorption by intervening ionized hydrogen and is not a property of the remnants themselves (e.g. KASSIM [18]).

References

1. V. Ginzburg, S. Syrovatski: Ann. Rev. Astr. and Ap. 3, 297 (1965).
2. J. Dickel, J. Eilek, E. Jones, S. Reynolds: Ap. J. Sup. 70, 497 (1989)
3. K. Weiler: In Supernova Remnants and Their X-ray Emission, ed. by J. Danziger and P. Gorenstein (Reidel, Dordrecht, 1983) p. 299.
4. D. Green: MN 209, 449 (1984).
5. D. Clark, F. Stephenson: The Historical Supernovae, (Pergamon Press, Oxford, 1977).
6. R. Arendt: Ph.D. Thesis, University of Illinois (1988).
7. R. Petre, A. Szymkowiak, F. Seward, R. Willingdale: Ap. J. 335, 215 (1988).
8. R. Manchester, J. Taylor: Pulsars, (Freeman, San Francisco, 1977).
9. J. Baars, R. Genzel, I. Pauliny-Toth, A. Witzel: Ast. Ap. 61, 99 (1977).
10. D. Milne, J. Caswell, R. Haynes: MN 191. 469 (1980).
11. R. Cowsik, S. Sarkar: MN 207, 745 (1984).
12. W. Erickson, M.J. Mahoney: Ap. J. 290, 596 (1985).
13. D. Milne, R.N. Manchester: Astron. Astrophys. 167, (1986).
14. D. Milne, J. Caswell, R. Haynes, M. Kesteven, K. Wellington, R. Roger, J. Bunton: Publ. Astron. Soc. Australia 6, 78 (1985).
15. W. Erickson, R. Perley: Ap. J. Lett. 200, L83 (1975).
16. L. De Noyer: A. J. 79, 1253 (1974).
17. A. Bridle: MN 136, 219 (1967).
18. N. Kassim: Ap. J. 347, in press, December (1989).

Observations of Molecular Clouds Associated with Supernova Remnants in the LMC

J. P. Hughes, L. Bronfman, & L. Nyman

I. Introduction

The Large Magellanic Cloud (LMC) has been systematically surveyed for CO line emission by COHEN *et al.* [1], who found a general correspondence between the CO emission and such Population I objects as supernova remnants (SNRs) and H II regions. However, because of the limited spatial resolution of the survey (8!8), a detailed association between individual objects and molecular cloud complexes was not possible. Guided by the survey results, we have undertaken a study of SNR-molecular cloud associations using the Swedish-ESO Submillimeter Telescope (SEST) at La Silla, Chile. The SEST has good spatial resolution: at the CO J = 2 - 1 line (230.5 GHz) the half power beam width is only about 22″. We have obtained data on two LMC SNRs: N49 and N132D, both of which have CO line emission in their vicinities.

We mapped the CO J = 2 - 1 line emission with a grid spacing of 20″ starting at the approximate center of each remnant. The telescope was operated in frequency switched mode with a 21 MHz frequency throw, corresponding to a velocity shift of about 27 km/s. The total frequency range sampled was 87 MHz (about 110 km/s in velocity). In most cases we integrated for 30 min., although for the regions of stronger emission near N132D we used 10 min. integrations. The system temperature varied from about 800 K to 1000 K during the two days of observations of N49, and varied from 1000 to 1250 K for the N132D observations (done on a single day). The data were folded, third order baselines were subtracted, and the CO emission was integrated over velocity for mapping purposes. We also integrated over a line-free region of the spectra in order to assess our signal to noise and establish the level of possible systematic error in our baseline fits. In a 10 km/s wide velocity region adjacent to the CO line the RMS velocity-integrated emission was ±0.086 K km/s for N49 and ±0.082 K km/s for N132D.

II. N49

The ***Einstein*** high resolution X-ray image of N49 is shown in Fig. 1 in a grey scale presentation along with contours of the velocity-integrated CO emission (W_{CO}). The lowest contour in the figure is 8.7 times the RMS noise level determined from integrating an adjacent 10 km/s velocity region, as described above. The average velocity (LSR) of this cloud is about 286 km/s, in excellent agreement with the velocity of the SNR: 286 ± 1 km/s (SHULL [2]). There is little difference in average velocity from point to point in the cloud (< 1 km/s), although the width (FWHM) of the line varies from about 3.5 km/s to almost 7 km/s. We note that the peak antenna temperature (uncorrected for beam efficiency) was 0.36 K.

The agreement in position and velocity between this molecular cloud and the SNR N49 strongly suggest that the objects are physically related. In addition, the cloud is located near the brightest region of X-ray emission, where the optical emission is also bright. Further study of our data (e.g., searches for high velocity CO emission) and comparisons with optical and infrared observations should allow us to investigate the relationship between the cloud and SNR.

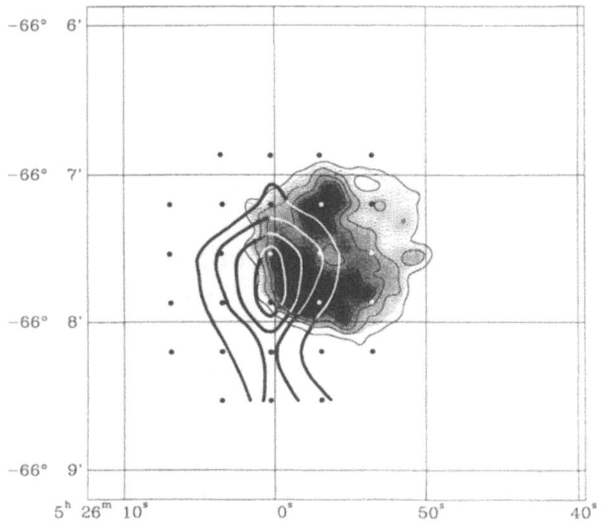

Figure 1. ***Einstein*** **X-ray image of N49 overlaid with contours of the velocity-integrated CO emission. The grid pattern for the CO observations is shown. The contour levels are 0.75, 1.0, 1.25, and 1.50 K km/s and the spectra were integrated from 280 km/s to 290 km/s.**

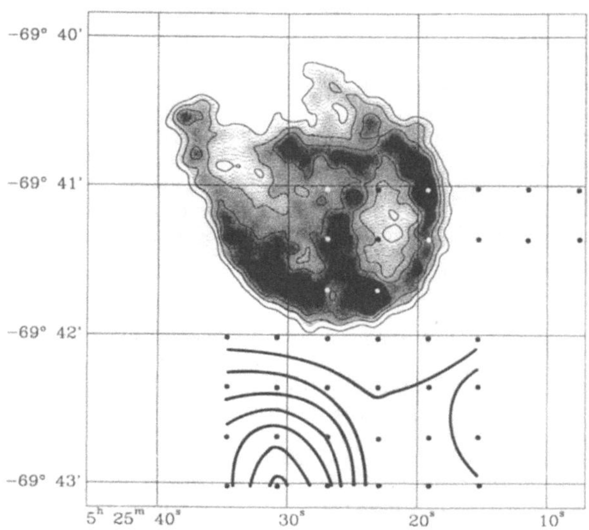

Figure 2. The same as Fig. 1, but for the SNR N132D. Here the contour levels are 3, 5, 7, 9, 11, 13, and 15 K km/s for a velocity integration range from 260 km/s to 270 km/s.

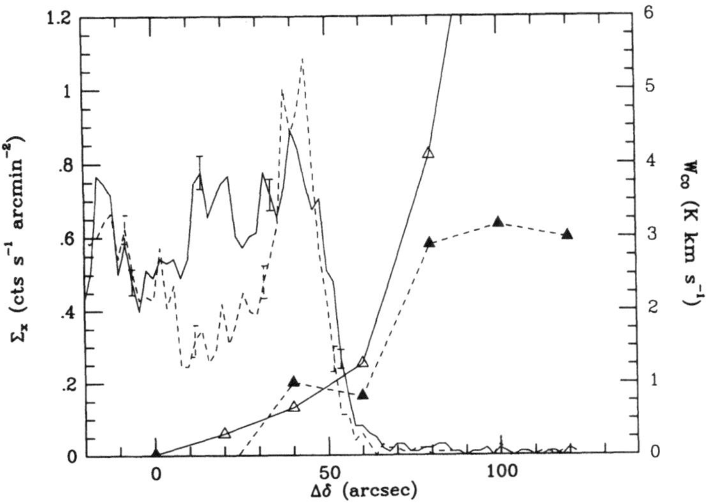

Figure 3. Comparison of X-ray surface brightness (curves on the left) with the velocity-integrated CO emission (curves rising to the right) for the SNR N132D. Two different strips through the remnant are shown.

III. N132D

Figure 2 shows the high resolution X-ray image of N132D and the contours of W_{CO}. The lowest contour in this figure is a factor of 37 times the RMS noise level and the peak antenna temperature is about 3 K. This molecular emission is significantly stronger than that in N49 by about a factor of ten. There is a strong velocity gradient across the CO emission region: the average velocity varies from 262 km/s in the east to about 268 km/s in the west. Although the uncertainty is large, it appears that the SNR lies at a different velocity (247 ± 16 km/s DANZIGER and DENNEFELD [3]). The velocity of the cloud and the SNR are both larger than that expected from the rotation curve of the LMC, which should be about 236 km/s at this point (DAME [4]).

Toward the south the match between the shape of the CO contours and the X-ray image is rather striking, even considering the incomplete CO coverage. This is shown in more detail in Fig. 3., where we plot the X-ray surface brightness and the W_{CO} emission for two slices through the SNR image. The slices correspond to $\alpha(1950) = 5^h\, 25^m\, 27^s$ (solid curves) and $\alpha(1950) = 5^h\, 25^m\, 23^s$ (dashed curves). The ordinate represents the distance in arcsec from declination value -69° 41′ 2″. The anticorrelation between X-ray and CO emission is apparent, although there is CO emission near the shell of N132D (near $\Delta\delta = 50''$ in Fig. 3) at about the same level as observed for N49. It is the presence of a rather bright molecular cloud complex to the south which is striking. Note, however, that the slices which were mapped to the west show little or no CO emission.

HUGHES [5] proposed an evolutionary scenario for N132D in which the precursor star blew a cavity in the surrounding interstellar medium (ISM) before it became a supernova, and that the SN blast wave has now propagated to the walls of the cavity. The model requires a gradient in the local ISM density, in the sense of increasing from the northeast to the southwest, in order to explain the morphology of the remnant. This requirement could be satisfied if the SN occurred at the edge of a molecular cloud, as the present observations seem to indicate. In this case a significant interaction should be occurring between the remnant and the molecular cloud.

We thank the SEST program committee for granting observation time to this project and we thank T. Dame, H. Tananbaum, and D. Harris for useful discussions and comments. This research was supported in part by Smithsonian Institution funds.

IV. References

1. R. S. Cohen, T. M. Dame, G. Garay, J. Montani, M. Rubio, and P. Thaddeus, *Ap. J. (Letters)*, **331**, L95, (1988).
2. P. Shull, *Ap. J.*, **275**, 611, (1983).
3. I. J. Danziger and M. Dennefeld, *Ap. J.*, **207**, 394, (1976).
4. T. M. Dame, private communication, (1989).
5. J. P. Hughes, *Ap. J.*, **314**, 103, (1987).

The Interaction of Supernova Remnant in the Early Phase with a Circumstellar Shell

Tatsuo Yoshida & Hitoshi Hanami

The observations in the ultraviolet(IUE), the near infrared(speckle interferometry), and the soft X-ray(Ginga) of SN1987A suggest circumstellar shells(CSSs) exist. The shells have been formed when the fast blue supergiant wind sweeps the red supergiant wind.

If the supernova remnant(SNR) hits the CSS, hydrodynamical instability may arise in the acceleration of the high-density shell by the lower-density shocked circumstellar matter. In this paper, we give the numerical results of our two dimensional calculations to investigate how the instability develops.

Model

We use the spherically symmetric self-similar solution(CHEVALIER [1]) as a model of SNR, assuming that the envelope of the ejecta and the circumstellar matter(CSM) have power-law density profiles as $\rho \propto r^{-7}$ and $\rho \propto r^{-2}$, respectively. We start calculations from 0.5 yr after the explosion. The parameters are as follows: the explosion energy is 10^{51}erg and the mass of the ejecta is 10 $M_{\odot}$. The ratio of the progenitor's mass loss rate to the wind velocity is $2.4\times10^{-6} M_{\odot} yr^{-1}/10\ kms^{-1}$ (YOSHIDA and HANAMI [2]).

The parameters of the shell are based on Masai's model (MASAI et al. [3]), which explains the flare in January 1988 of SN1987A. The shell and the CSM are assumed to be initially in pressure equilibrium. The surface of the shell is located at $r_c=4.1\times10^{16}$ cm with a density of $\rho_{cloud}=\delta\cdot\rho_{CSM}(r_c)$ and a thickness d of 5.4×10^{15} cm. The shell covers $\Omega/4\pi\sim15$ % of the spherical area. We assume that initially the surface of the shell has sinusoidal corrugations. The amplitude a_0 is 0.1d and the wavelength λ is $2\pi r_c/\Omega$, where Ω is the azimuthal wave number.

Numerical Results

Figure 1a and b show how the density distribution and the velocity field of the SNR and the CSS evolve at an interval of 1.0 yr for the case δ=20 and Ω=48. We find that the corrugations grow and finger-like structures are formed.

In order to see how the instability develops, we pursue the motion of the marker particles, which are set on the surface of the shell initially. Figure 2a shows the positions of the marker particles at evolutionary stages. The particles are connected by a line at each stage. Figure 2b shows the time evolution of the corrugation amplitude. We find that there are three periods, which have different growth rates of the amplitude.

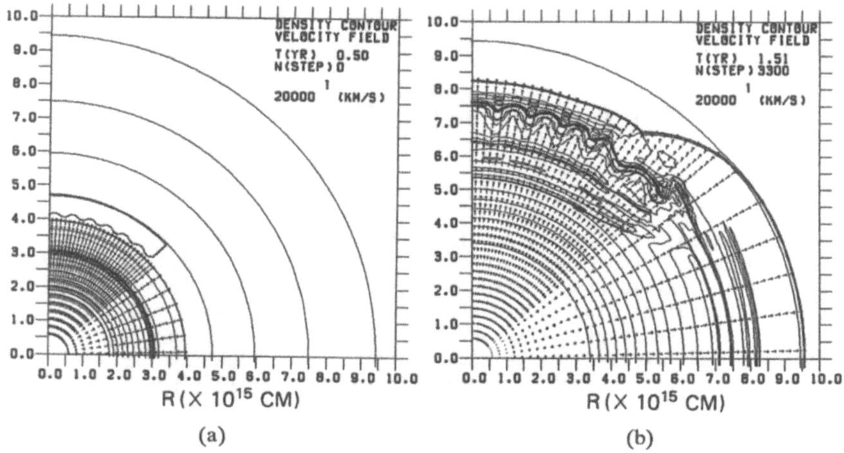

Figure 1. (a) Density contours and velocity vectors for the case δ=20 and ℓ=48 at the initial condition(t=0.5 yr). Lines of constant dehsity are logarithmically spaced with $\log\Delta\rho$=0.2. (b) The same as (a) except t=1.5 yr.

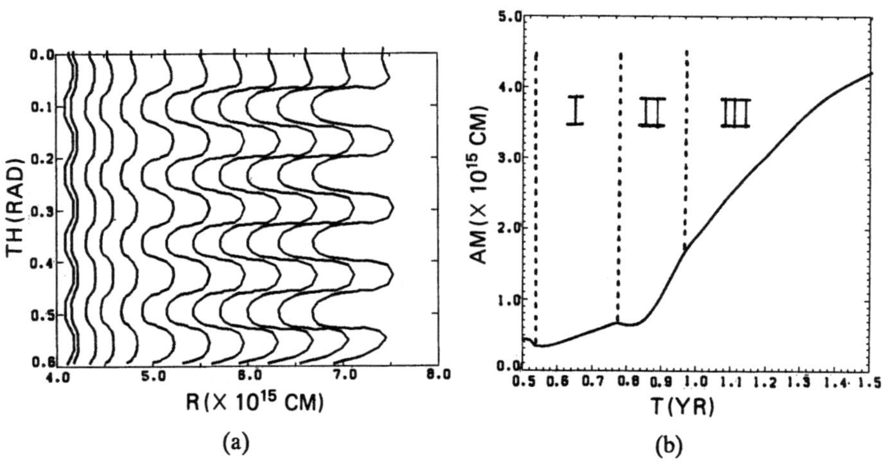

Figure 2. (a) The positions of the marker particles are shown. (b) Time evolution of the amplitude.

In the first period the amplitude grows linearly. Richtmyer [4] finds that, in systems which have been impulsively accelerated by a shock, the amplitude grows at a constant rate with time according to

$$a(t)/a(0)=1+\alpha t, \quad \alpha=\frac{2\pi}{\lambda}\left(\frac{\rho_{cl}-4\rho_0}{\rho_{cl}+4\rho_0}\right)\Delta v, \tag{1}$$

where Δv is the increment of the velocity imparted by the acceleration. If we know the value of Δv, we can estimate the value of α. Figure 3 shows the velocity history of the marker particle at a crest of corrugations. This figure shows after the particle is impulsively accelerated, the value of the particle velocity becomes nearly constant. Using this as the value of Δv, we can calculate the value of α. This calculated value of α is 1.6 times as large as one obtained from figure 2b. The discrepancy could be attributed to numerical diffusion.

In the second period, when the shock goes out of the dense shell, the growth rate increases rapidly. This is because rarefaction wave is formed and the shell compressed by the shock becomes subject to expansion. Figure 3 shows that the particle is accelerated again in this period. In the third period, the growth rate decreases slowly. However, the amplitude becomes comparable with the thickness of the shell since the growth rate in the second period is large.

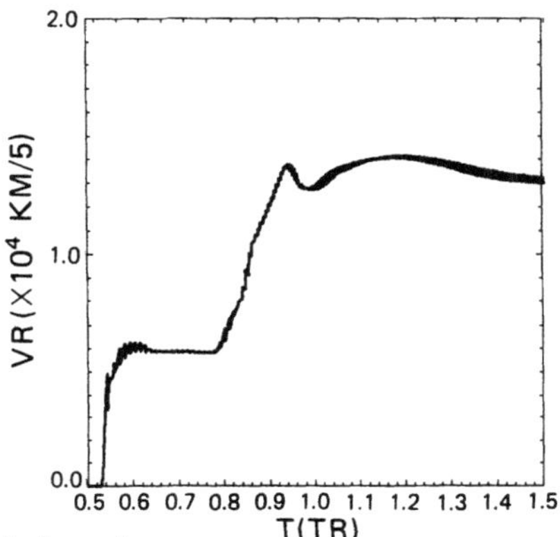

Figure 3. A velocity history of the marker particle at a crest of corrugations.

Acknowledgments

T.Y would like to thank Professor S.Sakashita, Dr.A.Habe, and Dr.T.Ebisuzaki for their valuable discussions T.Y. is grateful to the Nukazawa Memorial Foundation for financial support. The calculations were carried out on HITAC M-680H at the Center for Information Processing Education of Hokkaido University, HITAC S-820/80 at the Computing Center of Hokkaido University, FACOM VP200 at the Space Data Analysis Center, the Institute of Space and Astronautical Science, and FACOM VP50 at the Nobeyama Radio Observatory.

References

1. R. Chevalier: Ap.J. 258, 790(1982).
2. T. Yoshida, H. Hanami: Prog.Theor.Phys. 80, 83(1988).
3. K. Masai, S. Hayakawa, H. Inoue, H. Itoh, and K. Nomoto: Nature 335, 804(1988).
4. R. D. Richtmyer: Com.Pure.Appl.Math. 13, 297(1960)

Global Effects of Supernova Remnants on the Interstellar Medium

Christopher McKee

1. Introduction

The overall energetics of the interstellar medium (ISM) are governed largely by the outflow of energy from stars through the ISM to the intergalactic medium. This energy is in the form of radiation, winds, and supernova explosions; supernovae dominate the injection of kinetic energy into the ISM. Much of this energy is from massive stars, which are continually forming out of the ISM and which return most of their mass back to the ISM.

The ISM is observed to be highly inhomogeneous, with much of the mass concentrated in clouds—both atomic and molecular—which occupy a small fraction of the volume, whereas much of the volume is filled with warmer intercloud gas. Field, Goldsmith, and Habing (1969; hereafter FGH) provided a theoretical explanation of how such multiphase behavior is possible: the temperature dependence of the cooling rate of interstellar gas is such that cold ($T \sim 10^2$ K), neutral (H I) clouds can coexist in pressure equilibrium with warm ($T \sim 10^4$ K) intercloud H I only over a limited range of pressures, close to those observed in the ISM. Subsequently, Cox and Smith (1974) argued that supernova explosions in the Galaxy are sufficiently frequent that supernova remnants tend to overlap, forming a network of hot tunnels in the ISM with a filling factor of about 10%; numerical simulations of this model were carried out by Smith (1977). McKee and Ostriker (1977; hereafter MO) argued that the filling factor of the hot gas is substantially larger than this, so that it forms the background medium in which the two phases of the FGH model are embedded. The ISM thus consists of three phases: the cold neutral medium (CNM), taken to have a temperature of 80 K; the the warm medium at $T \sim 8000$ K, which has both neutral (WNM) and ionized (WIM) parts; and the hot ionized intercloud medium (HIM) at a calculated temperature $T \sim 5 \times 10^5$ K. The most obvious difference between the three–phase model and the two–phase model is that a large fraction of the volume of the ISM is predicted to be almost empty, filled with low-density, hot gas; as a result, mass exchange between the hot phase and the other phases, due to thermal evaporation and condensation, is crucial in determining the equilibrium. The thermal pressure in the ISM is determined dynamically by the evolution of supernova remnants (SNRs) in the ISM, permitting prediction of the distribution of the pressures in the ISM. The three–phase model remains controversial; in particular, Cox (1988) and Cox and Slavin (1989) have suggested that the filling factor of the hot gas is only about 10%. Determining the filling factor of the hot gas in the ISM is one of the critical observational problems in interstellar astrophysics (see Spitzer 1990).

Observations in the last decade have shown that the ISM extends much farther from the plane of the Galaxy than originally believed: Lockman (1984) has found a component of HI with a scale height of about 500 pc, Reynolds (1989*a*) has inferred a scale height for ionized gas of 1–2.5 kpc from observations of pulsar dispersion measures, and a number of investigators have studied halo gas far from the plane through its absorption lines (Savage 1987). The mass and scale height of the various components of the ISM have been reviewed by Bloemen (1987), Spitzer (1990), and McKee (1990*a*). A quantitative view of the vertical structure of the ISM is given in Table 1, taken from McKee (1990*a*). The densities are given in terms of $\bar{n}_{\rm H}(z)$, the volume averaged density of hydrogen nuclei as a function of distance from the Galactic plane, z. The surface densities Σ include the mass of helium, which we take to be 10% by number. References for the various entries are given in McKee (1990*a,b*).

Some comments on the Table are necessary. The division of the H I into three components is based on the work of Lockman (1984). Kulkarni and Heiles (1987) identify the component with the lowest scaleheight with the HI clouds (the CNM) and the remaining two components with the intercloud HI (which we label as $\rm WNM_1$ and $\rm WNM_2$). One of the components ($\rm WNM_2$) extends into the halo, as has been confirmed by ultraviolet observations of high latitude stars (Lockman, Hobbs, and Shull 1986). Ionized hydrogen near the Galactic plane is in two forms, photoionized (the WIM) and collisionally ionized (the HIM). The density and scale height of the WIM can be inferred from observations of pulsar dispersion measures (e.g., Kulkarni and Heiles 1987), which imply an electron distribution of the form given in the Table, with a scale height $H_w \simeq 1$ kpc. Allowing for the roughly 4% of the electrons contributed by helium (Mathis 1986), a correction more important in principle than in practice, gives the mean hydrogen density listed in Table 1. Observations of the WIM in emission lines (reviewed in Reynolds 1989*b*) also imply a thick disk of warm, ionized gas.

Table 1. Vertical Mass Distribution of the ISM

Component	$\bar{n}_{\rm H}(z)$ (cm^{-3})	Σ ($M_\odot$ pc^{-2})
$\rm H_2$	$0.54 \exp -\frac{1}{2}(z/60)^2$	2.80
HI: CNM	$0.39 \exp -\frac{1}{2}(z/135)^2$	4.54
$\rm WNM_1$	$0.093 \exp -\frac{1}{2}(z/254)^2$	2.04
$\rm WNM_2$	$0.053 \exp -(\|z\|/480)$	1.75
HII:[a] WIM	$0.96[0.015 \exp -(\|z\|/70) + 0.025 \exp -(\|z\|/H_w)]$	1.72/1.36
HIM	0 / $0.0015 \exp -(\|z\|/3000)$	0/0.31
Total		12.8

[a] Two cases are considered for the hot component of the ISM (HIM): (1) no HIM / (2) HIM based on MO. The second case accounts for some of the electrons observed in the dispersion measure of high altitude pulsars, thereby reducing the scale height of the WIM from 1 kpc for Case 1 to 780 pc for Case 2.

The characteristics of the HIM are more controversial, as indicated above. To gauge the effects of the HIM on the vertical structure of the ISM, two cases are considered: (1) The HIM is negligible, in the sense that its mean pressure is small compared to that of the other components. (2) The HIM is similar to that envisioned by MO, with a temperature of about 5×10^5 K, a density $n_{\rm H} \simeq 3 \times 10^{-3}$ cm^{-3}, and a filling factor f_h of about 0.5 in the plane. A scale height of $H_h = 3$ kpc for the mean density has been adopted.

The accuracy of this inventory of the mass distribution is difficult to assess, but it is consistent with the available data on the column densities and scale heights of H_2, HI, and HII; the total column density of all the components in this model is 13 $M_\odot$ pc^{-2}, the same as that quoted by Kulkarni and Heiles (1987). Furthermore, the total midplane density is 1.1 cm^{-3}, in good agreement with the value 1.2 cm^{-3} found from an analysis of extinction data by Spitzer (1978). The question to be resolved is: why is the scale height as large as it is? This question has been addressed observationally by Kulkarni and Fich (1985), who concluded that the H I is sufficiently turbulent that it can support itself; McKee (1990*a*) has shown that this conclusion applies to the WIM as well. The question then becomes, why is the ISM so turbulent? We shall argue that this turbulence is due to energy injection by supernovae, and that the remnants of the supernovae must essentially fill the ISM in order to explain the observed level of turbulence.

Before addressing these issues, a brief comment on superbubbles is in order. One of the major advances in our understanding of the ISM in the past decade is the realization that the clustering of supernovae in stellar associations can have a dramatic effect on the ISM. HI shells far too large to have been created by individual supernovae have been observed throughout the Galaxy, with large shells and supershells occurring primarily outside the solar circle (Heiles 1979) and vertical structures ("worms") occurring inside (Heiles 1984). A number of authors, beginning with Bruhweiler *et al.* (1980), have attributed these structures to energy injection by associations of massive stars, leading to the creation of superbubbles in the ISM. A discussion of this topic is beyond the scope of this paper; a thorough review has recently been given by Tenorio–Tagle and Bodenheimer (1988). It should be noted that a number of the theoretical models of superbubbles assumed that the gas disk of the Galaxy is thin, so that the energy could easily break out into the halo; however, the observations discussed in §II above show that the gas disk is in fact rather thick, so that most superbubbles should remain confined (Cox 1989). Observationally, superbubbles appear to occupy only about 10–20% of the area of the Galactic plane in the solar neighborhood (Heiles 1980), leaving most of the volume available for a two– or three–phase ISM.

2. Isothermal Models for the Vertical Structure of the ISM

In view of the inordinate complexity of the observed ISM, it is instructive to pull back and develop simple toy models to see what the ISM would be like if only a few processes were operating. Here we shall consider isothermal models for the vertical structure of the ISM. Let $C \equiv (P/\rho)^{1/2}$ be the isothermal sound speed for the ISM, where P is the total pressure and ρ the density. In order that the model not be too unrealistic, we shall include a magnetic pressure and a pressure due to cosmic rays in P. Let $g(z)$ be the magnitude of the z–component of the gravitational acceleration, and let $\bar{g}(z)$ be the mean value,

$$z\bar{g}(z) \equiv \int_0^z g(z')dz'. \tag{1}$$

Then the solution of the equation of hydrostatic equilibrium is

$$\rho(z) = \rho_0 \exp - \left[\frac{z\bar{g}(z)}{C^2}\right] . \tag{2}$$

If the gas disk is sufficiently thin and has a small fraction of the total mass, then the acceleration will be approximately a linear function of height, $g(z) \simeq zg_0'$, with the midplane gradient g_0' constant. In this case the density is a Gaussian,

$$\rho(z) = \rho_0 \exp -\frac{1}{2}\left(\frac{z^2}{H^2}\right), \tag{3}$$

where the scale height is

$$H = \frac{C}{g_0'^{1/2}} = 116 C_6 \quad \text{pc}, \tag{4}$$

with $C_6 \equiv C/(10^6 \text{ cm s}^{-1})$. The numerical value for the scale height is based on an acceleration gradient $g_0' = 2.4 \times 10^{-11}$ cm s^{-2} pc^{-1}; g_0' is proportional to the total density in the Galactic plane, and the adopted value is intermediate between those corresponding to densities of 0.185 $M_\odot$ pc^{-3} (Bahcall 1984) or 0.1 $M_\odot$ pc^{-3} (Kuijken and Gilmore 1989).

2.1 One Phase: The Dead ISM

We begin with the simplest case, in which the rate of energy injection into the ISM is so low that the gas is all cold, say at a temperature of 80 K. If we assume that the thermal, magnetic, and cosmic ray pressures are all equal, then the isothermal sound speed in the medium is 1.25 km s^{-1}, and the scale height of the gas is only about 15 pc! Such an ISM is lifeless. The scale height is small even compared to that of the molecular gas in the Galaxy, so this model can be rejected outright.

2.2 Two Phases: The Quiescent ISM

Consider now a higher heating rate, so that some of the gas is at a temperature of order 10^4 K. It is generally possible for interstellar gas to have two phases, one warm and one cold, coexisting at the same pressure: The warm phase can be in equilibrium for thermal pressures less than $P_{th,max}$, whereas the cold phase can be in equilibrium for thermal pressures greater than $P_{th,min}$; a two-phase medium is possible provided $P_{th,max} > P_{th} > P_{th,min}$ (FGH; Begelman and McKee 1990). A recent calculation of the heating and cooling of the interstellar gas including heating due to the ejection of photoelectrons from grains and the damping of hydromagnetic waves found $\tilde{P}_{th,max} \simeq 2000$ cm^{-3} K (we use $\tilde{P} \equiv P/k$ for convenience; 2000 cm^{-3} K corresponds to 2.8×10^{-13} dyne cm^{-2}) and $\tilde{P}_{th,min} \simeq 150$ cm^{-3} K (Ferriere, Zweibel, and Shull 1989).

Provided the thermal pressure in the midplane exceeds $P_{th,max}$, gas in the midplane would lie in a cold, thin disk, as in the one-phase model described above; the disk would be about 30 pc thick. At a height above the plane at which the thermal pressure has dropped below $P_{th,max}$, a two-phase structure would become possible, with cold clouds embedded in a warm intercloud medium. The scale height of the warm gas would be about 130 pc. Clouds could continue to exist up to heights somewhat greater than this, until the thermal pressure dropped below $P_{th,min}$; for

$\tilde{P}_{th,min} = 150$ cm^{-3} K, this height is about 300 pc. At greater heights, the gas would be homogeneous and warm. If not supported by turbulent motions or magnetic fields, clouds which formed in the two–phase zone would rain down on the cold disk.

This model is a substantial improvement over the one–phase model, but it is a far cry from reality. The possibility that clouds could exist in the two–phase zone above the cold disk could increase the scale height of the cold gas somewhat above 15 pc, but it would still be much less than the observed 135 pc. Warm gas is observed to have an effective scale height $\Sigma/2\rho_0$ of 430 pc (see Table 1), several times greater than the model predicts. This discrepancy is due to the fact that the model has no turbulence; it is quiescent, in contrast to the observed ISM. We conclude that isothermal models without injection of kinetic energy cannot account for the observed ISM.

3. The Three Phase ISM

The actual ISM is rent by stellar winds and supernova explosions, leading McCray and Snow (1979) to term it the "violent ISM". Supernovae are the dominant source of energy; they heat large volumes of the ISM and agitate it, inflating the ISM to its observed height. Cold clouds have a scale height much greater than 15 pc because of their turbulent motions. The scale height for the hot gas at a temperature of 5×10^5 K, the temperature calculated by MO, is about 3 kpc (this is higher than implied by equation [4] because the hot gas extends well above the stellar disk, so that $g \ll g_0' z$). If magnetic fields or cosmic rays contribute to the support of the hot gas, the scale height would be yet higher. The hot gas can confine clouds far from the plane, as envisaged by Spitzer (1956); absorption line observations indicate that these clouds extend to about 3 kpc from the plane. What is the filling factor f_h of this gas near the Galactic plane? More specifically, what is the filling factor of this gas outside superbubbles?

Consider that subset of supernovae which are effectively random in space and time; this includes supernovae from low–mass progenitors (Type Ia) and those supernovae from high–mass progenitors which are not in large associations. The rate of these supernovae per unit volume is denoted by S. Let $V(t)$ be the volume of an isolated SNR as a function of its age t. For a one–phase or two–phase ISM, this volume first increases as the remnant expands, and then decreases as the hot gas cools off. In a three–phase ISM, the remnants cannot be approximated as isolated; the SNR expansion is well defined, but thereafter the remnant merges into the hot medium. Let $dQ(t)$ be the probability that a given point is in a remnant of age between t and $t + dt$; in terms of the SN rate, we have $dQ(t) = SV(t)dt$. The expected number of SNRs younger than t encompassing a given point is then

$$Q(t) = S \int_0^t V(t')dt'. \tag{5}$$

The expected number of SNRs of *any* age which encompass a point is $Q(t \to \infty)$; we denote this by Q, without an argument. Q is sometimes termed the "porosity" of the medium (Cox and Smith 1974). Since $V(t)$ is not well-defined for an SNR in a three–phase medium after its expansion stops, Q is not precisely defined in such a medium, but it is large: any point may be regarded as being in an SNR of some age. In the absence of interactions among the SNRs, the probability of not being in any remnant is $\exp(-Q)$, and the probability of being in at least one SNR is $1 - \exp(-Q)$.

Observe that Q is directly proportional to the *volume in space-time*, or four–volume, occupied by an SNR, $\mathcal{V} \equiv \int V dt$. We can estimate this four–volume with a simple model in which the SNR expands as a power–law in time, $R \propto t^{\eta}$, until it reaches a radius R_m at time t_m; at this point, its expansion velocity v_m is a factor β times the ambient isothermal sound speed C_0,

$$v_m = \frac{\eta R_m}{t_m} \equiv \beta C_0. \tag{6}$$

If the filling factor of the hot gas is not large, the SNR will subsequently contract (the importance of this contraction was emphasized by Heiles [1987] and by B.–C. Koo, private communication). Under the assumption that this contraction occurs at a constant velocity over a time $t_{\rm con}$, the total four–volume of the SNR is

$$\mathcal{V} = \left(\frac{V_m t_m}{3\eta + 1} + \frac{V_m t_{\rm con}}{4}\right) \equiv q V_m t_m, \tag{7}$$

where $V_m = (4\pi R_m^3/3)$ and q is a numerical coefficient of order unity. If the contraction occurs at the velocity v_m, we have

$$q = \frac{1}{3\eta + 1} + \frac{1}{4\eta}. \tag{8}$$

This model is certainly over–simplified: it ignores the transition from expansion to contraction, which is accompanied by an overshoot in which the internal pressure drops below the ambient value (e.g., Ostriker and McKee 1988), and which would tend to increase Q; and it does not include the effects of the embedded clouds, which tend to destroy the remnant from within toward the end of its evolution, and which would tend to decrease Q.

Cioffi, McKee, and Bertschinger (1988) have studied the late evolution of spherically symmetric SNRs in a uniform medium. They found that the expansion can be approximated with $\eta = 0.3$ at late times. The maximum radius of the remnant is

$$R_m = 77.3 \left(\frac{E_{51}^{0.316}}{n_0^{0.153} \zeta_m^{0.051} \beta^{0.429} \tilde{P}_{04}^{0.214}}\right) \quad \text{pc}, \tag{9}$$

which is reached at a time

$$t_m = 2.97 \times 10^6 \left(\frac{E_{51}^{0.316} n_0^{0.348}}{\zeta_m^{0.051} \beta^{1.429} \tilde{P}_{04}^{0.714}}\right) \quad \text{yr}, \tag{10}$$

where n_0 is the ambient density of hydrogen nuclei, $\tilde{P}_{04} = \tilde{P}_0/(10^4 \text{ cm}^{-3} \text{ K})$, and ζ_m allows cooling rate to depend on the metallicity of the gas; we shall take $\zeta_m = 1$ in our numerical estimates. With $q = 1.36$ from equation (8), the SNR four–volume is

$$\mathcal{V} = 7.82 \times 10^{12} \left(\frac{E_{51}^{1.26}}{n_0^{0.110} \zeta_m^{0.204} \beta^{2.72} \tilde{P}_{04}^{1.36}}\right) \quad \text{pc}^3 \text{ yr}. \tag{11}$$

This result is quite close to the value found by MO over a decade ago, aside from the factor $q \simeq 1.36$. Note in particular that $\mathcal{V}$ is insensitive to the ambient density n_0. On the other hand, the four–volume is sensitive to the parameter β; for the case in which the interstellar magnetic field does not dominate the pressure, we assume that the SNR expansion stops when the velocity drops to the ambient isothermal sound speed, so that $\beta = 1$. For a two–phase ISM in which the mean intercloud density is 0.3 cm^{-3} and the ambient pressure is $\tilde{P}_{04} = 0.62$ (comprised of a thermal pressure $\tilde{P}_{th} = 3600$ cm^{-3} K and a magnetic pressure corresponding to a 3 μG field), the four–volume of an SNR is $\mathcal{V} = 1.7 \times 10^{13} E_{51}^{1.27}$ pc^3 yr. If the magnetic field is signficantly higher, say 5 μG as envisaged by Cox and Slavin (1989), then the four–volume is reduced by a factor of about 3 (McKee 1990*b*).

The porosity of the ISM is simply $Q = S\mathcal{V}$, the product of the SN rate per unit volume and the SNR four–volume. Evidence from historical SN and from the observed rate of massive star formation suggests that 0.022 yr^{-1} is a conservative estimate of the Galactic SN rate (McKee 1990*b*). This value is identical to van den Bergh's (1983) estimate of the Galactic SN rate, and it is consistent with the rate estimated from extragalactic SN by Tammann (1982) and by van den Bergh (1990) provided the Hubble constant is in the range 50–70 km s^{-1} Mpc^{-1}. Under the assumption that the SN rate per unit area is proportional to the surface density of disk stars (Ratnatunga and van den Bergh 1989), the SN rate per unit area at the solar circle is 2.6×10^{-11} pc^{-2} yr^{-1}.

To determine S, we must determine the fraction of the SN that are effectively random in space and time, and their scale height. According to Evans, van den Bergh, and McClure (1989), almost 90% of SN are of Types Ib or II, so we focus on them. These stars are believed to have massive progenitors. Humphreys and McElroy (1984) surveyed over 5000 OB stars, and found them to be nearly evenly divided between those in associations and those in the field; we therefore assume that about half of the supernovae are random. Heiles (1990) comes to the same conclusion based on an analysis of the size of an association required to create a superbubble. The observed scale height of OB stars is about 60 pc (e.g., Allen 1963). The scale height of these stars when they die might be somewhat larger; indeed, the scale height of the youngest pulsars is about 150 pc (Taylor and Manchester 1977). We shall adopt the latter value to be conservative; this gives

$$S = 0.5 \left(\frac{2.6 \times 10^{-11}\ \mathrm{pc}^{-2}\ \mathrm{yr}^{-1}}{300\ \mathrm{pc}} \right) = 4.4 \times 10^{-14} \quad \mathrm{pc}^{-3}\ \mathrm{yr}^{-1} \tag{12}$$

for the solar neighborhood. This is about half the nominal rate $S = 10^{-13}$ pc^{-3} yr^{-1} adopted by MO, primarily because we are excluding correlated SN from the rate. Note that the conversion from the SN rate per unit area to the SN rate per unit volume, S, is essentially the same for SN Ia as for the massive SN: for SN Ia, all the SN contribute to the porosity, but their scale height is larger (Heiles [1987] estimates 325 pc). Thus, if SN Ia make a larger contribution to the SN rate than estimated by Evans *et al.*, the value of S in equation (16) would be unaffected.

For the typical two–phase medium described below equation (11), the resulting porosity is $Q = S\mathcal{V} = 0.75$. For such a large value of Q, the ISM will be a three–phase medium rather than a two–phase medium, and interactions among the SNRs will tend to increase the filling factor of the hot gas above the value it would have in the absence of interactions (Smith 1977). On the other hand, in the highly magnetized ISM considered by Cox and Slavin (1989), the porosity is considerably smaller, $Q \simeq 0.23$ (assuming $E_{51} = 1$). It appears that additional evidence, both observational and theoretical, must be brought to bear in order to determine the filling factor of hot gas in the ISM; new theoretical evidence will be presented in §4.

4. Interstellar Turbulence and the Porosity of the ISM

The observations of the ISM described in §1 show that the scale height of the gas is several times greater than can be explained in one– or two–phase models of the ISM (§2). Observationally, the large scale height is associated with a high degree of turbulence in the ISM (Kulkarni and Fich 1985). Here we show that the level of turbulence in the ISM is directly associated with the porosity of the ISM (McKee 1990*a*).

Consider supernovae occurring randomly at a rate S per unit volume. Assume the porosity Q is small, so that interactions can be neglected and the SNRs expand into the warm intercloud medium. Furthermore, assume that thermal pressure forces dominate the evolution of the SNRs; the remnants are small compared to a scale height, so that gravity may be neglected, and the magnetic field does not dominate the pressure. After becoming radiative, the SNRs slow down until they merge with the ISM, carrying a radial momentum $(\rho_0 V_m)\beta C_0$ (see §3). This momentum is approximately conserved, and is concentrated in a thin shell which expands at about the sound speed (see Landau and Lifschitz 1959), which we take to be βC_0. The shell overlaps another shell at a radius R_{ov} (corresponding to a four–volume $\mathcal{V}_{ov}$) given by

$$S\mathcal{V}_{ov} = \frac{S}{4}\left(\frac{4\pi R_{ov}^3}{3}\right)\left(\frac{R_{ov}}{\beta C_0}\right) \equiv 1, \tag{13}$$

from equation (7); in this case, $\eta = 1$. We assume that the momentum is annihilated at R_{ov}. We obtain an upper limit on the turbulent pressure by assuming that the momentum flux is carried by the dynamic pressure, ρv^2. Approximating the shell as a delta function, we can express the momentum flux density as

$$\rho v^2 = \left(\frac{\rho_0 V_m \beta C_0}{4\pi r^2}\right)\delta\left(t - \frac{r}{\beta C_0}\right). \tag{14}$$

Averaging this over space and time yields

$$\int \rho v^2 dV dt \equiv \langle\rho v^2\rangle \mathcal{V}_{ov} = \frac{\langle\rho v^2\rangle}{S} = R_{ov}\rho_0 V_m \beta C_0. \tag{15}$$

Since

$$Q = S\mathcal{V} = qSV_m t_m = qSV_m\left(\frac{\eta R_m}{\beta C_0}\right) \tag{16}$$

from equations (6) and (7), we find

$$\frac{\langle\rho v^2\rangle}{\rho_0 C_0^2} = \frac{\beta^2}{q}\left(\frac{QR_{ov}}{\eta R_m}\right). \tag{17}$$

The turbulent pressure is proportional to the one–dimensional velocity dispersion, $P_{turb} = \langle\rho v^2\rangle/3$; eliminating R_{ov}/R_m with the aid of equations (13) and (16), we find

$$\frac{P_{turb}}{P_{th}} = \frac{4^{1/4}\beta^2}{3}\left(\frac{Q}{q\eta}\right)^{3/4} \simeq 0.9Q^{3/4}, \tag{18}$$

where the numerical evaluation is for the values used in §3. This estimate of the turbulent pressure does not include the contribution of the motions due to SNRs prior to merging with the ISM or during their subsequent collapse (§IV*a*), but one can show that this contribution is of higher order and does not alter the result in equation (18) when Q is small. We conclude that *the turbulent pressure is significant if and only if the porosity due to explosive energy injection is large.* Thus, provided the interstellar magnetic field is not too strong, the observation that the gas disk of the Galaxy is thick and highly turbulent implies that the porosity of the ISM is large. In their calculation of the turbulent velocity of the clouds, MO found $P_{turb} \simeq P_{th}$, consistent with the large porosity they inferred; at such a large Q, the simple estimate in equation (18) cannot be assumed to be quantitatively correct, however. It is unlikely that the conclusion that Q is large can be avoided by appealing to superbubbles to produce the turbulence: the fact that the ISM extends to great heights implies that most superbubbles are confined within the ISM (Cox 1989), so the relation between the turbulent pressure and the porosity given above should apply to them as well, at least approximately. Since superbubbles are observed to have a small filling factor (Heiles 1980), they cannot account for the observed level of interstellar turbulence.

What if the interstellar magnetic field is large? In that case, a fraction of the energy from the supernovae can be stored in the oscillations of the field (Cox 1988). However, if the interstellar field is large, it must be tangled on scales $\lesssim 100$ pc, since the mean field observed toward pulsars is $\lesssim 2.5\ \mu$G, and perhaps as low as 1.6 μG (Rand and Kulkarni 1989). Such a tangled field contributes a pressure of only 1/3 that of an ordered field, so it is difficult to see how such a field could inflate the ISM to the required height. Further observations of the Galactic magnetic field are required to clarify this point.

5. Concluding Remarks

It has long been recognized that the injection of energy by supernovae plays an essential role in the structure and dynamics of the ISM. Indeed, as the simple isothermal models developed in §2 demonstrated, one and two–phase models without substantial energy injection bear little resemblance to the observed ISM. Recent observations have shown that the ISM extends well above its natural scale height (even allowing for magnetic and cosmic ray pressure), and that this is accompanied by a high degree of turbulence. An enormous injection of kinetic energy is required to account for the observed turbulence; since supernovae are the dominant source of kinetic energy in the ISM, their remnants must essentially fill space if they are to supply the required turbulence (§4). Direct observational tests of this prediction, which underlies the three–phase model of the ISM, should be possible in the next decade as increasingly sophisticated and powerful instruments become available.

Acknowledgments: I wish to thank Stan Woosley for organizing such a stimulating conference and for inviting me to speak. The material in this article is taken from work reported on in McKee (1990*a,b*). My research is supported in part by NSF grant AST–8615177.

REFERENCES

Allen, C.W. 1963, *Astrophysical Quantities* (2d ed.;London: Athlone Press), p. 241.
Bahcall, J.N. 1984, *Ap. J.*, **276**, 169.
Begelman, M.C., and McKee, C.F. 1990, *Ap. J.*, in press.
Bloemen, J.B.G.M. 1987, *Ap. J.*, **322**, 694.

Bruhweiler, F.C., Gull, T.R., Kafatos, M., and Sofia, S. 1980, *Ap. J. (Letters)*, **238**, L27.
Cioffi, D.F., McKee, C.F., and Bertschinger, E. 1988, *Ap. J.*, **334**, 252.
Cox. 1988, in *Supernova Remnants and the Interstellar Medium*, ed.R.S. Roger and T.L. Landecker (Cambridge: Cambridge University Press), p. 73.
——. 1989, in *Structure and Dynamics of the Interstellar Medium*, eds. G. Tenorio–Tagle, M. Moles, and J. Melnick (Berlin: Springer–Verlag), in press.
Cox, D.P., and Slavin, J.D. 1989, in *EUV Astronomy*, eds.R.F. Malina and S. Bowyer (New York: Pergamon), in press.
Cox, D.P., and Smith, B.W. 1974, *Ap. J. (Letters)*, **189**, L105.
Evans, R., van den Bergh, S., and McClure, R.D. 1989, *Ap. J.*, **345**, 752.
Ferriere, K.M., Zweibel, E.G., and Shull, J.M. 1988, *Ap. J.*, **332**, 984.
Field, G.B., Goldsmith, D.W., and Habing, H.J. 1969, *Ap. J. (Letters)*, **155**, L149 (FGH).
Heiles, C. 1979, *Ap. J.*, **229**, 533.
——. 1980, *Ap. J*, **235**, 833.
——. 1984, *Ap. J. Suppl.*, **55**, 585.
——. 1987, *Ap. J.*, **315**, 555.
——. 1990, *Ap. J.*, in press.
Humphreys, R.M., and McElroy, D.B. 1984, *Ap. J.*, **284**, 565.
Kuijken, K., and Gilmore, G. 1989, *M.N.R.A.S.*, **239**, 651.
Kulkarni, S.R., and Fich, M. 1985, *Ap. J*, **289**, 792.
Kulkarni, S.R., and Heiles, C. 1987, in *Interstellar Processes*, ed.D. Hollenbach and H. Thronson (Dordrecht: Reidel), p. 87.
Landau, L.D., and Lifschitz, E.M. 1959, *Fluid Mechanics* (Reading: Addison–Wesley).
Lockman, F.J. 1984, *Ap. J.*, **283**, 90.
Lockman, F.J., Hobbs, L.M., and Shull, J.M. 1986, *Ap. J.*, **301**, 380.
Mathis, J.S. 1986, *Ap. J.*, **301**, 423.
McCray, R., and Snow, T.P. 1979, *Ann. Rev. Astr. Ap.*, **17**, 213.
McKee,C.F.. 1990*a*, *Ap. J.*, to be submitted.
——. 1990*b*, in *The Evolution of the Interstellar Medium*, ed. L. Blitz (San Francisco: Astronomical Society of the Pacific), in press.
McKee, C.F., and Ostriker, J.P. 1977, *Ap. J.*, **218**, 148 (MO).
Ostriker, J.P., and McKee, C.F. 1988, *Rev. Mod. Phys.*, **60**, 1.
Rand, R.J., and Kulkarni, S.R. 1989, *Ap. J.*, **343**, 760.
Ratnatunga, K.U., and van den Bergh, S. 1989, *Ap. J.*, **343**, 713.
Reynolds, R.J. 1989*a*, *Ap. J. (Letters)*, **339**, L29.
——. 1989*b*, in *Galactic and Extragalactic Background Radiation*, ed.S. Bowyer and C. Leinert, in press.
Savage, B.D. 1987, in *Interstellar Processes*, ed.D.J. Hollenbach and H. Thronson (Dordrecht: Reidel), p. 123.
Spitzer, L. 1956, *Ap.J.*, **124**, 20.
——. 1978, *Physical Processes in the Interstellar Medium* (NewYork: Wiley).
——. 1990, *Ann. Rev. Astr. Ap.*, **28**, in press.
Tammann, G. 1982, in *Supernovae: A Survey of Current Research*, ed. M.J. Rees and R. Stoneham (Reidel: Dordrecht), p 371.
Taylor, J.H., and Manchester, R.N. 1977, *Ap. J.*, **215**, 885.
Tenorio–Tagle, G., and Bodenheimer, P. 1988, *Ann. Rev. Astr. Ap.*, **26**, 145.
van den Bergh, S. 1983, *Pub. A.S.P.*, **95**, 388.
——. 1990, in *Supernovae*, ed.S. Woosley (Berlin: Springer–Verlag), in press.
van den Bergh, S., McClure, R.D., and Evans, R. 1987, *Ap. J.*, **323**, 44.

The Effect of Supernova Remnants on Interstellar Clouds

Richard I. Klein, Christopher F. McKee, & Philip Colella

INTRODUCTION

The interaction between supernova remnants (SNRs) and interstellar clouds in the galaxy is known to play a major role in determining the structure of the interstellar medium (ISM). We know that the ISM is highly inhomogeneous, consisting of both diffuse atomic clouds (T~100K) and dense molecular clouds (T~10K) surrounded by a low density warm ionized gas (T~10^4K) and by a very hot coronal gas (T~10^6K). Next to radiation directly from stars, supernova explosions represent the most important form of energy injection into the ISM; they determine the velocity of interstellar clouds, accelerate cosmic rays, and can compress clouds to gravitational instability, possibly spawning a new generation of star formation. The shock waves from supernova remnants can compress, accelerate, disrupt and render hydrodynamically unstable interstellar clouds, thereby ejecting mass back into the intercloud medium. Thus, while the interaction of the SNR blast wave with cloud inhomogeneities can clearly alter the appearance of the ISM, the cloud inhomogeneities can similarly have a profound effect on the structure of the SNR.

Recent observations of SNR of enhanced emission in the Balmer line filaments show evidence of cloud shock interactions for Tycho (Braun, 1988). Velusamy (1987) finds evidence of the remnant cloud interaction in his radio observations of W28 and W44 taken at 327 MHz. These observations clearly show the distortion of the radio shell as the remnant begins to wrap around a dense cloud. The observations of the SNR IC443 by Braun and Strom (1986) show the later evolution of the cloud shock interacting with the outer layers of the cloud stripped off at high velocity.

Given the importance of the interaction of the supernova shocks with clouds for understanding the structure and the dynamics of the ISM as well as the potential importance of the interaction as a means of triggering new star formation, the problem has been studied both analytically and numerically over the past decade. All of the previous work on this important problem leave unanswered several questions of key importance: What is the ultimate fate of clouds that have been impacted by SNR shocks? What is the total momentum delivered to the cloud? How much mass is lost from the cloud? What are the mechanisms by which clouds are disrupted and to what extend does disruption take place? How does cloud morphology scale with cloud density, shock Mach number and cloud size? Is the cloud driven to gravitational instability or is the cloud destroyed? What is the effect of the interstellar magnetic field on the evolution? What are the observable consequences of the interaction?

We have recently found (Klein, Colella and McKee, 1989a,b) that highly complex shock-shock interactions and instabilities and shear flow motions play a major role in determining the morphology of the cloud. To address these physical complexities, we have used the local adaptive mesh refinement techniques with second order Godonov methods for 2-D axisymmetry developed by Berger and Colella, 1989 (cf. Klein, Colella, and McKee, 1989a,b). We assume that the cloud and intercloud gas are both adiabatic, although we allow the cloud and intercloud medium to have different values of the adiabatic index γ.

From the point of view of being able to resolve detailed complex physical structures with reasonable amounts of supercomputer time and memory, the most important feature of our code is that it employs a dynamic regridding strategy known as local Adaptive Mesh Refinement (AMR) to dynamically refine the solution in regions of interest or excessive error.

This is effected by placing a finer grid over the region in question with the grid spacing reduced by some even factor (typically) in each spatial dimension. Multiple levels of grid refinement are possible with the maximum number of nested grids supplied as a parameter in the calculation. Typically our calculations employ two nested grids over the initial coarse grid.

CLOUD SIZE SCALES

As the SNR expands through the ISM, it drives a shock into any cloud it encounters. Assuming that these are strong shocks, the pressure behind the blast wave and the pressure behind the transmitted cloud shock are comparable, and one finds that (McKee and Cowie, 1975)

$$v_s \approx (\rho_i/\rho_c)^{1/2} v_b , \tag{1}$$

where v_s and v_b are the cloud shock and blast wave velocities and ρ_c and ρ_i the initial cloud and intercloud densities, respectively. Following McKee (1988), we define characteristic timescales for the cloud-shock interaction. Let $\chi \equiv \rho_c/\rho_i$ be the density contrast and assume that $\chi >> 1$. Assume that the cloud is a sphere with radius a at a distance R_b from the supernova explosion. The blast wave in the Sedov-Taylor phase will expand as $R_b \propto t^{2/5}$, so the age of the SNR is,

$$t \equiv \frac{dR_b}{dt} = \frac{2}{5}\frac{R_b}{v_b} . \tag{2}$$

The blast wave in the intercloud medium crosses the cloud in a time

$$t_{ic} \equiv \frac{2a}{v_b} , \tag{3}$$

whereas the cloud shock crushes the cloud in a time

$$t_{cc} \equiv \frac{a}{v_s} = \frac{\chi^{1/2} a}{v_b} . \tag{4}$$

The cloud crushing time t_{cc} is of the order of the sound crossing time in the crushed cloud; it is also about the timescale for the growth of large scale Rayleigh-Taylor instabilities. Finally, the cloud accelerates up to the velocity of the intercloud gas in a characteristic drag time t_d defined by $\rho_i v_b t_d = \rho_c a$, or

$$t_d = \frac{\chi a}{v_b} = \chi^{1/2} t_{cc} . \tag{5}$$

In this paper, we will consider only clouds that can be characterized as "small", so that the SNR does not evolve significantly during the time for the cloud to be crushed:

$$t > t_{cc} \Rightarrow a < \frac{0.4R}{\chi^{1/2}} . \tag{6}$$

Indeed, we shall focus on the case in which the cloud is "very small", so that $t >> t_d$, and $a << 0.4R/\chi$. In either case, we have $a << R$ so that the blast wave may be treated as a planar shock. In the opposite limit of a shock interaction with a large cloud, the SNR blast wave will undergo substantial weakening over the time it takes to cross the cloud. We expect substantial disruption for the small clouds, but only impulsive effects for large clouds.

CLOUD EVOLUTION

a. Cloud Crushing

Since there are no intrinsic scales in the problem, it is parameterized by the Mach number of the SNR blast wave M and the density ratio χ. Our calculations assumed 2-D axisymmetry for an inviscid fluid with no magnetic field. Two cases were considered for the cloud: $\gamma = 1.1$ and $\gamma = 5/3$. The intercloud gas was assumed to have $\gamma = 5/3$. Several calculations have been made for Mach numbers in the range 10-1000 and density ratios 10-400.

It is useful to follow the morphological evolution of the cloud through several cloud crushing times to obtain a sense of the different stages of development. We present the time-development of the isodensity contours of the cloud for the case γ (cloud) = γ (intercloud) = 5/3, $\chi=10$, $M=10$. At $t=0.84\ t_{cc}$ (Fig. 1), the transmitted shock is compressing the cloud from the front, secondary shocks have enveloped the sides of the cloud as the blast wave passes over the cloud, and a reflected bow shock moves upstream into the intercloud medium. The reflected shock becomes a standing bow shock and eventually a weak acoustic wave carrying away a small amount of energy from the supernova shock (Spitzer, 1982). At $t=1.05t_{cc}$ (Fig. 2) the blast wave behind the cloud reflects off the axis giving rise to a Mach reflected shock back into the cloud. Substantial flattening of the cloud is observed at $t=2.1t_{cc}$ from the strong shocks which have squeezed it like a vise. The pressure maximum on the nose of the cloud exceeds the pressure minimum on the sides and the cloud begins to expand laterally (Fig. 3). We note the growth of Richtmyer-Meshkov instabilities (Richtmyer, 1960) on the cloud nose which grow more slowly than the classic Rayleigh Taylor modes and evidence of Kelvin Helmholtz instabilities on the sides of the cloud.

b. Shear Flow and Vortex Production

At $3.78t_{cc}$ a prominent shear layer exists due to the motion of the cloud through the ICM. The shear produces copious vortex rings along the shear flow layer. The cloud consists of a distorted unstable axially flattened core component and a severely disrupted halo of cloud material. Over 70% of the original cloud mass is in small fragments which, in the absence of cooling, should merge with the intercloud medium. The unstable break up is dominated by large scale differential shear. At $t=9.7\ t_{cc}$, the cloud is completly destroyed (Fig. 4) and consists of several thousand fragments. At $4.2\ t_{cc}$ the strong supersonic vortex rings align along the shear flow layer produced in the dominant arm of cloud material that has been pulled from the main core of the cloud as well as along a second substantially fractured mass of cloud that has been fragmented from the arm. In Fig. 5 we show the associated flow field alongside of isodensity contours of the cloud and intercloud gas at $t=4.2\ t_{cc}$. It is clear that regions of strong circulation (high vorticity, numbered 1-5) are associated with positions along the shear flow layer where the cloud has undergone severe fragmentation. As vortex rings are formed in the shear layer and move away from the initial cloud are, the vortex rings are broken off. The process is called vortex shedding. It is suggestive of the possibility that the vorticity in the intercloud matter is acting to enhance the cloud break-up along the differential shear layer, thus acting as a mix-master aiding the development of the Kelvin-Helmholtz instabilities. This interesting possibility is worth further study.

The vorticity depends upon a baroclinic term which is the major source of vorticity in the cloud-shock interaction. The shock is curved as it interacts with the cloud surface and produces surfaces of constant pressure that are not coincident with surfaces of constant density at the interface of the cloud and intercloud matter. This gives a non-zero cross product of gradients. The vorticity in the ICM is greater than that in the cloud because of the higher velocities in the lower density material. Our calculations show that most of the vorticity remains concentrated near the cloud boundary, where it originated. An additional term that can be important is vortex diffusion. If the gas has a frictional force due to viscosity, F/ρ, it can be represented as $F/\rho = \nu\nabla^2 u$ where ν is the viscosity; then $\nabla \times (F/\rho) \sim \nu\nabla^2\omega$. This represents the diffusion of vorticity from regions of high to low concentration. It is proportional to the amount of numerical viscosity in the finite difference approximations. Given the importance of vorticity as a possible observational diagnostic of the remnant cloud

Figure 1

Figure 2

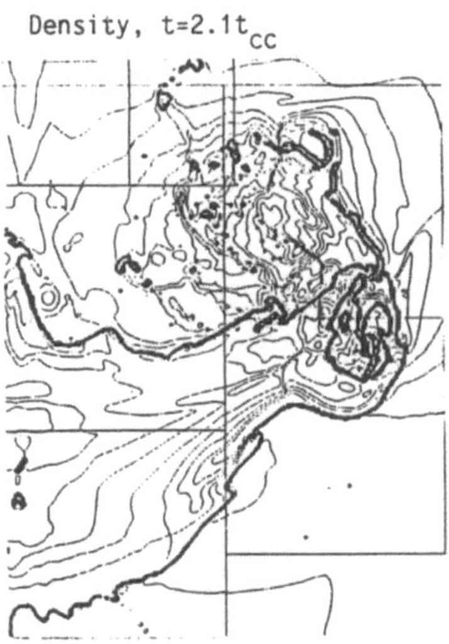

Figure 3

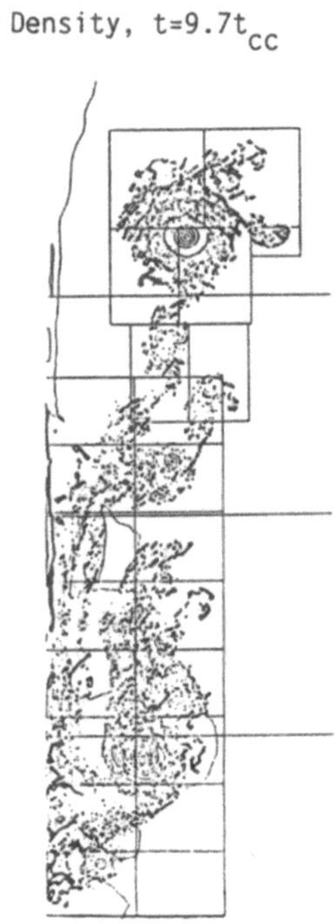

Figure 4

Figures 1-4 Isodensity contours of cloud and intercloud matter at different times.

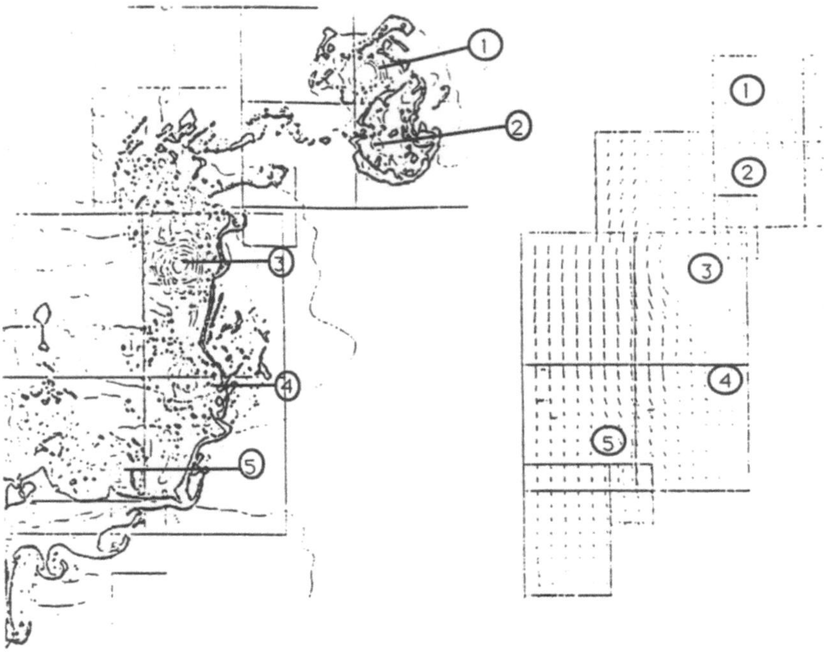

Figure 5 Isodensity contours (on left) at t=4.2 t_{cc}, flow field (on right). Numbers are sites of vorticity maximums.

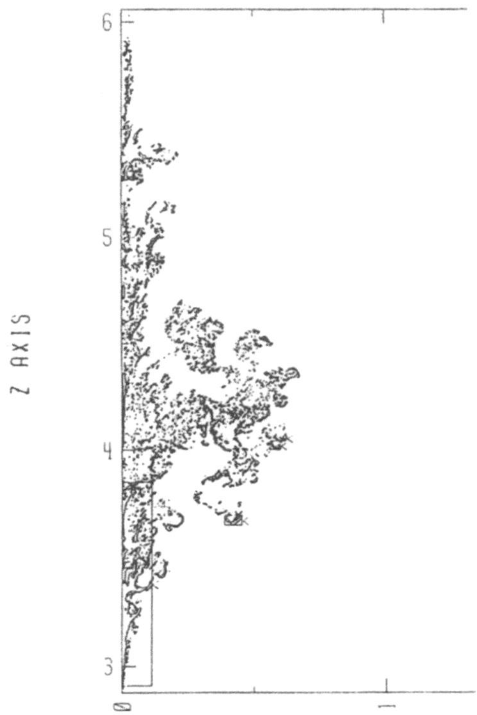

Figure 6 Isodensity contours for χ=100, M=100 at t=4.0 t_{cc}

interaction as well as its possible role in the cloud fragmentation, it is of great importance to demonstrate that numerical viscosity does not play a role in determining the amount of vorticity production. We have computed the time evolution of the cloud for four increasingly resolved initial grids, doubling the number of cells in both Δr and Δz with each increase in resolution. We have found that the time evolution of the vorticity for even the coarsest mesh tracks to a remarkable degree of accuracy the vorticity of the finest grid resolution, which is equivalent to a 7×10^6 zone calculation for a fixed grid method. This clearly establishes that numerical viscosity, which is proportional to grid resolution, does not affect the production of vorticity for the adaptive grid techniques we are using. This type of calculation is a powerful check on the conservation of vorticity.

Let us consider the characterization of the evolution of the interstellar cloud in more detail. In Table 1, we display the results of adiabatic calculations for three models in which $\gamma = 5/3$ in both the cloud and ICM. The calculations are done for two models (M=10 and 100) for density contrast χ =10 and one model (M=100) for density contrast 100. The first entry in the table is the time normalized to the intercloud crossing time. The second entry gives the time normalized to the cloud crushing time and the drag time, $t_d = \chi^{1/2}\, t_{cc}$. The next column is the sound speed behind the cloud shock normalized to the blast wave velocity. The shocked intercloud gas moves at a velocity (3/4) v_b relative to the cloud for $\gamma = 5/3$, so the next entry measures the ratio of the current cloud/intercloud relative velocity Δv to its initial value; in the frame of the shocked intercloud gas, this is a measure of cloud deceleration. The next column is a characterization of the cloud's aspect ratio in the radial and axial direction weighted by its half mass distribution. Here $r_{1/2}$ is the radial half-mass distance and $Z_{1/2}$ is the axial half-mass distance. The last column gives the radial $\dot{r}_{1/2}$ and axial $\dot{Z}_{1/2}$ expansion velocities of the cloud. These velocities are computed by using the half mass distance distributions at the two final times in the calculation.

Table 1

	t/t_{ic}	t/t_{cc} t/t_{drag}	c_c/v_b	$\frac{4}{3}(\Delta v / v_b)$	$r_{1/2}(t)/r_{1/2}(0)$ $Z_{1/2}(t)/Z_{1/2}(0)$	$\dot{r}_{1/2}/v_b$ $\dot{Z}_{1/2}/v_b$
χ=10						
M=10	6.7	4.2 1.3	0.18	0.16	1.8 3.2	~0.0 0.35
	15.3	9.66 3.0		0.074	2.38 5.69	~0.0 ≤0.045
M=100	6.7	4.2 1.3	0.18	0.14	2.0 2.6	~0.0 0.32
χ=100						
M=100	21.3	4.3 0.43	.056	0.25	3.7 8.4	~0.0 0.42

Several conclusions can be drawn from these results. Comparing the results at the same normalized "final" time $t = 4.2 t_{cc}$ for clouds of the same density $\chi = 10$, but subjected to blast waves of different Mach number, 10 and 100, we note that both clouds have decelerated to about 0.15 of their initial velocities. Thus, these clouds have almost stopped, leading to a small pressure differential between the front of the cloud surface and the sides so that there is little force driving further radial expansion; hence the clouds have a radial expansion velocity $\dot{r}_{1/2} \approx 0$. The strong shear flow in the cloud is still dominant, however, and both clouds are supersonically shearing apart at about the same axial expansion velocity $\dot{Z}_{1/2}$ of 3 times the cloud velocity. The physical extent of the stretching in both the radial and axial direction

$$\frac{r_{1/2}(t)}{r_{1/2}(0)}, \frac{Z_{1/2}(t)}{Z_{1/2}(0)}$$

is essentially the same for the two cases. The remarkable agreement of these features of the clouds and their similar morphological structure leads one to suspect that the cloud evolution may scale similarly with the Mach number of the SNR shock. This Mach scaling can be clearly seen if we scale the time, velocity and pressure as $t' = t/M$, $v' = vM$ and $P' = PM$. Substituting these scaled quantities into the Euler equations, we find that Euler equations are invariant under this transformation. Thus, we find that for fixed γ and density contrast χ, the morphological evolution is a function of t/t_{cc} only, in the limit of large M.

Clouds with greater density contrasts χ show greater expansion in both the radial and axial directions, as shown both by the results in Table 1 and by Fig. 6, which portrays the state of a shocked cloud with $\chi = 100$ at 4 t_{cc}. This follows from the fact that the characteristic expansion time for the cloud is the sound crossing time (which, as remarked above, is about t_{cc}), whereas the time for the cloud to decelerate is the drag time $t_d = \chi^{1/2} t_{cc}$. The lateral expansion of the cloud is due to the lower pressure on the sides of the cloud caused by the Venturi effect (Nittman et al. 1982). This pressure difference decays on the drag time; by the time shown in Fig. 6, this expansion has stopped. At $t = 4\ t_{cc}$, the axial expansion velocity is a substantial fraction of v_b for both $\chi = 10$ and $\chi = 100$; since t_{cc} is larger for $\chi = 100$, the length of the cloud is greater in this case. We expect the axial expansion of the cloud to stop within a few drag times. This has been verified for the $\chi = 10$ case, but not the $\chi = 100$ case.

c. <u>Cloud Fragmentation</u>

At late times (several t_{cc}) the clouds is turbulent with many fragments reduced to a foam on the scale of grid resolution. It is of great interest to follow the mass loss of the cloud as it fragments, and to understand how the fragmentation scales with varying cloud density. In Fig. 7, we show the mass of the cloud core as a function of time for clouds with density contrasts χ=10,100,400. The cloud core is defined to be the most massive cloud fragment. The mass loss vs time has been fitted with a exponential to determine the fragmentation time t_f, defined as the time for each cloud to be left with 1/e of its original mass. We find for χ=10 that the cloud fragments initially into two roughly equal mass fragments. The mass fragments then begin a series of further fragmentation stages into smaller pieces due to combined Rayleigh Taylor and Kelvin Helmholtz instabilities. In Fig. 4 we show isodensity contours of the cloud at t=9.67 t_{cc} where the cloud is completely destroyed. The final fate of this cloud consists of a quasi-static halo of fragments of which 50% of the mass resides in an axially elongated distribution stretched out 5-6 times its initial shape, and the rest of the mass resides in a multitude of fragments much less dispersed.

For clouds with χ>>10, the stripping process proceeds differently. For χ=400, the cloud fragments gradually, with a continuous erosion by loss of small fragments (cf. Fig. 7, Fig. 8). Since small fragments rapidly become comoving with the intercloud medium whereas the cloud core decelerates gradually, small fragments trail far behind the massive cloud core until the core itself is destroyed by Kelvin-Helmholtz instabilities as it drags through the intercloud medium. The cloud core mass at $t = 2t_{cc}$ (Fig. 8) is 26% of the original cloud and we see that the cloud has the distinct morphology of a dense cloud core trailed by a multitude of fragments in a narrow tail.

Our results show that clouds are fragmented in a time $t_f \sim (1.5 - 4)\ t_{cc}$ as χ ranges from 400 to 10; recall that t_{cc} is of order the Rayleigh Taylor timescale. The numerical coefficient is smaller for the higher density contrasts, presumably because the relative velocity of the cloud remains greater.

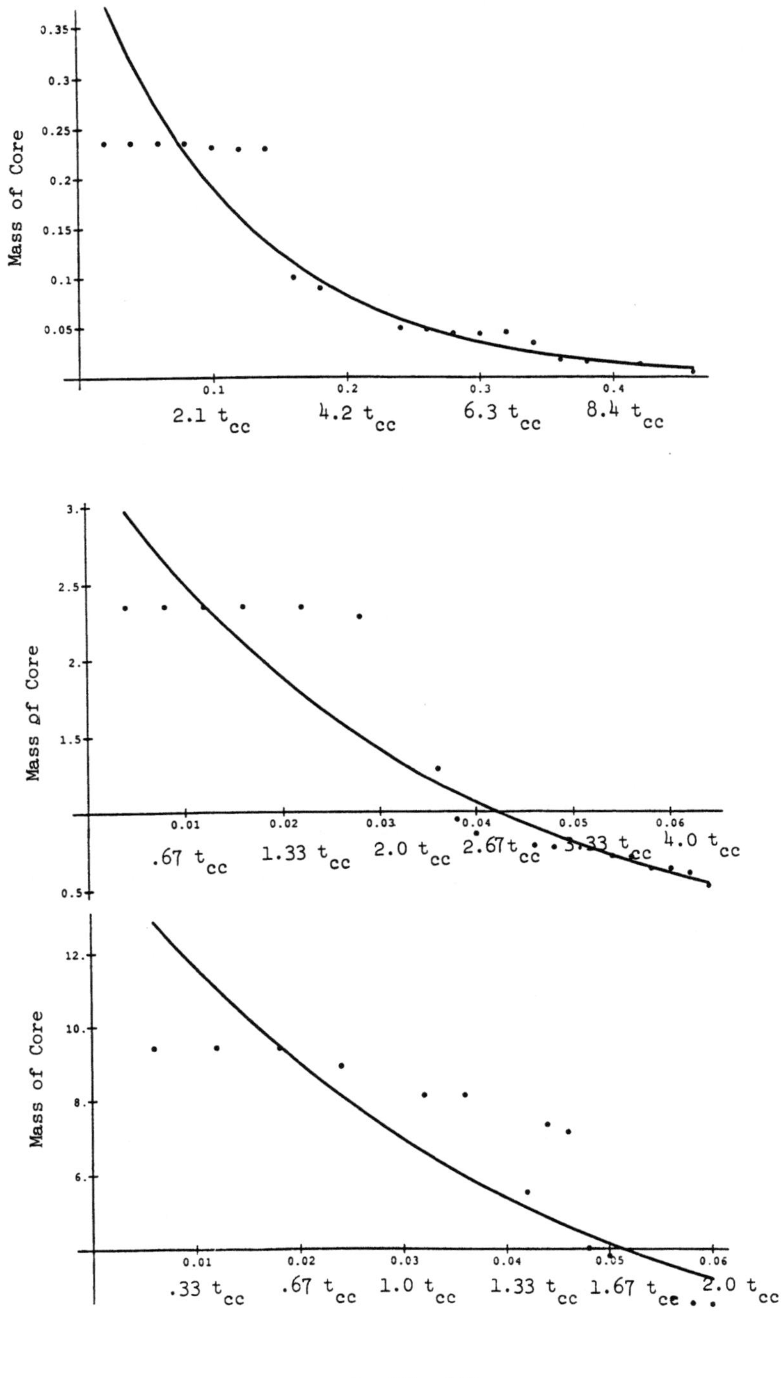

Figure 7 Core mass vs time for χ=10, 100, 400

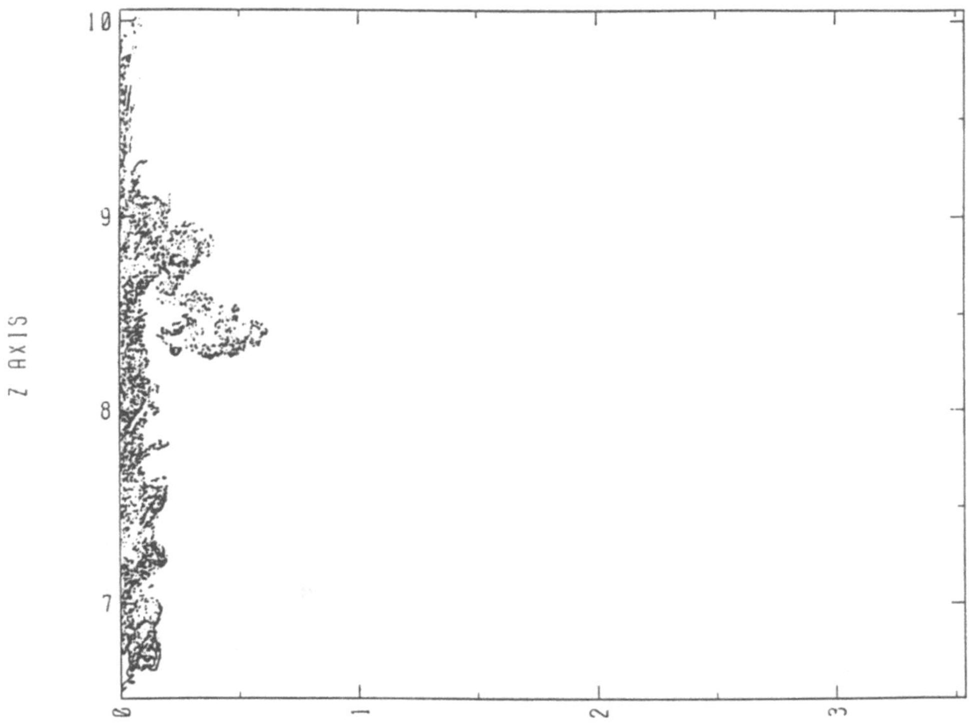

Figure 8 Isodensity contours for χ=400, M=100 at t=2.0 t_{cc}. Note morphology of cloud consisting of a dense "head" followed by a trail of several thousand fragments with an aspect ratio of 20 to 1.

This conclusion is consistent with those of Nittman et al. (1982), who concluded that the cloud would be destroyed in a time ~ 3 t_{cc} due to the combined effect of lateral expansion and strong fluid rotation behind the cloud. Because of the increase in the cross section of the cloud due to the lateral expansion, the time for the fragmented cloud to accelerate up to a velocity comparable to that of the shocked ICM is several times less than the initial drag time $t_d = \chi a/v_b$. For $\chi \lesssim 10^2$, the fragmentation time and the acceleration time are comparable; on this point, our conclusion differs from that of Nittman et al. In an analytic study of the related problem of the stripping of gas from a galaxy moving through an intracluster medium, Nulsen (1982) concluded that the stripping time is of the order of the drag time t_d; in the absence of gravitational effects (which he found to be generally small), our results indicate that stripping will occur substantially more rapidly.

We have performed calculations for several similar models for γ=1.1 in the cloud. This softer equation-of-state is more representative of clouds that are radiative, although it should be pointed out that truly radiative clouds can get rid of their stored energy efficiently, and we would expect substantially more shock compression than the models considered here. We note that these "radiative" clouds move substantially more rapidly than their γ=1.67 counterparts. These clouds are significantly more radially compressed, and thus experience far less drag than the γ=1.67 clouds. This can again be understood by consideration of the sound speed in these clouds. We find that the scaling of sound speed c_c with γ is such that $c_c(\gamma=1.1) << c_c(\gamma=5/3)$, so that these "radiative" clouds expand laterally more slowly. We note that the high density "radiative" cloud is still experiencing large supersonic axial shearing. As with the previous γ=5/3 models Mach scaling appears to be established.

OBSERVATIONAL CONSEQUENCES

An outcome of these calculations that may be potentially very important for observations of SNR is the discovery of the copious production of vortex rings distributed along the strong shear flow layer (Fig. 5). Approximating the rotation of these vortices by rigid body rotation, we can relate the vorticity ω in an individual vortex ring to the pressure differential across the vortex ΔP, and we find that $\omega = (8\Delta P/\rho)^{1/2}/r$. This appears to be an excellent approximation when compared to our detailed calculations. Those rings with large aspect ratio may be subject to non-axisymmetric instabilities and break up into yet smaller vortex structures (Saffman and Baker, 1979). "Fat" rings, with small aspect ratio, are likely to remain intact. Recent high resolution radio observations of the Cas A SNR (Tuffs 1986) have revealed several hundred intense compact radio emission peaks distributed throughout the remnant. We have demonstrated that strong shear flows associated with shock-cloud interactions result in the production of many supersonic vortex rings. These vortex rings can be expected to wind up ambient magnetic fields present in the interstellar clouds until equipartition between the energy in the field and the vortex is achieved. It is quite possible that the resulting intense wound up magnetic field and the associated betatron acceleration could account for the synchrotron emission of electrons, thus explaining the observations in Cas A. Equipartition magnetic fields are often involved in astrophysics to explain non-thermal emission; our results suggest that the fields may indeed reach equipartition, but only in a small fraction of the volume. Chevalier (1976) postulated the presence of turbulent vortices, acting as magnetic scattering centers in SNRs to explain particle acceleration by a second-order Fermi mechanism. We conjecture that the radio hot spots may indeed be indirect observational evidence of the presence of vortex rings produced behind the shocked clouds. The vortex rings would have low density and pressure at the center, thus appearing weak in optical, UV and x-ray emission.

Finally, the cloud morphology itself is a important signature. The clouds can be expected to be elongated structures with aspect ratios ~5-6, and multitudes of fragments trailing behind the cloud core. A possible example of this has been observed by Braun and Strom (1986).

CONCLUSIONS

We have performed second order accurate, high resolution local adaptive mesh refinement calculations of the interaction of a supernova shock with interstellar clouds. These extremely powerful hydrodynamic techniques have enabled us to calculate exceedingly complex flows much more rapidly and much more accurately, and much further in time than previous work with standard fixed grid hydrodynamics. We have followed the evolution of interstellar clouds well into the regime of fragmentation. Our calculations have demonstrated high accuracy with 80,000 grid cells in the cloud that would only be achievable with fixed grid high order accurate hydrodynamic schemes with >1,000,000 grid cells. We find:

1) Small non-radiative interstellar clouds are efficiently destroyed in a few cloud crushing times times by combined Rayleigh-Taylor and Kelvin Helmholtz instabilities dominated by large scale shear flow. Clouds that have the same density but are enveloped by strong shocks of differing Mach number exhibit scaling behavior in their morphological evolution.

2) Small clouds are highly fragmented by non-radiative shocks. Cloud fragments will most likely eventually feed their mass back into the ISM by thermal evaporation.

3) Small adiabatic clouds fragment to such an extent that it is unlikely that fragments large enough to become gravitationally unstable and form stars will survive. The cloud destruction proceeds more rapidly than the free fall time.

4) Clouds evolve toward a elongated structures with aspect ratios of five to six, consisting of multitudes of fragments.

5) Our calculations indicate the copious production of supersonic vortex rings. These vortex rings may be effective in winding up the ambient magnetic field in clouds, increasing the magnetic field strength and enhancing the synchrotron emission of cosmic ray electrons. This could explain the recent observations of numerous compact radio hot spots in Cas A.

In the future, we shall use adaptive mesh refinement hydrodynamic techniques to investigate a broad range of astrophysical gas dynamical phenomena.

ACKNOWLEDGEMENTS

The calculations presented in this paper were performed on the XMP416 and YMP832 at the Lawrence Livermore National Laboratory. This work was performed under the Auspices of the U.S. DOE by LLNL under Contract W-7405-Eng-48, supported in part by the Applied Mathematical Science Program of the office of energy research, and was also performed in part under the auspices of a special NASA astrophysics theory program which supports a joint Center for Star Formation Studies at NASA Ames Research Center, University of California, Berkeley, and University of California, Santa Cruz. The work of CFM is supported by NSF grant AST 86-15177.

BIBLIOGRAPHY

Berger, M. J., and Colella, P., 1989, to appear in J. Comp. Phys.

Braun, R. and Strom, R.G., 1986, Astronomy and Astrophysics 164, 193.

Braun, R., 1988, IAU Coll. 101, Supernova Remnants and the Interstellar Medium, Ed. R. S. Roger and I. L. Landecker, Cambridge Univ. Press, 227.

Chevalier, R. A., 1976, Ap.J., 207, 450.

Colella, P. and Woodward, P., 1984, J. Comp. Phys. 54, 174.

Glaz, H. M., Colella, P., Glass, I.I., Deschambault, R. L., 1985, Proc. Roy. Soc. Lond. A 398, 117.

Hornung, H., 1986, Ann. Rev. Fluid Mech., 18, 33.

Klein, R. I., Colella, P. and McKee, C. F., 1989, Ap. J. in preparation.

Klein, R. I., Colella, P., and McKee, C. F., 1989a, "The Physics of Compressible Turbulent Mixing International Workshop", Princeton University, Ed. W. Dannevik, 1989, Springer Verlag, New York Inc., Lecture Notes Series.

Klein, R. I., McKee, C. F. and Colella, P., Proceedings of the Astronomical Society of the Pacific, 100th Centennial, Berkeley, Calif., Ed. L. Blitz, 1989b.

McKee, C. F. and Cowie, L. L., 1975, Ap.J., 195, 715.

McKee, C. F., 1988, IAU Coll.101, Supernova Remnants and the Interstellar Medium, Ed. R. S. Roger and I. L. Landecker, Cambridge Univ. Press, 205.

Nulsen, P.E.J. 1982, M.N.R.A.S., 198, 1007.

Richtmyer, R.D., 1960, Comm. Pur. Appl. Math., 13, 297.

Saffman P. G., Baker, G. R., 1979, Ann. Rev. of Fluid Mech. 11, 95.

Spitzer, L., 1982, Ap.J., 262, 315.

Tuff, R. J., 1986, M.N.R.A.S., 219, 13.

Velusamy, T., 1988, IAU Coll 101, Supernova Remnants and the Interstellar Medium, Ed. R. S. Roger and I. L. Landecker, Cambridge Univ. Press, 265.

SECTION XIII

SUPERNOVA RATES, SEARCHES, AND USE AS STANDARD CANDLES

Galactic and Extragalactic Supernova Rates

Sidney van den Bergh

1. Introduction

Supernova explosions are rare events. Only 5 supernovae [Lupus (1006), Crab (1054), Tycho (1572), Kepler (1604) and Cas A (1658 ± 3)] are known with certainty to have occurred in the Galaxy during the last millenium. Other Galactic supernovae, no doubt, occurred during this period but escaped detection because of heavy obscuration.

Most extragalactic supernovae are discovered serendipitously, or during the course of large surveys of rather poorly-defined galaxy samples. Such data are, of course, ill-suited to the determination of supernova rates. Surveys designed to maximize the supernova discovery rate by observing rich clusters will be strongly biassed in favor of early-type galaxies. Some other recent surveys have concentrated on study of supernova-prone ScI galaxies. Finally serendipitous discoveries in spiral galaxies tend to be biassed in favor of pretty face-on spirals, which are photographed more frequently than esthetically less-pleasing edge-on spirals!

2. Extragalactic Supernova Rates

At the New Delhi IAU meeting the Rev. Robert EVANS [1] gave a lecture on amateur observations of supernovae. During the period 1980-84 Evans discovered a total of 10 supernovae. While listening to his talk I was struck by the thought that there must also be a large amount of useful information in the tens of thousands of observations of galaxies in which Evans did <u>not</u> find a supernova. Fortunately Evans kept a detailed record of all of his observations. A first attempt to extract information on supernova rates from this material is given in VAN DEN BERGH,

MCCLURE and EVANS [2], and MCCLURE, VAN DEN BERGH and EVANS [3]. These authors derive supernova rates from ~50000 observations of 748 Shapley-Ames (SANDAGE and TAMMANN [4]) galaxies, including 15 supernova discoveries, by Evans during the period 1980-85. More recently EVANS, VAN DEN BERGH and MCCLURE [5] have used ~75000 observations of 855 Shapley-Ames galaxies, during which 24 supernovae were discovered, to re-derive supernova rates for various types of galaxies. These results obviously suffer from small-number statistics, but have the advantage that they deal with a homogeneous, and well-observed, galaxy sample.

To determine supernova frequencies one first has to establish a "control time" (ZWICKY [6]) for each galaxy. In other words one needs to know the total period of time during which supernovae of a certain type would have been observable in a given galaxy. For any individual galaxy the control time will depend on (1) the distribution and frequency of the observations, (2) the limiting magnitude of the search, (3) the lightcurve shape and maximum magnitude for the supernova type considered, and (4) the distance of the target galaxy. Control time does not, however, depend on the numerical value of the Hubble parameter. This is so because both the distance of a galaxy, and the M(max) values for each type of supernova, depend on the adopted value of H. Within any class of supernovae the lightcurves of individual objects exhibit considerable variation. The mean shapes of the lightcurves adopted for each type of supernova are shown in VAN DEN BERGH, MCCLURE and EVANS [2]. EVANS, VAN DEN BERGH and MCCLURE [5] adopt the following values for the luminosity at maximum light for the three types of supernovae that are presently recognized:

Type Ia $M(max) = -19.79$,
Type Ib $M(max) = -18.19$,
Type II $M(max) = -17.8 + 0.65\,(M^{oi}_{B_T} + 20.9)$.

All of these relations assume $H = 50$ km s^{-1} Mpc. Since $(B - V)_{max} \approx 0.0$ the difference between $M_B(max)$ and $M_V(max)$ has been ignored. VAN DEN BERGH [7] has speculated that the apparently significant correlation ($r = 0.45 \pm 0.14$) between M(max) of SNII and the absorption-free, inclination corrected parent galaxy luminosity $M^{oi}_{B_T}$ (SANDAGE and TAMMANN [4], is due to a dependence of the M(max) values of individual SNII on the metallicities of their progenitors. Such a relationship might be understood by assuming that SNII progenitors in luminous galaxies have higher metallicities than do those that occur in less luminous systems. It was also assumed that all supernovae in Shapley-Ames galaxies suffered the same absorption $A^o + A^i$ (SANDAGE and TAMMANN [4]) as their parent galaxies. (This may result in a systematic underestimate of absorption for young massive supernovae in dusty star forming regions). Finally V(lim) ≈ 14.5 was adopted for Evans' 1980-85 observations with a 25cm telescope, and V(lim) ≈ 15.4 for his observations with a 41cm telescope during the period 1985-88. The latter assumption is, perhaps, somewhat suspect because no supernovae fainter than V = 14.5 were discovered by Evans during the last three years of his survey. This lack of discoveries of faint supernovae might well be due to the vagaries of small number statistics. Alternatively the completeness limit of Evans' visual search may have been overestimated, and the supernova frequency might therefore have been slightly underestimated by EVANS, VAN DEN BERGH and MCCLURE [5].

For an average Shapley-Ames galaxy these authors find the following rates:

SNIa $(0.28 \pm 0.10)h^2$ SNU,
SNIb $(0.27 \pm 0.15)h^2$ SNU,
SNII $(1.04 \pm 0.30)h^2$ SNU.

In these relations h = H/100 km s^{-1} Mpc, and 1 SNU is defined as one supernova per 10^{10} $L_B(\odot)$ of parent galaxy luminosity per century. The relatively large errors quoted for these rates are due to the fact that they are based on rather small number statistics. These results were derived from a total surveillance time of 12267 years for SNIa, 5091 years for SNIb and 4608 years for SNII. [For a galaxy of luminosity L_B the surveillance time is defined as the control time multipled by $L_B/1 \times 10^{10} L_B(\odot)$]. Supernova frequencies in galaxies of differing Hubble type, that were derived from Evans' survey, are collected in Table I. The high frequency derived for SNII in galaxies of types Sm + Im is entirely due to the fact that SN1987A occurred in the LMC during the survey period!

3. The Theoretical Galactic Supernova Rate

In the Shapley-Ames Catalog 528 galaxies have types earlier or equal to Sab, and 426 have types Sc or later. The Galaxy, which is generally

TABLE I. Extragalactic Supernova Rates

	No. observed			SN rate*		
Type	SNIa	SNIb	SNII	SNIa	SNIb	SNII
E+SO	2	0	0	$0.3h^2$	-	-
Sa	2	0	0	$0.6h^2$	-	-
Sab, Sb	2.5	1.5	6	$0.3h^2$	$0.6h^2$	$1.8h^2$
Sbc-Sd	2	2	5	$0.2h^2$	$0.4h^2$	$1.3h^2$
Sm+Im	0	0	1	-		$(31h^2)$

* Expressed in supernovae per century per $10^{10}L_B(\odot)$

believed to be of type Sbc, is therefore close to being an "average" Shapley-Ames galaxy. It follows that the supernova rates listed in §2 can, once the luminosity of the Galaxy is known, be used to calculate the Galactic supernova rate. Unfortunately the luminosity of our own Milky Way system is still quite uncertain. Recent estimates for the luminosity of the Galaxy range from $1.6 \times 10^{10}L_B(\odot)$ [$M_B = -20.1$] (DE VAUCOULEURS and PENCE [8]) to $3.9 \times 10^{10}L_B(\odot)$ [$M_B = -21.1$] (TAMMANN [9]). In a review of presently available data, DE VAUCOULEURS [10] derives a face-on absolute magnitude $M_B^o = -20.2 \pm 0.15$ for the Galaxy. A correction $A_B \approx 0.3$ mag has to be applied to this value to transform it to the dust-free magnitude $M_{B_T}^{oi}$ of SANDAGE and TAMMANN [4]. A value $M_{B_T}^{oi} = -20.5 \pm 0.2$ corresponds to a blue luminosity $(2.2 \pm 0.4) \times 10^{10}$ $L_B(\odot)$. This result is consistent with $L_B = (1.8 \pm 0.3) \times 10^{10}L_B(\odot)$ that VAN DER KRUIT [11] recently obtained for the Galactic disk. Adding a spheroid luminosity of $(1.5 \pm 0.5) \times 10^9L_B(\odot)$ yields a total blue luminosity of $(1.95 \pm 0.3) \times 10^{10}L_B(\odot)$ for the Galaxy. Both this determination, and that by de Vaucouleurs, are based on the assumption that the Sun is located at a distance $R_o = 8.5$ kpc from the Galactic center. For other Galactocentric distances the Galactic luminosity (and supernova rate!) scales approximately as $(R/8.5)^2$. In the subsequent discussion it will be assumed that $L_B = (2.0 \pm 0.3) \times 10^{10}L_B(\odot)$. With this value the Galactic supernova rates become: SNIa $(0.56 \pm 0.22)h^2$, SNIb $(0.54 \pm 0.31)h^2$ and SNII $(2.08 \pm 0.68)h^2$ per century. These values correspond to rates of 0.6, 0.5 and 2.1 per century if H = 100 km s^{-1} Mpc^{-1}, and to 0.14, 0.14 and 0.5 per century if H = 50 km s^{-1} Mpc^{-1} for SNIa, SNIb and SNII, respectively. The total Galactic supernova rate is therefore predicted to be $(3.18 \pm 0.78)h^2$ per century; corresponding to 0.8, 1.8 and 3.2 per century for Hubble parameters of 50, 75 and 100 km s^{-1} Mpc^{-1}, respectively.

VAN DEN BERGH [12] has reviewed the evidence on masses of supernova progenitors. He concludes that the progenitors of SNIb are stars with masses greater than those of the progenitors of SNII. If the hypothesis that SNIb have very massive progenitors is correct then the total Galactic core collapse rate of supernovae with massive progenitors (SNIb + SNII) is $(2.62 \pm 0.75)h^2$ per century.

Major fluctuations in the Galactic star formation rate probably occur on a timescale >1 x 10^8 years. Since the progenitors of core-collapse supernovae have main sequence lifetimes <1 x 10^8 years it is (neglecting effects of mass-transfer in close binaries!) safe to assume that the birthrate of these objects is approximately equal to their death rate. The total rate at which massive stars die in the Galactic disk is therefore equal to the rate at which they form, which is

$$N(\mathcal{M}_{min}) = 2\pi \int_{\mathcal{M}_{min}}^{\mathcal{M}_{max}} \int_{R_{min}}^{R_{max}} \psi(\mathcal{M}) \frac{\sigma(R)}{\sigma(R_o)} RdRd \quad . \tag{1}$$

In Eqn. (1) $\mathcal{M}_{min}$ is the minimum mass of stars that are still capable of becoming supernovae by core collapses and $\mathcal{M}_{max}$ is the largest observed stellar mass. $\psi(\mathcal{M})$ is the present stellar birthrate function (initial mass function), integrated perpendicular to the Galactic plane, for stars in the range $\mathcal{M}$ to $\mathcal{M} + d\mathcal{M}$. In Eqn. (1) $\sigma(R)/\sigma(R_o)$ is the (azimuthally averaged) stellar surface density, as a function of Galactocentric radius R, normalized to its value at the solar radius. Adopting the $\psi(\mathcal{M})$ relation found by SCALO [13], and assuming ψ is the same throughout the Galactic disk, RATNATUNGA and VAN DEN BERGH [14], find that the core-collapse supernova frequency is approximated (to an accuracy of ~5 percent) by the relation

$$N(\mathcal{M}_{min}) \approx A\, 10^{-1.91\log \mathcal{M}_{min} - 1.3 \pm 0.3} . \tag{2}$$

With the assumption of an exponential radial profile, the integral of the surface density over radius is

$$A(\mathrm{kpc}^2) = 2\pi\, H_R^2 \exp(R_o/H_R). \tag{3}$$

Eqn. (3) gives the area that the star forming Galactic disk would have if it were of uniform surface density equal to that in the solar neighborhood. Adopting H_R = 4 kpc and R_o = 8.5 kpc Eqn. (3) yields an equivalent surface area of 842 kpc^2 for the Galactic disk. The regions beyond 15 kpc and interior to 3 kpc, in which there is probably not much current star formation, contribute only 11% and 17%, respectively, to the integral over radius. Assuming a scale-length H_R = 8 kpc would yield an equivalent area A = 1163 kpc^2. Adopting A = 1000 ± 250 kpc^2 one finds $1.0^{+1.5}_{-0.6}$ stellar collapses per century for $M_{min} = 8\ M_\odot$ and $2.2^{+2.7}_{-1.6}$ stellar collapses per century for $M_{min} = 5\ M_\odot$. Clearly supernova frequencies in the range one per 45 years to one per 100 years are disappointing to those who hope to observe neutrinos from Galactic corecollapse supernovae! The supernova rates quoted above are consistent with a rate of $(2.62 \pm 0.75)h^2$ Galactic core collapse supernovae per century that was derived from Evans' supernova survey. Equating the extragalactic rate of SNIb + SNII to the rate of Galactic core collapse events yields $0.34 < h < 1.15$ for $M_{min} = 8\ M_\odot$ and $0.42 < h < 1.62$ for $M_{min} = 5\ M_\odot$.

4. Observational Estimates of Galactic Supernova Rates

KATGERT and OORT [15] and TAMMANN [9] have used the distribution of nearby historical supernovae to derive the Galactic supernova rate. Such estimates of supernova frequencies are, of course, wrought with uncertainty because of incompleteness in the historical record, uncertainties in supernova distances, and the vagaries of small-number statistics. From the data on recent supernovae Tammann obtains a rather high rate of 6.3 ± 2.5 supernovae per century.

According to CLARK and STEPHENSON [16] three supernovae (the Crab, 3C58 and Tycho) have occured in the region with $100° < \ell < 260°$ during the last two millennia, yielding an observed rate of 0.15 ± 0.09 per century. From the fact that 16 out of 20 radio SNR's in this longitude zone suffer so little obscuration that they can be seen optically (VAN DEN BERGH [17]) it follows that historical supernova discoveries in the Galactic anti-center direction are probably ~80% complete. [As seen from

Beijing the region with $\ell < 210°$, $b = 0°$ culminates above 10° altitude now, and 15° above the horizon in the first century AD]. The corrected supernova rate in the anti-center direction is therefore ~0.19 ± 0.11 per century. The Galaxy contains 46 SNR's with a surface brightness $>3.0 \times 10^{-20}$ W m^{-2} Hz^{-1} Sr^{-1} at 408 MHz; of these 4 are located in the anti-center region with $100° < \ell < 260°$. The total Galactic supernova rate may therefore be estimated to be $0.19 \times 46/4 \approx 2$ per century.

The currently favored scenario for the formation of SNIa (WOOSLEY and WEAVER [18]) involves an accreting white dwarf that is pushed over the Chandrasekhar limit. [But see HACHISU, KATO and SAIO [19] for an alternate view in which novae and SNIa are alternate evolutionary paths for stars that are initially embedded in a common envelope]. Classical novae are also believed to be due to accretion onto a white dwarf in a close binary system. It therefore seems reasonable to assume that novae and SNIa occur in similar stellar populations. The _ratio_ of the number of novae to the number of SNIa might therefore be expected to be a constant. From observations of 8 novae in Virgo cluster ellipticals by PRITCHET and VAN DEN BERGH [20] the nova rate in E galaxies is 18 ± 6.5 per year per $10^{10} L_B(\odot)$ of parent galaxy luminosity. This value assumes a Virgo distance modulus $m - M = 31.5$, corresponding to a distance of 20 Mpc. According to TAMMANN [15] the _supernova_ rate in E galaxies is $(0.88 \pm 0.24)h^2$ per $10^{12} L_B(\odot)$ per year. With h = 0.67 km s^{-1} Mpc^{-1} (VAN DEN BERGH [21]) this yields a supernova rate of 0.4 ± 0.1 per $10^{12} L_B(\odot)$ per year. The nova to SNIa ratio is therefore $(18 \pm 6.5) \times 100/0.4 \pm 0.1 \approx 4500 \pm 1800$. Note that this ratio is independent of the Hubble parameter, but that it does depend slightly on the adopted infall velocity into the Virgo cluster.

According to CAPACCIOLI _et al._ [22] the nova rate in M31 is 29 ± 4 per year, while that in M33 is found to be 4 ± 2 per year (DELLA VALLE [23]. From these data CAPACCIOLI _et al._ [22] derive a Galactic nova rate of 15 ± 5 per year. It then follows that the Galactic SNIa rate is $100 \times (15 \pm 5)/(4500 \pm 1800) = 0.33 \pm 0.17$ per century. According to EVANS, VAN DEN BERGH and MCCLURE [5] ~18% of all supernovae in the Shapley-Ames Catalog are of type Ia. If the same ratio applies to the Galaxy then the expected Galactic supernova rate is ~1.8 per century. This value is in satisfactory agreement with the other estimates of Galactic supernova frequency that have been given above.

5. Summary and Conclusions

(a) Evans' discovery of 24 supernovae in Shapley-Ames galaxies during the period 1980-1988 are used to derive the following supernova rates in an average Shapley-Ames galaxy:

SNIa (0.28 ± 0.10)h^2 SNU,
SNIb (0.27 ± 0.15)h^2 SNU, and
SNII (1.04 ± 0.30)h^2 SNU.

In these relations h = H/100 km s^{-1} Mpc, and one SNU is one supernova per century per $10^{10} L_B(\odot)$ of parent galaxy luminosity. The total supernova rate in an average Shapley-Ames galaxy is therefore 1.59 ± 0.35 SN per century for h = 1.0 and 0.40 ± 0.09 SNU for h = 0.5. Of these supernovae ~82 percent are of the core-collapse type which have massive progenitors.

(b) From the extragalactic supernova rate and the assumption that the Galaxy is an "average" Shapley-Ames galaxy with L_B = (2.0 ± 0.3) x $10^{10} L_B(\odot)$ the following Galactic supernova rates are obtained: SNIa (0.56 ± 0.22)h^2, SNIb (0.54 ± 0.31)h^2 and SNII (2.08 ± 0.68)h^2. The corresponding total supernova rates in the Galaxy are 3.2 ± 0.8 for h = 1.0, and 0.8 ± 0.2 for h = 0.5.

(c) From SCALO's [13] mass spectrum of star formation RATNATUNGA and VAN DEN BERGH [14] find a rate of $1.0^{+1.5}_{-0.6}$ supernovae per century in the Galaxy if the minimum mass for core-collapse is 8 $M_\odot$. For a minimum core-collapse mass of 5 $M_\odot$ the Galactic rate of supernovae with massive progenitors is found to be $2.2^{+2.7}_{-1.6}$. These figures assume that the Galactic disk has an equivalent area, with a star formation rate equal to that near the Sun, of 1000 ± 250 kpc.

(d) From a comparison of the fraction of all supernova remnants in the anti-center direction with the total number of Galactic SNR's, with the historical rate of optical supernovae in the anti-center region, a rate of ~2 Galactic supernovae per century is derived.

(e) Scaling from the nova rate in M31 and M33, and adopting N(novae)/N(SNIa) = 4500 ± 1800, yields a SNIa rate of 0.33 ± 0.17 per century. If the Galaxy is a typical Shapley-Ames object then ~18% of all supernovae are of type Ia. It then follows that the total Galactic supernova rate is ~1.8 per century.

References

1. R. Evans: Highlights of Astronomy **7**, 579 (1986).
2. S. van den Bergh, R.D. McClure and R. Evans: Ap.J. **323**, 44 (1987).
3. R.D. McClure, S. van den Bergh and R. Evans: Pub. Dom. Ap. Obs. **16**, 281 (1987).
4. A. Sandage and G.A. Tammann: A Revised Shapley-Ames Catalog of Bright Galaxies (Washington, Carnegie Institution) (1981).
5. R. Evans, S. van den Bergh and R.D. McClure: Ap.J. **343**, xxx (1989).
6. F. Zwicky: Ap.J. **96**, 28 (1942).
7. S. van den Bergh: A.J. **96**, 701 (1988a).
8. G. de Vaucouleurs and W.D. Pence: A.J. **83**, 1163 (1978).
9. G.A. Tammann: in Supernovae: A Survey of Current Research, eds. M.J. Rees and R.J. Stoneham (Dordrecht, Reidel) p.371 (1982).
10. G. de Vaucouleurs: Ap.J. **268**, 451 (1983).
11. P.C. van der Kruit: A.Ap. **157**, 230 (1986).
12. S. van den Bergh: Ap.J. **327**, 156 (1988b).
13. J.S. Scalo: Fund. Cosmic Phys. **11**, 1 (1986).
14. K.U. Ratnatunga and S. van den Bergh: Ap.J. **343**, xxx (1989).
15. P. Katgert and J.H. Oort: Bull. Astr. Inst. Netherlands **19**, 239 (1967).
16. D.H. Clark and F.R. Stephenson: in Supernovae: A Survey of Current Research, eds. M.J. Rees and R.J. Stoneham (Dordrecht, Reidel) p.355 (1982).
17. S. van den Bergh: in IAU Symposium No. 101, Supernova Remnants and Their X-Ray Emission, eds. I.J. Danziger and P. Gorenstein (Dordrecht, Reidel) p.597 (1983).
18. S.E. Woosley and T.A. Weaver: Ann. Rev. A. Ap. **24**, 205 (1986).
19. I. Hachisu, M. Kato and H. Saio: Ap.J. (Letters) **342**, L19 (1989).
20. C.J. Pritchet and S. van den Bergh: Ap.J. **318**, 507 (1987).
21. S. van den Bergh: A.Ap.Rev. **1**, xxx (1989).
22. M. Capaccioli, M. Della Valle, M. D'Onofrio and L. Rosino: A.J. **97**, 1627 (1989).
23. M. Della Valle: in The Extragalactic Distance Scale, A.S.P. Conference Proceedings Vol. 4, eds. S. van den Bergh and C.J. Pritchet (Provo, Brigham Young University Press) p.73 (1988).

The Asiago Supernova Catalog

Roberto Barbon, Enrico Cappellaro, &
Massimo Turatto

The large number of supernovae (SNe) discovered in the last years, i.e. since the publication of the Revised Supernova Catalogue [1] which contained all SNe discovered up to 1983, led us to publish a new version including all SNe discovered up to 1988 December 31. The number of listed supernovae amounts to 661, of which 267 have been classified.

The Catalogue is intended as a quick reference for statistical studies, so only the main data relative to all extragalactic SNe and their parent galaxies are presented.

Besides the inclusion of the newly discovered supernovae, a great effort has been devoted to update the information relative to all objects already included in the 1984 edition of the Catalogue by collecting new data either from the literature, private communications, further inspection of archival material and by new observations. The Catalogue has also been cross-checked with all major extragalactic files in order to improve data on parent galaxies.

Some differences in the data presentation have been suggested us by the experience of investigators in the field, in order to improve the quality of the information.

The main differences with respect of the past version are as follows:

- asterisks preceding the SN designation denote now all supernovae in that same parent galaxy;

- morphological types of the parent galaxies include *Sm* subtypes. Intermediate spirals, once classified SAB, have been put in either class by careful searching, through galaxy catalogues, for a more accurate classification. This will help when doing statistics with small numbers. Finally, peculiarity has been also marked;

- anonymous galaxies are coded as in the RC2 [2] for easier identification;

- when available, the integrated B magnitudes of the parent galaxies are given, instead of the m_{pg} from ZWICKY *et al.* [3] which are not of high accuracy and, moreover, present systematic errors expecially for low latitude fields;

- supernova types include the recent subclasses Ia and Ib, whereas type III, IV and V have been kept mostly for historical reasons, although they are, very likely, spectroscopically similar to type II;

- to better discriminate between *epoch of maximum* and *date of discovery*, an asterisk marks this latter one for poorly studied supernovae;

-name(s) of the discoverer(s) is given.

The symbol "-" in the first column indicates that some modification with respect to the previous version, either correction or implementation, has been applied to that data line.

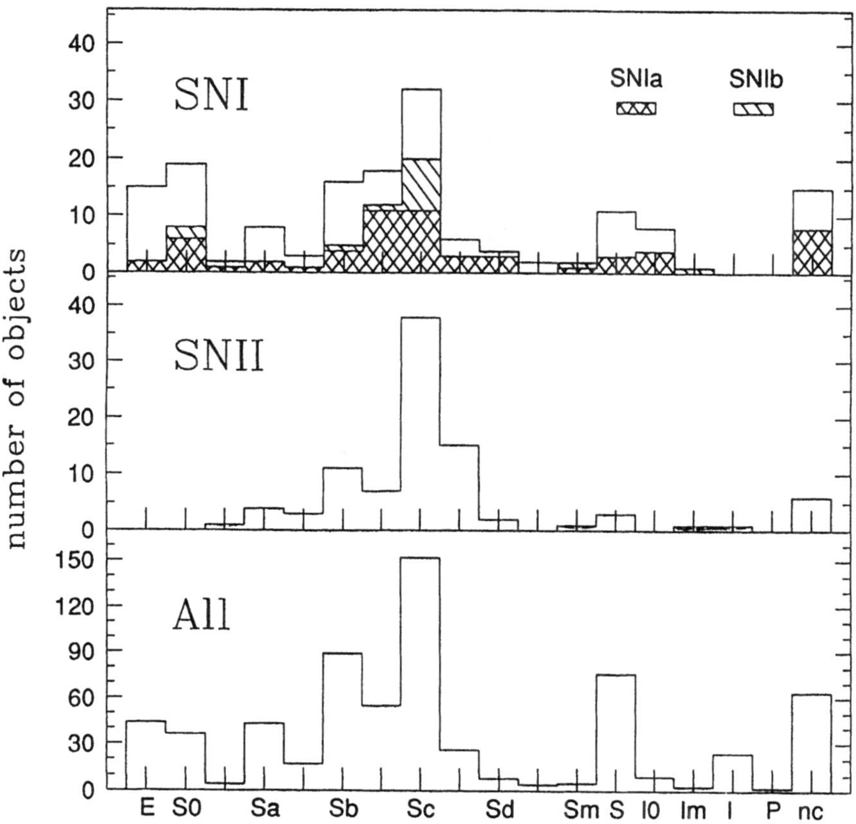

Figure 1: Frequency distribution of supernovae according to the morphological types of their parent galaxies. The bottom panel includes classified (255) and unclassified (406) objects.

The meaning of the different columns in Tables is similar to that of the previous edition [1].

Figure 1 shows the frequency distributions of the various types of classified supernovae with respect to the different morphological types of their parent galaxies, *nc* meaning parent galaxies without classification. In the Figure, normal and barred spirals have been binned together.

In Figure 2, the productivity of the worldwide SN search since 1970 is shown, where the shaded region refers to classified SNe. From this Figure the efforts of the observers to provide full information on each new object is outstanding, thus revealing the growing interest in such field of research.

Table 1 gives an example of the data presentation. The whole Catalogue will appear in a forthcoming issue of Astronomy and Astrophysics Supplement Series.

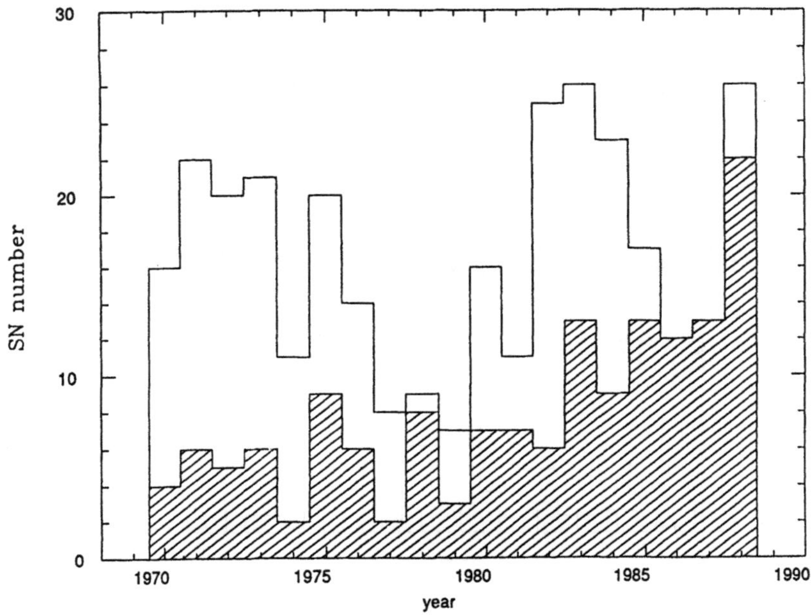

Figure 2: Productivity of the worldwide SN search since 1975. The dashed area represents the number of classified objects.

References

1. Barbon,R., Cappellaro,E., Ciatti,F., Turatto,M., Kowal,C.T.: 1984, *Astron. Astrophys. Suppl. Ser.* **58**, 735
2. de Vaucouleurs,G., de Vaucouleurs,A., Corwin,H.G.Jr.: 1976, *Second Reference Catalogue of bright galaxies* (RC2) (The University of Texas Press, Austin)
3. Zwicky,F., Herzog,E., Karpowicz,M., Kowal,C.T., Wild,P.: 1960-1968, *Catalogue of Galaxies and Clusters of Galaxies* (California Institute of Technology, Pasadena) Vol. 1-6

Table 1. The Asiago Supernova Catalogue by date.

SN	Galaxy	α (1950.0) h	m	δ °	'	type	GALAXY DATA Lc	inc	V_{hel}	mag	log D	SUPERNOVA DATA offset		mag	type	date	discoverer
- 1885 A	NGC 224	00	40.0	+41	00	Sb	2	69	-299	B 4.4	3.25	15W	4S	V 5.9	I	Aug21	Hartwig (S And)
- 1895 A	NGC4424	12	24.7	+09	42	Sa p	5:	59	439	B12.3	1.57	75E	11S	12.5*		Mar *	Wolf (VW Vir)
- * 1895 B	NGC5253	13	37.1	−31	23	I0 p		–	403	B11.0	1.60	16E	23N	8.0	I	Jul 6	Fleming (Z Cen)
- 1901 A	NGC2535	08	08.2	+25	21	Sc p	1:	54	4135	B13.1	1.47	19E	7N	14.7*		Jan *	Reinmuth
- * 1901 B	NGC4321	12	20.4	+16	06	Sc	1	27	1568	B10.1	1.84	110W	4N	15.6*	I	Mar	Curtis
- 1907 A	NGC4674	12	43.5	−08	23	SB0		67	1301	14.5	1.28	10W	11N	13.5:	I	May10	Luyten
- * 1909 A	NGC5457	14	01.5	+54	36	Sc	1	12	266	B 8.2	2.43	620W	408N	B13.5	Pec	Feb:	Wolf (SS Uma)
- * 1912 A	NGC2841	09	18.5	+51	11	Sb	1	62	631	B10.1	1.91	50W	20N	13.0*		Feb *	Pease, Curtis
- * 1914 A	NGC4321	12	20.4	+16	06	Sc	1	27	1568	B10.1	1.84	24E	111S	15.7*		Mar *	Curtis
- 1915 A	NGC4527	12	31.6	+02	56	Sb	3	69	1738	B11.3	1.80	44E	8S	15.5*		Mar *	Curtis
- * 1917 A	NGC6946	20	33.8	+59	59	Scd	1	27	46	B 9.6	2.04	37W	105S	14.6*	II	Jul	Ritchey
- 1919 A	NGC4486	12	28.3	+12	40	E0 p		–	1258	B 9.6	1.86	15W	100N	12.3	I	Feb26	Balanowsky
- 1920 A	NGC2608	08	32.2	+28	39	SBb:	3	50	2119	B12.8	1.40	19W	5N	11.8*	I	Jan	Wolf
- * 1921 A	NGC4038	11	59.3	−18	35	SBm p		46	1650	B11.3	1.41					Mar *	Hubble, Duncan
- * 1921 B	NGC3184	10	15.3	+41	40	Scd	3	12	418	B10.4	1.84	32E	160S	13.5	II:	Apr:	Zwicky,Hubble
- * 1921 C	NGC3184	10	15.3	+41	40	Scd	3	12	418	B10.4	1.84	79E	236S	11.0	I	Dec11	Jones
- * 1923 A	NGC5236	13	34.3	−29	37	SBc	2	24	506	B 8.2	2.05	109E	58N	B12.4:	II	Feb20:	Lampland
- * 1926 A	NGC4303	12	19.4	+04	45	Sc	1	24	1566	B10.2	1.78	11W	69N	14.2	II	May 5	Wolf, Reinmuth
- * 1926 B	NGC6181	16	30.1	+19	56	Sc	1	60	2158	B12.5	1.41	0	48N	14.8*	-	Jun *	Van Maanen
1934 A	IC4719	18	29.0	−56	46	I		–		B14.6		6E	13S	13.6:		Oct11	Boyd
1935 A	IC4652	17	22.0	−59	41	Sbc								15.0*		Jun *	Boyd, Huruhata
- 1935 B?	NGC3115	10	02.7	−07	29	S0		68	698	B10.1	1.92	36W	60N			Apr *	Samaha
- 1935 C	NGC1511	03	59.3	−67	46	Sab		27	1525	B12.1	1.52	55E	8S	12.5:		Sep19:	Boyce
- 1936 A	NGC4273	12	17.4	+05	37	SBc	3	47	2378	B12.4	1.36	0	29N	14.9	II	Jan26	Hubble, Moore
- 1936 B	M+02-04-29	01	18.4	+15	26	SBc:		41		15.6	1.03			14.0*		Sep *	Zwicky
- * 1937 A	NGC4157	12	08.6	+50	47	Sbc	3:	76	916	B11.6	1.84	42E	42N	15.5	II	Jan19	Zwicky
- 1937 B	M-04-52-18	22	07.8	−22	54	Sbc		17	9337	14.8	1.28	29E	31S	15.3*		Aug *	Zwicky
- 1937 C	IC4182	13	03.5	+37	52	Sm	8	21	225	B11.5	1.76	30E	40N	8.4	Ia	Aug24	Zwicky
- 1937 D	NGC1003	02	36.1	+40	39	Scd	5	67	585	B12.1	1.73	48E	1S	12.9	Ia	Sep18	Zwicky
- 1937 E	NGC1482	03	52.4	−20	39	S0		42	1655	14.4	1.19	24W	51N	15.0*	I:	Dec:	Zwicky
- * 1937 F	NGC3184	10	15.3	+41	40	Scd	3	12	418	B10.4	1.84	5E	149S	13.7	II	Dec15	Zwicky,Jones
- 1938 A	M+06-06-68	02	34.6	+34	13	SBa	5:	45	4800	B14.2	1.17	8W	28S	15.2:	I:	Nov	Zwicky
- 1938 B	NGC2672	08	46.5	+19	16	E1		–	4223	B12.6	1.41			15.5*			Wachmann
1938 C	Anon1313+25	13	13.7	+25	26	I:		–				31E	20S	17.7*		May *	Klein
- 1939 A	NGC4636	12	40.3	+02	58	E1		–	979	B10.5	1.79	26W	20N	12.6	I	Jan25	Zwicky
- 1939 B	NGC4621	12	39.5	+11	55	E4		–	424	B10.8	1.71	0	53S	12.0	I	May 1	Zwicky
- * 1939 C	NGC6946	20	33.8	+59	59	Scd	1	27	46	B 9.6	2.04	215W	24N	13.4*	I:	Jun:	Zwicky
- 1939 D	NGC 321	00	54.9	−05	18	SBcd	3:	46	5775	14.5		9W	11N	16.0*		Nov *	Zwicky
- 1940 A	NGC5907	15	14.6	+56	30	Sc	3:	82	535	B11.0	2.09	137E	310S	14.3	II	Feb22	Johnson
- * 1940 B	NGC4725	12	48.0	+25	46	Sab p	1	44	1114	B10.0	2.04	95E	118N	13.1	II	May 8	Johnson
- 1940 C	IC1099	15	05.6	+56	42	Sbc	3	32		15.0	1.15			16.3*	II:	Apr:	Zwicky

Views from the OCA Schmidt Telescope

Christian Pollas

Fifteen SNe were found with this 90/152/360 cm telescope in the South-East of France near Grasse [1] and announced in Circulars of UAI. The rate of discoveries has been seven a year for the last two years. Most were between 17 and 20 magnitude, i.e. faint SNe with z up to 0.05. Since early 1987 the Palomar and the OCA Schmidt Telescopes have together provided fifty percent of the SNe production.

This is not new, as Zwicky and Kowal, Wischnjewsky more recently [2] produced record-numbers of faint SNe. Multiple discoveries were predicted and observed by Mnatskanian in 1968 [3]. We confirmed this, with two then three SNe on one plate. But today many observations are possible on large telescopes or with CCD, allowing classification and discoveries of some peculiarities. We especially need spectral observations on our faint objects,the position of which we can rapidly provide by means of electronic mail.

But our research is not a specific SN program and is not exhaustive for all the galaxies in our view. These discoveries have been made in the run of other regular astronomical programs on our Schmidt telescope. The method used is only an ocular comparison between the images of galaxies on our plates and on the Palomar Observatory Sky Survey prints. Sometimes direct examination of brighter galaxies of 14-15 magnitude (between one and 50 galaxies on each plate) provides 1 or 2 SN/y. 1984 P, 1987 J, 1988 A, 1988 Z, 1989 J were found this way. Other times a binocular scanning provides a localisation of galaxies with star image nearby that are compared in a second step with the red Poss print then with the blue if necessary. This second method may produce from 100 to 800 of candidate galaxies (up to eighteenth magnitude) for further examination. Any big schmidt can do it!

Since July 1987, 13000 to 20000 galaxies provided 7 detected SNe on our plates. Our annual plate production is only two hundred, with 0 to a thousand of examinable galaxies by plate. This work can take quite a long time: between 0 and 8 hours per plate. Nevertheless only a few stars around galaxies are examined perhaps to 7000 on a plate. It is fast in regards of the very high number of objects present to 22 magnitude on a 5dx5d field.

The most often used emulsion is Kodak Technical Pan Plate 15301 which has a very high signal to noise ratio. Its spectral sensitivity extends from UV to 7000Å including an interesting but dangerous detection of H alpha . It includes all the images that may be faint enough to be visible on only one of either the blue or red POSS prints which are used for comparison. The best resolution and detection quality of this 15301 plate drive us to abandon sometimes up to 3 faint possible SNe on a plate. Collaboration work has started with French astronomers on imagery and photometry with 2m-T and 1m-T of Pic du Midi and 1.2m-T of OHP, and the development of this working relationship between spectroscopics observers and theoricians is expected [4],[5].

Locally, we hope to replace expensive plates by film and to build a videoblink [6]. A specific Sne research program on a circumpolar field or on a chosen cluster of galaxies should be planned.

We will continue production of accurate positions of faint Sne and nearby stars in the field of observational cooperations as made with A. Filippenko and as suggested at this workshop.

References

1 J-L. Heudier: The INAG Schmidt Telescope in Proceedings of the IAU Colloquium 48 Modern Astrometry (1978).

2 R. Barbon, E. Cappellaro, M. Turatto: The Asiago Supernovae Catalogue A&A (1989) in press.

3 R.G.Mnatsakanian: IBVS 785 (1973).

4 C.Pollas, J-L.Heudier: Journal des Astronomes Français 32 (1988).

5 C.Pollas, R.Moschkovich, P.Prugniel: JAF 35 (1989).

6 A.Maury: Observatoire de la Cote d'Azur Technical report (1988).

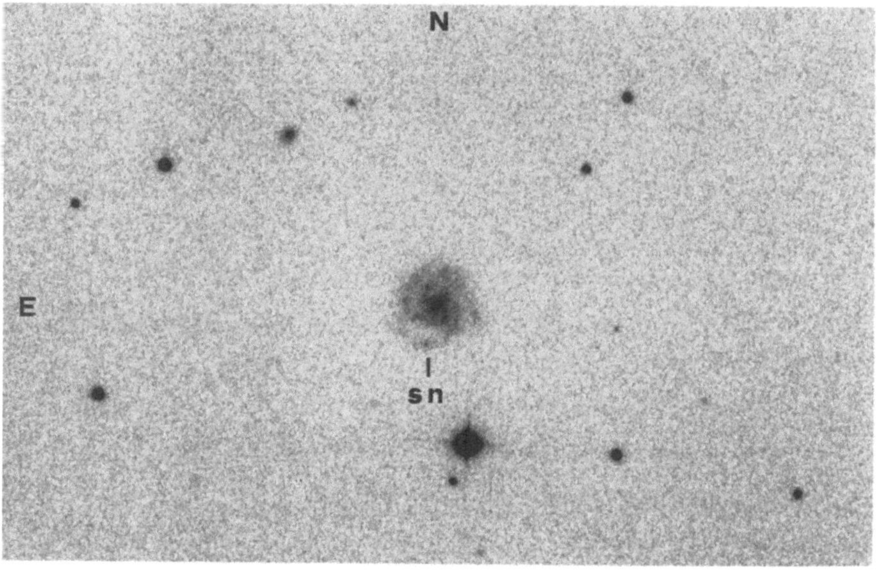

Fig. 1. Supernova 1989 H in MCG +6-30-64 discovered on a February 7 plate at Bj=20. The image obtained on 1989 March 12 shows the object in the 15301 emulsion spectral band, probably fainter than the 22 magnitude. It is superimposed on a blue region visible on the POSS print. Offset from nucleus 3.6" E and 16.3" S . Exposure: 40 mn. ©OCA.

Progress and New Directions for the Berkeley Supernova Search

Saul Perlmutter, Heidi J. Marvin, Richard A. Muller, Carlton R. Pennypacker, Timothy P. Sasseen, Craig K. Smith, & Li-Ping Wang

1. Review of Real-Time Automated Search

The Berkeley Automated Supernova Search is a systematic search for supernovae in a known galaxy set. The primary goals of our experiment are to discover close supernovae early in their light curves (before maximum light) and to discover a large number of supernovae which can be used for statistical purposes. To aid in the study of supernova explosions and their classification, we are conducting VRI photometry on supernovae discovered in our data set (including those found by other observers). The supernova discoveries themselves will eventually be used to determine the supernova rates for various supernova types, galaxy types, and positions within the galaxy. With this in mind, we keep an extensive data base of all observations, their results, and characteristics of each galaxy.

The automated search system consists of a 30" Ritchey Chretien reflecting telescope, an RCA CCD, and a network of computers and software. The software generates galaxy observation lists, controls telescope motion and data acquisition, compares the galaxy image with a previous image of the same galaxy, archives images, stores the results of its analysis in a database, and notifies scientists at LBL of the possible supernova candidates. Apart from maintenance and constant improvements, the search requires one person to start up and shut down the telescope every night and one person to check the search results for supernovae every morning.

The automated search allows more prompt reporting of supernovae and less human effort than the previous generation "semi-automated" supernova search which utilized human scanners to search for supernovae in subtractions of a reference image from a recent image. The software searches for supernovae as the images are taken. Then, if a candidate is detected, a second image is taken immediately to rule out the possibility that the candidate is a cosmic ray. If the potential supernova is still present, a third image is taken one hour later to verify that it is not an asteroid (which would move in the space of an hour). Therefore, we have three images of a supernova to examine the following morning (Figure 1) and can report the discovery to the IAU that very day. Using these procedures, the Berkeley team has never falsely reported a supernova. (See PERLMUTTER *et al.* [1] for a more detailed description of our search).

2. Current Progress and Recent Discoveries

This second generation search is currently capable of imaging about 300-600 fields per night to a limiting detection threshold of 17th magnitude. In the eighteen months it has been in operation since January 1987, the automated search has accumulated 960 galaxy years of observations (unnormalized for galaxy luminosity) and has added four supernova discoveries to the list of four found previously by the semi-automated search. The more recent discoveries were SN 1988H, 1988L, 1989A, and 1989L (PERLMUTTER and

PENNYPACKER [2,3,4,5]. The last two of these were discovered at or before maximum light. During this same period, two additional supernovae were discovered in our data set (Table 1 gives the complete list of supernova discoveries in our galaxy set). One of them, 1989B, was "re-discovered" at a later time by our automated search. The second, 1989M, was not discovered due to an inadequate reference picture of the galaxy. Although this

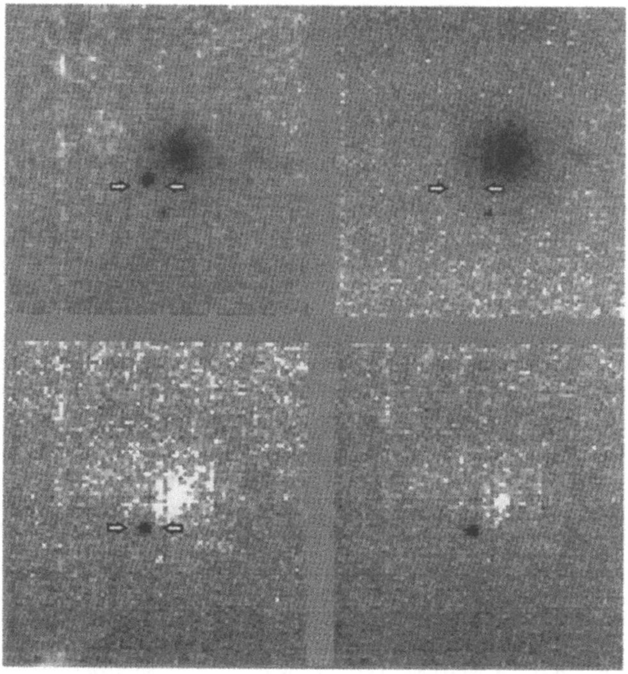

Figure 1. Sample morning report printout. This is the discovery printout for supernova 1989A. This shows the January 19 image (top left), the reference picture (top right), the subtraction used by the supernova search to identify the supernova (bottom left), and a version of the subtraction used to aid the eye in verifying supernovae (bottom right). The arrows point to the spot where the software has found the supernova.

represents a real inefficiency in our search, events such as this one will not ultimately effect our statistics because these inefficiencies will be noted when we characterize our reference image set for statistical analyses (and also eliminated whenever possible). From these supernovae, we find approximate rates consistent with the recent supernova rates of VAN DEN BURGH, MCCLURE, and EVANS [6,7] (type II supernovae must be calculated with a lower effective number of galaxy years due to its dimmer absolute magnitude).

An automated filter wheel is now incorporated into our search, and was used to collect VRI photometry of 1989A, 1989B, 1989L, and 1989M. These data will be published when the supernovae have faded and low noise reference images can be taken in three colors. We intend to continue collecting such follow-up photometry on all discoveries in our galaxy set. In addition, it is our standing policy to notify observers immediately upon our detection of a supernova event so that early spectra can be taken and the supernova can be typed. For one of these supernovae, 1989A, we are organizing a collaboration to publish the spectra and light curve.

Table 1.
Supernovae Discovered
in the Galaxy Set of the Berkeley Supernova Search

Supernova	Galaxy	Discovery Observation	Discovery Magnitude	Type	Discovered By Us	Follow-up Photometry
1986I	M99	May 17	14	II	Y	Y
1986N	NGC 1667	Dec 11	15	Ia	Y	N
1986O	NGC 2227	Dec 24	14	Ia	Y	N
1987K	NGC 4651	July 28	~15	II/I	Y	N
1988H	NGC 5878	Mar 3.5	15.5	II	Y	N
1988L	NGC 5480	May 3.4	16.5	Ia	Y	N
1989A	NGC 3687	Jan 19.5	15.3	Ia	Y	Y
1989B	NGC 3627	Jan 30.5	13	Ia	N	Y
1989L	NGC 7339	Jun 1.4	16	II	Y	Y
1989M	NGC 4579	Jun 28.8	12.2	Ia	N	Y

3. New Telescope

We plan to increase the sensitivity and number of images per night by moving our supernova search to new site. We are considering sites with improved seeing and much lower sky background than our present Leuschner Observatory, which is located only 12 miles into the Berkeley hills. For this purpose, we have purchased a 30" Ritchey Chretien (f/6.75) telescope which has been delivered to us at Lawrence Berkeley Laboratories and is being fitted with our own motion controller and search hardware.

Currently, our processing time is about 70 seconds per image. With a new computer, we can reduce the processing time by a factor of about three. However, we will have to move to a site with better conditions to reduce the imaging time. At a new site with 1" seeing and 21st magnitude sky background, we will cut our time in half while finding supernovae three magnitudes fainter. In addition, we will operate many more nights due to better weather. We are in the process of further automating our observatory so that it can be operated at a remote site.

4. Deep Supernova Search

One of the original motivations for our search for supernovae was to measure the deacceleration parameter of the universe, q_0, and thus determine whether the universe is open, flat, or closed. It appears that it may now be possible to achieve this, with the apparent constancy of the peak absolute magnitude of type Ia supernovae (LEIBUNDGUT [8]; CADONAU, SANDAGE, and TAMMANN [9]) and the recent discovery of a distant type Ia supernova by a Danish group at ESO (NORGAARD-NIELSEN *et al.* [10]). We have set up a collaboration with Warrick Couch and Brian Boyle at the AAO and Shane Burns at Harvey Mudd to attempt to find a statistically significant number of distant supernovae. Thus while we are expanding our search for nearby supernovae, we are also beginning an entirely new search for supernovae of redshifts out to about z=0.3.

The deep search will utilize the prime focus of the AAT 3.9 meter telescope in Coonabarrabran, Australia. We have installed and are testing f/1 reducing optics and a Thomson 1024x1024 pixel CCD, which allow us to image a 16'x16' field with a pixel scale of 1 arcsec/pixel. With 5-10 minute exposures, we should be able to detect supernovae to about 23rd magnitude. With 100-200 interesting galaxies per field and one type Ia supernova every 300-400 galaxy years, we should discover about one supernova per night of observation.

This work has been supported in part by the National Science Foundation, the Ann and Gordon Getty Foundation, and the U.S. Department of Energy under contract number DE-AC03-76SF00098

REFERENCES

1. Perlmutter, S., Crawford, F.S., Muller, R.A., Pennypacker, C.R., Sasseen, T.S., Smith, C.K., Treffers, R., and Williams, R., "The Status of Berkeley's Real-Time Supernova Search," *Instrumentation for Ground-Based Optical Astronomy*, ed. L. B. Robinson, New York: Springer-Verlag, p. 674-680 (1988).

2. Perlmutter, S., and Pennypacker, C.R., IAU CIRCULAR #4560, March 8, 1988.

3. Perlmutter, S., and Pennypacker, C.R., IAU CIRCULAR #4590, May 5, 1988.

4. Perlmutter, S., and Pennypacker, C.R., IAU CIRCULAR #4721, January 24, 1989.

5. Pennypacker, C.R., and Perlmutter, S., IAU CIRCULAR #4791, June 3, 1989.

6. Evans, R., van den Bergh, S., and McClure, R.D., "Revised supernova rates in Shapely-Ames galaxies," submitted to *Astrophysical Journal*, (1989).

7. van den Burgh, S., McClure, R.D., and Evans, R., *Astrophysical Journal*, **323**, 44-53 (1987).

8. Leibundgut, B., "Supernovae Ia as standard candles," this volume.

9. Cadonau, R., Sandage, A., Tammann, G.A., "Type I supernovae as standard candles," in*Supernovae as Distance Indicators*, ed. N. Bartel, Berlin:Springer, p.151-165, (1985).

10. Nørgaard-Nielsen, H.U., Hansen, L., Jørgensen, H.E., Salamanca, A.A., Ellis, R.S., Couch, W.J., *Nature,* **339**, 523-525, (1989).

Searching for Supernovae in Starburst Galaxies

Michael Richmond & Alexei V. Filippenko

1. Introduction

A number of explanations have been suggested to account for the strong emission lines seen in "starburst galaxies," the most successful of which postulates that a short-lived, but intense, period of star formation yields the high-energy photons needed to excite the lines. Several different measurements (e.g., the luminosity of a hydrogen Balmer line or the far-infrared continuum) can be used to estimate the number of O and early-B main-sequence stars in such galaxies. We are currently monitoring nearby starburst galaxies to look for the many supernovae (SNe) that ought to be produced in these star-forming regions. Within the next year or so, we will be able to check quantitatively whether most of the massive stars end their lives as normal SNe, subluminous SNe, or some other objects such as black holes. If the observed SN rate is too low, it is even possible that vigorous star formation is *not* responsible for the energetic phenomena in some of these peculiar galaxies.

2. Expectations: a Supernova Per Year?

Based on extensive visual observations by Robert Evans in Australia, the supernova rate in "normal" spiral galaxies has been estimated as roughly $1.8\,h^2$ SNU for all types of SNe combined (VAN DEN BERGH *et al.* [1]). Here, $h = \mathrm{H}_0/(100 \text{ km s}^{-1} \text{ Mpc}^{-1})$, and a SNU (SuperNova Unit) is defined as one supernova per $10^{10}\,L_B(\odot)$ per century (TAMMANN [2]). More recent work by VAN DEN BERGH [3], reported at this meeting, refined this rate to $(1.59 \pm 0.35)\,h^2$ SNU, where about two-thirds of the SNe are of type II and one-sixth each of types Ia and Ib. For our Milky Way Galaxy, with an assumed $L_B = 2 \times 10^{10}\,L_B(\odot)$, the total rate is only about two SNe per century ($h = 0.75$, used throughout this paper).

But there are many relatively nearby galaxies whose predicted SN rates are much higher, per unit blue luminosity, than in normal galaxies. In particular, the observed phenomena in "starburst" galaxies, which are characterized by a bright, compact nucleus and spectra containing strong, narrow lines of H I, [O III], [N II], and other species, can be explained by a vigorous, recent burst of star formation. A population of many hot, young stars emits high-energy photons that ionize the interstellar medium and produce spectra very similar to those observed. Because so many massive stars are formed, however, many SNe should follow some $10^6 - 10^7$ years after the onset of activity. In such galaxies, the SN rate can be one or two orders of magnitude greater than in "normal" galaxies.

For example, (WEEDMAN *et al.* [4]) studied the galaxy NGC 7714 at many wavelengths and found that there must be 10^4 supernova remnants within the inner 280 pc in order to explain the observed X-ray and radio fluxes. Since a typical remnant lifetime is of order 10^4 years, they estimated that the SN rate in NGC 7714 must be about *one per year* in the inner regions alone! This agrees with the rate they calculated from the number of massive stars needed to produce the measured, dereddened luminosity of Hα in NGC 7714. Not all starburst galaxies are so active, but SN rates of $0.2 - 0.5$ yr^{-1} must be fairly common. A set of 150 such galaxies, observed for a year, might yield $30 - 75$ SNe *if* obscuration is negligible.

3. The Sample

Our sample comes primarily from two sources: the list of starburst galactic nuclei of BALZANO [5], and a group of *IRAS*-bright galaxies (SOIFER *et al.* [6]). Balzano selected galaxies with strong blue and UV continua from the lists of Markarian. All Markarian galaxies are fainter than magnitude 13, so she added a number of bright galaxies from the *Second Reference Catalog* (DE VAUCOULEURS *et al.* [7]) to fill the gap at small distances. She chose galaxies with a bright, stellar or semi-stellar nucleus; those with diffuse inner regions were excluded. She took spectra of her final sample of 102 galactic nuclei through either a circular (4″ diameter) or a rectangular (8″ × 3″) aperture. The Hα fluxes, together with the galaxy redshifts, yield intrinsic luminosities of 10^{40} to 6×10^{42} ergs s^{-1} in Hα, from which she concluded that the galaxies were in the process of forming $10^7 - 10^9\,M_\odot$ of massive stars.

The *IRAS* sample consists of all extragalactic *IRAS* sources with a 60 μm flux of more than 5.4 Jy, a Galactic latitude of $|b| > 30°$, and declinations generally larger than $-15°$. For a total of 324 such sources, with identifications and redshifts, SOIFER *et al.* [6] fit a single-temperature Planck function to the 60 μm and 100 μm fluxes and compute a total far-infrared (FIR) luminosity. Most of the galaxies are very luminous in the far IR, with the peak of the sample distribution at $L_{FIR} \approx 10^{10.5}\,L_\odot$. According to BECKLIN [8], at luminosities greater than $10^{10}\,L_\odot$ it is likely that star formation is the dominant form of energy generation in IR-bright galaxies; moreover, a study of the 56 galaxies shared by both samples (DEUTSCH and WILLNER [9]) showed that the Hα luminosity correlates well with the *IRAS* FIR luminosity. Thus, the *IRAS*-bright galaxies may be expected to have high SN rates as well. On the other hand, considerable quantities of dust are present in most of these galaxies, making it more likely that the SNe will be heavily obscured.

It is significant that in all six of the *IRAS* galaxies that had previous IR photometry through a 50″ circular aperture, 1.3 to 5 times more FIR emission was detected through the larger $[(1.5' - 3') \times 5']$ effective aperture of *IRAS*. This implies that in these galaxies, and probably in many of our sample, there are comparable amounts of star formation occurring both near the nucleus and well outside it — an important point for our search.

The total sample we chose from these sources is somewhat heterogeneous, but it does consist of several complete subsets. All of the galaxy redshifts (z) are relatively low, so normal SNe should easily be detected in our survey. From the Markarian galaxies of BALZANO [5], we took all galaxies with an *observed* Hβ luminosity greater than 10^{40} ergs s^{-1} and $z \leq 0.025$; in addition, some galaxies with $z \leq 0.03$ were included. Comparison with the NGC 7714 calculation of WEEDMAN *et al.* [4] indicates that these galaxies should have an *observed* SN rate of 0.03 yr^{-1}. Their *intrinsic* SN rate, determined from the extinction-corrected Hβ luminosity, is ~ 0.2 yr^{-1}. From the *IRAS* sample, we included all galaxies with $L_{FIR} \geq 10^{10.5}\,L_\odot$ and

$z \leq 0.0125$, as well as all galaxies with $L_{FIR} \geq 10^{11.0} L_{\odot}$ and $z \leq 0.03$. Once again, the intrinsic SN rates are predicted to be high ($\gtrsim 0.5$ yr^{-1}). We also added a small number of somewhat less luminous galaxies in areas of the sky which were sparsely populated by galaxies, so that throughout the year there would always be objects to examine at reasonable airmass. This was done primarily in the range $16^h \lesssim \alpha \lesssim 23^h$, where the Milky Way blocks extragalactic views. We ended up with a total of about 150 galaxies, whose distributions with distance and absolute blue magnitude can be seen in Figure 1.

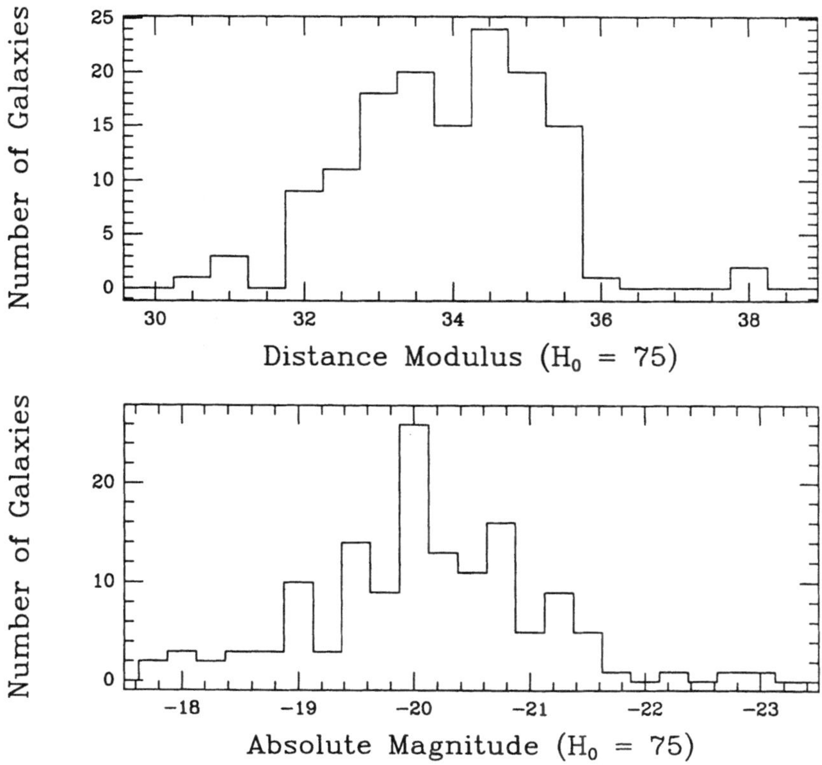

Figure 1: Starburst galaxies in the current survey ($H_0 = 75$ km s^{-1} Mpc^{-1}).

Using the SN rates of VAN DEN BERGH [3] for *normal* galaxies, we can calculate the *minimum* number of SNe we expect to see per unit time in our sample of *starburst* galaxies. If we adopt the *observed* blue luminosities as our basis, then extinction by dust in the parent galaxy should not seriously affect our calculations: in general, if the blue photons from hot stars can reach us, so ought the blue photons from SNe. Simply binning our sample by absolute blue magnitude and adding up the rates from each bin should provide a rough estimate of the minimum *observable* SN rate. We find that our sample should yield about $160\, h^2$ SNe per century, so we may expect to see about one SN per year, assuming all 150 galaxies are observed throughout the year (i.e., 150 galaxy-years of coverage). If starburst galaxies produce *more* SNe per unit blue luminosity than normal galaxies, then our observed rate should be significantly higher. This is true even if some of the SNe are partially obscured, because most of our images go several (and in some cases, many) magnitudes deeper than the expected unobscured magnitude of a SN at maximum brightness.

4. The Search Method

At present, our search for SNe is a simple one. We use the 1-m Anna Nickel reflector at Lick Observatory to take short (two to seven minute) CCD exposures of each galaxy. During the first eight months we used no filters, hoping to gain an extra magnitude, but recently we have begun taking all images through the "Spinrad red" filter (similar to the one described by DJORGOVSKI [10]). We find that the sky is so bright in the shorter visual wavelengths, due to nearby San Jose and to scattered moonlight, that the benefits of being able to perform accurate photometry more than make up for the little signal lost with the filter in place.

We attempt to take an image of each galaxy that is well placed in the sky once every two weeks, near the night of quarter moon, although the region in right ascension from 11^h to 14^h is so crowded that for some galaxies a full month elapses between images. Immediately after each picture is taken, the frame is flattened and displayed on a computer screen in the telescope control room. The observer then compares, by eye, the image with a hardcopy version of a previous picture of the same galaxy. If he notices a difference, he takes one or more additional images before the end of the night so that we can distinguish defects and slowly-moving objects from SN candidates. Several asteroids have been found in this manner, but usually we have too little data to calculate their orbits.

We also take regular pictures, through a number of wide-band filters, of any SNe in the sky which are bright enough to remain visible for an extended period of time — generally those which reach sixteenth magnitude or brighter. The resulting light curves can be used in conjunction with spectra obtained during other observing programs to study the physical processes at work in SNe. Typically, there are only three or four visible on any given night, so we don't lose much time from our search.

Although our visual comparison is less sophisticated than computer analysis of CCD images, we have found that it is adequate to detect SNe which occur outside the nucleus of galaxies in our sample. Preliminary tests indicate that a trained comparator can distinguish an object of magnitude $m \lesssim 17.5$ within $5'' - 10''$ of the nucleus (but not within $3''$), $m \lesssim 18 - 18.5$ in the disk of a spiral galaxy, and $m \lesssim 19$ in a typical arm. The limit of detection in "the field" of our exposures ranges between $m = 19.5$ and $m = 21$. Since almost all our galaxies have distance moduli $\lesssim 36$, and about half have $m - M \lesssim 34$, a typical SN will exceed our detection limit of $m \lesssim 17.5$ for at least twenty days around maximum, and usually for several months (see Table 3 of VAN DEN BERGH *et al.* [1]). We are likely to miss a SN only if it is superposed directly on the galactic nucleus (but see §6). As pointed out above, however, certain galaxies in our sample, especially the *IRAS*-bright galaxies, experience star formation over an extended region. With this simple approach, we therefore hope to find at least some of the SNe that may be produced in starburst galaxies.

5. Preliminary Results

Up through September 1989, we have been searching for ten months, during which we observed 21 clear (or relatively clear) nights. We have a total of 691 images. To get a rough idea of how many galaxy-years we have covered, we subtract the initial image of each galaxy, leaving about 540 images, and estimate that each of those images covers two weeks of time. This is likely to be an underestimate, since we can detect a majority of SNe for much longer after maximum; on the other hand, most images of a given galaxy are taken just two weeks after the previous image. We therefore have about twenty galaxy-years of images. With an assumed average rate of one SN every three years for our galaxies, and neglecting obscuration, to date we may expect to have found seven SNe associated with starbursts, including the *nuclear* starbursts that we have not yet thoroughly checked. The most likely expected number, if some dust extinction is included, is probably $1-2$ SNe. If the VAN DEN BERGH [3] rate for normal galaxies is used (§3), the corresponding estimate is only $0.5\,h^2$ SNe, or about 0.3 SNe ($h = 0.75$).

Our current SN count is two (see Appendix) — SN 1988ab (RICHMOND [11]) and SN 1988ac (RICHMOND [12]). However, SN 1988ac (which breaks the record of 28 SNe found in 1954!) should probably not be included in our starburst statistics; it occurred in NGC 3995, a *companion* to one of our sample starburst galaxies (NGC 3994) that just happened to fall within our CCD field of view. It can also be argued that SN 1988ab does not affect our *starburst* statistics, making the actual starburst-SN count equal to zero. SN 1988ab was located at the outer edge of a spiral arm about 33″ from the nucleus of NGC 762 (Mrk 1012), a galaxy from the list of BALZANO [5] whose starburst characteristics are strongly confined to its *nucleus*.

So, while it is still too early to make any definitive statements about the frequency of SNe in starburst galaxies, we suspect that the rates may be lower than theoretical predictions. We plan to extend this project for at least another year, by the end of which we should accumulate over fifty galaxy-years of observations. If we continue to see very few SNe associated with starbursts during this interval, we may be able to significantly constrain models of the end states of stellar evolution. Some of the SNe in such galaxies might be subluminous, or perhaps certain types of massive stars can end their lives as black holes. At the current time, though, a plausible explanation for our low observed rate may be the concentration of SNe in the central few *arcseconds* of each galaxy, since many of our galaxies have luminous starbursts in the *nuclear* region.

6. Future Prospects

To detect SNe which occur in the central parts of galaxies and are hidden by the glare of bright nuclei, we must perform a more quantitative analysis of our CCD images. Accurate photometry of the innermost 5″ of each galaxy could identify normal SNe in many of our galaxies. For example, WEEDMAN *et al.* [4] calculated that a SN going off in the magnitude 13.8 nucleus of NGC 7714 would increase its brightness by about 10%, which could easily be seen. Our images, about 4′ on a side, usually contain two or more reasonably bright ($m \lesssim 18$) stars as well as the target galaxy; these stars may serve as relative calibrators so that we can derive useful differential photometric measurements of the galactic nuclei, even on nights which are not completely cloud-free. This requires a substantial amount of work, however, and it will take a number of months to go through the backlog of data.

Another possibility for improving our search is the acquisition of a dedicated telescope. There is a good chance that within the next year we may have a fully-automated observatory with a 0.8-m telescope, imaging CCD, and autoguider. This observatory will open and close its dome, schedule observations, find galaxies, and guide long exposures all by itself. We plan to reach mag 21 in deep images, through standard *UBVRI* filters. Such an instrument will free human operators the chore (and expense) of travelling to the mountain and spending long hours watching a TV screen. It will also take images every clear night, extending our coverage of galaxies and making much more frequent observations; this should increase our chances of finding SNe before they reach maximum light. Finally, and most importantly, it will allow us to *monitor* SNe, and many other types of variable or ephemeral objects (e.g., Seyfert nuclei, quasars, faint variable stars, novae), on an almost nightly basis. Indeed, discovering new SNe is not the major goal of our group; we wish instead to complement the efforts of the existing Berkeley Automated Supernova Search Team (PERLMUTTER *et al.* [13]). By making long-term follow-up observations of all available SNe, we will construct the accurate light curves badly needed for a more thorough understanding of the processes which lead these massive stars to explode, light up the sky, and then slowly, quietly, fade away.

This work is partially funded by a Presidential Young Investigator Award (NSF grant AST–8957063) to AVF, as well as by the Center for Particle Astrophysics, an NSF Science and Technology Center operated by the University of California at Berkeley under Cooperative Agreement AST–8809616. Support from the California Space Institute (grant CS–41–88) is also appreciated. Lick Observatory receives partial funding from NSF through Core Block grant AST–8614510. We thank our colleagues at U.C. Berkeley, especially Joseph Shields, for obtaining some of the data used in our survey.

Appendix: Supernovae Discovered Thus Far

SN 1988ab was detected at mag 15.6 (red) in an image obtained on 4 December 1988 UT, the very first night devoted to this search. Through a stroke of bad luck, we did not recognize it as a SN for over eight months. Since it appeared in our *initial* image of NGC 762, which was adopted as the comparison standard, its subsequent appearance in an image taken on 3 January 1989 UT did not attract our attention, despite its considerably fainter magnitude of 17.9 (unfiltered). Unfortunately, shortly thereafter the galaxy became unobservable due to its position in the sky, and the next image was not obtained until August 1989. Although we quickly realized that the "missing" star was a SN, by that time it had faded below our detection limit.

Our other discovery, SN 1988ac, also has no spectroscopic classification. We have determined that SN 1988ac brightened to at least 16.5 mag (unfiltered) sometime between 15 December and 30 December 1988 UT. Unfortunately, we only noticed it in October 1989, while going back through all our images during the course of reductions. Because it was superposed on a bright H II region, it was difficult to detect in images taken long after maximum. A spectrum of the H II region, obtained with the 3-m Shane reflector at Lick Observatory, also did not reveal the old SN.

References

1. S. van den Bergh, R. D. McClure, and R. Evans: *Ap. J.*, **323**, 44 (1987).
2. G. A. Tammann: In *Supernovae: A Survey of Current Research*, ed. by M. J. Rees and R. J. Stoneham (Reidel, Dordrecht 1982), p. 371.
3. S. van den Bergh: these *Proceedings* (1990).
4. D. W. Weedman, *et al.*: *Ap. J.*, **248**, 105 (1981).
5. V. A. Balzano: *Ap. J.*, **268**, 602 (1983).
6. B. T. Soifer, *et al.*: *Ap. J.*, **320**, 238 (1987).
7. G. de Vaucouleurs, A. de Vaucouleurs, and H. G. Corwin: *Reference Catalog of Bright Galaxies*, 2nd ed. (University of Texas Press, Austin 1977).
8. E. E. Becklin: In *Star Formation Galaxies*, ed. by C. J. Persson (U.S. Government Printing Office, Washington D.C. 1987).
9. L. K. Deutsch and S. P. Willner: *Ap. J. (Letters)*, **306**, L11 (1986).
10. S. Djorgovski: *Pub. A. S. P.*, **97**, 1119 (1985).
11. M. Richmond: *IAU Circular* No. 4836 (1989).
12. M. Richmond: *IAU Circular* No. 4900 (1989).
13. S. Perlmutter, *et al.*: In *Instrumentation for Ground-Based Optical Astronomy*, ed. by L. B. Robinson (Springer-Verlag, New York 1988), p. 67.

The Value of the Hubble Constant Through Novae and Supernovae

Massimo Della Valle, Massimo Capaccioli, Enrico Cappellaro, & Massimo Turatto

The reliability of novae as standard candles has been questioned on several occasions, particularly in relation to the determination of the distance of M31 (COHEN [1]). This problem has been recently re-discussed by CAPACCIOLI et al. [2]; they found, through a new calibration of the galactic maximum magnitude *vs.* rate of decline relationship (=MMRD), a distance modulus of M31 in agreement with the values given by other indicators. We apply here these results to the sample of novae discovered by PRITCHET and VAD DEN BERGH [3] in Virgo to determine the distance of the Virgo cluster.

On the same time, the distance of the Coma cluster in units of the Virgo has been derived by comparing the average magnitudes at maximum of selected samples of SNI-a. Finally, we give a value of the Hubble constant of $H_\circ = 59 \pm 13$ $[kms^{-1}Mpc^{-1}]$, nearly independent of the assumption on the values of the Virgo–infall correction and of the mean galactocentric redshift of Virgo cluster. In the following we give a short account of the different steps involved in this analysis whereas, the complete discussion will be given elsewhere (CAPACCIOLI et al. [4]).

I. The Distance of the Virgo Cluster

The best fit (Fig. 1) between the new MMRD calibrated through M31 novae and seven novae discovered in Virgo cluster [3], yields a differential distance modulus $(m_{Vir} - m_{M31})_\circ = 7.0 \pm 0.35$; with $(m - M)_\circ^{M31} = 24.30 \pm 0.20$ the distance modulus for the Virgo cluster is $(m - M)_\circ = 31.30 \pm 0.40$.

In principle, using this number, we can soon derive the Hubble constant; however, because of the present uncertainties on both the infall correction toward Virgo $\Delta V \simeq 100 \div 400$ kms^{-1} (DRESSLER et al. [5]) and the average recession velocity of this cluster $\langle V \rangle \simeq 900 \div 1200$ kms^{-1} (KRAAN–KORTEWEG [6], HUCHRA [7]), the global error on the value of $H_\circ$ would be rather closer to 35% than to 20%, this latter figure results taking into account the only uncertainty on the distance modulus.

II. The Distance Virgo–Coma from SNI-a

To avoid the afore mentioned uncertainty we decided to determine the relative distance Virgo–Coma using SNI-a as distance indicators. The selected supernovae are listed in Table 1. Epoch (col. 5) and magnitude of maximum light (cols. 6 and 7) of each SN result from a new critical analysis of the original observations and by best fitting the revised light curves to the average light curve for type I SNe (BARBON et al.[8]).

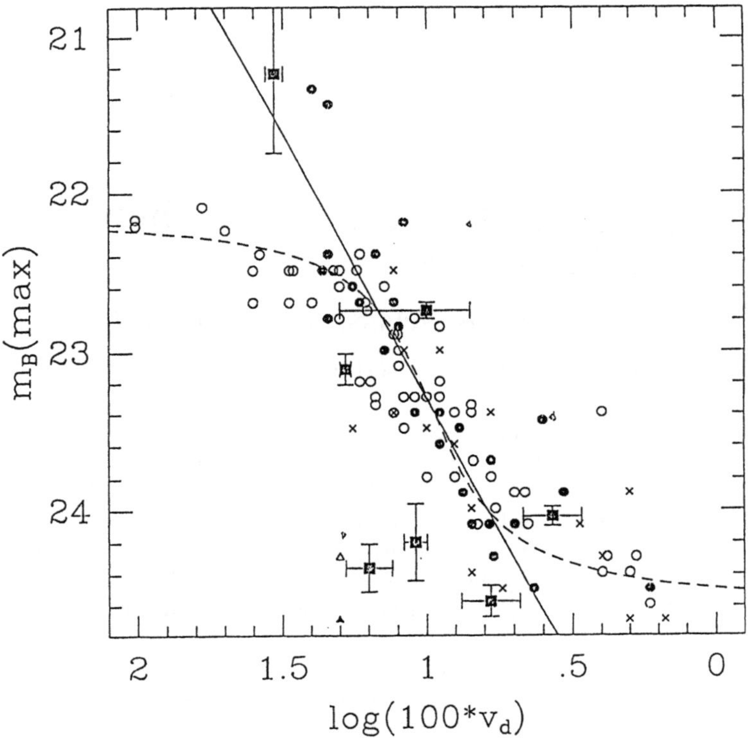

Figure 1: Maximum magnitude versus rate of decline for seven novae (*filled squares*) discovered in two Virgo E galaxies (NGC 4365 and NGC 4472). Smaller symbols are M31 novae, the dark circles refer to objects with observed maxima. The magnitudes have been corrected for color term, internal and galactic extinction, and reduced to the *adopted distance* of the Virgo cluster ($(m_{V_{ir}} - m_{M31})_o = +7.0$). The *dashed* curve reproduces equation 2 of CAPACCIOLI et al. [2], with the zero point adjusted to give the best fit of Virgo novae. The *solid* line is the linear fit of the central part of the MMRD.

Table 1: Photometric data for type I SNe

SN		Parent Galaxy		SN Photometry				
iden.	type	iden.	type	J.D.	B_{max}	V_{max}	$(B-V)_{max}$	$E(B-V)$
				VIRGO				
1919 A	I	NGC 4486	E	22016	12.25 ± 0.15 (pg)			
1939 A	I	NGC 4636	E	29289	12.60 ± 0.10 (pg)	12.25 ± 0.15 (pv)	0.29 ± 0.15	0.33
1957 B	I	NGC 4374	E	35967	12.20 ± 0.10 (pg)			
1960 R	I	NGC 4382	S0 P	37288	11.90 ± 0.20		0.25 ± 0.30	0.29
1961 H	I–a	NGC 4564	E_2	37428	11.80 ± 0.10			
1963 I	I	NGC 4178	SBdm	38155	13.30 ± 0.20		0.50 ± 0.30	0.54
1965 I	I–a	NGC 4753	I0	38929	12.50 ± 0.10	12.40 ± 0.15	0.10 ± 0.18	0.14
1981 B	I–a	NGC 4536	Sbc	44673	12.00 ± 0.05	11.90 ± 0.05	0.10 ± 0.07	0.14
1983 G	I–a	NGC 4753	I0	45431	12.85 ± 0.10	12.65 ± 0.10	0.20 ± 0.14	0.24
1984 A	I–a	NGC 4419	SBa	45716	12.45 ± 0.10	12.15 ± 0.10	0.30 ± 0.14	0.34
				COMA				
1961 D	I	M+05-30-101	E	37305	16.00 ± 0.30 (pg)	16.30 ± 0.30 (pv)	-0.01 ± 0.31	0.00
1962 A	I	M+05-31-102	S0	37685	15.20 ± 0.20 (pg)	15.40 ± 0.15 (pv)	0.08 ± 0.21	0.00
1963 C	I	M+05-31-32	E	38056	15.60 ± 0.15 (pg)	16.00 ± 0.10 (pv)	-0.09 ± 0.15	0.00
1963 M	I:	M+05-31-35	Sa	38195	15.70 ± 0.30 (pg)	15.90 ± 0.30 (pv)	0.08 ± 0.36	0.11
1973 F	I	NGC 4944	S0	41782	15.70 ± 0.15	15.80 ± 0.15	-0.10 ± 0.21	0.00

When necessary, maximum magnitudes were reduced to the B, V photometric bands using standard relations (see [4] for references).

In order to evaluate the global extinction, we make the assumption that all SNeI–a have the same intrinsic color at maximum and we adopt as reference $(B-V)_\circ = -0.02$, corresponding to the average color of the 5 SNeI–a discovered in early–type galaxies of Coma. The observed color and the derived $E(B-V)$ for each SN are listed in Table 1, cols 8 and 9 respectively. The extinction is then computed from the relation $A_B = R_B \times E(B-V)$ where the value of R_B is not well established [4]. Assuming $R_B = 4$, we derive a $(m-M)_{Coma} - (m-M)_{Virgo} = 4.1 \pm 0.3$, $(m-M)_\circ^{Coma} = 35.4 \pm 0.5$ and $\langle M_B \rangle SNI - a = -19.7 \pm 0.5$.

III. The Value of the $H_\circ$

The mean galactocentric radial velocity for the Coma cluster, reduced to the motion of the centroid of the Local Group is $\langle V \rangle = 6890$ $[kms^{-1}]$ [5], with a correction of 240 $[kms^{-1}]$ to account for a Virgocentric infall velocity of $\simeq 250$ $[kms^{-1}]$, the cosmological expansion velocity turns out $V_\circ = 7130 \pm 200$ $[kms^{-1}]$.

The Hubble constant results $H_\circ = 59 \pm 13$ $[kms^{-1}Mpc^{-1}]$.

We like to stress the fàct that this determination of $H_\circ$ does not depend from the assumptions on the values of the Virgo–infall and mean galactocentric redshift of Virgo cluster, and the attached error ($\sim 20\%$) reflects the present uncertainty on the distance modulus of Coma cluster (~ 0.5 mag).

References

1. Cohen, J.G. 1985, *Astrophys. J.*, **292**, 90.
2. Capaccioli, M., Della Valle, M., D'Onofrio, M., and Rosino, L. 1989, *Astron. J.*, **97**, 1622.
3. Pritchet, C.J., and van den Bergh, S. 1987*a*, *Astrophys. J.*, **318**, 507.
4. Capaccioli, M., Cappellaro, E., Della Valle, M., D'Onofrio, M., Rosino, L., Turatto, M. 1989, *Astrop J.*, in press.
5. Dressler, A., Lynden–Bell, D., Burstein, D., Davies, R., Faber, S.M., Terlevich, R., and Wegner, G. 1987, *Astrophys. J.*, **313**, 42.
6. Kraan–Korteweg, R.C. 1981, *Astron. Astrophys.*, **104**, 280.
7. Huchra, J.P. 1985, in "*The Virgo Cluster*", O.G. Richter and B. Binggeli eds., ESO: Garching, p. 181.
8. Barbon, R., Ciatti, F. and Rosino, L. 1973, *Astron. Ap.*, **25**, 241.

From the Expansion of SN 1987A to the Expansion of the Universe

Robert V. Wagoner

1. The Expanding Photosphere Method of Distance Determination

Sixty-three years ago Walter BAADE [1] proposed a method for directly determining the distances of oscillating stars. But, as Robert Kirshner has emphasized, it was Leonard Searle who realized that an extension of this method could be applied to supernovae. His direction led to the pioneering work of BRANCH and PATCHETT [2] and KIRSHNER and KWAN [3]. As more sophisticated atmospheric models have been employed, the power as well as the limitations of this method have become more apparent. For a review, see WAGONER [4]; for other recent results see the contributions of Höflich (denoted by H) and Eastman and Kirshner (denoted by EK) in these proceedings, as well as HÖFLICH [5], CHILUKURI and WAGONER [6] (denoted by CW), and EASTMAN and KIRSHNER [7].

We will summarize in Section 5 how well this method has passed its most critical test, afforded by the unique opportunity of SN 1987A (whose properties have been authoritatively reviewed by ARNETT et al. [8]). However, it must be remembered that for our ultimate task of mapping the Hubble flow, we will not have the luxury of a known progenitor and a flood of photons. Thus it is our philosophy that the detailed technique employed to extract the distance must be as model-independent as possible, with the minimum number of assumptions. The implementation of this philosophy is one of the focuses of this contribution.

In the next Section, we will argue that the early recombination phase is the optimum era within which to apply the method, with the only major assumption being spherical symmetry. The expansion has become homologous, with the matter velocity $v = r/t = const.$, with t the time since the matter was last accelerated (t_0). The density can then be specified in terms of conditions at some fiducial time t_* as

$$\rho = \rho_* \left(\frac{v}{v_*}\right)^{-\gamma(v)} \left(\frac{t}{t_*}\right)^{-3} . \tag{1}$$

Since by this epoch the photosphere can be considered as quasi-static, the second input is the depth-independent luminosity $L(t)$; or equivalently, the effective temperature at some fiducial photospheric radius, $T_e(r_p)$. The remaining quantities to be specified are then t_0 and the effective heavy-element abundance Z_* (which controls the line opacity).

With this input, a radiative transfer program can then output the spectral luminosity $L_\nu(t)$. Comparing a catalog of such spectra with the shape of the observed continuum can then determine the model parameters $T_e(r_p)$ (mainly from the optical region), ρ_* (mainly from the Balmer jump), and Z_* (mainly from the UV region). The line profiles determine the remaining effective model parameters $v(r_p)$ $[= r_p/t \Rightarrow r_p]$ and $\gamma(r_p)$, if the photosphere is sufficiently sharp. Once we have thereby determined the model appropriate to each observation of the spectral flux $f_\nu(t)$, the distance can be obtained from $L_\nu(t)$ if the redshift of the parent galaxy has been measured. We will address the two major problems, the effects of extinction and asymmetry, in Sections 3 and 4.

2. Advantages of the Early Recombination Phase

We have chosen to employ this method of distance determination during this particular phase for the following reasons.

- As shown by CW, the effective thickness of the photosphere $\delta r(\nu)$ is reduced by the abrupt drop in the electron density at a temperature $T \approx 6000$ K. This sharpening reduces the dependence of the computed spectrum on the density profile function $\gamma(v)$. In addition, the weaker P-Cygni line profiles will have sharper mimima, allowing a more accurate determination of the velocity $v(r_p)$.

- The complications introduced by the evolution of the atmosphere during the time it takes a photon to traverse the photosphere (EK) become negligible after $t \approx (\delta r)\tau_m/c$, typically a few days. [$\tau_m \sim 10$ is the maximum (frequency-dependent) thermalization optical depth.]

- For the luminous Type II supernovae that will be observed in the Hubble flow, this phase begins a few weeks after the explosion. Thus the uncertainty in the explosion time t_o affects the determination of r_p less than if earlier epochs were employed.

- As shown in Fig. 1 (from PEREZ and WAGONER [9]; see also KARP et al. [10]), the effective continuum (scattering) opacity produced by the $\sim 10^5$ Doppler-broadened lines is an important contribution mainly at wavelengths $\lambda < 4000$ Å. Thus the effects of the uncertainties in this opacity (incorporated partially in our parameter Z_*) are reduced because the fraction of the underlying flux at these wavelengths has become small at the recombination temperature.

- The emitted flux is less affected by (large or small scale) deviations from spherical symmetry during recombination because the photospheric temperature evolves relatively slowly.

An observational disadvantage of this era is obviously the reduced luminosity. The theoretical disadvantage is a more difficult radiative transfer problem, due to the near degeneracy of the equations of radiative and statistical equilibrium (mainly for the first excited state of hydrogen) if complete linearization is employed to achieve convergence.

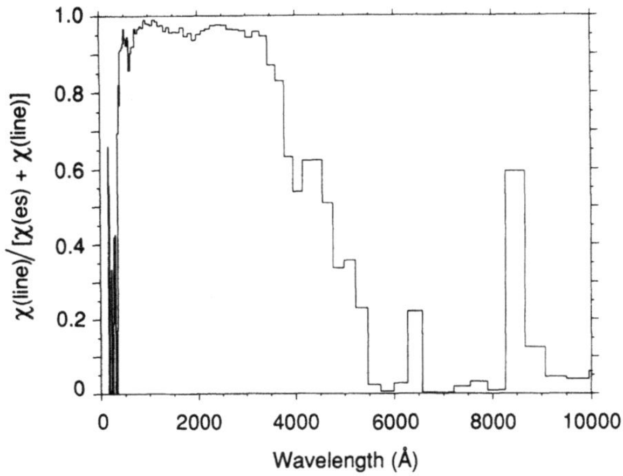

Figure 1. The ratio of the effective continuum opacity produced by lines to the total (electron plus line) scattering opacity is averaged over wavelength bands $\Delta \log \lambda = 0.02$. The parameters are $r/v = 10$ days, $\log \rho\,(\mathrm{g\ cm^{-3}}) = -13.5$, and $T = 6000$ K. The line list was provided by R. Kurucz.

3. Reducing the Effects of Extinction

From the existing data on our galaxy, the Large Magellenic Cloud, and the Small Magellenic Cloud [11], there is no evidence that the wavelength dependence of interstellar extinction $A_\lambda = E_{B-V}\psi(\lambda)$ is not universal for those wavelengths $\lambda > 2600$ Å at which there is significant flux during recombination . Adopting this assumption of universality then allows us to employ the three-frequency indices

$$\Phi_{ABC} = 2.5\left\{\log\left[\frac{\Phi(\nu_B)}{\Phi(\nu_A)}\right] - \left[\frac{\psi(\nu_B) - \psi(\nu_A)}{\psi(\nu_C) - \psi(\nu_A)}\right]\log\left[\frac{\Phi(\nu_C)}{\Phi(\nu_A)}\right]\right\}, \tag{2}$$

where Φ can be the observed flux f_ν or the emitted luminosity L_ν (simply generalized to broad bands). By employing an appropriate variety of such indices in comparing the observed and computed spectra, the values of the photospheric model parameters can be obtained in a manner independent of the value of the amount of extinction E_{B-V}. Of course, a Doppler correction of the form $\Delta \log L_\nu = (v/c)F(d \log L_\nu/d \log \nu)$ must be applied to the spectrum obtained in the comoving frame (CW).

The effect of the unknown extinction on the distance obtained is then minimized if L_ν/f_ν is employed at as long a wavelength as is feasible ($\lambda_0 \geq 3\,\mu$m).

4. Effects of Asymmetry

Supernova 1987A has provided us with the first observational evidence for deviations from spherical symmetry during the early expansion of the ejecta. However, as we shall see, there is no evidence that the matter that matters (near the photosphere during the early recombination phase: March 1–14, 1987) had significant asymmetries. Nevertheless, we shall illustrate via a simple model how large-scale deviations would affect the determination of distance, and how this might be detected.

The most relevant observation during this epoch is the March 7 polarization scan across the Hα line [12]. The only feature, representing a decrease in polarization of 0.2%, is seen at velocities approximately twice that of the matter at the photosphere. Since the otherwise constant level of polarization (0.8%) could be of interstellar origin, as indicated by nearby stars, the amount of asymmetry *at this epoch* could be small. The next scan, on May 5, shows much greater variations in polarization.

The other indication of large-scale asymmetry is the reported resolution of the supernova image by speckle interferometry [13]. Although an axis ratio of about 3/2 is inferred (especially in the Hα filter), the earliest observations were three months after the epoch of interest. Thus the claimed asymmetry referred to matter which was well below the photosphere during March 1–14. Although the position of the major axis is close to that of the polarization, radiative transfer effects in the line-producing region above the photosphere could enhance these observational consequences above the level expected from the photospheric asymmetry (H).

Evidence for smaller-scale asymmetries (mixing and clumping) come from the X-ray and γ-ray observations, as summarized by Nomoto in these proceedings (and [8]). However, the observational evidence again comes from much later epochs. In addition, the shells that appear to be Rayleigh-Taylor unstable or to be stirred by hot bubbles from Ni^{56} decay were well below the photosphere during the early recombination epoch [14,15].

It should be noted, moreover, that the more luminous Type II supernovae (which are thought to be produced by red supergiants) will be the ones used to determine cosmological distances. As CHEVALIER and SOKER [16] have pointed out, there are three reasons why these progenitors should produce more spherical supernovae: a) Their flatter density profile tends to reduce the asymmetry of the propagating shock; b) being more extended, they should have smaller rotational distortions; c) the probability of a companion close enough to significantly affect the flow or tidally distort the mantle remains small.

Nevertheless, we shall illustrate how asymmetry could affect this method of distance determination via a model (first employed by SHAPIRO and SUTHERLAND [17]) which incorporates the dominant factor. As shown in Fig. 2, we consider a sharp spheroidal supernova photosphere, but note that its matter velocity $\vec{v}(\theta)$ has become essentially radial by the epochs of interest. The 'radius' that is determined from the line profiles is then

$$R \equiv v_n t = (\vec{v} \cdot \hat{k})\, t \quad . \tag{3}$$

The actual luminosity per unit solid angle directed toward the observer, and that which he would infer if he assumed that the photosphere was spherical, are given by

$$dL_\nu/d\Omega = \int I_\nu(\mu)\,\mu\,dA \qquad (\mu = \hat{n}\cdot\hat{k} > 0) \quad , \tag{4}$$

$$(dL_\nu/d\Omega)_s = R^2 F_\nu \quad . \tag{5}$$

(Here $\hat{n}$ is the unit normal to the photosphere.) We shall take the ratio of flux F_ν to specific intensity I_ν to be that for pure scattering (a good approximation), and the flux to be constant over the surface (for the reason indicated earlier).

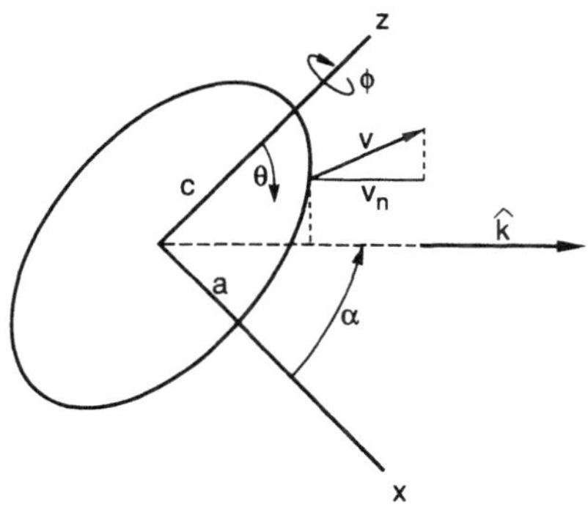

Figure 2. A radially expanding spheroidal photosphere with axis of symmetry (in the z direction) inclined at an angle $\pi/2 - \alpha$ from the direction $\hat{k}$ of the observer. The minima of weak P-Cygni lines are shifted by the velocity v_n.

The ratio of the actual distance to the 'spherical distance' of the supernova is then

$$Q(a/c,\alpha) = \left[\frac{dL_\nu/d\Omega}{(dL_\nu/d\Omega)_s}\right]^{\frac{1}{2}} \quad . \tag{6}$$

This quantity is shown as a function of viewing angle α for various axis ratios a/c of prolate spheroids in Fig. 3a and oblate spheroids in Fig. 4a. We have also computed the mean and the r.m.s. deviation of Q for random viewing, shown in Figs. 3b and 4b.

From these results, we conclude that although the distance of a very distorted supernova can be significantly over- or underestimated, the average from many supernovae is remarkably accurate. On the other hand, the scatter ΔQ_{RMS} about the mean of N supernovae is a measure of the average degree of distortion. Finally, we note that the requirement that $L_\nu(t)/f_\nu(t)$ remain constant throughout the era of interest could eliminate very distorted supernovae from consideration, in addition to constraining the other parameters of the model atmosphere.

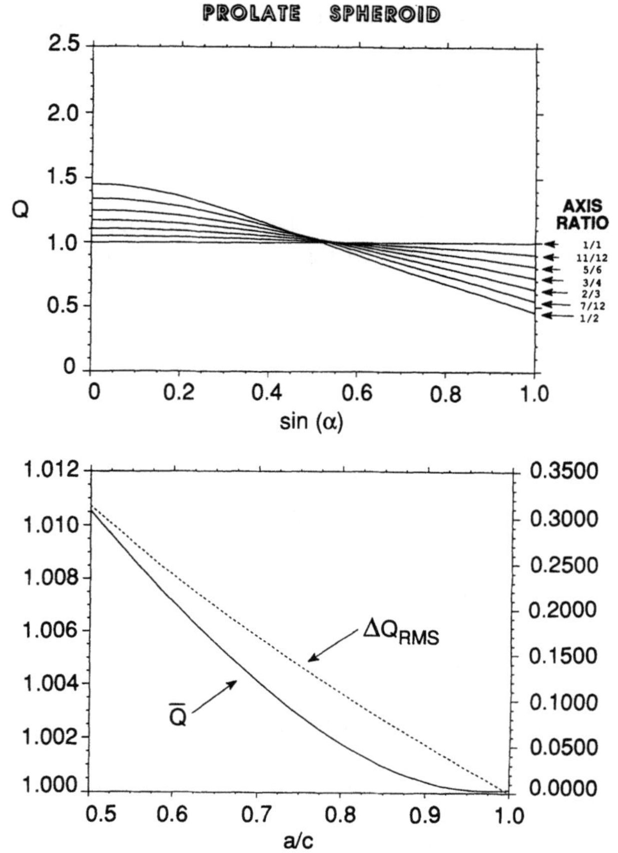

(a)

(b)

Figure 3. a) The dependence of the distance factor Q (6) on the orientation of prolate supernovae is shown for various axis ratios a/c. b) The average (left axis) and r.m.s. deviation (right axis) of the distance factor over random orientations.

5. The Distances of Supernova 1987A and the Large Magellenic Cloud

The supernova of February 23, 1987 has provided as one of its many gifts the first critical test of the expanding photosphere method of distance determination. Until then, no Type II supernova in a galaxy whose distance could be estimated by other methods had been observed with sufficient spectral range and line resolution to allow this method to be reliably employed.

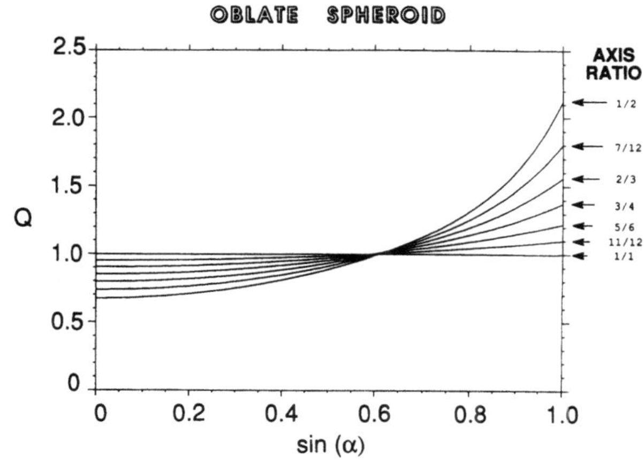

(a)

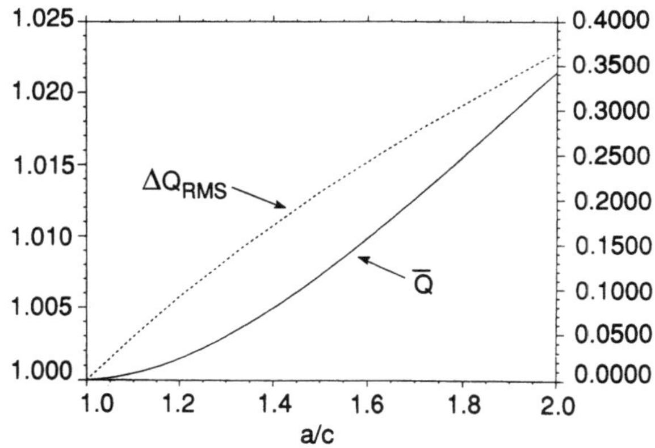

(b)

Figure 4. Same as Figure 3, for oblate spheroids.

Although BRANCH [18] obtained a distance of 55 ± 5 kpc (distance modulus 18.70 ± 0.20) by assuming that SN1987A emitted a blackbody spectrum at visible wavelengths, there have been three independent determinations from physically consistent models of the atmosphere. HÖFLICH [5] employed a nonLTE, static, extended atmosphere code which has since been enlarged (H) to include the blanketing effects of many lines (in LTE). He obtained good fits to many of the optical line profiles. EASTMAN and KIRSHNER [7] have applied a nonstatic code to the epoch 2–10 days after explosion, also obtaining good fits to many lines (see also EK). But in both cases, the fit to the UV continuum suffers from an incomplete line list.

CW employed a nonLTE, static, plane-parallel continuum code which incorporated the approximate line opacities of KARP et al. [10]. A comparison of their best-fitting model (corresponding to the values of the parameters indicated) with the March 1, 1987 spectrum is shown in Fig. 5. A blackbody spectrum characterized by the temperature $T_c = 6850$ K obtained by fitting the observed spectrum longward of the B band [19] is shown for comparison. Note the expected dilution of the flux produced by the dominance of scattering over absorption [20], as also indicated by the lower effective temperature $T_e = 5750$ K. Subsequently, a fit to the March 6 spectrum has been obtained, producing the same distance. In addition, a correction for extension has been derived by computing the increase in the radius of the photosphere (CW) at the flux comparison wavelength $\lambda_0 = 3.16\,\mu$m (due to a larger absorptive fraction of the opacity) over that at optical wavelengths, where it is determined by the line profiles. This correction is inversely proportional to the density profile parameter γ, as expected.

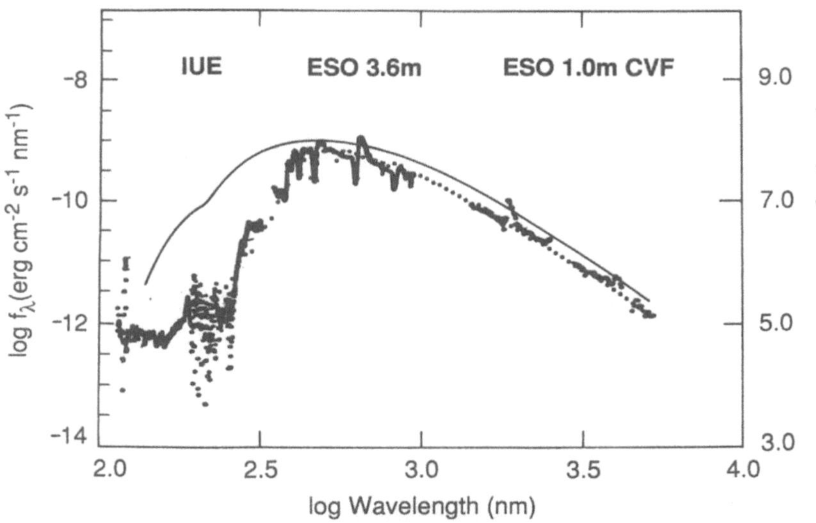

Figure 5. A composite spectrum of SN 1987A observed on March 1, 1987 [22] is compared with the best-fitting model of CW (small dots) and a blackbody of $T_c = 6850$ K [19]. An extinction corresponding to $E_{B-V} = 0.19$ has been applied to both emitted fluxes F_ν. The parameters of the CW model are $T_e = 5750$ K, $\log \rho_p(\text{g cm}^{-3}) = -11.8$, and $Z_* = 0.01$.

A survey of the distances to the LMC obtained from a variety of methods has been included in a recent review of Cepheids as distance indicators [21]. In Table

1, we compare the determinations of the distance modulus $m - M$ of the supernova with that of the LMC. This comparison is probably reasonable because the LMC is believed to be flattened roughly along our line of sight.

Table 1. Comparison of distances.

Supernova 1987A	$m - M$
Chilukuri and Wagoner (1988)	$18.20 \pm 0.20 + 0.7/\gamma$
Höflich (1988)	18.41 ± 0.18
Eastman and Kirshner (1989)	18.45 ± 0.22
Large Magellenic Cloud	$m - M$
Cepheids and other variables	18.47 ± 0.15
Old clusters	$18.42 \pm 0.2?$
Intermediate-age clusters	18.35 ± 0.35
Young clusters	18.45 ± 0.25

If we adopt the value $\gamma \cong 7 - 11$ obtained by fitting the line shapes (EK), the three supernova distances are remarkably consistent with a value that agrees with those to the LMC. This agreement provides strong evidence that this new method of distance determination is reliable. In particular, the results presented in the previous Section indicate that the photosphere of this supernova did not have a large-scale distortion during the first two weeks, unless our direction of viewing (or possibly the distribution of flux) was fortuitous.

However, before we venture into the Hubble flow, progress in three directions would reduce the uncertainties that will arise. It would be desirable to obtain more frequent polarization observations on another nearby Type II supernova, hopefully of the more luminous variety. This could give us important constraints on the degree of large-scale asymmetry. Secondly, more sensitive infrared detectors are needed (eventually in space) because of the importance of such wavelengths, as indicated above. Finally, we (and others) plan to include more of the relevant physics in our models of the atmosphere and improve the convergence of our radiative transfer codes.

6. Acknowledgments

Important contributions were made by Chris Perez in the production of Figs. 1, 3, and 4; and by Robert Kurucz, who provided his new line list. This work was supported in part by NSF (grant PHY 86-03273) and by NASA (grant NAGW-299).

REFERENCES

1. W. Baade: Astr. Nach. 228, 359 (1926).
2. D. Branch, B. Patchett: M. N. R. A. S. 161, 71 (1973).
3. R. P. Kirshner, J. Kwan: Ap. J. 193, 27 (1974).
4. R. V. Wagoner: In Theory and Observational Limits in Cosmology, ed. by W. R. Stoeger, S.J. (Vatican Observatory, 1987, p. 345).
5. P. Höflich: In Atmospheric Diagnostics of Stellar Evolution, I.A.U. Colloquium 108, ed. by K. Nomoto (Springer-Verlag, Berlin, Heidelberg, 1988, p. 288).
6. M. Chilukuri, R. V. Wagoner: In Atmospheric Diagnostics of Stellar Evolution, I.A.U. Colloquium 108, ed. by K. Nomoto (Springer-Verlag, Berlin, Heidelberg, 1988, p. 295).
7. R. G. Eastman, R. P. Kirshner: Ap. J., in press (1989).
8. W. D. Arnett, J. N. Bahcall, R. P. Kirshner, S. E. Woosley: Ann. Rev. Astron. Ap. 27, 629 (1989).
9. C. Perez, R. V. Wagoner: in preparation (1989).
10. A. H. Karp, G. Lasher, K. L. Chan, E. E. Salpeter: Ap. J. 214, 161 (1977).
11. G. C. Clayton, P. G. Martin: Ap. J. 288, 558 (1985).
12. M. Cropper, J. Bailey, J. McCowage, R. D. Cannon, W. J. Couch, J. R. Walsh, J. O. Strade, F. Freeman: M. N. R. A. S. 231, 695 (1988).
13. C. Papaliolios, M. Karovska, L. Koechlin, P. Nisenson, C. Standley, S. Heathcote: Nature 338, 565 (1989).
14. W. D. Arnett, B. A. Fryxell, E. Müller: Ap. J. (Letters), in press (1989).
15. T. Ebisuzaki, T. Shigeyama, K. Nomoto: Ap. J., submitted (1989).
16. R. A. Chevalier, N. Soker: Ap. J. 341, 867 (1989).
17. P. R. Shapiro, P. G. Sutherland: Ap. J. 263, 902 (1982).
18. D. Branch: Ap. J. (Letters) 320, L23 (1987).
19. J. W. Menzies et al.: M. N. R. A. S. 227, 39P (1987).
20. R. V. Wagoner: Ap. J. (Letters) 250, L65 (1981).
21. M. W. Feast, A. R. Walker: Ann. Rev. Astron. Ap. 25, 345 (1987).
22. I. J. Danziger, R. A. E. Fosbury, D. Alloin, S. Christiani, J. Dachs, C. Gouiffes, B. Jarvis, K. C. Sahu: Astron. Ap. 177, L13 (1987).

Supernovae Ia as Standard Candles

Bruno Leibundgut

1. Introduction

The use of Supernovae (SNe) as a mean of distance measurements was proposed already in the early times of supernova research. An extensive study by Baade (1938) of all available data at that time illustrates how SNe were used to infer cosmological parameters.

Although the recognition of various subtypes of SNe (Minkowski 1964, Kirshner et al. 1973, Oke and Searle 1974, Wheeler and Levreault 1985, Uomoto and Kirshner 1985) hampers the use of SNe In general for distance determinations, the possibility remains of using SNe of type II as "custom yardsticks" (Kirshner and Kwan 1974, Höflich 1987, Wagoner 1988, Eastman and Kirshner 1989) and SNe of type Ia as standard candles (Tammann 1982, Leibundgut 1988, Leibundgut and Tammann 1989). The second method needs, of course, a good calibration of SNe Ia as objects with equal, if not identical, evolution of their light emission. Spectroscopic studies of SNe Ia have shown differences between individual events (Branch et al. 1988), but the photometric observations have exhibited astonishing uniformity (Leibundgut 1988). The exceptional case of the well studied SN 1986G in the peculiar galaxy NGC 5128 (Cen A; Phillips et al. 1987, Frogel et al. 1987, see also Canal et al. 1988) poses a strong challenge to the significance of standard candles for SNe Ia. We would like to understand what caused the differences, for instance in the infrared light curves of SN 1986G compared to standard SNe Ia like SNe 1972E, 1981B, 1980N (Leibundgut 1988) and the dispersion of the expansion velocities in SNe Ia (Branch et al. 1988), before we really may rely on distances from SNe. The little knowledge on extinction in external galaxies complicates accurate determinations even further, but as will be shown below, it might still be acceptable to neglect this contaminating effect for most SNe Ia.

We will first demonstrate the photometric uniformity of SNe Ia and then outline their possible uses for cosmology. The observations of SN 1988U (Norgaard-Nielsen et al. 1989) provide a first test of the predictions.

2. Evidence for Photometric Uniformity of SNe Ia

2.1 Light Curves

The light curves of SNe I exhibit, unlike those of SNe II, a much closer resemblance for individual events (Barbon et al. 1973, 1979). The normalization of data sets of different SNe to analyze the photometric light curves, however, suffers from the very inhomogeneous composition of the available observations and the photometric errors involved. The improvement on the photometry was demonstrated by Cadonau et al. (1985; their Fig. 1) who compared photographic, mostly old observations, with those in B (from more recent SNe). A much better definition of the light curves was found for the B data, although a wider sample of SNe I was considered. Some of the remaining dispersion can be attributed to SNe Ib which by themselves show a larger dispersion (Ensman and Woosley 1988, Leibundgut 1988).

Thus we concentrate on SNe Ia with sufficient observations to determine good light curves in the optical (UBV) and the near infrared (JHK) which are used to find the intrinsic deviations, if at all present, between the light curves. The only SNe Ia with suitable observations were SNe 1972E, 1980N, 1981B, and 1981D. In Table 1 we list the number of photometric observations as well as the scatter about a templet light curve as defined by Cadonau (1986; cf. Leibundgut 1988). The dispersion around a standard light curve from 5 days before until 110 days past B maximum are exceedingly small. The errors in the near infrared are dominated by SN 1972E. The more recent SNe show again very small deviations from the templet curves (Elias et al. 1981, 1985, Leibundgut 1988). We want to stress the need of more observations in the infrared to improve the available data. Especially the J filter in which a strong Fe absorption trough is observed (Graham et al. 1986, Frogel et al. 1987) might be sensible to differences between individual SNe.

Table 1: Standard SNe Ia

	1972E	1980N	1981B	1981D	all SNe
galaxy	NGC 5253	NGC 1316	NGC 4536	NGC 1316	
galaxy type	I0	Sa_{pec}	Sc	Sa_{pec}	
N_U	47	32	31	9	119
N_B	47	44	82	9	182
N_V	55	45	85	9	194
N_J	5	13	15	7	40
N_H	9	14	15	7	45
N_K	19	10	14	7	50
σ_U	0.12	0.18	0.19	0.19	0.16
σ_B	0.08	0.06	0.09	0.13	0.08
σ_U	0.09	0.07	0.11	0.08	0.10
σ_J	0.24	0.20	0.10	0.10	0.16
σ_H	0.06	0.08	0.05	0.04	0.06
σ_K	0.26	0.11	0.13	0.03	0.18

The accurate repetition of fiducial light curves in six filter over the entire optical and the near infrared wavelength range is a strong indication that we might consider SNe Ia as standard candles.

2.2 Colors at Maximum

Although extinction in galaxies is unknown colors of SNe Ia may show a small dispersion. To test the uniformity of SNe Ia colors the magnitudes at maximum in each observed filter were inter- or extrapolated for 24 SNe I by means of the templet light curves. The photometric data were taken Cadonau and Leibundgut (1989). The optical measurements were corrected only for foreground absorption (Sandage and Tammann 1987); no corrections were applied to the infrared data. Any intrinsic absorption in the parent galaxies was assumed to be negligible. In spite of this simplification the resulting dispersions are very small (Table 2). It is probably significant that the scatter decreases with increasing wavelength, i.e. with decreasing influence of any internal reddening. The infrared colors are given at 10 days past the B maximum because no observations at earlier phases are known for SNe Ia. This evaluation of a standard character of SNe Ia has the advantage of being independent of distances (Leibundgut and Tammann 1989).

Table 2: Colors of SNe Ia at maximum

color	mean	σ	N(SNe)
$(U\text{-}B)^{max}$	-0.26	0.21	13
$(B\text{-}V)^{max}$	0.02	0.22	24
$(B\text{-}H)^{(10)}$	-0.53	0.21	6
$(J\text{-}H)^{(10)}$	0.92	0.12	6
$(H\text{-}K)^{(10)}$	0.05	0.04	6

2.3 Uniformity of the Peak Luminosity

To compare the brightness at maximum of individual SNe one is forced to use at least relative distances and hence an appropriate Virgo infall model as long as not enough SNe Ia are observed in distant galaxies out in the Hubble flow. To check the dispersion of SNe Ia at maximum, however, the SNe I in Virgo offer the unique opportunity to compare a set of SNe at essentially the same distance - neglecting the depth of the cluster (Leibundgut and Tammann 1989). Only six SNe are suited for this test (SNe 1957B, 1960F, 1960R, 1961H, 1981B, and 1984A). The mean apparent magnitude at maximum in B is 11.91 with a dispersion of only $0.^m19$ (Leibundgut and Tammann 1989) probably enlarged by internal absorption. The new SNe Ia in Virgo (e.g. SN 1989M) will hopefully add more data.

The sample of SNe Ia in Virgo yields only a value for m_B^{max} due to the poor observations in other filters. To widen the set and extend the study to more filters we used a model for the virgocentric infall (cf. Schechter 1980, Aaronson et al. 1982, Kraan-Korteweg 1986) to determine relative distances of the parent galaxies. The distances were taken from Kraan-Korteweg (1986) and are based on an infall velocity of 220 km s^{-1}. All SNe magnitudes were reduced to the Virgo distance and in Table 3 we give the mean apparent magnitudes, the standard deviations and the number of SNe used in deriving these numbers. The values for the infrared bands refer to the phase 10 days past the B maximum. The scatters are once more extremely small regarding that in this sample the uncertainties in distances add to the partially poor photometry and the untreated internal extinction. This derivation does **not** depend on absolute distances, i.e. H_0, nor the Virgo distance (i.e. absolute magnitudes of SNe Ia; see below).

Table 3: Apparent magnitudes of SNe Ia at Virgo distance

filter	$<m_{Virgo}>$	σ	N
U^{max}	11.79	0.42	12
B^{max}	12.06	0.32	22
V^{max}	12.04	0.34	18
$J^{(10)}$	13.98	0.22	5
$H^{(10)}$	13.05	0.15	5
$K^{(10)}$	13.01	0.15	5

The photometric properties of SNe Ia are thus very uniform. The light curves in six filters are highly repetitive, the colors at maximum are identical within the uncertainties and the absolute peak magnitudes show quite small scatter which could well be due to mainly observational errors of the photometry. SNe Ia are hence standard candles to a high degree and suitable for distance determinations and fine tools for the measurement of cosmological parameters.

3. Use of SNe Ia as Standard Candles

With a sufficiently large sample of maximum magnitudes and galaxy redshifts it is possible to measure the Hubble constant H_0 *and* the deceleration parameter q_0 (we assume Λ=0 here) for Friedmann models of the universe. It will, however, be essential to obtain high quality photometry at peak light for many SNe Ia in galaxies within the Hubble flow (for H_0) and at high redshifts ($z \geq 0.5$; for q_0). In the following we will explore these possibilities and describe furthermore a test for the nature of cosmological redshifts (Leibundgut 1989).

3.1 H_0 from SNe Ia

The calibration of H_0 with SNe Ia is not new. Values determined have been mostly low ($H_0 \leq 70$ km Mpc^{-1} s^{-1} (Branch 1988, Tammann 1982, Tammann and Leibundgut 1989). The method described here rests on the SNe Ia in Virgo (cf. section 2.3) and the Hubble diagram of SNe Ia at maximum (Fig. 1). The velocities are corrected to the centroid of the Local Group, but *not* for the infall into Virgo. This causes the relatively large scatter at low velocities in addition to the reported effect of absorption in certain galaxies (Branch 1989). Model dependent velocities improve the picture considerably (Tammann and Leibundgut 1989). Thus only SNe with $v_0 \geq 2000$ km s^{-1} were considered. The determination of H_0 is now possible through the best fit of SNe Ia in the Hubble diagram. Using the formula

$$\log H_0 = 5 \log v_0 - m + 25 + M \tag{1}$$

we can calculate H_0 by averaging over all SNe with well observed maximum provided the absolute magnitude is known. Given a distance modulus to the Virgo cluster of $(m\text{-}M)_0$=31.70±0.09 (Tammann 1988, Leibundgut and Tammann 1989) an absolute magnitude of M_B^{max}=-19.79±0.12 is derived for SNe Ia. This together with the nine SNe with $v_0 \geq 2000$ km s^{-1} yields a value of H_0=54±1.5. The error includes the uncertainty in the Virgo distance modulus, the scatter of the SNe Ia in Virgo, and the scatter of the distant SNe about the Hubble line in the diagram.

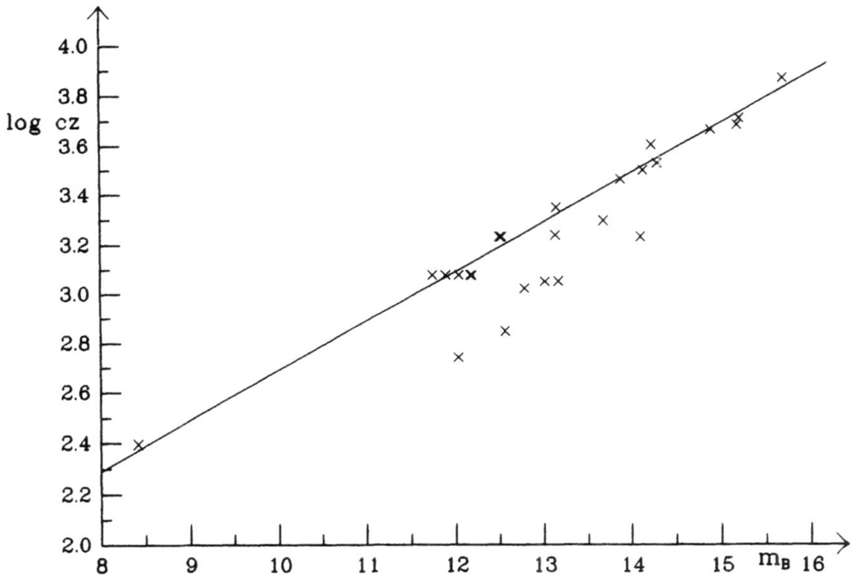

Figure 1: The Hubble diagram of SNe Ia at maximum

3.2 The Use of SNe Ia to Measure q_0

SNe Ia might open a new way to q_0. The small dispersion in absolute magnitude will allow to confine the range of q_0 (and Ω_0) to theoretically interesting values ($0 < q_0 < 1/2$). With a few SNe Ia with well observed maxima at redshifts of ~0.5 a distinction between these two extreme cases of Friedmann universes should be feasible.

The effects of redshift on the apparent magnitudes at maximum, i.e. the K-corrections, were studied for this purpose for the V filter (Leibundgut 1989). Figure 2 shows the expected Hubble diagram for SNe Ia at maximum for q_0=0 and q_0=1/2. As a comparison also the uncorrected line for an empty universe (q_0=0) is drawn. The K_V-corrections are negative for small z due to the blue flux shifted into the V passband and the blue colors of SNe Ia at maximum. This effect is stopped as soon as the ultraviolet flux is redshifted into the V and the flux drops significantly ($z \geq 0.6$) which drives the curves across the uncorrected predictions. In other words SNe Ia appear brighter at small redshifts (0.1<z<0.6) than "local" SNe Ia, but fainter beyond. This is a prediction due to the fact that they *are* standard candles! If SNe Ia do not follow this predictions then they are either effected by evolutionary effects and/or are not suitable as cosmological probes.

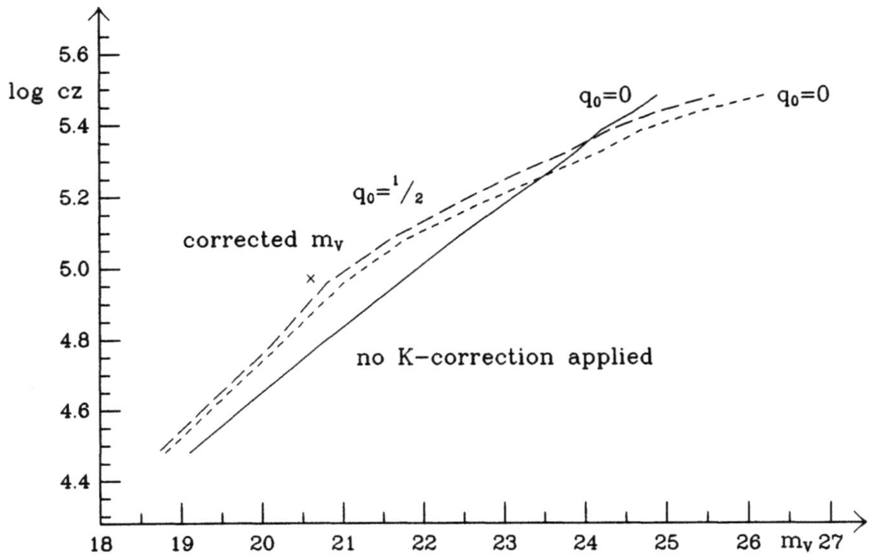

Figure 2: The predicted Hubble diagram for SNe Ia. The cross marks SN 1988U.

The observations of SN 1988U in the Abell cluster AC118 (Norgaard-Nielsen et al. 1989) at a redshift of z=0.31 unfortunately did not cover the maximum of the supernova. Although it supposedly was of type Ia (due to spectroscopic evidence) the lack of the maximum definition precludes an accurate test with this SN. Nevertheless we tried to fit the observations on a templet of an expected V curve of SNe Ia at z=0.3 (Leibundgut 1989; cf. next section). The inferred maximum value is shown in Fig. 2. The uncertainties in the determination are very large, for both the maximum estimate - which is $0.^m9$ brighter than derived by Norgaard-Nielsen et al. - and the curves based on an assumed M_V^{max}=-19.8. The validity of the corrections, however, is supported by this measurement. With more SNe and well defined maxima this potentially will be a route to measure q_0.

3.3 A Test to the Nature of Cosmological Redshifts

The identical photometric evolution in time provides a further clue to the nature of our universe. The explosion of a SN starts a clock, the ticks of which we measure relative to local clocks by comparing the shapes of the light curves. To illustrate this point the expected curve of SNe Ia at z=0.3 is drawn in Fig. 3 with and without the correction for time dilation (Leibundgut 1989). The differences are obvious and easily measurable as they are $\approx 0.^m6$ at 20 days past maximum or ~30% in time to fade by the same amount. Thus SNe Ia provide an easy test to probe the nature of cosmological redshifts, i.e. a differentiation between expansion versus tired-light models. This method is much easier to perform than the well known test using the surface brightness within linear diameters of galaxies (Sandage 1988).

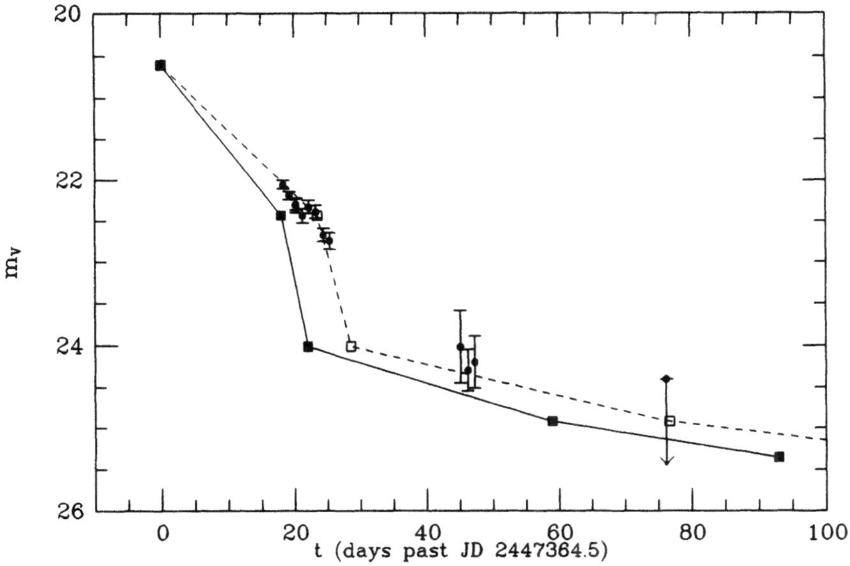

Figure 3: Comparison of the observations of SN 1988U and the predicted light curves.

The observations of SN 1988U allow us to carry out this test. The slope of the light curve after maximum clearly distinguishes between the models ($0.^m12$/day for expanding universes versus $0.^m15$/day for non-expanding models). The observations indicate a slope of $\sim 0.^m10$/day and clearly favor the curve corrected for time dilation (Fig. 3) and hence give evidence that the cosmological redshifts are due to an expansion of the universe and are not caused by tired light or equivalent models. To our knowledge this is the first clear prove of this assumption.

4. Conclusions

The conjecture of SNe Ia being standard candles is based on several independent determinations. The few well observed SNe Ia with light curves in six filters (UBVJHK) are identical to within the photometric errors (Leibundgut 1988) over the entire observed range (i.e. in time and in wavelengths), the colors at maximum are very similar for many SNe Ia and, in regard of the contamination by internal reddening, they are probably identical (Leibundgut and Tammann 1989), and finally their absolute magnitudes are identical to within $0.^m1$ - $0.^m2$ in all six filters (Leibundgut and Tammann 1989, Leibundgut 1988).

Nevertheless a few question marks remain in this picture. SN 1986G (Phillips et al. 1987, Frogel et al. 1987), one of the closest SN Ia, clearly did not follow the standard photometric evolution of SNe Ia (cf. Leibundgut 1988). How much this is caused by the very strong extinction in Cen A is unclear, but we certainly do not understand this SN in the context of normal SNe Ia. Also the spectroscopic differences, e.g. expansion velocities (Branch et al.1988) contrast with the photometric uniformity of SNe Ia.

The confined behaviour of most SNe Ia requires a narrow explosion scenario, like the proposed deflagration or detonation of white dwarfs (Nomoto et al. 1984, Woosley and Weaver 1986).

The small dispersion in absolute magnitudes makes SNe Ia the best standard candles known. As cosmological probes they have several advantages over first-ranked cluster galaxies, e.g.

- SNe are point sources which facilitates the photometry considerably. The contribution by background galaxy light can be handled by careful reduction of the observations.
- no obvious luminosity evolution is present. The physics of an exploding white dwarf are set by natural constants and even chemical changes should not alter the explosion energies by much.
- no dynamical evolution, like the one of galaxies in clusters (mergers, interaction, etc.), introduces a secular term.

To check on the latter two assumptions it is imperative to observe many SNe Ia at various redshifts. The close correspondence of the predictions and the observations of SN 1988U indicate that such effects, if present, are probably small. This supernova also supports the fundamental assumption of the expansion of the universe, demonstrating the potential they bear as standard candles, not only for distance measurements.

Hence, although SNe Ia are transient objects, they provide us with a superb tool to determine the cosmological relevant parameters H_0 and, in the forseeable future, q_0.

Acknowledgement: It is a pleasure to thank Prof.Dr. G.A. Tammann for many enlightening discussions on the cosmological implications of SNe Ia. I would like to thank the Swiss Society for Astronomy and Astrophysics for travel support and the Swiss National Science Foundation which partially financed the research.

References

Aaronson, M., Huchra, J., Schechter, P.L., Tully, R.B.: 1982, Astrophys.J. **258**,64

Baade, W.: 1938, Astrophys.J. **88**,285

Barbon, R., Ciatti, F., Rosino, L.: 1973, Astron. Astrophys. **25**,241

Barbon, R., Ciatti, F., Rosino, L.: 1979, Astron. Astrophys. **72**,287

Branch, D., Drucker, W., Jeffery, D.J.: 1988, Astrophys.J. **330**,L117

Branch, D.: 1988, in *The Extragalactic Distance Scale*, eds. S. van den Bergh and C.J. Pritchet, San Francisco: Astron.Soc.Pacific,p146

Branch, D.: 1989, this workshop

Cadonau, R.: 1986, Ph.D. Thesis, University of Basel

Cadonau, R., Leibundgut, B.: 1989, submitted to Astron.Astrophys.Suppl.Series

Cadonau, R., Sandage, A., Tammann, G.A.: 1985, in *Supernovae as Distance Indicators*, ed. N.Bartel, Berlin:Springer, p.151

Canal, R., Isern, J., Lopez, R.: 1988, Astrophys.J. **330**,L113

Eastman, R.G., Kirshner, R.P.: 1989, Astrophys.J., in press
Ensman, L.M., Woosley, S.E.: 1988, Astrophys.J **333**,754
Elias, J.H, Frogel, J.A., Hackwell, J.A., Persson, S.E.: 1981, Astrophys.J **251**,L13
Elias, J.H., Matthews, K., Neugebauer, G., Persson, S.E.: 1985, Astrophys.J. **296**,379
Frogel, J.A., Gregory, B., Kawara, K., Laney, D., Phillips, M.M., Terndrup, D., Vrba, F., Whitford, A.E.: 1987, Astrophys.J. **315**,L129
Graham, J.R., Meikle, W.S.P, Allen, D.A., Longmore, A.J., Williams, P.M.: 1986, Mon.Not.R.Astr.Soc **218**,93
Höflich, P.: 1987, in *Nuclear Astrophysics*, eds. W. Hillebrandt, R. Kuhfuss, E. Müller, J.W. Truran, Berlin:Springer,p.307
Kirshner, R.P., Oke, J.B., Penston, M.V., Searle, L.: 1973, Astrophys.J. **185**,303
Kirshner, R.P., Kwan, J.: 1974, Astrophys.J. **193**,27
Kraan-Korteweg, R.C.: 1986, Astron.Astrophys.Suppl.Ser. **66**,255
Leibundgut, B.: 1988, Ph.D. Thesis, University of Basel
Leibundgut, B.: 1989, Astron.Astrophys., in press
Leibundgut, B., Tammann, G.A.: 1989, submitted to Astron.Astrophys.
Minkowski, R.: 1964, *Ann.Rev.Astron.Astrophys.* **2**,247
Nomoto, K., Thielemann, F.-K., Yokoi, K.: 1984, Astrophys.J. **286**,644
Norgaard-Nielsen, H.U., Hansen, L., Jorgensen, H.E., Salamanca, A.A., Ellis, R.S., Couch, W.J.: 1989, Nature **339**,523
Oke, J.B., Searle, L.: 1974, *Ann.Rev.Astron.Astrophys.* **12**,315
Oke, J.B., Sandage, A.: 1968, Astrophys.J. **154**,21
Phillips, M.M. et al.: 1987, Publ.Astron.Soc.Pac. **99**,592
Sandage, A.: 1988, *Ann.Rev.Astron.Astrophys.* **26**,561
Sandage, A., Tammann, G.A.: 1987, *A Revised Shapley-Ames Catalog of Bright Galaxies*, 2nd edition, Washington: Carnegie Institution of Washington
Schechter, P.L.: 1980, Astron.J. **85**,801
Tammann, G.A.: 1982, in *Supernovae: A Survey of Current Research*, eds. M.J. Rees and R.J. Stoneham, Dordrecht:Reidel,p.371
Tammann, G.A.: 1988, in *The Extragalactic Distance Scale*, eds. S.van den Bergh and C.J. Pritchet, San Francisco:Astron.Soc.Pacific,p.282
Tammann, G.A., Leibundgut, B.: 1989, in preparation
Uomoto, A., Kirshner, R.P.: 1985, Astron.Astrophys. **149**,L7
Wheeler, J.C., Levreault, R.: 1985, Astrophys.J. **294**,L17
Woosley, S.E.,Weaver, T.A.: 1986, *Ann.Rev.Astron.Astrophys.* **24**,205

An Estimate of the Distance to M100 in the Virgo Cluster via VLBI of SN1979C: Updates and Prospects

Norbert Bartel

1. METHOD

A VLBI determination, at time t after the explosion, of the angular radius, Θ, of a supernova's radiosphere, coupled with an optical-spectroscopic determination of the radial expansion velocity v_{shock}, at time t_0 after the explosion, of the supernova's shockfront allows estimates to be made of the distance, D, to the supernova's host galaxy and of Hubble's constant (BARTEL [1]; BARTEL *et al.* [2]; see also BARTEL [3], for a comparison of this method with other methods (e.g., BRANCH [4]) that use supernovae as distance indicators):

$$D = \frac{v_{\mathrm{shock}}\, t\, \mu\, \eta}{\Theta\, m\, \kappa} \left[\frac{t}{t_0}\right]^{m-1} \tag{1}$$

$$H_0 = \frac{v_{\mathrm{redshift}} + v_{\mathrm{infall}}}{D} \quad . \tag{2}$$

The parameter m is a measure of the deceleration (or acceleration) of Θ, with $\Theta \propto t^m$, η is the ratio between the expansion velocity of the radiosphere and that of the shockfront, and κ the model dependence of Θ. We define the parameter κ so that $\kappa = 1.0$ for a supernova having a shell-like brightness distribution with a shell thickness of $\sim 15\%$. The parameter μ describes the effect of any deviation from spherical symmetry on the distance determination. For spherically symmetric SNe, $\mu = 1$. The parameter v_{redshift} denotes the redshift velocity of the host galaxy, and v_{infall} its correction due to local peculiar velocities. For a uniformly expanding, spherically symmetric supernova shell eq. (1) becomes simply: $D = v_{\mathrm{shock}} \cdot t\eta/\Theta$.

The goal is to determine observationally each of the parameters in eq. (1) and thereby to estimate a host galaxy's distance and H_0. Since 1982, we have made VLBI observations of SN1979C in the galaxy M100 in the Virgo cluster at several epochs and combined the results with those from optical spectroscopic observations. In the remainder, I will describe the results and uncertainties and indicate prospects for more accurate determinations of M100's distance via additional VLBI observations.

2. PARAMETER DETERMINATIONS

For SN1979C, three of the six parameters in eq. (1) have been determined and a useful lower bound on a fourth parameter can be estimated from our VLBI observations of SN1987A. From VLBI observations of SN1979C, we obtained $\Theta = 0.42 \pm 0.03$ at $t = 3.69$ yr and $m = 1.03 \pm 0.15$ (BARTEL, this volume). From optical spectroscopic observations (BRANCH *et al.* [5]), the parameter v_{shock} can be estimated.

Since the largest unambiguous value for the expansion velocity is $\sim 12000 \pm 200$ km s^{-1}, determined from the blue edge of the absorption trough of the NaD line profile 0.24 yr after the explosion, and since the blue edge of the Hα emission profile, which is equivalent to 10500 km s^{-1}, remains constant until the last date of Branch *et al.*'s spectroscopic observations, 0.65 yr after the explosion, we take $v_{\rm shock} = 12000 \pm 200$ km s^{-1} at $t_0 = 0.65$ yr.

A lower bound on the parameter η can be estimated from our VLBI observations of SN1987A (BARTEL *et al.* [6]; JAUNCEY *et al.* [7]; SHAPIRO *et al.* [8]), if the physical processes responsible for the radio emission from SN1987A are typical for SNe in general. Are these processes typical for SNe in general? In Fig. 1, I plot the spectral luminosity at 1.5 GHz of SNe and young SNRs versus the time since their explosions. Since the spectral luminosity of SN1987A is two to four orders of magnitude smaller than the spectral luminosities of other recent SNe, and since its peak was reached at least an order of magnitude faster than for the other SNe, it was speculated that the prompt radio burst was perhaps a precursor of, and quite distinct from, a more prominent outburst. Such a precursor would have been undetectable for more distant supernovae and, moreover, because of its likely occurrence well before maximum light, unobservable from other SNe discovered well after the epoch of explosion.

However, two and a half years have passed without SN1987A's flaring up again at radio frequencies (0.8–8 GHz). Further, the circumstellar interaction model was quite successful in fitting the evolution of the prompt burst's light curve and spectrum. Also, the shape of the prompt burst's light curve and the evolution of its spectrum resembled those of other supernovae in detail. A fit of the circumstellar interaction model to the radio flux densities observed at four frequencies (STOREY and MANCHESTER [19]; CHEVALIER and FRANSSON [20]) suggested a density of the circumstellar medium two to three orders of magnitude lower than the corresponding densities for SN1979C and SN1980K. This lower density is consistent with the progenitor of SN1987A having been a blue supergiant (wind velocity ~ 500 km s^{-1}), in contrast to the progenitors of SN1979C and SN1980K, believed to have been red supergiants (wind velocities ~ 10 km s^{-1}) (CHEVALIER and FRANSSON [20]).

Thus, it appears now that the relatively low radio luminosity and the early turn-on of radio emission were caused by the relatively low density of the circumstellar medium encountered by the SN1987A shockfront. Further, it appears that the physical processes responsible for the radio emission were indeed typical for SNe in general and that a lower bound on the parameter η can be estimated from our VLBI observations of SN1987A. The lower bounds on the size and velocity of the radio shell relative to, respectively, the radii and velocities inferred from the absorption minimum and the blue edge of the Hα-profile imply that $\eta > 0.9$ for the values for SN1979C (see, e.g., BARTEL, this volume). The upper bound on η is model dependent. The circumstellar interaction model predicts $\eta = 1.2$ for $0.9 \lesssim m < 1$ (CHEVALIER [21]). Based on these values, we assume $\eta = 1.0^{+0.2}_{-0.1}$.

Since we have not yet determined the brightness distribution of SN1979C, any estimate of the SN's angular radius is somewhat model-dependent. With $\kappa = 1.0$ for an optically thin, shell-like brightness distribution with a shell thickness of $\sim 15\%$ of the shell radius, κ can vary between ~ 0.8 for the distribution as a ring and ~ 1.3 for it as an optically thin uniform sphere. The variation of κ as a function of the shell thickness itself is smaller: $\pm 4\%$ for a thickness between 25% and $\lesssim 5\%$ (e.g., MARSCHER [22]). There are reasons to believe that the morphology of SN1979C is indeed shell-like. First, $\sim 90\%$ of all known SNRs with discernible morphology have a shell-like structure (GREEN [23]). Second, SNRs with steep radio spectra like that of SN1979C are shells, in contrast to SNRs with flat spectra like that of the centrally powered Crab nebula, which are plerions (WEILER and SRAMEK [24]). Since, further, a ring of emission is most likely unphysical for SNe, we adopt $\kappa = 1.00^{+0.15}_{-0.05}$, which allows for a morphology between that of a thin shell of emission and that of a composite between a shell and a plerion. Such a value for κ is also consistent with the

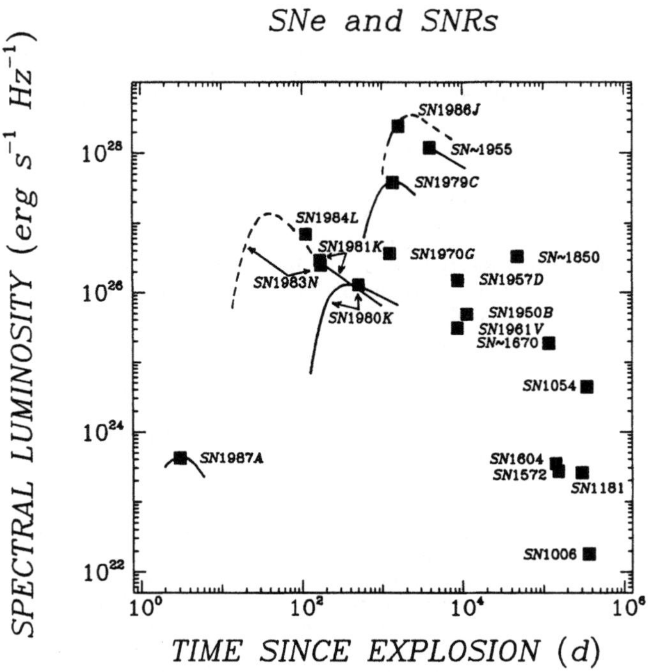

Fig. 1. The spectral luminosity at 1.5 GHz and, for SN1987A only, at 1.4 GHz, vs. the age of radio supernovae and young supernova remnants. The squares indicate the largest measured spectral luminosities of the corresponding sources, the solid curves represent all measured spectral luminosities, and the dashed curve predicts spectral luminosities according to CHEVALIER [9]. The time at which the flux density of a SN reaches maximum is frequency-dependent, typically threefold shorter at 8 than at 1.5 GHz. Note that SN~1955 (SNR41.9+58) is the brightest source among more than 20 hot spots in M82, which are most likely also SNe or SNRs. Their spectral luminosities are a few times to hundredfold smaller than the spectral luminosity of SN~1955, and their ages range from a few years to a few hundred years (UNGER *et al.* [10]; KRONBERG, BIERMANN, and SCHWAB [11]; KRONBERG and SRAMEK [12]; BARTEL *et al.* [13]). Their data are not plotted here. The data for SN1987A are from TURTLE *et al.* [14], for SN1986J from RUPEN *et al.* [15], BARTEL *et al.* [6], and WEILER, this volume, for SN1961V from COWAN, HENRY, and BRANCH [16], for SN~1955 (SNR41.9+58) in M82 from KRONBERG, BIERMANN, and SCHWAB [11] and BARTEL *et al.* [13], and for SN~1850 in NGC4449 from DE BRUYN [17]. The data from all other sources are from WEILER *et al.* [18] and references therein.

circumstellar interaction model for SNe (CHEVALIER [25] and, e.g., LUNDQVIST and FRANSSON [26]).

As to the parameter μ, we have not yet been able to determine with useful accuracy how much, if at all, the brightness distribution of SN1979C deviates from circular symmetry. The relatively small declination of the source of $\delta = 16°$ and the locations of the four telescopes used in the observations caused the u-v coverage to be rather elongated (see Fig. 2) and limited our sensitivity in modeling the SN in two dimensions. For now, we therefore have to rely on the assumption that SN1979C is approximately circularly symmetric and adopt $\mu = 1.0 \pm 0.15$.

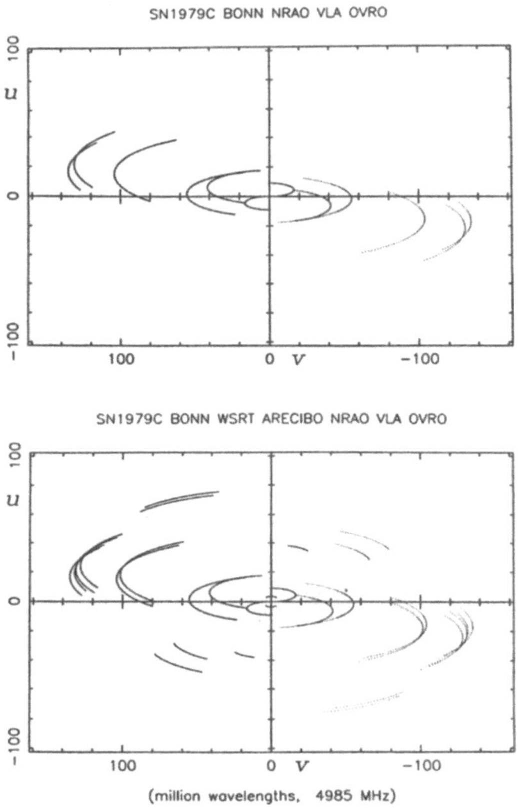

Fig. 2. The upper part of the figure shows the u-v coverage when SN1979C is observed with a four-station array, as during the first three sessions of our monitoring program. The four stations are: the 100-m antenna near Bonn, West Germany; the 43-m antenna in Green Bank, WV; the 27 × 25-m VLA near Socorro, NM; and the 40-m antenna near Big Pine, CA. The lower part shows the equivalent coverage when the array is enlarged by the 305-m antenna in Arecibo, PR, and the 14 × 25-m array at Westerbork, The Netherlands. The dotted curves show reflected tracks.

With these values and uncertainties, we get

$$D_{\mathrm{M100}} = 22^{+7}_{-6} \ \mathrm{Mpc} \quad .$$

For estimating H_0, we take $v_0 + v_{\mathrm{infall}} = 1250 \pm 150$ km s^{-1} for the Virgo cluster center (see 12 references in Bartel *et al.* 1985). If we assume that the distance to M100 is within 10% of the distance to this center, since M100 is within 5° of the center and has a redshift consistent with other Virgo cluster members, we get

$$H_0 = 60 \pm 20 \ \mathrm{km\ s^{-1}\ Mpc} \quad ,$$

with the uncertainties of both estimates representing 1–2σ. A more complete error analysis is pending.

3. PROSPECTS

We anticipate making more and more extended VLBI observations of SN1979C to determine the parameters m, κ, and μ more accurately. In the spring of 1990, we will add the upgraded Arecibo telescope and the phased Westerbork array to our presently east–west oriented VLBI array of four antennas (Owens Valley Radio Telescope, VLA, Green Bank telescope, and Effelsberg telescope). Each of these additional antennas will add significantly to the sensitivity of our array and counterbalance to some degree the combined effects of SN1979C's decreasing flux density and increasing size on the rising difficulty of measuring the SN's angular size. Apart from determining the parameter m more accurately, we may also be able to distinguish whether the emission region has the morphology of a ring, an optically thin shell, or an optically thin uniform sphere; thereby we can also obtain an estimate of the parameter κ independent of our present educated guess. Further, the inclusion of Arecibo into the array will broaden the u-v coverage twofold (see Fig. 2) and should allow a useful estimate to be made on the bounds on μ.

4. CONCLUSIONS

VLBI observations of SN1979C in the galaxy M100 in the Virgo cluster were made to determine the angular radius Θ of the SN's radiosphere at several epochs and its deceleration parameter m. VLBI observations of SN1987A revealed a useful lower bound on the ratio η between the size of the radiosphere and that of the Hα line-forming region. Combined with a) a model-dependent upper bound on η, b) an educated guess of the possible range of values κ for the morphologies of SN1979C, and c) the assumption of approximate spherical symmetry of SN1979C, the distance to M100 could be determined with an uncertainty of about 30%.

Anticipated new observations could allow a more accurate measurement of the deceleration parameter m, and, for the first time, determinations of SN1979C's morphology and of a bound on any deviation of circular symmetry of its brightness distribution. With such determinations, the uncertainty of M100's distance could perhaps be limited to about 20%.

5. ACKNOWLEDGMENT

This research was supported in part by the NSF under grant No. AST-8902087.

6. REFERENCES

1. Bartel, N. 1985, in *Supernovae as Distance Indicators*, Lecture Notes in Physics, ed. N. Bartel (Springer–Verlag, Berlin), **224**, 107.
2. Bartel, N., Rogers, A. E. E., Shapiro, I. I., Gorenstein, M. V., Gwinn, C. R., Marcaide, J. M., and Weiler, K. W. 1985, *Nature*, **318**, 25.
3. Bartel, N. 1986, in *Highlights of Astronomy*, ed. J. P. Swings (Reidel, Dordrecht), **7**, 655.
4. Branch, D. 1985, in *Supernovae as Distance Indicators*, Lecture Notes in Physics, ed. N. Bartel (Springer–Verlag, Berlin), **224**, 138.
5. Branch, D., Falk, S. W., McCall, M. L., Rybski, P., Uomoto, A., and Wills, B. J. 1981, *Ap. J.*, **244**, 780.
6. Bartel, N. *et al.* 1988, in *Supernova 1987A in the Large Magellanic Cloud*, eds. M. Kafatos and A. Michalitsianos (Cambridge Univ. Press, Cambridge), p. 81.
7. Jauncey, D. L. *et al.* 1988, *Nature*, **334**, 412.
8. Shapiro, I. I. *et al.* 1988, in *IAU Symposium 129, The Impact of VLBI on Astrophysics and Geophysics*, eds. M. J. Reid and J. M. Moran (Reidel, Dordrecht), p. 185.

9. Chevalier, R. A. 1984, *Ap. J. (Letters)*, **285**, L63.
10. Unger, S. W., Pedlar, A., Axon, D. J., Wilkinson, P. N., and Appleton, P. N. 1984, *M. N. R. A. S.*, **211**, 783.
11. Kronberg, P. P., Biermann, P., and Schwab, F. R. 1985, *Ap. J.*, **291**, 693.
12. Kronberg, P. P., and Sramek, R. A. 1985, *Science*, **277**, 28.
13. Bartel, N. *et al.* 1987, *Ap. J.*, **323**, 505.
14. Turtle, A. J. *et al.* 1987, *Nature*, **327**, 38.
15. Rupen, M. P., van Gorkom, J. H., Knapp, G. R., Gunn, J. E., and Schneider, D. P. 1987, *A. J.*, **94**, 61.
16. Cowan, J. J., Henry, R. B. C., and Branch, D. *Ap. J.*, **329**, 116.
17. de Bruyn, A. G. 1983, *Astr. Ap.*, **119**, 301.
18. Weiler, K. W., Sramek, R. A., Panagia, N., van der Hulst, J. M., and Salvati, M. 1986, *Ap. J.*, **301**, 790.
19. Storey, M. C. and Manchester, R. N. 1987, *Nature*, **329**, 421.
20. Chevalier, R. A. and Fransson, C. 1987, *Nature*, **328**, 44.
21. Chevalier, R. A. 1985, in *Supernova as Distance Indicators*, Lecture Notes in Physics, ed. N. Bartel (Springer–Verlag, Berlin), **224**, 123.
22. Marscher, A. P. 1985, in *Supernova as Distance Indicators*, Lecture Notes in Physics, ed. N. Bartel (Springer–Verlag, Berlin), **224**, 130.
23. Green, D. A. 1984, *M. N. R. A. S.*, **209**, 449.
24. Weiler, K. W., and Sramek, R. A. 1988, *Ann. Rev. Astr. Ap.*, **26**, 295.
25. Chevalier, R. A. 1982, *Ap. J.*, **259**, 302.
26. Lundqvist, P. and Fransson, C. 1988, *Astr. Ap.*, **192**, 221.

Hard X-ray Radiation from Supernova 1987A. The Results of Kvant Module in 1987–1989

R. A. Sunyaev, A. S. Kaniovsky, V. V. Efremov, S. A. Grebenev, A. V. Kuznetsov, J. Englhauser, S. Doebereiner, W. Pietsch, C. Reppin, J. Truemper, E. Kendziorra, M. Maisack, B. Mony, & R. Staubert

For the first time during two years observations of the Supernova 1987A in 1989 June the Roentgen observatory aboard the Mir-Kvant module was not able to detect its hard X-ray radiation during current series of observations. Radiation flux in energy band 45-105 keV decreased more than 8.5 times in comparison with maximal flux detected in 1988 January.

The results obtained during two years of the Supernova 1987A observations were given in the papers (Sunyaev et al., 1987a,b, 1988, 1989). By the present time we have succeeded in calibration of the third and fourth detectors of the HEXE device using the results of the Crab Nebula observation. These detectors have lower energy resolution in comparison with the first and the second ones. Therefore we reprocessed all the obtained data about SN1987A hard X-rays using the results of all four detectors of the HEXE device once more that is increased data significance and decreased statistical errors.

Paper not presented at Workshop; Late manuscript accepted by editor

In Fig.1a,b spectra of SN1987A hard X-ray radiation are presented. They were obtained in seven series of intense observations carried out by the Roentgen observatory during two years. The spectra demonstrate an increase of the flux from 1987 August to 1988 January. This increase is connected with rapid decreasing of the envelope transparence. From 1988 January to 1989 June a continuous decline of the flux is observed which is mainly connected with decreasing of ^{56}Co amount in the envelope. Already in 1988 September a strong change of the spectral shape was detected. This is explained by decreasing of the envelope optical thickness with respect to Thomson scattering. The number of successive scatterings experienced by majority of photons became insufficient to decrease energy of the ^{56}Co decay gamma-photons due to multiple recoil effect up to value $h\nu \leq 50$ keV.

Note the sharp cutoff of flux at the energies below 20 keV in 1987 August and 1988 January was connected with photo-absorption by heavy elements. At that time the photon diffusion in the envelope accompanied by the recoil effect moved majority of photons in the band $h\nu \leq 20$ keV where the photoabsorption dominated.

In Fig.2,3 light curves of the SN1987A emission in three energy bands: 15-45, 45-105, 105-200 keV (see also Table 1) are presented. Note here especially the point corresponding to observations on June 16, 1987 when the hard X-ray flux in the 45-105 keV spectral channel was detected at four standard deviation level (first it was noticed by Englhauser et al., 1989).

The light curves (Fig.3) testified to hard X-ray flux from the supernova changed smoothly in accordance with predictions of the model of this radiation appearance due to radioactive cobalt decay in opaque envelope. For two years of observations we have not been able to observe neither traces of a shock wave generated due to collision between expanding envelope of the supernova and a stellar wind emitted by the presupernova on a red giant stage of the evolution nor traces of X-ray radiation of the stellar remnant - a young pulsar or an accreting object, nor any manifestations of emission connected with cosmic rays.

Mixing of Radioactive Elements in the Expanding Envelope. Early detection of the SN1987A hard X-ray radiation by the Ginga satellite and the Kvant module (Dotani et al., 1987, Sunyaev et al., 1987a,b) was the first evidence of a radioactive ^{56}Co strong mixing over the envelope volume (Ito et al., 1987, Ebisuzaki and Shibazaki, 1988, Grebenev and Sunyaev, 1988, Pinto and Woosley, 1988). At present this conclusion is confirmed by direct observations of a velocity dispersion of infrared lines of the iron and cobalt ions(Ericson et al., 1988) and also by a broad spectral width of the ^{56}Co direct escape gamma-lines (Matz et al., 1988, Rester et al., 1989). These direct observations testify to presence of radioactive cobalt in the envelope layers having expansion velocities from 400 up to 3000 km/s. This would be impossible if a strongest mixing of envelope material due to generation of the Rayleigh-Taylor instability was not occured (Hachisu et al., 1989, Arnett et al., 1989).

The supernova hard X-ray light curve gives a possibility to estimate the distribution of radioactive cobalt over the envelope using the simplest assumptions. The ^{56}Co distribution (mass-fraction) consistent with the observed light curve is shown in Fig.4 by crosses (vertical line of a cross corresponds to error at one standard deviation level). The problem of reconstruction of cobalt distribution over the envelope is

considerably simpler if cobalt radial distribution is searched as a superposition of two Gaussians: narrow one localized near the envelope centre and an extensive one with broader ^{56}Co distribution over the envelope. The regions of ^{56}Co distribution consistent with the observed light curve in this simple model are also presented in Fig.4.

It is obvious that two different approaches give quite close results. About 60% of cobalt is in central region of the envelope having low velocities. And about 40% is mixed over all the envelope volume. Note that the data of only two hard energy bands 45-200 keV were used during the cobalt distribution reconstruction. The flux at lower energies strongly depends on photoabsorption in the envelope but the photoabsorption efficiency strongly depends on degree of the cobalt mixing.

All the calculations the results of which were presented above and will be discussed below were carried out on the basis of the velocity and density distribution model resulted from hydrodynamics simulations by Arnett (1988).

In Fig.3 it is shown how the accepted model of cobalt distribution coincides with the observed X-ray light curve. Deviations are maximal at the beginning of the supernova X-ray observation in a soft 15-45 keV band. This points out at more strong photoabsorption in comparison with photoabsorption in the used model. It may be connected with the enhanced cobalt concentration in outer envelope layers. High X-ray and gamma-ray radiation from these layers appeared at early stages of the envelope expansion before the beginning of the Roentgen observatory systematical observations.

Abundance of ^{57}Co. By the beginning of the second year after the explosion a ^{57}Co isotope will be able to become an important energy source in the supernova envelope as it decays

3.5 times moreslowly than ^{56}Co. The simulations of explosive nucleosynthesis (Woosley et al., 1986, Hashimoto et al., 1989) predicted a ratio of ^{57}Co/^{56}Co abundances 2 times exceeding the ratio of ^{57}Fe/^{56}Fe abundances at the Earth. There are two ways to define the abundance of ^{57}Co in the SN1987A envelope: the first, by direct determination of flux in the ^{57}Co lines of 12? and 136 keV in the supernova spectrum and the second one, by determination of a ^{57}Co photon portion in the X-ray continuous spectrum in the 45-105 keV energy band. We means the photons emitted in the ^{57}Co lines 122 and 136 keV but undergoing to multiple scatterings in the envelope and decreasing their energy due to recoil effect. Because of relatively low energy resolution of Phoswich detectors the HEXE device aboard Mir-Kvant module gave considerably better results when the second method was used.

The results presented below depend on an accepted envelope model (the Arnett's model (1988) is used) and on a cobalt distribution over the envelope (the distribution presented in Fig.4 is used). It is also supposed that ^{57}Co is distributed similarly to ^{56}Co.

For the whole period from 1988 September to 1989 June the Roentgen observatory has not detected a statistically significant enhancement of X-ray luminosity in the 45-105 kev energy band over the model predictions in which the whole observed flux is connected with the ^{56}Co decay. The upper limits at three standard deviation level for the ratio of ^{57}Co/^{56}Co relative abundance in the supernova envelope to the Earth's ^{57}Fe/^{56}Fe relative abundance were equal to 2.4 in 1988 September, 3.3 in 1988 December and 1.8 in 1989 June at the accepted assumptions. Note that the ratio of ^{57}Fe/^{56}Fe abundances at the Earth is 0.024 (Cameron, 1986). All the data obtained from 1988 September to 1989 June allowed us to obtain a limit at three standard deviation level for a portion of ^{57}Co decay photons in the light curve of the SN1987A hard X-ray radiation. This limit corresponds to the $^{57}Co/^{56}Co$ abundance in 1.5 times exceeding the Earth's $^{57}Fe/^{56}Fe$ relative abundance. The observations in 1989 May-June gave upper limit on the cobalt 122 keV line flux $3.9 \cdot 10^{-4}$ photons$\cdot cm^{-2}s^{-1}$ (at 3σ level). This limit corresponds in frames of the model being discussed to the $^{57}Co/^{56}Co$ abundance 6 times exceeding the Earth's abundance of $^{57}Fe/^{56}Fe$.

The data obtained in 1989 May-June give also a possibility to set up an upper limit on a fraction of the ^{22}Na and ^{44}Ti radioactive photons in the X-ray 45-105 keV flux from SN1987A in 830 days after the explosion. The corresponding upper limits at three standard deviation level on mass of ^{22}Na and ^{44}Ti contained in the envelope at the moment of explosion are $1.3 \cdot 10^{-3} M_{\odot}$ and $9 \cdot 10^{-3} M_{\odot}$. These limits exceed the amount of ^{44}Ti, $M_{44} \sim 1.2 \cdot 10^{-4} M_{\odot}$, and ^{22}Na, $M_{22} \sim 3 \cdot 10^{-5} M_{\odot}$, predicted by Hashimoto et al. (1989) and Woosley et al. (1986) on basis of the explosive nucleosynthesis calculations at one order of magnitude.

Limits on the Stellar Remnant Luminosity. The observations of the Roentgen observatory in 1989 May-June set up strong restrictions on X-ray luminosity of a stellar remnant produced during the explosion L_X(1-6 keV) $\leq 3.6 \cdot 10^{36}$, L_X(6-15 keV) $\leq 5.4 \cdot 10^{36}$ and L_X(15-105 keV) $\leq 1.35 \cdot 10^{37}$ erg/s for the assumed distance 55 kpc (Sunyaev R.A. et al., 1990, Table 1). At that time a Thomson optical depth of the envelope yet exceeded 3-4, and a X-ray spectrum of the remnant was considerably distorted by a photoabsorption and a compton scattering. The absorbed energy went on the envelope heating and were reemitted in the infrared, submillimeter and optical bands. The measurements by Bouchet et al. (1990) showed that emission of a dust in the envelope had a black body spectrum with $T \approx 160$K and the supernova bolometric luminosity in 1030 days after the explosion were equal to $(2.0 \pm 0.1) \cdot 10^{38}$ erg/s. Using the data of Monte-Carlo calculations, the upper limit on the hard X-ray flux in the 15-105 keV band obtained by the HEXE device on the 830th day and the information on the envelope emission at low frequencies rather interesting restrictions on an intrinsic spectrum of the stellar remnant may be obtained. For example, assuming that the remnant (pulsar) has a power law spectrum in 1-1000 keV energy band, $I_\nu \sim \nu^{-\alpha}$ [photons$\cdot$cm^{-2}s^{-1}keV^{-1}], and using the 3σ upper limit presented above for the X-ray 15-105 keV flux escaping the envelope on the 830th day we obtain the 3σ upper limit on the pulsar luminosity in the 1-1000 keV energy band $L_B \le 2.4 \cdot 10^{38}$ and $\le 4.4 \cdot 10^{38}$ erg/s for two spectral indexes $\alpha = 1.5$ and 2.1. In neglecting the pulsar spindown and its luminosity changing we may find the upper limit on the energy absorbed in the envelope that is its low frequency luminosity on the 1100th day $L_{IR} \le 1.0 \cdot 10^{38}$ and $\le 3.5 \cdot 10^{38}$erg/s for $\alpha = 1.5$ and 2.1 correspondingly. It is clear that in case when the spectral index is 1.5 such a spectrum is not able to give the observed low frequency luminosity of the envelope. It may be easily shown that any spectrum with $\alpha \le 1.75$ does not coincide with the data obtained by Bouchet et al. (1990).

The presented example shows a possibility for using our data to obtain restrictions on parameters of a pulsar hidden inside the expanding envelope. In the case when an accreting object is situated in the envelope centre our estimates are less definite. Nevertheless such an analysis with using the

HEXE data presented above for the object with spectrum similar to the spectrum of the wellknown source Cygnus X-1 in a low state (Sunyaev and Truemper, 1979) also demonstrates the impossibility to satisfy the low frequency data. If the infrared radiation is the result of dust reprocessing of a central object hard emission the X-ray spectrum of the stellar remnant should be soft enough.

Another way to explain the excess of infrared radiation detected by Bouchet et al. (1990) is that radioactive isotopes ^{57}Co, ^{22}Na and ^{44}Ti are more abundant in the envelope than it was assumed. Their hard radioactive emission transforms in the opaque envelope into the low frequency emission which was observed. As the excess luminosity on the 1100th day after the explosion was equal to $2\cdot10^{38}$erg/s it was necessary that about $4\cdot10^{-2}M_\odot$ of ^{57}Co (that is the $^{57}Co/^{56}Co$ ratio exceeded the Earth's $^{57}Fe/^{56}Fe$ ratio about 22 times), or $9\cdot10^{-4}M_\odot$ of ^{22}Na, or $9\cdot10^{-3}M_\odot$ of ^{44}Ti were hidden insidè the envelope. Comparing these values with the HEXE upper limits for ^{57}Co, ^{22}Na and ^{44}Ti abundances, $M_{57} \leq 2.8\cdot10^{-3}M_\odot$, $M_{22} \leq 1.3\cdot10^{-3}M_\odot$ and $M_{44} \leq 9\cdot10^{-3}M_\odot$, we come to conclusion that the assumption about the radioactive nature of excess has failed in the case of ^{57}Co and is unlikely in the case of ^{44}Ti and ^{22}Na.

Discussion. All the data obtained by the four HEXE detectors in 1987 August - 1988 January confirm the identification of the hard X-ray source in the Large Magellanic Cloud having an unusual spectrum (see Fig.5) with SN1987A. The upper limits at 3σ level obtained during this localization on the X-ray fluxes from LMC X-1 and 50-millisecond pulsar PSR 0540-693 are presented in Table 2.

In 1989 May-June the Kvant module did not detect a statistically confident signal from the SN1987A region in spite of other X-ray sources LMC X-1 and PSR 0540-693 having been in the HEXE field of view. Taking into account the deviations between the direction of the telescope axis and directions on these X-ray sources an efficiency of the flux detecting from different sources differed. The upper limits on the fluxes from SN1987A and other sources are presented in Fig.1,6 and also in Table 3.

The weakness of the hard X-ray flux from LMC X-1 in 1989 May-June (Sunyaev et al., 1989) and the closeness of the upper limit on the LMC X-1 hard emission obtained by the HEXE device to limits obtained during the HEAO A2 (Wait and Marshall, 1984) and HEAO A4 (Matteson and Peterson, 1987) experiments testify to a small portion of the LMC X-1 flux to the flux detected by the Kvant module during 1987 August - 1988 April. Note that the flux detected in 1988 January exceeds the upper limits obtained in 1989 May-June about one order of magnitude.

Note in conclusion that the supernova light curve in 45-105 keV energy band did not show a single sharp statistically confident burst similar to the burst observed by the Ginga satellite in 1988 January in the softer energy band. In hard X-rays the light curve was smooth as it was expected for the light curve of the source connected with radioactive decay.

The authors are grateful to V.D.Blagov, V.M.Loznikov, V.G.Rodin, A.M.Prudkoglyd, the team headed by Yu.P.Semenov and the cosmonauts working aboard the Mir space station for the observatory control.

REFERENCES

Arnett W.D.//Astrophys.J., 1988, V.331, P.377.

Arnett W.D., Fryxell B.A. and *Muller E.*//Astrophys.J. Letters, 1989, 341, L63.

Bouchet P., Danziger I.J. and *Lucy L.B.*//IAU Circ., N°4933, 1990.

Cameron A.J.W.//Nuclear Astrophysics (ed. Barnes C.A., Clayton D.D. and Schramm D.N.), Moscow, Mir, 1986, P.33.

Clark D.H., Tuohy I.R., Long K.S. et al.//Astrophys.J, 1982, V.255, P.440.

Dotani T., Hayashida K., Inoue H. et al.//Nature, 1987, V.330, P.230.

Ebisuzaki T. and *Shibazaki N.*//Astrophys.J. Letters, 1988, V.327, P.L5.

Englhauser J., Doebereiner S., Pietsch E. et al.//23d ESLAB Symp. Proc., 1989.

Erickson E.F., Haas M.R., Colgan S.W.J. et al. //Astrophys.J. Letters, 1988, V.330, P.139.

Grebenev S.A. and *Sunyaev R.A.*//Soviet Astron. Letters, 1988, V.14, P.675.

Hachisu I., Matsuda T., Nomoto K. and *Shigeyama T.*// Astrophys.J. Letters, 1990, in press.

Hashimoto M., Nomoto K. and *Shigeyama T.*//Astron. Astrophys., 1989, V.20, P. L5.

Itoh M., Kumagai S., Shigeyama T. et al.// Nature, 1987, V.330, P.233.

Matteson J.L. and *Peterson L.E.*//1987, private communication.

Matz S.M., Share G.H., Leising M.D. et al.//Nature, 1988, V.331, P.416.

Pinto P.A. and *Woosley S.E.*//Astrophys.J, 1988, V.329, P.820.

Rester A.S., Coldwell R.L., Dunnam F.E. et al.//Astrophys.J. Letters, 1989, V.342, P.L71.

Seward F.D., Harnden F.R. and Helfand D.J.//Astrophys.J. Letters, 1984, V.287, P.L19.

Sunyaev R.A., Kaniovsky A.S., Efremov V.V. et al.//Nature, 1987a, V.330, P.227.

Sunyaev R.A., Kaniovsky A.S., Efremov V.V. et al.//Soviet Astron. Letters, 1987b, V.13, P.1027.

Sunyaev R.A., Efremov V.V., Kaniovsky A.S. et al.//Soviet Astron. Letters, 1988, V.14, P.579.

Sunyaev R.A., Kaniovsky A.S., Efremov V.V. et al.//Soviet Astron. Letters, 1989, V.15, P.291.

Sunyaev R.A., Gilfanov M.R., Churazov E.M. et al.//Soviet Astron. Letters, 1990, V.16, in press.

Sunyaev R.A. and Truemper J.//Nature, 1979, V.279, P.506.

White N.E. and Marshall F.E.//Astrophys.J., 1984, V.281, P.354.

Woosley S.E. and Weaver T.A.//Nuclear Astrophysics (ed. Barnes C.A., Clayton D.D. and Schramm D.N.), Moscow, Mir, 1986, P.359.

Table 1. The SN1987A X-ray flux evolution in accordance with da a of the HEXE device observations in 1987-1989.

Day since the outburst	Fluxes and 1σ errors $[10^{-6} phot \cdot cm^{-2} s^{-1} keV^{-1}]$ in the energy bands					
	15 - 45 keV		45 - 105 keV		105 - 200 keV	
143.9 - 144.1	2.	34.	51.	12.	18.	17.
169. - 182.	68.2	5.7	46.7	2.1	21.3	3.3
186. - 204.	96.7	8.5	50.9	3.5	26.0	5.2
231. - 247.	83.1	8.3	57.4	3.3	24.3	7.2
258. - 266.	85.	10.	55.2	4.4	12.8	9.7
291. - 309.	97.	11.	66.1	5.6	30.4	8.9
328. - 343.	100.	10.	62.6	3.5	25.1	4.8
413. - 426.	63.	9.5	51.1	3.8	27.9	6.1
444. - 446.	46.	27.	24.	12.	24.0	16.
559. - 569.	17.2	9.6	20.5	4.1	7.2	6.5
590. - 599.	21.4	6.1	14.5	3.0	8.2	4.7
630. - 648.	12.2	4.1	10.6	2.6	14.5	7.0
820. - 840.	8.2	7.0	3.8	2.9	-2.8	4.2

Table 2. The upper limits on the X-ray fluxes (in $photons \cdot cm^{-2} s^{-1} keV^{-1}$) from LMC X-1 and PSR 0540-693 at three standart deviation level in accordance with the HEXE data obtained in 1987 August - 1988 January. They were reconstructed during localization used a number of offset observations.

Source	15-45 keV	45-105 keV
LMC X-1	$5.5 \cdot 10^{-5}$	$8.0 \cdot 10^{-6}$
PSR 0540-693	$2.9 \cdot 10^{-5}$	$6.8 \cdot 10^{-6}$

Table 3. The average efficiencies of the observatory Roentgen pointing and the corresponding upper limits at three standard deviation level on the X-ray fluxes (in photons·$cm^{-2}s^{-1}keV^{-1}$) from the LMC sources observed by the HEXE device in 1989 May-June.

Source	Efficiency	15-45 keV	45-105 keV
SN1987A	52 %	$2.1\cdot10^{-5}$	$8.6\cdot10^{-6}$
LMC X-1	32 %	$8.5\cdot10^{-5}$	$1.2\cdot10^{-5}$
PSR 0540-693	45 %	$3.4\cdot10^{-5}$	$9.5\cdot10^{-6}$

Figure 1. The SN1987A X-ray spectra obtained by the Roentgen observatory in 1987 August (1) and October-November (2), 1987 December-1988 January (3), 1988 April (4), September-October (5) and November (6), 1989 May-June (7) (diamonds and crosses- the HEXE and Pulsar X-1 data respectively, crosses marked by circles - the TTM telescope upper limits). The errors correspond to one standard deviation, the upper limits - to three standard deviations (in the last graph the HEXE upper limits are shown by triangles). Results of the Monte Carlo calculations carried out according to the envelope model accepted in the present paper are presented by solid lines (time after the outburst is shown near each curve). In graphs 5,6,7 a ^{56}Co portion in the total ^{56}Co and ^{57}Co emission is shown by dotted lines. The relative abundance of $^{57}Co/^{56}Co$ is equal to two-abundance of $^{57}Fe/^{56}Fe$ at the Earth.

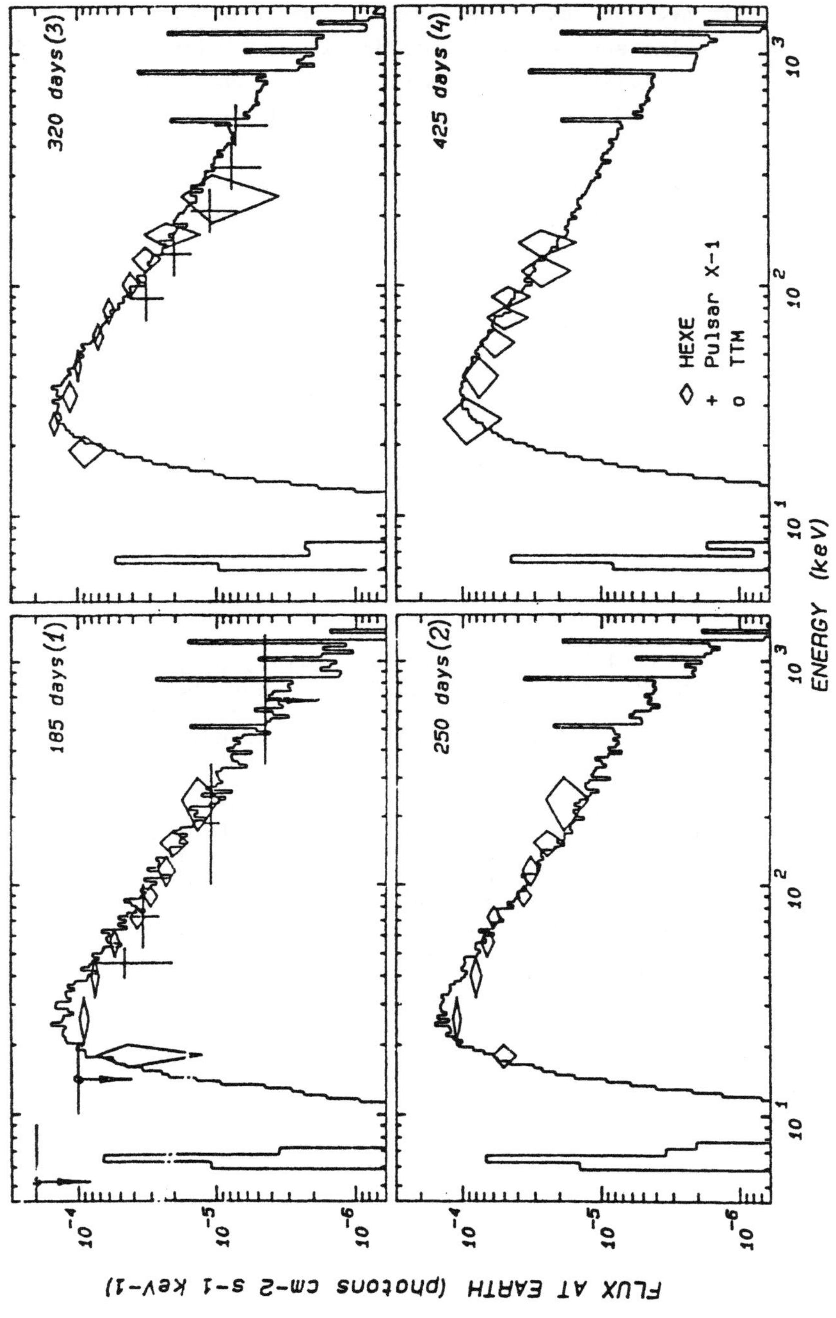
185 days (1)
250 days (2)
320 days (3)
425 days (4)
◇ HEXE
+ Pulsar X-1
o TTM
FLUX AT EARTH (photons cm-2 s-1 keV-1)
ENERGY (keV)

Fig. 1a

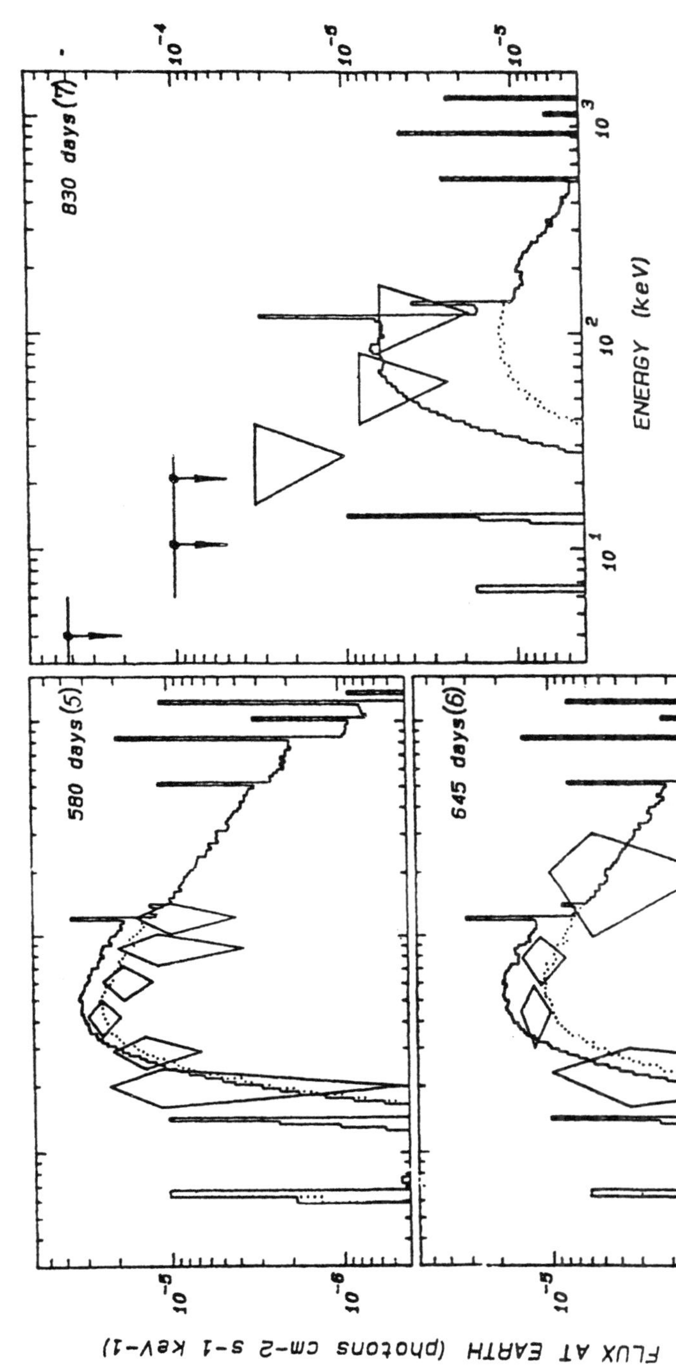
580 days (5)
645 days (6)
830 days (7)
ENERGY (keV)
FLUX AT EARTH (photons cm-2 s-1 keV-1)

Fig. 1b

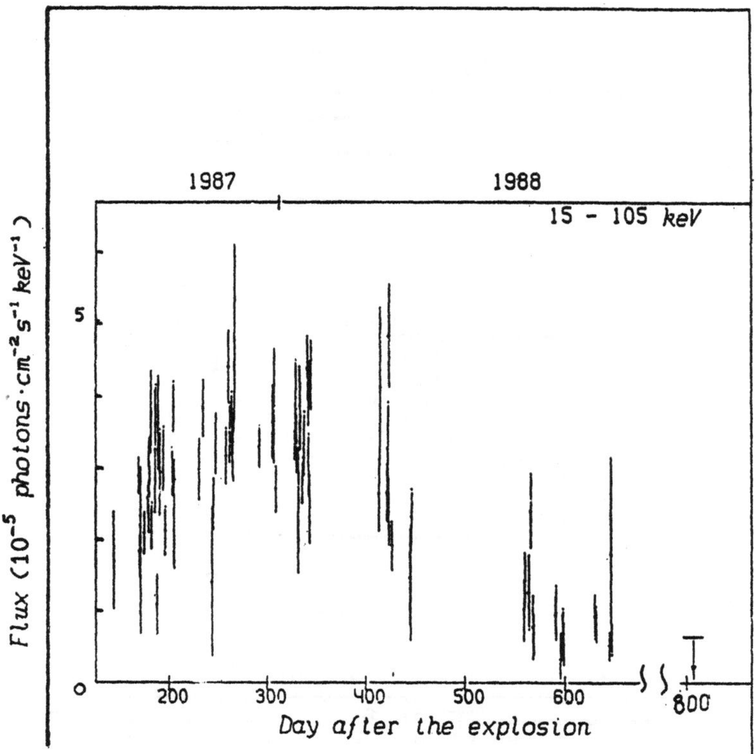

Figure 2. The SN1987A X-ray 15-105 keV flux as a function of time according to the HEXE observations. Each point corresponds to one observational day, the errors correspond to one standard deviation. Three sigma upper limit on the flux observed in 1989 May-June is presented.

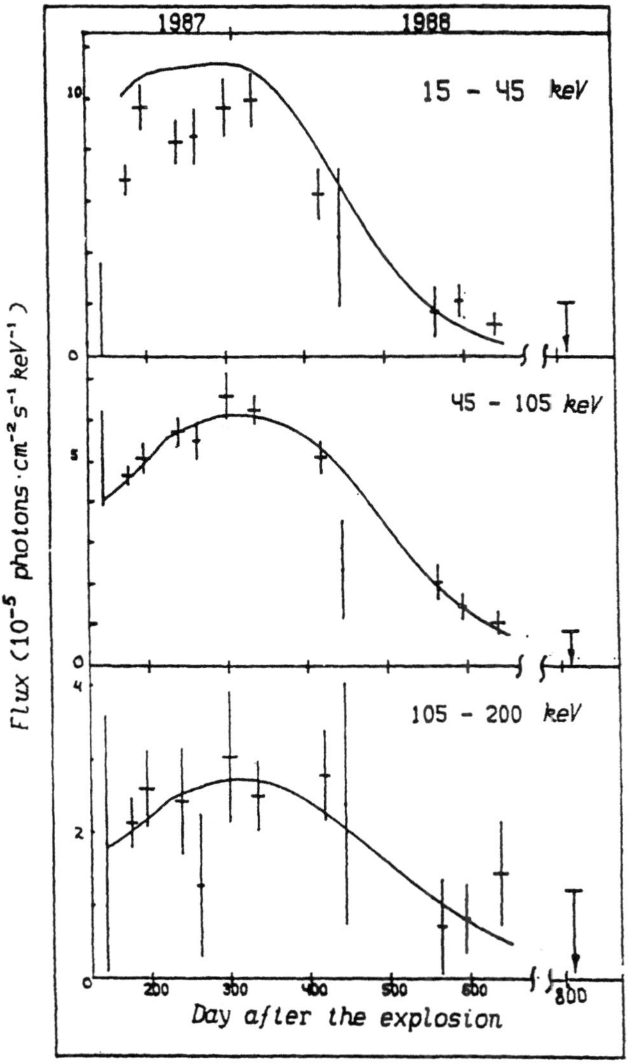

Figure 3. The SN1987A X-ray fluxes as functions of time according to the HEXE data in three colors 15-45, 45-105 and 105-200 keV. Each point presents data averaged over long period of observations. The errors correspond to one standard deviation. The results of Monte-Carlo simulations carried out for the envelope model accepted in the present paper are shown by solid lines.

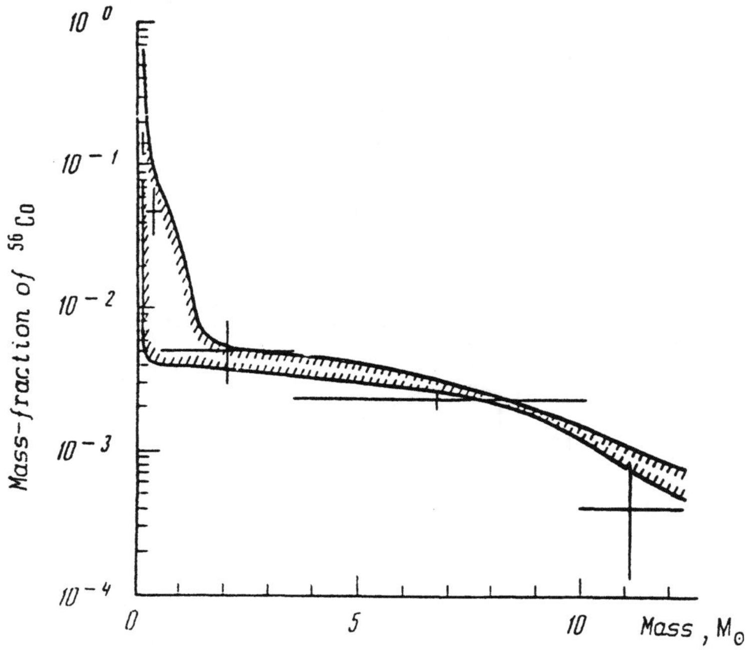

Figure 4. Region of the most probable ^{56}Co distribution (mass fraction) over the SN1987A envelope which gives a possibility to simulate the observed X-ray light curves of the source. The results of two independent approaches are presented. In the first approach it is assumed that cobalt is uniformly distributed over five spherical layers of the envelope (crosses, errors of the cobalt concentration in each layer are given at one sigma level). In the second approach the 67% confidence level region of the distribution described by superposition of two Gaussians is obtained. It is clear that both approaches give similar results.

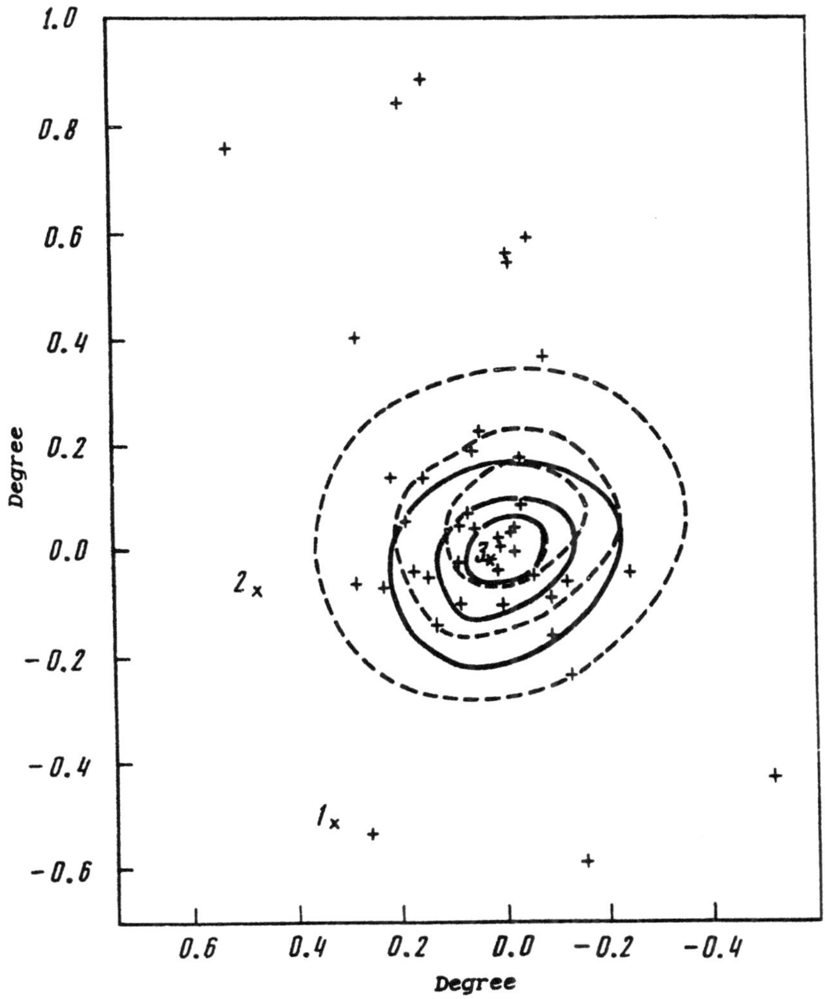

Figure 5. Localization of the hard X-ray source in the Large Magellanic Cloud according to the HEXE data obtained from 1987 August to 1988 January. The contours of 67%, 99%, and 99.9% significance for the energy bands 15-45 (Dotted lines) and 45-105 keV (solid lines) are presented. Positions of the sources LMC X-1 (1), PSR 0540-693 (2), SN1987A (3) are marked. Pointings of the HEXE device in different series of observations are shown by small crosses.

Figure 6. (a) Spectrum of LMC X-1 according to the HEAO A2 experiment (crosses) (Wait et al., 1984). The upper limits on the hard X-ray flux from this source according to the HEXE data. The analytical approximation of the TTM instrument data is shown by a solid line (Sunyaev et al., 1990). (b) Spectrum of PSR 0540-693 (a power law approximation) in accordance with the observatory Einstein (Clark et al., 1982, Seward et al., 1984) and the upper limits on the hard X-ray flux according to the HEXE data. The spectrum obtained by the TTM instrument (Sunyaev et al., 1990) coincides with the presented power law approximation within the limits of experimental data errors.

Figures 6a and 6b on next page

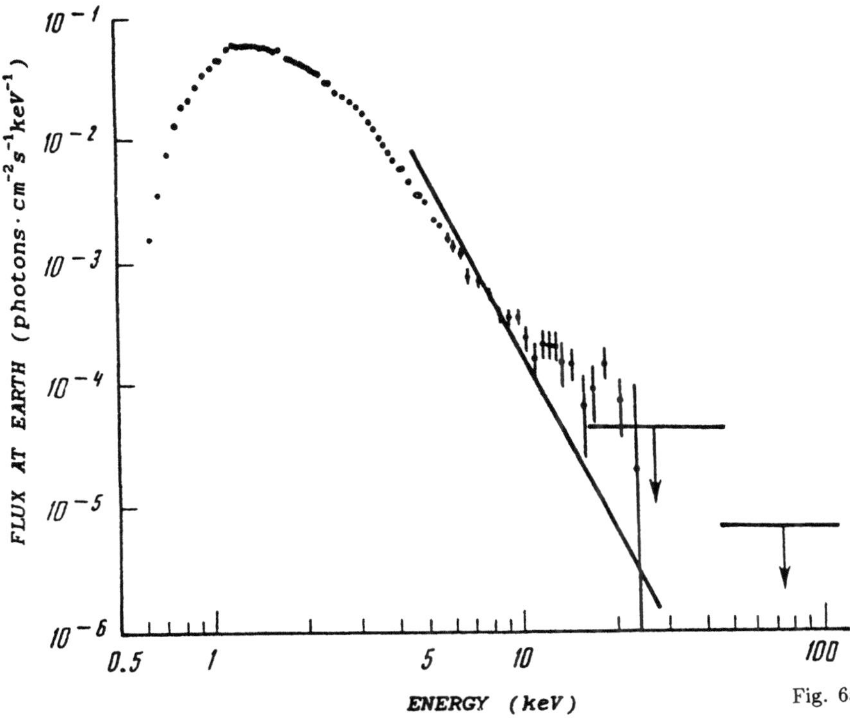
FLUX AT EARTH (photons·cm^{-2}s^{-1}keV^{-1})
10^{-1}
10^{-2}
10^{-3}
10^{-4}
10^{-5}
10^{-6}
0.5
1
5
10
100
ENERGY (keV)

Fig. 6a

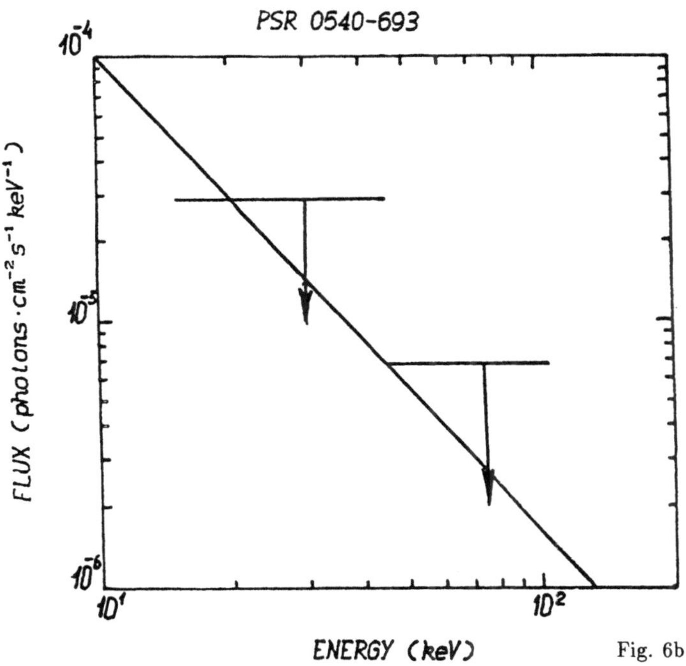
PSR 0540-693
FLUX (photons·cm^{-2}s^{-1}keV^{-1})
10^{-4}
10^{-5}
10^{-6}
10^{1}
10^{2}
ENERGY (keV)

Fig. 6b

POSTSCRIPT

The "Supernova Song," such as it is, as performed by Woosley at the Conference Dinner.

D D_7 A D D_7 A A_7
Su-per nova, su-per nova; Super-no-va over you.

D D_7
I just can't believe it,

G G_7
But before Reverand Evans sees it,

D A A_7 D G D A_7
Gotta get this telegram off to the IAU.

Supernova, supernova, your models are best it's plain to see.
You've got a great big Cray,
And it does, what you say,
In three dimensions and non-LTE.

Supernova, supernova, will we ever know you?
You live so far away,
Beyond the Milky Way.
I bet you've still got a secret or two.

And if you're not tired by this point repeat the first verse. The tune is very much like the song "Freedom" (with apologies to Odetta).

MIX
Papier aus verantwortungsvollen Quellen
Paper from responsible sources
FSC® C105338

If you have any concerns about our products,
you can contact us on
ProductSafety@springernature.com

In case Publisher is established outside the EU,
the EU authorized representative is:
Springer Nature Customer Service Center GmbH
Europaplatz 3, 69115 Heidelberg, Germany

Printed by Libri Plureos GmbH
in Hamburg, Germany